# SEDIMENTATION IN CONTINENTAL RIFTS

*Edited by:*


ROBIN W. RENAUT
*Department of Geological Sciences, University of Saskatchewan, Saskatoon, Saskatchewan S7N 5E2, Canada*

GAIL M. ASHLEY
*Department of Geological Sciences, Rutgers University, Piscataway, New Jersey 08854-8066, U.S.A.*





*Laura J. Crossey, Editor of Special Publications*
*SEPM Special Publication Number 73*

*Tulsa, Oklahoma, U.S.A.* *October, 2002*

ISBN 1-56576-082-4



Printed in the United States of America

# SEDIMENTATION IN CONTINENTAL RIFTS

Robin W. Renaut and Gail M. Ashley, Editors

# CONTENTS

## Introduction



## Tectonic, Volcanic, and Hydrological Controls on Rift Sedimentation



## Fluvial and Alluvial Depositional Systems in Rift Basins

## Lacustrine, Spring, and Wetland Depositional Systems in Rift Basins

# Introduction

# RIFT SEDIMENTATION

GAIL M. ASHLEY
*Department of Geological Sciences, Rutgers University, Piscataway, New Jersey 08854-8066, U.S.A.*
*e-mail: gmashley@rci.rutgers.edu*
AND
ROBIN W. RENAUT
*Department of Geological Sciences, University of Saskatchewan, Saskatoon, Saskatchewan S7N 5E2, Canada*
*e-mail: robin.renaut@usask.ca*

## INTRODUCTION

Continental rift basins have long been of interest to sedimentologists. Of all the terrestrial depositional settings, rift basins typically provide the greatest accommodation space, and consequently have some of the longest records of continental sedimentation. These records are a product of a complex interplay between several factors. These include geological structure and tectonic activity, volcanism, climate and its temporal variability, hydrology, hydrogeology, biology, and time (Fig. 1). The lithological records in rifts, which are naturally dominated by fluvial and lacustrine deposits, have become increasingly prominent in recent years because of their potential for studying long-term climatic changes and for testing hypotheses of orbital forcing (e.g., Olsen, 1986; Johnson and Odada, 1996). More recently, the continuing quest for the paleontological and cultural records of human origins that are entombed in the sedimentary rocks of the East African rift has raised further questions on the tempo of climatic change, changing paleolandscapes, and the environmental stresses that might have affected human evolution (Vrba et al., 1995; Andrews and Banham, 1999). Rapid burial of thick sedimentary fills, high geothermal gradients, and consequent early maturation of lacustrine organic matter, much of which is sapropelic, have made rift basins attractive targets for petroleum exploration (e.g., Robbins, 1983; Katz, 1990; Lambiase, 1990, 1995). Indeed, most of our understanding of rift-basin geometry has resulted from seismic profiling sponsored by oil companies (Scholz et al., 1990; Lezzar et al., 1996; Morley, 1999).

A conference at the Geological Society of London in 1984 brought together a group of researchers working mainly in the East African Rift, and spawned the publication "Sedimentation in the African Rifts", edited by Frostick et al. (1986). While relaxing one evening on the shore of Lake Turkana in the summer of 2000, we discussed the possibility of assembling a similar volume, seeing that almost fifteen years had passed since that book was published, and many new ideas had evolved. This collection of nineteen papers is the result of that discussion.

### *Volume Contributions*

This volume presents results of some of the varied studies on rift sedimentation that have been undertaken in the past few years. The coverage of topics is certainly not comprehensive, but it reflects the diversity of studies that are currently taking place in rift basins, both modern and ancient. Unlike the papers in Frostick et al. (1986), which had a bias toward lacustrine environments, some of the lesser-known rift settings are highlighted in this volume, which includes studies beyond the East African Rift.

The volume opens with four papers that consider the tectonic development of rift basins as repositories for sediments and sedimentary rocks. *Ebinger et al.* examine rift development at the crustal scale, emphasizing vertical motions during extension and the creation of basins for sediment accumulation. *Morley* provides a lengthy discussion of the tectonic settings of different types of extensional basin worldwide, and shows the relationships between these basins and their styles of volcanism, rift topography, and patterns of sedimentation. *Withjack et al.* document the different structural facets of rift basins, the evolution of rifts, and their relationship to the stratigraphic record of rift basins. *Hinderer and Einsele* take a different approach, and model the development of rift-basin fills quantitatively by considering them as linked denudation–accumulation systems.

Paleosols can record a wealth of paleoclimate information. *Driese and Mora* present exciting new information from pedogenic carbonates of Upper Triassic rift sediments from the Durham Basin, North Carolina, U.S.A. The $\delta^{13}C$ values show a shift from wetter to drier conditions associated with elevated atmospheric levels of carbon dioxide.

Paleoclimate signatures can also be gleaned from the sedimentary records of lakes. *Yemane et al.* interpret temperatures and rainfall from an ~ 700-m-thick freshwater lake sequence in the Permian mid-Zambezi basin of southern Africa. They show that the climate was moist and cool in the interior of Gondwana at 50–60° S. Their findings are in stark contrast to earlier numerical models that had found severe aridity and extreme seasonal temperatures at this latitude.

Pedogenic mud aggregates are sand-size particles composed of clay and silt and are important vehicle for transportation of fine-grained sediment in rift basins and other environments. *Gierlowski-Kordesch and Gibling* have examined their characteristics and development in the Mesozoic Hartford rift basin of Connecticut, U.S.A. They show that the clasts are reworked from floodplains during high-discharge events.

Lake-margin environments in steep-sided rift basins are not flat featureless plains like those in some other tectonic settings. Studies in the rift basins in northern Tanzania by *Deocampo* reveal the surprising variety of subenvironments and lithofacies that are present in lake-margin settings. He also documents the impact of hippos on sedimentation in these low-gradient environments.

The Okavango in northern Botswana has previously been called an inland delta, a wetland, and an alluvial fan. *McCarthy et al.* present a synthesis of two decades of biological, sedimentological, and geochemical studies into a depositional model of this fascinating modern system. Sedimentation history is linked to incipient rifting in the distal end of the rift valley.

*Mack et al.* provide an unusual three-dimensional glimpse into basin-fill architecture in the Rio Grande Rift in New Mexico, U.S.A. Tectonic tilting of the Neogene Palomas Basin likely triggered toe cutting of the basin-margin alluvial fans by the axial river as it shifted laterally.

Sedimentation in Continental Rifts
SEPM Special Publication No. 73, Copyright © 2002
SEPM (Society for Sedimentary Geology), ISBN 1-56576-082-4, p. 3–7.

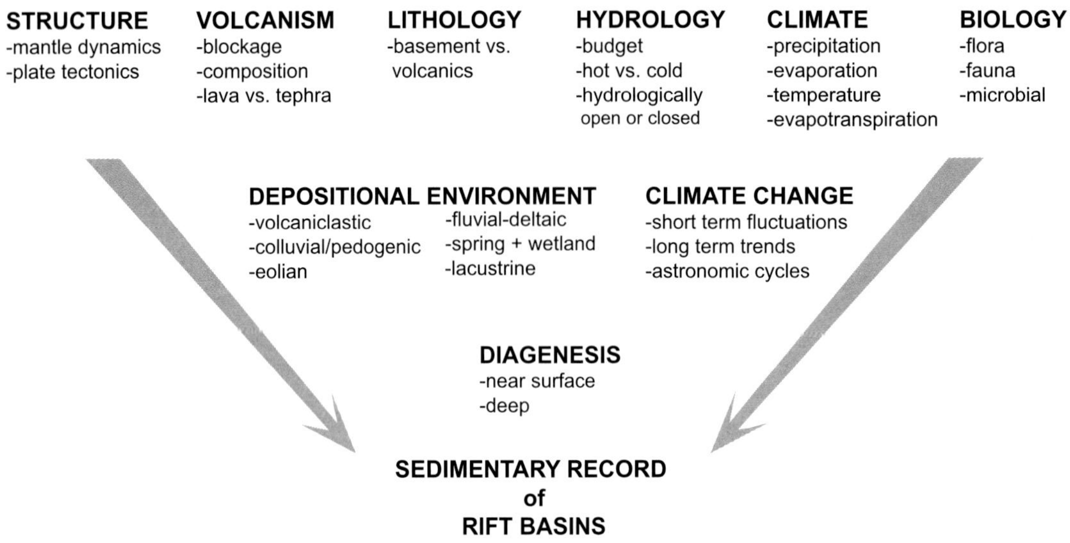


FIG. 1.—Factors that determine the nature of the sedimentary record in continental rift settings.

Long-term sedimentary sequences in rifts record the complex interplay among tectonics, climate, and volcanism, and allow the examination of sedimentation at a range of temporal and spatial scales. *Behrensmeyer et al.* analyze a million-year-long, well dated record from the Olorgesailie Basin in the southern Kenya Rift, and show clearly that these three driving mechanisms operate at different time scales. *Ashley and Hay* use a high-resolution record (20 km wide, 12 m thick, 100,000 years long) from the Olduvai basin of northern Tanzania to decipher sedimentation patterns during the early Pleistocene. A large groundwater-fed wetland was identified, and it appears to have been the only perennial freshwater resource in this otherwise semiarid rift-margin environment. Hominids, as well as other fauna, used this valuable resource.

With the extensive faulting, volcanism, and high regional heat flow in rift basins, it is no surprise that thermal springs are a common feature of modern rift systems. Their sedimentary record, however, is sparse or difficult to recognize. *Renaut et al.* describe fossil hot-spring deposits from the Turkana Basin of northern Kenya and discuss some of the impacts of hydrothermal fluids on rift sedimentation and diagenesis.

Lacustrine sediments form major parts of many rift-basin fills and have been the subject of several other recent volumes (e.g., Johnson and Odada, 1996; several papers in Gierlowski-Kordesch and Kelts, 2000). Lake waters in rifts, particularly those with volcanic rocks in their catchments, commonly have waters rich in silica and nutrients. Consequently, diatoms can thrive. *Owen* provides an overview of the controls of diatom sedimentation in rift lakes and includes a new classification of lacustrine diatomaceous sediments, based on study of many lake basins in East Africa. *Johnson* examines the budget of biogenic silica in lakes Malawi and Victoria and shows that shifts in abundance of biogenic silica through time can be related to climatic forcing.

*Yuretich and Ervin* use clay mineralogy of sediments in lakes Malawi and Turkana as indicators of paleoenvironmental change. They show the critical importance of determining whether the clays are detrital or authigenic before making inferences on their paleoclimatic significance.

Carbonates are a common component of lacustrine rift fills, but their origins and paleoenvironmental significance are not always clear. Using a combination of sedimentological and geochemical evidence, *de Wet et al.* examine the origins of limestones in the Mesozoic rift-basin fills of eastern North America. They show that the limestones, which have several origins (both as lacustrine and spring deposits), formed at times when the siliciclastic influx to the basins was reduced, and when there was an increased supply of calcium carbonate, including detrital components.

Lake Magadi, in the Kenya Rift, is one of the best-known lakes in Africa, and to limnogeologists it is famous for its trona deposits, authigenic zeolites, magadiite, and associated cherts. Two papers consider the cherts at Lake Magadi. *Behr* demonstrates that much of the chert at Magadi may have originated by microbial precipitation of silica, and did not have a magadiite precursor, as originally proposed by Eugster (1967). Relatively little chert at Magadi may be Magadi-type chert. *Kerrich et al.* have examined the trace-element geochemistry of cherts at Magadi, Olduvai, and other places in the rift, and show that alkaline lake cherts have a unique geochemical signature that will help their future recognition in the geological record.

What trends can we see from this selection of papers? If this selection is a representative sampling of current research, then clearly sedimentology in rift basins has progressed beyond basic facies description and the general relationships between sedimentary facies, structures, and fill geometry. Detailed local studies and the increasing application of geochemical methods (e.g., Hay and Kyser, 2001) are being used to advance our understanding of sedimentary processes and products in rifts. Future plans for the East African Rift include deep drilling of large rift lakes with the aim of recovering long paleoclimatic records that can be linked to those interpreted from deep-sea cores.

## *Editors' Perspectives*

There remain, however, many gaps in our understanding of rift sedimentology. From our perspective these include (1) the role of bedrock composition in affecting processes and the nature of deposits; (2) the character of soil development in volcanically active settings; (3) the importance of fine-grained eolian sediments; (4) the impact of colluvial processes on the sedimentary record; and (5) the significance of groundwater-fed freshwater wetlands, to name just a few.

### *Bedrock Composition*

Except for eolian dust and pyroclastic air fall, weathering of the bedrock in the drainage basin provides the detrital sediment that accumulates in continental rifts. In an influential paper, Yuretich (1982) contrasted the eastern branch of the East African Rift, where volcanic rocks are abundant, with the western branch, which is dominated by metamorphic basement rocks. He showed that the differences in bedrock have a major impact on the ionic composition of the basin waters, rift-basin fills, and lacustrine sedimentation. Bedrock lithology also affects siliciclastic sedimentation in colluvial, fluvial, and pedogenic environments in continental rifts. Weathering of quartz-poor lavas yields feldspathic silts, clay minerals, and gravels, whereas volcanic rock fragments (VRFs) and K-feldspar typically dominate the sand fraction. Feldspars break down rapidly to silt and fine sand during transportation, whereas VRFs originally deposited as sands are prone to early diagenetic alteration to mudstone. In contrast, metamorphic basement gneisses, schists, and quartzites typically produce plentiful quartz sand under the same environmental conditions (Fig. 2A, B).

These basic mineralogical differences should affect the sediment supply to the fluvial system in rifts, and consequently the nature of rift fills. Models that predict the distribution of sandstone hydrocarbon reservoirs in rift basins, for example, may be unsuitable in volcanic catchments. There is much scope for research into the role of bedrock composition in siliciclastic sedimentation in rift basins, besides its well-known controls on hydrochemistry.

### *Soil Development*

Paleosols may contain a wealth of environmental and climate data, but the record is time-averaged and episodic influx of pyroclastic material to the surface typically leads to immature and cumulative soils (Bestland and Retallack, 1997; Ashley and Driese, 2000) (Fig. 2C). We encourage the integration of pedogenic analysis with the more traditional lithologic analysis to provide a more complete understanding of environmental conditions in rift basins.

### *Eolian Dust*

The tendency has been for researchers on rift sedimentology to focus on fluvial and lacustrine environments. There is a general lack of appreciation of the importance of wind-blown sediments in rifts, except for the more obvious sand dunes (e.g., Olsen, 1986; Hubert and Mertz, 1984; Smoot, 1991). Nevertheless, the weathering of volcanic rocks produces copious fine-grained sediment, and eolian dust (loess) should be an important component of the sedimentary record (Fig. 2D). Understandably, it is difficult to recognize loess when mixed with fine-grained fluvial and lacustrine deposits. However, eolian deposits can indicate a negative hydrologic budget (more evaporation than precipitation); the ability to differentiate a wind-blown fraction from the water-transported fraction has important paleoenvironmental connotations.

### *Colluvial Facies*

Colluvium covers large areas of rift systems. Slopes, however, represent zones of sediment transport, and their deposits may not be preserved (Fig. 2E, F; Frostick and Reid, 1982; Reid and Frostick, 1986). Because hill slopes cover large portions of the rift region, equivalent in surface area to sites of accumulation (i.e., floodplains, eolian flats, deltas, and lakes), we can expect that a portion of the colluvial sediments will eventually be preserved in the geological record. For a more complete picture of the rift environment, there is a clear need for the development of colluvial facies models.

### *Freshwater Wetlands*

Springs and groundwater-fed wetlands (Thompson and Hamilton, 1983) are common in rifts, particularly associated with faults and bedrock fracture systems (Fig. 2G, H). They are an important part of rift ecology, but their preservation potential is low. Springs, particularly cool springs, may not leave a mineral record (Deocampo and Ashley, 1999), and much of the record from wetlands is likely to oxidize with time. There is a major knowledge gap in our understanding of the processes that ultimately lead to the geological record of springs and wetlands, particularly in arid settings.

These are just a few of many aspects of rift sedimentology that appear to be ripe for research. This list is far from exhaustive; every sedimentologist working in modern rifts and paleorifts could undoubtedly add many other topics. It is clear that there will be promising new avenues for investigation in the sedimentology of continental rifts for many years to come.

## REFERENCES


Andrews, P., and Banham, P., eds., 1999, Late Cenozoic Environments and Hominid Evolution; A Tribute to Bill Bishop: Geological Society, London, 276 p.

Ashley, G.M., and Driese, S.G., 2000, Paleopedology and paleohydrology of a volcaniclastic paleosol interval: implications for early Pleistocene stratigraphy and paleoclimate record, Olduvai Gorge, Tanzania: Journal of Sedimentary Research, v. 70, p. 1065–1080.

Bestland, E.A., and Retallack, G.J., 1993, Volcanically influenced calcareous paleosols from the Miocene Kiahera Formation, Rusinga Island, Kenya: Geological Society of London, Journal, v. 150, p. 293–310.

Deocampo, D.M., and Ashley, G.M., 1999, Siliceous islands in a carbonate sea: Modern and Pleistocene records of spring-fed wetlands in Ngorongoro Crater and Olduvai Gorge, Tanzania: Journal of Sedimentary Research, v. 69, p. 974–979.

Eugster, H.P., 1967, Hydrous sodium silicates from Lake Magadi, Kenya. Precursors of bedded chert: Science, v. 157, p. 1177-1180.

Frostick, L.E., and Reid, I., 1982, Alluvial processes, mass wasting and slope evolution in arid environments: Zeitschrift für Geomorphologie, Supplement Band 44, p. 53–67.

Frostick, L.E., Renaut, R.W., Reid, I., and Tiercelin, J.-J., 1986, Sedimentation in the African Rifts: Geological Society of London, Special Publication 25, 382 p.

Gierlowski-Kordesch, E.H., and Kelts, K.R., eds., 2000, Lake Basins through Space and Time: American Association of Petroleum Geologists, Studies in Geology no. 46, 648 p.

Hay, R.L., and Kyser, T.K., 2001, Chemical sedimentology and paleoenvironmental history of Lake Olduvai, a Pliocene lake in northern Tanzania: Geological Society of America, Bulletin, v. 113, p. 1505–1521.

Hubert, J.F., and Mertz, K.A., 1984, Eolian sandstones of the Upper Triassic–Lower Jurassic red beds of the Fundy Basin, Nova Scotia: Journal of Sedimentary Petrology, v. 54, p. 798–810.

Johnson, T.C., and Odada, E.O., eds., 1996, The Limnology, Climatology and Paleoclimatology of the East African Lakes: Amsterdam, Gordon & Breach, 664 p.

A
B
C
15 cm
D
E
F
C
LS
G
H

KATZ, B.J., ED., 1990, Lacustrine Basin Exploration—Case Studies and Modern Analogs: American Association of Petroleum Geologists, Memoir 50, 340 p.

LAMBIASE, J.J., 1990, A model for tectonic control of lacustrine stratigraphic sequences in continental rift basins, *in* Katz, B.J., ed., Lacustrine Basin Exploration—Case Studies and Modern Analogs: American Association of Petroleum Geologists, Memoir 50, p. 265–276.

LAMBIASE, J.J., ED., 1995, Hydrocarbon Habitat in Rift Basins: Geological Society of London, Special Publication 80, 381 p.

LEZZAR, K.E., TIERCELIN, J.-J., DE BATIST, M., COHEN, A.S., BANDORA, T., VAN RENSBERGEN, P., LE TURDU, C., MIFUNDU, W., AND KLERKX, J., 1996, New seismic stratigraphy and late Tertiary history of the North Tanganyika Basin, East African Rift system, deduced from multichannel and high resolution reflection seismic data and piston core evidence: Basin Research, v. 8, 1–28.

MORLEY, C.K., ED., 1999, Geoscience of Rift Systems—Evolution of East Africa: American Association of Petroleum Geologists, Studies in Geology, no. 44, 250 p.

OLSEN, P.E., 1986, A 40-million year lake record of early Mesozoic orbital climatic forcing: Science, v. 234, p. 842–848.

REID, I., AND FROSTICK, L.E., 1986, Slope processes, sediment derivation and landform evolution in a rift basin northern Kenya, *in* Frostick, L.E., Renaut, R.W., Reid, I., and Tiercelin, J.-J., eds., Sedimentation in the African Rifts: Geological Society of London, Special Publication 25, p. 99–111.

ROBBINS, E.I., 1983, Accumulation of fossil fuels and metallic minerals in active and ancient rift systems: Tectonophysics, v. 94, p. 633–658.

SCHOLZ, C.A., ROSENDAHL, B.R., AND SCOTT, D.L., 1990, Development of coarse-grained facies in lacustrine rift basins: examples from East Africa: Geology, v. 18, p. 140–144.

SMOOT, J.P., 1991, Sedimentary facies and depositional environments of Early Mesozoic Newark Supergroup basins, eastern North America: Palaeogeography, Palaeoclimatology, Palaeoecology, v. 84, p. 369–423.

THOMPSON, K., AND HAMILTON, A.C., 1983, Peatlands and swamps of the African continent, *in* Gore, A.J.P., Ecosystems of the World, Volume 4B: Mires: Swamp, Bog, Fen, and Moor, Regional Studies: Amsterdam, Elsevier, p. 331–373.

VRBA, E.S., DENTON, G.H., PARTRIDGE, I.C., AND BURCKLE, L.H., EDS., 1995, Paleoclimate and Evolution with Emphasis on Human Origins: New Haven, Connecticut, Yale University Press, p. 262–288.

YURETICH, R.F., 1982, Possible influences upon lake developments in the East Africa rift valleys: Journal of Geology, v. 90, p. 329–337.

## ACKNOWLEDGMENTS

The contents of this volume are a reflection not only of the researchers who contributed their results and ideas but also of the reviewers who helped to evaluate the data and interpretations presented. The editors of this volume gratefully acknowledge the time and effort given by the reviewers of the manuscripts in this compilation:

| | |
|---|---|
| Philip Barker | Victoria Hover |
| Charles Bristow | Jacqueline Huntoon |
| Paul Buchheim | Joseph Lambiase |
| Cathy Busby | Peter LeTourneau |
| Angeles Bustillo | Andrew D. Miall |
| Lluis Cabrera | Richard Murray |
| Joel Casanova | Gregory Nadon |
| Henry Chafetz | David Peacock |
| Amy Clifton | Vickie A. Pedone |
| Isabelle Cojan | Gregory Retallack |
| Richard Collier | Robert Riding |
| Nancye Dawers | Gary Smith |
| Grant Ferris | Charlotte Schreiber |
| Robert Gawthorpe | Kathryn Schubel |
| Margaret Gucchione | Frederick Schwab |
| Li Guo | Carl Swisher |
| Sanjeev Gupta | Torbjörn Törnqvist |
| Anthony Hall | Bruno Vendeville |
| Kurt Haberyan | Giday Woldegabriel |
| Robert Hecky | |

Thanks are also expressed to Robert Dalrymple, former SEPM Special Publications Editor, for his enthusiastic support of this project from the beginning. We are grateful to SEPM Special Publications Editor Laura Crossey, Kris Farnsworth (SEPM Publications Coordinator), and to Robert Clarke for his meticulous help with production. We are especially grateful to John Southard for technical editing, wise advice, and his exceptional patience.

←

FIG. 2 (opposite page).—Environments and topics for future research in rift basins. **A)** Ephemeral stream floored by loose quartzose sand, near Koobi Fora, Lake Turkana basin. The catchment contains basement rocks that have provided the sand. **B)** Ephemeral stream near Lake Baringo, Kenya, floored by lava gravel and fine sands and silts, composed mainly of feldspars derived from weathering of volcanic rocks. Coarse sand is rare because of the paucity of quartz. **C)** Paleosol development in tuffaceous claystones (1.75 Ma) at toe of pyroclastic fan, Olduvai basin, Tanzania (see Ashley and Driese, 2000). Prismatic structure, peds, and carbonate rhizoliths typify soil-forming processes in arid regions; light-colored sediments reflect zeolitization from saline, alkaline soil waters. Sample pit is 10 cm wide. **D)** Dust storm on the Sandai Plain north of Lake Bogoria, Kenya. The dust is mainly K-feldspar but also contains efflorescent $Na–CO_3$ salts. **E)** Vegetated colluvial slope on Ngorongoro Volcanic Highland, East African Rift, northern Tanzania. Although these surfaces tend to function as sediment transport zones, some colluvial deposits may end up in the geological record but not be recognized. **F)** Quaternary colluvial gravels and sands (C), several meters thick, disconformably overlie lacustrine sediments (LS) of the Miocene Tambach Formation, Kerio Escarpment road near Tambach, central Kenya Rift. **G)** A cool, freshwater spring discharging at base of slope yields a small perennial stream that supports lush wetland vegetation within the arid rift valley, Lake Natron, northern Tanzania. **H)** Loboi Swamp, a large (12 $km^2$) groundwater-fed wetland, located between Lakes Bogoria and Baringo, Kenya Rift Valley. A band of *Papyrus* in topographic low in mid-ground is bracketed by *Typha* vegetation on adjacent higher ground.

# CAUSES AND CONSEQUENCES OF LITHOSPHERIC EXTENSION: THE UPS AND DOWNS OF CONTINENTAL RIFTS

CYNTHIA EBINGER
*Department of Geology, Royal Holloway University of London, Egham, TW20 0EX, U.K.*
*e-mail: c.ebinger@gl.rhul.ac.uk*
AND
CAROLE PETIT AND EVGENE BUROV
*Laboratoire Tectonique ESA 7072, Université de Paris VI, Place Jussieu, Paris, Cedex 05, France*

**Abstract:** In this synthesis of current models for the initiation and evolution of continental rift zones, we discuss how the thermo-mechanical properties of the continental lithosphere produce topographic relief and control rift architecture. Our aim is to identify the key processes and parameters controlling the geometry and kinematics of continental rift basins, thereby providing a theoretical framework for fault arrays within rift basins. Case studies of the East African and Baikal rifts are used to illustrate rifting processes and the importance of lithospheric rheology in determining basin form.

On the basis of theory and observations, we suggest a rift classification scheme based on the lithospheric strength at the time of rifting. At the length scale of individual rift basins, lithospheric rheology also plays an important role. Lengths of border faults in cratonic parts of the East African and Baikal rifts are 80–120 km, whereas border faults that form in younger lithosphere or collapsing orogens are on the order of 30–60 km, indicating that the dimensions of rift basins are also controlled by the mechanical properties of the lithosphere. Variations in lithospheric thickness between continental domains probably have corresponding variations in thickness of the mechanical lithosphere. Once created, these lateral variations in rheology clearly will concentrate stress along terrane boundaries, suggesting that terrane boundaries are susceptible to later tectonic episodes. These results indicate that we can develop predictive models of the regional morphology and along-axis segmentation of active and ancient rift basins based on a relatively small data base: short, narrow basins with narrow uplifted flanks will form in young, hot, weak, lithosphere and be characterized by < 15 km seismogenic layer thickness / effective elastic thickness; long, wide, deep basins with broad uplifted flanks will develop in old, cold, strong continental lithosphere with seismogenic layer thickness / effective elastic thickness ≥ 30 km.

## INTRODUCTION

Tectonic processes in continental rift zones produce topographic relief and magmatism at length scales from $10^0$ to $10^6$ m (e.g., Ruppel, 1995). At the largest scale (> 1000 km), intracontinental plateaus and sag basins reflect thermal processes within the mantle lithosphere and asthenosphere. Magmatic processes may also contribute to topographic relief across a zone much broader than the rift valleys: volcanic construction across a 2000-km-wide zone may occur within a flood basalt province, and magmatic underplating may be widespread (e.g., White and McKenzie, 1989). At the scale of extensional basin systems (100–1000 km), density changes induced by the geometry of lithospheric necking produce vertical crustal movements and lateral pressure gradients that augment or counter far-field plate stresses. Magmatic underplating and explosive felsic volcanism may also occur at this scale. At the scale of individual basins and their uplifted flanks (< 100 km), normal and oblique-slip fault arrays produce a half-graben or full-graben basin morphology. Individual shield volcanoes, zones of intense dike injection, explosive felsic volcanism, and erosion modify the basin morphology as well as the density structure across the rift zone.

In this synthesis of current models for the initiation and evolution of continental rift zones at a regional scale, we discuss how the thermo-mechanical properties of the continental lithosphere produce topographic relief and control rift architecture. These lithospheric deformation processes are modulated by asthenospheric processes up until the breakup stage, when asthenospheric processes begin to dominate (e.g., Boutilier and Keen, 1999). The primary aim of this study is to identify the key processes and parameters controlling the geometry and kinematics of continental rift basins, thereby providing a theoretical framework for fault arrays in rift basins (e.g., Morley, this volume; Schlische, this volume). We emphasize factors controlling vertical and horizontal crustal movements through time and their characteristic length scales, attempting to simplify classification schemes for rift basins. This organization is intended to allow one to predict the response of the lithosphere to extension, given the conditions specific to individual rift basins. We commence with a summary of lithospheric and asthenospheric processes affecting the morphology of the rift basin and flanks, and then present case studies of two tectonically active rift systems: East Africa and Baikal.

## HORIZONTAL AND VERTICAL CRUSTAL MOVEMENTS

Isostasy, as well as asthenospheric and fault processes, governs the ups and downs of continental rift basins and flanks, and operates on a variety of length scales (e.g., McKenzie, 1978; Weissel and Karner, 1989). The process of lithospheric extension produces lateral variations in density structure across a continental rift zone, which can be considered as contributions from different levels (e.g., Fig. 1). These density changes, referenced to a normal lithosphere, are compensated isostatically on time scales of $10^3$ to $10^5$ years (e.g., Weissel and Karner, 1989). This is very slow in comparison to the time scales of coseismic and postseismic deformation, which are measured over periods < 500 years (e.g., Calais et al., 2002).

Take the case of a single border fault defining an asymmetric rift basin (Fig. 1). Crustal thinning, or necking, causes both an upwarp of the crust–mantle interface, as well as a basin at the surface which will be filled by water, air, sediments, and/or volcanic strata. The replacement of light crust with denser mantle lithosphere beneath the Moho upwarping tends to pull

Sedimentation in Continental Rifts
SEPM Special Publication No. 73, Copyright © 2002
SEPM (Society for Sedimentary Geology), ISBN 1-56576-082-4, p. 11–23.

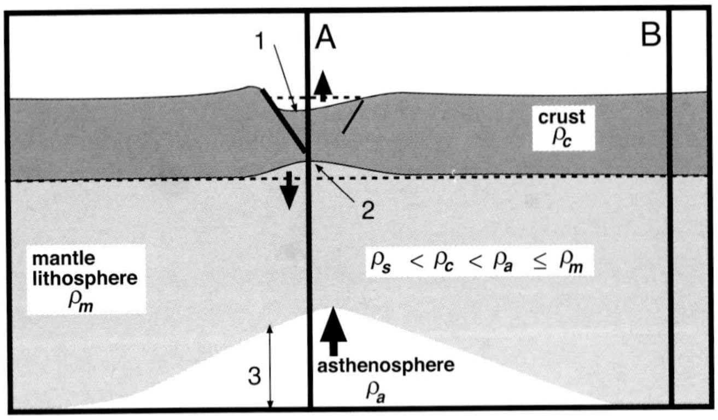


FIG. 1.—Schematic diagram illustrating lateral variations in density induced by the process of lithospheric extension. The magnitude of forces induced by these changes depends upon the geometry and density of sedimentary strata ($\rho_s$), crust ($\rho_c$), and upper mantle ($\rho_m$), as well as the degree of asthenospheric ($\rho_a$) and lithospheric heating. If in isostatic equilibrium, the weight of column B (reference lithosphere) equals that of column A (rift). 1. Replace crust by air or water or sediments = buoyancy force. 2. Replace crust by mantle lithosphere = downward force. 3. Replace mantle lithosphere by asthenosphere = buoyancy force.

the lithospheric section down, whereas the lighter basin fill produces the opposite effect. The amplitude of isostatic compensation for the fault-controlled basin subsidence depends on the geometry of crustal thinning and the density of the rift infill (e.g., Fig. 1).

The flexural rigidity, or mechanical strength, of the lithosphere (discussed below) determines the length scale and distribution of the rebound to crustal stretching (e.g., Weissel and Karner, 1989; Braun and Beaumont, 1989). If the plate has no strength (Airy isostasy), then the isostatic rebound occurs only across the basin, and there will be no uplifted rift flanks (e.g., McKenzie, 1978) (Fig. 2). The rebound of a strong plate will be distributed across a broader region than for a weak plate (e.g., Fig. 2). Small-scale convection induced by the steep gradients at the lithosphere–asthenosphere boundary beneath the rift margins may also contribute to rift flank uplift, but this thermal effect decays with time (e.g., Buck, 1986). The flexural uplift along the rift flanks does not decay with time, but it is subject to the competing process of erosion.

The density of the crust and upper mantle decreases with increasing temperature, and so lithospheric thinning and heating cause a reduction in density. Although the decrease in density at any one depth is small, it is the depth-integrated density contrast between stretched and unstretched lithosphere that determines the uplift; the depth extent may be 100 km or more in cratonic lithosphere. The subsequent uplift compensates the warpings of density contrasts and occurs over a broader region than the rift basin; the magnitude and shape of the uplift depends on the geometry of the mantle lithospheric thinning (e.g., Royden and Keen, 1980). This thermal uplift is transient; subsequent cooling of the heated lithosphere returns it to its original thickness, producing the post-rift thermal subsidence (e.g., McKenzie, 1978). The conduction of heat through the lithosphere is slow, and the thermal effects on the crust and sedimentary infill require millions of years to be observed at the surface (e.g., Nyblade et al., 1990; Lavier and Steckler, 1997).

In addition to isostatic compensation for density changes occurring in the extended lithosphere, there may be dynamic compensation for asthenospheric density variations in areas of convective upwellings or mantle plumes. Density changes in the convecting asthenosphere cause large-scale (> 1000 km) changes in continental topography, with amplitudes on the order of hundreds of meters to perhaps 1000 m (e.g., Lithgow-Bertelloni and Silver, 1998; Gurnis et al., 2000). This dynamic topography is compensated at great depth in the convecting asthenosphere. Dynamic, or non-isostatic, compensation may also occur at the crustal scale (e.g., Burov and Cloetingh, 1997). Finally, phase changes and serpentinization of mafic material may also contribute to vertical crustal movements in some rift locales (e.g., Louden and Chian, 1999).

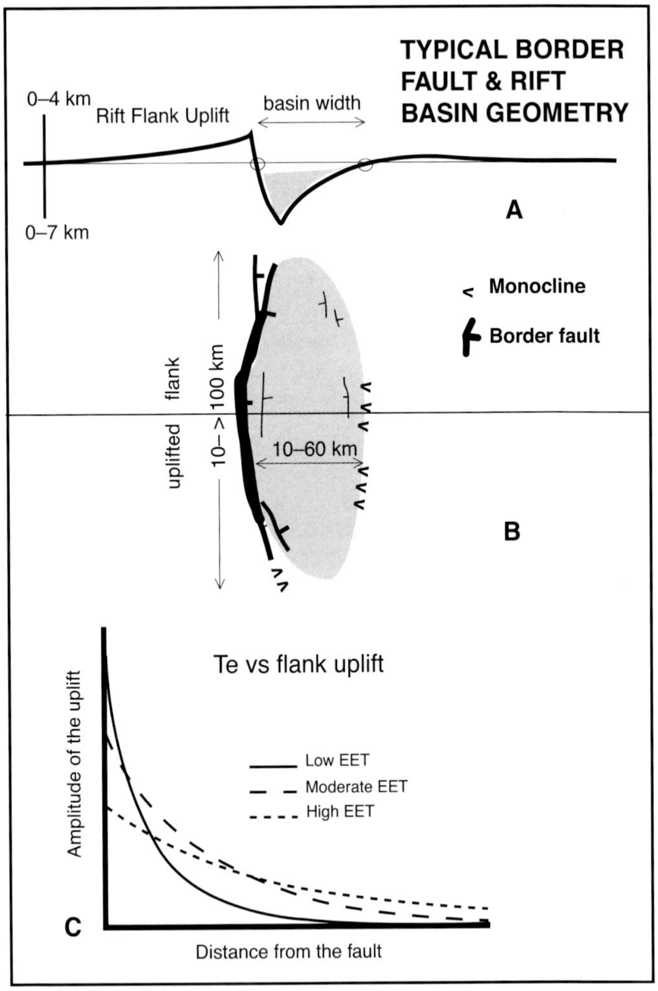


FIG. 2.—Generalized form of a tectonically active continental rift basin showing border faults in cross-section (A) and plan form (B). **A)** Most youthful single-phase rift basins are half-graben, with nearly all of the deformation accommodated by strain along the border fault. Flexural isostatic compensation leads to uplift of the rift flanks, with the breadth and height of the flanks dependent upon *Te*. The lower-relief margin may show minor faults, or be a monocline. **B)** Border faults show maximum displacements near their centers, with decreasing throws toward their tips. The approximate line of cross section A is shown by the bold line. The depression is shown by shading (after Ebinger et al., 1999).

## *Rheology*

The rheology of the lithological layers constituting the continental lithosphere is perhaps the single most important parameter controlling the response of the lithosphere to extension. Yet, the composition and hydration state of the lower continental crust and upper mantle are poorly determined in most continental regions. Experimental studies of crustal materials (quartz, feldspar, diabase) and upper mantle (olivine) have been extrapolated over seven orders of magnitude to the slow rates of geological processes, adding additional uncertainty (e.g., Kirby and Kronenberg, 1987). Nonetheless, numerical models that incorporate a multilayered rheology constrained by the rock-mechanics data reproduce the features of rift basins, suggesting that the extrapolation is reasonable (e.g., Burov and Diament, 1995; Buck et al., 1999) (Fig. 3A).

Numerical models of the cumulative yield strength of the compositionally and rheologically layered continental lithosphere have shown that its deformation behavior is governed principally by its thermal state and crustal composition, and, to a lesser but significant extent, by strain rate and plate curvature (e.g., Kusznir and Karner, 1985; Kusznir and Park, 1987; Burov and Diament, 1995; Bassi, 1995; Hopper and Buck, 1996). The yield-strength envelope of continental lithosphere places bounds on the elastic stresses the lithosphere can support (Fig. 3). The brittle stress increases linearly with depth, with the slope of the line depending on pore-fluid pressure, friction, and weight of the overlying rocks (Brace and Kohlstedt, 1980) (Fig. 3B). At high temperatures, ductile flow may occur in the lower crust and mantle lithosphere; the stress required to initiate flow depends on composition and temperature (e.g., Kusznir and Park, 1986; Hopper and Buck, 1996). Variations of the crustal thickness and composition lead to mechanical coupling (mafic and/or thin crust) or decoupling (quartz-dominated and/or thick crust) between resistant crust and mantle layers (e.g., Burov and Cloetingh, 1997; Lavier and Steckler, 1997).

Recent investigations of the behavior of the lower continental crust in extension suggest that the rheology and probably the composition of the lower crust is highly variable, leading to differences in the breadth of the rift zone, as well as the morphology of basins. In areas with high geothermal gradients and quartzofeldspathic composition the lower crust is too weak to sustain shear stresses and undergoes viscous flow (e.g., Hopper and Buck, 1996; Kruse et al., 1997) (Fig. 3A). In cratonic areas with very low geothermal gradients and diabase lower crust, the lower crust may behave in a brittle manner throughout its depth extent (e.g., Déverchère et al., 1993; Foster and Jackson, 1998).

Braun and Beaumont (1989) simulated lithospheric extension, or necking, and examined the effects of a two-layered rheological model on vertical crustal movements across rift zones. They showed that the distribution of thinning and the resultant vertical crustal movements are strongly affected by the position of the layer with maximum strength. This *necking depth* is the level where no vertical movements would occur prior to isostatic compensation. There may be net upward or downward deflections of the basin and flank system, depending upon lithospheric rheology (e.g., Braun and Beaumont, 1989; Watts and Stewart, 1998). In application to multilayered models (e.g., Fig. 4), each layer may have a necking level, and necking depth becomes ambiguous or even misleading.

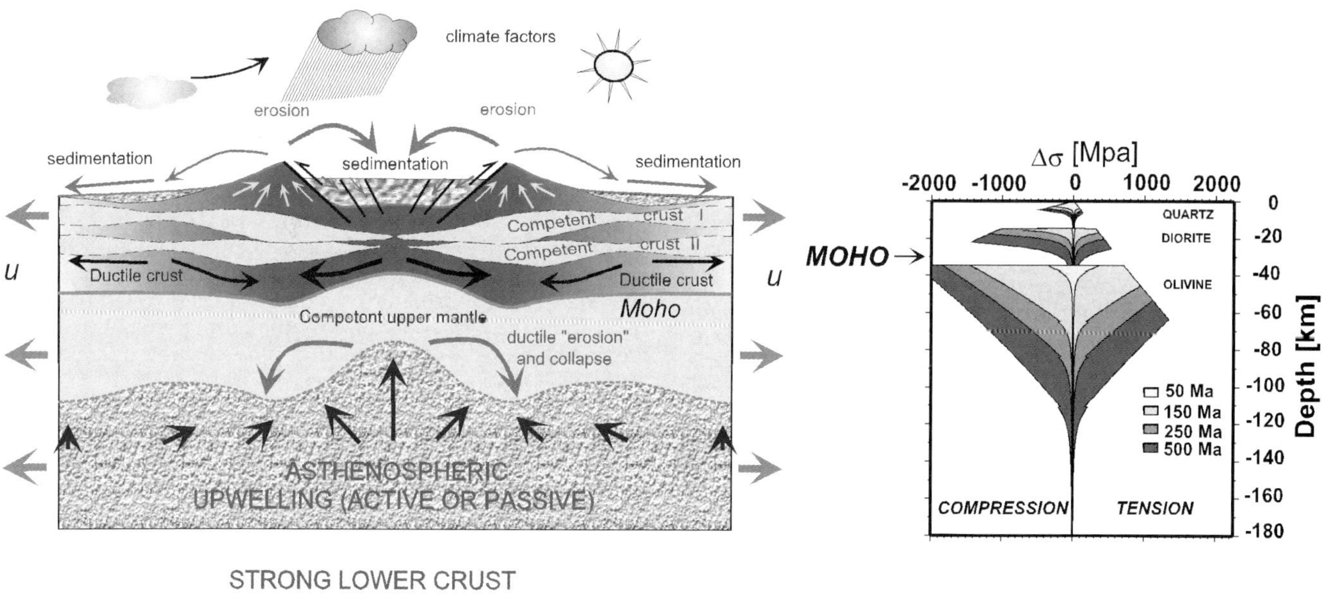


FIG. 3.—Generalized model of lithospheric extension processes illustrating the effects of rheological layering (after Burov and Cloetingh, 1997, and Burov and Poliakov, 2001). Sedimentary loads derived from erosion of rift flanks accumulate in the rift basin. The stronger parts of the crust and mantle lithosphere are deflected downward and to higher geothermal gradients by the weight of this load. The resultant elevation in temperatures leads to a reduction in plate strength beneath the load, but there is a time lag between extension and response to the loading. Strong (50–70%) variations in the thickness of the crustal brittle layers shown in the figure result not only from thinning and heating but also from plastic yielding under flexural and tensional stresses. Note that the brittle layers are flexurally weakened not only below the basin but also immediately outside the rift flanks (Burov and Diament, 1995; Burov and Cloetingh, 1997). The lower crust may flow from beneath the basin towards the shoulders, enhancing uplift. Right-hand panel shows yield-stress envelope for continental lithosphere for a number of thermal ages: hotter lithosphere is considerably weaker than old, cold, lithosphere. Δσ is deviatoric stress.

In addition, one must remember that the lithosphere can show mechanical anisotropy, caused by inherited, "frozen" olivine fabric of the upper mantle (Vauchez et al., 1997). Frequently, rift zones develop parallel to the preexisting orogenic fabric of continental plates, suggesting that mechanical strength is lower perpendicular to inherited fabric.

### *Horizontal Forces Induced by Rifting*

Lateral variations in density lead to horizontal pressure gradients that augment or counter the far-field, or plate-driving, forces acting across the rift zones (e.g., Fig. 1). In many rift zones, particularly collapsing orogenic belts, the positive topographic relief and comparatively lower-density upper mantle cause a horizontal deviatoric stress. Where the buoyancy forces caused by lithospheric heating and crustal thinning, combined with far-field stresses, are larger than the yield strength of the lithosphere, rifts form (e.g., Buck, 1991).

The deposition of sedimentary rocks and the accumulation of water above thinned continental crust produces a load on the lithosphere (e.g., Figs. 2, 3). Although this load is less dense than the original crustal material, isostatic compensation generally occurs more rapidly than does the infilling of the basin by sedimentary strata. The sedimentary strata, therefore, are a positive load. The weight of the basin fill may lead to lateral flow of material beneath the basin, effectively smoothing the relief at the base of the crust and delocalizing strain (e.g., Hopper and Buck, 1996) (Fig. 3). Thus, lithospheric rheology plays a critical role at the scale of rift zones, as well as in the evolution of discrete basins. An alternative but equivalent approach is to consider the *gravitational potential energy*, which is the difference between the depth-integrated weight of a lithospheric column in the rift and in a reference column (e.g., England and Houseman, 1986) (Fig. 1).

### *Erosion and Deposition*

Erosion and deposition in rift systems compete with and augment aspects of the extensional processes described above. First, erosion of the rift flanks and intrabasinal highs selectively thins the competent, brittle upper crust, effectively weakening the lithosphere (e.g., Burov and Cloetingh, 1997) (Figs. 3, 4). This erosion occurs over time periods of millions of years, and involves both the erosional retreat of the border fault, as well as drainage away from the rift along the uplifted rift shoulders (e.g., van der Beek et al., 1998). Isostatic rebound in response to the denudation restores the topographic relief to slightly less than its elevation prior to erosion, thereby maintaining the form of the rift shoulders through time. Where the lower crust is weak and deforms by ductile flow, the rebound to erosion along the rift flanks may be compensated by the outward flow of lower crust from beneath the rift, accelerating subsidence (and infilling). The effect is to amplify the effects of crustal thinning across the rift and to decrease the strength of the lithosphere (Burov and Cloetingh, 1997). If lower-crust flow has occurred, then estimates of pre-rift and syn-rift crustal thickness do not represent the total extension across the basin.

The accumulation of thick sedimentary sequences in rift basins tends to reduce elastic plate thickness, for several reasons. The top of crystalline basement beneath the rift is deeper than adjacent crust, and therefore experiences higher geothermal gradients; the sedimentary strata insulate the crust; and the lighter load of the sediments exerts a stress on the plate (e.g., Lavier and Steckler, 1997). Inelastic flexure under the accumulating sedimentary load results in zones of localized yielding (Burov and Cloetingh, 1997).

### *Modification by Magmatic Processes*

We briefly highlight the consequences of magmatism on vertical crustal movements in rift zones. In plume provinces, hot, low-density material rises diapirically through the mantle and interacts with the base of the lithosphere, forming a broad head that may partially melt to produce flood basalts and radial dikes (e.g., Ribe and Christensen, 1994). This buoyant material displaces denser, normal asthenosphere and heats the lithosphere, leading to regional uplift, lithospheric thinning, and pressure-release magmatism (e.g., White and McKenzie, 1989). Production of flood-basalt melt requires mantle potential temperatures of 1400–1600°C and lithospheric thinning to ca. 90 km (or preexisting lithospheric thickness no more than) to allow adiabatic decompression melting. Extension by dike injection may locally increase density in the rift, leading to subsidence.

Silicic volcanism in plume and nonplume rifts reveals the importance of high-level magma chambers and crustal melting in rift zones. These magma chambers may decrease the density of the crust, leading to local uplift. The intrusion of magmatic material may increase the density at shallow levels in the rift zone, leading to a net downward force; magmatic underplating effectively thickens the crust, leading to a buoyancy force that may be regional (e.g., McKenzie, 1984) or local (e.g., Birt et al., 1997).

## CONSTRAINTS ON RHEOLOGY

### *Seismogenic Layer Thickness*

The *seismogenic thickness* is the depth extent of earthquakes in a particular area averaged over periods of years. Scholz (1988) equates the seismogenic thickness with the thickness of the brittle upper crust, allowing for a strong uppermost mantle beneath a weak lower crust. We can see that effective elastic thickness and seismogenic layer thickness should show similar regional variations, but the actual values may differ depending on lithospheric rheology.

### *Effective Elastic Thickness*

Forward, or direct, models provide important insights into the critical parameters controlling the general form of rift basins, but they are difficult to compare with detailed observations from rift basins because they require the above parameters, which are poorly known in most areas. An alternative approach to forward modeling is to estimate the integrated yield strength of the continental lithosphere: flexural rigidity, or an effective elastic plate thickness, $Te$. In the case of mechanical coupling, the lithosphere behaves as a single layer whose effective elastic thickness is close to the depth of the mechanical lithosphere. In the case of mechanical decoupling, the lithosphere behaves as two horizontal elastic leaves separated by a ductile layer, which causes a drastic decrease (up to 50%) in $Te$ (Burov and Diament, 1995). The effective elastic thickness, therefore, does not correspond to a particular geological boundary in the continents.

Gravity anomalies and topographic relief provide strong constraints on the deformation behavior of the continental lithosphere, allowing us to differentiate between different rift models. Flexure and/or displacement of the surface and crust–mantle boundaries, where large density contrasts exist, in response to horizontal and vertical forces, produce measurable gravity-field variations, which are commonly expressed in free air or Bouguer gravity anomalies.

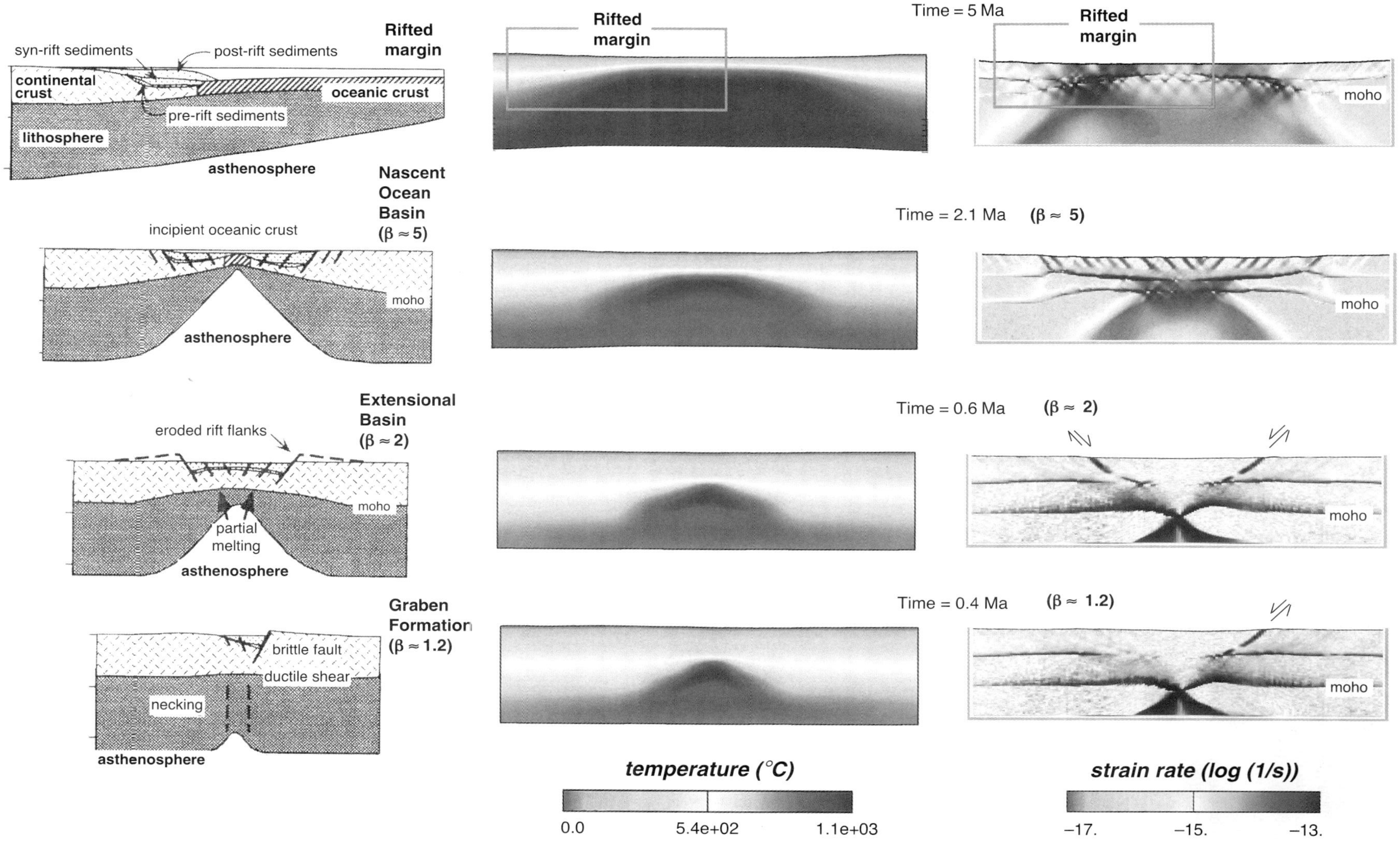


FIG. 4.—Numerical simulation of the rift to breakup evolution of a continental rift system assuming a rheologically layered lithosphere and faulting (adopted from Burov and Poliakov, 2001). The left-hand panels show a qualitative summary of rift evolution (after Salveson, 1978). The central panels show the thermal structure of the stretched and unstretched lithosphere, which has a thermotectonic age of 250 Ma. The right-hand panels show strain rate localization (active faults and shear zones) produced by numerical simulations of a lithosphere of intermediate age. The middle diagram shows evolution of the temperature field during the simulations. Note the strongly two-dimensional character of the temperature field and the sharp localization at the top during the earlier stages of rift evolution. These relations result from the nonlinear rheology used (see Fig. 3). Note also that the patterns of strain rate and temperature shown here correspond to different stages of the synrift phase. The geometry of faults and shear zones corresponds to those that were active in the corresponding time snapshot. Some of these faults and especially shear zones may be later deactivated and healed and not seen on current seismic data. Note: In the post-rift stage, gravity and temperature driven stresses may change the fault patterns (Burov and Poliakov, 2001). This effect (i.e., post-rift compression) was included in the computations but is not shown. β is extension factor.

Forward models and spectral analyses of free-air or Bouguer gravity and topography have been used to estimate *Te* in a number of rifts and along passive continental margins, with *Te* estimates ranging from 0 km (McKenzie, 1978) to > 30 km (e.g., Ruppel, 1995; Simons et al., 2000; Petit and Ebinger, 2000). The wide range of estimates in part reflects assumptions made on the compensation depth and rheology of the lower crust (e.g., Simons et al., 2000; Petit and Ebinger, 2000).

Contrary to current theory, Maggi et al. (2000) cite the absence of well located earthquakes in the upper mantle as evidence for a weak upper mantle. Maggi et al. (2000) suggest that the presence of small melt fractions in the upper mantle blocked by an impermeable granulitic lower crust make the cratonic mantle lithosphere weak. The Maggi et al. (2000) argument is provocative, in that evidence for a strong upper mantle is indirectly determined: forward and inverse estimates of elastic plate thickness in rifts and orogens that are greater than crustal thickness (e.g., Cloetingh and Burov, 1996; Petit and Ebinger, 2000); studies of heat flow and S-wave propagation that indicate a high-viscosity upper mantle (e.g., Nyblade et al., 1990; Ritsema et al., 1998; Simons et al., 2000); the preservation of high-density upper crustal loads (e.g., batholiths) that require a high-viscosity upper mantle (e.g., Burov et al., 1998); and upper-mantle earthquakes (e.g., Déverchère et al., 1993).

We consider the absence of deep earthquakes as an argument either (1) in favor of high brittle yield strength of the upper mantle, or (2) in favor of a more complex rheological behavior at depth (e.g., Ranalli, 1995). If we extrapolate Byerlee's law to mantle pressures, tectonically induced stress never exceeds brittle yield stress limits in the upper mantle (Byerlee, 1978). In-as-much as it is currently impossible to test the applicability of Byerlee's law at great depth, it is difficult to draw conclusions from the presence or absence of upper-mantle earthquakes in continental rifts. Further considerations are the long-term chemical and mechanical properties of cratonic lithosphere, which would be difficult to achieve if the upper mantle has no strength.

## CURRENT RIFT MODELS

Mantle plume or convective upwellings drive extension in classic "active" rifting models, with regional uplift and synrift volcanism diagnostic features (e.g., White and McKenzie, 1989). "Passive" rifts are driven by far-field stresses (e.g., Sengör and Burke, 1978). These rifts form at or near sealevel, and have little or no volcanism during the synrift stage. As more precise geochronological techniques become available for rift studies, we find that uplift–subsidence patterns and timing of volcanism do not necessarily fit these models of the driving mechanism, and exceptions abound (e.g., Ruppel, 1995; Menzies et al., 2000). As discussed by Huismans et al. (2001), passive extension may induce small-scale convection (active rifting), and active processes may promote passive extension in nearby regions. In addition, neither "active" nor "passive" modes includes the large category of continental rift basins that developed on collapsing orogenic belts (e.g., Aegean, Basin and Range). We instead recommend the classification on the basis of mechanical properties of the plate through time, discussed below.

A wide variety of structural styles have been observed in active and ancient rift basins worldwide. At the largest scale, the distribution and style of strain in rift zones is controlled by lithospheric rheology (e.g., England, 1983; Zuber and Parmentier, 1986). Buck (1991) demonstrates that three modes of extension (discrete rifts, diffuse rifts, low-angle detachments or core complexes) may develop for the same plate driving forces, depending on the initial thermal and compositional properties of the lithosphere undergoing extension. *Discrete rifts* comprise a narrow zone of deformation where basins are deep and broad. These rifts form where the mechanical lithosphere is cold, thick, and strong (e.g., Buck, 1991). A number of models have been proposed to explain broad provinces of subparallel extensional basins, *diffuse rifts*, including strain hardening at low strain rates (England, 1983) and lithospheric boudinage (e.g., Zuber and Parmentier, 1986). Expanding on work by Hopper and Buck (1996), Buck et al. (1999) analyze the development of faults and necking, and show that some wide rifts develop where the crust is thick and/or the lithosphere is thin, independently of strain rate. The major difference between diffuse rifts and narrow rifts is that thicker crust leads to greater buoyancy forces for the same geothermal gradient. As rifting approaches continental breakup the zone of active extension should localize, and diffuse rifts should evolve to narrow rifts (e.g., Hopper and Buck, 1996). Important to note is that only a small combination of parameters allow one to delocalize crust and mantle lithospheric thinning (Buck et al., 1999), suggesting that conceptual models of simple lithospheric shear (e.g., Wernicke, 1985) are unlikely to occur. Diffuse rifts, therefore, include collapsing orogenic belts where crustal thicknesses are anomalously large (> 50 km) but where geothermal gradients may be high. Clearly, wide rifts can also be produced in other more complicated settings, for example, where the zone of rifting migrates or jumps (e.g., Morley, 1994).

Core complexes represent a small class of rift basins, and models for their creation are debated. Large offsets along low-angle detachments expose mid-crustal material, and the low fault dips commonly observed are believed by many to reflect flexural rotation of the footwall and overlying strata following denudation along the fault (e.g., Forsyth, 1992). Focal mechanism solutions showing low-angle faults from the Woodlark basin (Abers et al., 1997) and the Gulf of Corinth (Rietbrock et al., 1996) suggest that some low-angle faults are active, contrary to rock-mechanics studies (e.g., Anderson, 1951). Lavier et al. (1999) simulate the development of faults in a brittle layer with different values of cohesion, and find that large offsets along a single detachment can develop only where the fault zone is very weak. More recently, Burov and Poliakov (2001) simulate extension of a rheologically layered lithosphere and observe the development of shear zones (faults). Their models reproduce diffuse and discrete rifts with combinations of weak–strong upper crust, lower crust, and upper mantle layers (e.g., Fig. 4), indicating the necessity of coupled rheological and surface-processes models. Chéry (2001) suggests that extensional fault systems in high-strain rifts may involve high-angle normal faults in the upper crust and low-angle decollements in the lower crust that are exhumed in core complexes. After the initial graben-forming faults develop, the low-angle decollements develop at mid-crustal levels owing to a combination of high differential stress, stress rotation, and high pore pressures (Chéry, 2001). These geologically realistic but necessarily complex models indicate that lithospheric rheology and thermal state largely control the distribution of strain across the plate boundary.

The mechanical properties of the lithosphere also play a critical role in the three-dimensional geometry of discrete, fault-controlled rift basins, which are the components of narrow, diffuse, and core-complex rifts. Observations from continental rifts reveal that border faults control basin geometry during the early stages of rifting (e.g., Bosworth, 1985; Ebinger et al., 1999) and that they accommodate most of the strain across the rift (e.g., Morley, 1988). Basins, which are commonly asymmetric, are bounded along one or both sides by border faults and are flanked by broad uplifts of amplitude 500–3000 m (Fig. 2). The breadth and height of the flanks depend on lithospheric strength and the density of the infilling material, with volcanic infill pulling down the flanks (e.g., Weissel and Karner, 1989) (Fig. 2). Ebinger et al. (1999) compare border-fault lengths with *Te* and seismogenic layer thickness in a number of rifts worldwide, and find that the fault length increases linearly with either parameter. The length scales of the along-axis segmentation of continental rifts reflects the mechanical properties of the lithosphere; longer faults and deeper basins develop where elastic / seismogenic layer thickness is greater (e.g., Hayward and Ebinger, 1996; Foster and Jackson, 1998; Lavier et al., 1999).

## CASE STUDIES OF THE EAST AFRICA AND BAIKAL RIFT

We now compare models and observations from the East African and Baikal rift systems, and discuss the general implications for sedimentary sequences in continental rifts. Our aim is to illustrate the general concepts discussed above through case studies of wide and diffuse rifts in East Africa and Baikal.

### *Constraints on East African Model*

The East African rift system extends over 3000 km of the African continent, and comprises incipient basins and new oceanic spreading centers (Fig. 5). The rift system north of Lake Turkana has been affected by one or more mantle plumes since Eocene time (e.g., George et al., 1998; Ebinger and Sleep, 1998), and its morphology is strongly influenced by Eocene to Recent magmatism. We concentrate on the largely fault-controlled and isostatically controlled basins lying on the broad East African plateau, where magmatic construction and probably magmatic intrusion has been minor (Fig. 5) to illustrate vertical and horizontal crustal movements in rift zones.

The East African rift system south of Lake Turkana developed in Archean to late Precambrian lithosphere during Miocene–Recent times (e.g., Ebinger et al., 1989; Morley et al., 1994). The Eastern rift, and probably the Western rift, youngs to the south where basins developed during the past 5 Myr (e.g., Foster et al., 1997). The Western and Eastern rift systems transect the > 1000-m-high East African plateau with the Archean Tanzania craton in its center (Figure 5A). Analyses of the long-wavelength gravity field over Africa suggest that more than half of the plateau uplift is dynamically supported by asthenospheric processes (Ebinger et al., 1989; Lithgow-Bertelloni and Silver, 1998). The Eastern rift, barring the Eyasi basin, lies in a late Precambrian orogenic belt (Fig. 5). Rift basins in East Africa are unusually broad and long: basin widths are 40–70 km, and lengths (excluding multiphase rift basins) are > 80 km, indicating that the elastic lithosphere is thick beneath these young basins (e.g., Jackson and Blenkinsop, 1997; Scholz and Contreras, 1998).

The high-velocity, mantle lithospheric root beneath the Archean craton that continues to depths of 200 km or more (Ritsema et al., 1998) probably does not contribute to the observed gravity anomaly because the low-temperature effects of the root are balanced by its depleted composition (e.g., Doin et al., 1996). The seismic results are very similar to estimates of lithospheric thickness of > 170 km or more provided by petrographic and geochemical analyses of Cenozoic mantle (Chesley et al., 2000). Heat-flow data from the Archean craton and surrounding Early Proterozoic (Western rift) and Late Proterozoic (Eastern rift) orogenic belts show little or no elevation relative to global means, arguing against significant lithospheric thinning in Cenozoic time (Nyblade et al., 1990).

Estimates of crustal extension across the central Eastern rift south of Lake Turkana, where the East African rift overprints Cretaceous–Paleogene rifts, from seismic-refraction, receiver-function, and fault studies show that crustal thinning is less than 10% (e.g., Birt et al., 1997). Extension across the Western rift is estimated at less than 20%, on the basis of analyses of fault patterns (e.g., Morley, 1994). Strain rate in the East African rift system is estimated from extensional velocities ~ 6 mm / yr in East Africa (Chu and Gordon, 1999). Estimates of erosional denudation along the rift flanks from analyses of apatite fission-track data are small (< 200 m), indicating that the observed topographic relief is the response of the lithosphere to tectonic processes (van der Beek et al., 1998; Noble, 1996).

### *Constraints on Baikal Model*

The Baikal rift zone is currently developing in the Asian plate, closely following a Paleozoic suture delineating the Archean Siberian craton (Logatchev and Zorin, 1992) (Fig. 5B). This suture was formed by accretion of various terranes against the craton during Paleozoic and Mesozoic orogenic episodes (e.g., Melnikov et al., 1994; Delvaux et al., 1995). The deepest rift basins, in the southern and central parts of the rift, are occupied by Lake Baikal and are largely superimposed on the suture zone bounding the Siberian platform (Figure 5B). The northeastern part of the rift comprises several discontinuous basins and is located in the Mesozoic Sayan–Baikal orogenic belt. Thus, the Baikal rift system encompasses basins in the "narrow rift mode" as well as basins in the "diffuse rift mode".

Analyses of sedimentary strata and structural data indicate that shallow depressions and small mountain ranges started to develop in the Late Oligocene–Early Miocene under a transtensional stress regime associated with low subsidence rates. The main stage of rift basin development started in the late Pliocene and is characterized by increasing subsidence and uplift rates, and a stable extensional stress regime (Hutchinson et al., 1992; Petit et al., 1996; Delvaux et al., 1997). Detailed studies of local and teleseismic activity show that earthquakes can occur in the crust down to 30 km and possibly in the upper mantle down to the depth of 40 km (Déverchère et al., 1991; Déverchère et al., 1993). The broad Central Baikal basin is bounded by the ~ 120 km-long Primorskiy border fault (Agar and Klitgord, 1995). Border fault lengths south of Lake Baikal (diffuse rift) are roughly half that length (e.g., Petit et al., 1996).

Heat-flow data depict a cold geotherm (40–60 mW / m$^2$) with only locally elevated values linked to hydrothermal circulation in the rift basins (Lysak, 1992). Analyses of mantle xenoliths (Ionov et al., 1995) and forward models of gravity data provide little evidence for lithospheric thinning or weakening, arguing against dynamic mantle upwelling (Ruppel et al., 1993; Burov et al., 1994; van der Beek, 1997). These forward models reveal great strength of the lithosphere beneath Baikal, with *Te* ranging from 45 to 60 km.

Recent global high-resolution tomographic models (e.g., Grand et al., 1997; van der Hilst et al., 1997) show that the seismic velocity structure beneath the Baikal and East African rifts is quite different. For instance, the East African rift is underlain by a narrow, large negative velocity anomaly that extends throughout the upper and lower mantle (e.g., Keller et al., 1994), while nothing comparable is observed beneath the Baikal area. Seismic velocity structure beneath the two cratons, however, is very similar: both show deep, high-velocity roots, arguing against significant lithospheric heating during Cenozoic time (Petit et al., 1997; Ritsema et al., 1998).

### *Thermo-Mechanical Models of East Africa and Baikal Rifts*

We prepared profiles of gravity and topography data across the East African rift system, choosing profiles that crossed the rifts, Archean cratons, and younger orogenic belts (Fig. 5; see also

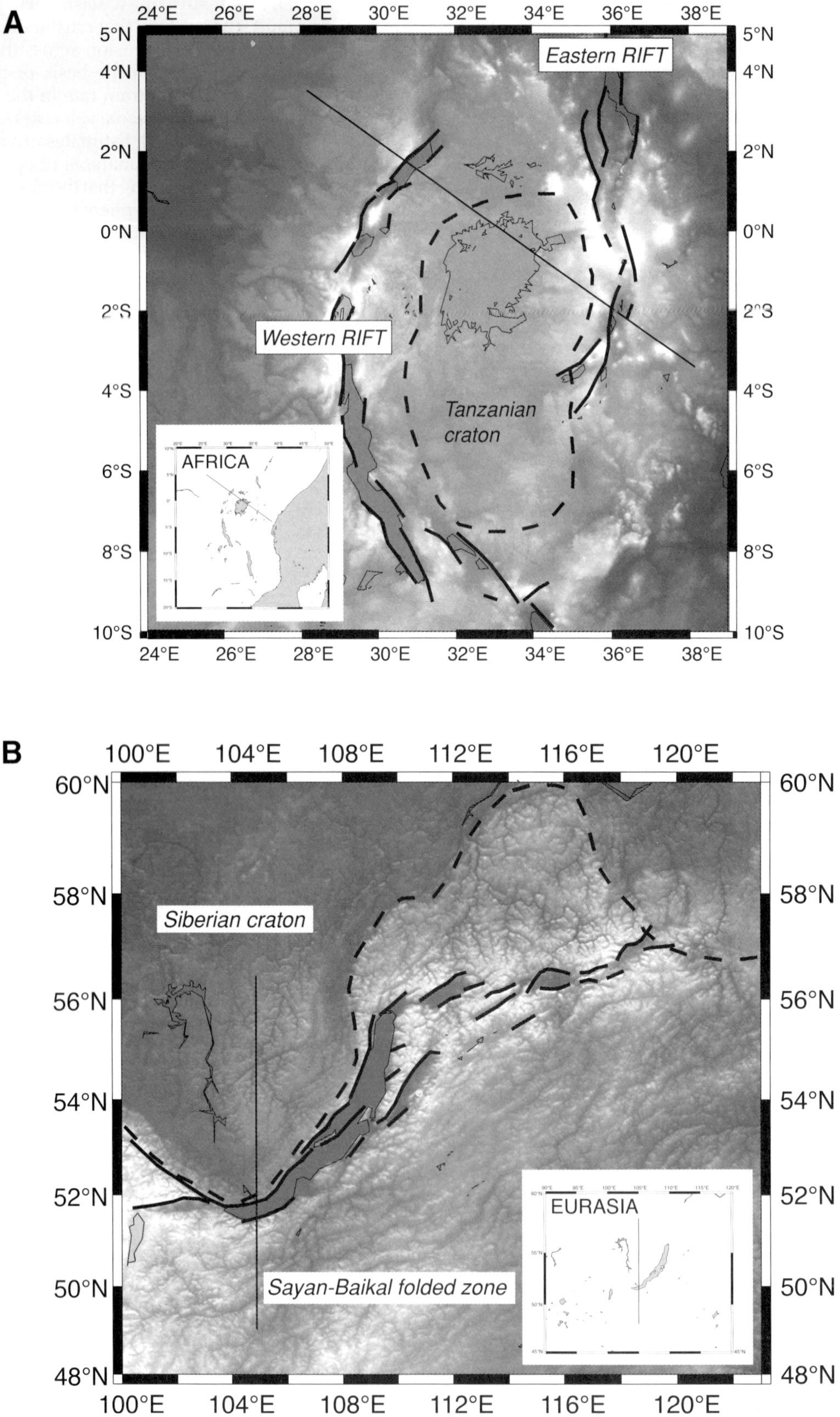


FIG. 5.—Tectonic setting of the **A)** East African rift and **B)** Baikal rift. Thick lines represent the main active faults. Profiles shown in Figure 6 are indicated.

Petit and Ebinger, 2000). The East African rift profile (Figs. 5, 6) crosses the central Tanganyika rift, the Archean Tanzania craton, and the Pan-African orogenic belt on the eastern seaboard of Africa. It crosses the central Western rift and the youthful southern part of the Eastern rift. The absence of a faulted rift valley along the eastern side of the craton allows us to examine the inherent differences in lithospheric rheology related to thermal age of the lithosphere. The Baikal profile crosses the Archean craton, the south-central basins of the Lake Baikal, and Phanerozoic compressional belts. Important to note is that these profiles have had the effects of deep mantle dynamic processes removed prior to modeling (e.g., Ebinger et al., 1989).

Neither models that assume Airy isostasy nor models that assume a laterally constant lithospheric rheology can explain the observed gravity anomalies (Petit and Ebinger, 2000). Considering the plate tectonic settings of these two locales, we instead assume a laterally varying rheology and compare model and observed gravity and topography curves across these representative profiles of the two rift systems. A complete discussion of these results is found in Burov and Diament (1995) and Petit and Ebinger (2000).

We match the observed Bouguer anomaly by allowing deflection of the crust–mantle interface resulting from regional compensation of the surface topography acting as a vertical, horizontally distributed load and boundary forces, or plate-driving forces, produced by the interaction between lithospheric plates. Observational constraints are used to estimate the far-field and density-driven forces.

Plate bending may lead to inelastic (brittle or ductile) deformation where bending stresses exceed the yielding limits. In other words, *Te* depends on deviatoric stresses caused by plate bending, so there is an interdependence between local strength and plate structure. Plate discontinuities can be introduced in the models to simulate major weaknesses where flexure causes significant vertical differential movements affecting the whole crust and the uppermost mantle. Deflections of the lithosphere–asthenosphere boundary contribute little to the gravity anomaly, inasmuch as the density contrast is $< 1/6$ that of the crust–mantle boundary. The vertical forces applied at the plate ends simulate the combined effects of sediment and water loading, crustal thinning, denudation along the fault, and frictional resistance between the two plates (Burov et al., 1994). They are applied on both sides of plate discontinuities in order to reproduce the plate flexure, as indicated by the Bouguer gravity anomaly. The model output is the deflection of the Moho, which is used to calculate a synthetic Bouguer gravity anomaly. Important to note is that deflection of the Moho does not necessarily mirror that of the surface as in an elastic model, because of possible crust–mantle decoupling and lower crustal flow (e.g., Figs. 3, 4).

Predicted profiles of gravity anomalies across the Baikal and East African rifts that include lateral variations in plate strength match the observed Bouguer gravity anomalies, indicating that most of the short-wavelength topography is supported by upward flexure of the lithosphere in response to denudation along the border fault. Features at length scales < 10 km are uncompensated and are probably related to intrabasinal structures not considered in these flexural models. As further tests of the model, the Moho deflections are ~ 3 km for the East African and Baikal rift basins, consistent with seismic observations of 3–4 km of crustal thinning across the southern part of the Eastern rift (e.g., Birt et al., 1997). The implications for vertical forces on the plates in the East African and Baikal rift valleys are in good agreement with those obtained from simple uniform elastic plate models in the Western rift (e.g., Ebinger et al., 1989) and the Baikal rift (van der Beek, 1997). These authors showed that flexural rebound accounts for most of the intermediate- to short-wavelength topographic relief.

Conversely, the forward model of the south-central Baikal rift shows that the high topography encountered south of the Siberian craton is associated with downward plate flexure and crustal thickening, while upward flexure occurs on the margin of the Siberian craton (Figs. 5, 6). This illustrates the need to model simultaneously the topography and gravity, because similar topographic profiles (Figures 6A and 6B) may in fact be associated with completely different modes of compensation.

The effective elastic thickness corresponding to the resulting yield stress across the rift (Figure 6) shows that the cratons appear strong and undeformed in both rifts. The lithosphere of the Archean craton is significantly stronger than that beneath the extensional provinces: *Te* is ~ 60 km across the uplifted plateau between the Western and Eastern rift systems and ~ 45 km beneath the Precambrian orogenic belts surrounding the craton (Figure 6). Our results argue against significant heating or thinning of the old, cold lithosphere, regardless of whether extension is driven by mantle upwelling or far-field plate kinematic processes. In both the Baikal and East African rifts, the major part of the deformation is concentrated along the borders of the craton, i.e., where our models show the greatest rheological contrast; the inherent lateral variations in density structure across the craton boundary would make it susceptible to deformation, and could explain the along-axis propagation of the Eastern rift on or near the suture zone (e.g., Birt et al., 1997).

## IMPLICATIONS FOR RIFT-BASIN MORPHOLOGY

With regard to the traditional classification of rift basins, recent studies demonstrate that the passive vs. active method of rift classification may be misleading: the morphology of "narrow rift basins" in East Africa and Baikal are, to first order, very similar, yet the East African rift is a classic plume-related rift, and Baikal is a passive rift driven by the Himalayan orogeny. The similarities are found in those parts of the rift that cross or come near cold, strong, thick, Archean lithosphere. Thus, at length scales of 10–500 km, lithospheric strength at the time of rifting is the critical factor determining rift structure, rather than the rifting mechanism. Petit and Ebinger (2000) show that the conditions of topography genesis may be highly variable even along the length of a single rift system, as has been suggested in orogenic belts (Stewart and Watts, 1997). Clearly, subcrustal processes may also account for a non-negligible part of the short-wavelength rift topography, even in the East African rift, where flexural rebound seems to predominate.

At the length scale of individual rift basins, lithospheric rheology also plays an important role. Border-fault lengths in cratonic parts of the East African and Baikal rifts are 80–120 km, whereas border faults that form in younger lithosphere or collapsing orogens are on the order of 30–60 km (e.g., Jackson and Blenkinsop, 1997; Ebinger et al., 1999), indicating that the dimensions of rift basins are also controlled by the mechanical properties of the lithosphere. The large basin widths, the long fault segments, and the great depths of teleseismic and local earthquakes (up to 35 km) all indicate a strong crust in some places of the East African rift (e.g., Nyblade and Langston, 1995; Foster and Jackson, 1998). Clearly, though, as one approaches continental breakup and strain localizes to a narrow zone, asthenospheric processes predominate (e.g., Hayward and Ebinger, 1996; Boutilier and Keen, 1999).

Variations in lithospheric thickness among continental domains probably have corresponding variations in thickness of the mechanical lithosphere, which may control the location and style of deformation in response to horizontal and vertical forces on the

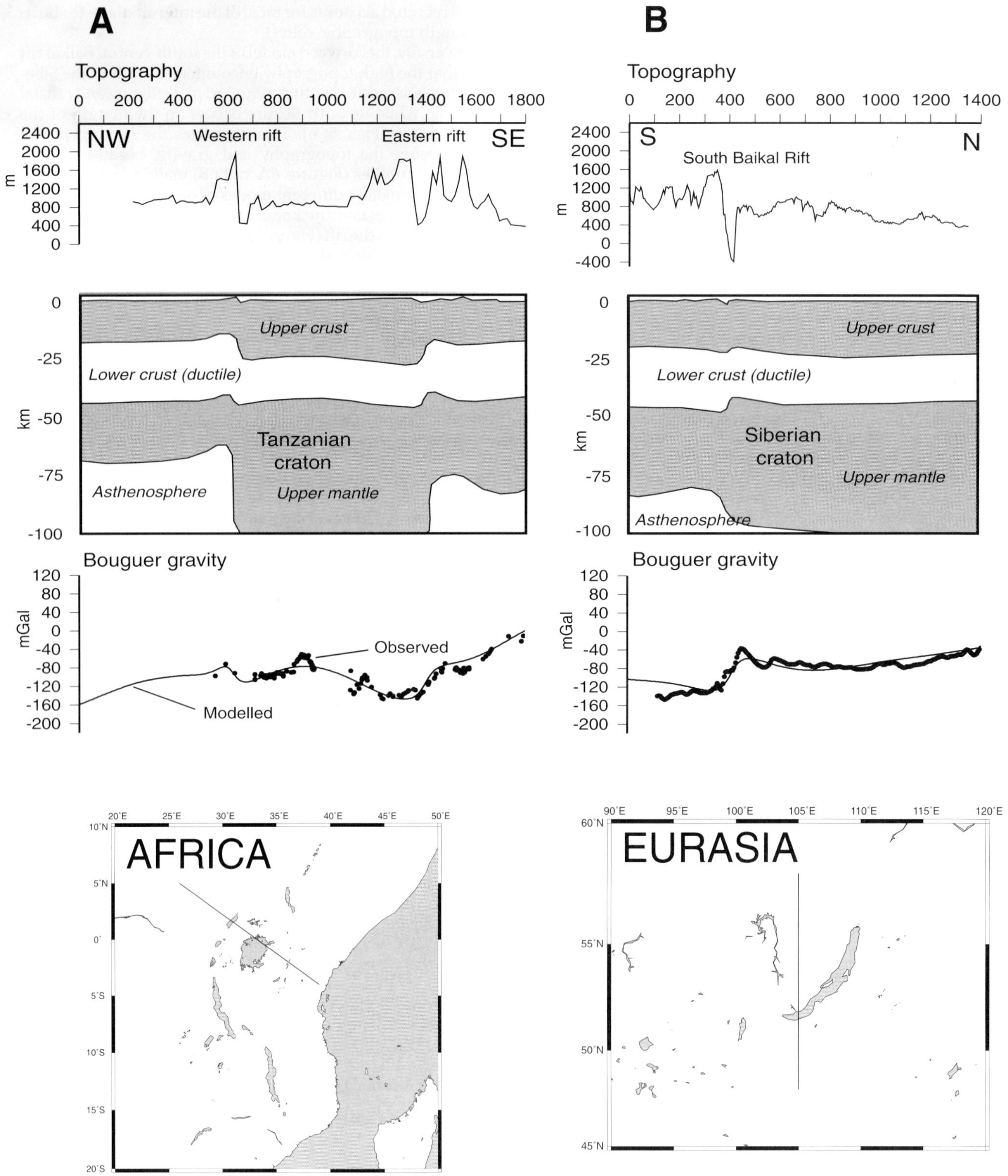


FIG. 6.—Layered discontinuous plate models of **A)** the East African and **B)** the South Baikal gravity profiles. Strength variations are represented as in Burov and Cloetingh (1997): white and dark areas correspond to maximum and minimum strength, respectively.

plates. Once created, these lateral variations in rheology clearly concentrate stress along terrane boundaries (e.g., Petit and Ebinger, 2000), suggesting that terrane boundaries are susceptible to later tectonic episodes.

These results indicate that we can develop predictive models of the regional morphology and along-axis segmentation of active and ancient rift basins based on a relatively small data base: a) thermal age of lithosphere or heat flow; b) seismogenic layer thickness; c) maximum length and/or displacement of normal faults in a basin array; d) effective elastic plate thickness during the synrift stage; and e) gravity anomalies (for active basins). Short, narrow basins with narrow uplifted flanks form in young, hot, weak, lithosphere and are characterized by < 15 km seismogenic layer thickness and / or effective elastic thickness; long, wide, deep basins with broad uplifted flanks develop in old, cold, strong continental lithosphere with seismogenic layer thickness / $Te > 30$ km (e.g., Fig. 2). Clearly, basin width also depends upon the stretching factor.

The along-axis segmentation of the rift is, therefore, a function of the mechanical properties of the lithosphere, with intrabasinal highs or rift offsets occurring less frequently in a strong lithosphere than in weak lithosphere. With these results one can predict the original shapes of the flanks and the spacing of the along-axis segmentation, both of which exert strong controls on drainage-basin size and sediment pathways to the rift basins (e.g., Morley, this volume).

## ACKNOWLEDGMENTS

Reviews by C.K. Morley, G. Nichols, and R. Gloaguen improved the manuscript. This project was supported by National Environment Research Council grant NER/T/S/2000/00647 awarded to CJE.

## REFERENCES

Abers, G., Mutter, C., and Fand, J., 1997, Shallow dips of normal faults during rapid extension: Earthquakes in the Woodlark–d'Entrecastaux rift system, Papua–New Guinea: Journal of Geophysical Research, v. 102, p. 15,301–15,317.

Agar, S., and Klitgord, K., 1995, Rift flank segmentation, basin initiation, and propagation: A neotectonic example from Lake Baikal: Geological Society of London, Journal, v. 152, p. 849–860.

Anderson, E., 1951, The Dynamics of Faulting: London, Oliver & Boyd, 206 p.

Bassi, G., 1995, Relative importance of strain rate and rheology for the mode of continental extension: Geophysical Journal International, v. 122, p. 195–210.

Birt, C., Maguire, P.K.H., Khan, M.A., Thybo, H., Keller, G., and Patel, J., 1997, The influence of pre-existing structures on the evolution of the southern Kenya rift valley: Evidence from seismic and gravity studies: Tectonophysics, v. 278, p. 211–242.

Bosworth, W., 1985, Geometry of propagating continental rifts: Nature, v. 316, p. 625–627.

Boutilier, R., and Keen, C., 1999, Small-scale convection and divergent plate boundaries: Journal of Geophysical Research, v. 104, p. 7389–7403.

Brace, W., and Kohlstedt, D., 1980, Limits on lithospheric stress imposed by laboratory experiments: Journal of Geophysical Research, v. 85, p. 6248–6252.

Braun, J., and Beaumont, C., 1989, A physical explanation of the relation between flank uplifts and the break-up unconformity at rifted continental margins: Geology, v. 17, p. 760–764.

Buck, W.R., 1986, Small-scale convection induced by passive margin rifting: The cause of uplift of rift shoulders: Earth and Planetary Science Letters, v. 77, p. 363–372.

Buck, W.R., 1991, Modes of continental extension: Journal of Geophysical Research, v. 96, p. 20,161–20,178.

Buck, W.R., Lavier, L., and Poliakov, A., 1999, How to make a rift wide: Royal Society [London], Philosophical Transactions, A, v. 357, p. 671–693.

Burov, E.B., and Diament, M., 1995, The effective elastic thickness (Te) of continental lithosphere: What does it really mean?: Journal of Geophysical Research, v. 100 , p. 3905–3927.

Burov, E.B., and Cloetingh, S., 1997, Erosion and rift dynamics: New thermomechanical aspects of post-rift evolution of extensional basins: Earth and Planetary Science Letters, v. 150, p. 7–26.

Burov, E., and Poliakov, A., 2001, Erosion and rheology controls on syn- and post-rift evolution: Verifying old and new ideas using fully coupled numerical models: Journal of Geophysical Research, v. 106, p. 16,461–16,482.

Burov, E.B., Jaupart, C., and Mareschal, J.-C., 1998, Large-scale crustal heterogeneities and lithospheric strength in cratons: Earth and Planetary Science Letters, v. 164, p. 205–219.

Burov, E.B., Houdry, F., Diament, M., and Deverchère, J., 1994, A broken plate beneath the North Baikal rift zone revealed by gravity modelling: Geophysical Research Letters, v. 21, p. 129–132.

Byerlee, J.D., 1978, Friction of rocks: Pure and Applied Geophysics, v. 116, p. 615–626.

Calais, E., Vergnolle, M., and Déverchère, J., 2002, Are post-seismic effects of the M = 8 Bolnay earthquake still influencing GPS velocities in the Mongolia–Baikal area?: Geophysical Journal International, v. 149, p. 157–168.

Chéry, J., 2001, Core complex mechanics: From the Gulf of Corinth to the Snake Range: Geology, v. 29, p. 439–442.

Chesley, J., Rudnick, R., and Lee, C.-T., 1999, Re–Os systematics of mantle xenoliths from the East African rift: age, structure, and history of the Tanzania craton: Geochimica et Cosmochimica Acta, 1999, v. 63, p. 1203–1217.

Chu, C., and Gordon, R., 1999, Evidence for motion between Nubia and Somalia along the Southwest Indian Ridge: Nature, v. 398, p. 64–67.

Cloetingh, S., and Burov, E., 1996, Thermomechanical structure of European lithosphere: constraints from rheological profiles and EET estimates: Geophysical Journal International, v. 124, p. 695–723.

Delvaux, D., Melnikov, A., and Ermikov, V., 1995, Paleostress reconstruction and geodynamics of the Baikal region, Central Asia, I, Paleozoic and Mesozoic pre-rift evolution: Tectonophysics, v. 252, p. 61–101.

Delvaux, D., Moeys, R., Stapel, G., Petit, C., Levi, K., Miroshnichenko, A., Ruzhich, V., and Sankov, V., 1997, Paleostress reconstruction and geodynamics of the Baikal region, Central Asia. Part II: Cenozoic tectonic stress and fault kinematics: Tectonophysics, v. 282, p. 1–38.

Déverchère, J., Houdry, F., Diament, M., Solonenko, N.V., and Solonenko, A.V., 1991, Evidence for a seismogenic upper mantle and lower crust in the Baikal rift: Geophysical Research Letters, v. 18, p. 1099–1102.

Déverchère, J., Houdry, F., Solonenko, N.V., Solonenko, A.V., and Sankov, V.A., 1993, Seismicity, active faults and stress field of the north Muya region, Baikal rift: New insights on the rheology of extended continental lithosphere: Journal of Geophysical Research, v. 98, p. 19,895–19,912.

Doin, C., Fleitout, L., and McKenzie, D., 1996, Geoid anomalies and the structure of continental and oceanic lithospheres: Journal of Geophysical Research, v. 101, p. 16,119–16,135.

Ebinger, C.J., and Sleep, N.H., 1998, Cenozoic magmatism throughout east Africa resulting from impact of a single plume: Nature, v. 395, p. 788–791.

Ebinger, C.J., Bechtel, T.D., Forsyth, D.W., and Bowin, C.O., 1989, Effective elastic plate thickness beneath the East African and Afar plateaus and dynamic compensation of the uplifts: Journal of Geophysical Research, v. 94, p. 2893–2901.

Ebinger, C.J., Jackson, J.A., Foster, A.N., and Hayward, N.J., 1999, Extensional basin geometry and the elastic lithosphere: Royal Society [London], Philosophical Transactions, A., v. 357, p. 741–765.

ENGLAND, P., 1983, Constraints on extension of continental lithosphere: Journal of Geophysical Research, v. 88, p. 1145–1152.

ENGLAND, P., AND HOUSEMAN, G., 1986, Finite strain calculations of continental deformation 2: Comparison with the India–Asia collision zone: Journal of Geophysical Research, v. 91, p. 3664–3676.

FORSYTH, D.W., 1992, Finite extension and low angle normal faulting: Geology, v. 20, p. 27–30.

FOSTER, A.N., AND JACKSON, J.A., 1998, Source parameters of large African earthquakes: implications for crustal rheology and regional kinematics: Geophysical Journal International, v. 134, p. 422–448.

FOSTER, A.N., EBINGER, C.J., MBEDE, E., AND REX, D., 1997, Tectonic development of the northern Tanzanian sector of the East African rift system: Geological Society of London, Journal, v. 154, p. 689–700.

GEORGE, R., ROGERS, N., AND KELLEY, S.P., 1998, Earliest magmatism in Ethiopia: Evidence for two mantle plumes in one flood basalt province: Geology, v. 26, p. 923–926.

GRAND, S., VAN DER HILST, R., AND WIDIYANTORO, S., 1997, Global seismic tomography: a snapshot of convection in the Earth: GSA Today, v. 7, no. 4, p. 1–7.

GURNIS, M., MITROVICA, J., RITSEMA, J., AND VAN HEIJST, H-J., 2000, Constraining mantle density structure using geological evidence of surface uplift rates: The case of the African superplume: $G^3$, v. 1.

HAYWARD, N., AND EBINGER, C.J., 1996, Variations in the along-axis segmentation of the Afar rift system: Tectonics, v. 15, p. 244–257.

HOPPER, J.R., AND BUCK, W.R., 1996, The effect of lower crustal flow on continental extension and passive margin formation: Journal of Geophysical Research, v. 101, p. 20,175–20,194.

HUISMANS, R., PODLADCHIKOV, Y.Y., AND CLOETINGH, S.A., 2001, The transition from passive to active rifting: Relative importance of asthenospheric doming and passive extension of the lithosphere: Journal of Geophysical Research, v. 106, p. 11,271–11,292.

HUTCHINSON, D.R., GOLMSHTOK, A.J., ZONENSHAIN, L.P., MOORE, T.C., SCHOLZ, C.A., AND KLITGORD, K.D., 1992, Depositional and tectonic framework of the rift basins of Lake Baikal from multichannel seismic data: Geology, v. 20, p. 589–592.

IONOV, D.A., O'REILLY, S.Y., AND ASHCHEPKOV, I.V., 1995, Feldspar-bearing lherzolite xenoliths in alkali basalts from Hamar–Daban, southern Baikal region, Russia: Contributions to Mineralogy and Petrology, v. 122, p. 174–190.

JACKSON, J.A., AND BLENKINSOP, T., 1997, The Bilila–Mtakataka fault in Malawi: an active, 100 km long normal fault segment in thick seismogenic crust: Tectonics, v. 16, p. 137–150.

KELLER, R., MECHIE, J., BRAILE, L., MOONEY, W., AND PRODEHL, C., 1994, Seismic structure of the uppermost mantle beneath the Kenya rift: Tectonophysics, v. 236, p. 201–210.

KIRBY, S.H., AND KRONENBERG, A.K., 1987, Rheology of the lithosphere: Selected topics: Reviews of Geophysics, v. 25, p. 1219–1244.

KRUSE, S.E., ZHENGONG, J.L., NAAR, D.F., AND DUNCAN, R.A., 1997, Effective elastic thickness of the lithosphere along the Easter seamount chain: Journal of Geophysical Research, v. 102, p. 27,305–27,317.

KUSZNIR, N., AND KARNER, G., 1985, Dependence of the flexural rigidity of the continental lithosphere on rheology and temperature: Nature, v. 316, p. 138–142.

KUSZNIR, N., AND PARK, R., 1987, The extensional strength of the continental lithosphere: its dependence on geothermal gradient, crustal composition, and thickness, *in* Coward, M., Dewey, J., and Hancock, P., eds., Continental Extensional Tectonics: Geological Society of London, Special Publication 28, p. 35–42.

LAVIER, L., AND STECKLER, M., 1997, The effect of sedimentary cover on the flexural strength of continental lithosphere: Nature, v. 389, 476–479.

LAVIER, L., BUCK, W.R., AND POLIAKOV, A.N.B., 1999, Self-consistent rolling-hinge model for the evolution of large-offset low-angle normal faults: Geology, v. 27, p. 1127–1130.

LITHGOW-BERTELLONI, C., AND SILVER, P. , 1998, Dynamic topography, plate driving forces, and the African superswell: Nature, v. 395, p. 269–272.

LOGATCHEV, N.A., AND ZORIN, Y.A., 1992, Baikal rift zone: Structure and geodynamics: Tectonophysics, v. 208, p. 273–286.

LOUDEN, K., AND CHIAN, D., 1999, The deep structure of non-volcanic rifted continental margins: Royal Society [London], Philosophical Transactions, A, v. 357, p. 767–804.

LYSAK, S.V., 1992, Heat flow variations in continental rifts: Tectonophysics, v. 208, p. 309–323.

MCKENZIE, D., 1978, Some remarks on the formation of sedimentary basins: Earth and Planetary Science Letters, v. 40, p. 25–32.

MCKENZIE, D., 1984, A possible mechanism for epeirogenic uplift: Nature, v. 307, p. 616–618.

MAGGI, A., JACKSON, J., MCKENZIE, D., AND PRIESTLEY, K., 2000, Earthquake focal depths, effective elastic thickness, and the strength of continental lithosphere: Geology, v. 28, p. 495–498.

MELNIKOV, A.I., MAZUKABZOV, A.M., SKLYAROV, E.V., AND VASILIEV, E.P., 1994, Baikal rift basement: structure and tectonic evolution, Centres de Recherche Exploration et Production Elf Aquitaine, Bulletin, v. 18, p. 99–122.

MENZIES, M., EBINGER, C., AND KLEMPERER, S., 2000, Volcanic rifted margins: GSA Today, v. 10, no. 8, p. 8–11.

MORLEY, C., 1988, Variable extension in Lake Tanganyika: Tectonics, v. 7, p. 785–801.

MORLEY, C.K., 1994, Interaction of deep and shallow processes in the evolution of the Kenya rift, Tectonophysics, v. 236, p. 81–91.

NOBLE, W., 1997, Post Pan-African tectonic evolution of eastern Africa: an apatite fission track study: Unpublished PhD dissertation, LaTrobe University, Melbourne, Australia, 310 p.

NYBLADE, A.A., AND LANGSTON, C.A., 1995, East African earthquakes below 20 km depth and their implications for crustal structure: Geophysical Journal International, v. 121, p. 49–62.

NYBLADE, A., POLLACK, H. JONES, D., PODMORE, F., AND MUSHAYANDEBVU, M., 1990, Terrestrial heat flow in East and Southern Africa: Journal of Geophysical Research, v. 95, p. 17,371–17,384.

PETIT, C., AND EBINGER, C., 2000, Flexure and mechanical behavior of cratonic lithosphere: Gravity models of the East African and Baikal rifts: Journal of Geophysical Research, v. 105, p. 19,151–19,162.

PETIT, C., DÉVERCHÈRE, J., HOUDRY, F., SANKOV, V.A., MELNIKOVA, V.I., AND DELVAUX, D., 1996, Present-day stress field changes along the Baikal rift and tectonic implications: Tectonics, v. 15, p. 1171–1191.

PETIT, C., BUROV, E., AND DÉVERCHÈRE, J., 1997, On the structure and mechanical behavior of the extending lithosphere in the Baikal rift from gravity modeling: Earth and Planetary Science Letters, v. 149, p. 29–42.

RANALLI, G., 1995, Rheology of the Earth, 2nd Edition: London, Chapman & Hall, 413 p.

RIBE, N., AND CHRISTENSEN, U. 1994, Three-dimensional modeling of plume–lithosphere interaction: Journal of Geophysical Research, v. 99, p. 669–682.

RIETBROCK, A., TIBERI, C., SCHERBAUM, F., AND LYON-CAEN, H., 1996, Seismic slip on a low angle normal fault in the Gulf of Corinth: Evidence from high-resolution cluster analysis of microearthquakes: Geophysical Research Letters, v. 23, p. 1817–1820.

RITSEMA, J., NYBLADE, A., OWENS, T., LANGSTON, C., AND VAN DECAR, J., 1998, Upper mantle seismic velocity structure beneath Tanzania: implications for the stability of cratonic roots: Journal of Geophysical Research, v. 103, p. 21,200-21,214.

ROYDEN, L., AND KEEN, C., 1980, Rifting processes and thermal evolution of the continental margin of eastern Canada determined from subsidence curves: Earth and Planetary Science Letters, v. 51, p. 343–361.

RUPPEL, C., 1995, Extensional processes in continental lithosphere: Journal of Geophysical Research, v. 100, p. 24,187–24,215.

RUPPEL, C., KOGAN, M.G., AND MCNUTT, M.K., 1993, Implications of new gravity data for Baikal rift zone structure: Geophysical Research Letters, v. 20, p. 1635–1638.

SALVESEN, J., 1978, Variations in the geology of rift basins: a tectonic model: Conference Proceedings, Los Alamos Laboratory, Los Alamos, New Mexico, v. 7487, p. 82–86.

SCHOLZ, C., 1988, The brittle–plastic transition and the depth of seismic faulting: Geologische Rundschau, v. 77, p. 319–328.

SCHOLZ, C., AND CONTRERAS, J., 1998, Mechanics of continental rift architecture: Geology, v. 26, p. 967–970.

SENGÖR, C., AND BURKE, K., 1978, Relative timing of rifting and volcanism on Earth and its tectonic implications: Geophysical Research Letters, v. 5, p. 419–421.

SIMONS, F., ZUBER, M., AND KORENAGA, J., 2000, Isostatic response of the Australian lithosphere: Estimation of effective elastic thickness and anisotropy using multitaper spectral analysis: Journal of Geophysical Research, v. 105, p. 19,163–19,184.

STEWART, J., AND WATTS, A.B., 1997, Spatial variations of flexural rigidity at mountain ranges: Journal of Geophysical Research, v. 102, p. 5327–5352.

VAN DER BEEK, P., 1997, Flank uplift and topography at the Central Baikal Rift (SE Siberia): A test of kinematic models for continental extension: Tectonics, v. 16, p. 122–136.

VAN DER BEEK, P., MBEDE, E., ANDRIESSEN, P., AND DELVAUX, D., 1998, Denudation history of the Malawi and Rukwa rift flanks from apatite fission track thermochronology: Journal of African Earth Sciences, v. 26, p. 363–385.

VAN DER HILST, R., WIDIYANTORO, S., AND ENGDAHL, E.R., 1997, Evidence for deep mantle circulation from global tomography: Nature, v. 386, p. 578–584.

VAUCHEZ, A., BARRUOL, G., AND TOMMASI, A., 1997, Why do continents break-up parallel to ancient orogenic belts: Terra Nova, v. 9, p. 62–66.

WATTS, A.B., AND STEWART, J., 1998, Gravity anomalies and segmentation of the continental margin offshore west Africa: Earth and Planetary Science Letters, v. 156, p. 239–252.

WEISSEL, J.K., AND KARNER, G.D., 1989, Flexural uplift of rift flanks due to mechanical unloading of the lithosphere under extension: Journal of Geophysical Research, v. 94, p. 13,919–13,950.

WERNICKE, B., 1985, Uniform-sense normal simple shear of the continental lithosphere: Canadian Journal of Earth Sciences, v. 22, p. 108–125.

WHITE, R., AND MCKENZIE, D., 1989, Magmatism at rift zones: the generation of volcanic passive continental margins and flood basalts: Journal of Geophysical Research, v. 94, p. 7685–7729.

ZUBER, M., AND PARMENTIER, M., 1986, Lithospheric necking: A dynamic model for rift morphology: Earth and Planetary Science Letters, v. 77, p. 373–383.

# TECTONIC SETTINGS OF CONTINENTAL EXTENSIONAL PROVINCES AND THEIR IMPACT ON SEDIMENTATION AND HYDROCARBON PROSPECTIVITY

CHRIS K. MORLEY

*Department of Geology and Petroleum Geology, Kings College, Aberdeen University, Meston Building, Aberdeen, AB24 3UE, Scotland, U.K.*
*present address: Department of Petroleum Geoscience, University of Brunei, Darussalam, Bandar, Seri Begawan, Brunei Darussalam*

ABSTRACT: Extensional basins can be classified according to plate-tectonic setting and driving mechanism into: (1) regional continental extension, (2) extension at the tips of permanently or temporarily abandoned spreading centers, (3) in the upper plate of volcanic arc–oceanic crust subduction zones, (4) soft collision zones, (5) post-orogenic wide rifts (Cordilleran type), (6) Himalayan-type extension; and (7) in response to far-field orogenic stresses. These settings display differences in mantle geometry, volcanic history, fault geometry and evolution, generation of lithospheric thermal anomalies, rift topography, post-rift history, sediment source characteristics, sediment pathways, and timing and intensity of inversion tectonics, which can impact hydrocarbon prospectivity. In convergent settings subduction rollback and slab-suction forces are very important mechanisms for generating extension, creating the conditions for non-uniform lithospheric thinning, and very characteristic thick (4–6 km), rapidly formed (10–15 Myr) thermal subsidence basins. The enhanced gravity potential of overthickened crust appears to be an important driving mechanism for extension in orogenic belts, but it is usually accompanied by other mechanisms (escape tectonics, subduction rollback, propagating spreading centers) for extension to occur.

## INTRODUCTION

This paper reviews the different structural and tectonic characteristics of rift systems, stressing how these characteristics affect syn-rift and post-rift sedimentation and hydrocarbon prospectivity. There have been a wide range of attempts to categorize rifts. One divided driving forces into active and passive rifts, implying that rifts form because of either lithospheric (passive rifts) or mantle plume–related asthenosphere (active rifts) stresses. Another is based on plate-tectonic setting, e.g., whether the extensional basins form on a continental interior, along passive margins, or in back-arc settings (Kingston et al., 1983). However, widespread extension associated with orogenic belts, and especially metamorphic core complexes, had not been described when Kingston et al. (1983) was published. More recently crustal extension has been classified according to structural style into narrow, wide, and metamorphic core complex as a means to investigate the inferred lithospheric stress and strain conditions that cause these structural styles (Kusznir and Park, 1987; Buck, 1991; Brun, 1999).

When considering how variations in volcanism, structural style, surrounding topography, sediment source area, and post-rift tectonic history have affected deposition and hydrocarbon systems in extensional basins, it is necessary to look at a wider range of extensional settings than those in the broad classifications outlined above. Hydrocarbon systems will not work if one of any of the key components (reservoir, seal, trap, trap timing, migration, source rock) fail. Hence if one of the factors above (or others) repeatedly impact any of the key components in a particular tectonic setting, there will be a significant reduction in hydrocarbon prospectivity. The classification used in this paper divides rifts into seven categories, defined according to tectonic setting and driving mechanism.

## TECTONIC SETTINGS

The tectonic settings of extensional basins include: (1) aulacogens or triple junctions to spreading centers, (2) passive margins, (3) back arc, (4) collapse of mountain belts (Dewey, 1988), and (5) in the upper (Lake Baikal) or lower (Rhine Graben) plates of collisional orogenic belts developed in response to far-field orogenic stresses (e.g., Molnar and Tapponnier, 1975). A selection of rifts from these settings is illustrated in Figure 1.

The tectonic settings used here are slightly different from the traditional settings above. In particular, the driving mechanisms for extension are emphasized. Also, the post-rift history (in particular thermal subsidence and tectonic inversion) is critical to how hydrocarbon systems develop (e.g., Morley, 1999c) and commonly varies in a predictable way depending on tectonic setting. Justification for the divisions briefly described below forms the main body of the text.

Aulacogens are defined as long-lived, deeply subsiding sedimentary troughs (failed rifts) that extend at high angles from the margins towards the interiors of cratons. They are generally associated with triple junctions (Dickinson, 1974; Burke, 1976). In this paper three rift settings are defined from those which could be classified as aulacogens: failed regional continental extension, rifts at abandoned spreading tips, and extensional basins in the upper plate of subduction zones.

Failed regional continental extension defines rifts formed by single or repeated attempts to break up large continental landmasses. In particular, repeated attempts to break up Pangea have led to many crosscutting, long, narrow, linear branching rift systems. Rifts found along passive margins originated as regional continental rifts. Failed regional continental extension is meant to emphasize the extensive branching networks of rifts that develop in response to both regional litho-spheric stresses (slab pull, ridge push) and buoyancy forces associated with mantle plumes (e.g., Zoback, 1992; Bott, 1992, 1993). Excluded from the category are (alucogen-type) extensional basins formed in the upper plate of subduction zones. This is because forces related to subduction rollback and slab suction (e.g., Bott, 1993) create extensional basins with very characteristic crustal and mantle geometries, and post-rift histories that differ from failed regional continental rifts. Rifts formed at the tips of permanently or temporarily abandoned spreading centers are placed in a different category from regional continental rifts because of the characteristics of their structural setting.

A wide range of extensional basins are associated with regions of plate convergence, from areas of subduction of oceanic crust beneath island arcs, to major collisional belts. Variations in the style of extension superimposed on orogenic belts coincides quite neatly with variations in the type of plate interaction. Hence the tectonic settings are divided into: extensional basins associated with subduction zones of oceanic crust–island arc or oceanic crust–continental crust; soft continental collisions (Royden and

Sedimentation in Continental Rifts
SEPM Special Publication No. 73, Copyright © 2002
SEPM (Society for Sedimentary Geology), ISBN 1-56576-082-4, p. 25–55.

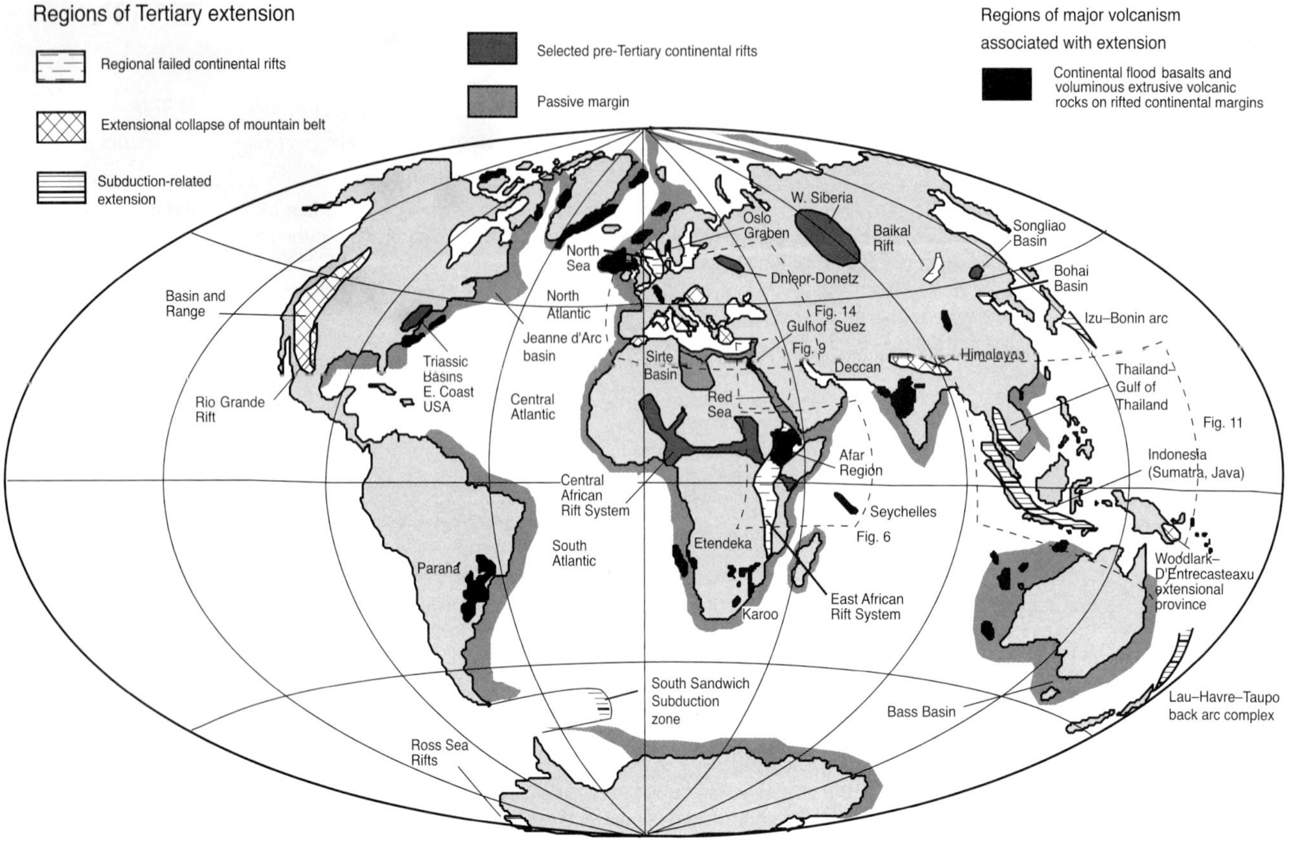


FIG. 1.—World map showing regions of extension, and major volcanic provinces associated with extensional tectonics (modified from White and McKenzie, 1989, and Twiss and Moores, 1992).

Burchfiel, 1989); post-orogenic wide rifts (Cordilleran type?); and hard collisions (Himalayan type). Driving forces for extension include subduction rollback and buoyancy forces related to enhanced gravity potential (e.g., Dewey, 1980; England and Houseman, 1989). Back-arc basin is a term avoided in the main classification because extension caused by subduction rollback can occur in a wide range of locations in the upper plate, not just the back-arc. The final category of rifts related to collisions are those formed in the foreland and hinterland of orogenic belts by far-field stresses from orogenic belts (Molnar and Tapponnier, 1975).

## FAILED REGIONAL CONTINENTAL EXTENSION

### *Introduction*

Many continental rifts are precursors to (e.g., European and African Triassic rifts; Late Jurassic North Sea rifts), or failed arms associated with (e.g., East African rift (EAR)), continental breakup (e.g., Guiraud, 1998; Dore et al., 1999). Such rift systems tend to be thousands of kilometers long, narrow to wide (30–200 km wide) systems of half grabens and full grabens (e.g., Rosendahl, 1987; Nelson et al., 1992). They form the classic rift systems which are long, linear, zig-zagging and branching features that often involve three rift trends (triple junctions). The best modern example is the 3500-km-long East African rift system, a failed rift system which branches off the "successful" Red Sea and Gulf of Aden rifts, where oceanic crust is actively being generated (Figs. 1, 6; e.g., Girdler, 1991).

### *Structural Styles*

Typically major continental rift systems are dominated by high-angle fault sets (45–70°) which, for relatively small amounts of displacement, can give rise to large amounts of subsidence; commonly basin fill is up to 6–7 km thick (Fig. 2; e.g., Rosendahl, 1987; Morley, 1989; Banks et al., 1995). Occasionally lower-angle faults (down to about 20° dip) can also occur, but these are comparatively rare (Morley, 1999b). The amount of crustal extension is typically low in failed branches ($\beta < 1.5$; e.g., North Sea, Roberts et al., 1993; Kenya Rift, Morley, 1994; Anza Graben, Morley et al., 1999b; Western branch East African Rift, Ebinger, 1989; Luangwa Valley Karoo rifts, Zambia, Banks et al., 1995). Continental rifts lining passive margins can exhibit much higher extension ($\beta < 3$–$4$) (e.g., Brun and Beslier, 1996).

Preexisting fabrics exert a strong control on the overall rift geometry, causing obliquely extending segments, transfer zones, jumps and gaps in the location of the rift, splays, and curved and zig-zag geometries (Figs. 3, 4; Morley et al., 1990; Nelson et al., 1992). Large tilted fault blocks commonly develop on rift margins (e.g., Shetland Platform, North Sea; e.g., Lee and Hwang, 1993), and associated sedimentary units are frequently modified by detachments and slumps (e.g., Rouby et al., 1996; Underhill et al., 1997)

Magmatism can significantly affect structure style. In the Kenya Rift during the Pliocene older half grabens were abandoned and replaced by a narrow (10 km wide) swarm of minor faults focused along a line of Plio-Pleistocene volcanoes (Baker

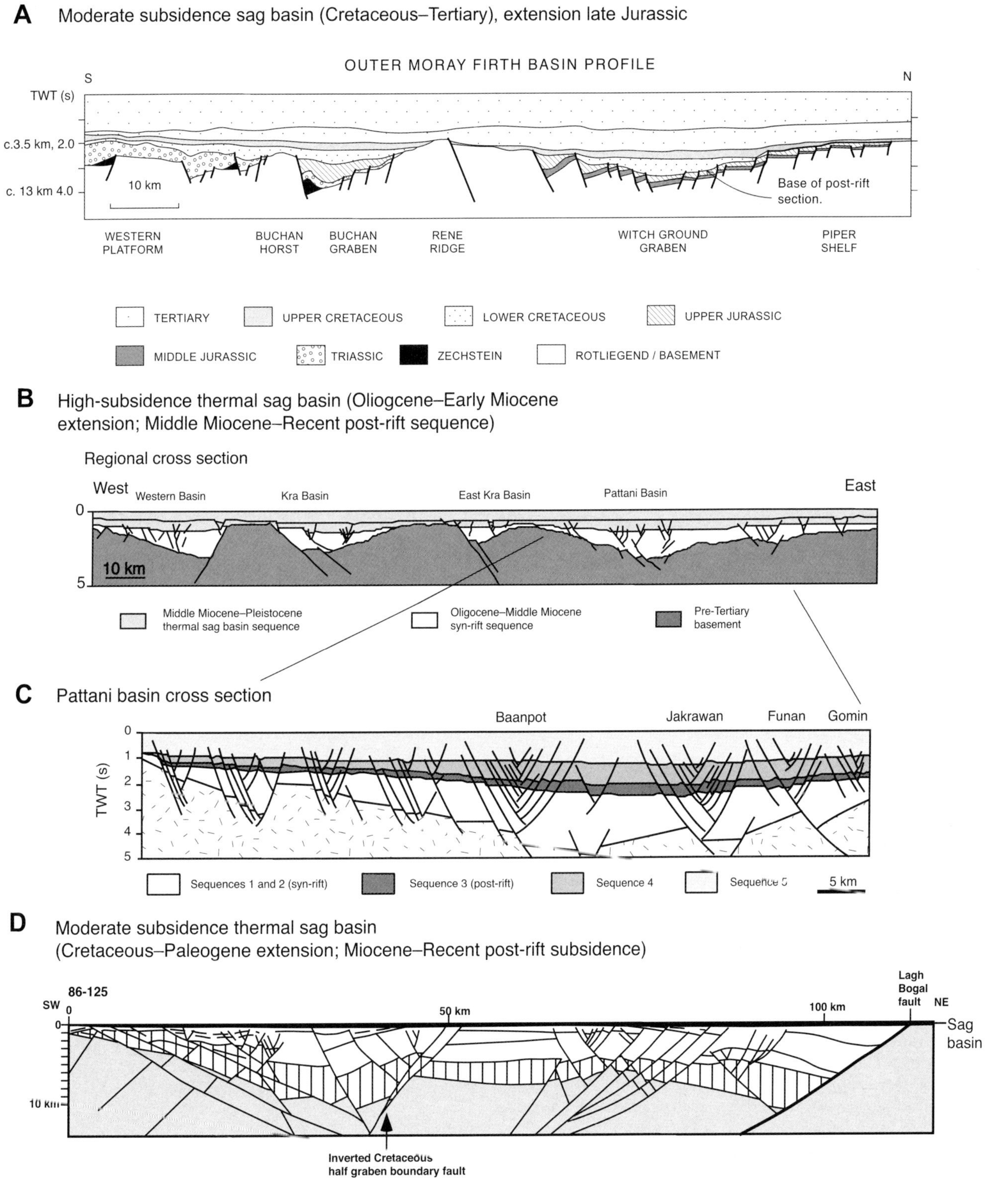


FIG. 2.—Cross sections of marine and continental post-rift basins illustrating some key differences between marine and non-marine sag basins. **A)** Outer Moray Firth basin profile (marine rift; redrawn from Errant et al., 1999) showing considerable topography on the base of the post-rift section and onlap of the Cretaceous and Tertiary section onto major highs. **B, C)** Cross section through the northern Gulf of Thailand and detail of the Pattani basin (from Watcharanantakul and Morley, 2001); note how little onlap there is of the high areas (continental setting) compared with the marine setting of the Outer Moray Firth Basin. **D)** Cross section through the Anza Graben (continental rift; Morley et al., 1999b). Note that the sag basin is the same age as the Pattani Basin, and there is more upper-crustal extension in the Anza Graben than the Pattani Basin, yet the Anza Graben basin is much thinner.

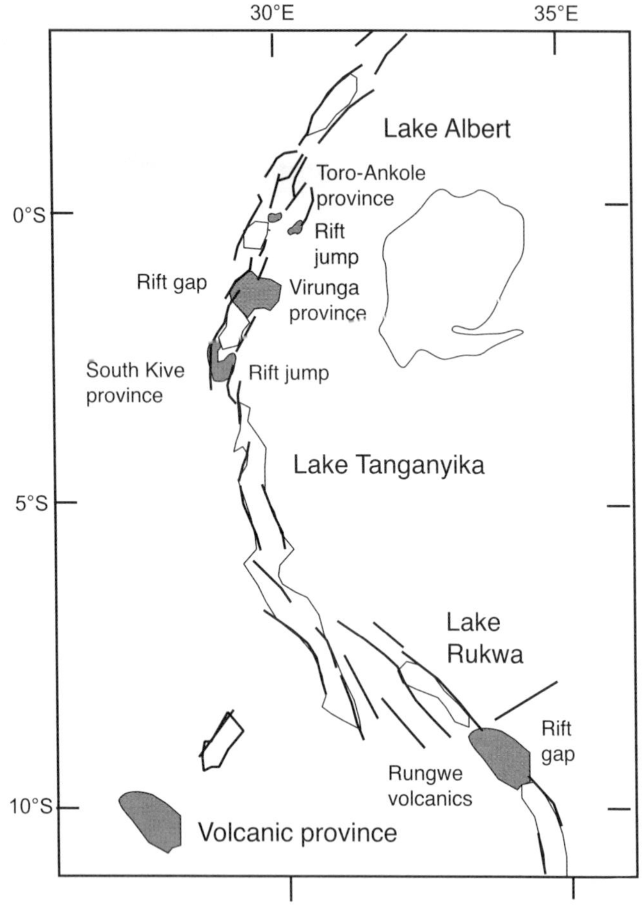


FIG. 3.—Map of volcanics and faults and basins, western branch EAR, illustrating major changes in rift segment geometry coinciding with igneous provinces. Simplified from Ebinger (1989).

and Wohlenberg, 1971). Similar structural styles are found in the Ethiopian Rift (Hayward and Ebinger, 1996). These fault swarms appear to reflect extreme weakening of the upper crust by magma and consequent reduction in thickness of the brittle layer and reduced effective elastic thickness (Fig. 5; Mohr, 1987; Ebinger et al., 1999).

In the Western Branch of the EAR isolated volcanic centers clearly are associated with breaks in the rift segment geometry (Fig. 3). The volcanic centers developed early in the rift evolution (e.g., Ebinger, 1989) and hence appear to be the cause of major rift transfer zones as marked by gaps and jumps in rift segment geometry (Fig. 3; Nelson et al., 1992). It is uncertain whether the volcanic centers represent changes in extensional strain from brittle faults to magmatic intrusions, or areas of decreased strain, or both.

### *Sedimentation and Syn-Rift Topography*

Rifts associated with attempted continental breakup tend to lie within relatively cold, normal-thickness continental crust (e.g., Buck, 1991) or cold overthickened crust of old orogenic belts (Vink et al., 1984). Doming due to an underlying mantle plume typically causes uplifts up to 1 km high over a region 1000+ km in diameter (Ebinger et al., 1989; White and McKenzie, 1989). Shorter-wavelength uplifts can result from magmatic underplating

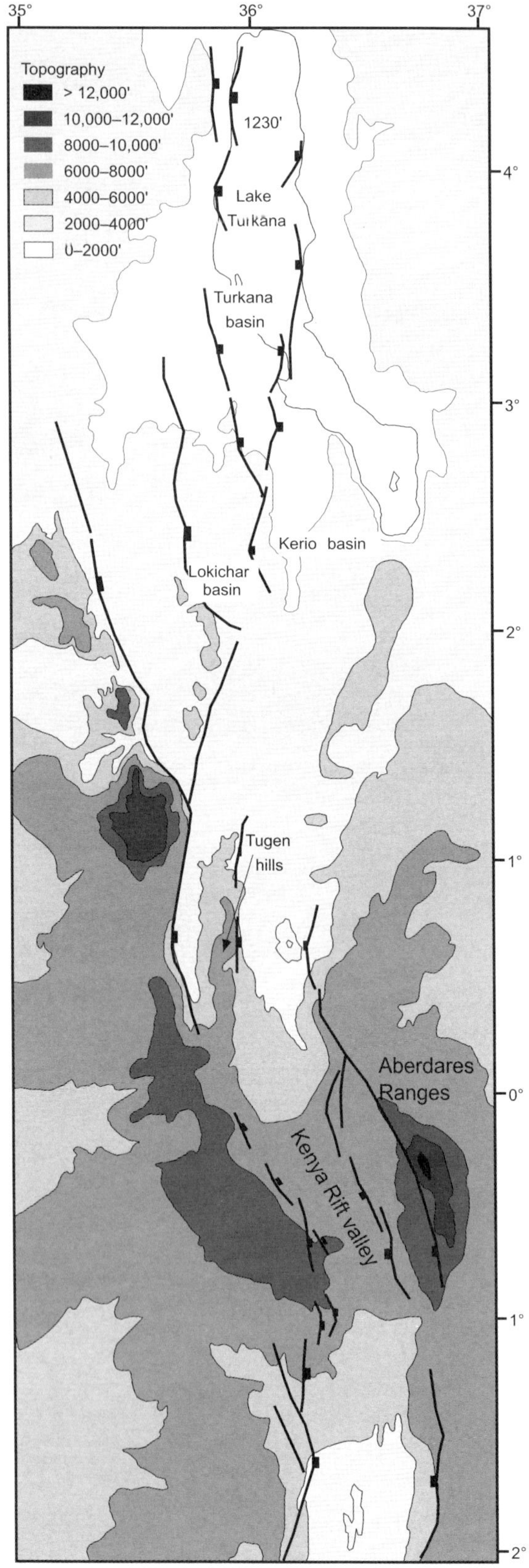


FIG. 4.—Topographic map of the Kenya rift, East Africa, illustrating the rapid lateral variations in elevation in rifts, and the strong structural control by major boundary faults on topography.

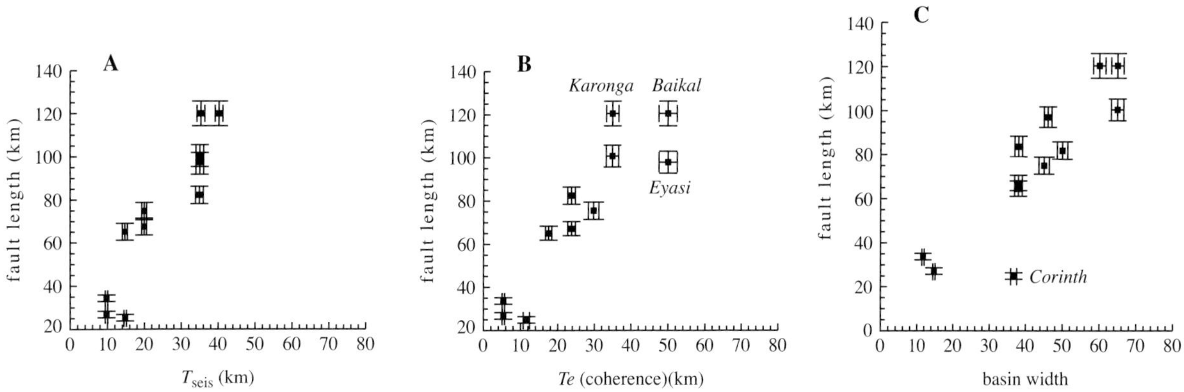


FIG. 5.—Variations in border-fault length in extensional provinces and **A)** seismogenic layer thickness (Tseis); **B)** effective elastic thickness (Te); and **C)** basin width. The diagrams show a linear increase in fault length as Te, Tseis, and basin width increase. The extensional province with the highest regional extension is the Corinth basins (Aegean). Redrawn from Ebinger et al. (1999).

and isostatic uplift in the footwalls of normal faults. Volcanoes also create substantial topography (e.g., Mounts Elgon, Kenya, and Kilimanjaro, Kenya rift).

Flank uplift along the Kenya rift is spectacular, with over 2 km difference in elevation between the rift floor and the Aberdare ranges of the rift shoulder (Fig. 4). The Tanganyika and Albert rifts also display maximum uplifts of about 2 km (between the uplifted shoulders and the lake surface) some 10–15 km horizontal distance into the footwall from border fault systems. Such topography produces a wide vertical range of climatic conditions over short horizontal distances; in Kenya the rift floor is hot savanna, the Aberdare ranges are covered by almost temperate forests, and Mt Kenya and Mt Kilimanjaro are capped by snow.

Rift topography exerts profound regional to local-scale effects on drainage. In general the domal and flank uplifts promote drainage away from the rifts (e.g., Cohen, 1990). However, the rift structure at a variety of scales is not continuous, but has gaps, often created by transfer zones, especially relay ramps (Griffiths, 1980; Crossley, 1984; Morley et al., 1990), which enable rivers draining off the flanks to enter the rift and form axial drainage along the rift (Cohen, 1990). Propagating fault systems can impact the development of fans, and aerial and subaqueous drainage systems (e.g., Gupta et al., 1999).

The largest high-angle boundary faults in rifts typically display maximum footwall uplifts of 1–2 km (Fig. 4). Maximum hanging-wall displacements are often in the range of 6–8 km, and permit enormous accumulation of syn-rift sediments over short periods of time (5–15 Myr; e.g., Rosendahl, 1987; Hendrie et al., 1994). Commonly subsidence rate exceeds sediment supply, and consequently large lakes develop in fault-bounded depressions, e.g., Lakes Malawi, Rukwa, Tanganyika, and Albert, EAR; (Rosendahl, 1987; Ebinger, 1989). The strong control exerted by major boundary faults on depositional environment occurs both in continental (e.g., Morley, 1989; Cohen, 1990; Lambiase and Bosworth, 1995) and marine (e.g., Ravnas and Steel, 1998) settings.

## *Volcanism*

Continental rifts range from nonmagmatic (e.g., the Cretaceous–Paleogene Central African rift system; Schull, 1988; Genik, 1992) to strongly magmatic where volcanic and volcaniclastic rocks dominate the basin fill (e.g., central and southern Kenya rift, Ethiopian rift; Baker et al., 1972). Extension of the lithosphere causes crustal thinning and uplift of the geotherms under the rift (McKenzie, 1978). The resulting reduction of pressure while temperature remains constant can lead to the generation of melt (adiabatic decompression melting; McKenzie and Bickle, 1988). For normal-thickness continental crust, extension of about $\beta = 2$–3 is required to generate significant quantities of melt. Consequently for most failed continental rifts, where typical values of $\beta$ lie between 1 and 2, passive rifting is unlikely to cause much volcanism (e.g., Hendrie et al., 1994).

The eastern branch of the EAR is the most magmatically rich continental rift system in the world, with the volume of igneous rocks estimated around 1 million $km^3$ (combining estimates from Baker et al., 1971, Williams, 1978, and Morley et al., 1992). Volcanism began early in the tectonic development of the Kenya rift, in places preceding the onset of significant extension in the upper crust, inconsistent with adiabatic decompression melting. Instead timing and magma geochemistry point to melt generation at the head of a mantle plume at depths between 40 and 80 km (e.g., Latin et al., 1993; Hendrie et al., 1994). Separate plumes underlie the Kenya and Ethiopian rifts, but considerably more volcanism is associated with the Ethiopian rift (Fig. 6; Baker et al., 1972).

Figure 7 schematically illustrates some key features of how crustal structure and mantle plume development interact, on the basis of the East African rift (Zeyen et al., 1997). Some key points are as follows. (1) The plume head flattened against the base of the lithosphere and spread laterally to preferentially thin the lithosphere beneath the mobile belt. (2) Volcanism closest to the plume pipe starts first in the eastern rift branch, and later spreads to the western branch. (3) Thicker Archean cratonic crust between the two branches contains little extension or volcanism. Both branches are focused on the weaker crust of Proterozoic mobile belts. (4) Heat flows are higher in the eastern branch and the brittle–ductile transition is higher. Such conditions, particularly when igneous intrusions are present, lead to reduced effective elastic thicknesses and impact fault geometry and reduce basin size (Fig. 5). (5) Non-uniform thinning of the crust is due to replacement of the lithospheric mantle by the asthenospheric mantle by both uplift of the asthenosphere and mineral phase changes within the lithosphere. It is not due to vertical variations in the amount of extension.

## *Rift Structural Evolution*

The largest scale of rift evolution is the succession of failed rift basins that marks repeated attempts at continental breakup,

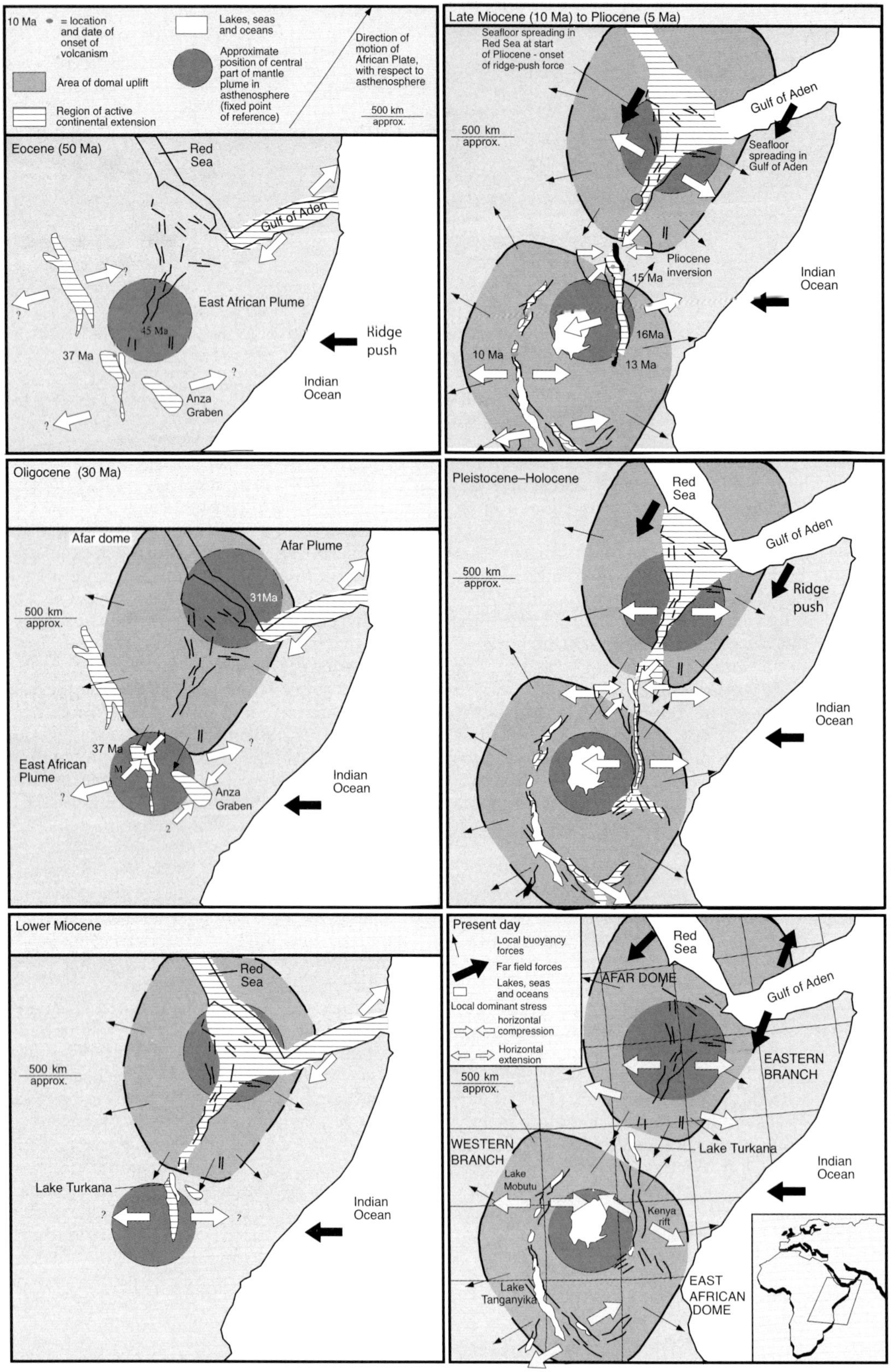


FIG. 6.—Tectonic evolution of the East African Rift System. The estimated location of two mantle-plume centers in the asthenosphere are shown as light gray circles, their position is fixed, and East Africa is moved with respect to the plumes following the path shown in George et al. (1998). The East African geological map is modified from Ebinger et al. (1989) and Ebinger and Ibrahim (1994). Stress evolution is mostly from Ring et al. (1993), Bosworth et al. (1992), and Morley et al. (1999c).

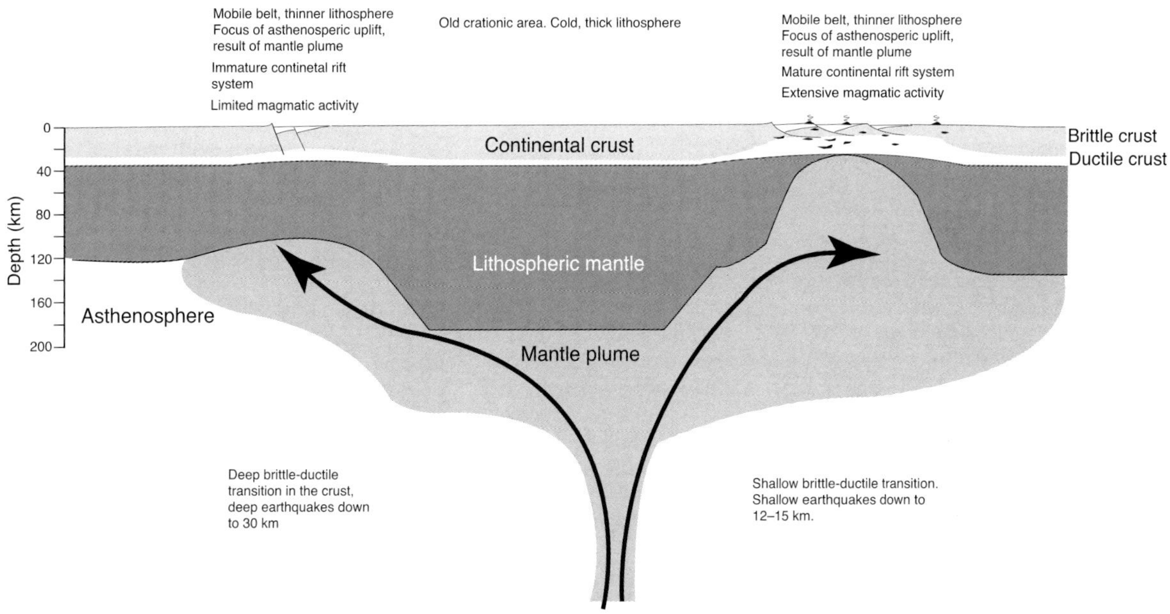


FIG. 7.—Cross section through the lithosphere and upper asthenosphere of a failed regional continental rift system, based on the East African Rift system (modified from Zeyen et al., 1997).

of which the supercontinent Pangea is an outstanding example. The result is seen in Europe and Africa where Permo-Triassic, Cretaceous–Paleogene, and late Tertiary rift systems (EAR) either cross cut each other or are reactivated at different times (e.g., Guiraud, 1998; Dore et al., 1999). The Jurassic period in Africa is very distinctive, showing no good examples of continental rifts in the interior of Africa (except the inverted rift forming the Atlas mountains). The Jurassic rifts do, however, commonly follow older Karoo (late Paleozoic) rift trends along what is now the margin of Africa. The breakup of Pangea began in the Early Jurassic, at about 180–160 Ma, in the Central Atlantic rift zone on the plume-affected, overthickened crust of the Appalachian Mountains (Sheridan, 1983; Emery and Uchupi, 1984; Vink et al., 1984).

Some failed rift systems are composed of one extensional event, which typically lasts 10–15 Myr (e.g., Western Branch, EAR, Ebinger, 1989; Oslo Graben, Pallesen, 1993; Jurassic North Sea rifts, e.g., Rattey and Hayward, 1993). Such rifts are characterized by a relatively simple rift structure and little migration of fault activity with time. Longer-lived rift systems (lasting some 20–100 Myr) tend to be composed of several separate extensional events lasting about 10–15 Myr each, e.g., Cretaceous–Paleogene rifts in Sudan (Schull, 1988), the Cretaceous–Paleogene Anza Graben (Morley et al., 1999b), Oligocene–Recent Kenya Rift, Turkana area (Morley et al., 1999a). These rifts show superimposed or migrating basins of different age. Some rifts even show alternating rift and sag basin sequences (e.g., Sudan Rift, Schull, 1988). Other areas of extension are composed of multiple rifting events widely separated in time; for example, the Rockall Trough, off the northwest coast of Ireland, is composed of Triassic, Late Jurassic, and Early Cretaceous rifting events, where extension became more focused to the basin center with time, and parts of older rifts form perched basins on the margins of the youngest rift (e.g., Shannon et al., 1999). Such areas of repeated rifting can have highly attenuated crust, stretched by a β factor of 3–4.

Rift systems evolve in a variety of ways. They can propagate both along strike and laterally (Fig. 6; Baker and Wohlenberg, 1971; Morley, 1994). Depocenters can remain fixed for long periods of time (e.g., Rukwa rift, Morley, 1999a) or migrate as faults propagate and link (e.g., Schlische and Anders, 1996; Morley, 1999a). In some cases the location of active extensional faults follows changes in the location of active volcanism, suggesting a magmatic control on structural evolution; the Kenya rift is probably the best documented example of this behavior (Baker and Wohlenberg, 1971; Morley, 1994).

## *Driving Mechanism(s)*

Major continental rift systems are thought to form either in response to lithospheric stresses such as slab pull and ridge push or in response to buoyancy forces associated with mantle plumes and to crustal and lithospheric thinning and thickening (e.g., Bott, 1992, 1993). Results of the world stress map have been interpreted in different ways: the predominance of compressional intraplate stresses led Zoback (1992) to suggest that buoyancy stresses are of similar magnitude to plate driving forces. The East African rift illustrates the problem of understanding plate driving forces. Africa is surrounded by mid-ocean ridges, which place the continent under compression (Zoback, 1992). Consequently it is difficult to envision how plate driving forces have created extension in East Africa, hence Zoback (1992) appeals to buoyancy forces related to plume-driven domal uplifts (Fig. 6). However, the East African rift did not evolve in a stable stress field; extensional stresses have rotated with time (e.g., Bosworth et al., 1992), and there have been at least three phases of inversion (Fig. 6; Morley et al., 1999c). Bosworth et al. (1992) have argued that this evolution is not consistent with simple buoyancy-driven extension.

The number of continental rifts of different age that crosscut each other is a remarkable feature of Pangea, and may be related to its size. It is possible that the insulating effect of the large Pangean continental mass trapped heat, reducing the heat flow, and possibly raised the upper-mantle temperature by as much as 200°C (Anderson, 1982). This may have led to thermal and chemical effects similar to mantle plumes being mimicked by upper-mantle convection (Anderson, 1982). The combination of this instability, the effects of actual mantle plumes, plus lithospheric stresses led to repeated large-scale rifting events that either failed to break up the continent or resulted in the successful splitting off of continental fragments from the Permian to the present day.

### *Evolution from Extension to Thermal Subsidence*

Heat flow is highly variable in regional continental rifts, ranging between about 50 $mWm^2$ on the rift flanks up to 110 $mWm^2$ within the rift (e.g., Kuznir and Park, 1987). Local variations are commonly associated with fluid migration along faults and the presence of subsurface magma chambers. Regional variations in heat flow depend largely upon the amount of lithospheric extension (passive rifting; McKenzie, 1978) and the presence or absence of mantle-plume-related heating. Post-rift subsidence is a consequence of the time-dependent decay of perturbed syn-rift lithospheric geotherms and consequent density changes (McKenzie, 1978). Regional continental rift systems are the best candidates of all rift types for generating the classic thermal subsidence post-rift basins. Modeling makes no differentiation between initiation of the post-rift stage in a marine and in a continental environment, but geologically the difference is an important one.

Typically in marine rifts the covering of water protects much of the syn-rift topography, although structural highs near sea level are sites of erosion or nondeposition (Fig. 2). In areas of deeper water, topography is slowly smoothed out by sedimentation, which continues without a break from syn-rift to post-rift, infilling the remnant hanging-wall lows (Prosser, 1993). For example, following late Jurassic extension in the North Sea the marine sag basin remained persistently fine grained throughout the Cretaceous, with episodic influxes of turbidite sands into the deeper-water areas (e.g., Rattey and Hayward, 1993). The low topography associated with sag basins reduces coarse clastic input and is conducive to carbonate deposition, particularly on remnant syn-rift highs. This may happen at a very early stage in the history of the sag basin when structural highs are slightly submerged, as is the case in the Sirt basin (Parsons et al., 1980; Spring and Hansen, 1998).

The syn-rift to post-rift transition for continental deposition differs considerably from the marine setting, because no covering of water protects the syn-rift topography. As strain rate decreases, subsidence becomes reduced on the major boundary faults and coarse clastic fluviodeltaic deposits tend to form the latest syn-rift infill, replacing earlier fine-grained lacustrine deposits (see Lambiase and Bosworth, 1995, for a review). Following infilling of the basins the high parts of the rift topography continue to be eroded. The old rift-basin lows will have a thin veneer of post-rift sediment but be predominantly regions of sediment bypass, e.g., the west Turkana basins in Northern Kenya today (Morley et al., 1999a). Unlike marine rift–sag transitions, where considerable submarine rift topography is preserved for tens of millions of years after the end of rifting, sag basins in a continental setting tend to develop on highly eroded and smoothed syn-rift topography (Fig. 2B, C). Note also the very thin post-rift section in the Anza Graben (Fig. 2C) despite over 25 Myr having passed since extension stopped. Often inversion occurs towards the end of extension, and thermal subsidence occurs post-inversion (Fig. 2C; Morley et al., 1999c).

After several millions of years thermal subsidence may begin to affect the rifts. In the case of the Lokichar basin in northern Kenya, thermal subsidence has yet to develop a noticeable sag basin even 10 Myr after the area stopped extending. In continental basins the first post-rift basin fill tends to be highly sand-prone, fluviodeltaic deposits (e.g., Anza Graben, Kenya; Morley et al., 1999b). As the remnant rift topography becomes reduced and the sag basin broadens, more differentiation of the sedimentary environments occurs, and in well developed, large sags there may be an overall fining-upwards sequence as lacustrine conditions are established in the center of the basin. For basins located close to a seaway, sag-basin subsidence (possibly coupled with a sea-level highstand) may lead to a marine incursion transgressing the early fluviodeltaic sediments, again resulting in a gross fining-upwards trend.

The amount of thermal subsidence is generally in the order of 3–4 km over 70–100 Myr (e.g., Cretaceous–Tertiary of the North Sea; White and McKenzie, 1988, Roberts et al., 1993; Fig. 2). This reflects the generally low amounts of extension ($\beta < 2$) and the passive nature of the extension.

### *Examples*

(1) Permo-Triassic rift systems: Dnepr–Donets, Oslo Graben; (2) Triassic rift systems: North Sea, Karroo rifts Africa, East Coast USA; (3) Jurassic rifts: North Sea, Atlas Mountains (inverted); (4) Cretaceous Central African Rift system; (5) Tertiary: East African Rift system.

### *Hydrocarbon Prospectivity*

Rifts associated with failed continental breakup are commonly highly prospective (e.g., North Sea, Central African Rift systems, Dnepr–Donets basin). The presence of the thermal sag basin is particularly important to prospectivity by causing subsidence and widespread maturity of syn-rift source rocks, providing a top seal to rift structures, especially tilted fault blocks, and by forcing lateral migration of hydrocarbons (e.g., Morley, 1999c). Failed continental rifts offer good opportunities to develop deep-water marine and lacustrine shales, with good source-rock potential. There are usually opportunities for sands to enter the rift system axially, from flexural margins and via transfer zones. However, parts of the rift system are likely to have topographic configurations that restrict sand entry. Prospectivity is greatly reduced when large volumes of volcanics occur in the rift system (e.g., Ethiopian Rift, Kenya Rift, Oslo Graben) and in rifts that have not developed thermal-subsidence basins or had them eroded (Morley, 1999c). Tectonic inversion seems remarkably common in this rift setting for a regional extensional setting. It may be that forces such as ridge push and arising from differential drag on the base of the lithosphere are responsible for much of the inversion (e.g., Withjack et al., 1998).

## TEMPORARY AND PERMANENTLY ABANDONED PROPAGATING RIFT TIPS

### *Introduction*

Associated with the development of back-arc and oceanic spreading centers is the formation of continental rifts ahead of propagating spreading centers (PSCs). Most commonly oceanic spreading centers propagate gradually and progressively through the continental rifts, which become inactive and consequently line passive continental margins. Continental rifts associated with PSCs tend to have structural geometries similar to failed regional continental rifts (see section on Failed Regional Continental Extension, this article), but instead of lying within continental masses, they lie along passive margins, en echelon to, or along strike from active spreading centers (Figs. 8, 9, 11). One

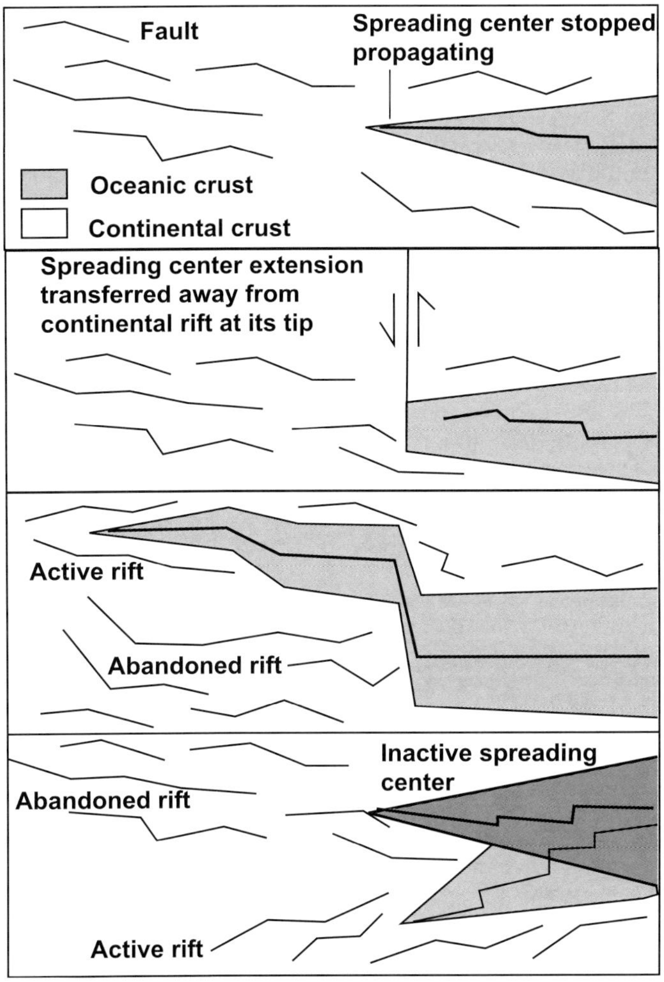


FIG. 8.—Schematic illustrations showing how different types of abandonment or stalling of propagating spreading center tips may develop.

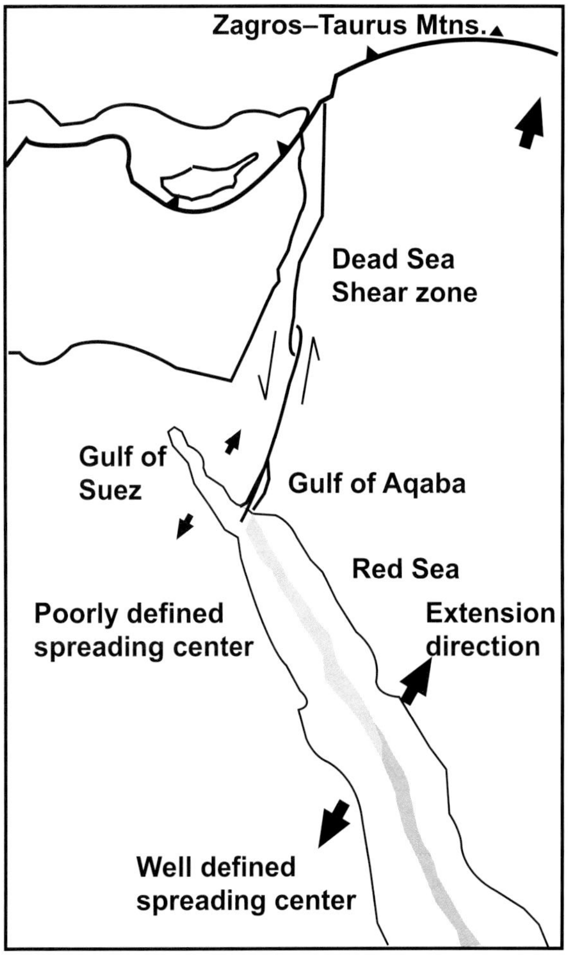


FIG. 9.—Regional setting of the Gulf of Suez rift at the tip of the Red Sea spreading center. From the late Miocene onwards extension from the Red Sea province has largely been transferred to the Taurus–Zagros collision zone, via the Dead Sea–Gulf of Aqaba shear zone, leaving the Gulf of Suez rift abandoned. Hence the dramatic decrease in width of the Red Sea passing into the Gulf of Suez. (Modified from Joffe and Garfunkel, 1987).

difference is that they can be broader regions of extension, and areas of higher strain (e.g., Brun and Bessler, 1996). Consequently low-angle extensional detachment faults and strike-slip transfer faults may be more important than in the setting discussed above (e.g., Lister et al., 1986; Tankard and Welsink, 1989; Brun and Bessler, 1996). Also the transition to oceanic crust requires evolution from extension by faulting to extension that is accommodated mostly by igneous intrusions (e.g., Parson and Wright, 1996).

### *Patterns of Development of Propagating Spreading Centers*

In the simplest cases, spreading centers propagate along strike into continental or island-arc crust that has undergone some degree of extension. The rifted area may extend along strike from the tip of the propagating spreading center some 1000–1500 km (e.g., Parson and Wright, 1996). For the Lau–Havre–Taupo back-arc basin in the Pacific, the highly rifted crust near the spreading center tip has undergone at least 120 km extension, with extension rates ranging between 15 and 45 mm $yr^{-1}$ (Parson and Wright, 1996). Extension occurred by both faulting and magmatic intrusion. Along strike, where oceanic crust is being created, full spreading rates are about 90 mm $yr^{-1}$.

Under certain circumstances continental rifts formed at the tip of a spreading center become permanently abandoned, because either (Fig. 8): (1) seafloor spreading stopped, or ceased propagating (e.g., the extensive Tertiary rifts southwest of the South China Sea spreading center; Northwestern Atlantic spreading center, e.g., Dore et al., 1999); or (2) the spreading center took a path away from the rift (e.g., Reconcavo basin, Brazil, Milani and Davison, 1988; Gulf of Suez rift at the tip of the Red Sea spreading center, Bonatti, 1985; Patton et al., 1994; Sorvestnaget Basin, Barents Sea, Dore et al., 1999) or became reoriented (e.g., Woodlark Basin spreading center, Goodliffe et al., 1997).

In the case of the Gulf of Suez, the Aqaba–Dead Sea shear zone tranferred about 107 km extensional displacement in the Red Sea northeastwards into a collisional plate margin (e.g., Fig. 9; Joffe and Garfunkel, 1987). Northwest of the shear zone the Gulf of Suez forms the abandoned tip of the PSC. The Gulf of Suez displays a relatively rapid decrease in strain over a strike distance of about 200 km from 34–36 km extension near the Aqaba–Dead

Sea shear zone (β = 1.56–1.62, Bosworth, 1995) to β = 1.1–1.2 in the northwestern Gulf (Patton et al., 1994). Sea-floor spreading in the Red Sea was pulsed, with late Oligocene–early Miocene spreading, a hiatus during the late Miocene, and rejuvenation of spreading in the Plio-Pleistocene (Girdler, 1991). Such pulsed spreading would temporarily stall propagation of rifting, which can sometimes be identified in the geological record. For example, as the South Atlantic spreading center developed during the Cretaceous, south of the Walvis Ridge continental rifting ended in the early Aptian and was replaced by seafloor spreading, while north of the ridge the end of continental extension was delayed until the late Aptian (Fig. 10; Duval et al., 1992; Clemson et al., 1997).

The late Jurassic–Tertiary history of seafloor spreading in the North Atlantic area illustrates how varied the propagation of spreading centers can be (e.g., Dore et al., 1999, their fig. 6). In the late Jurassic the broad Jeanne D'Arc basin formed at the stalled Central Atlantic spreading center tip. In the mid-Cretaceous, extension in the Rockall trough lay at the tip of the North Atlantic spreading center. Then in the Paleogene–Early Eocene the spreading center propagated northwards between Greenland and Canada; subsequently the branch became abandoned, leaving continental rifts at its tip offshore central west Greenland along the Davis Strait. During the Oligocene–Miocene, spreading switched to the northeastern Atlantic between Greenland and northwest Europe along a line of superimposed Triassic, Jurassic, and Cretaceous failed regional continental rifts and seafloor spreading–related Tertiary rifts. The switch in spreading-center locations was probably caused by mantle-plume activity in the northeastern Atlantic area (as reviewed in Dore et al., 1999).

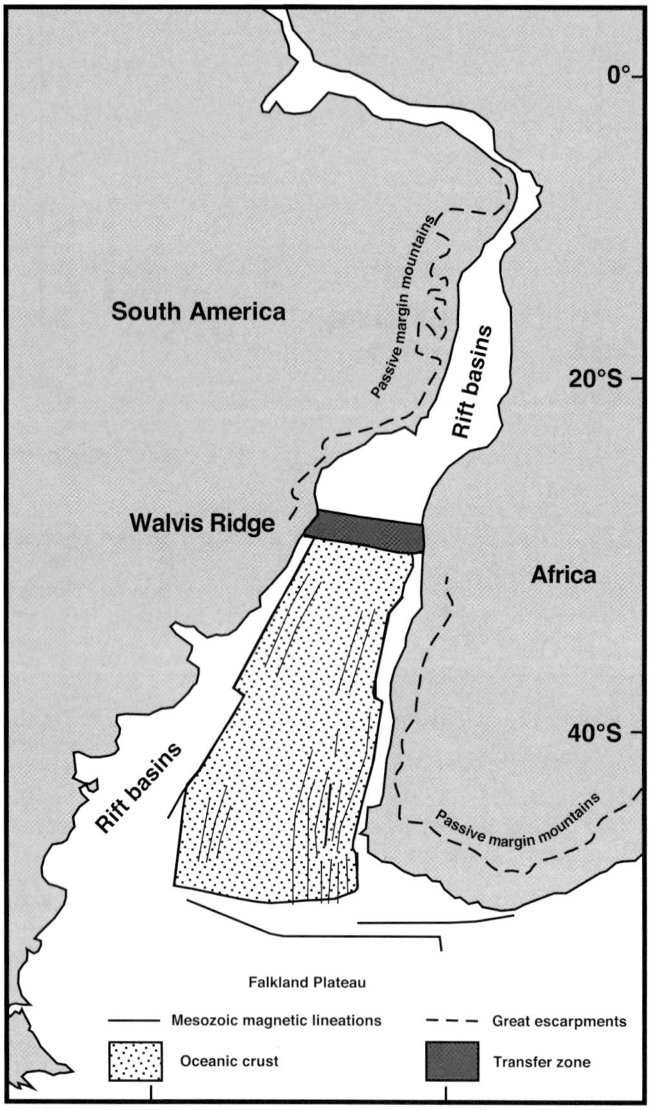


FIG. 10.—Blockage of the propagating South Atlantic rift at the Walvis Ridge (Aptian; from Etheridge et al., 1989).

## *Structural Styles*

Rift systems associated with failed continental breakup tend to be over 1000 km long, whereas those at permanently abandoned propagating tips tend to be less than 1500 km long. Rifts at PSCs can be narrow (e.g., Gulf of Suez, 50–80 km wide), or stubby broad rifts like the Sirt basin (400 km wide) and the Jurassic rifts (including the Jeanne D'arc basin) east of Newfoundland (200–300 km wide; Tankard and Welsink,1989).

Extension at stalled PSCs is relatively low-strain-rate deformation dominated by faulting. Low strain rates probably cause increased penetration of water into the lower crust and mantle, leading to advective cooling of the lower crust and upper mantle, and serpentinization of the mantle (Louden and Chian, 1999). Serpentinized mantle appears to develop predominantly in the transition zone between continental rifting and sea-floor spreading (Louden and Chian, 1999). Louden and Chian (1999) propose that cooling and serpentinization may significantly decrease lithospheric strength, leading to the formation of wide zones of extension where development of serpentinized mantle is greatest. This may favor the development of low-angle detachments on specific segments of passive margins where serpentinized mantle is best developed (e.g., Reston, 1996).

Although most rifts at PSCs are typical continental-style rifts, metamorphic core complexes are sometimes developed. Metamorphic core complexes occur where a spreading center has propagated into the weak, hot, thickened continental crust of an orogenic belt (e.g., Woodlark basin–D'Entrecasteaux Islands, Papua New Guinea; Hill et al., 1992; Mutter et al., 1996; Fig. 11).

## *Volcanic History*

Major oceanic spreading centers are associated with volcanism generated by adiabatic decompression melting, which requires large beta factors (McKenzie and Bickle, 1988). Spreading centers located over plumes produce flood-basalt provinces (White and McKenzie, 1989), but thermal weakening of the lithosphere by the plume means that stalled rifts are unlikely to develop. Consequently there tends to only be minor volcanic activity associated with rifts at stalled PSC tips (e.g., Gulf of Suez, Patton et al., 1994; Reconcavo basins, Ghignone and Andrade, 1970). Along strike, however, there may be extensive plume-related volcanic provinces. Thermal perturbations at the PSC may lead to localized volcanism (e.g., Late Tertiary volcanism, southeastern Vietnam; Wang et al., 2001). Propagating back-arc basins have both arc volcanism related to subduction of oceanic crust and adiabatic decompression melting at the spreading center.

## *Driving Mechanism(s)*

Slab pull and ridge push, locally aided by mantle plume–related forces, are the main driving mechanisms.

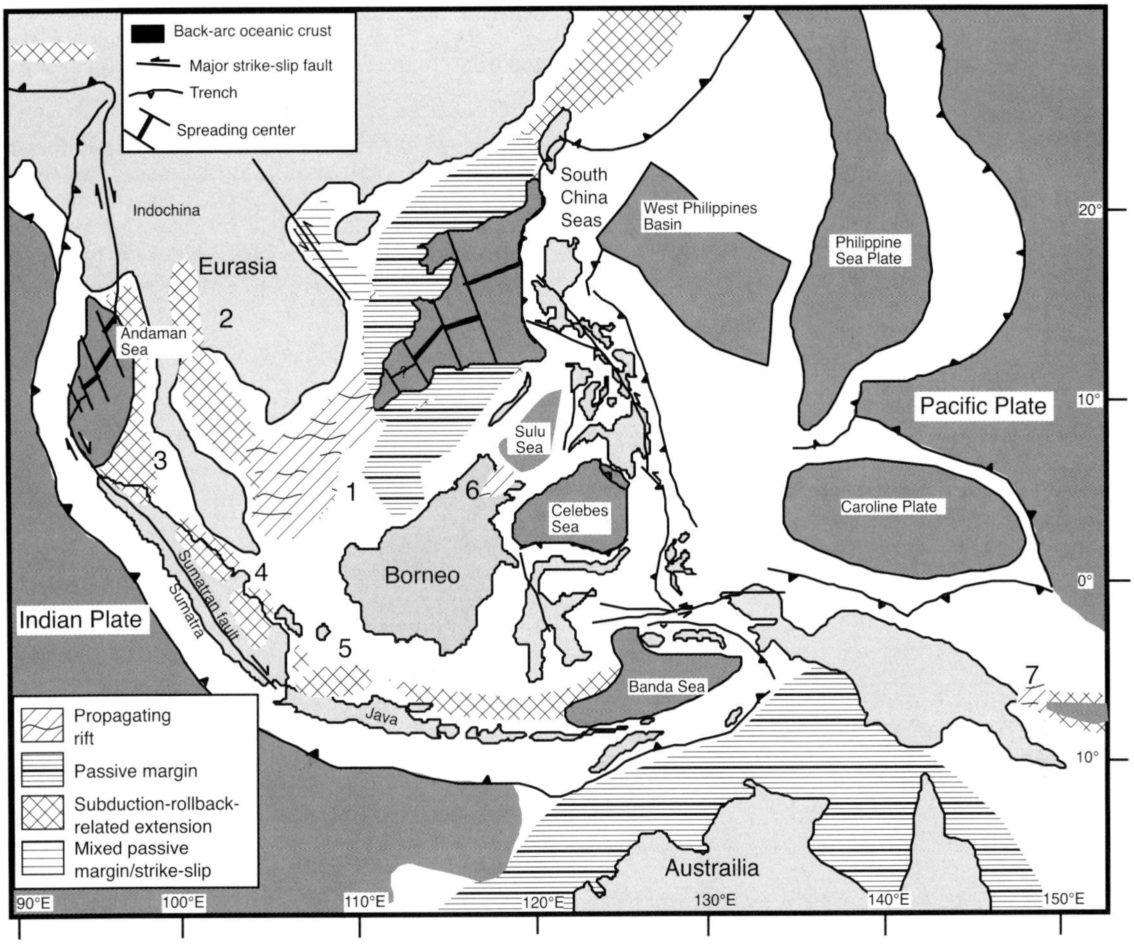


FIG. 11.—Map of extensional provinces in South East Asia. 1) Nam Con Song, West Natuna basins, 2) Thailand rift basins, 3) Andaman Sea, 4) Sumatra rifts, 5) Java rifts, 6) propagating tip of Sulu Sea rift, Borneo, 7) propagating tip of Woodlark basin spreading center and D'Entrecastinaux Islands metamorphic core complex.

### *Sedimentation Style*

The timing of spreading-center propagation with respect to the timing of continental extension is critical for whether marine or continental sedimentation occurs, and this factor is highly variable. Rifts along the west coast of Africa are predominantly continental, indicating that the PSC was some distance away when the rifts developed. Southwest of the South China Sea spreading center, the rifts (Cuu Lon, Nam Con Son, weat Natuna basins, e.g., Ginger et al., 1993; Matthews et al., 1997; Fig. 11) are also filled predominantly by continental deposits. Marine conditions were established only in the post-rift phase. Only during the initial stages of extension, however, does the Gulf of Suez display continental sedimentation; most of the syn-rift section is marine (e.g., Patton et al., 1994).

### *Evolution from Extension to Thermal Subsidence*

Rifts formed at temporarily stalled PSCs are usually covered later by passive-margin sedimentary rocks. If the spreading center jumped a considerable distance from the abandoned continental rift, however, then the rift may lie beyond the region of thermal subsidence, and on the uplifted passive-margin flanks. The Reconcavo–Tucano rift, Brazil (Fig. 1) is such an example: it has no associated post-rift subsidence (e.g., Ghignone and Andrade, 1970). Abandoned rifts ahead of PSCs usually undergo typical thermal subsidence to form a sag basin (e.g., West Natuna basin, Ginger et al., 1993; Matthews et al., 1997). Rifts close to the spreading center, however, may experience a prolonged transition to thermal subsidence because of interaction with the extensional forces related to the spreading center.

The Gulf of Suez is an example of an episodic, prolonged transition to post-rift subsidence. As summarized by Patton et al. (1994), the Gulf of Suez rift began extending during the Oligocene. A marked phase of generally reduced subsidence rates and structural reorganization occurred after a widespread unconformity of middle Miocene age called the mid Rudeis event. The origin of this event is not completely certain, but following Steckler and ten Brink (1986) many workers favor motion along the Aqaba–Dead Sea wrench system as replacing much of the extension in the Gulf of Suez. Such a mechanism could give rise to an abrupt cessation of extension followed by post-rift subsidence. Indeed Wescott et al. (1998) chose to define this event as the start of the post-rift sequence. In the Gulf of Suez, however, it is remarkably difficult to define whether post-rift subsidence has begun even today.

Some three and a half million years after the mid Rudies event, at the top of the Kareem Formation is a pronounced unconformity, where a marked slowing of subsidence rates is recorded, followed by the deposition of evaporites. Patton et al. (1994) identify this event as the time when the Gulf of Suez was abandoned as a site of active extension for about 7 Myr. Extension was later renewed in the central and southern Gulf of Suez from the latest Miocene to the present (see review in Patton et al., 1994; McClay et al., 1998). Consequently, McClay et al. (1998) have classified the entire sequence from the Oligocene to the present day as syn-rift.

The problem with classifying the start of the post-rift sequence in the Gulf of Suez centers around the assumption, inherited from the passive-rift model, that there is on a geological time scale an abrupt change from syn-rift to post-rift subsidence. The last 16 Myr history (post–mid Rudeis event) of the Gulf of Suez has elements of both post-rift and syn-rift subsidence and must be regarded as an extended transitional event that has yet to end. The topography around the Gulf of Suez is still very pronounced, so that even if extension stopped today a considerable period of time would elapse before thermal subsidence created a sag basin wider than the rift basin.

The presence of marine gulfs created at temporarily (e.g., West Africa north of the Walvis ridge, Fig. 10) or permanently (e.g., Gulf of Suez) stalled PSCs creates an opportunity for evaporite deposition for a geologically short period of time, e.g., the Aptian (6–7 Myr) in Angola and the early Tortonian (1–2 Myr) in the Gulf of Suez (Duval et al., 1992; Patton et al., 1994). Economically the presence of evaporites is important as both a seal and for creating traps for hydrocarbons (e.g., Duval et al., 1992).

### *Inversion*

The occurrence and intensity of inversion tectonics is highly variable from very little (none?) in the Gulf of Suez and the Reconcavo basin, to spectacular examples in the West Natuna Basin (Ginger et al., 1993). In the northwestern Atlantic area, the onset of seafloor spreading between Greenland and Europe placed northern Europe under enhanced horizontal northwest–southeast oriented regional compressive stresses, by ridge-push forces from the morthwestern Atlantic spreading center to the northwest and Alpine compressional stresses to the south (e.g., Dore et al., 1999). The result was Oligocene–Miocene inversion of many older extensional structures along the Atlantic margin.

### *Examples*

Nam Con Song, Cuu Long–West Natuna basins, South China Sea; Gulf of Suez; Woodlark Ridge–Papua New Guinea; North Borneo–Sulu Sea; Reconcavo Basin, Brazil; Voring basin, northwest European margin; possibly Sirt Basin, Libya; Falklands Plateau; East Coast Jurassic rift basins around Newfoundland; West Africa–South America north of Walvis Ridge; Valencia Trough, Mediterranean; Bass Basin, Australia.

### *Hydrocarbon Prospectivity*

In terms of hydrocarbon prospectivity, the propagating-rift setting is similar in many ways to the failed-continental-rift setting. The main differences are that the prospective rift trends are shorter and commonly broader, and may pass laterally into passive margins or ultimately become part of a passive margin. Typically the post-rift section is marine or mixed continental and marine. Areas of stalled rifting may be hydrocarbon sweet spots on passive margins, for example the Jeanne D'Arc basin along the east coast of North America. As noted above, the presence of evaporites is important as a top seal in several examples of propagating rifts. Commonly the post-rift section acts as a good seal; in the Gulf of Suez the excellent evaporite seal means that all the large fields are confined to the syn-rift and pre-rift section, predominantly in tilted-fault-block plays (Patton et al., 1994). In the Sirt Basin most of the largest traps are concentrated in the pre-rift or post-rift section along horst blocks trends (platforms) flanking the rift basins, the broad nature of the rift gives rise to several productive platform trends (e.g., Parsons et al., 1980; Spring and Hansen, 1998).

## EXTENSIONAL BASINS ASSOCIATED WITH SUBDUCTION ROLLBACK (OCEANIC CRUST–ISLAND ARC OR OCEANIC CRUST–CONTINENTAL CRUST)

### *Introduction*

The breakup of Pangea around its African core represents very widespread rifting events and propagation of large oceanic spreading centers. Subduction rollback in a volcanic-arc setting typically produces back-arc extension, as exemplified by numerous parts of Southeast Asia (Figs. 11, 12). It is a very different kind of continental extension from the breakup of Pangea. Back-arc extension can be developed at continental margins where a long, narrow region of continental crust is rifted away from the main continental region, to leave a rhomboidal or wedge-shaped marine basin floored by oceanic crust. The Sulu Sea is such an example (Fig. 11). The isolated, U-shaped South Sandwich subduction zone (Fig. 1) is a more extreme example of back-arc seafloor spreading. There are also numerous failed back-arc continental rifts; particularly well developed examples are found along the northern parts of Java and Sumatra, where the region of back-arc continental extension lies about 300–800 km from the trench. Use of the term back-arc in the section heading was avoided because although rifting associated with subduction rollback generally occurs in the back-arc setting (e.g., Java, Sumatra, Andaman Sea, Lau–Hau–Taupo back-arc basin; Parson and Wright, 1996), more rarely subduction rollback may generate extension in the fore-arc (e.g., Alexander Island, Antarctica; Macdonald et al., 1999), volcanic arc (Sumisu Rift, Izu–Bonin Arc; Taylor et al., 1991), or 1000 km or more away from the trench (e.g., Bohai Basin; Allen et al., 1997; possibly the Thailand–Gulf of Thailand; Watcharanantakul and Morley, 2000).

The consistent location of extensional basins in the upper plate of subduction zones led Karig (1971) to propose that the descending slab is in some way responsible for causing extension. Steepening of the subducting plate with time causes the subduction zone and trench to migrate backwards or roll back in the direction of the (main) spreading center (Fig. 12). This situation leads to extension in the overlying plate when the rollback rate exceeds the convergence rate of the two plates. Hence despite the overall convergence of the two plates, locally extension can dominate. Four mechanisms which could drive subduction rollback are as follows: (1) The descending slab steepens with time under the influence of gravity (Dewey, 1980). (2) Relative rates of motion of the converging plates change (Hynes and Mott, 1985). (3) Approximately westward global motion of the lithosphere with respect to the underling mantle creates flow against west-dipping subduction zones, making them steep and prone to subduction rollback (Doglioni, 1992). (4) Slab suction forces may drive slab rollback, or cause extension independently of slab rollback (Bott, 1993).

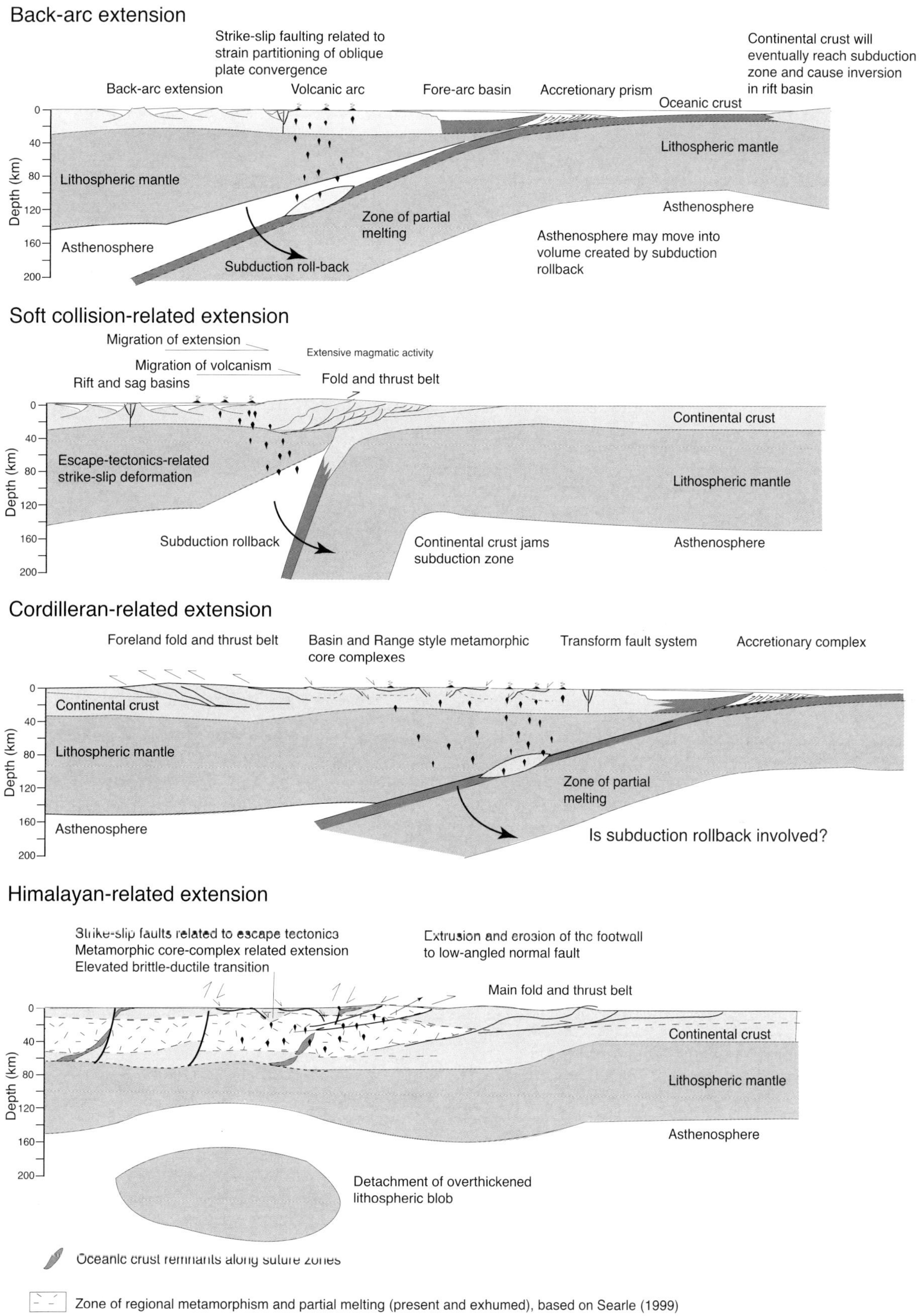


FIG. 12.—Cross sections illustrating different types of extension associated with convergent plate-tectonic settings.

### *Structural Style*

Typical back-arc rift basins are similar to continental rifts in the general size and distribution of half-graben and full-graben building blocks. Their length is controlled by the subduction-zone geometry, and more locally preexisting fabrics can exert a strong control on basin orientation. In Central Sumatra, for example, the trench and volcanic arc trends are northwest–southeast but the dominant trends in the rift basins are commonly north–south (Williams and Eubank, 1995). Instead of the long-linear trends seen in failed continental rifts, rifts in a back-arc setting are commonly composed of shorter, wider segments (Williams and Eubank, 1995).

Extension can also occur along volcanic arcs, where a number of significant differences between extension on volcanic arcs and continental crust exist, as discussed by Taylor et al. (1991) for the Sumisu Bonin arc. (1) The typical preexisting fabrics of continental crust do not exist for volcanic arcs. Instead volcanic centers tend to control the location of rifts, and rifts are better developed between major volcanic centers, probably because igneous intrusions accommodate extension around the sites of the volcanic centers. (2) Volcanic-arc lithosphere is very weak (effective elastic thickness of about 3 km); consequently the rift tends to be narrower (20–50 km wide) and be more densely faulted than continental rifts. Many faults detach just a few kilometers below the seafloor. (3) Volcanic rocks are intruded along faults. (4) Very high sedimentation rates (up to 6 m/1000 yr) keep pace with fault-related subsidence, resulting in a smooth basin floor.

### *Inversion*

The location of extensional basins in the upper plate of subduction zones makes the likelihood of inversion affecting the basins the highest for any type of rift system. Island arcs are commonly cut through by arc-parallel strike-slip faults developed because of strain partitioning along regions of oblique plate convergence (e.g., McCaffry, 1992). The interaction of strike-slip and extensional geometries can produce complex syn-rift and post-rift structural relationships, and lead to inversion (e.g., Ombilin Basin, Sumatra; Williams et al., 1995). The post-rift history is also likely to include episodes of compression or strike-slip deformation as various elements of thickened crust, including continents and microcontinents, reach the subduction zone. In Sumatra and Java, for example, rift basins have undergone post-rift inversion (e.g., Eubank and Makki, 1981; Letouzey, 1990) due to the impingement of the Australian continental crust on the island-arc system (Fig. 11; e.g., Hall, 1996). Collision beyond the region of back-arc extension can cause inversion related to escape tectonics. For example, the rift basins of Thailand may lie in a highly oblique back-arc setting; they experienced inversion as a result of northwest–southeast to north–south compressional stresses related to the Himalayan orogenic belt, which lies to the northwest (Huchon et al., 1994; Morley, 2001).

### *Effects of Subduction Rollback on Volcanism and Heat Flow*

Subduction rollback, in addition to causing extension, also affects the thermal structure of the lithosphere. As the slab steepens, extra volume is created above the slab, which can be filled either by lateral flow of the lithosphere (Watcharanantakul and Morley, 2000) or by emplacement of the asthenosphere immediately above the slab (Fig.12; Tatsumi and Kimura, 1991; Doglioni, 1992).

Tatsumi and Kimura (1991) discussed the effects on magmas produced by cooling of a wedge of hot asthenosphere emplaced above a subducting slab. They noted that as cooling progresses the effects are (1) magma becomes more alkaline and (2) the volcanic front migrates towards the back-arc side with time. The initial emplacement of the asthenospheric wedge produced a short pulse of high-magnesium andesites, which require anomalously high-temperature conditions to develop.

Differential or non-uniform stretching could be caused by subduction rollback. If the slab steepens as it rolls back, then the farther (i.e., deeper) the slab is from the hinge line, the greater the horizontal component of extension (Fig. 13). For example, if, with a hinge line at the trench, a slab dipping at 15° steepens to 17° then at 40 km depth (about at the Moho) this would result in extension of 19 km, whereas at the base of the lithosphere (120 km) extension would be 55 km (Watcharanantakul and Morley, 2000). This mechanism requires coupling of the overlying plate lithosphere with the slab as it rolls back. It can help explain some phenomena in the upper crust, such as a rapid subsidence in short-lived thermal subsidence basins, and high heat flows (Watcharanantakul and Morley, 2000).

### *Thermal Subsidence*

Heat flow and geothermal gradients in rifts developed in the upper plate of subduction zones tends to be high because of the non-uniform lithospheric thinning caused by subduction rollback and emplacement of asthenosphere above the subducting slab. For example: in the central Sumatran rifts, the average geothermal gradient is 6°C/100 m (Williams and Eubank, 1995); in the Malay Basin the maximum heat flow is 105 $mWm^2$ (Mazlan, 1997); in the Pattani Basin, the maximum geothermal gradient is 6°C/100 m and heat flow is 101 $mWm^2$ (Bustin and Chonchawalit, 1995). Most back-arc basins seem to display normal to relatively rapid thermal subsidence rates compared with failed regional continental rifts. Given the high heat flows recorded during the thermal subsidence phases of these rifts today, rapid thermal subsidence is to be expected, but only when the basins are relatively close to the upwelling asthenospheric. For example, in the Bohai basin, which lies about 1000 km from the trench, post-rift subsidence for about the last 25 Myr has a comparatively modest maximum thickness of 2 km (Allen et al., 1997). Conversely, the Malay Basin has undergone thermal subsidence for the last 25 Myr, and the post-rift basin reaches a maximum thickness of 8 km, whereas the 10 Ma old thermal sag of the Pattani basin reaches a thickness of 4 km (Mazlan, 1997; Watcharanantakul and Morley, 2000).

### *Examples*

Java and Sumatran rifts, Indonesia (Williams and Eubank, 1995); Andaman Sea–Thailand rift basins? (Morley, 2001); Bohai Basin (Allen et al., 1997); Sumisu Rift, Izu–Bonin Arc; Lau–Hau–Taupo back-arc basin, Parson and Wright (1996).

### *Hydrocarbon Prospectivity*

Young back-arc rifts have characteristics of hydrocarbon prospectivity similar to those of regional continental rifts with thermal sag basins, and have proven productive in Java, Sumatra, the Bohai Basin, and the Gulf of Thailand. A number of factors can impact prospectivity: (1) Because the rift basin is close to a volcanic arc, volcanism may either fill it with volcanic and volcaniclastic rocks, or reservoir sands may be of poor quality because of volcanic sediment source areas. (2) The back-arc

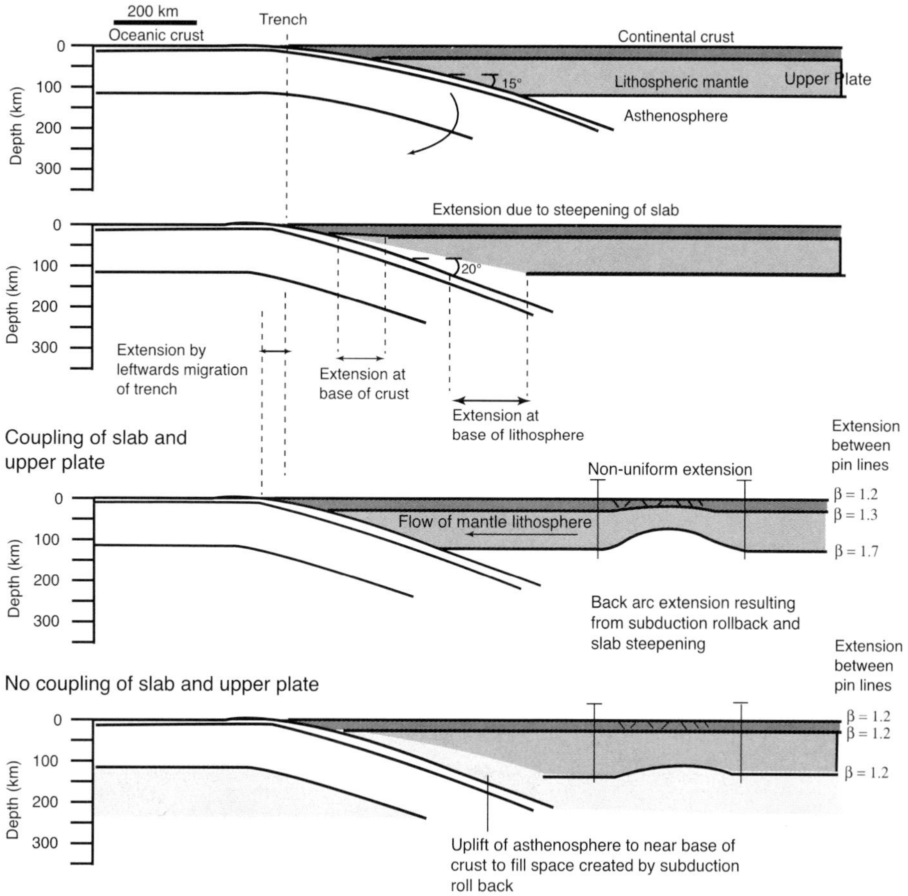


FIG. 13.—Schematic illustration of how subduction rollback and slab steepening could lead to non-uniform extension of the lithosphere, or uplift of the asthenosphere above the descending slab. (After Watcharanantakul and Morley, 2000.)

setting is prone to being involved later in continental collisions, hence the rift basins may be severely inverted and traps destroyed. This is particularly a problem for older basins. Conversely, in Java and Sumatra, late Tertiary inversion created major traps (anticlines called Sunda folds) in the post-rift section and form the largest oil fields in the basins (Williams and Eubank, 1995). The generally high heat flows in these basins is usually advantageous and allows extensive maturation of source rocks even in comparatively thin syn-rift and post-rift sequences.

## "SOFT" CONTINENTAL COLLISIONS

### *Introduction*

Zones of soft collision between continents occur where subduction rollback has inhibited major uplift, erosion and exposure of a high-grade metamorphic core in the hinterland of the thrust belt (Fig. 12; e.g., Royden and Burchfiel, 1989; Royden, 1993). Instead of the interior of the thrust belt being high and mountainous, it is low-lying and flat, or even covered by marine conditions, because of the presence of a thermal sag basin, as exemplified by the Rif–Betic mountains and the Alboran Sea, Apennines and the Tyrrhenean Sea, Corsica–Sardinia and the Liguro–Provençal basin, and the Carpathians–Pannonian basin in the Mediterranean region (Fig. 14). The low-lying interiors underwent extension and were subsequently covered by a thermal sag basin. The key characteristic of this diverse group of extensional basins is the development of a sag basin contemporaneous with the later stages of thrusting (e.g., Morley, 1993). Yet it is difficult to make generalizations because the main variables that affect the interior zone of extension are profound. They include:

1. The timing of extension with respect to crustal thickening. Some thrust belts may have undergone extension relatively early and crustal thickening of the internal zones (prior to extensional thinning) was minimal (e.g., Carpathians; Morley, 1993). Other thrust belts have undergone crustal thickening prior to extensional thinning (e.g., Rif–Betic Cordillera–Alboran Sea; Platt and Vissers, 1989; Zeck et al., 1992; Longeran and White, 1997; Platt et al., 1998).
2. The geometry of the collisional margins is highly variable, ranging from straight (e.g., the Apennines), through arcuate (e.g., the Carpathians), to U-shaped (the Rif–Betic Cordillera) (Fig. 14). Margin geometry affects the shape and extent of the extensional province and the development of strike-slip faults associated with escape tectonics. The geometry and magnitude of displacement on strike-slip faults is highly variable from region to region. In the Pannonian basin (Carpathians), for example, numerous strike-slip faults are superimposed on the extensional belt (e.g., Ratschbacher et al., 1991), whereas in the Apennines there are comparatively few.

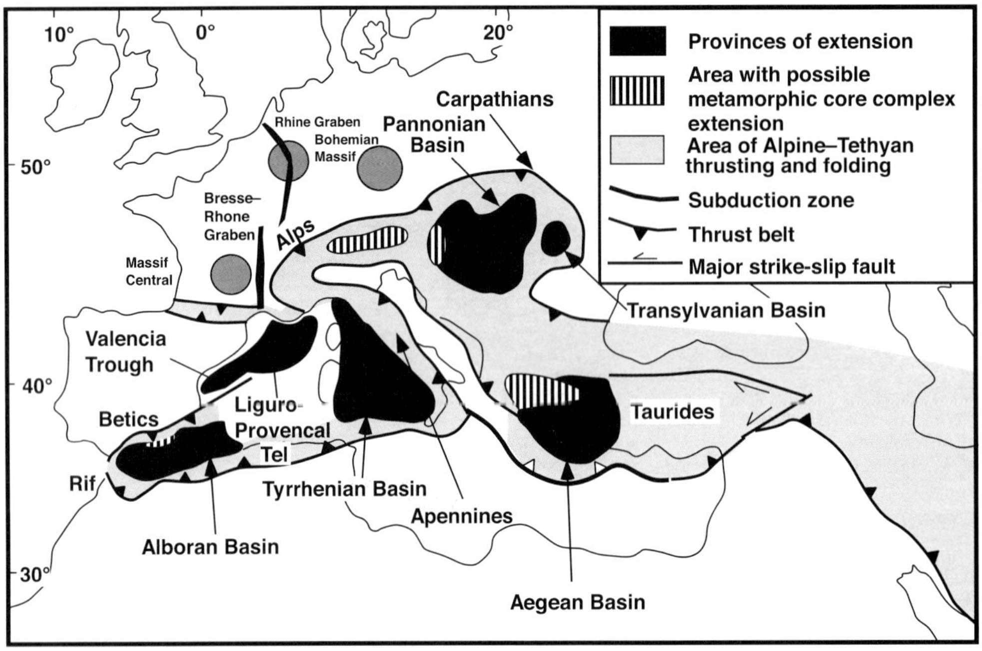


FIG. 14.—Map showing the locations of regions of extension associated with the Tertiary Alpine thrust belts of Europe and North Africa. (Modified from Dewey, 1988.)

### *Structural Styles*

Extensional basins have recognizable half-graben shapes, but the basin fill tends to be much thinner than in well developed rift basins in the East African rift, for example. Syn-rift basin fill is commonly just one or two kilometers thick. Rift basins may also be quite narrow, between 10 and 20 km wide. Seismic sections across the basins tend to show numerous horst-and-graben or half-graben structures, with considerable syn-rift topography draped by the post-rift section.

Extensional structures are superimposed on a fold and thrust belt; consequently many major extensional faults in this setting are low-angle because they reactivate older thrusts (negative inversion) (e.g., Pascucci et al., 1999). Hence, extensional imbricate structures are common. Extensional basins may be superimposed on older thrust-top sedimentary basins (Pascucci et al., 1999).

### *Syn-Rift Topography*

Today the topography of the Carpathians–Pannonian basin, Rif–Betics–Alboran Sea and Apennines–Tyrrhenian Sea is one of a highly subsided thrust-belt hinterland flanked on one or more sides by high mountains. Hence the borders of the rift system are controlled by thrust-related fringing mountains, which provide enormous amounts of sediment to fill the hinterland basins. Within the rift, uplift mechanisms include fault-related footwall uplift and variable uplift related to magmatic underplating and strike-slip faults (escape tectonics).

In the Alboran Sea and adjacent Betic Cordillera there is a complex history of unroofing, and exposure of high-grade metamorphic rocks near the end of rifting (e.g., Platt et al., 1998; Azanon and Crespo-Blanc, 2000). Crustal thickening in parts of the orogen began during the Eocene (e.g., Platt and Vissers, 1989) followed relatively early in the orogenic history (late Oligocene to Early Miocene) by extensional collapse of the hinterland contemporaneous with thrusting in the external zones. The region of the present Alboran Sea appears to have undergone rapid elevation of crust from depths of 30–40 km to the near surface (Zeck et al., 1992; Platt et al., 1998). The present-day thickness of the crust beneath the Alboran Sea is about 18 km, which, added to the thickness of eroded material, suggests that the original crustal thickness may have been about 50–60 km (Platt et al., 1998).

Thermal subsidence removes much of the syn-rift topography and creates a low, flat interior basin. Where other thrust-belt configurations block the basin from the sea, nonmarine conditions may be established (e.g., Pannonian basin, Carpathians; Nemcok et. al., 1998). More commonly, marine conditions are established during thermal subsidence (e.g., Tyrhennian Sea, Alboran Sea).

### *Volcanic History*

Subduction-related volcanism accompanies the formation of extensional and sag basins, although the igneous history varies from region to region. Volcanism tends to evolve from calc-alkaline to alkaline with time (e.g., Bellon, 1981; Rehault et al., 1985; Maillard and Mauffret, 1993). Volcanism in some regions, such as the Apennines, migrated towards the foreland with time accompanied by a similar migration in extension (Serri et al., 1993). Extension may be sufficiently intense to generate back-arc oceanic crust (e.g., Kastens and Mascle, 1990). Szabo et al. (1992) discuss a typical example of the types of volcanism exhibited in subduction-rollback settings. They identified three causes of volcanism in the Carpathian–Pannonian region: (1) Early Miocene acidic volcanism associated with right lateral shear along the Mid Hungarian shear zone; (2) Miocene–Pliocene calc-alkaline volcanism in the inner western and eastern Carpathians, about 100–150 km behind the thrust front, related to subduction of an oceanic slab; and (3) Plio-Pleistocene alkaline basaltic volcanism related to cooling of asthenosphere that had upwelled above the slab.

### *Driving Mechanism(s)*

Subduction rollback appears to be the main driving force for soft-collision thrust belts (e.g., Royden, 1993; Morley, 1993); this is particularly well established for the Carpathians (Tomek and Hall, 1993; Nemcok et al., 1998) and the Apennines (e.g., Amato et al., 1993; Mariucci et al., 1999). For the Rif–Betic Cordillera–Alboran Sea area, however, a variety of mechanisms have been proposed, including enhanced gravitational potential due to detachment of a lithospheric blob (see Lonergan and White, 1997, for a review). The subduction-rollback mechanism seems to be the only possible mechanism for thrust belts that developed interior rift basins prior to significant crustal thickening, such as much of the Carpathians–Pannonian basin. The Rif–Betic Cordillera does appear to have been thickened prior to extension, and modeled P–T paths for exhumed high-grade rocks from ODPS Site 976 indicate that lithospheric delamination (or thinning) occurred (Platt et al., 1998). Hence a combination of lithosphere detachment–delamination and subduction rollback may have driven deformation in the Rif–Betic Cordillera (Morley, 1993; Lonergan and White, 1997). Other extensional provinces in the Mediterranean have had gravitational collapse (Corsica–Ligurian Sea; Jolivet et al., 1991; Laubscher et al., 1992); delamination (Tyrrhenian Sea; Kastens and Mascle, 1988) and subduction rollback (Doglioni et al., 1997; Carminati et al., 1998) proposed as the main driving mechanisms for extension.

A problem with gravity collapse as the primary driving mechanism concerns space for extension to occur. Most extensional collapse occurs during thrusting, when regionally compressional stresses are dominant, not at the end of thrusting. The ductile lower crust can forcefully flow away from the region of thickened crust into regions of thinner crust, causing elevation changes as it does so (e.g., Liu et al., 2000; McKenzie et al., 2000). For associated large-scale brittle extension of the upper crust to occur, however, the resulting expanding area must be accommodated by relatively rigid movements of brittle adjacent continental blocks to accommodate it. Gravity collapse may be one driving mechanism, but it requires another mechanism to create space for the brittle upper-crustal extension, unless the magnitudes of the regional stresses are smaller than the stresses generated by gravity potential and space can be forcefully created (e.g., England and Houseman, 1989). Small regional stresses cannot, however, be the explanation for external-zone thrusting contemporaneous with extensional collapse.

Today the crust of the Rif–Betic and Carpathian thrust belts is significantly thicker than the internal regions of extension (reviews in Lonergan and White, 1997; Nemcok et al., 1998), so gravity collapse should be acting in the direction opposite to the thrust belts towards the extensional basins. Indeed, Bada et al. (2001) have proposed that horizontal stresses originating from the gravity potential of the thickened thrust belt crust are important for causing post-rift inversion in the Pannonian Basin. Extension in the Pannonian Basin and the Alboran Sea has proceeded beyond crustal thickness appropriate for gravity collapse. The extensional basins are areas of highly thinned crust, with low topography that clearly are no longer sites of overthickened, gravitationally unstable crust (Morley, 1993). One way to create space for initial gravity collapse, and to continue extension after loss of gravitational potential, is subduction rollback. For these reasons, arguments for a single driving mechanism seem to have unnecessarily polarized views about how extensional basins in thrust belts have developed. Commonly gravity collapse needs to be linked with space-creating mechanisms such as subduction rollback (Carpathians, Apennines), extrusion tectonics (Alps, Himalayas) and the tip of a propagating spreading center (D'Entrecastinaux Islands–Woodlark spreading center).

The effects of overthickened crust on the western side of the Pannonian Basin may have come into play in the transition zone from the Eastern Alps to the main part of the Pannonian Basin. Tari et al. (1992) noted that much of the Pannonian Basin displays relatively high-angle fault–bounded extensional basins characterized by 20–40% extension. To the west (Danube Basins, Great Hungarian Plain), however, seismic reflection data indicates the presence of metamorphic core complexes, possibly representing 200% extension. Tari et al. (1992) proposed that the core complexes developed when thickened crust built from Cretaceous Alpine nappes collapsed during the late Oligocene–early Miocene during a phase of east-directed escape tectonics.

### *Thermal Subsidence*

Extension and subduction rollback can elevate the asthenosphere to the base of the crust (Figs. 11, 12). The resulting highly elevated geothermal gradient would begin to decay once rollback ceased. Regionally high heat flows are common in this setting. Perhaps the best documented example is the Pannonian Basin, where values cluster between 80 and 100 $mWm^2$, with a maximum of 115 $mWm^2$ (Dövényi and Horváth, 1988). In the southern Tyrrhenean Sea, where back-arc oceanic crust has developed, heat flow is on the order of 160 $mWm^2$ (Pasquale et al., 1999).

Very thick sag-basin sequences (4–6 km) develop over very short periods of time (10 Myr) in the Pannonian Basin. This would appear to require the decay of very high geothermal gradients, and high sediment loading due to the large volume of eroded material from the surrounding mountains. The rift–sag transition associated with the Pannonian Basin defies the logic of a basin-fill pattern typically associated with thermal subsidence, where the cessation of extension produces an overall shallowing-upward trend across the syn-rift to post-rift transition (e.g., Lambiase and Bosworth, 1995). The syn-rift to post-rift transition in the Pannonian Basin is marked by a deepening of lacustrine conditions (Bérczi, 1988; Mattick et al., 1988). This observation suggests that thermal subsidence was underway prior to the end of low-strain-rate extension.

In addition to the unusual deepening event across the syn-rift to post-rift transition, there are two other characteristics of the Pannonian basin atypical of classic rifts. First, the extensional structures are, in general, of relatively small displacement and are associated with relatively thin syn-rift deposits (e.g., Horváth and Pogácsás, 1988; Mattick et al., 1988; Dank, 1988). Secondly, the post-rift section is remarkably thick, in places exceeding 5 km (Mattick et al., 1988), and it developed only in about the last 10 Myr. Backstripping of the post-rift section yields high estimates for crustal stretching ($\beta$ between 3 and 5; Sclater et al., 1980). The observed crustal extension indicates that these values are unrealistic, hence extra mantle lithospheric thinning has been suggested by Horváth and Royden (1981), Royden et al. (1982), and Royden and Dövényi (1988). Later publications suggest that subduction rollback may have caused the non-uniform lithospheric thinning (Figs. 12 , 13; Royden, 1993; Morley, 1993).

Catastrophic loss of the mantle lithosphere beneath the Alboran Sea in the late Oligocene early Miocene occurred either by delamination, subduction rollback, or detachment of a lithospheric blob (e.g., Platt and Vissers, 1989; Frizon de Lamotte et al., 1991; Morley, 1993; Lonergan and White, 1997; Platt et al., 1998). Although the mechanism is still a current source of controversy, there is little doubt that the consequence of emplacing hot asthenospheric mantle near the base of the crust caused uplift and erosion around the Rif–Betic orogen and the emplacement of peridotite bodies within the crust (e.g., Zeck et al., 1992). Following an erosional hiatus, thermal subsidence began in the Alboran Sea in the middle Miocene (if not the late early Miocene) (Morley, 1993). Hence active mantle processes in the Alboran Sea had a very marked effect on the nature of the syn-rift to post-rift transition. The subsequent thermal sag basin is very thick for its age (about 6 km since the middle Miocene; Morley, 1993), and it is associated with shale diapirs, strike-slip deformation in places, and inversion or thrust tectonics.

### *Structural Evolution*

Subduction rollback tends to produce a migration in the timing of extension and volcanics: both tend to propagate towards the trench with time. For example, extension in the Tyrrhenian Sea has propagated eastwards towards the trench with time, the youngest extensional faults lie onshore in Italy crosscutting older thrusts in the Apennines, and the onset of post-rift subsidence also youngs eastwards (Pascucci et al., 1999).

### *Inversion*

Given the structural setting, inversion during the thermal subsidence phase is likely to be widespread and common in these basins. Inversion may occur because of strike-slip deformation related to escape tectonics, far-field compressional stresses, and/or the local gravitational potential of thickened crust (e.g., Bada et al., 2001). Inversion has been noted in the Alboran Sea by Chalouan et al. (1997) and in the Pannonian Basin by Tari et al. (1992).

### *Examples*

Rif–Betics–Alboran Sea; Gulf of Lyon–Corsica–Sardinia; Tyrrhenean Sea–Apennines; Carpathians–Pannonian basins. Possible examples: Caspian Sea–Alborz mountains?, Triassic rifts, Khorat Plateau, Thailand?.

### *Hydrocarbon Prospectivity*

Most soft-collision rifts are predominantly marine, or began as continental rifts, and then, as subsidence progressed, evolved to marine conditions. The Pannonian is the exception (being filled by continental deposits) and is notably the most significant hydrocarbon province in this setting. High heat flows mean that comprehensive maturation of source rocks is likely, but overmaturation can be a problem. Marine basins tend to be gas prone; some, notably the Alboran Sea, have a very thick post-rift sequence with deformation by shale diapirs. Reservoirs tend to be turbidite sandstones sealed in deep marine shales. Inversion during the post-rift phase is common, and is associated with compression and strike slip.

## POST-OROGENIC WIDE RIFTS (CORDILLERAN TYPE?)

### *Introduction*

The two main examples of post-orogenic wide rifts are the Basin and Range of the western United States and the Aegean, which Buck (1991) and Brun (1999) classify simply as wide rifts. The term post-orogenic wide rift is used here to denote the link between extension and earlier crustal thickening. Wide rifting is too general a term because broad rifts, like the Sirt Basin in Libya, and broad zones of extension on passsive margins, developed in initially normal-thickness crust might mistakenly be included. Brun (1999) states that "wide rifting occurs in previously thickened domains where the dominantly ductile behavior of the lithosphere makes it able to spread under its own weight". A key characteristic is that extension follows cessation of a previous period of crustal thickening by more than 20 Myr. This time delay permitted thermal relaxation to weaken the lithosphere considerably (Sonder et al., 1987; Brun, 1999). The timing of extension post-thrusting is a marked difference from extension in soft collision and hard collision orogenic belts, as discussed in the sections dealing with these topics.

Another term considered to head this category was Cordilleran-type extension, because both the Basin and Range and the Aegean are formed on Cordilleran- or Andean-type margins. It may be that only this setting creates the right lithospheric conditions for massive post-orogenic extensional collapse. These margins are characterized by subduction of oceanic crust under continental crust (e.g., Taboada et al., 2000), whose key features passing towards the continent are: a subduction zone–accretionary prism complex, a volcanic arc, a continental interior–verging fold and thrust belt, and a foredeep basin (Fig. 12). There can be considerable modification to this basic pattern, as is seen passing along the west coasts of North and South America. The western margin of South America displays little evidence for the strike-slip deformation and extensional collapse that characterizes California, but it does display significant changes in subduction-zone geometry (e.g., Taboada et al., 2000). In contrast, the west coast of North America has been subject to accretion of numerous microplates, subduction of a spreading center, and their subsequent dismemberment by strike-slip deformation, plus the Cordilleran mountain belt has undergone massive extensional collapse to form the Basin and Range province (Fig. 15; e.g., Coney, 1987).

### *Structural styles*

The Basin and Range province represents an extreme state of extension in continental crust. It is a very large and diffuse extensional province in the western USA (Fig. 15). Its size (1500 km long and up to 1000 km wide) and the absence of a discrete, well defined, narrow zone of rifting set the region apart from typical continental rift systems. The province lies east of the Cascade volcanic arc, and it avoids the Colorado Plateau and the Rocky Mountains.

The present-day topographic pattern that gives the Basin and Range its name is a series of linear ranges spaced 20–35 km apart, separated by basins 10–20 km wide filled with Tertiary and Quaternary continental sediments (Fig. 15). Many are fault-bounded ridges, giving a horst-and-graben pattern to the region. The seismically active faults appear to be planar, and they penetrate to about 16 km depth and dip around 45° (Stein and Barrientos, 1985). Regionally the extension direction is WNW–ESE (Zoback and Zoback, 1980). Extension on the modern, apparently high-angle fault systems is estimated to be up to 23% (Zoback et al., 1981). This latest system of faults, formed during the last 10–17 Myr, contrasts strongly with the low-angle normal faults or detachments faults of the older extensional phase (50–10 Ma), which may have accommodated about 60–100% extension or 200–300 km (e.g., Sonder et al., 1987 ; Wernicke et al., 1988). Extension is distributed unevenly, with large-magnitude extension domains some 100–150 km wide separated by regions of little extension (Fig. 15; Gans, 1987; Wernicke et al., 1988).

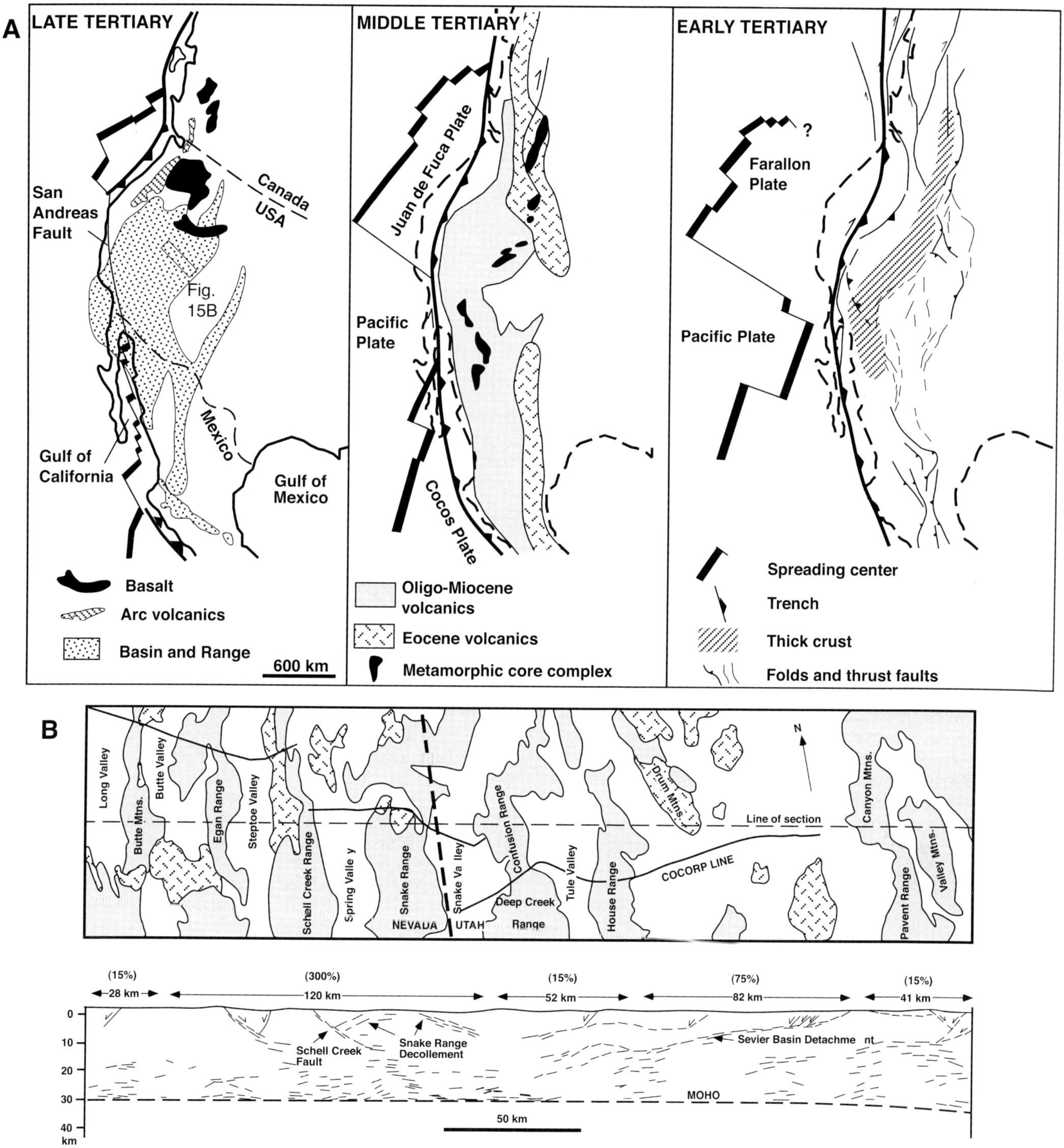


FIG. 15.—**A)** Geological evolution of the western coast of North America during the Tertiary (redrawn from Coney, 1987). The early Tertiary marked the end of compressional deformation in the Rocky Mountains. In the Middle Tertiary extensive volcanism accompanied extensional collapse in the hinterland of the old thrust belt. Subduction of part of the Pacific plate spreading center led to the formation of a major strike-slip system on the west coast, which includes the San Andreas fault. By the late Tertiary, extension in the Basin and Range had switched from the metamorphic-core-complex style to high-angle normal faults. **B)** Strip map in the vicinity of the COCORP 40° N transect illustrating the Basin and Range morphology, which comprises alternating low-lying flat (basin) areas (white) and hilly areas (ranges, gray), and line drawing of the COCORP seismic line illustrating the regional structural geometry of the Basin and Range (redrawn from Gans, 1987).

### *Volcanic History*

The volcanic history of the Basin and Range is complex. There has been extensive subduction-related volcanism (e.g., Armstrong and Ward, 1991). In addition there are also the effects of mantle plumes and large-scale crustal extension on magma generation to be considered (e.g., Gans, 1997; Parsons et al., 1998). The problem of determining the origins of volcanism is summed up by Gans (1997). He commented, "Though most of the early to middle Cenozoic magmatic activity is calc–alkaline in character and occurred at a time of continued subduction beneath western North America, its complex space–time distribution is in striking contrast to the linear volcanic chains typical of modern Andean type continental arcs, and it is difficult to assess the role of subduction versus other processes in generating these melts". Although igneous intrusions probably helped trigger extension, extension in turn may have dispersed magma into smaller bodies and increased their interaction with meteoric water, relatively enhancing the volume of syn-extensional plutonism and inhibiting concomitant volcanism (Gans and Bohrson, 1998).

### *Driving Mechanisms*

Metamorphic core complexes occur in regions of thickened crust, where the overall plate setting is convergent. This may seem an unlikely setting for the largest extensional provinces on the planet. The subduction-rollback mechanism appealed to in the back-arc setting may (e.g., ten Veen and Postma, 1999) or may not (Sachpazi et al., 2000) be appropriate for extensive metamorphic-core-complex provinces like the Basin and Range and the Aegean. Hence the driving mechanism for extension has been sought in the gravitational potential of thickened crust (e.g., England and Houseman, 1989; Buck, 1991). In particular, if convective removal of the mantle lithosphere occurs there will be a resulting rapid increase in potential energy in the overlying lithosphere (e.g., Platt and England, 1994). However, differences in potential energy alone may not generate significant extension (Lonergan and White, 1997). It also appears necessary for the lower crust to be hot and readily undergo ductile flow; such a trigger may be the emplacement of igneous intrusions (e.g., Gans et al., 1989).

Deep crustal seismic images across the Basin and Range have revealed the absence of either a topographic low above the core complexes or Moho relief beneath them (Fig. 15; McCarthy and Thompson, 1988). This suggests that the crust is too weak to support any substantial topography in the vicinity of core complexes (Gans, 1987; Buck, 1988). Extension in the upper crust does not result in rift-valley-like pure-shear thinning of the lower crust over a relatively narrow zone. Instead, lower crustal material apparently flows in from the surrounding region.

Both the Basin and Range and the Aegean are affected by extensive igneous activity. Extensional activity in the Basin and Range crudely follows the shifting patterns of magmatic activity (Fig. 15), but the importance of the linkage between magmatic activity and extension remains a topic of considerable debate (e.g., Gans et al., 1989; Axen et al., 1993; Gans, 1997). It does seem appealing, however, to require a combination of crustal thickening and magmatic thermal weakening to generate large metamorphic core complexes.

The Aegean Sea (Fig. 14) has recently been recognized as a Basin and Range–type province, with metamorphic core complexes (Dinter and Royden, 1993), driven by gravitational collapse of thickened continental crust (Davies et al., 1997; Gautier et al., 1999). However, other mechanisms may also have helped drive extension, including: escape tectonics associated with the extrusion of Anatolia (Le Pichon et al., 1995; Armijo et al., 1996), which affected the northwestern part of the Aegean Sea; slab detachment (Spakman et al., 1988), and subduction rollback (ten Veen and Postma, 1999).

### *Evolution from Extension to Thermal Subsidence*

The Basin and Range province is still an active extensional province, so it is difficult to know whether thermal subsidence will occur. There is a suggestion that post-extension lower crustal flow causes uplift and hence inhibits sag-basin formation (McKenzie et al., 2000). If thermal subsidence does occur, it will affect an enormous area. There are no well established older analogues with which to make a comparison, although in terms of size and juxtaposition with a thrust belt (Urals) the West Siberia basin might be appropriate (e.g., Tull, 1997; Knapp et al., 1998). However, the deep structure of this basin is poorly known. One other possible older example is the 450 km x 1000 km Songliao Basin, in northeastern China (Yang et al., 1985; Hu et al., 1998). Perhaps the best analogues for the future development of the Basin and Range are the basins with thick sediment fills and rapid thermal subsidence like the Pannonian and Alboran Sea basins discussed earlier.

### *Examples*

Basin and Range, U.S.A.; Aegean Sea, Mediterranean. Questionable examples: West Siberian Basin, Songliao Basin.

### *Hydrocarbon Prospectivity*

The young extensional basins tend to be too structurally complex, and produce basins that are too laterally restricted, to contain large hydrocarbon fields. A few small fields have been found in the Basin and Range. Prospectivity may be improved considerably if a thermal-subsidence basin in present. If the West Siberian and Songliao basins are underlain by Basin and Range–type tectonics, then the model they provide involves (1) source rocks from syn-rift and post-rift sources, (2) source, reservoir, and seal within the post-rift section, (3) traps largely restricted to the post-rift as drape over syn-rift structural highs, inversion anticlines due to reactivation of syn-rift faults, and stratigraphic traps. The very large area of the thermal-subsidence basins gives room for the presence of subtle traps of very large area, large source-rock-kitchen areas, and marked stratigraphic variations within a marine or lacustrine setting.

## "HARD" CONTINENT COLLISIONS

### *Introduction*

Himalayan-type collisions (e.g., Mattauer, 1986) are characterized by the collision of two major continental masses, with resultant crustal thickening, uplift, and erosion. Unlike soft-collision mountain belts, in Himalayan-type collisions the interior of the orogenic belt is highly uplifted and reveals high-grade metamorphic rocks and old suture zones, and may expose lower crust and mantle rocks (Fig. 12). Studies in several Alpine mountain belts have revealed the presence of rift basins and metamorphic core complexes in the internal zones, e.g., the Eastern Alps (Selverstone, 1988; Ratschbacher et al., 1991; Tari et al., 1992), the Norwegian Caledonides (e.g., Andersen, 1998), and the Tibetan Plateau (Chen et al., 1990). Geographically, however, the core complexes are not as extensive as those in the Basin and Range and Aegean.

Extension in the Tibetan Plateau area differs from the Basin and Range in several ways: (1) Extension is sub-perpendicular to the mountain belt in the Tibetan Plateau (Fig. 16; e.g., Cogan et al., 1998) instead of subparallel in the latter; this probably reflects a fundmental change in the boundary conditions of deformation (e.g., Liu et al., 2000). (2) Extension occurs at a much earlier stage in the orogeny, during thrusting. (3) The amount of extension is much greater in the Basin and Range, which now is composed of relatively thin crust (30–40 km thick; Fig. 15) whereas the crustal thickness under the Tibetan Plateau is about 70 km (Fig. 12). Metamorphic core complexes reported from the Central Alps (Selverstone, 1988) are also associated with extension subparallel to the regional strike of the mountain belt and may have been caused by eastward expulsion of crustal blocks (escape tectonics) during orogenic convergence (Ratschbacher et al., 1991). In the Norwegian Caledonides too there are indications that escape tectonics were associated with the collision (e.g., Lyberis and Manby, 1999), to which extension might be linked.

The Tibetan plateau is the result of crustal thickening caused by the collision between India and Asia. During most of the Tertiary north–south shortening on east–west striking faults dominated the deformation of the plateau (Fig. 12; England and Houseman, 1989). It is commonly stated that over the last five million years, extension on north–south trending faults has affected the area. Coleman and Hodges (1995), however, demonstrated that some north–south trending extensional faults have been active since 14 Ma. These faults appear to be linked with eastwards flowage of the thickened ductile crust away from the collision zone. Earthquakes have shown that a mixture of extensional and strike-slip faults are active, with both north–south convergence and east–west extension occurring. Overall the crust is undergoing a small amount of net thinning (Chen and Molnar, 1983). There are several possible causes for this change, including: (1) reduced compressional stresses by reduction of the convergence rate between India and Asia, and (2) reduced strength of the Asian continental lithosphere. England and Houseman (1989) argued that reducing the strength of the Asian lithosphere would only increase compressional strain in the Tibetan Plateau region, because, regionally, compressional stresses are still dominant. They proposed that the thermal evolution of thickened continental lithosphere triggered extension by the detachment of a thermally unstable blob of lower-lithosphere mantle. This convective instability created a change in elevation. England and Houseman (1989) calculated that a change in elevation for the Tibetan Plateau of 1.5 to 3 km would raise the potential energy of the lithosphere (overlying the zone of detachment) sufficiently to replace north–south compression with extension.

The hot thickened crust of the Himalayan collision zone has actually undergone two types of extension, thrust-belt parallel and thrust-belt perpendicular (Fig. 16). This section concentrates on the thrust-belt-perpendicular extension because it generated rift basins. However, the thrust-belt-parallel system is very important to the development of the Himalayas and provides an interesting solution to the volume problem associated with extension during overall plate convergence. The thrust-belt-parallel extensional system forms the South Tibetan Detachment system, and is seen near the crest of the Himalayas and actually is exposed cutting through Mt. Everest (Fig. 16; see review in Searle, 1999). The hanging wall did not move much to the east relative to the underlying Indian Plate. Instead the hot,

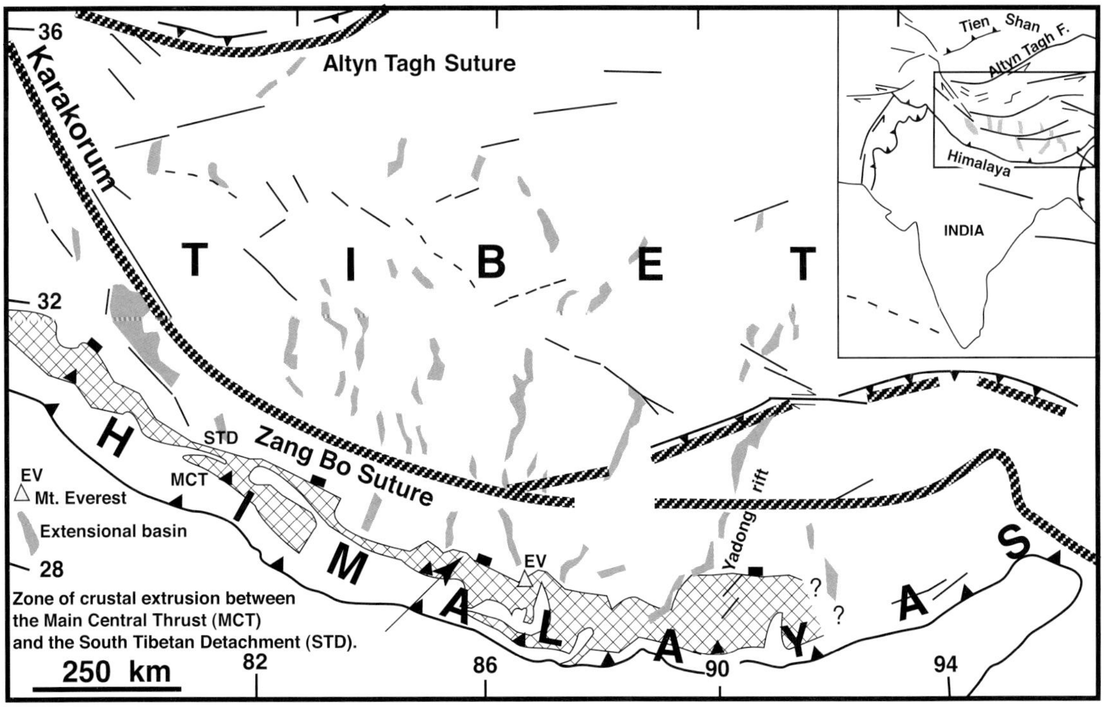


FIG. 16.—Two Tertiary extension systems in the Himalayas: A late Oligocene–Miocene thrust-belt-parallel extrusion system, and a thrust-belt-suborthogonal system of Miocene–Recent grabens in Tibet. The map is modified from Tapponnier et al. (1986); extent of the Greater Himalayan belt, between the Main Central Thrust and the South Tibetan Detachment, is from Wu et al. (1998) and Hodges (2000).

ductile footwall (Greater Himalayan Belt) appears to have been extruded towards the surface, and was continuously eroded, creating a conveyer-belt mechanism by which crustal shortening and heating led to granite production. The granites were intruded below the detachment, exhumed to the surface in the footwall of the detachment/hanging wall of the Main Central Thrust, and eroded away (Fig. 12; e.g., Burchfiel et al., 1992; Wu et al., 1998; Searle, 1999). This system operated in the Late Oligocene to the Pliocene.

### *Structural Styles*

The only rift in Tibet described from subsurface data is the Yadong–Gulu Rift. It is different in dimensions from typical failed continental rifts. The Yadong–Gulu Rift is about 500 km long, and the entire faulted region is some 20–30 km wide (Fig. 16). Basin fill is up to 1.5 km thick, but more typically only a few hundred meters thick. Individual half grabens are only about 12 km wide (one-third to one-quarter the width typically found in failed regional continental rifts). Remarkably for such apparently small rifts, the footwalls are composed of intensely sheared rocks forming ductile deformation zones up to 150 m wide (Pan and Kidd, 1992). The footwalls display rocks from mid-crust levels (Harrison et al., 1995; Cogan et al., 1998). Hence, the apparently small-displacement grabens (e.g., Armijo et al., 1986) may actually represent proto-detachment (core complex) faults that have exhumed their footwalls from depths of about 17 km (Cogan et al., 1998). It is thought that the middle crust at depths of 15–20 km is ductile and helped promote the extension by ductile flow (Cogan et al., 1998). Further evidence for ductile flow comes from rift uplift geometry (see below).

### *Syn-Extension Topography*

The location of rifts in the interiors of Himalayan-type thrust belts means that they typically are associated with high elevations and rugged terraine. Uplifts directly associated with rifts are of short wavelength (30–40 km), particularly for crust that is 70 km thick (Masek et al., 1994). Maximum flank uplift is about 1 km, for narrow rift systems some 10–15 km across. This contrasts strongly with rifts developed in cold, normal-thickness (30–40 km) crust (e.g., East African Rift, Baikal Rift), which are typically 50 km wide and have flank uplifts of around 1–2 km (e.g., Masek et al., 1994; Ebinger et al., 1999). For the Tibetan rifts Masek et al. (1994) calculated that the topography required low flexural rigidity ($D = 2\text{–}6 \times 10^{26}$ Nm), crust with low (2–4 km) elastic thickness, and isostatic compensation within low-viscosity ductile lower crust. This situation is similar to the Basin and Range.

### *Volcanic History*

Some syn-orogenic intrusives, in particular granites, do accompany the development of Himalayan-type orogenic belts. However, the degree of intrusion is much less than in Andean-type margins, and the impact of volcanism on extensional fault development is difficult to separate from the presence of extensive hot, thick, ductile crust under the Tibetan plateau resulting from orogenic contraction (e.g., Searle, 1999).

### *Evolution from Extension to Thermal Subsidence*

Extensional thinning does not seem to have much impact on offsetting the effects of crustal thickening due to collision; hence the classic cause of thermal subsidence (McKenzie, 1978) is not applicable to Himalayan-type thrust belts. The location of the rifts in the uplifted mountainous interior is also not conducive to the formation of thermal sag basins (too much topography to erode).

### *Inversion*

The potential for inversion of extensional basins would seem to be high because of the overall compressional settings of the basins. However, inversion has not been widely reported for Himalayan-type settings; this is to be expected given the immaturity of the extensional structures.

### *Examples*

Central–Eastern Alps; Tibetan Plateau, Himalayas; Devonian basins, Norwegian Caledonides.

### *Hydrocarbon Prospectivity*

Rifts in Himalayan settings appear to have little hydrocarbon prospectivity. The basins are small, and are associated with considerable uplift, due both to thrusting and to the development of metamorphic core complexes. They do not appear to develop thermal sag basins, the absence of which is a major problem for prospectivity (Morley, 1999c). There is also a high potential for inversion to occur.

## EXTENSION ASSOCIATED WITH FAR-FIELD COLLISIONAL STRESSES

### *Introduction*

Some rifts in continental plates lie in the foreland or hinterland of synchronous major thrust belts and do not fit into any of the tectonic settings discussed above. The West Europe Cenozoic rift in the foreland of the Alps and Lake Baikal in the hinterland of the Himalayas are the best known examples. Molnar and Tapponnier (1975) proposed that far-field stresses from the orogenic belts gave rise to such rifts. As reviewed in Petit et al. (1996), subsequent work supports this theory for the Baikal Rift. Extension on the West Europe Cenozoic rift was triggered by the arrival of a mantle plume in the Early Tertiary in the Massif Central–Eifel–Westerwald area (Fig. 14; Granet et al., 1995) but was controlled by the far-field crustal stress field resulting from the Alpine orogeny and from ridge push from the Mid-Atlantic Ridge (Zeyen et al., 1997).

### *Structural Styles*

The Rhine Graben and the Baikal Rift display structural styles similar to those involved in failed continental breakup. They are built from half grabens and full grabens 30–80 km wide, bounded by high-angle normal faults. Syn-rift sediments can be thick (up to about 7 km), and as in Lake Tanganyika, similar deep lacustrine conditions (up to 1.5 km water depth) are present in Lake Baikal (Hutchinson et al., 1992; Mats, 1993). The West Europe Cenozoic rift and the Baikal Rift tend to be shorter than rifts involved in failed continental breakup. The Baikal Rift is about 1000 km long. Its southwestern end terminates in a strike-slip fault regime (e.g., Hutchinson et al., 1992; Petit et al., 1996). The West Europe Cenozoic rift is composed of three segments, which form the Eger, Rhine Graben, and Massif Central grabens. Maximum extension is about 17 km (Brun et al., 1992). Neither end of the Baikal and Rhine rifts ends at a contemporaneous passive margin or failed rift system, unlike failed continental rifts (Jurassic North Sea rift, Central African rift, and East African rift systems).

Hutchinson et al. (1992) compared the structural styles of the Baikal and East African rifts and noted some differences. In particular, the lengths of basins in the Baikal Rift tend to be longer (160–240 km) than in the East African rift (80–160 km). Flexural margins in the Baikal Rift are not as well developed as in East Africa, and the boundary fault systems in the Baikal Rift mostly form a series of synthetic transfer zones, with no alternations of polarity. Hutchinson et al. (1992) proposed that differences in the driving mechanisms of the two rift systems might be the reason. There is also the possibility, however, that preexisting fabrics influenced the boundary fault systems to dip consistently in one direction in the Baikal Rift. Secondly, the length of the faults might reflect the linkage of initially separate boundary faults. Ebinger et al. (1999) noted that both the Baikal and East African rift-boundary fault systems were of similar dimensions and were typical for extensional deformation of normal-thickness, relatively cold continental lithosphere (Fig. 5).

In general, the characteristics of uplift, volcanism, and thermal subsidence are similar to the failed regional continental rifts (see section of same name on page 26). The close relationship between the rifts and orogenic belt–related stresses means that there is a strong likelihood of inversion during and after rifting. There is also a more intimate relationship between strike-slip faults and extension in some far-field-stress settings than in regional failed continental rifts. The Baikal Rift is linked with strike-slip faults (e.g., Delvaux et al., 1997). Farther to the east, extension in graben systems in the Ordos region of China shows marked strain partioning between NE–SW striking graben-bounding extensional faults and synchronous, extrusion-related NNW–SSW and NNW–SSE striking strike-slip faults (Zhang et al., 1998). Both the Baikal and the Ordos rifts show early (Paleogene–early Miocene) phases of slow rifting, and accelerated extension late in their history (late Miocene–Quaternary), which may be an effect of changing lithospheric conditions or variations in orogeny-related stresses (Delvaux et al., 1997; Zhang et al., 1998).

### *Examples*

Rhine Graben, Baikal Rift, Ordos rifts.

### *Hydrocarbon Prospectivity*

Rifts like the Baikal and Rhine grabens are similar to failed continental rifts in terms of syn-rift architecture, and hence prospectivity. High rates and amounts of subsidence are associated with these rifts. Some rifts in the Ordos region have a remarkably thick syn-rift sequence (up to 12 km; Zhang et al., 1998). Hence there is plenty of opportunity for maturation of source rocks, and for the development of deep syn-rift lacustrine conditions, which are favorable for source-rock development. Neither the Baikal Graben nor the Rhine Graben is associated with thermal sag basins, which significantly impacts prospectivity in a negative way (Morley, 1999c). However, there is no reason why older examples of this type of rift, with sag basins, should not be prospective for hydrocarbons in a way similar to regional continental rifts. Being generally shorter, more discontinuous rift systems, the fairways for making hydrocarbon plays will be more limited than in regional continental rifts.

## EXTENSION ASSOCIATED WITH STRIKE SLIP

In many rifts there is a close association of extensional structures and strike-slip features. The subject is too big to cover in this paper, but it is necessary to acknowledge some of the ways in which strike-slip deformation and extension interact, including the following:

1. Old strike-slip faults oblique to the regional extension direction become reactivated passively by extensional faults. This is a very common feature in rifts, but where detailed information is scarce it is common for major active throughgoing strike-slip faults to be incorrectly interpreted (e.g., Gulf of Thailand, North Sea).

2. Preexisting fabrics result in extensional structures forming oblique to the regional extension direction. Transtensional rift basins are developed (e.g., Rukwa Rift, East African rift system; Witch Ground Graben, North Sea).

3. Oblique fabrics in high-strain areas of extension commonly become sites for strike-slip faults to develop which link offset areas of extension. Such strike-slip faults are transfer faults. The offsets may evolve into transform faults if a spreading center develops. Extension can also be transferred from a spreading center to a plate boundary (e.g., Red Sea–Dead Sea shear zone–Taurus/Zagros Mountains).

4. In mountain belts escape tectonics results in the synchronous formation of strike-slip faults and extensional structures (including metamorphic core complexes) (e.g., Himalayas, Eastern Alps, Aegean Sea, Pannonian Basin, Alboran Sea).

5. Pull-apart basins formed at restraining bends or left-stepping, left lateral, and right-stepping, right-lateral en echelon fault segments along strike-slip fault zones (e.g., Dead Sea Rift).

6. In oblique-convergence volcanic arcs, strain partitioning may lead to arc-parallel major strike-slip faults superimposed on regions of back-arc extension (e.g., Sumatra).

## DISTRIBUTION OF HYDROCARBONS IN RIFT SYSTEMS

Morley (1999c) concluded that rifts capped by thermal sag basins were the most productive of all rift systems. Marine settings were somewhat better than continental deposition. In this paper rifts with sag basins occur in a variety of tectonic settings, and some further distinctions can be made. (1) Marine deposition in failed regional continental rifts has produced the North Sea rifts. Most rifts in this setting are dominated by continental deposits and are remarkably unproductive, with the Sudan rifts being the most productive example (ca. 1–2 billion barrels of reserves). (2) All the rifts with large reserves, capped by thermal sag basins, filled with continental sedimentary rocks, are found in convergent settings. These include: Sumatra and Java Tertiary rifts; Thailand rift basins; Pannonian Basin; and Bohai Basin. This category probably also includes the West Siberia Basin and the Songliao Basin, although their tectonic origins are more uncertain. (3) Permanently or temporarily stalled rift systems are particularly good hydrocarbon provinces. They include: the Gulf of Suez, the Angola passive margin, the Newfoundland Jurassic rifts, the Sirt Basin, the West Natuna Basin, and the Reconcavo Basin.

## DISCUSSION AND SUMMARY

This review sets the stage for a volume largely concerned with sedimentation in extensional basins. Consequently the emphasis in discussing the different tectonic settings focuses on how structural style, basin evolution, basin geometry, uplift, and volcanism vary between the different settings. These variations may affect drainage patterns, pathways of sands into rifts, the basin fill, sediment source areas, and the lateral extent of source rocks and sealing facies. The differences are summarized in Table 1. A few important differences are highlighted below.

Attempts have been made to classify rifts on the basis of morphology (e.g., narrow, wide, and core complex) and to relate the morphology to crustal type (e.g., normal-thickness crust, hot crust, hot thick crust) and strain rate (e.g., Kusznir and Park, 1987; Buck, 1991). These are important parameters which control fault geometry in rifts, another aspect of which is the elastic thickness of the crust (e.g., Ebinger et al., 1999). Normal-thickness, cold continental crust tends to be extended along high-angle faults, which create long, deep basins. Extension of hot, normal-thickness crust is usually associated with igneous intrusions and produces swarms of minor faults and relatively narrow, small half grabens. Extension of hot, thickened crust tends to include the development of metamorphic core complexes in a variety of tectonic environments, including: propagating spreading centers encountering orogenic belts (e.g., Papua New Guinea); escape tectonics in Himalayan-type thrust belts (Eastern Alps; Himalayas); and gravity collapse (subduction rollback?) of hot, thick, intruded orogenic continental crust in the Basin and Range and the Aegean Sea.

Defining the mechanical significance of wide rifts is more difficult. Strain hardening can produce wide rifts in normal-thickness crust (e.g., Kusznir and Park, 1987), as can the initial lithospheric strength conditions plus the effects of dominant local crustal buoyancy forces (Buck, 1991). Classification of rifts into narrow and wide types (e.g., Buck, 1991; Brun, 1999) simplifies the discussion of how variations in lithospheric strength affect structural style. However, these oversimplifications ignore the variety and combinations of driving mechanisms that can cause rifting, especially subduction rollback. For example, wide rifts can form in response to subduction rollback and laterally migrating spreading centers.

Extension in convergent orogenic belts is divided into four types (Table 1), but the categories require some flexibility. Considerable diversity of rift styles and geometries exists within these settings, in part because of the irregular geometries of the colliding continental margins. The Rif–Betic Cordillera is placed within the soft collision category but is perhaps better viewed as a transition between soft collisions and Himalayan-type collision because it shares some of the characteristics of both.

A common problem is the assumption that extension in all mountain belts has a similar origin (i.e., gravitational collapse) and that the variations between different thrust belts can be explained in terms of maturity of the process. Different plate-tectonic settings produce different types of collision, different amounts of volcanic intrusions, and varying degrees of extension-driving mechanisms, in particular: (1) subduction-driven extension, and (2) enhanced gravitational potential due to crustal thickening, delamination, or detachment of the lithosphere. Enhanced gravitational potential can usually be viewed only as a partial driving mechanism; generally it does not account for how space is created for brittle upper crustal extension to occur (note the exception of the thrust-belt-parallel extension in the Himalayas; e.g., Burchfiel et al., 1992). For example, in Himalayan-type thrust belts the aspect of escape tectonics is an important factor associated with the north–south trending rift basins. In Papua New Guinea metamorphic core complexes in a thrust belt were generated at the tip of a propagating spreading center, thus demonstrating that when a mechanism for creating extra volume is present, extensional collapse can occur. Enhanced gravity potential cannot explain how extensional thinning of the crust to less than the thickness of adjacent (non-extended) crust has occurred. Hence, in the soft-collision thrust belts, subduction-driven extension is probably an equally important or dominant mechanism. Post-orogenic wide rifting in a Cordilleran setting appears to be an example where buoyancy forces are the dominant cause of extension. The (comparatively) free margin of the continent against subducting oceanic crust in a Cordilleran-type setting would permit gravity-driven extending continental crust to increase area by moving over oceanic crust.

Morley (1999c) discussed how the post-extension history of extensional basins is very important for determining their hydrocarbon prospectivity, and concluded that rift basins capped by thermal-sag basins were the most productive types of rifts, followed by passive margins. In this paper rifts with thermal-sag basins fall into several categories and serve to highlight some of the key differences in post-rift evolution between tectonic settings. Such differences include the following.

1. Marine basins tend to preserve considerably more syn-rift topography, progressively infilled by the sag-basin fill, than continental rifts. In continental rifts erosion tends to remove the syn-rift topography before thermal subsidence commences.

2. Compared with regional continental rifts, rifts at abandoned propagating spreading centers may undergo a much longer transition from syn-rift to post rift, lasting many millions of years (e.g., the Gulf of Suez). The proximity of abandoned propagating spreading centers to marine conditions commonly makes them sites for late rift to post-rift evaporite deposition, which can have a significant impact on hydrocarbon prospectivity in terms of migration pathways, and seal and trap formation. Typical amounts of subsidence are ca. 4 km in 60–100 Myr.

3. Thermal sag basins associated with subduction rollback may accumulate to great thicknesses (> 4 km) in a short time (10–15 Myr) associated with non-uniform thinning of the lithosphere. There may also be lateral shifts in the onset of thermal subsidence due to the lateral migration of extension following trench migration.

4. Subduction in oceanic crust–volcanic arc settings inevitably leads to collisions with continents and continental fragments. Consequently thermal-subsidence basins will be affected by tectonic inversion. Hence the timing of creation and destruction of hydrocarbon traps is very sensitive in this setting.

5. Very broad sag basins, particularly in continental settings, associated post-orogenic wide rifts (e.g., Songliao and West Siberia basins?), certain types of stalled propagating rift tips (e.g., Sirte Basin, Libya); and soft collisions (e.g., Pannonian Basin) permit greater differentiation of sedimentary environments compared with tectonic settings that favor narrow rifts (such as regional continental extension).

## ACKNOWLEDGMENTS

This paper has drawn from too many sources to make individual acknowledgments possible, but work and conversations with colleagues in Amoco, Elf, and Brunei and Aberdeen Universities over the years has been of great help in piecing this paper together. B. Vendeville and A. Clifton are thanked for constructive reviews.

TABLE 1.—Summary of the characteristics of the rift basin classification presented in this paper.

| Type of rift | Volcanic history | Thermal history | Shape/ dimensions | Evolution to thermal subsidence | Typical sediment source material | Thermal subsidence rate | Rift sediment fill | Syn-rift topography | Inversion | Structural style | Examples |
|---|---|---|---|---|---|---|---|---|---|---|---|
| Failed regional continental rifts | Variable from zero to extensive; plume-related, minor adiabatic DCM | Variable, passive to active | Long (1000s of km) narrow (30–150 km) | Gradual, often several-million-year time gap between the two in continental settings | Older continental sedimentary rocks, crystalline basement. Igneous, if rift is volcanically active | Average (i.e. 3–4 km over 70–100 Myr). | Continental to marine | Uplift related to mantle plume doming, footwall uplift, and magmatic underplating | Absent to strong; one to multiple phases. Causes (ridge push, gravity collapse, far-field stresses) | Mostly high-angle normal faults typically $\beta < 2$; strike-slip as transfer structures | East African Rift system, Central African rift system, North Sea |
| Propagating rift at tip of spreading center | Variable from zero to extensive; adiabatic-decomp. In back-arc setting, slab related volc. | Variable, passive to active | up to ca. 1500 km long, narrow (30–150 km) | Gradual, with pulses of extension continuing into the thermal subsidence phase | Older contiental sedimentary rocks, crystalline basement. Igenous, if rift is volcanically active | Average (i.e. 3–4 km over 70–100 Myr). Thermal subsidence either as sag basin or on passive margin | Early continental fill, with transition later to marine conditions. Commonly early post-rift marked by evaporite deposition | Uplift related to mantle plume doming, footwall uplift, and magmatic underplating | Absent to strong | Low to high-angle nomal faults, wide range on β. Strike-slip as transfer structures Occasional metamorphic core complex. | Can be a failed rift (e.g., Gulf of Suez) or passive margin, e.g., W. Africa north of Walvis Ridge, Papua New Guinea, N. Borneo |
| Upper-plate subduction zones-oceanic crust–island arc, oceanic crust–continental crust | Nearby extensive volcanism (island arc)-subduction related | Variable, passive to active | Bundles of relative short (hundreds of km) narrow (30–150 km) rifts following trace of volcanic arc | Relatively rapid, often transition marked by passage from continental to marine sedimentation | Older continental sedimentary rocks, crystalline basement. Commonly igneous material from volcanic arc | Rapid (i.e. 3–4 km in 10–15 Myr) | Usually continental but can also be marine or mixed continental and marine. | Footwall uplift, and magmatic underplating. Generally relatively subdued | Liable to undergo strong inversion events due to subduction-related collisions | Low to high-angle nomal faults, typically $\beta < 2$. Extension may be mixed with strike-slip deformation | Sumatra, Java, Bohai Basin, Thailand rift basins, Sumisu Rift, Lau–Hau–Taupo basin |
| Soft continental collision | Moderate; derived from subducted slab | High geothermal gradients | Broad oval–circular or linear regions 500–1000 km in extent | Either rapid deepening (Pannonian Basin) or uplift and erosion then subsidence (Alboran Sea) | Older continental sedimentary rocks, crystalline basement. | Rapid (i.e. 3–4 km in 10–15 Myr) | Continental and/or marine | Low area of rifting in center of fringing orogenic-belt mountains | Often extensive strike-slip related deformation during and after rifting. Also compression. | Low to high-angle nomal faults, typically $\beta < 2$. Extension may be mixed with strike-slip deformation | Carpathians–Pannonian basin Rif–Betic–Alboran Sea. Apennines–Tyrrhenean Sea. Caspian Sea?? |
| Post-orogenic extension (Cordilleran type?) | Intense; extensive batholith complexes | High | Up to very large, diffuse area, 1000 km+ wide and long | No sag basin? Possible big ones: Songliao, W. Siberia | Older continental sedimentary rocks, crystalline basement. Igneous | Uncertain. Possibly rapid | Continental and/or marine | Rift super-imposed on highly uplifted mountainous topography | None to weak | Low to high-angle nomal faults, typically $\beta > 2$. Extensive metamorphic core complexes | Basin and Range Aegean Songlia basin? W. Siberia basin? |
| Hard continental collision (Himalayan type) | Zero to moderate; crustal melting | Moderate | Variable shape, generally short | No sag basin | Older continental sedimentary rocks, crystalline basement. | Crustal thickness and elevation too great | Continental Due to elevation; strong possibility of glacially related sediments | Rift superimposed on highly uplifted mountainous topography | Unknown, potentially considerable | Low to high-angle nomal faults, typically $\beta < 2$. In places metamorphic core complex | Himalayas–Tibet Alps, Norwegian Caledonides–Devonian basins |
| Collision belt far-field stress | Zero to minor; plume-related | Generally moderate | Long (up to 1000 km) narrow 50–150 km) | No sag basin? | Older continental sedimentary rocks, crystalline basement. | Examples not known, probably moderate | Continental | Uplift related to mantle plume doming, footwall uplift | None to moderate | High-angle nomal faults, typically $\beta < 2$. May be strongly associated with strike-slip faults | Rhine Graben Baikal Rift Ordos Rifts |

## REFERENCES


Allen, M.B., Macdonald, D.I.M., Xun, Z., Vincent, S.J., and Brouet-Menzies, C., 1997, Early Cenozoic two-phase extension and late Cenozoic thermal subsidence and inversion of the Bohai Basin, northern China: Marine and Petroleum Geology, v. 14, p. 951–972.

Amato, A., Alessandrini, B., Cimini, G.B., Frepoli, A., and Selvaggi, G., 1993, Active and remnant subducted slabs beneath Italy: Evidence from seismic tomography and seismicity: Annali di Geofisica, v. 36, p. 201–214.

Anderson, D.L., 1982, Hotspots, polar wander, Mesozoic convection and the geoid: Nature, v. 297, p. 391–393.

Andersen, T.B., 1998, Extensional tectonics in the Caledonides of southern Norway, an overview: Tectonophysics, v. 285, p. 333–351.

Armijo, R., Meyer, B., King, G.C.P., Rigo, A., and Papanastassiou, D., 1996, Quaternary evolution of the Corinth Rift and its implications for the Late Cenozoic evolution of the Aegean: Geophysical Journal International, v. 126, p. 11–53.

Armijo, R., Tapponnier, P., Mercier, J.L., and Tong-Lin, H., 1986, Quaternary extension in southern Tibet: field observations and tectonic implications: Journal of Geophysical Research, v. 94, p. 13,803–13,872.

Armstrong, R.L., and Ward, P., 1991, Evolving geographic patterns of Cenozoic magmatism in the North American Cordillera: The temporal and spatial association of magmatism and metamorphic core complexes: Journal of Geophysical Research, v. 96, p. 13201–13224.

Axen, G.J., Taylor, W.J., and Bartley, J.M., 1993, Space–time patterns and tectonic controls of Tertiary extension and magmatism in the Great Basin of the western United States: Geological Society of America, Bulletin, v. 105, p. 56–76.

Azanon, J.-M., and Crespo-Blanc, A., 2000, Exhumation during a continental collision inferred from the tectonometamorphic evolution of the Alpujarride Complex in the central Betics (Alboran Domain, SE Spain): Tectonics, v. 19, p. 549–565.

Bada, G., Horváth, F., Cloetingh, S., Coblentz, D.D., and Toth, T., 2001, Role of topography-induced gravitational stresses in basin inversion: The case study of the Pannonian basin: Tectonics, v. 20, p. 343–363.

Baker, B.H., and Wohlenberg, J., 1971, Structure and evolution of the Kenya Rift Valley: Nature, v. 229, p. 538–542.

Baker, B.H., Williams, L.A.J., and Fitch, F.J., 1971, Sequence and geochronology of the Kenya rift volcanics: Tectonophysics, v. 11, p. 191–215.

Baker, B.H., Mohr, P.A., and Williams, L.A.L., 1972, Geology of the Eastern Rift System of Africa: Geological Society of America, Special Paper 136, 67 p.

Banks, N.L., Bardwell, K.A., and Musiwa, S., 1995, Karoo Rift basins of the Luangwa Valley, Zambia, *in* Lambiase, J.J., ed., Hydrocarbon Habitat in Rift Basins: Geological Society of London, Special Publication 80, p. 285–295.

Bellon, H., 1981, Chronologie radiometrique (K–Ar) des maniefstations magmatiques autout de la Mediterranée occidentale entre 33 et 1 Ma., *in* Wezel, F.C., ed., Sedimentary Basins of Mediterranean Margins: Bologna, Italy, Tecnoprint, p. 341–360.

Bérczi, I, 1988, Preliminary sedimentological investigation of a Neogene depression in the Great Hungarian Plain, *in* Royden, L.H., and Horváth, F., eds., The Pannonian Basin: American Association of Petroleum Geologists, Memoir 45, p. 107–116.

Bonatti, E., 1985, Punctiform initiation of seafloor spreading in the Red Sea during transition from a continental to oceanic rift: Nature, v. 316, p. 33–37.

Bosworth, W., 1995, A high-strain rift model for the southern Gulf of Suez (Egypt), *in* Lambiase, J.J., ed., Hydrocarbon Habitat in Rift Basins: Geological Society of London, Special Publication 80, p. 75–102.

Bosworth, W., Strecker, M.R., and Blisniuk, P.M., 1992, Integration of East African paleostress and present-day stress data: Implications for continental stress field dynamics: Journal of Geophysical Research, v. 97, p. 11,851–11,856.

Bott, M.H.P., 1992, The stress regime associated with continental breakup, *in* Storey, B.C., Alabaster, T., and Pankhurst, R.J., eds., Magmatism and the Causes of Continental Breakup: Geological Society of London, Special Publication 68, p. 125–136.

Bott, M.H.P., 1993, Modelling the plate-driving mechanism: Geological Society of London, Journal, v. 150, p. 941–951.

Brun, J.P., 1999, Narrow rifts versus wide rifts: inferences for the mechanics of rifting from laboratory experiments: Royal Society [London], Philosophical Transactions, v. 357, p. 695–712.

Brun, J.P., and Beslier, M.O., 1996, Mantle exhumation at passive margins: Earth and Planetary Science Letters, v. 142, p. 161–173.

Brun, J.P., Gutscher, M.A., and DEKORP-ECORS teams, 1992, Deep crustal structure of the Rhine Graben from DEKORP-ECORS seismic reflection data: A summary: Tectonophysics, v. 208, p. 139–147.

Buck, R.W., 1988. Flexural rotation of normal faults: Tectonics, v. 7, p. 959–973.

Buck, W.R., 1991, Modes of continental lithospheric extension: Journal of Geophysical Research, v. 96, p. 20,161–20,178.

Bustin, R.M., and Chonchawalit, A., 1995, Formation and tectonic evolution of the Pattani Basin, Gulf of Thailand: International Geology Review, v. 37, p. 866–892.

Burchfiel, B.C., Chou, Z., Hodges, K.V., Liu, Y., Royen, L., Deng, C., and Xu, J., 1992, The South Tibetan Detachment System, Himalayan Orogen: extension contemporaneous with and parallel to shortening in a collisional mountain belt: Geological Society of America, Special Paper 269, 41 p.

Burke, K., 1976, Development of graben associated with the intial rupture of the Atlantic ocean: Tectonophysics, v. 36, p. 93–112.

Carminati, E., Wortel, M.J.R., Spakman, W., and Sabadini, R., 1998, The role of slab detachment processes in the opening of the western–central Mediterranean basins: some geological and geophysical evidence: Earth and Planetary Science Letters, v. 160, p. 651–665.

Chalouan, A., Saji, R., Michard, A., and Bally, A.W., 1997, Neogene tectonic evolution of the Southwestern Alboran Basin, as inferred from seismic data off Morocco: American Association of Petroleum Geologists, Bulletin, v. 81, p. 1161–1184.

Chen, W.P., and Molnar, P., 1983, Focal depths of intra-continental and intraplate lithosphere: Journal of Geological Research, v. 88, p. 4183–4214.8

Clemson, J., Cartwright, J., and Booth, J., 1997, Structural segmentation and the influence of basement structure on the Namibian passive margin: Geological Society of London, Journal, v. 154, p. 477–482.

Cogan, M.J., Nelson, K.D., Kidd, W.S.F., Wu, C., and Project INDEPTH Team, 1998, Shallow structure of the Yadong–Gulu rift, southern Tibet, from refraction analysis of Project INDEPTH common midpoint data: Tectonics, v. 17, p. 46–61.

Cohen, A.S., 1990. Tectono-stratigraphic model for sedimentation in Lake Tanganyika, Africa, *in* Katz, B.J. ed., Lacustrine Basin Exploration: Case Studies and Modern Analogs: American Association of Petroleum Geologists, Memoir 50, p. 137–150.

Coney, P.J., 1987, The regional tectonic setting and possible causes of Cenozoic extension in the North American Cordillera, *in* Coward, M.P., Dewey, J.F., and Hancock, P.L., eds., Continental Extensional Tectonics: Geological Society of London, Special Publication 28, p. 177–186.

Crossley, R., 1984, Controls on sedimentation in the Malawi rift valley, Central Africa: Sedimentary Geology, v. 40, p. 33–50.

Dank, V., 1988, Petroleum geology of the Pannonian basin, Hungary: an overview, *in* Royden, L.H., and Horváth, F., eds., The Pannonian Basin: American Association of Petroleum Geologists, Memoir 45, p. 319–332.

Davies, R., England, P., Parsons, B., Billiris, H., Paradissis, D., and Veis, G., 1997, Tectonic strain of Greece in the interval, 1892–1992: Journal of Geophysical Research, v. 102, p. 24,571–24,588.

Delvaux, D., Moeys, R., Stapel, G., Petit, C., Levi, K., Miroschnichenko, A., Suzhich, V., and San'kov, V., 1997, Paleostress reconstructions and geodynamics of the Baikal region, central Asia, Part 2. Cenozoic rifting: Tectonophysics, v. 282, p. 1–38.

Dewey, J.F., 1980, Episodicity, sequence and style at convergent plate boundaries, *in* Geological Association of Canada, Special Paper 20, p. 553–574.

Dewey, J.F., 1988, Extensional collapse of orogens: Tectonics, v. 7, p. 1123–1139.

Dickinson, W.R., 1974, Plate tectonics and sedimentation, *in* Dickinson, W.R., ed., Tectonics and Sedimentation: Society of Economic Paleontologists and Mineralogists, Special Publication 22, p. 1–27.

Dinter, D.A., and Royden, L., 1993, Late Cenozoic extension in northeastern Greece: Strymon Valley detachment system and Rhodope metamorphic core complex: Geology, v. 21, p. 45–48.

Doglioni, C., 1992, Main differences between thrust belts: Terra Nova, v. 4, p. 152–164.

Doglioni, C., Gueguen, E., Sabat, F., and Fernandez, M., 1997, The western Mediterranean extensional basins and the Alpine orogen: Terra Nova, v. 9, p. 109–112.

Dore, A.G., Lundin, E.R., Jensen, L.N., Birkeland, O., Elaissen, P.E., and Fichler, C., 1999, Principal tectonic events in the evolution of the northwest European Atlantic margin, *in* Fleet, A.J., and Boldy, S.A.R., eds., Petroleum Geology of Northwest Europe: Proceedings of the 5th Conference: Geological Society of London, p. 41–61.

Dövényi, P., and Horváth, F., 1988, A review of temperature, thermal conductivity and heat flow data from the Pannonian basin, *in* Royden, L.H., and Horváth, F., eds., The Pannonian Basin: American Association of Petroleum Geologists, Memoir 45, p. 195–234.

Duval, B., Cramez, C., and Jackson, M.P.A., 1992, Raft tectonics in the Kwanza Basin, Angola: Marine and Petroleum Geology, v. 9, p. 389–404.

Ebinger, C.J., 1989, Tectonic development of the western branch of the East African Rift system: Geological Society of America, Bulletin, v. 101, p. 885–903.

Ebinger, C.J., and Ibrahim, A., 1994, Multiple episodes of rifting in Central and East Africa: A re-evaluation of gravity data: Geologische Rundschau, v. 83, p. 689–702.

Ebinger, C.J., Bechtel, T.D., Forsyth, D.W., and Bowin, C.O., 1989, Effective elastic thickness beneath the East African and Afar Plateaux and dynamic compensation of the uplifts: Journal of Geophysical Research, v. 94, p. 2883–2901.

Ebinger, C.J., Jackson, J.A., Forster, A.N., and Hayward, N.J., 1999, Extensional basin geometry and the elastic lithosphere: Royal Society [London], Philosophical Transactions, v. 357, p. 741–765.

Emery, K.O., and Uchupi, E., 1984, The Geology of the Atlantic Ocean: New York, Springer-Verlag, 1050 p.

England, P.C., and Houseman, G., 1989, Extension during continental convergence, with application to the Tibetan Plateau: Journal of Geophysical Research, v. 94, p. 17,561–17,579.

Errant, D., Thomas, G.M., and Wall, G.R.T., 1999, The evolution of the Central North Sea Rift, *in* Fleet, A.J., and Boldy, S.A.R., eds., Petroleum Geology of Northwestern Europe, Proceedings of the 5th Conference: Geological Society of London, p. 63–82.

Etheridge, M.A., Symonds, P.A., and Lister, G.S., 1989, Applications of the detachment model to reconstructions of conjugate passive margins, *in* Tankard, A.J., and Balkwill, H.R., eds., Extensional Tectonics and Stratigraphy of the North Atlantic Margins: American Association of Petroleum Geologists, Memoir 46, p. 23–40.

Eubank R.T., and Makki, A., 1981, Structural geology of the Central Sumatra back-arc Basin: Indonesian Petroleum Association, 10th Ann. Convention, Proceedings, p. 153–196.

Frizon de Lamotte, D., Andrieux, J., and Guezou, J.C., 1991, Cinematique des chevauchements Neogène dans l'arc Betico–Rifain: Discussion sur les modéles géodynamiques: Société Géologique de France, Bulletin, v. 8, p. 337–344.

Gans, P., 1987, An open system, two-layer crustal stretching model for the eastern Great Basin: Tectonics, v. 6, p. 1–12.

Gans, P.B., 1997, Large-magnitude Oligo-Miocene extension in southern Sonora: Implications for the tectonic evolution of northwest Mexico: Tectonics, v. 16, p. 388–408.

Gans, P.B., and Bohrson, W.A., 1998, Suppression of volcanism during rapid extension in the Basin and Range province, United States: Science, v. 279, p. 66–68.

Gans, P.B., Mahood, G.A., and Schermer, E., 1989, Synextensional magmatism in the Basin and Range province: A case study from the eastern Great Basin: Geological Society of America, Special Paper 233, 53 p.

Gautier, P., Brun, J-P., Moriceau, R., Sokoutis, D., Martinod, J., and Jolivet, L., 1999, Timing, kinematics and cause of Aegean extension: a scenario based on a comparison with simple analogue experiments: Tectonophysics, v. 315, p. 31–72.

George, R., Rogers, N., and Kelley, S., 1998, Earliest magmatism in Ethiopia: Evidence for two mantle plumes in one flood basalt province: Geology, v. 26, p. 923–926.

Ghignone, J.I., and Andrade, G., 1970, General geology and major oil fields of Reconcavo basin, Brazil, *in* Halbouty, M.E., ed., Geology of Giant Petroleum Fields: American Association of Petroleum Geologists, Memoir 14, p. 337–358.

Genik, G.J., 1992. Regional framework, structural and petroleum aspects of rift basins in Niger, Chad, and the Central African Republic (C.A.R.): Centre du Recherches Exploration Production Elf Aquitaine, Bulletin, v. 13, p. 39–53.

Ginger, D.C., Ardjakusumah, W.O., Hedley, R.J., and Pothecary, J., 1993, Inversion history of the West Natuna Basin: Examples from the Cumi-Cumi PSC: Indonesian Petroleum Association, 22nd Annual Conference, Proceedings, p. 635–658.

Girdler, R.W., 1991, The Afro–Arabian rift systems—an overview: Tectonophysics, v. 197, p. 139–153.

Goodliffe, A.M., Taylor, B., Martinez, F., Hey, R., Maeda, K., and Ohno, K, 1997, Synchronous reorientation of the Woodlark Basin Spreading Center: Earth and Planetary Science Letters, v. 146, p. 233–242.

Granet, M., Stoll. G., Dorel, J., Achauer, U., Poupinet, G., and Fuchs, K., 1995, Massive Central (France): new constraints on the geodynamical evolution from teleseismic tomography: Geophysical Journal International, v. 121, p. 33–48.

Griffiths, P.S., 1980, Box-fault systems and ramps: a typical association of structures form the eastern shoulder of the Kenya rift: Geological Magazine, v. 117, p. 579–586.

Guiraud, R., 1998, Mesozoic rifting and basin inversion along the northern African Tethyian margin: an overview, *in* MacGregor, D.S., Moody, R.T.J., and Clarke-Lowes, D.D., eds, Petroleum Geology of North Africa: Geological Society of London, Special Publication 132, p. 217–229.

Gupta, S., Underhill, J.R., Sharp, I.R., and Gawthorpe, R.L., 1999, Role of fault interactions in controlling synrift sediment dispersal patterns: Miocene, Abu Alaqa Group, Suez Rift, Sinai, Egypt: Basin Research, v. 11, p. 167–189.

Hall, R., 1996, Reconstructing Cenozoic SE Asia, *in* Hall, R., and Blundell, D.J., eds.,Tectonic Evolution of Southeast Asia: Geological Society of London, Special Publication 106, p. 203–224.

Harrison, T.M., Copeland, P., Kidd, W.S.F., and Lovera, O.M., 1995, Activation of the Nyainqentanglha Shear Zone: Implications for uplift of the southern Tibet Plateau: Tectonics, v. 14, p. 658–676.

Hayward, N., and Ebinger, C., 1996, Variations in along-axis segmentation of the Afar Rift system: Tectonics, v. 15, p. 244–257.

Hendrie, D.B., Kusznir, N.J., Morley, C.K., and Ebinger, C.J., 1994, Cenozoic extension in northern Kenya: A quantitative model of rift basin development in the Turkana region: Tectonophysics, v. 236, p. 409–438.

Hill, E.J., Baldwin, S.L., and Lister, G.S., 1992, Unroofing of active metamorphic core complexes in the D'Entrecasteaux Islands, Papua New Guinea: Geology, v. 20, p. 907–910.

Hodges, K.V., 2000, Tectonics of the Himalaya and southern Tibet from two perspectives: Geological Society of America, Bulletin, v. 112, p. 324–350.

Horváth, F., and Pogacasas, G., 1988, Contribution of seismic reflection data to chronostratigraphy of the Pannonian basin, *in* Royden, L.H., and Horváth, F., eds., The Pannonian Basin: American Association of Petroleum Geologists, Memoir 45, p. 97–106.

Horváth, F., and Royden, L.H., 1981, Mechanism for formation of the intra-Carpathians basins: a review: Earth Evolution Science, v. 1, p. 307–316.

Huchon, P., Le Pichon, X., and Rangin, C., 1994, Indo-China Peninsular and the collision of India and Eurasia: Geology, v. 22, p. 27–30.

Hutchinson, D.R., Golmshtok, A.J., Zonenshain, L.P., Moore, T.C., Scholz, C.A., and Klitgord, K.D., 1992, Depositional and tectonic framework of the rift basins of Lake Baikal from multi-channel seismic data: Geology, v. 20, p. 589–592.

Hutchison, C., 1996, Geological evolution of South–East Asia: Geological Society of Malaysia, Kuala Lumpur, 368 p.

Hu, W., Cai, C., Wu, Z., and Li, J., 1998, Structural style and its relation to hydrocarbon exploration in the Songliao basins, northeast China: Marine and Petroleum Geology, v. 15, p. 41–55.

Hynes, A., and Mott, J., 1985, On the causes of back-arc spreading: Geology, v. 13, p. 387–389.

Joffe, S., and Garfunkel, Z., 1987, Plate tectonics of the circum Red Sea—A re-evaluation: Tectonophysics, v. 141, p. 15–22.

Jolivet, M., Daniel, J.M., and Fournier, M., 1991, Geometry and kinematics of extension in Alpine Corsica: Earth and Planetary Science Letters, v. 104, p. 278–291.

Karig, D.E., 1971, Origin and development of margin basins in the western Pacific: Journal of Geophysical Research, v. 76, p. 2542–2561.

Kastens, K., and Mascle, M., 1988, ODP Leg 107 in the Tyrrhenian Sea: insights into passive margin and back-arc evolution: Geological Society of America, Bulletin, v. 100, p. 1140–1156.

Kastens, K., and Mascle, M., 1990, The geological evolution of the Tyrrhenian Sea: An introduction to the scientific results of ODP Leg 107: Proceedings of the Ocean Drilling Program, Scientific Results, v. 107, p. 3–26.

Kingston, D.R., Dishroon, C.P., and Williams P.A., 1983, Global basin classification: American Association of Petroleum Geologists, Bulletin, v. 67, p. 2175–2193.

Knapp, J.H., Diaconescu, C.C., Bader, M.A., Sokolov, V.B., Kashubin, S.N., and Rybalka, A.V., 1998, Seismic reflection fabrics of continental collision and post-orogenic extension in the Middle Urals, Central Russia: Tectonophysics, v. 288, p. 115–126.

Kusznir, N.J., and Park, R.G., 1987, The extensional strength of the continental lithosphere: its dependence on geothermal gradient, and crustal composition and thickness, *in* Coward, M.P., Dewey, J.F., and Hancock, P.L., eds., Continental Extensional Tectonics: Geological Society of London, Special Publication 28, p. 35–52.

Lambiase, J.J., and Bosworth, W., 1995, Structural controls on sedimentation in continental rifts, *in* Lambiase, J.J., ed., Hydrocarbon Habitat in Rift Basins: Geological Society of London, Special Publication 80, p. 117–144.

Latin, D., Norry, M.J., and Tarzey, R.J.E., 1993, Magmatism in the Gregory Rift, East Africa: Evidence for melt generation by a plume: Journal of Petrology, v. 34, p. 1007–1027.

Laubscher, H., Biella, G.C., Cassinis, R., Gelati, R., Lozej, A., Scarascia, S., and Tabacco, I., 1992, The collisional knot in Liguria: Geologishe Rundschau, v. 81, p. 275–289.

Lee, M.J., and Hwang, Y.J., 1993, Tectonic evolution and structural styles of the East Shetland Basin, *in* Parker, J.R., ed., Petroleum Geology of Northwest Europe: Proceedings of the 4th Conference: Geological Society of London, p. 1137–1149.

LePichon, X., Chamot-Rooke, N., Lallemant, S., Noomen, R., and Veis, G., 1995, Geodetic determination of the kinematics of central Greece with respect to Europe: implications for eastern Mediterranean tectonics: Journal of Geophysical Research, v. 100, p. 12675–12690.

Letouzey, J., 1990, Fault reactivation, inversion and fold-thrust belt, *in* Letouzey, J., ed. Petroleum and Tectonics in Mobile Belts: Paris, Editions Technip, p. 101–128.

Lister, G.S., Banga, G., and Feenstra, A., 1984, Metamorphic core complexes of Cordilleran type in the Cyclades, Aegean Sea, Greece: Geology, v. 12, p. 221–225.

Lister, G.S., Etheridge, M.A., and Symonds, P.A., 1986, Detachment faulting and the evolution of passive contiental margins: Geology, v. 12, p. 246–250.

Liu, M., Shen, Y., and Yang, Y., 2000, Gravitational collapse of orogenic crust: A preliminary three-dimensional finite element study: Journal of Geophysical Research, v. 105, p. 3159–3173.

Lonergan, L., and White, N., 1997, Origin of the Betic–Rif mountain belt: Tectonics, v. 16, p. 504–522.

Louden, K.E., and Chian, D., 1999, The deep structure of non-volcanic rifted continental margins: Royal Society [London], Philosophical Transactions, v. 357, p. 767–804.

Lyberis, N., and Manby, G., 1999, Continental collision and lateral escape deformation in the lower and upper crust: An example from Caledonide Svalbard: Tectonics, v. 18, p. 40–63.

McKenzie, D., Nimmo, F., and Jackson, J.A., 2000, Characteristics and consequences of flow in the lower crust: Journal of Geophysical Research, v. 105, p. 11,029–11,046.

Macdonald, D.I.M., Leat, P.T., Doubleday, P.A., and Kelly, S.R.A., 1999, On the origin of fore-arc basins: new evidence of formation by rifting form the Jurassic of Alexander Island, Antarctica: Terra Nova, v. 11, p. 186–193.

Mariucci, M.T., Amato, A., and Montone, P., 1999, Recent tectonic evolution and present stress in the Northern Apennines (Italy): Tectonics, v. 18, p. 108–118.

Masek, J.G., Isacks, B.L., Fielding, E.J., and Browaeys, J., 1994, Rift flank uplift in Tibet: Evidence for a viscous lower crust: Tectonics, v. 13, p. 659–667.

Mats, V.D., 1993, The structure and development of the Baikal rift depression: Earth-Science Reviews, v. 34, p. 81–118.

Mattauer, M., 1986, Intracontinental subduction, crust–mantle decollement and crust-stacking wedge in the Himalayas and other collision belts, *in* Coward, M.P., and Ries, A.C., eds., Collision Tectonics: Geological Society of London, Special Publication 18, p. 37–50.

Matthews, S.J., Fraser, A.J., Lowe, S., Todd, S.P., and Peel, F.J., 1997, Structure, stratigraphy and petroleum geology of the SE Nam Con Son basin, offshore Vietnam, *in* Fraser, A.J., Matthews, S.J., and Murphy, R.W., eds., Petroleum Geology of S.E. Asia: Geological Society of London, Special Publication 126, p. 89–106.

Mattick, R.E., Phillips, R.L., and Rumpler, J., 1988, Seismic stratigraphy and depositional framework of sedimentary rocks in the Pannonian basin in southeastern Hungary, *in* Royden, L.H., and Horváth, F., eds., The Pannonian Basin: American Association of Petroleum Geologists, Memoir 45, p. 117–146.

Mazlan B. Hj. Madon, 1997, Analysis of tectonic subsidence and heat flow in the Malay Basin (offshore Peninsula Malaysia): Geological Society of Malaysia, Bulletin, v. 41, p. 127–138.

McCaffry, R., 1992, Oblique plate convergence, slip vectors and forearc deformation: Journal of Geophysical Research, v. 97, p. 8905–8915.

McCarthy, J., and Thompson, G.A., 1988, Seismic imaging of extended crust with emphasis on the western United States: Geological Society of America, Bulletin, v. 100, p. 1361–1374.

McClay, K.R., Nichols, G.J., Khalil, S.M., Darwish, M., and Bosworth, W., 1998, Extensional tectonics and sedimentation, eastern Gulf of Suez, Egypt, *in* Purser, B.H., and Bosence, D.W.J., eds., Sedimentation and Tectonics of Rift Basins: Red Sea–Gulf of Aden: London, Chapman & Hall, p. 223–238.

McKenzie, D.P., 1978, Some remarks on the development of sedimentary basins: Earth and Planetary Science Letters, v. 40, p. 25–32.

McKenzie, D.P., and Bickle, M.J., 1988, The volume and composition of melt generated by extension of the lithosphere: Journal of Petrology, v. 29, p. 625–667.

Milani, E.J., and Davison, I., 1988, Basement control and transfer tectonics in the Reconcavo–Tucano–Jatoba rift, Northeast Brazil: Tectonophysics, v. 154, p. 41–71.

Mohr, P.A., 1987, Structural style of continental rifting in Ethiopia: reverse decollements: Eos, Transactions of the American Geophysical Union, v. 68, p. 729–730.

Molnar, P., and Tapponnier, P., 1975, Cenozoic tectonics of Asia: Effects of a continental collision: Science, v. 189, p. 419–426.

Morley, C.K., 1989, Extension, detachments and sedimentation in continental rifts (with particular reference to East Africa): Tectonics, v. 8, p. 1175–1192.

Morley, C.K., 1993, A possible delamination origin for hinterland basins to the Rif–Betic cordillera and Carpathians: Tectonophysics, v. 226, p. 359–376.

Morley, C.K., 1994, Interaction of deep and shallow processes in the evolution of the Kenya rift: Tectonophysics, v. 236, p. 81–91.

Morley, C.K., 1999a, Patterns of displacement along large normal faults: Implications for basin evolution and fault propagation, based on examples from East Africa: American Association of Petroleum Geologists, Bulletin, v. 83, p. 613–634.

Morley, C.K., 1999b, Marked along-strike variations in dip of normal faults—the Lokichar fault, N. Kenya rift: a possible cause for metamorphic core complexes: Journal of Structural Geology, v. 21, p. 479–492.

Morley, C.K., 1999c, Comparison of hydrocarbon prospectivity in rift systems, *in* Morley, C.K., ed., Geoscience of Rift Systems—Evolution of East Africa: American Association of Petroleum Geologists, Studies in Geology, no. 44, p. 233–242.

Morley, C.K., 2001, Combined escape tectonics and subduction rollback–back arc extension: a model for the evolution of Tertiary rift basins in Thailand, Malaysia and Laos: Geological Society of London, Journal, v. 158, p. 461–474.

Morley, C.K., Nelson, R.A., Patton, T.L., and Munn, S.G., 1990, Transfer zones in the East African rift system and their relevance to hydrocarbon exploration in rifts: American Association of Petroleum Geologists, Bulletin, v. 74, p. 1234–1253.

Morley, C.K., Wescott W.A., Stone D.M., Harper R.M., Wigger S.T., and Karanja, F.M., 1992, Tectonic evolution of the northern Kenya rift: Geological Society of London, Journal, v. 149, p. 333–348.

Morley, C.K., Wescott, W.A., Stone, D.M., Harper, R.M., Wigger, S.T., Day, R.A., and Karanja, F.M., 1999a, Geology and geophysics of the Western Turkana basins, Kenya, *in* Morley, C.K., ed., Geoscience of Rift Systems—Evolution of East Africa: American Association of Petroleum Geologists, Studies in Geology, no. 44, p. 19–54.

Morley, C.K., Day, R.A., Lauck, R., Bosher, R., Stone, D.M., Wigger, S.T., Wescott, W.A., Haun, D., Bassett, N., and Bosworth, W., 1999b, Geology and Geophysics of the Anza Graben, *in* Morley, C.K., ed., Geoscience of Rift Systems—Evolution of East Africa: American Association of Petroleum Geologists, Studies in Geology, no. 44, p. 67–90.

Morley, C.K., Harper, R.M., and Wigger, S.T., 1999c, Tectonic inversion in East Africa, *in* Morley, C.K., ed., Geoscience of Rift Systems—Evolution of East Africa. American Association of Petroleum Geologists, Studies in Geology, no. 44, p. 193–210.

Mutter, J.C., Mutter, C.Z., and Fang, J., 1996, Analogies to oceanic behaviour and the continental breakup of the western Woodlark basin: Nature, v. 380, p. 333–336.

Nelson, R.A., Patton, T.L., and Morley, C.K., 1992, Rift-segment interaction and its relation to hydrocarbon exploration in continental rift systems: American Association of Petroleum Geologists, Bulletin, v. 76, p. 1153–1169.

Nemcok, M., Pospisil, L., Lexa, J., and Donelick, R.A., 1998, Tertiary subduction and slab break-off model of the Carpathian–Pannonian region: Tectonophysics, v. 295, p. 307–340.

Pallesen S., 1993, Crustal extension in the Oslo Graben, SE Norway: a method incorporating magmatism and erosion: Tectonophysics, v. 221, p. 155–172.

Pan, Y., and Kidd, W.S.F., 1992, Nyainqentanglha Shear Zone: A late Miocene extensional detachment in the southern Tibet Plateau: Geology, v. 20, p. 775–778.

Parson, L.M., and Wright, I.C., 1996, The Lau–Havre–Taupo back-arc basin: A southward-propagating, multi-stage evolution from rifting to spreading: Tectonophysics, v. 263, p. 1–22.

Parsons, M.G., Zagaar, A.M., and Curry, J.J., 1980, Hydrocarbon occurrences in the Sirte basin, Libya, *in* Maill, A.D., ed., Facts and Principles of World Petroleum Occurrence: Canadian Society of Petroleum Geologists, Memoir 6, p. 723–732.

Parsons, T., Thompson, G.A., and Smith, R.P., 1998, More than one way to stretch: a tectonic model for extension along the plume track of the Yellowstone hotspot and adjacent Basin and Range Province: Tectonics, v. 17, p. 221–234.

Pascucci, V., Merlini, S., and Martini, I.P., 1999, Seismic stratigraphy of the Miocene–Pleistocene sedimentary basins of the Northern Tyrrhenian Sea and western Tuscany (Italy): Basin Research, v. 11, p. 337–356.

Pasquale, V., Verdoya, M., and Chiozzi, P., 1999, Thermal state and deep earthquakes in the southern Tyrrhinean: Tectonophysics, v. 306, p. 435–448.

Patton, T.L., Moustafa, A.R., Nelson, R.A., and Abdine, A.S., 1994, Tectonic evolution and structural setting of the Suez Rift, *in* Landon, S.M., ed., Interior Rift Basins: American Association of Petroleum Geologists, Memoir 59, p. 7–55.

Petit, C., Deverchere, J., Houdry, F., Sankov, V.A., Melnikova, V.I., and Delvaux, D., 1996, Present-day stress field changes along the Baikal rift and tectonic implications: Tectonics, v. 15, p. 1171–1191.

Platt, J.P., and England, P.C., 1994, Convective removal of lithosphere beneath mountain belts: Thermal and mechanical consequences: American Journal of Science, v. 294, p. 307–336.

Platt, J.P., and Vissers, R.L.M., 1989, Extensional collapse of thickened continental lithosphere: A working hypothesis for the Alboran Sea and the Gibraltar Arc: Geology, v. 17, p. 540–540.

Platt, J.P., Soto, J.-I., Whitehouse, M.J., Hurford, A.J., and Kelley, S.P., 1998, Thermal evolution, rate of exhumation, and tectonic significance of metamorphic rocks from the floor of the Alboran extensional basin, western Mediterranean: Tectonics, v. 17, p. 671–689.

Prosser, S., 1993, Rift-related linked depositional systems and their seismic expression, *in* Williams, G.D., and Dobb, A., eds., Tectonics and Seismic Sequence Stratigraphy: Geological Society of London, Special Publication 71, p. 117–144.

Ratschbacher, L., Frisch, W., Linzer, H-G., and Merle, O., 1991, Lateral extrusion in the eastern Alps, II: structural analysis: Tectonics, v. 10, p. 257–271.

Rattey, R.P., and Hayward, A.B., 1993, Sequence stratigraphy of a failed rift system: the Middle Jurassic to Early Cretaceous basin evolution of the Central and North Sea, *in* Parker, J.R., ed., Petroleum Geology of Northwest Europe, Proceedings of the 4th Conference: Geological Society of London, p. 215–249.

Ravnas, R., and Steel, R.J., 1998, Architecture of marine rift-basin successions: American Association of Petroleum Geologists, Bulletin, v. 82, p. 110–146.

Rehault, J.-P., Boillot, G., and Mauffret, A., 1985, The Western Mediterranean Basin, *in* Wezel, F.C., ed., Geological Evolution of the Mediterranean Basin: Berlin, Springer-Verlag, p. 101–129.

Reston, T.J., 1996, The S reflector west of Galicia: The seismic signature of a detachment fault: Geophysical Journal International, v. 127, p. 230–244.

RING, U., BETZLER, C., AND DELAVAUX, D., 1993, Normal vs. strike-slip faulting during rift development in East Africa: The Malawi rift: Geology, v. 20, p. 1015–1018.

ROBERTS, A.M., YIELDING, G., KUSZNIR, N.J., WALKER, I., AND DORN-LOPEZ, D., 1993, Mesozoic extension in the North Sea: constraints from flexural backstripping, forward modelling and fault populations, *in* Parker, J.R., ed., Petroleum Geology of Northwest Europe: Geological Society of London, p. 1123–1136.

ROSENDAHL, B.R., 1987, Architecture of continental rifts with special reference to East Africa: Annual Review of Earth and Planetary Sciences, v. 15, p. 445–503.

ROUBY, D., FOSSEN, H., AND COBBOLD, P.R., 1996, Extension, displacement and block rotation in the larger Gullfaks area, Northern North Sea: Determined from map view restoration: American Association of Petroleum Geologists, Bulletin, v. 80, p. 875–890.

ROYDEN, L., 1993, The tectonic expression slab pull at continental convergent boundaries: Tectonics, v. 12, p. 303–325.

ROYDEN, L., AND BURCHFIEL, B.C., 1989, Are systematic variations in thrust belt style related to plate boundary processes? (The Western Alps versus the Carpathians): Tectonics, v. 8, p. 51–61.

ROYDEN, L., AND DÖVÉNYI, P., 1988, Variations in extensional style with depth across the Pannonian basin system, *in* Royden, L.H., and Horváth, F., eds., The Pannonian Basin: American Association of Petroleum Geologists, Memoir 45, p. 235–256.

ROYDEN, L., HORVÁTH, F., AND BURCHFIEL, B.C., 1982, Transform faulting, extension and subsidence in the Carpathian–Pannonian region: Geological Society of America, Bulletin, v. 73, p. 717–725.

SACHPAZI, M., HIRN, A., CLEMENT, C., HASINGER, F., LAIGLE, M., KISSLING, E., CHARVIS, P., HELLO, Y., LEPINE, J.-C., SAPIN, M., AND ANSORGE, J., 2000, Western Hellenic subduction and Cephalonia Transform: local earthquakes and plate transport and strain: Tectonophysics, v. 319, p. 301–319

SCHLISCHE, R.W., AND ANDERS, M.H., 1996, Stratigraphic effects and tectonic implications of the growth of normal faults and extensional basins, *in* Beratan, K.K., ed., Reconstructing the Structural History of Basin and Range Extension Using Sedimentology and Stratigraphy: Geological Society of America, Special Paper 303, p. 183–203.

SCHULL, T.J., 1988, Rift basins of interior Sudan: petroleum exploration and discovery: American Association of Petroleum Geologists, Bulletin, v. 72, p. 1128–1142.

SCLATER, J.G., ROYDEN, L., HORVÁTH, F., BURCHFIEL, B.C., SEMKEN, S., AND STEGENA, L., 1980, The formation of the intra-Carpathian basin as determined from subsidence data: Earth and Planetary Science Letters, v. 51, p. 139–162.

SEARLE, M.P., 1999, Extensional and compressional faults in the Everest–Lhotse massif, Khumbu Himalaya, Nepal: Geological Society of London, Journal, v. 156, p. 227–240.

SELVERSTONE, J., 1988, Evidence for east–west crustal extension in the Eastern Alps: Implications for the unroofing history of the Tauern Window: Tectonics, v. 7, p. 87–105.

SERRI, G., INNOCENTI, F., AND MANETTI, P., 1993, Geochemical and petrological evidence of the subduction of delaminated Adriatic continental lithosphere in the genesis of the Neogene–Quaternary magmatism of central Italy: Tectonophysics, v. 223, p. 117–147.

SHANNON, P.M., JACOB, A.W.B., O'REILLY, B.M., HAUSER, F., READMAN, P.W., AND MAKRIS, J., 1999, Structural setting, geological development and basin modelling in the Rockall Trough, *in* Fleet, A.J., and Boldy, S.A.R., eds., Petroleum Geology of Northwest Europe, Proceedings of the 5th Conference: Geological Society of London, p. 421–431.

SHERIDAN, R.E., 1983, Phenomena of pulsation tectonics related to the breakup of the eastern North American continental margin: Tectonophysics, v. 94, p. 169–185.

SONDER, L.J., ENGLAND, P.C., WERNICKE, B.P., AND CHRISTIANSEN, R.L., 1987, A physical model for Cenozoic extension of western North America, *in* Coward, M.P., Dewey, J.F., and Hancock, P.L., eds., Continental Extensional Tectonics: Geological Society of London, p. 187–201.

SPAKMAN, W., WORTEL, M.J.R., AND VLAAR, N.J., 1988, The Hellenic subduction zone: a tomographic image and its geodynamic implications: Geophysical Research Letters, v. 15, p. 60–63.

SPRING, D., AND HANSEN, O.P., 1998, The influence of platform morphology and sea level on the development of a carbonate sequence: the Harash Formation, Eastern Sirt Basin, Libya, *in* MacGregor, D.S., Moody, R.T.J., and Clarke-Lowes, D.D., eds., Petroleum Geology of North Africa: Geological Society of London, Special Publication 132, p. 335–353.

STECKLER, M.S., AND TEN BRINK, U.S., 1986, Lithospheric strength variations as a control on new plate boundaries: examples from the northern Red Sea region: Earth and Planetary Science Letters, v. 79, p. 120–132.

STEIN, R.S., AND BARRIENTOS, S.E., 1985, Planar high angled faulting in the Basin and Range: geodetic analysis of the, 1983 Borah Peak Idaho, earthquake: Journal of Geophysical Research, v. 90, p. 11,355–11,366.

SZABO, C., HARANGI, S., AND CSONTOS, L., 1992, Review of Neogene and Quaternary volcanism of the Carpathian–Pannonian region: Tectonophysics, v. 208, p. 243–256.

TABOADA, A., RIVERA, L.A., FUENZALIDA, A., CISTERNAS, A., PHILIP, H., BIJWAARD, H., OLAYA, J., AND RIVERA, C., 2000, Geodynamics of the northern Andes: subductions and intracontinental deformation (Colombia): Tectonics, v. 19, p. 787–813.

TANKARD, A.J., AND WELSINK, H.J., 1989, Mesozoic extension and styles of basin formation in Atlantic Canada, *in* Tankard, A.J., and Balkwill, H.R., eds., Extensional Tectonics and Stratigraphy of the North Atlantic Margins: American Association of Petroleum Geologists, Memoir 46, p. 175–195.

TAPPONNIER, P., PELTZER, G., AND ARMIJO, R., 1986, On the mechanics of collision between India and Asia, *in* Coward, M.P., and Ries, A.C., eds., Collision Tectonics: Geological Society of London, Special Publication 19, p. 115–157.

TARI, G., HORVÁTH, F., AND RUMPLER, J., 1992, Styles of extension in the Pannonian Basin: Tectonophysics, v. 208, p. 203–219.

TATSUMI, Y., AND KIMURA, N., 1991, Backarc extension versus continental breakup: petrological aspects for active rifting: Tectonophysics, v. 197, p. 127–137.

TAYLOR, B., KLAUS, A., BROWN, G.R., MOORE, G.F., OKAMURA, Y., AND MURAKAMI, F., 1991, Structural development of Sumisu Rift, Izu–Bonin arc: Journal of Geophysical Research, v. 96, p. 16,113–16,129.

TEN VEEN, J.H., AND POSTMA, G., 1999, Rollback controlled vertical movements of outer-arc basins of the Hellenic subduction zone (Crete, Greece): Basin Research, v. 11, p. 243–266.

TOMEK, C., AND HALL, J., 1993, Subducted continental margin imaged in the Carpathians of Czechoslovakia: Geology, v. 21, p. 535–538.

TULL, S.J., 1997, The diversity of hydrocarbon habitat in Russia: Petroleum Geoscience, v. 3, p. 315–325.

TWISS, R.J., AND MOORES, E.M., 1992, Structural Geology: New York, Freeman,

UNDERHILL, J.R., SAWYER, M.J., HODGSON, P., SHALLCROSS, M.D., AND GAWTHORPE, R.L., 1997, Implications of fault scarp degredation for Brent Group prospectivity, Ninian Field, Northern North Sea: American Association of Petroleum Geologists, Bulletin, v. 81, p. 999–1022.

VINK, G.E., MORGAN, W.J., AND ZHAO, W.-L., 1984, Preferential rifting of continents: a source of displaced terranes: Journal of Geophysical Research, v. 89, p. 10,072–10,076.

WANG, J.-H., YIN, A., HARRISON, GROVE, M., ZHANG, Y.-Q., AND XIE, G.-H., 2001, A tectonic model for Cenozoic igneous activities in the eastern Indo–Asian collision zone: Earth and Planetary Science Letters, v. 188, p. 123–133.

WATCHARANANTAKUL, R., AND MORLEY, C.K., 2000, Syn-rift and post-rift modelling of the Pattani Basin, Thailand: Evidence for a ramp-flat detachment: Marine and Petroleum Geology, v. 17, p. 937–958.

WERNICKE, B.P., AXEN, G.J., AND SNOW, J.K., 1988, Basin and Range extensional tectonics at the latitude of Las Vegas, Nevada: Geological Society of America, Bulletin, v. 100, p. 1738–1757.

WESCOTT, W.A., KREBS, W.N., NUMMEDAL, D., AND DOLSON, J.C., 1998, Miocene sequence stratigraphy, chronostratigraphic correlation and subsurface exploration and production analogues, Gulf of Suez: outcrop to subsurface: American Association of Petroleum Geologists, Hedberg Conference, Field Seminar Notes.

WHITE, R., AND MCKENZIE, D., 1988, Formation of the steer's head geometry of sedimentary basins by differential stretching of the crust and mantle: Geology, v. 16, p. 250–253.

WHITE, R., AND MCKENZIE, D.,1989, Magmatism at rift zones: The generation of volcanic continental margins and flood basalts: Journal of Geophysical Research, v. 94, p. 7685–7729.

WILLIAMS, L.A.J., 1978, The volcanological development of the Kenya rift, *in* Neumann, E., and Ramberg, I. B., eds., Petrology and geochemistry of continental rifts: Dordrecht, The Netherlands, D. Riedel, p. 101–121.

WILLIAMS, H.H., AND EUBANK, R.T., 1995, Hydrocarbon habitat in the rift graben of the Central Sumatra Basin, Indonesia, *in* Lambiase, J.J. ed., Hydrocarbon Habitat of Rift Basins: Geological Society of London, Special Publication 80, p. 331–371.

WILLIAMS, H.H., FOWLER, M., AND EUBANK, R.T., 1995, Characteristics of selected Palaeogene and Cretaceous lacustrine source basins of Southeast Asia, *in* Lambiase, J.J. ed., Hydrocarbon Habitat of Rift Basins: Geological Society of London, Special Publication 80, p. 241–282.

WITHJACK, M.O., SCHLISCHE, R.W., AND OLSEN, P.E., 1998, Diachronous rifting, drifting and inversion on the passive margin of central eastern North America: an analog for other passive margins: American Association of Petroleum Geologists, Bulletin, v. 82, p. 817–835.

WU, C., NELSON, K.D., WORTMAN, G., SAMSON, S., YUE, Y., LI, J., KIDD, W.S.F., AND EDWARDS, M.A., 1998, Yadong cross-structure and South Tibetan Detachment in the east central Himalaya (89°–90°E): Tectonics, v. 17, p. 28–45.

YANG, W., LI, Y., AND GAO, R., 1985, Formation and evolution of nonmarine petroleum in Songliao basin, China: American Association of Petroleum Geologists, Bulletin, v. 69, p. 1112–1122.

ZECK, H.P., MONIE, P., VILLA, I.M., AND HANSEN, B.T., 1992, Very high rates of cooling and uplift in the Alpine belt of the Betic Cordilleras, southern Spain: Geology, v. 20, p. 79–82.

ZEYEN, H., VOLKER, F., WEHRLE, V., FUCHS, K., SOBOLEV, S.V., AND ALTHERR, R., 1997, Styles of continental rifting: crust–mantle detachment and mantle plumes: Tectonophysics, v. 278, p. 329–352.

ZHANG, Y.Q., MERCIER, J.L., AND VERGELY, I., 1988, Extension in the graben systems around the Ordos (China), and its contribution to the extrusion tectonics of south China with respect to Gobi–Mongolia: Tectonophysics, v. 285, p. 41–75.

ZOBACK, M.L., ANDERSON, R.E., AND THOMPSON, G.A., 1981, Cainozoic evolution of the state of stress and style of tectonism of the Basin and Range province of the western United States: Royal Society [London], Philosophical Transactions, v. A300, p. 407–434.

ZOBACK, M.L., 1992. First- and second-order patterns of stress in the lithosphere: the world stress map project: Journal of Geophysical Research, v. 97, p. 11,703–11,728.

ZOBACK, M.L., AND ZOBACK, M., 1980, State of stress in the coterminus United States: Journal of Geophysical Research, v. 85, p. 6157–6173.

# RIFT-BASIN STRUCTURE AND ITS INFLUENCE ON SEDIMENTARY SYSTEMS

MARTHA OLIVER WITHJACK AND ROY W. SCHLISCHE
*Department of Geological Sciences, Rutgers University, Piscataway, New Jersey 08854-8066, U.S.A.*
*e-mail: drmeow3@rci.rutgers.edu; schlisch@rci.rutgers.edu*

AND

PAUL E. OLSEN
*Lamont-Doherty Earth Observatory of Columbia University, Palisades, New York 10964, U.S.A.*
*e-mail: polsen@ldeo.columbia.edu*

ABSTRACT: Rift basins are complex features defined by several large-scale structural components including faulted margins, the border faults of the faulted margins, the uplifted flanks of the faulted margins, hinged margins, deep troughs, surrounding platforms, and large-scale transfer zones. Moderate- to small-scale structures also develop within rift basins. These include: basement-involved and detached normal faults; strike-slip and reverse faults; and extensional fault-displacement, fault-propagation, forced, and fault-bend folds.

Four factors strongly influence the structural styles of rift basins: the mechanical behavior of the prerift and synrift packages, the tectonic activity before rifting, the obliquity of rifting, and the tectonic activity after rifting. On the basis of these factors, we have defined a standard rift basin and four end-member variations. Most rift basins have attributes of the standard rift basin and / or one or more of the end-member variations. The standard rift basin is characterized by moderately to steeply dipping basement-involved normal faults that strike roughly perpendicular to the direction of maximum extension. Type 1 rift basins, with salt or thick shale in the prerift and / or synrift packages, are characterized by extensional forced folds above basement-involved normal faults and detached normal faults with associated fault-bend folds. In Type 2 rift basins, contractional activity before rifting produced low-angle thrust faults in the prerift strata and / or crystalline basement. The reactivation of these contractional structures during rifting created the low-angle normal faults characteristic of Type 2 rift basins. In Type 3 rift basins, preexisting zones of weakness in the prerift strata and / or crystalline basement strike obliquely to the direction of maximum extension, leading to oblique rifting. Type 3 rift basins are characterized by faults with strike-slip, normal, and oblique-slip displacement and with multiple trends. Contractional activity followed rifting in Type 4 rift basins. These inverted rift basins are affected by late-formed contractional structures including normal faults reactivated with reverse displacement, newly formed reverse faults, and contractional fault-bend and fault-propagation folds.

Structures within rift basins affect depositional patterns by creating sites of uplift and erosion, by controlling pathways of sediment transport, and by defining the accommodation space for sediment deposition and preservation. The relationships among basin capacity (structurally controlled), sediment supply, and water supply determine the primary depositional regime in nonmarine rift basins, fluvial or lacustrine. Changes in basin capacity resulting from the growth of a rift basin may yield a tripartite stratigraphy (fluvial, deep lacustrine, and shallow lacustrine–fluvial) common to many nonmarine rift basins.

## INTRODUCTION

Knowledge of continental rifting has increased steadily during the past several decades. Field geologists have more precisely defined the timing and style of deformation in many exposed rift basins (e.g., the Newark basin of eastern North America, the Upper Rhine rift basin of central Europe, and Kenya basin of East Africa). Simultaneously, seismic interpreters, using 2D and 3D data acquired during hydrocarbon exploration and production activities, have identified and mapped many of the world's subsurface rift basins (e.g., the Viking and Central rift basins of the North Sea, the Jeanne d'Arc rift basin of the Canadian Grand Banks, and the Dampier rift basin on the Northwest Shelf of Australia). Additionally, experimentalists have used scaled laboratory models to simulate rift-basin development under a variety of geologic conditions. Together, these field, seismic, and experimental approaches have yielded valuable information about the structural styles and depositional patterns of rift basins.

Most previous reviews of rift-basin structures have focused on one particular rift system using one, or possibly two, of the above approaches (e.g., Rosendahl, 1987; Schlische, 1993; Morley, 1995, 1999b). In this paper, we examine rift systems from around the world, using information provided by field, seismic, and experimental approaches. This broad perspective allows us to describe the variety of structures associated with continental rifting and to infer the influence of these structural styles on the depositional patterns within them. Because of space limitations, we have left out or abbreviated many important topics related to continental rifting. Readers interested in tectonic-scale aspects of rifting and continental extension should consult the review papers by Roberts and Yielding (1994), Leeder (1995), and Ruppel (1995). Also, Peacock et al. (2000) presented a useful compilation of the voluminous nomenclature related to normal faults, rift basins, and extensional tectonics.

## RIFT BASINS AND RIFT SYSTEMS

Rift basins are elongate crustal depressions bounded on one or both sides by basement-involved normal faults (i.e., faults that cut the crystalline basement) (Figs. 1 and 2). These extensional features are up to several kilometers deep, tens of kilometers wide, and hundreds of kilometers long. Rift systems are collections of stepping, intersecting, and / or parallel rift basins (e.g., Nelson et al., 1992) (Fig. 3). For example, the Bresse and Upper Rhine rift basins of southern France and Germany are right-stepping rift basins that form part of the Tertiary rift system of central Europe (e.g., Ziegler, 1992) (Fig. 3B). The Culpeper and Taylorsville rift basins of the southeastern United States are parallel rift basins that form part of the Mesozoic rift system of eastern North America (e.g., Withjack et al., 1998) (Fig. 3C). Although many rift systems are associated with continental breakup and passive-margin development (e.g., the Mesozoic rift system of eastern North America), rift systems can form in a variety of tectonic settings. For example, the Tertiary rift system

Sedimentation in Continental Rifts
SEPM Special Publication No. 73, Copyright © 2002
SEPM (Society for Sedimentary Geology), ISBN 1-56576-082-4, p. 57–81.

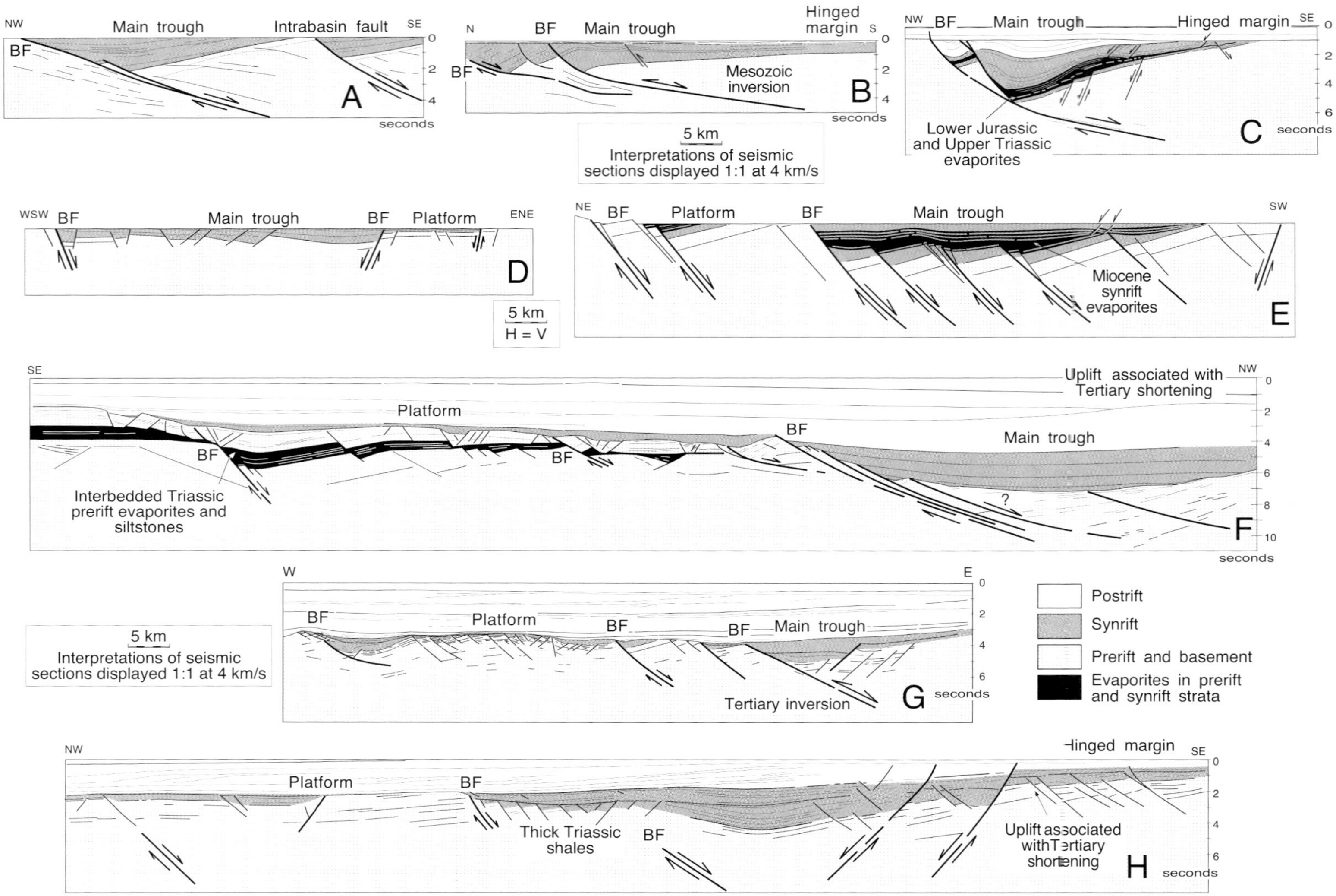


FIG. 1—Cross sections of rift basins. All of the sections have the same horizontal scale, and are shown with the faulted margin and border faults (BF) on the left. The geologic cross sections are displayed without vertical exaggeration. The seismic sections are displayed without vertical exaggeration assuming a velocity of 4 km / s; the vertical axes are in seconds of two-way travel time. **A)** Line drawing of seismic line NB-1 through the Mesozoic Newark rift basin, eastern United States (after Withjack et al., 1998). Figure 2A gives the line location. **B)** Line drawing of seismic line 81-47 through the Mesozoic Fundy rift basin, southeastern Canada (after Withjack et al., 1998). **C)** Line drawing of seismic line 85-4A through the Mesozoic Jeanne d'Arc rift basin, southeastern Canada (after Withjack and Callaway, 2000). **D)** Geologic cross section through the Tertiary Upper Rhine rift basin, Germany (after Sittler, 1969). Figure 2B gives the section location. **E)** Geologic cross section through the Tertiary Suez rift basin, Egypt (after Colletta et al., 1988). Figure 2C give the section location. **F)** Line drawing of a segment of seismic line GMNR94-310 through the Mesozoic Vøring rift basin of offshore Norway (after Withjack and Ingebrigtsen, 1999; Withjack and Callaway, 2000). **G)** Composite line drawing of several seismic lines through the Mesozoic Viking rift basin, northern North Sea. **H)** Composite line drawing of several seismic lines through the Mesozoic Dampier rift basin of the Northwest Shelf, offshore Australia (after Withjack and Eisenstadt, 1999; Schlische et al., 2002). Figure 2D gives the line location.

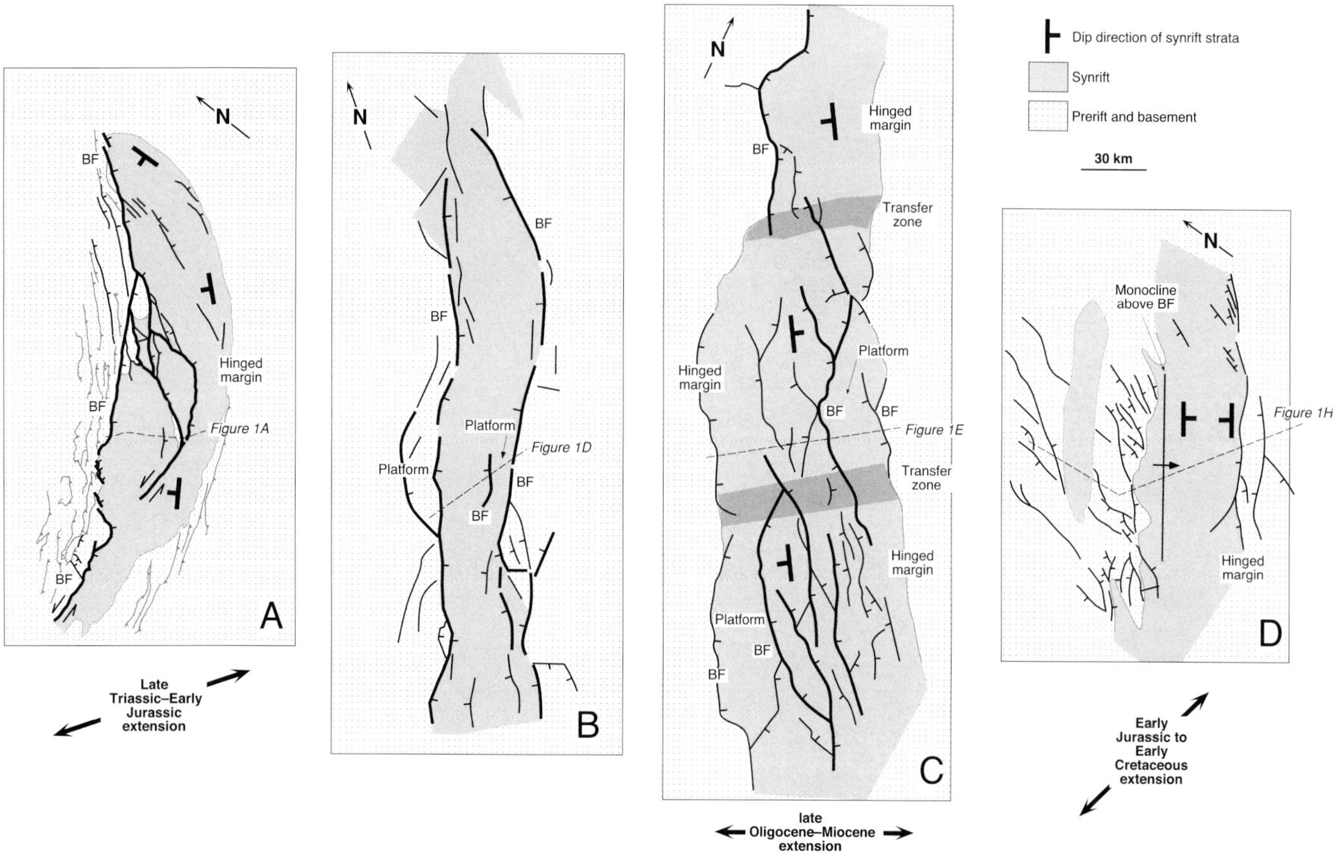


FIG. 2.—Map views of rift basins. All maps are shown at the same scale, and are oriented such that the rift axis is vertical. Thin black lines are normal faults with more than several hundred meters of displacement. Thick black lines are normal faults with several kilometers of displacement. BF is border fault. **A)** Mesozoic Newark rift basin, eastern United States (after Schlische, 1992). The strike of the rift-basin border faults is subparallel to the strike of the Paleozoic reverse faults (gray barbed lines). **B)** Tertiary Upper Rhine rift basin, Germany (after Breyer, 1974; Illies and Greiner, 1978). **C)** Tertiary Suez rift basin, Egypt (after Patton et al., 1994). **D)** Mesozoic Dampier rift basin of Northwest shelf, offshore Australia (after Withjack and Eisenstadt, 1999; Schlische et al., 2002). The strike of many of the normal faults on the northwestern margin of the basin is oblique to the trend of the rift axis.

of central Europe developed adjacent to the active Alpine collision zone (e.g., Ziegler, 1992) (Fig. 3B), whereas the Mesozoic rift system of eastern North America formed within a Paleozoic orogenic belt (e.g., Withjack et al., 1998) (Fig. 3C).

## LARGE-SCALE STRUCTURAL COMPONENTS OF RIFT BASINS

Rift basins are complex features defined by several large-scale structural components. These components include faulted margins, the border faults of the faulted margins, the uplifted flanks of the faulted margins, hinged margins, deep troughs, large-scale intrabasin fault blocks, and large-scale transfer zones (Figs. 1 and 2).

### *Faulted Margins*

Basement-involved normal faults with displacements of several kilometers and lengths of several tens of kilometers are a fundamental part of every faulted margin. These faults are commonly called border faults, and their spatial arrangement varies considerably (Fig. 4). Some border faults have a stepping geometry and similar dip directions, forming asymmetric rift basins with a faulted margin and a hinged margin (Fig. 4A). For example, the northwestern faulted margin of the asymmetric Newark rift basin is composed of closely spaced, right-stepping, southeast-dipping, basement-involved normal faults (e.g., Withjack et al., 1998) (Figs. 1A and 2A). Some border faults have a stepping geometry and convergent dip directions, causing the faulted and hinged margins to shift from side to side of the rift basin (Fig. 4B). For example, the faulted and hinged margins shift from side to side in the Suez rift basin (e.g., Moustafa, 1976; Thiebaud and Robson, 1979; Colletta et al., 1988) (Fig. 2C) and in many of the rift basins of the East African rift system (e.g., Rosendahl et al., 1986; Ebinger et al., 1987; Morley et al., 1990). Some border faults are parallel, overlapping, and have similar dip directions (Fig. 4C). The fault blocks between these border faults are called platform structures in the Upper Rhine rift basin (Breyer, 1974) (Figs. 1D and 2B), stepfault platforms in the Kenya rift basin (Baker and Wohlenberg, 1971), border zones in the Suez rift basin (Garfunkel and Bartov, 1977) (Figs. 1E and 2C), benches in the Rio Grande rift basin (Kelley, 1979), and terraces in the Vøring rift basin of offshore Norway (Blystad et al., 1995) (Fig. 1F). Some border faults are parallel, overlapping, and have con-

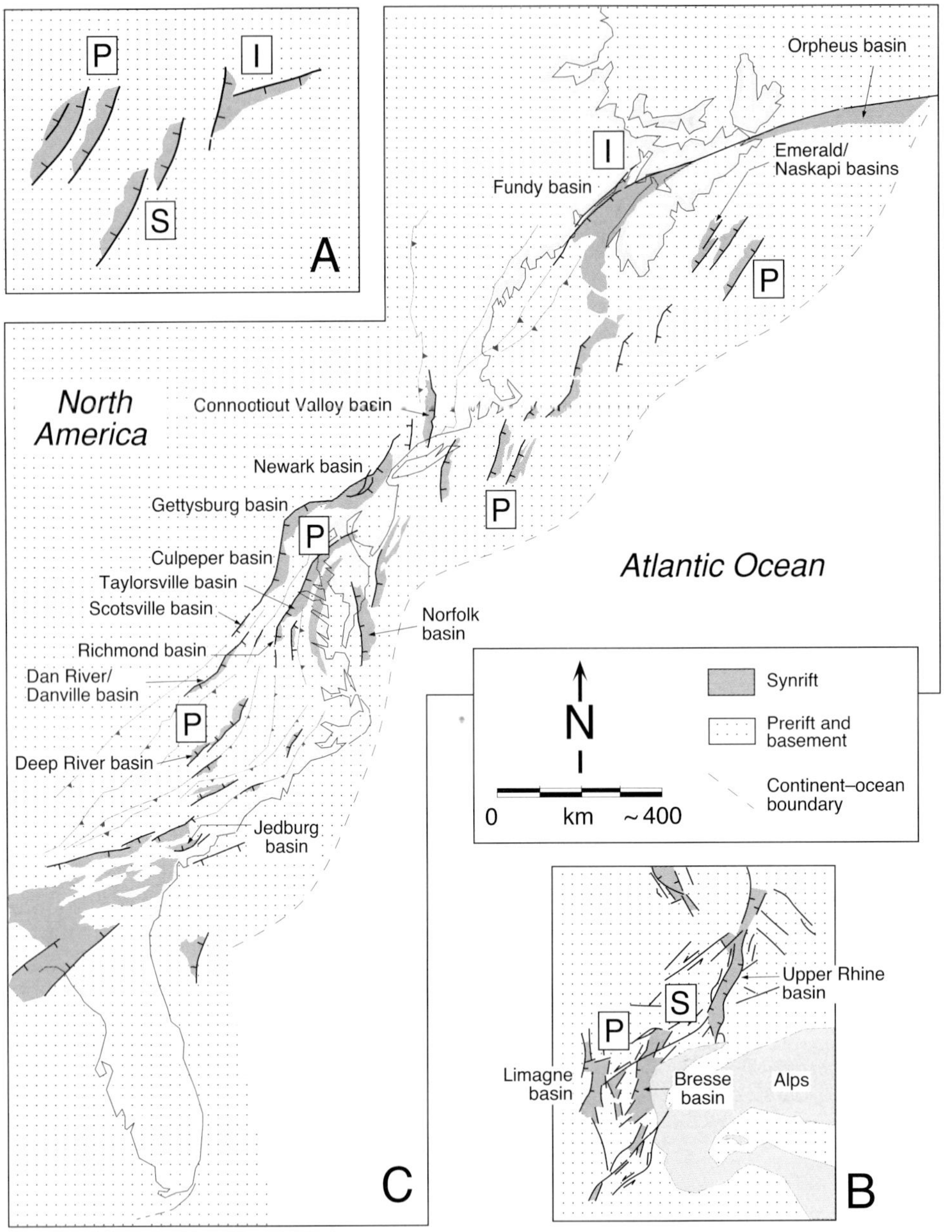


FIG. 3.—Map views of rift systems. Maps B and C are shown at the same scale. **A)** Schematic drawing of parallel (P), stepping (S), and intersecting (I) rift basins. **B)** Tertiary rift system of central Europe, showing a right-stepping arrangement of the Upper Rhine and Bresse rift basins and a parallel arrangement of the Bresse and Ligmagne rift basins (after Ziegler, 1992). Note the proximity of the coeval Alpine collision zone. **C)** Mesozoic rift system of eastern North America, showing parallel and intersecting rift basins (after Withjack et al., 1998). The rift system is now part of the Atlantic passive margin. It formed within a Paleozoic orogenic belt. The strike of the rift-basin border faults is subparallel to the strike of the Paleozoic contractional structures (gray, barbed lines).

vergent dip directions, forming symmetric rift basins with two faulted margins and no hinged margin (Fig. 4D). For example, the western and eastern faulted margins of the symmetric Upper Rhine rift basin are composed of east- and west-dipping border faults, respectively (e.g., Sittler, 1969) (Figs. 1D and 2B). Finally, some border faults intersect each other (Fig. 4E). For example, the northwestern faulted margins of the Fundy and Chignecto rift basins are composed of northeast-striking, basement-involved normal faults. An intersecting east-striking, basement-involved fault with both normal and strike-slip components of displacement bounds the adjoining Minas rift basin (Withjack et al., 1995b) (see Figs. 7C and 7D).

The secondary deformation associated with border faults also varies considerably. In many rift basins, folds form in the sedimentary cover above the border faults. For example, monoclinal flexures occur above the border faults of the southern Upper Rhine rift basin (Laubscher, 1982; Maurin, 1995) (see Fig. 12E) and in the Haltenbanken area of offshore Norway (Fig. 1F). Evaporites in the prerift package facilitated the development of these folds by decoupling the shallow strata from the deep,

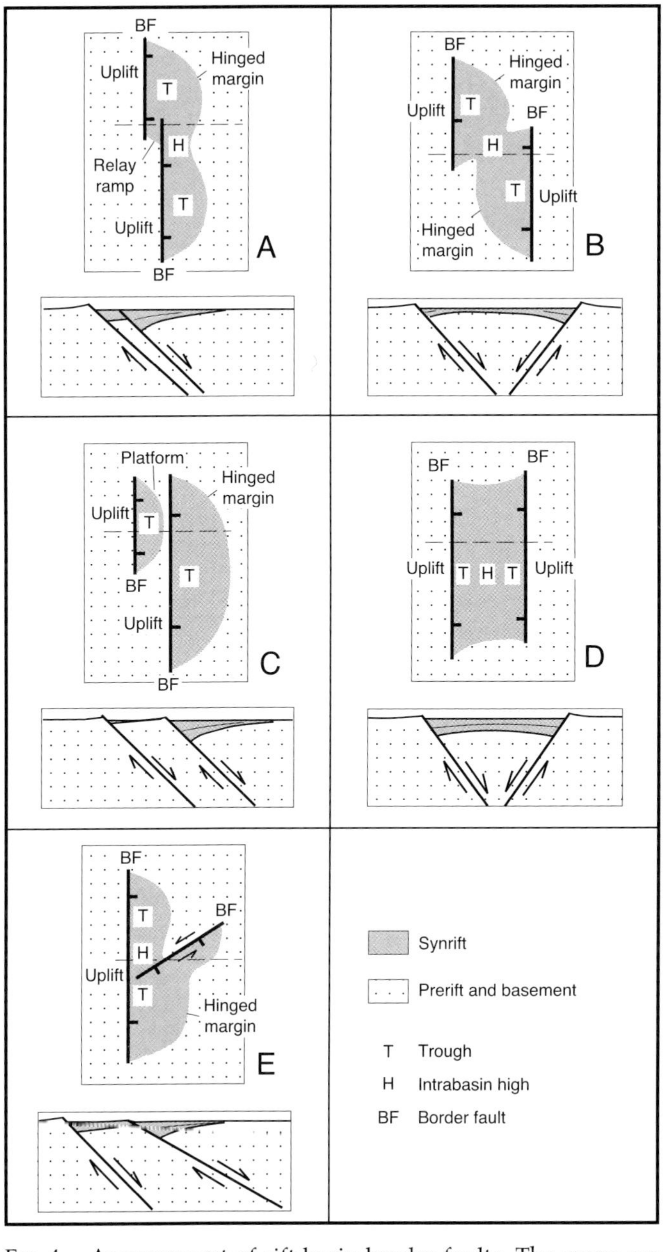


FIG. 4.—Arrangement of rift-basin border faults. The maps are displayed at half the size of the cross sections. **A)** Border faults have a stepping geometry and similar dip directions, forming an asymmetric rift basin with a faulted margin and a hinged margin. **B)** Border faults have a stepping geometry and convergent dip directions, causing the faulted and hinged margins to shift from side to side of the rift basin. **C)** Border faults are parallel, overlapping, and have similar dip directions, forming a platform. **D)** Border faults are parallel, overlapping, and have convergent dip directions, forming a symmetric rift basin with two faulted margins and no hinged margin. **E)** Border faults intersect.

faulted strata and basement. In some rift basins, the border faults are oblique-slip faults with both normal and strike-slip components of displacement. In these rift basins, many of the secondary normal faults strike obliquely to the trend of the border faults. For example, the trend of the faulted margin (and the deep-seated border faults) of the Dampier rift basin is northeast–southwest. Most of the secondary normal faults in the sedimentary cover, however, strike north–south (Withjack and Eisenstadt, 1999) (Fig. 2D).

### *Flanks and Troughs*

The footwall of each border fault is generally uplifted, producing an elevated rift flank (Figs. 4 and 5A) (e.g., Zandt and Owens, 1980; Jackson and McKenzie, 1983; King and Ellis, 1990; Anders and Schlische, 1994). Similarly, the hanging wall of each border fault is generally depressed, producing a trough (Figs. 4 and 5A). The magnitude of the footwall uplift and hanging-wall subsidence is greatest near the border fault and decreases away from the border fault into the footwall and hanging-wall blocks, respectively (e.g., Barnett et al., 1987) (Fig. 5B). It also decreases toward the tips of the border fault (e.g., Jackson and McKenzie, 1983; Anders and Schlische, 1994) (Fig. 5C). The ratio of footwall uplift to hanging-wall subsidence varies significantly, ranging from 1:1 to 1:10 (e.g., Jackson and McKenzie, 1983; Stein et al., 1988; Anders and Schlische, 1994).

Most rift basins are composed of several troughs separated by intrabasin highs (Fig. 4). The arrangement of the troughs and

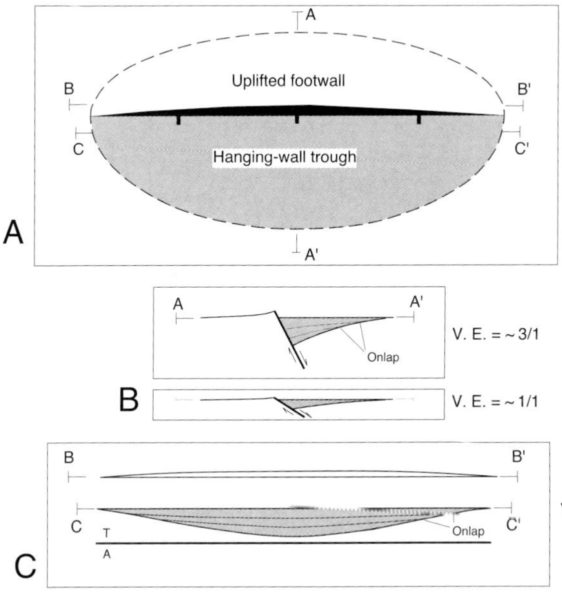


FIG. 5.—Footwall uplift and hanging-wall subsidence associated with border faults. **A)** Map view of a single border fault. The black line is a fault scarp; the line thickness indicates the displacement magnitude. **B)** Traverse section A–A' (with and without vertical exaggeration). The magnitude of the footwall uplift and hanging-wall subsidence decreases away from the border fault. Sedimentary units thicken toward the fault and progressively onlap the prerift units. **C)** Longitudinal sections. Longitudinal section B–B' shows that footwall uplift decreases toward the fault tips. Longitudinal section C–C' shows that hanging-wall subsidence decreases toward the fault tips. Sedimentary units thicken toward the center of the basin and progressively onlap the prerift units. The black line beneath the basin fill is a strike view of the border fault. T is motion toward the reader; A is motion away from the reader. Modified from Gibson et al. (1989) and Schlische (1995).

highs depends on the arrangement of the border faults. If the border faults have a stepping geometry and the same dip direction, then a series of adjacent troughs and highs develops on the faulted margin of the rift basin (Fig. 4A). If the border faults have a stepping geometry and convergent dip directions, then a series of offset troughs separated by intrabasin highs develops on opposing sides of the rift basin (Fig. 4B). If the border faults are parallel, overlapping, and have the same dip direction, then a series of parallel troughs and highs develops on the faulted margin of the rift basin (Fig. 4C).

### *Large-Scale Intrabasin Fault Blocks*

Most rift basins are dissected by large-scale intrabasin faults (i.e., faults with displacements of more than several hundred meters and lengths of tens of kilometers) (Figs. 1 and 2). These intrabasin faults can strike parallel or obliquely to the trend of the border faults, and they can have normal and / or strike-slip components of displacement. Experimental studies suggest that the orientation and displacement on these intrabasin faults is controlled, in part, by the obliquity of the maximum extension direction relative to the rift-basin trend (Withjack and Jamison, 1986; Tron and Brun, 1991; Clifton et al., 2000) (Fig. 6). For orthogonal extension, intrabasin faults are likely to strike parallel to the rift trend and perpendicular to the maximum extension direction and have normal displacements (Fig. 6A). For oblique extension, intrabasin faults are likely to strike obliquely to the rift trend and subperpendicular to the maximum extension direction and have mostly normal displacement (Figs. 6B and 6C).

### *Large-Scale Transfer Zones*

We define large-scale transfer zones as zones that accommodate the transfer of extension from one set of large-scale normal faults to another set of large-scale normal faults. Large-scale transfer zones affect most rift basins (e.g., Rosendahl, 1987; Morley et al., 1990; Faulds and Varga, 1998) (Fig. 7). In some cases, the deformation is focused, consisting one or two strike-slip or oblique-slip faults. In other cases, the deformation is diffuse, consisting of flexures and numerous normal, strike-slip, and oblique-slip faults. Experimental models (e.g., Clifton et al., 2000) and field studies (e.g., Moustafa, 1997) suggest that the deformation within large-scale transfer zones is more likely to be focused if the magnitude of extension is large, the sedimentary cover is thin, and/or the sedimentary cover contains limited salt or shale. Thus, a transfer zone can narrow through time (as the magnitude of extension increases) or along strike (as the thickness of the sedimentary cover decreases).

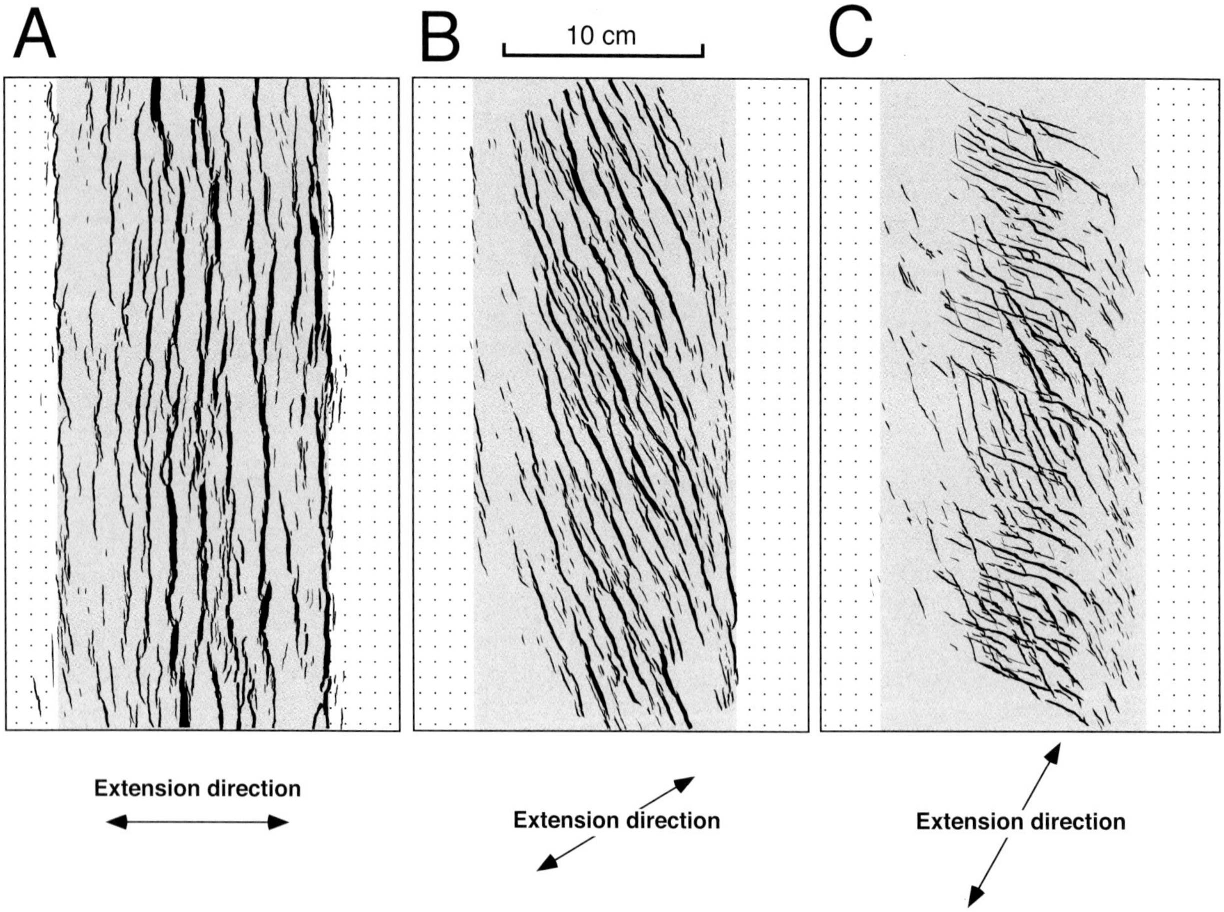


FIG. 6.—Surface fault patterns for orthogonal and oblique rifting based on experimental (clay) models (after Clifton et al., 2000). The rift axis is vertical. Gray areas are extended regions; dotted areas are regions with no extension. **A)** Orthogonal extension with the displacement direction perpendicular to the rift axis. **B)** Oblique extension with the displacement direction 60° from the rift axis. **C)** Oblique extension with the displacement direction 30° from the rift axis.

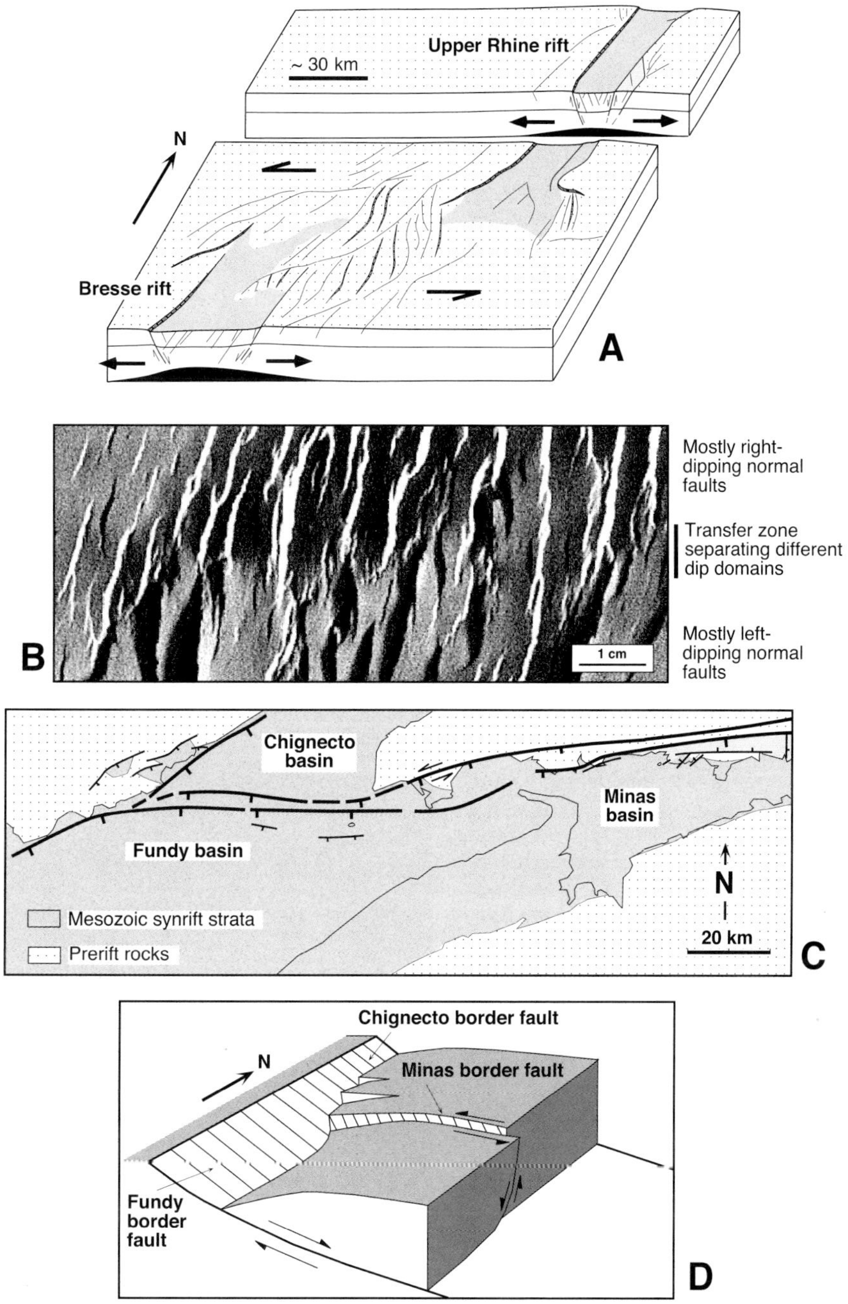


FIG. 7.—Large-scale transfer zones. **A)** Block diagram of a transfer zone that links laterally offset zones of extension. This broad transfer zone, composed of numerous normal, strike-slip, and oblique-slip faults, links the northern end of the Bresse rift basin with the southern end of the Upper Rhine rift basin (after Illies, 1977). **B)** Map view of a model of distributed extension in homogeneous clay. A transfer zone developed between regions with normal faults with different dip directions and locations (after Clifton et al., 2000). **C)** Map view of a transfer zone that separates regions with different amounts of extension. An east-trending oblique-slip fault zone forms the northern border of Minas rift basin. This zone separates the Chignecto rift basin (less extension) from the Fundy rift basin (more extension) (after Withjack et al., 1995b). **D)** Schematic block diagram showing the geometries of the border faults in the Fundy, Chignecto, and Minas rift basins.

One type of large-scale transfer zone connects stepped zones of extension. The intrabasin highs between stepping border faults with convergent dip directions are examples of this type of transfer zone (Fig. 4B). A much larger-scale example of this type of transfer zone is the zone of distributed deformation that links the northern end of the Bresse rift basin with the southern end of the Upper Rhine rift basin (Contini and Theobald, 1974; Angelier and Bergerat, 1983; Fig. 7A). A second type of large-scale transfer zone separates regions in which the normal faults have different dip directions and locations. Examples include the zones that bound the three dip provinces of the Suez rift basin (Moustafa, 1976; Thiebaud and Robson, 1979; Colletta et al., 1988; Fig. 2C). In each of these dip provinces, the dip directions of the normal faults and strata differ from those in the adjacent dip province(s). Experimental models of distributed extension with a homogeneous modeling material have produced similar transfer zones (Clifton et al., 2000) (Fig. 7B). A third type of large-scale transfer zone separates regions with different amounts of extension. For example, the east-striking oblique-slip fault zone on the northern margin of the Minas rift basin is a transfer zone that separates the Fundy rift basin from the less-extended Chignecto rift basin (Figs. 7C and 7D).

## FAULTS

### *Normal Faults*

Normal faults are the most common structures within rift basins. Many normal faults are deep-seated, involving the crystalline basement (Fig. 8A). Others detach into salt or shale above the basement (Fig. 8B). Examples of basement-involved normal faults include the border faults of the Newark, Fundy, Jeanne d'Arc, and Suez rift basins (Figs. 1A–C, E). Examples of detached normal faults include the small-scale normal faults that die out within salt within the Jeanne d'Arc, Suez, and and Vøring rift basins (Figs. 1C, E, and F).

In cross section, the geometries of normal faults vary considerably (Fig. 1). Some normal faults are planar, whereas other normal faults are listric. Some normal faults have steep dips (e.g., most normal faults within the Upper Rhine and Suez rift basins), whereas other normal faults have gentle to moderate dips (e.g., the border faults of the southern Newark, Fundy, and Jeanne d'Arc rift basins). Several processes can produce low-angle normal faults. Reactivation of gently dipping zones of weakness creates low-angle normal faults. For example, the Mesozoic reactivation of Paleozoic thrust faults produced many of the low-angle normal faults in the Newark and Fundy rift basins (Plint and van de Poll, 1984; Ratcliffe et al., 1986; Withjack et al., 1995b). "Domino-style" rotation of fault blocks increases stratal dips and decreases fault dips (e.g., Proffett, 1977). For example, 30° of "domino-style" rotation increases stratal dips from 0° to 30° and decreases fault dips from 60° to 30°. Compaction commonly alters the cross-sectional shapes of normal faults. Compaction causes fault dips to decrease as burial depth increases (Roux, 1979; Davison, 1987; Xiao and Suppe, 1989).

In map view, most normal faults have a sinuous appearance and trend subperpendicular to the direction of maximum extension (Fig. 2). Numerical and experimental models suggest that this characteristic map pattern forms during fault-population evolution (Cowie et al., 1995; Cowie, 1998; Marchal et al., 1998; Ackermann et al., 2001). During the early stages of normal faulting, numerous minor normal faults form; these trend subperpendicular to the direction of maximum extension (Fig. 9A). They are isolated, "lens-shaped" structures with maximum displacements near their centers and no displacement at their ends. Examples include centimeter-scale normal faults in the Danville rift basin (Schlische et al., 1996) and meter- to kilometer-scale minor normal faults in the Volcanic Tableland, California, of the Basin-and-Range province (Dawers et al., 1993) and the Kenya rift basin (Chapman et al., 1978) (see Fig. 11). During subsequent stages of normal faulting, fault displacements and lengths increase (Figs. 9B and C). Some of the minor normal faults that formed during the early stage coalesce, forming sinuous normal faults with greater displacements and lengths. This linkage process continues as extension increases, producing a few major, through-going

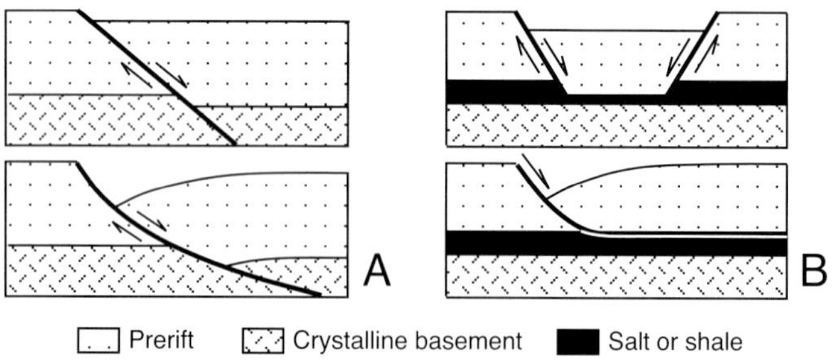


FIG. 8.—Schematic cross sections of **A)** basement-involved and **B)** detached normal faults. Either type of normal fault can be planar or listric.

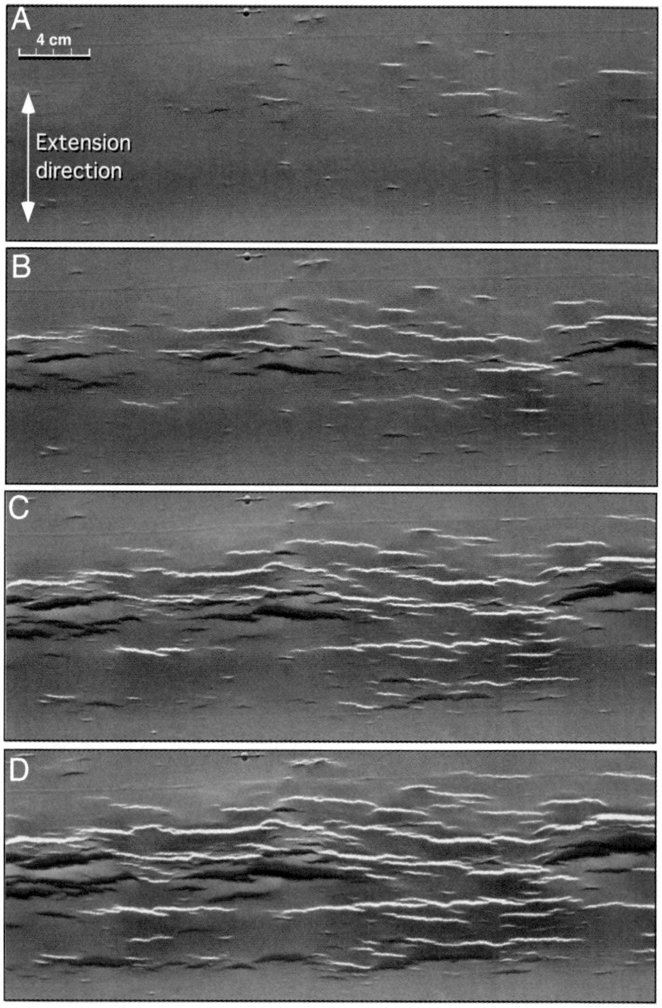


FIG. 9.—Development of normal faults in map view. Photos show faults cutting the upper surface of a homogeneous clay layer deformed above a uniformly extended rubber sheet. The amount of extension of the rubber sheet is **A)** 1.5 cm, **B)** 1.75 cm, **C)** 2.0 cm, and **D)** 2.25 cm. The bright faults dip toward the light source (bottom of image), and the dark faults dip away from the light source. The majority of the faults begin as isolated features. Linkage of adjacent faults results in complex fault traces.

normal faults (Fig. 9D). Many of the faults not involved in the linkage process either stop growing entirely or grow at a slower rate. Because major normal faults are composed of linked minor normal faults, fault displacements, trends, and dips can vary considerably along strike (Fig. 9D).

### *Strike-Slip, Oblique-Slip, and Reverse Faults*

Strike-slip, oblique-slip, and reverse faults also form within rift basins. Strike-slip and oblique-slip faults are most common in large-scale transfer zones (e.g., Rosendahl, 1987) and in rift basins produced by oblique extension (e.g., Withjack and Jamison, 1986) (Figs. 6B, 6C, and 7). Experimental models show that reverse faults can develop as steeply dipping normal faults propagate to the surface (Fig. 10A; Horsfield, 1977; Vendeville, 1987; Withjack and Callaway, 2000). Reverse faults in the Suez rift basin might be analogous to those observed in these models (Fig. 10B; Patton, 1984; Gawthorpe et al., 1997). Some reverse faults in rift basins are actually rotated normal faults. For example, strata and normal faults rotated up to 70° during the past 30 million years in the Lemitar Mountains of the Rio Grande rift basin (Chamberlin, 1983) (Fig. 10C). This large rotation caused some initially high-angle, west-dipping normal faults to become high-angle, east-dipping reverse faults. Finally, reverse faults are common in inverted rift basins, (e.g., Cooper and Williams, 1989). These reverse faults form after rifting, and many are reactivated normal faults. For example, normal faults reactivated with reverse displacement exist in the Fundy rift basin (Fig. 1B), the Viking rift basin (Fig. 1G), and the Sunda arc rift basins (Fig. 10D).

## FOLDS

Three types of folds are genetically associated with normal faulting: fault-displacement folds, fault-propagation folds, and fault-bend folds (e.g., Figs. 11–14; Hamblin, 1965; Withjack and Drickman Pollock, 1984; Schlische, 1995; Janecke et al., 1998). As discussed below, the geometry referred to in the literature as "normal-drag" folding (i.e., the hanging-wall beds dip away from the normal fault) is generally produced by fault-propagation folding and/or fault-bend folding. The geometry referred to in the literature as "reverse-drag" folding (i.e., the hanging-wall

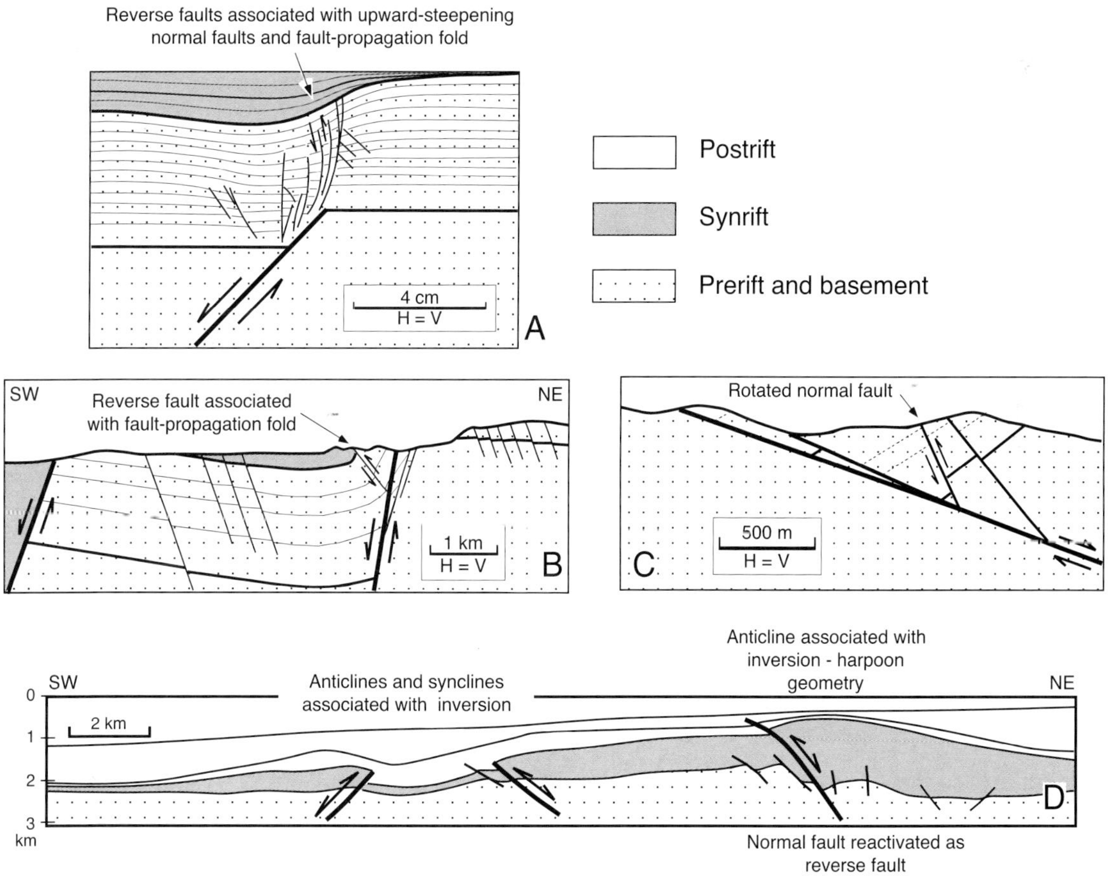


FIG. 10.—Reverse faults associated with rifting. **A)** Line drawing of a clay model showing steeply dipping reverse faults associated with an upward-propagating normal fault (after Withjack and Callaway, 2000). **B)** Example of a reverse fault associated with fault-propagation folding in the Suez rift basin (after Gawthorpe et al., 1997). This reverse fault may be analogous to those in the experimental models. **C)** Rotated normal fault from the Lemitar Mountains of the Rio Grande rift basin (after Chamberlin, 1983). Strata and normal faults rotated up to 70°, causing some initially high-angle, west-dipping normal faults to become high-angle, east-dipping reverse faults. **D)** Normal faults reactivated with reverse displacement in inverted rift basins from the Sunda arc. During basin inversion, some normal faults became reverse faults, producing synclines and anticlines with harpoon geometries (after Letouzey, 1990).

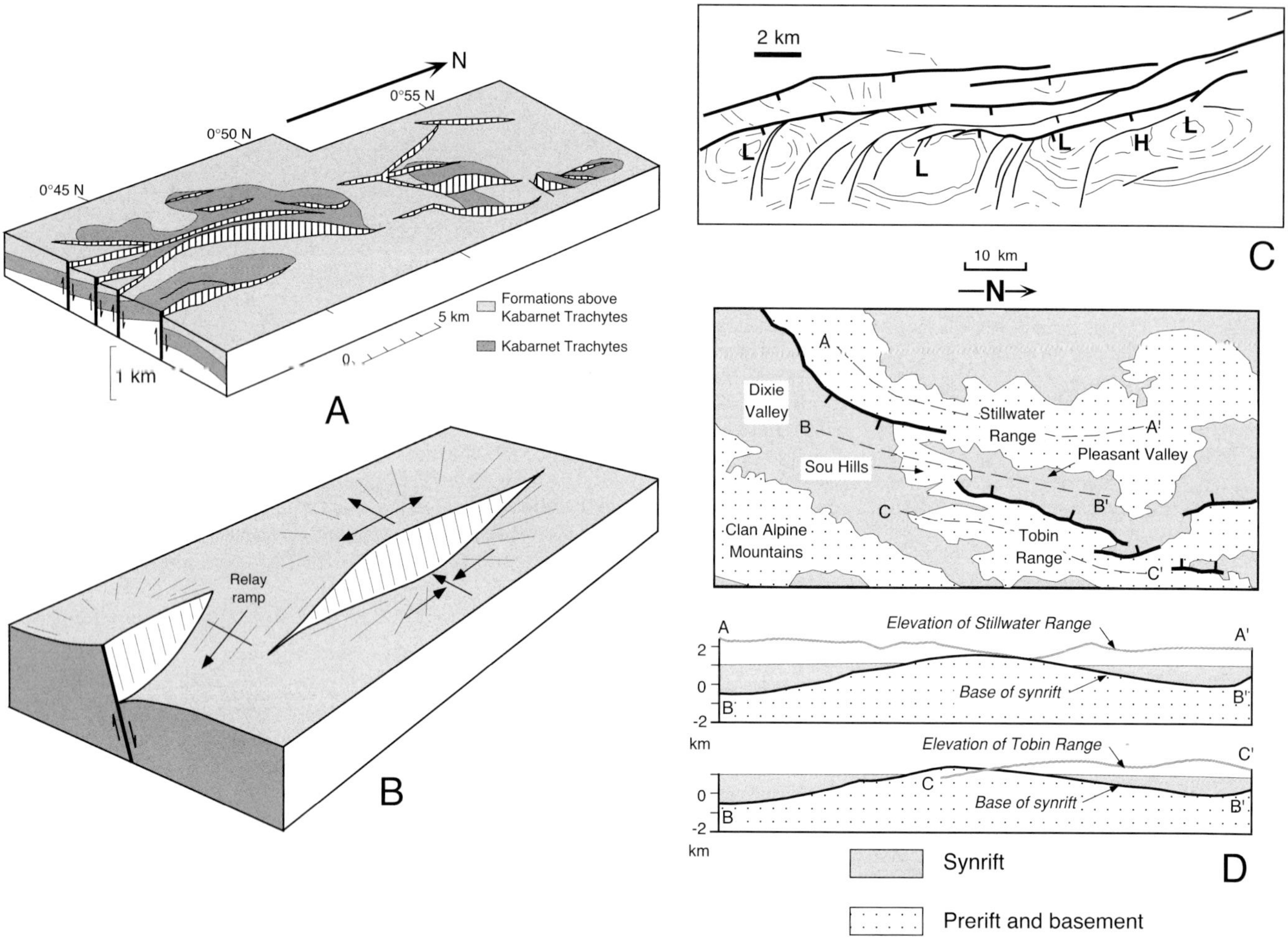


FIG. 11.—Examples of fault-displacement folds. **A)** Fault-displacement folds in the Kenya rift basin (after Chapman et al., 1978). Footwall uplift is greatest where fault displacement is greatest, exposing the underlying Kabarnet Trachytes. **B)** Fault-displacement folds and relay ramp associated with stepping normal faults. **C)** Line drawing of a time slice from 3D seismic survey from southeast Asia showing relative highs (H) and lows (L) in hanging-wall of normal faults. **D)** Fault-displacement folds associated with normal faults in the Basin and Range rift system (after Zhang et al., 1991; Jackson and Leeder, 1994). *Top:* Geologic map of the Dixie Valley–Pleasant Valley fault system of Nevada. *Bottom:* Longitudinal profiles of footwall uplift and hanging-wall subsidence along three transects indicated in the upper panel. The Sou Hills is an interbasin high associated with footwall elevation lows in the Stillwater and Tobin Ranges.

beds dip toward the normal fault) is generally produced by fault-displacement folding and/or fault-bend folding.

### *Fault-Displacement Folds*

We define fault-displacement folds as flexures produced by changes in fault displacement (Fig. 11). The axes of some fault-displacement folds are subparallel to the associated fault; they form because the displacement decreases with increasing distance from a normal fault (Fig. 5B). This displacement geometry leads to monoclinal upfolding in the footwall block (i.e., footwall uplift) and monoclinal downfolding (i.e., "reverse-drag") in the hanging-wall block (e.g., Zandt and Owens, 1980; Barnett et al., 1987). The axes of other fault-displacement folds are subperpendicular to the associated fault; they form because the displacement varies along the strike of a normal fault (Walsh and Watterson, 1987) (Fig. 5C). Several factors produce displacement variations along strike: fault segmentation during the early stages of faulting, linkage during the later stages of faulting, and/or fault-surface irregularities. These fault-displacement folds include the hanging-wall synclines and footwall anticlines that form where fault displacements are greatest (Fig. 11B) and the footwall synclines and hanging-wall anticlines that form where fault displacements are least. As shown below, fault-displacement folding produces three-dimensional culminations and depressions in the footwalls and hanging walls of many normal faults.

Fault-displacement folds form at all scales. At the small scale, folds in the footwalls and hanging walls of normal faults in the Kenya rift basin of East Africa are produced by fault-displacement folding (Chapman et al., 1978; Fig. 11A). Relay ramps (e.g., Goguel, 1962; Kelley, 1979; Larsen, 1988; Peacock and Sanderson, 1991, 1994) are fault-displacement folds that form between stepping normal faults that dip in the same direction. These folds connect the hanging-wall block of one fault with the footwall block of the other fault (Fig. 11B). Fault-

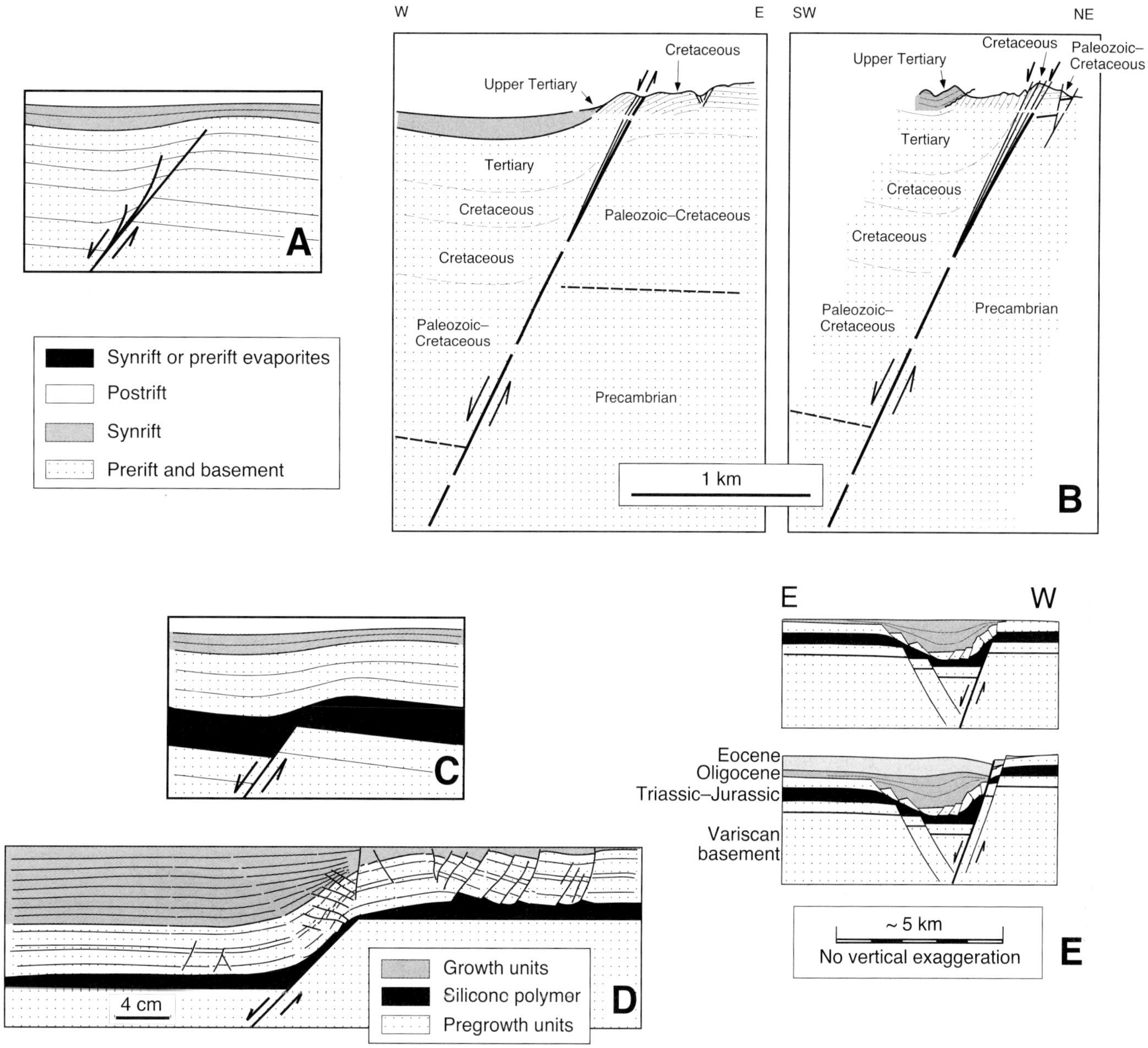


FIG. 12.—Extensional fault-propagation and forced folds. **A)** Sketch of an extensional fault-propagation fold with smooth upward transition from faulting to folding. **B)** Extensional fault-propagation folds from the Suez rift basin (after Patton, 1984; Withjack et al., 1990). **C)** Sketch of an extensional forced fold with abrupt upward transition from faulting to folding. Presence of salt facilitates development of extensional forced folds. **D)** Line drawing of clay–putty model of an extensional forced fold (after Withjack and Callaway, 2000). The black area is silicone putty deposited before faulting. The dotted pattern below the putty represents aluminum blocks; the dotted pattern above the putty is clay deposited before faulting. The gray area is clay deposited during faulting. **E)** An extensional forced fold from the eastern margin of the Rhine rift basin (after Maurin, 1995). *Top*: A deep-seated normal fault has not propagated through the sedimentary section. *Bottom*: A deep-seated normal fault has propagated through the sedimentary section.

displacement folds are evident on time slices from 3D seismic surveys, appearing as anticlinal highs and synclinal lows in the footwalls and hanging walls of normal faults (Fig. 11C). The residual intrabasin highs between stepping border faults with convergent dip directions are large-scale fault-displacement folds (as well as transfer zones) (Figs. 4D and 11D). Finally, at the largest scale, the uplifted flank and the depressed trough of a rift basin are produced by fault-displacement folding associated with the border fault (Fig. 5).

### *Extensional Fault-Propagation Folds and Forced Folds*

Folds commonly develop above upward-propagating normal faults. If the vertical transition from fault- to fold-dominated

(e.g., Pat
sional fa
those fro
1990; Wi
Exam
eastern a
12E; Lau
13A and
Haltenba
Withjack
area, the
Cretaceo
rate fold
basemen
have up
tached g
of these
sional fo
multi-la
and Call

Exter
hanging
1992). Si
remains
fault inf
wall (Fi
dipping
geometr
wall of t
from th
normal
dipping
forms in
The fold
drag" f
this latt
faults co
fault be
Withjac
Vari
fault-be
are para
perpend
resembl
cline ha
"bump"
basin (F
bounda

Fou
basins:
ages, th
and the
we hav
variatio
of struc
has litt
preexis
baseme
activity



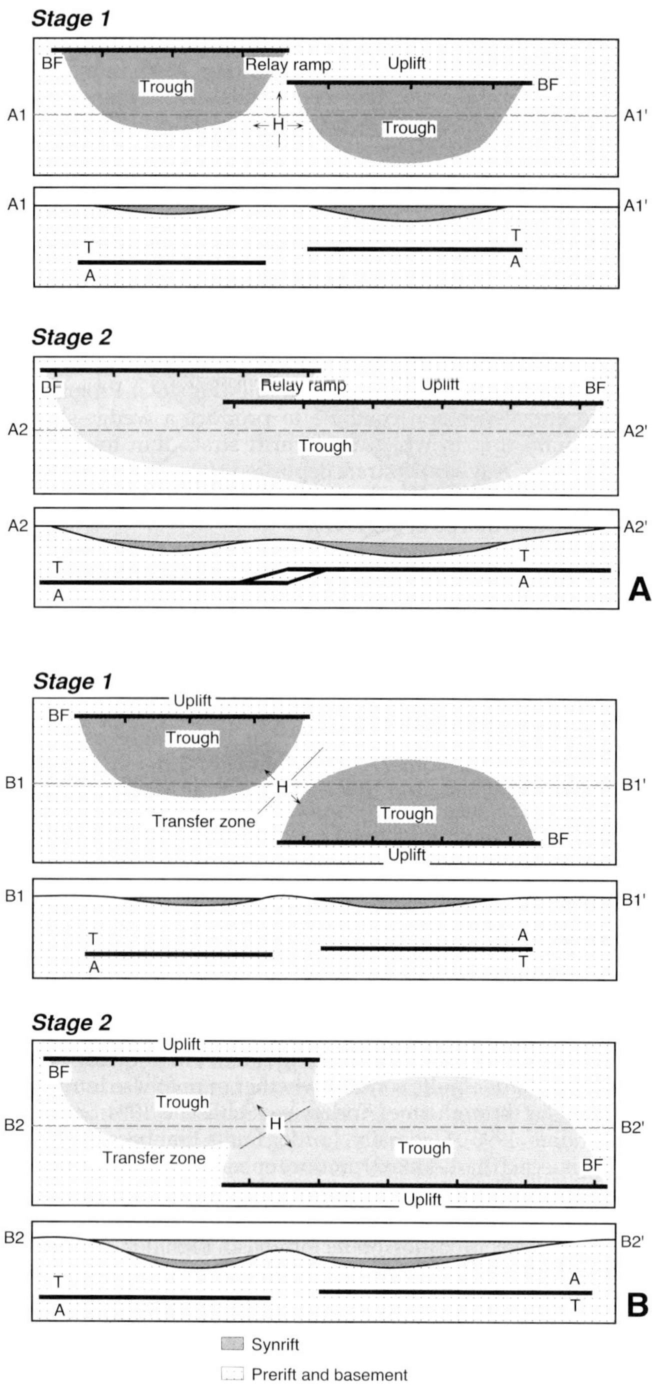


FIG. 19.—Possible geometries of synrift sedimentary units in longitudinal cross sections. Maps (top diagrams in A and B) show the location of cross sections A–A' and B–B', respectively. **A)** Rift basins with stepping border faults (BF) with similar dip directions. **B)** Rift basin with stepping border faults with convergent dip directions shown in two stages. The black line beneath the basin fill in the cross sections is a strike view of the border fault. T is motion toward the reader; A is motion away from the reader. See text for discussion.

Gawthorpe, 1987). Footwall uplift generally produces topographic surfaces that slope away from, and drainages that flow away from, the rift basin. Thus, except for the circumstances listed below, the eroded footwall supplies little sediment directly to the adjacent rift basin. (1) Small drainage systems can exist on the fault-line scarp associated with the border-fault system (Cohen, 1990). These drainage systems can supply limited sediment to the rift basin. (2) Streams can enter the rift basin from the footwall region between border faults where footwall uplift is minimal. This includes relay ramps between stepping border faults with similar dip directions (e.g., Gawthorpe and Hurst, 1993; Childs et al., 1995; Lambiase and Bosworth, 1995). In the Newark rift basin, conglomeratic alluvial-fan facies along the border-fault system are preferentially associated with such relay ramps (e.g., Schlische, 1992). Streams may also enter at transfer zones associated with stepping border faults with convergent dip directions (e.g., Morley et al., 1990; Childs et al., 1995; Lambiase and Bosworth, 1995). (3) Streams can flow across the footwall block if large antecedent drainages are able to erode through the uplifting footwall blocks. For example, in the Connecticut Valley rift basin in eastern North America, streams entered the basin from the footwall when fault-displacement rates were low and from the hanging wall when displacement rates were higher (e.g., Olsen, 1997). The topography on the hinged margin, if present, slopes toward the rift basin, and many streams enter the basin along this route (e.g., Leeder and Gawthorpe, 1987; Cohen, 1990). Many streams also enter a rift basin axially (e.g., Cohen, 1990; Lambiase, 1990; Lambiase and Bosworth, 1995), including streams that flow around the footwall uplift.

### *Influence of Intrabasin Faults and Folds*

Intrabasin faults and folds also influence the depositional patterns of rift basins. Fault-displacement folds locally influence sedimentation, with synrift units thickening in the synclinal lows and thinning on the anticlinal highs in the footwalls and hanging walls of intrabasin normal faults (e.g., Figs. 11A and 11C). Schlische (1995) described examples from the Newark rift basin in eastern North America, and Dawers and Underhill (2000) and McLeod et al. (2000) described examples from the depocenter adjacent to the Statfjord fault in the Viking rift basin of the northern North Sea. Fault-propagation folding and forced folding also locally affect depositional patterns. For example, fault-propagation folding has influenced depositional patterns in the Suez rift basin (Gawthorpe et al., 1997; Gupta et al., 1999; Sharp et al., 2000) and in the Haltenbanken area of the Vøring rift basin (e.g., Figs. 13C and 13D). Progressive folding and deposition combine to produce a wedge-shaped sedimentary unit in which the synrift strata thin toward the underlying intrabasin faults.

### *Depositional Environments of Nonmarine Rift Basins*

A wide variety of depositional environments are present in nonmarine rift basins, but two depositional regimes predominate: fluvial and lacustrine. Fluvial depositional systems require a slope, whereas lacustrine (ponded water) depositional systems require that the basin outlet be perched above the depositional surface. The relationships among incremental accommodation space (the amount of accommodation space added to a basin for a given time increment), sediment supply, and water supply determine which depositional system predominates in a given rift basin (Schlische and Olsen, 1990; Schlische, 1991; Schlische and Anders, 1996; Olsen, 1997; Carroll and Bohacs, 1999) (Fig. 20). In cases where the sediment supply exceeds the incremental accommodation space, fluvial deposition predominates (Fig. 20A).

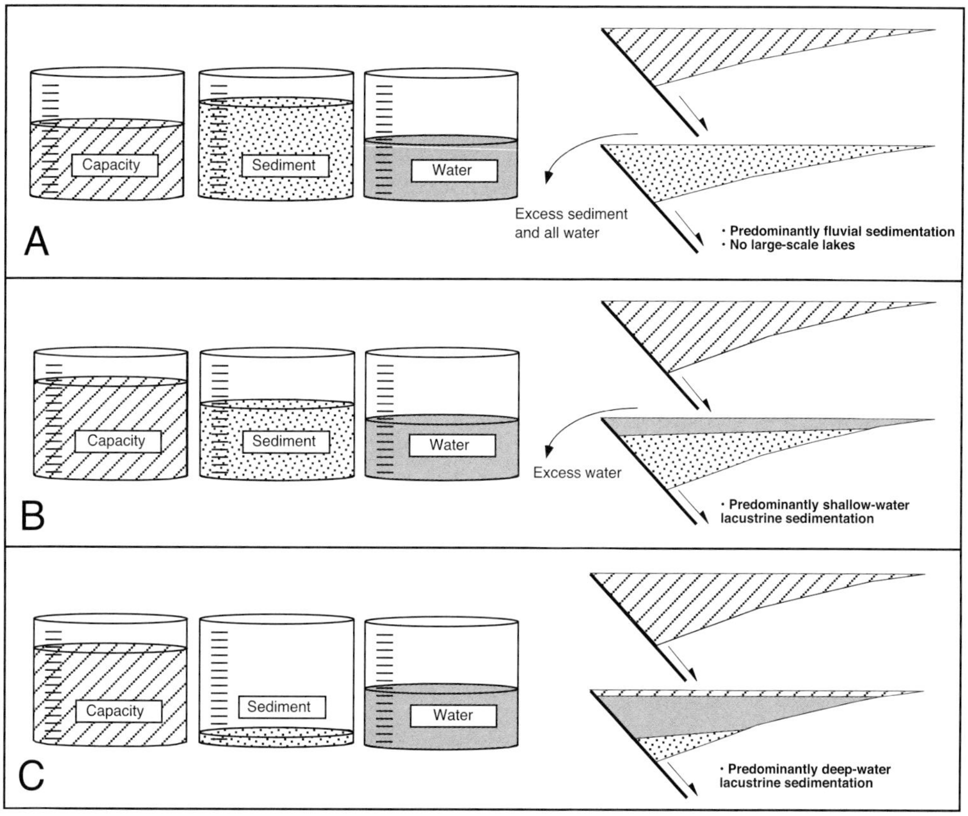


FIG. 20.—Relationships among basin capacity (incremental accommodation space), sediment supply, and water supply in nonmarine rift basins. **A)** Sediment supply exceeds basin capacity. **B)** Basin capacity exceeds sediment supply and water supply exceeds excess basin capacity. **C)** Basin capacity greatly exceeds sediment supply and water supply is less than excess basin capacity. See text for discussion.

In cases where the incremental accommodation space exceeds the sediment supply, lacustrine deposition predominates (although deltaic and fluvial deposition occurs around the margins of the lake). The relationship between the available supply of water and the excess capacity (the difference between the incremental accommodation space and the sediment supply) of the basin determines the hydrological conditions in the lake. If the water supply exceeds the excess capacity, the lake is hydrologically open (Fig. 20B). If the excess capacity exceeds the water supply, the lake is hydrologically closed (Fig. 20C). Given these relationships, the depth of the lake is an unreliable recorder of climate. For example, if the excess capacity of the basin is very small, the lake depth is controlled by the distance between the depositional surface and the lowest outlet. Thus, even if the climate is very wet, the lake will still be shallow.

Many nonmarine rift basins from a wide variety of climatic regimes and tectonic settings share a similar synrift stratigraphic architecture (e.g., Lambiase, 1990; Schlische and Olsen, 1990). The succession generally begins with a fluvial unit. This is overlain by a lacustrine unit that demonstrates a rapid deepening-upward interval to a lake-highstand interval. The deep-water lacustrine interval is succeeded by a gradually upward-shoaling lacustrine unit that is commonly capped by a fluvial unit. This succession is known as a tripartite stratigraphy. Lambiase (1990) observed that some continental rift basins contain multiple stratigraphic successions, consisting of all or part of the tripartite stratigraphy. For the Mesozoic rift basins of eastern North America, the individual stratigraphic successions are bounded by unconformities, particularly along the margins of the basins, separating the synrift deposits into tectonostratigraphic packages (Olsen, 1997; Olsen et al., 2000). Some dominantly marine rift basins (e.g., in Crete; Postma and Drinia, 1993) display similar stratigraphic successions. In these cases, marine deposits take the place of lacustrine deposits, but the water depth varies in a manner similar to that of continental rift basins.

Given the relationships among incremental accommodation space, sediment supply, and water supply, several mechanisms can produce the major stratigraphic transitions within the tripartite succession. The fluvial–lacustrine transition may result from an increase in incremental accommodation space and/or a decrease in sediment supply. The shallow-water to deep-water lacustrine transition may result from an increase in incremental accommodation space, a decrease in the sediment supply, or an increase in the water supply. The deep-water to shallow-water lacustrine transition may result from a decrease or increase in incremental accommodation space (depending on the geometry of the excess accommodation space in the basin), an increase in the sediment supply, and/or a decrease in the water supply. The lacustrine–fluvial transition may result from a decrease in incremental accommodation space or an increase in sediment supply. Because three of four of the stratigraphic transitions may result from an increase in incremental accommodation space and because accommodation space is structurally controlled (e.g., Leeder and Gawthorpe, 1987; Gibson et al., 1989), the structural evolution of rift basins most strongly influences the stratigraphic architecture (Lambiase, 1990; Schlische and Olsen, 1990; Schlische, 1991; Contreras et al., 1997).

### *Basin-Filling Models for Rift Basins*

The basin-filling model of Schlische (1991) emphasized the role of fault growth in an asymmetric rift basin with a single border fault (Fig. 21). As discussed previously, faults generally increase in length as fault displacement increases. Thus, as displacement on the border fault accrues, the rift basin generally increases in depth, length, and width. Thus, incremental accommodation space within the basin increases through time. If sedimentation keeps pace with incremental accommodation space, then the basin is always filled to its lowest outlet with sediment. Strata progressively onlap the hanging-wall hinged margin as the basin widens and lengthens through time. It is unlikely, however, that sedimentation always keeps pace with incremental accommodation space. If the sediment-supply rate and the water-supply rate are constant and if the incremental accommodation space progressively increases, then the rift basin undergoes three of the four major stratigraphic transitions contained within the tripartite stratigraphy. The fourth major stratigraphic transition, the lacustrine–fluvial transition, cannot be reproduced in a basin-filling model that requires a progressive increase in incremental accommodation space. Accommodation space, however, cannot increase indefinitely. Once extension slows considerably or ceases, the basin gradually infills with sediment, with fluvial conditions returning when sediment supply exceeds incremental accommodation space. Multiple extensional pulses may produce multiple synrift tectonostratigraphic packages separated by unconformities if the time between pulses is large (Olsen, 1997).

The quantitative basin-filling models of Contreras et al. (1997) predicted how accumulation rates vary with location and through time. These predictions are in broad agreement with actual accumulation rates from the Newark rift basin (Olsen, 1997). The qualitative basin-filling models of Lambiase (1990) and Lambiase and Bosworth (1995) emphasized the role of footwall uplift and transfer zones. They argued that deep lakes form only when high relief is produced by significant footwall uplift and the formation of transfer zones between stepping borders faults with convergent dip directions. They also pointed out that depocenters bounded by transfer zones exhibit different stratigraphies, particularly if a large axial supply of sediment exists. In this case, the depocenter located nearest the axial supply is the first to become completely infilled, with succeeding depocenters in the sediment-transport direction being progressively infilled. Smoot (1991) presented a qualitative basin-filling model that is most appropriate for the initial fluvial deposits and immediately succeeding lacustrine deposits. He hypothesized that reductions in fluvial gradients and clogging of the basin's outlets contribute to the fluvial–lacustrine transition. This model accounts for a change from initial braided-stream deposits to meandering-stream deposits observed in the basal fluvial sequences of many Mesozoic rift basins in eastern North America.

Gupta et al. (1998) and Cowie et al. (2000) presented a semiquantitative model to explain the large-scale structural and stratigraphic evolution of rift basins. They noted that many marine rift basins appear to undergo a period of accelerated subsidence (associated with deeper-water marine sedimentation) several million years after the initiation of rifting. Their basin-evolution model, based on the results of numerical models of fault-population evolution, showed that certain normal faults undergo periods of accelerated displacement, even though the regional extensional strain rate is constant (Cowie et al., 1995; Cowie, 1998; Cowie et al., 2000). They proposed that the period of increased subsidence and deeper-water marine sedimentation observed in many marine rift basins correlates with the period of strain localization on faults within the border-fault system. Applying their model to nonmarine rift basins, the period of widely distributed faulting would correspond to fluvial sedimentation, whereas deep-water lacustrine sedimentation would correspond to the phase of strain localization.

### *Influence of Plate-Tectonic Movements on Sedimentary Systems*

The forgoing discussion has highlighted how the structure within rift basins and rift systems influences sedimentary systems. Tectonics at the plate-tectonic scale also influences sedimentary systems. Plate movement can cause a rift basin to drift from one climatic zone to another. For example, the Mesozoic rift basins of eastern North America gradually drifted northward during Late Triassic and Early Jurassic time (e.g., Olsen et al., 2000; Olsen, 1997; Olsen and Kent, 2000). Relatively humid facies were deposited when a basin was located near the paleoequator, whereas relatively arid facies were deposited when a basin was located at more northern latitudes. Given the structural control on the tripartite stratigraphic architecture of synrift strata discussed above, it is important to draw a distinction between humid and arid facies deposited under different climatic regimes versus fluvial and lacustrine strata deposited under different hydrologic conditions (Olsen et al., 2000). The combinations of various facies (e.g., arid fluvial, humid fluvial, arid lacustrine, humid lacustrine) are produced by the creation of accommodation space by border faults, the balance between accommodation space and sediment infilling, and changes in climatic regime controlled by the drift of tectonic plates.

## SUMMARY

1. Rift basins are complex features defined by several large-scale structural components including faulted margins, the border faults of the faulted margins, the uplifted flanks of the faulted margins, hinged margins, deep troughs, large-scale intrabasin fault blocks, and large-scale transfer zones. A variety of moderate- to small-scale structures also develop within rift basins. These structures include normal faults (basement-involved and detached), strike-slip and reverse faults, and extensional folds (fault-displacement, fault-propagation, forced, and fault-bend). All of these structures, regardless of their size, affect the depositional patterns of rift basins by creating sites of uplift and erosion, by influencing pathways of sediment transport, and by defining the accommodation space for sediment deposition and preservation.

2. To provide a framework for understanding the variety of structural styles in rift basins, we have defined a standard rift basin and four end-member variations. Most rift basins have attributes of the standard rift basin and one or more of the end-member variations. The standard rift basin has little salt or shale in the prerift or synrift packages. Few preexisting zones of weakness in the prerift strata or crystalline basement were reactivated during rifting, and little tectonic activity followed rifting. The standard rift basin is characterized by moderately to steeply dipping basement-involved normal faults that strike roughly perpendicular to the direction of maximum extension. Type 1 rift basins have salt or thick shale in the prerift or synrift packages. They are characterized by extensional forced folds above basement-in-

volved normal faults and detached normal faults with associated fault-bend folds. Contractional activity preceded rifting in Type 2 rift basins. Many of the normal faults in these rift basins are thrust faults reactivated with normal displacement and, thus, are low-angle. Type 3 rift basins are produced by oblique rifting. They are characterized by faults with strike-slip, normal, and oblique-slip displacement that are parallel and oblique to the rift trend. One or more contractional events followed rifting in Type 4 rift basins. These inverted rift basins are affected by late-formed contractional structures including normal faults reactivated as reverse faults, newly formed reverse faults, and contractional fault-bend and fault-propagation folds.

3. Most rift basins are composed of several troughs separated by intrabasin highs. The arrangement of the troughs and highs depends on the arrangement of the rift-basin border faults. If the border faults are stepping and have similar dip directions, then a series of adjacent troughs and highs develops on the faulted margin of the rift basin. These intrabasin highs are commonly ephemeral because the stepping border faults become hard linked. If the border faults are stepping and have

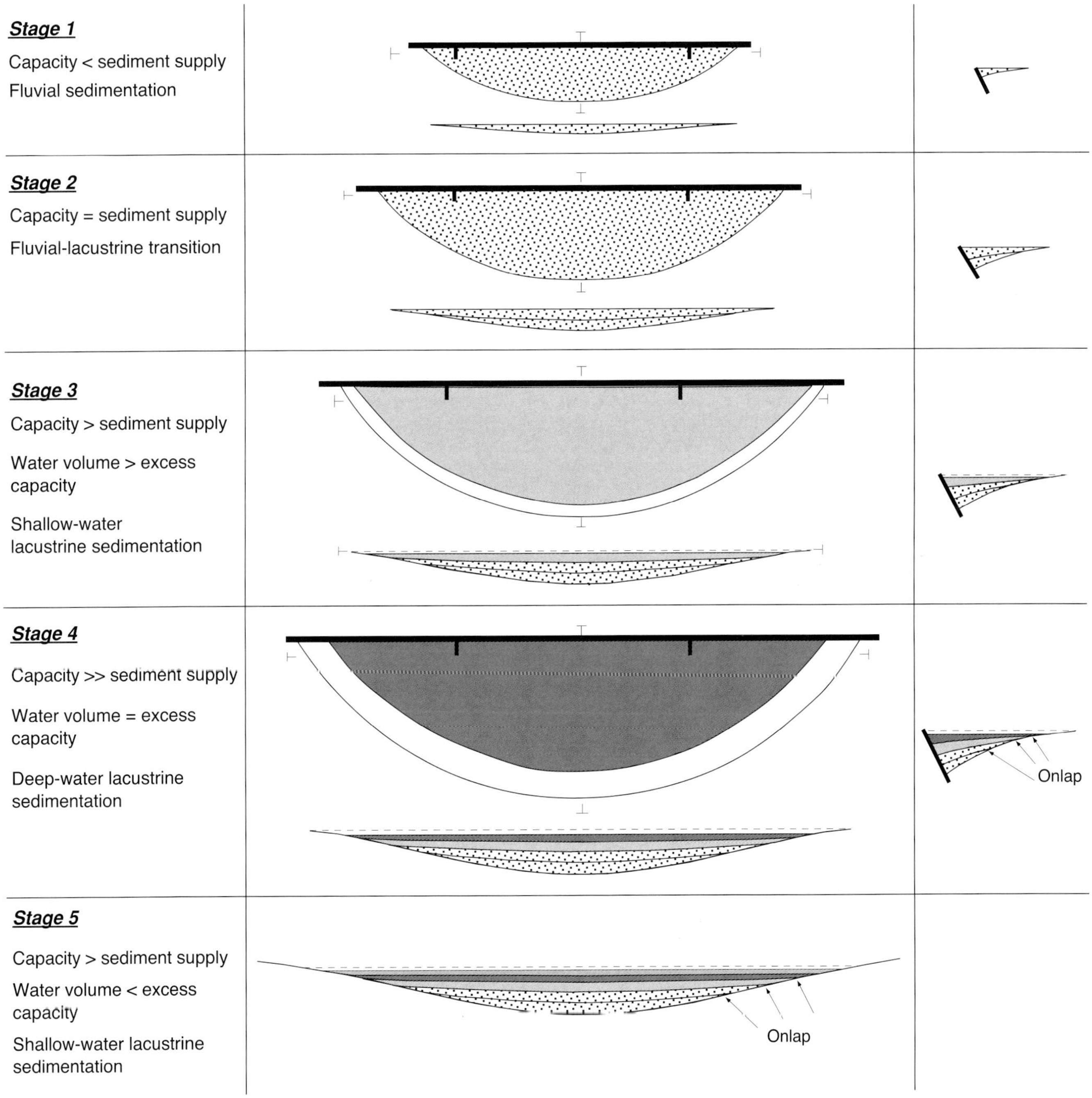


FIG. 21.—Simple filling model for a growing asymmetric rift basin shown in map view (stages 1–4), longitudinal cross section (stages 1–5), and transverse cross sections (stages 1–4). Dashed lines represent lake level. Modified from Schlische and Anders (1996).

convergent dip directions, then a series of offset troughs separated by highs develops on opposing sides of the rift basin. Generally, synrift units are thin to absent on these intrabasin highs throughout synrift sedimentation.

4. In cross sections perpendicular to the border-fault system, rift basins exhibit a variety of structural and depositional patterns. In many rift basins, troughs form that deepen toward the border faults. Progressive faulting and infilling combine to produce a wedge-shaped unit in which the synrift strata thicken toward the border faults. In other rift basins, platforms develop between parallel, overlapping border faults that dip in the same direction. These platforms have an intermediate elevation relative to the rift flank and trough and, consequently, contain a much thinner sequence of synrift strata than the trough. In most Type 1 rift basins, an extensional forced fold forms above the border faults. In these rift basins, troughs form that deepen away from, rather than toward, the border faults. Progressive folding and deposition combine to produce a wedge-shaped sedimentary unit in which the synrift strata thin toward the border faults.

5. Many nonmarine rift basins share a similar synrift stratigraphic architecture known as a tripartite stratigraphy. The succession begins with a fluvial unit. This is overlain by a lacustrine unit that demonstrates a rapid deepening-upward interval to a lake-highstand interval. This is succeeded by a gradually upward-shoaling lacustrine unit that is commonly capped by a fluvial unit. These stratigraphic transitions result from changes in the relative balance between sediment supply, water supply, and accommodation space, the last of which is structurally controlled. Fluvial sedimentation occurs when sediment supply exceeds accommodation space. Lacustrine sedimentation results when accommodation space exceeds sediment supply. Hydrologically open lakes result when the supply of water exceeds the basin's excess capacity (accommodation space minus sediment supply). Hydrologically closed lakes result when the supply of water is less than the excess capacity.

6. A simple rift basin increases in depth, length, and width as displacement on its border-fault system accrues. Thus, incremental accommodation space increases through time. If sedimentation keeps pace with incremental accommodation space, then the basin is always filled to its lowest outlet with sediment. If the sediment-supply rate and the water-supply rate are constant and if the incremental accommodation space progressively increases, then the rift basin may undergo three of the four major stratigraphic transitions contained within the tripartite stratigraphy. The final lacustrine-fluvial transition requires that faulting slow considerably or cease. As a result, the basin gradually infills with sediment, with fluvial conditions returning when sediment supply exceeds incremental accommodation space.

## ACKNOWLEDGMENTS

We thank Gail Ashley for her encouragement to write this paper and our colleagues, Rolf Ackermann and Amy Clifton, for providing their experimental insights about normal-fault development. We also thank Nancye Dawers and David Peacock for their careful and thoughtful reviews of the manuscript. Some of the research presented here was supported by Mobil Technology Company and National Science Foundation grants EAR-9017785 and EAR-9706199 to Schlische.

## REFERENCES

Ackermann, R.V., Schlische, R.W., and Withjack, M.O., 2001, The geometric and statistical evolution of normal fault systems: an experimental study of the effects of mechanical layer thickness on scaling laws: Journal of Structural Geology, v. 23, p. 1803–1819.

Anders, M.H., and Schlische, R.W., 1994, Overlapping faults, intrabasin highs, and the growth of normal faults: Journal of Geology, v. 102, p. 165–179.

Angelier, J., and Bergerat, F., 1983, Systemés de contrainte et extension intracontinentale: Bull. Centres Recherches Exploration–Production Elf-Aquitaine, v. 7, p. 137–147.

Baker, B.H., and Wohlenberg, J., 1971, Structure and evolution of the Kenya rift valley: Nature, v. 229, p. 538–542.

Barnett, J.A.M., Mortimer, J., Rippon, J.H., Walsh, J.J., and Watterson, J., 1987, Displacement geometry in the volume containing a single normal fault: American Association of Petroleum Geologists, Bulletin, v. 71, p. 925–937.

Barr, D., 1987, Structural/stratigraphic models for extensional basins of half-graben type: Journal of Structural Geology, v. 9, p. 491–500.

Blair, T.C., and Bilodeau, W.L., 1988, Development of tectonic cyclothems in rift, pull-apart, and foreland basins—sedimentary response to episodic tectonism: Geology, v. 16, p. 517–520.

Blystad, P., Brekke, H., Faerseth, R.B., Larsen, B.T., Skogseid, J., and Tørudbakken, B., 1995, Structural Elements of the Norwegian Continental Shelf, Part II—The Norwegian Sea Region: Norwegian Petroleum Directorate, Bulletin 8, 45 p.

Braun, J., and Beaumont, C., 1989, A physical explanation of the relation between flank uplifts and the breakup unconformity at rifted continental margins: Geology, v. 17, p. 760–764.

Breyer, F., 1974, Structure and development of the southern part of the Rhine graben according to geological and geophysical observations, *in* Illies, J.H., and Fuchs, K., eds., Approaches to Taphrogenesis: Inter-Union Commission on Geodynamics, Scientific Report 8, p. 145–155.

Brodie, J., and White, N.J., 1994, Sedimentary basin inversion caused by igneous underplating: Northwest European continental shelf: Geology, v. 22, p. 147–150.

Brown, R.N., 1980, History of exploration and discovery of Morgan, Ramadan, and July oilfields, Gulf of Suez, Egypt, *in* Maill, A.D., ed., Facts and Principles of World Petroleum Occurrence: Canadian Society of Petroleum Geologists, Memoir 6, p. 733–764.

Buck, W.R., 1986, Small-scale convection induced by passive rifting—the cause for uplift of rift shoulders: Earth and Planetary Science Letters, v. 77, p. 362–372.

Carroll, A.R., and Bohacs, K.M., 1999, Stratigraphic classification of ancient lakes—balancing tectonic and climatic controls: Geology, v. 27, p. 99–102.

Chamberlin, R.N., 1983, Cenozoic domino-style crustal extension in the Lemitar Mountains, New Mexico: a summary: New Mexico Geological Society 34th Field Conference, Guidebook, Socorro Region II, p. 111–118.

Chapman, G.R., Lippard, S.J., and Martyn, J.E., 1978, The stratigraphy and structure of the Kamasia Range, Kenya Rift Valley: Geological Society of London, Journal, v. 135, p. 265–281.

Childs, C., Watterson, J., and Walsh, J.J., 1995, Fault overlap zones within developing normal fault systems: Geological Society of London, Journal, v. 152, p. 535–549.

Clifton, A.E., Schlische, R.W., Withjack, M.O., and Ackermann, R.V., 2000, Influence of rift obliquity on fault-population systematics: results of clay modeling experiments: Journal of Structural Geology, v. 22, p. 1491–1509.

Coffield, D.Q., and Schamel, S., 1989, Surface expression of an accommodation zone within the Gulf of Suez rift, Egypt: Geology, v. 17, p. 76–79.

Cohen, A.S., 1990, Tectono-stratigraphic model for sedimentation in Lake Tanganyika, Africa, *in* Katz, B.J., ed., Lacustrine Basin Exploration—

Case Studies and Modern Analogs: American Association of Petroleum Geologists, Memoir 50, p. 137–150.

Colletta, B., Le Quellec, P., Letouzey, J., and Moretti, I., 1988, Longitudinal evolution of the Suez structure (Egypt): Tectonophysics, v. 153, p. 221–233.

Contini, D., and Theobald, N., 1974, Relations entre le Fosse Rhenan et le Fosse de la Soane, tectonique des régions sous-vosgiennes et préjurassiennes, *in* Illies, J.H. and Fuchs, K., eds., Approaches to Taphrogenesis: Inter-Union Commission on Geodynamics, Scientific Report 8, p. 310–321.

Contreras, J., Scholz, C.H., and King, G.C.P., 1997, A general model of rift basin evolution: constraints of first order stratigraphic observations: Journal of Geophysical Research, v. 102, p. 7673–7690.

Cooper, M.A., and Williams, G.D., eds., 1989, Inversion Tectonics: Geological Society of London, Special Publication 44, 375 p.

Cowie, P.A., 1998, A healing–reloading feedback control on the growth rate of seismogenic faults: Journal of Structural Geology, v. 20, p. 1075–1088.

Cowie, P.A., Gupta, S., and Dawers, N.H., 2000, Implications of fault array evolution for synrift depocentre development: insights from a numerical fault growth model: Basin Research, v. 12, p. 241–261.

Cowie, P.A., Sornette, D., and Vanneste, C., 1995, Multifractal scaling properties of a growing fault population: Geophysical Journal International, v. 122, p. 457–469.

Davison, I., 1987, Normal fault geometry related to sediment compaction and burial: Journal of Structural Geology, v. 9, p. 393–401.

Dawers, N.H., Anders, M.H., and Scholz, C.H., 1993, Growth of normal faults—displacement-length scaling: Geology, v. 21, p. 1107–1110.

Dawers, N.H., and Underhill, J.R., 2000, The role of fault interaction and linkage in controlling synrift stratigraphic sequences—Late Jurassic, Statfjord East area, northern North Sea: American Association of Petroleum Geologists, Bulletin, v. 84, p. 45–64.

de Voogd, B., Keen, C.E., and Kay, W.A., 1990, Fault reactivation during Mesozoic extension in eastern offshore Canada: Tectonophysics, v. 173, p. 567–580.

Ebinger, C.J., Rosendahl, B.R., and Reynolds, D.J., 1987, Tectonic model of the Malawi rift, Africa: Tectonophysics, v. 141, p. 215–235.

Falvey, D.A., 1974, The development of continental margins in plate tectonic theory: APEA Journal, v. 14, p. 95–106.

Faulds, J.E., and Varga, R.J., 1998, The role of accommodation zones and transfer zones in the regional segmentation of extended terranes, *in* Faulds, J.E., and Stewart, J.H., eds., Accommodation Zones and Transfer Zones: The Regional Segmentation of the Basin and Range Province: Geological Society of America, Special Paper 323, p. 1–45.

Garfunkel, Z., and Bartov, Y., 1977, The tectonics of the Suez rift: Geological Survey of Israel, Bulletin 71, 44 p.

Gawarecki, S.L., and Coffield, D.Q., 1990, Mechanical response to stratigraphic inhomogeneity during extension, and surface expression of buried structures, Gulf of Suez, Egypt: Tectonophysics, v. 174, p.7–20.

Gawthorpe, R.L., and Hurst, J.M., 1993, Transfer zones in extensional basins—their structural style and influence on drainage development and stratigraphy: Geological Society of London, Journal, v. 150, p. 1137–1152.

Gawthorpe, R.L., and Leeder, M.R., 2000, Tectono-sedimentary evolution of active extensional basins: Basin Research, v. 12, p. 195–218.

Gawthorpe, R.L., Sharp, I., Underhill, J.R., and Gupta, S., 1997, Linked sequence stratigraphic and structural evolution of propagating normal faults: Geology, v. 25, p. 795–798.

Gibson, J.R., Walsh, J.J., and Watterson, J., 1989, Modelling of bed contours and cross-sections adjacent to planar normal faults: Journal of Structural Geology, v. 11, p. 317–328.

Goguel, J., 1962, Tectonics (translated from French edition, 1952, by Thalman, H.E.): San Francisco, Freeman, 384 p.

Gupta, S., Cowie, P.A., Dawers, N.H., and Underhill, J.R., 1998, A mechanism to explain rift-basin subsidence and stratigraphic patterns through fault-array evolution: Geology, v. 26, p. 595–598.

Gupta, S., Underhill, J.R., Sharp, I.R., and Gawthorpe, R.L., 1999, Role of fault interactions in controlling synrift sediment dispersal patterns—Miocene, Abu Alaqa Group, Suez rift, Sinai, Egypt: Basin Research, v. 11, p. 167–189.

Hamblin, W.K., 1965, Origin of "reverse drag" on the downthrown sides of normal faults: Geological Society of America, Bulletin, v. 76, p. 1145–1164.

Horsfield, W.T., 1977, An experimental approach to basement-controlled faulting: Geologie en Mijnbouw, v. 56, p. 363–370.

Hutchinson, D.R., Klitgord, K.D., and Detrick, R.S., 1986, Rift basins of the Long Island Platform: Geological Society of America, Bulletin, v. 97, p. 688–702.

Illies, J.H., 1977, Ancient and recent rifting in the Rhinegraben: Geologie en Mijnbouw, v. 56, p. 329–350.

Illies, J.H., and Greiner, G., 1978, Rhinegraben and Alpine system: Geological Society of America, Bulletin, v. 89, p. 770–782.

Jackson, J., and Leeder, M., 1994, Drainage systems and the development of normal faults—an example from Pleasant Valley, Nevada: Journal of Structural Geology, v. 16, p. 1041–1059.

Jackson, J., and McKenzie, D., 1983, The geometrical evolution of normal fault systems: Journal of Structural Geology, v. 5, p. 471–482.

Janecke, S.U., Vanderburg, C.J., and Blankenau, J.J., 1998, Geometry, mechanisms and significance of extensional folds from examples in the Rocky Mountain Basin and Range province, U.S.A.: Journal of Structural Geology, v. 20, p. 841–856.

Keen, C.E., 1985, The dynamics of rifting: deformation of the lithosphere by active and passive driving mechanisms: Geophysical Journal of the Royal Astronomical Society, v. 80, p. 95–120.

Keen, C.E., Boutilier, R., DeVoogd, B., Mudford, B., and Enachescu, M.E., 1987, Crustal geometry and extensional models for the Grand Banks, eastern Canada: constraints from deep seismic reflection data, *in* Beaumont, C., and Tankard, A., eds., Sedimentary Basins and Basin-Forming Mechanisms: Canadian Society of Petroleum Geologists, Memoir 12, p. 101–115.

Kelley, V., 1979, Tectonics, middle Rio Grande rift, New Mexico, *in* Riecker, R.E., ed., Rio Grande Rift: Tectonics and Magmatism: Washington, D.C., American Geophysical Union, p. 57–70.

King, G., and Ellis, M., 1990, The origin of large local uplift in extensional regions: Nature, v. 348, p. 689–693.

Lambiase, J.J., 1990, A model for tectonic control of lacustrine stratigraphic sequences in continental rift basins, *in* Katz, B.J., ed., Lacustrine Basin Exploration—Case Studies and Modern Analogs: American Association of Petroleum Geologists, Memoir 50, p. 265–276.

Lambiase, J.J., and Bosworth, W., 1995, Structural controls on sedimentation in continental rifts, *in* Lambiase, J.J., ed., Hydrocarbon Habitat in Rift Basins: Geological Society of London, Special Publication 80, p. 117–144.

Larsen, P.-H., 1988, Relay structures in a Lower Permian basement-involved extension system, East Greenland: Journal of Structural Geology, v. 10, p. 3–8.

Laubscher, H.P., 1982, Die Südostecke des Rheingrabens—ein kinematisches und dynamisches Problem: Eclogae Geologicae Helvetiae, v. 75, p. 101–116.

Leeder, M.R., 1995, Continental rifts and proto-oceanic rift troughs, *in* Busby, C.J. and Ingersoll, R.V., eds., Tectonics of Sedimentary Basins: Cambridge, Massachusetts, Blackwell, p. 119–148.

Leeder, M.R., and Gawthorpe, R.L., 1987, Sedimentary models for extensional tilt-block/half-graben basins, *in* Coward, M.P., Dewey, J.F., and Hancock, P.L., eds., Continental Extensional Tectonics: Geological Society of London, Special Publication 28, p. 139–152.

Letouzey, J., 1990, Fault reactivation, inversion, and fold-thrust belt, *in* Letouzey, J., ed., Petroleum and Tectonics in Mobile Belts: Paris, IFP Editions Technip, p. 101–128

Marchal, D., Guiraud, M., Rives, T., and Van den Driessche, J., 1998, Space and time propagation processes of normal faults, *in* Jones, G., Fisher, Q.J., and Knipe, R.J., eds., Faulting, Fault Sealing, and Fluid Flow in

Hydrocarbon Reservoirs: Geological Society of London, Special Publication 147, p. 51–70.

MAURIN, J.-C., 1995, Drapage et décollement des séries jurassiques sur la faille de détachement majeure du rift rhénan sud: implications sur la géométrie des dépôts syn-rifts oligocenes: Académie des Sciences [Paris], Comptes Rendus, v. 321, p. 1025–1032.

MCLEOD, A.E., DAWERS, N.H., AND UNDERHILL, J.R., 2000, The propagation and linkage of normal faults—Insights from the Strathsey–Brent–Statfjord fault array, northern North Sea: Basin Research, v. 12, p. 263–284.

MORLEY, C.K., 1995, Developments in the structural geology of rifts over the last decade and their impact on hydrocarbon exploration, *in* Lambiase, J.J., ed., Hydrocarbon Habitat in Rift Basins: Geological Society of London, Special Publication 80, p. 1–32.

MORLEY, C.K., 1999a, Patterns of displacement along large normal faults—implications for basin evolution and fault propagation, based on examples from East Africa: American Association of Petroleum Geologists, Bulletin, v. 83, p. 613–634.

MORLEY, C.K., ED. 1999b, Geoscience of Rift Systems—Evolution of East Africa: American Association of Petroleum Geologists, Studies in Geology 44, 250 p.

MORLEY, C.K., NELSON, R.A., PATTON, T.L., AND MUNN, S.G., 1990, Transfer zones in the East African rift system and their relevance to hydrocarbon exploration in rifts: American Association of Petroleum Geologists, Bulletin, v. 74, p. 1234–1253.

MOUSTAFA, A.M., 1976, Block faulting in the Gulf of Suez: Proceedings of the 5th Exploration Seminar, Egyptian General Petroleum Organization, Cairo.

MOUSTAFA, A.R., 1996, Internal structure and deformation of an accommodation zone in the northern part of the Suez rift: Journal of Structural Geology, v. 18, p. 93–108.

MOUSTAFA, A.R., 1997, Controls on the development and evolution of transfer zones—the influence of basement structure and sedimentary thickness in the Suez rift and Red Sea: Journal of Structural Geology, v. 19, p. 755–768.

NELSON, R.A., PATTON, T.L., AND MORLEY, C.K., 1992, Rift-segment interaction and its relation to hydrocarbon exploration in continental rift systems: American Association of Petroleum Geologists, Bulletin, v. 76, p. 1153–1169.

OLSEN, P.E., 1997, Stratigraphic record of the early Mesozoic breakup of Pangea in the Laurasia–Gondwana rift system: Annual Review of Earth and Planetary Sciences, v. 25, p. 337–401.

OLSEN, P.E. AND KENT, D.V., 2000, High resolution early Mesozoic Pangean climatic transect in lacustrine environments, *in* Bachmann, G., and Lerche, I., eds., Epicontinental Triassic, Volume 3, Zentralblatt für Geologie und Paläontologie, VIII, p. 1475–1496.

OLSEN, P.E., KENT, D.V., FOWELL, S.J., SCHLISCHE, R.W., WITHJACK, M.O., AND LETOURNEAU, P.M., 2000, Implications of a comparison of the stratigraphy and depositional environments of the Argana (Morocco) and Fundy (Nova Scotia, Canada) Permian–Jurassic basins, *in* Oujidi, M. and Et-Touhami, M., eds., Le Permien et le Trias du Maroc, Actes de la Premièr Réunion su Groupe Marocain du Permien et du Trias: Oujda, Hilal Impression, p. 165–183.

PATTON, T., 1984, Normal fault and fold development in sedimentary rocks above a pre-existing basement normal fault [unpublished Ph.D. thesis]: Texas A&M University, College Station, 164 p.

PATTON, T.L., MOUSTAFA, A.R., NELSON, R.A., AND ABDINE, S.A., 1994, Tectonic evolution and structural setting of the Suez rift, *in* Landon, S.M., ed., Interior Rift Basins: American Association of Petroleum Geologists, Memoir 59, p. 9–55.

PEACOCK, D.C.P., AND SANDERSON, D.J., 1991, Displacements, segment linkage and relay ramps in normal fault zones: Journal of Structural Geology, v. 13, p. 721–733.

PEACOCK, D.C.P., AND SANDERSON, D.J., 1994, Geometry and development of relay ramps in normal fault systems: American Association of Petroleum Geologists, Bulletin, v. 78, p. 147–165.

PEACOCK, D.C.P., KNIPE, R.J., AND SANDERSON, D.J., 2000, Glossary of normal faults: Journal of Structural Geology, v. 22, p. 291–306.

PLINT, A.G., AND VAN DE POLL, H.W., 1984, Structural and sedimentary history of the Quaco Head area, southern New Brunswick: Canadian Journal of Earth Sciences, v. 21, p. 753–761.

POSTMA, G., AND DRINIA, H., 1993, Architecture and sedimentary facies evolution of a marine, expanding half-graben (Crete, Late Miocene): Basin Research, v. 5, p. 103–124.

PROFFETT, J.M., JR., 1977, Cenozoic geology of the Yerington district, Nevada, and implications for the nature and origin of Basin and Range faulting: Geological Society of America, Bulletin, v. 88, p. 247–266.

RATCLIFFE, N.M., BURTON, W.C., D'ANGELO, R.M., AND COSTAIN, J.K., 1986, Low-angle extensional faulting, reactivated mylonites, and seismic reflection geometry of the Newark Basin margin in eastern Pennsylvania: Geology, v. 14, p. 766–770.

ROBERTS, A., AND YIELDING, G., 1994, Continental extensional tectonics, *in* Hancock, P.L., ed., Continental Deformation: New York, Pergamon, p. 223–250.

ROBSON, D.A., 1971, The structure of the Gulf of Suez (Clysmic) rift, with special reference to the eastern side: Geological Society of London, Journal, v. 127, p. 247–276.

ROSENDAHL, B.R., 1987, Architecture of continental rifts with special reference to East Africa: Annual Review of Earth and Planetary Sciences, v. 15, p. 445–503.

ROSENDAHL, B.R., REYNOLDS, D.J., LORBER, P.M., BURGESS, C.F., MCGILL, J., SCOTT, D., LAMBIASE, J.J., AND DERSKEN, S.J., 1986, Structural expressions of rifting: lessons from Lake Tanganyika, Africa, *in* Frostick, L.E., Renault R.W., Reid, I., and Tiercelin, J.-J., eds., Sedimentation in the African Rifts: Geological Society of London, Special Publication 25, p. 29–43.

ROUX, W.F., 1979, The development of growth fault structures: American Association of Petroleum Geologists, Structural Geology Course Notes, 33 p.

RUPPEL, C., 1995, Extensional processes in continental lithosphere: Journal of Geophysical Research, v. 100, p. 24,187–24,215.

SCHLISCHE, R.W., 1990, Aspects of the structural and stratigraphic evolution of early Mesozoic rift basins of eastern North America [unpublished Ph.D. thesis]: Columbia University, New York, 479 p.

SCHLISCHE, R.W., 1991, Half-graben basin filling models: New constraints on continental extensional basin evolution: Basin Research, v. 3, p. 123–141.

SCHLISCHE, R.W., 1992, Structural and stratigraphic development of the Newark extensional basin, eastern North America—implications for the growth of the basin and its bounding structures: Geological Society of America, Bulletin, v. 104, p. 1246–1263.

SCHLISCHE, R.W., 1993, Anatomy and evolution of the Triassic–Jurassic continental rift system, eastern North America: Tectonics, v. 12, p. 1026–1042.

SCHLISCHE, R.W., 1995, Geometry and origin of fault-related folds in extensional settings: American Association of Petroleum Geologists, Bulletin, v. 79, p. 1661–1678.

SCHLISCHE, R.W., AND ANDERS, M.H., 1996, Stratigraphic effects and tectonic implications of the growth of normal faults and extensional basins, *in* Beratan, K.K., ed., Reconstructing the Structural History of Basin and Range Extension Using Sedimentology and Stratigraphy: Geological Society of America, Special Paper 303, p. 183–203.

SCHLISCHE, R.W., AND OLSEN, P.E., 1990, Quantitative filling model for continental extensional basins with applications to early Mesozoic rifts of eastern North America: Journal of Geology, v. 98, p. 135–155.

SCHLISCHE, R.W., YOUNG, S.S., ACKERMANN, R.V., AND GUPTA, A., 1996, Geometry and scaling relations of a population of very small rift-related normal faults: Geology, v. 24, p. 683–686.

SCHLISCHE, R.W., WITHJACK, M.O., AND EISENSTADT, G., 2002, An experimental study of the secondary fault patterns produced by oblique-slip normal faulting: American Association of Petroleum Geologists, Bulletin, v. 86, p. 885–906.

SHARP, I.R., GAWTHORPE, R.L., UNDERHILL, J.R., AND GUPTA, S., 2000, Fault-propagation folding in extensional settings—examples of structural style and synrift sedimentary response from the Suez rift, Sinai, Egypt: Geological Society of America, Bulletin, v. 112, p. 1877–1899.

SITTLER, C., 1969, The sedimentary trough of the Rhine Graben: Tectonophysics, v. 8, p. 543–560.

SMOOT, J.P., 1991, Sedimentary facies and depositional environments of early Mesozoic Newark Supergroup basins, eastern North America: Palaeogeography, Palaeoclimatology, Palaeoecology, v. 84, p. 369–423.

STECKLER, M.S., 1981, The thermal and mechanical evolution of Atlantic-type continental margins [unpublished Ph.D. dissertation]: Columbia University, New York, 261 p.

STECKLER, M.S., 1985, Uplift and extension at the Gulf of Suez: implications of induced mantle convection: Nature, v. 317, p. 135–139.

STEIN, R.S., KING, G.C.P., AND RUNDLE, J.B., 1988, The growth of geological structures by repeated earthquakes; 2. field examples of continental dip-slip faults: Journal of Geophysical Research, v. 93, p. 13,319–13,331.

TANKARD, A.J., AND WELSINK, H.J., 1989, Mesozoic extension and styles of basin formation in Atlantic Canada, *in* Tankard, A.J., and Balkwill, H.R., eds., Extensional Tectonics and Stratigraphy of the North Atlantic Margins: American Association of Petroleum Geologists, Memoir 46, p. 175–195.

THIEBAUD, C.E., AND ROBSON, D.A., 1979, The geology of the area between Wadi Warden and Wadi Gharandal, east Clysmic rift, Sinai, Egypt: Journal of Petroleum Geology, v. 1, p. 63–75.

THIEBAUD, C.E., AND ROBSON, D.A., 1981, The geology of Asl oilfield, western Sinai, Egypt: Journal of Petroleum Geology, v. 4, p. 77–87.

TRON, V., AND BRUN, J.-P., 1991, Experiments on oblique rifting in brittle–ductile systems: Tectonophysics, v. 188, p. 71–84.

VENDEVILLE, B., 1987, Champs de failles et tectonique en extension: modélisation expérimentale [unpublished Ph.D. dissertation]: Université de Rennes, France, 395 p.

WALSH, J.J., AND WATTERSON, J., 1987, Distributions of cumulative displacement and seismic slip on a single normal fault surface: Journal of Structural Geology, v. 9, p. 1039–1046.

WITHJACK, M.O., AND CALLAWAY, J.S., 2000, Active normal faulting beneath a salt layer—an experimental study of deformation in the cover sequence: American Association of Petroleum Geologists, Bulletin, v. 84, p. 627–651.

WITHJACK, M.O., AND DRICKMAN POLLOCK, D.J., 1984, Synthetic seismic-reflection profiles and rift-related structures: American Association of Petroleum Geologists, Bulletin, v. 68, p. 1160–1178.

WITHJACK, M.O., AND EISENSTADT, G., 1999, Structural history of the Northwest Shelf, Australia—an integrated geological, geophysical and experimental approach (abstract): American Association of Petroleum Geologists, Program and Abstracts, v. 8, p. A151.

WITHJACK, M.O., AND INGEBRIGTSEN, M., 1999, Basement-involved and detached extensional structures—examples from the Norwegian Sea (abstract): American Association of Petroleum Geologists, Program and Abstracts, p. 509.

WITHJACK, M.O., AND JAMISON, W.R., 1986, Deformation produced by oblique rifting: Tectonophysics, v. 126, p. 99–124.

WITHJACK, M.O., AND PETERSON, E., 1993, Prediction of normal-fault geometries—a sensitivity analysis: American Association of Petroleum Geologists, Bulletin, v. 77, p. 1860–1873.

WITHJACK, M.O., MEISLING, K.E., AND RUSSELL, L.R., 1989, Forced folding and basement-detached normal faulting in the Haltenbanken area, offshore Norway, *in* Tankard, A.J., and Balkwill, H.R., eds., Extensional Tectonics and Stratigraphy of the North Atlantic Margins: American Association of Petroleum Geologists, Memoir 46, p. 567–575.

WITHJACK, M.O., OLSON, J., AND PETERSON, E., 1990, Experimental models of extensional forced folds: American Association of Petroleum Geologists, Bulletin, v. 74, p. 1038–1054.

WITHJACK, M.O., ISLAM, Q., AND LAPOINTE, P., 1995a, Normal faults and their hanging-wall deformation—an experimental study: American Association of Petroleum Geologists, Bulletin, v. 79, p. 1–18.

WITHJACK, M.O., OLSEN, P.E., AND SCHLISCHE, R.W., 1995b, Tectonic evolution of the Fundy basin, Canada: Evidence of extension and shortening during passive-margin development: Tectonics, v. 14, p. 390–405.

WITHJACK, M.O., SCHLISCHE, R.W., AND OLSEN, P.E., 1998, Diachronous rifting, drifting, and inversion on the passive margin of central eastern North America—an analog for other passive margins: American Association of Petroleum Geologists Bulletin, v. 82, p. 817–835.

XIAO, H., AND SUPPE, J., 1992, Origin of rollover: American Association of Petroleum Geologists, Bulletin, v. 76, p. 509–529.

ZANDT, G., AND OWENS, T.J., 1980, Crustal flexure associated with normal faulting and implications for seismicity along the Wasatch front, Utah: Seismological Society of America, Bulletin, v. 70, p. 1501–1520.

ZHANG, P., SLEMMONS, D.B., AND MAO, F., 1991, Geometric pattern, rupture termination and fault segmentation of the Dixie Valley–Pleasant Valley active normal fault system, Nevada, U.S.A.: Journal of Structural Geology, v. 13, p. 165–176.

ZIEGLER, P.A., 1992, European Cenozoic rift system: Tectonophysics, v. 208, p. 91–111.

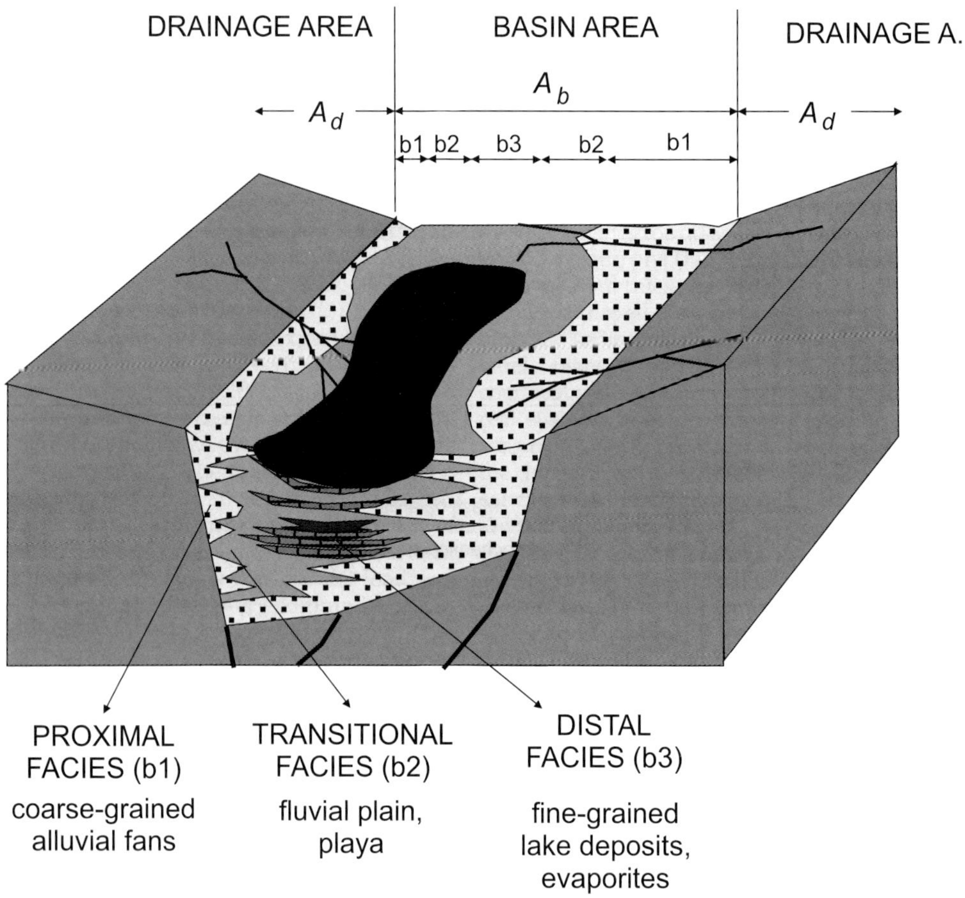


FIG. 4.—Vertical and lateral distribution of facies architecture of lacustrine rift basins used for two-dimensional modeling based on the DA concept.

mentary units. If continuous seismic transects from the basin margin to the center are available and the mineralogical and chemical composition of the sediments is known (e.g., from bore holes), accumulated sediment masses within particular stratigraphic boundaries can be quantified. To calculate sedimentation rates and convert them into denudation rates, also the time interval of accumulation has to be known. In some basins, radiometric ages derived from tephra layers yield good stratigraphic markers (e.g., Great Salt Lake basin; Williams, 1994). A higher time resolution, however, is achieved only by indirect methods, e.g., frequency analysis of sedimentary cycles (Glenn and Kelts, 1991; Matthews and Perlmutter, 1994). Despite great progress on this field, uncertainty remains, if these cycles represent Milankovich cycles caused by quasi-periodic elements of the earth's motion about the sun.

Sequence stratigraphy is a powerful tool to predict facies patterns and correlate stratigraphic units. As shown by, e.g., Gawthorpe and Leeder (2000), sedimentary architecture and hence sequence stratigraphic framework is controlled largely by the impact of fault growth processes. Combining the DA concept with sequence stratigraphy and coupled tectonosedimentary models opens a wide field of applications and offers better control on modeling of both landscape erosion and sediment supply to basins (Fig. 5). However, application of this combined concept is often hampered by lack of sedimentologic, stratigraphic, and geomorphologic data.

## CASE STUDIES OF RIFT BASINS

### *Locations*

To demonstrate our modeling approach, we selected two modern closed-basin lakes with a good data base, the Qinghai lake, China, and Lake Turkana, East Africa. Both lake basins developed in an extensional tectonic regime since the Late Tertiary. They contrast especially with respect to the lithology in their drainage areas. The DA concept is used to reconstruct and simulate the long-term evolution (million years) of both lake basins. The climate-controlled short-term evolution (thousands of years to a hundred thousand years) is simulated by a steady-state mass-balance approach and the integrated geochemical model of Yan (1998) and Yan et al. (2002).

**Qinghai lake** lies in a rift depression in the northeastern Tibet–Qinghai Plateau, China (Lanzhou Institute of Geology, 1979). The drainage area is composed mainly of carbonate-bearing sedimentary rocks. The lake has existed since the early Pleistocene and was hydrologically mostly open during the middle Pleistocene. At the end of the late Pleistocene (about 15 ka) the lake became topographically closed. At present, the salinity is 14.1 g/l and a cold and semiarid continental climate prevails with less than 400 mm/yr precipitation over the lake surface (Lister et al., 1991). In the early Holocene, the climate was more humid than at present. Precipitation in the early Holocene was 500–650 mm/

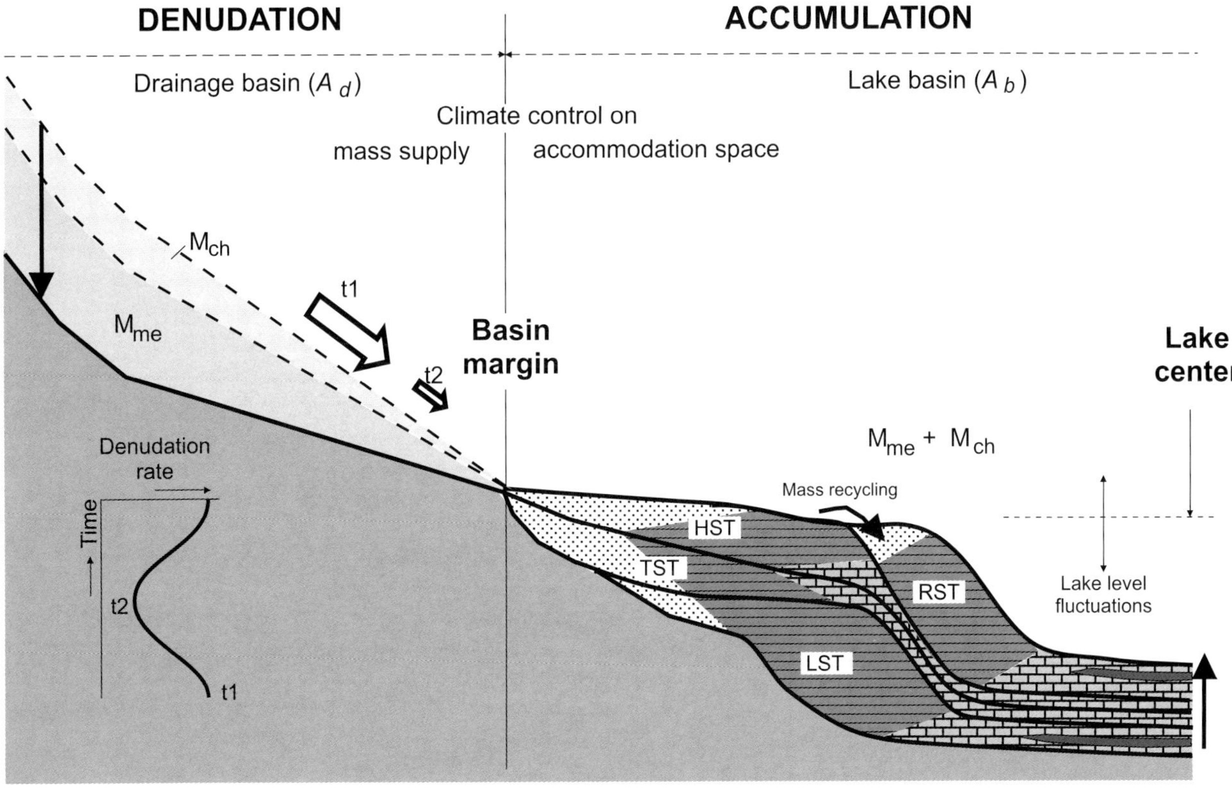


FIG. 5.—Sequence stratigraphy in rift basins linked with the DA concept. $M_{me}$ = mechanical mass, $M_{ch}$ = chemical mass, LST = lowstand systems tract, TST = transgressive systems tract, HST = highstand systems tract, RST = regressive systems tract.

yr and during the late Holocene it fluctuated between 400 and 500 mm / yr (Shi et al., 1993).

**Lake Turkana** is the largest water body in the eastern branch of the East African Rift system. The drainage area consists almost totally of silicate rocks (granites, metamorphic rocks, and volcanic rocks). The oldest lacustrine deposits date back to the Miocene (Savage and Williamson, 1978). In the Plio-Pleistocene, the lake varied between fresh open-lake and brackish closed-lake conditions (Cerling, 1979). Between 10 and 3.5 ka, the lake level reached a highstand (+ 80 m) and developed an overflow to the Nile river system (Owen et al., 1982). Since 3.5 ka, Lake Turkana has been topographically closed and its water level has fluctuated between 20 and 50 m above the present level (Johnson et al., 1991). The present-day lake water is brackish with about 2.5 g / l of total dissolved solids (Yuretich, 1979) . The modern climate in the Ethiopian Highland is humid, subtropical to tropical, and the lake area is semiarid with a high evaporation rate (Cerling, 1986).

## *One-Dimensional Modeling of the Long-term Evolution (ca. 12 million years)*

Figure 6 shows evolutionary diagrams of rift sedimentation for the Qinghai and Turkana basins since the Pliocene and projection into the next eight million years. The accommodation space for sediments is generated by the rates of subsidence and crustal extension. The long-term subsidence rates are estimated from sediment thickness and are assumed to be linear within the time intervals. This is a minimum estimate because temporally undersupplied conditions have prevailed. The subsidence rate in the Qinghai basin doubled from the Pliocene (ca. 100 mm / kyr) to the Quaternary (ca. 200 mm / kyr). In the Turkana rift subsidence seems to have been relatively constant. Since the Pliocene, about 1000 m of sediments has accumulated in the Turkana basin, which implies a subsidence rate of almost 200 mm / kyr (Cerling, 1994). Present extension rates in the East African rift are in the order of 6 mm / yr, according to remote sensing by satellites. The widths of both the Turkana and Qinghai rifts are ca. 40 to 80 km. Rift formation of the Turkana basin started somewhere in the Miocene (Savage and Williamson, 1978). Estimating a lifetime of ten million years, crustal extension rates would be in the order of 4 to 8 mm / yr and thus in good agreement with modern observations. Extension of the Qinghai basin presumably did not start before the Pliocene, which would imply twice the rates of the Turkana rift (Lanzhou Institute of Geology, 1979). Because dating of basal sediments in the Qinghai basin is uncertain and modern measurements are lacking, however, we used the same extension rate of 6 mm / yr in our model as for the Turkana rift.

Holocene total denudation rates are ca. 40 and 65 mm / kyr for the drainage areas of the Qinghai and Turkana basins derived from lacustrine sedimentation rates and present river loads (Yan, 1998). Mechanical denudation accounts for 18 and 60 mm / kyr, respectively. In spite of a relatively low terrigenous sediment supply, the combination of a high $A_d/A_b$ ratio and a low subsidence rate initially led to oversupplied conditions for the Qinghai basin. This is confirmed by fluvial deposits of Pliocene age at the base of the rift infill. During the Pleistocene, lakes formed in the basin because there was: (1) a shift towards undersupplied conditions, (2) total subsidence accelerated, and (3) the $A_d/A_b$ ratio progressively declined. Without subsidence, complete infill of

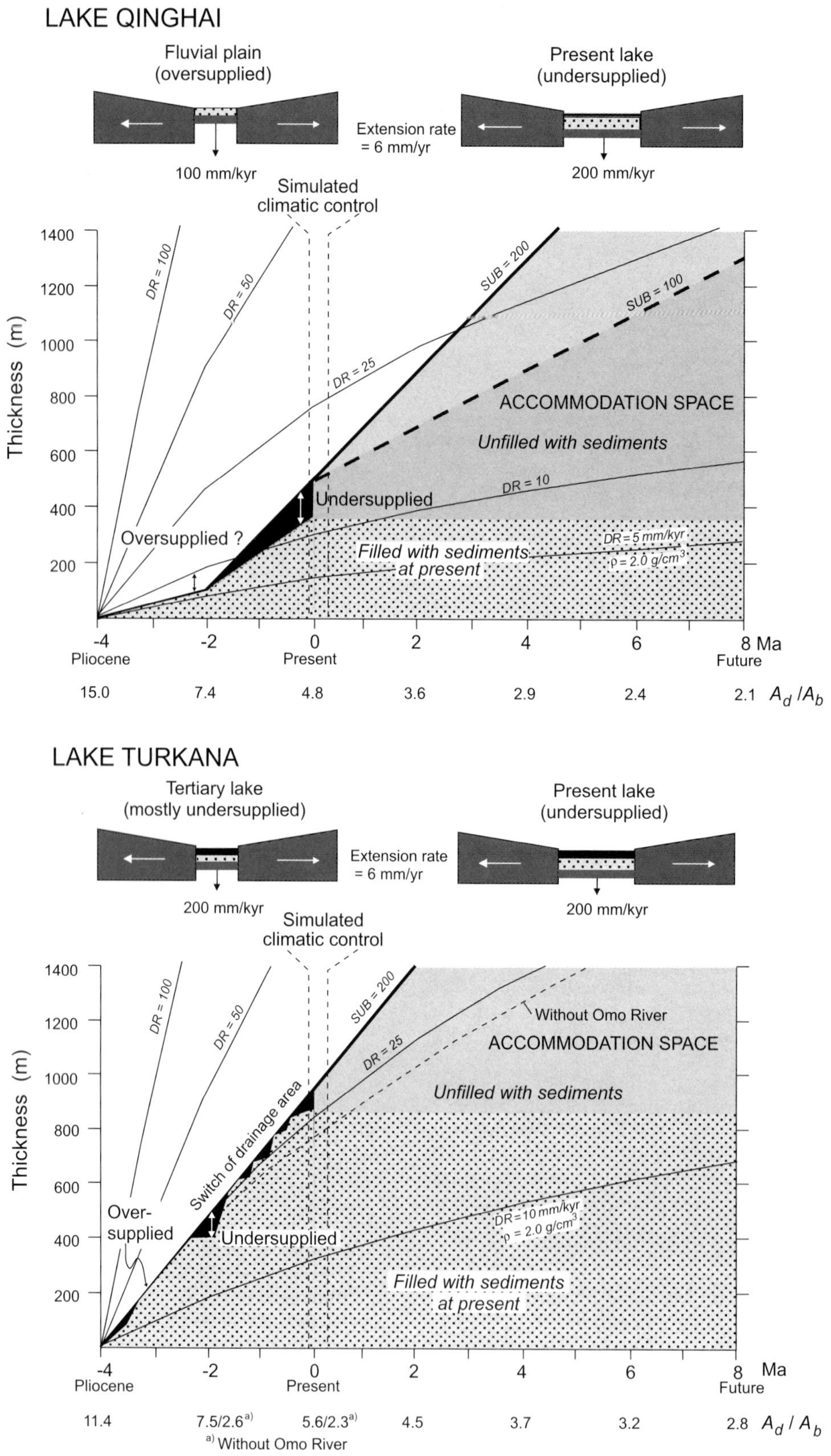


FIG. 6.—Evolutionary diagram of sediment filling of the Qinghai and Turkana basins from the Pliocene to the present and projection into the future. The curves of total denudation rates relate to a sediment density of 2.0 g/cm$^3$ (due to porosity) and the changing ratio of drainage area to basin area ($A_d/A_b$ ratio) as indicated below. Rates of extension and subsidence that generate accommodation space are estimated according to field data (see text).

the present lake up to its outflow level would take around 300 kyr (lifetime at steady state; Yan, 1998).

A total denudation rate of 60 mm / kyr and a total subsidence rate of 200 mm / kyr would have led to permanent oversupplied conditions until present for the Turkana rift. However, the first indications of lake formation date back to the Miocene, and a lake was also temporarily established during the Plio-Pleistocene (Savage and Williamson, 1978; Cerling, 1994). Possible explanations are tectonic pulses with periods of accelerated subsidence and / or periods of reduced denudation in the drainage area. In the Turkana rift, however, volcanic activity with lava flows damming and reorganizing large parts of the drainage basin and thus a repeated change of the $A_d / A_b$ ratio largely controlled the sediment flux into the rift basin. The major tributary to Lake Turkana is the Omo River, which drains the relatively humid Ethiopian Highland in the north and delivers most of the terrigenous sediments and dissolved load (85% inflow; Cerling, 1986). Between 4.1 and 2.3 Ma the Omo River flowed out of the basin to the Indian Ocean. Fluvial deposits with some lacustrine intercalations indicate predominantly oversupplied conditions. At about 2.3 Ma, a lava flow blocked the Omo River in the south and a deep freshwater lake was established for several hundred thousand years. Since 1.9 Ma, the high sediment supply of the Omo River repeatedly led to oversupplied conditions and hence to an overflow of the Omo River to the Nile River (Cerling, 1994). In consequence, the $A_d / A_b$ ratio switched from ca. 7 to 2.5 during that time. Using the present total denudation rate, the lifetime of Lake Turkana would be not more than 100,000 years under steady-state conditions. Including subsidence may increase the estimate to 130,000 years (sedimentation rate is ca. 790 mm / kyr with a mean sediment density of 2.0 g / cm$^3$, subsidence rate is ca. 200 mm / kyr).

### *One-Dimensional Modeling of the Short-term Evolution (ca. 300,000 years)*

Figures 7 and 8 show the results of inverse and forward dynamic modeling of late Quaternary to future sediment accumulation in lakes Qinghai and Turkana. The modeling starts at 40 ka because the evolution of climate and lake-level fluctuations is relatively well known for both rift lakes since that time. The future evolution was modeled for three scenarios (S1 to S3) by considering different mechanical and chemical denudation rates, brine evolution paths, and lake-level fluctuations. Accommodation space relates to the present level of basin overflow and is shown for steady-state and dynamic conditions with respect to basin subsidence. The model output of the three scenarios is presented as cumulative sediment thickness vs. time (Fig. 7) and as sedimentation rates vs. time (Fig. 8). The latter are calculated for the $A_d/A_b$ ratio (at outflow level) as well as for the $A_d/A_l$ ratio (at present lake level). Obviously, evaporitic sedimentation is restricted to the lake area. Terrigenous sedimentation, however, may start at the basin margin with the formation of alluvial fans and a fluvial plain. These fluvial deposits may be recycled in the case of a falling lake level if enough transport capacity by water is available, leading to a gap in the stratigraphic record. Further treatment of this problem is provided by a two-dimensional modeling approach (see below). In the one-dimensional model, however, intermittent sediment storage is neglected and the fluvial plain between the basin margin and the shoreline of the shrinking lake is considered as a region of bypassing of terrigenous and chemical masses.

During the late Quaternary, Qinghai lake evolved from an open lake into a closed lake with a short-lived return to open conditions (Figs. 7A, 8A). Together with these climate-controlled lake-level changes, mechanical and chemical denudation rates, as well as evaporitic conditions, also changed. Closure time was not long enough, however, to reach lake-water saturation with respect to sulfates and halites and hence only calcite and magnesite precipitated. Modeling of the future sedimention under a humid climate (Scenario S1, open lake) predicts dominance of clastic deposits with minor formation of carbonate and magnesite. The scenario at steady-state conditions (S2) leads to a permanent increase of solutes in the lake water and thus to the formation of gaylussite (Na–Ca carbonate) after 20,000 years, and mirabilite (Na sulfate) after 80,000 years. Formation of chlorides starts after 200,000 years. Continued increase of aridity (S3) causes a lowering of the lake level; therefore, lake-water saturation with respect to sulfates and chlorides is reached earlier, at 75,000 and 150,000 years, respectively. Increasing precipitation of evaporative minerals together with a decreasing input of terrigenous sediments from the drainage area (reduced transport capacities of rivers) increasingly dominate over mechanical sedimentation. The total sedimentation rate over the shrinking lake area increases by about 50%. The total infill rate of the Qinghai basin, however, is reduced if compared to scenarios S1 and S2.

Lake Turkana (Figs. 7B, 8B) experienced a strong lowering of lake level during the late Pleistocene, contemporaneous with other East African rift lakes (Johnson et al., 1991). From the early to the middle Holocene the East African monsoon reached maximum intensity and the lake became open. The last 3500 years are characterized by increasing aridity and continuous lake-level fall. Sedimentation in Lake Turkana is dominated by clastics because of the large input of terrigenous material and the low salinity of the lake water. The low concentration of dissolved solids in inflow water prevented saturation of the lake water with respect to sulfates and chlorides during the late Pleistocene lowstand. In the future, little sulfides are formed under scenario S3 after about 115,000 years. For steady-state conditions (S2) chemical sedimentation remains restricted to silica and carbonates over at least 120,000 years. Temporal change of sedimentation rates in Lake Turkana is much more affected by the variability of mechanical denudation and lake-level fluctuations compared to Qinghai lake (Fig. 8B). Although the input of terrigenous sediments and thus the mean sedimentation rate over the basin is markedly reduced under arid conditions, the sedimentation rate in the lake increases. This illustrates that in real systems 2D or even 3D stratigraphic information is needed in order to avoid misleading interpretations from, e.g., single wells with respect to changes of mass supply and denudation in the hinterland.

### *Two-Dimensional Modeling Based on Sequence Stratigraphy*

Accurate two-dimensional modeling of the two model lakes has not been possible so far because of the lack of data. However, Figure 9 shows an idealized temporal evolution of sedimentation rates for three settings from proximal to distal (b1 to b3) in the Qinghai basin (based on scenario S3 of Fig. 8A). The resulting sedimentation rates of the two-dimensional modeling deviate markedly from the mean sedimentation rates gained from the one-dimensional model (Fig. 9, left). However, they give the same mean values when they are weighted by their portion of the basin area ($A_{b1}$ to $A_{b3}$; Fig. 4). If only one section (b1, b2, or b3) is known, a quantitative reconstruction of the entire basin sedimentation is not possible. To model natural rifts as denudation–accumulation systems a sequence stratigraphic framework coupled with fault-growth models is required. So far, few studies have combined field analysis of stratigraphic geometries with numerical simulations.

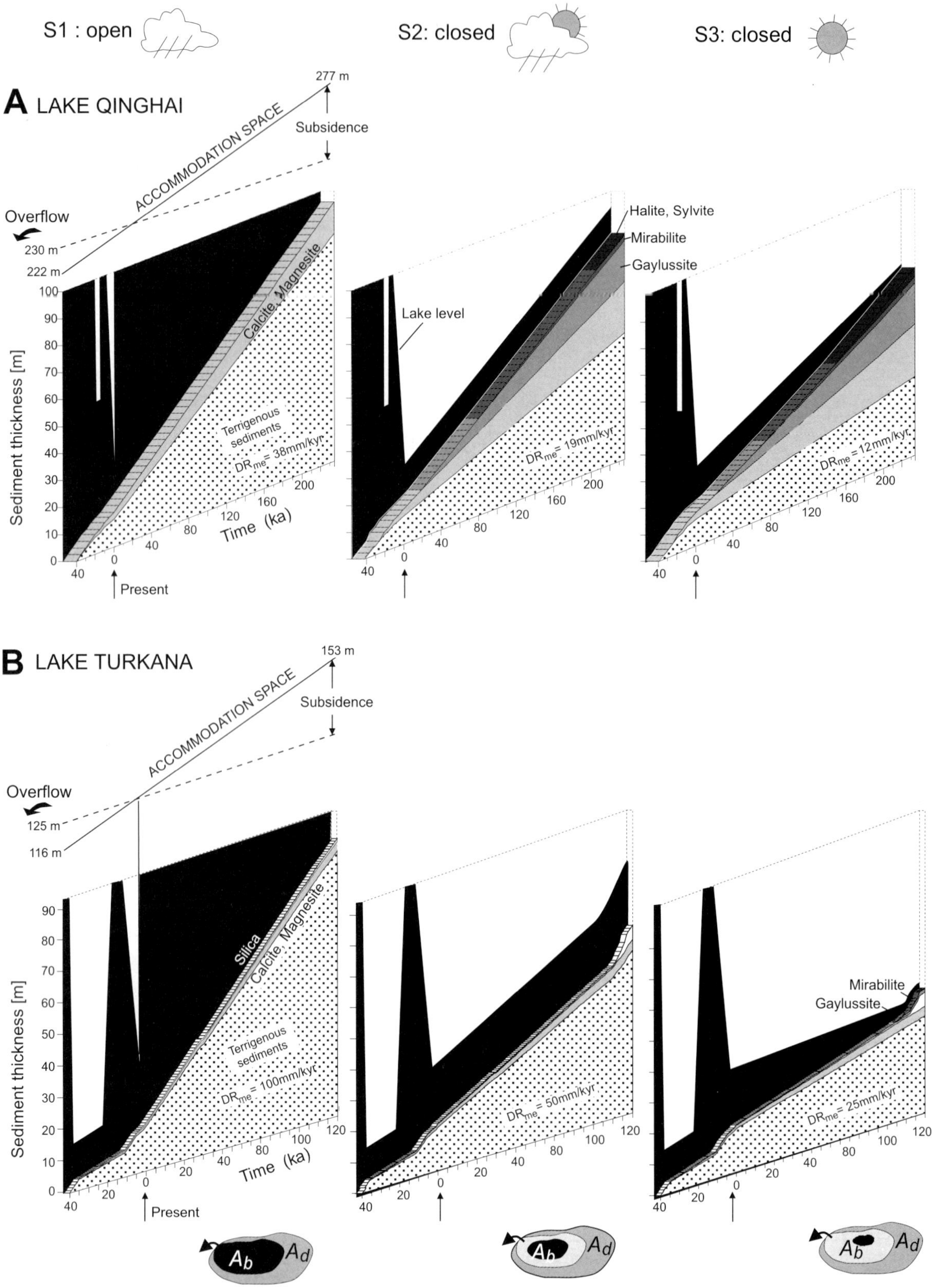


FIG. 7.—Results of one-dimensional modeling of the mechanical and chemical sediment filling of **A)** Qinghai and **B)** Turkana basins from the late Quaternary to the present and projection into the future using three climatic scenarios (see Fig. 3). Mechanical sedimentation is modeled by the DA concept; chemical sedimentation is modeled by an integrated geochemical computer program (Yan, 1998; Yan et al., 2002). Note the different time scales of forward modeling which relate to the potential lifetime of the lake basin.

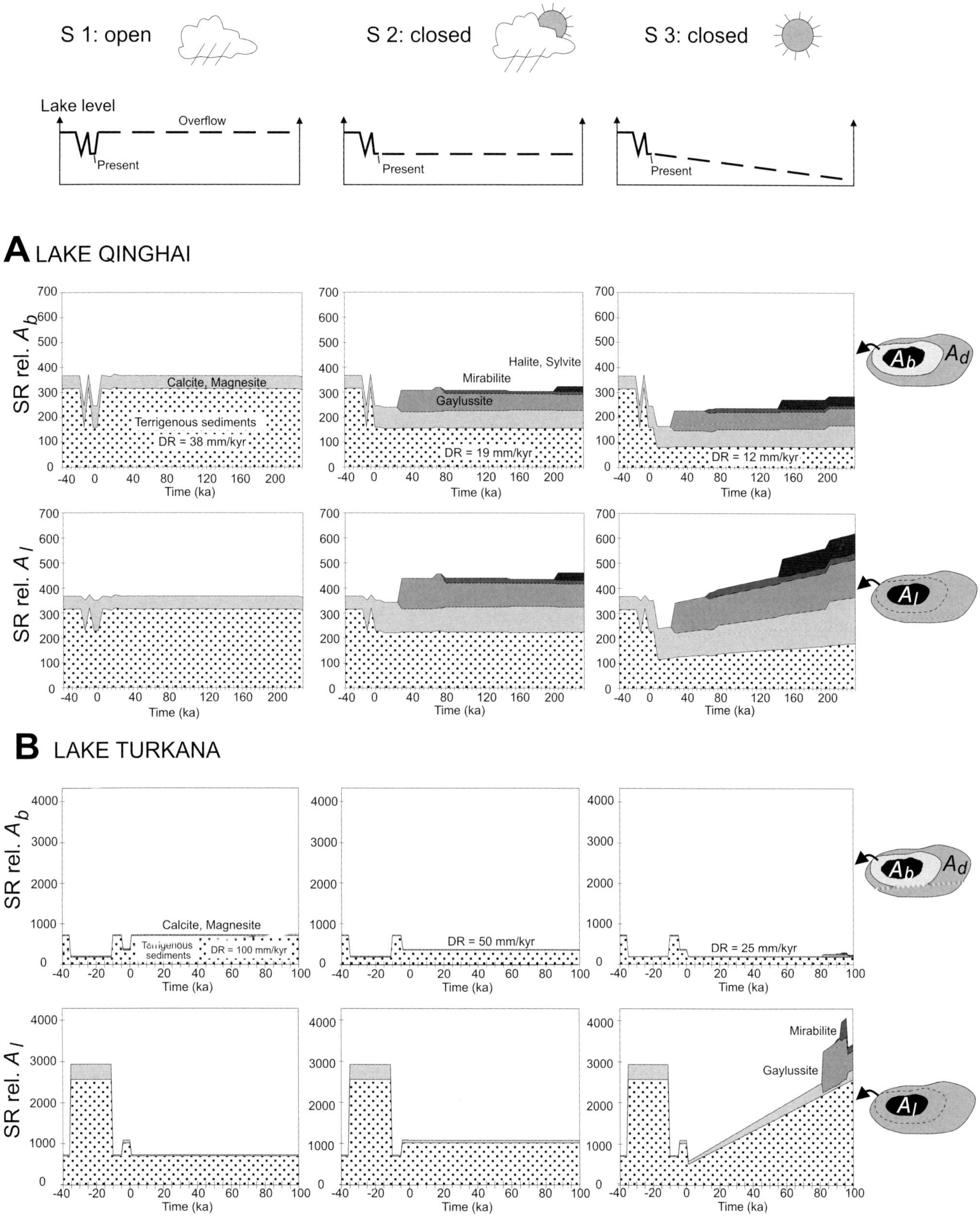


FIG. 8.—Results of one-dimensional modeling of the mechanical and chemical sedimentation rates of **A)** Qinghai and **B)** Turkana basins from the late Quaternary to the present and projection into the future using three climatic scenarios (see Fig. 3). Sedimentation rates relate to both the basin area, $A_b$, and the lake area, $A_l$. Mechanical sedimentation rates are modeled by the DA concept; chemical sedimentation rates are modeled by an integrated geochemical computer program (Yan, 1998; Yan et al., 2002). Note the different time scales of forward modeling which relate to the potential lifetime of the lake basin.

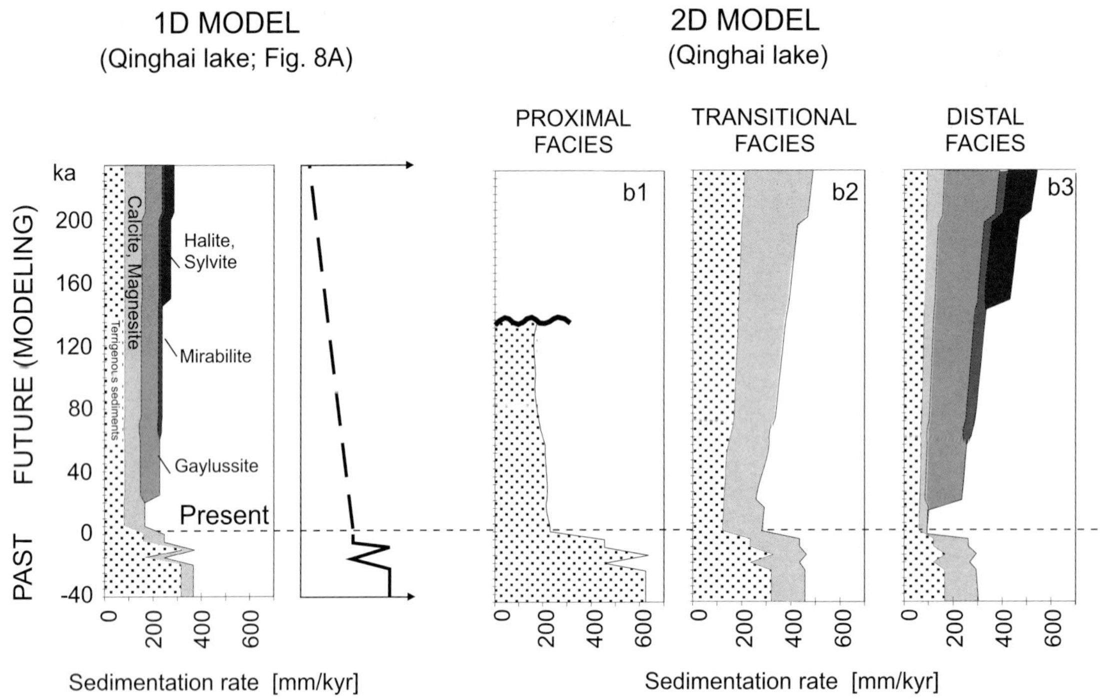


FIG. 9.—Conceptual comparison of the temporal evolution of sedimentation rates for proximal to distal settings (b1 to b3; see Fig. 4) in the Qinghai basin derived from two-dimensional modeling with the model output of scenario S3 derived from one-dimensional modeling (see Fig. 8A).

## DISCUSSION AND CONCLUSIONS

In spite of local complexities of stratigraphic sequences and facies associations in rift basins, general patterns can be identified that are related to the tectonic and climatic evolution of the rift. Commonly, rifts can be considered as sedimentologically closed systems. Hence, the DA concept can be successfully applied and provide a quantitative tool to link mechanical and chemical mass supply from the hinterland with sediment accumulation within the rift basin. Simplistic one-dimensional models using the DA concept help to better understand the evolution of more complex basins in nature and to identify the influence of various controlling factors. In the case of continuous extension, mechanical sedimentation rates tend to decrease during rift evolution. One-dimensional models demonstrate the importance of decreasing morphogenic ratios and decreasing denudation rates in the hinterland for this trend. In addition, facies associations in rift basins are controlled largely by subsidence rates. By comparing sedimentation rates with subsidence rates, the shift from oversupplied conditions during the initial rift stage to undersupplied conditions during an advanced rift stage with lacustrine and/or marine deposits can be predicted. In the case of ceasing subsidence and extension, rift basins may return to oversupplied conditions (failed rift). Other rift basins may develop into oceanic basins with a further decrease of mechanical sedimentation rates (drift stage).

In principle, chemical sedimentation rates in rift basins can also be predicted by the DA concept if the basin is hydrologically closed. The drainage area accounts for the supply of dissolved masses to the rift basin. However, dissolved solids are not immediately removed from solution but need a specific saturation stage of the lake water. This delays chemical sedimentation rates relative to chemical weathering rates by about one thousand to several hundred thousand years. In addition, the onset of specific evaporative minerals may strongly change the chemical sedimentation rate. Using an integrated geochemical computer model based on thermodynamic equilibria, we also could predict temporal changes in chemical sedimentation rates. The application of the one-dimensional mechanical and chemical models to two natural rift basins demonstrate the usefulness of this approach.

So far, we have not treated two-dimensional models in great detail. To carry out inverse or forward modeling for specific parts of the basin, numerous sophisticated models and concepts of lacustrine sequence stratigraphy already exist (Liro, 1993; Paola, 2000). However, they do not include the dynamics of the source area and its evolution with time. Hence, we propose the combination of the DA concept based on mass balancing with the concept of sequence stratigraphy and in a largely fault-controlled tectonosedimentary regime to better interpret the role of the hinterland in the basin-fill history and, vice versa, to use the sedimentary record for quantification of paleo–denudation rates with a higher time resolution. However, we see little point in trying to model exactly a particular stratigraphic architecture in a particular part of a rift basin, rather to explore external forcing of rift infill.

Our main results of applying the DA concept and geochemical modeling to rift basins are:

1. The decrease of mechanical sedimentation rate in rift basins with continuous extension with time is predicted by the DA concept by the increase of the $A_d/A_b$ ratio and the decrease of mechanical denudation rates due to smoothing of the relief during a more advanced stage of the rift.

2. Sedimentologically open or closed conditions, e.g., fluvial versus lacustrine facies associations, additionally are dependent

on subsidence rates. The timing of the shift from fluvial to lacustrine sedimentation or vice versa can be predicted well by plotting curves of sedimentation rates against subsidence rates.

3. Chemical sedimentation in rifts is influenced predominantly by climatic conditions and the lithology of the drainage area.

4. Whereas large portions of dissolved mass are lost under hydrologically open conditions, removal of dissolved solids by the formation of evaporitic minerals in closed basins lags behind chemical mass input by about one thousand to several hundred thousand years (temporal disequilibrium). If saturation is reached for specific evaporitic minerals, a marked change in chemical sedimentation rate may occur.

5. To predict types of evaporitic minerals and chemical sedimentation rates of closed rift basins, an integrated geochemical computer model has been applied. It reproduces successfully the late Pleistocene to recent evolution of lakes Qinghai and Turkana. Forward modeling of both lake basins has been carried out for three climate scenarios. The models predict the mineral sequence formed and the time intervals under various climatic conditions. Under steady-state conditions, Qinghai lake (carbonate-bearing sedimentary rocks in the drainage area) reaches saturation with respect to sulfates and halites after about 80,000 and 200,000 years, respectively, whereas Lake Turkana (silicate rocks in the drainage area) remains in the carbonate stage until 120,000 years.

## ACKNOWLEDGMENTS

We thank Robin Renaut and Gail Ashley for inviting us to contribute to this special publication. Comments and suggestions made by Oliver Kempf as well as the reviews by Joseph J. Lambiase and Cathy Busby helped us to improve the manuscript. This work has been supported by the German Research Foundation (SFB 275/TP D2).

## REFERENCES

Bethke, C.M., 1996, Geochemical Reaction Modelling; Concepts and Applications: New York, Oxford University Press, 390 p.

Cerling, T.E., 1979, Paleochemistry of Plio-Pleistocene Lake Turkana, Kenya: Palaeogeography, Palaeoclimatology, Palaeoecology, v. 27, p. 247–285.

Cerling, T.E., 1986, A mass balance approach to basin sedimentation—constraints on the recent history of the Turkana Basin: Palaeogeography, Palaeoclimatology, Palaeoecology, v. 54, p. 63–86.

Cerling, T.E., 1994, Lake Turkana and its precursors in the Turkana Basin, East Africa (Kenya, Ethiopia), *in* Kelts, K., and Gierlowski-Kordesch, E., eds., Global Geological Record of Lake Basins: World and Regional Geology: Cambridge, U.K., Cambridge University Press, p. 341–343.

Einsele, G., 2000, Sedimentary Basins; Evolution, Facies, and Sediment Budget: Heidelberg, Berlin, Springer-Verlag, 800 p.

Einsele, G., and Hinderer, M., 1998, Quantifying denudation and sediment-accumulation systems (open and closed lakes): basic concepts and first results: Palaeogeography, Palaeoclimatology, Palaeoecology, v. 140, p. 7–21.

Lanzhou Institute of Geology, 1979, Qinghai Lake monograph of the 1961 expedition: Beijing, Science Press, p. 1–175.

Gawthorpe, R.L., and Leeder, M.R., 2000, Tectono-sedimentary evolution of active extensional basins: Basin Research, v. 12, p. 195–218.

Gierlowski-Kordesch, E., and Kelts, K., 1994, Global Geological Record of Lake Basins; World and Regional Geology: Cambridge, U.K., Cambridge University Press, 427 p.

Glenn, C.R., and Kelts, K., 1991, Sedimentary rhythms in lake deposits, *in* Einsele, G., Ricken, W., and Seilacher, A., eds., Cycles and Events in Stratigraphy: Berlin, Heidelberg, Springer-Verlag, p. 188–221.

Haberyan, K.A., and Hecky, R.E., 1987, The late Pleistocene and Holocene stratigraphy and palaeolimnology of lakes Kivu and Tanganyika: Palaeogeography, Palaeoclimatology, Palaeoecology, v. 61, p. 169–197.

Hardie, L.A., and Eugster, H.P., 1970, The evolution of closed-basin lakes, *in* Morgan, B.A., ed., 15th. Anniversary Symposia: Mineralogical Society of America, Special Paper 3, p. 273–290.

Hardie, L.A., Smoot, J.P., and Eugster, H.P., 1978, Saline lakes and their deposits: a sedimentological approach, *in* Matter, A., and Tucker, E., eds., Modern and Ancient Lake Sediments: International Association of Sedimentologists, Special Publication 2, p. 7–41.

Herczeg, A.L., and Lyons, W.B., 1991, A chemical model for the evolution of Australian sodium chloride lake brines: Palaeogeography, Palaeoclimatology, Palaeoecology, v. 84, p. 43–53.

Hermanrud, C., 1993, Basin modelling techniques: an overview, *in* Dore, A.G., Augustson, J.-H., Hermanrud, C., Stewart, D.J., and Sylta-Oyvind, S.O., eds., Basin Modelling: Advances and Applications: Norwegian Petroleum Society (NPF), Special Publication no. 3, Amsterdam, Elsevier, p. 1–34.

Hinderer, M., and Einsele, G., 2001, The world's large lake basins as denudation–accumulation systems and implications for their lifetimes: Journal of Paleolimnology, v. 26, p. 355–372.

Johnson, T.C., Halfman, J.D., and Showers, W.J., 1991, Paleoclimate of the past 4000 years at Lake Turkana, Kenya, based on the isotopic composition of authigenic calcite: Palaeogeography, Palaeoclimatology, Palaeoecology, v. 85, p. 189–198.

Lambiase, J.J., 1990, A model for tectonic control of lacustrine stratigraphic sequences in continental rift basins, *in* Katz, B.J., ed., Lacustrine Basin Exploration: American Association of Petroleum Geologists, Memoir 50, p. 265–276.

Liro, L.M., 1993, Sequence stratigraphy of a lacustrine system: Upper Fort Union Formation (Paleocene), Wind River Basin, Wyoming, U.S.A., *in* Weimer, P., and Posamentier, H.W., eds., Siliciclastic Sequence Stratigraphy: American Association of Petroleum Geologists, Memoir 58, p. 317–333.

Lister, G.S., Kelts, K., Zao, C.K., Yu, J.-Q., and Niessen, F., 1991, Lake Qinghai, China: closed-basin lake levels and the oxygen isotope record for ostracoda since the latest Pleistocene: Palaeogeography, Palaeoclimatology, Palaeoecology, v. 84, p. 141–162.

Ludwig, W., and Probst, J.-L., 1998, River sediment discharge to the Oceans: present-day controls and global budgets: American Journal of Science, v. 298, p. 265–295.

Matthews, M.D., and Perlmutter, M.A., 1994, Global cyclostratigraphy: an application to the Eocene Green River Basin, *in* de Boer, P.L., and Smith, D.G., eds., Orbital Forcing and Cyclic Sequences: International Association of Sedimentologists, Special Publication 19, p. 459–481.

Meybeck, M., 1987, Global chemical weathering of surficial rocks estimated from river dissolved loads: American Journal of Science, v. 287, p. 401–428.

Miall, A.D., 1997, The Geology of Stratigraphic Sequences: Berlin, Springer-Verlag, 433 p.

Milliman, J.D., and Syvitski, J.P.M., 1992, Geomorphic/tectonic control of sediment discharge to the ocean: The importance of small mountainous rivers: Journal of Geology, v. 100, p. 525–544.

Owen, R.B., Barthelme, J.W., Renaut, R.W., and Vincens, A., 1982, Palaeolimnology and archaeology of Holocene deposits north-east of Lake Turkana, Kenya: Nature, v. 298, p. 523–528.

Paola, C., 2000, Quantitative models of sedimentary basin filling: Sedimentology, v. 47, p. 121–178.

Plummer, L.N., Parkhurst, D.L., Fleming, G.W., and Dunkle, S.A., 1988, PHRQPITZ, a computer program incorporating Pitzer's equations for calculation of geochemical reactions in brines: U.S. Geological Survey, Water-Resources Investigations Report, v. 88-4153, 309 p.

READING, H.G., 1996, Sedimentary Environments; Processes, Facies and Stratigraphy: Oxford, U.K., Blackwell Science, 688 p.

SAVAGE, J.E.G., AND WILLIAMSON, P.G., 1978, The early history of the Turkana depression, *in* Bishop, W.W., ed., Geological Background to Fossil Man: Edinburgh, Scottish Academic Press, p. 375–394.

SCHLISCHE, R.W., 1991, Half-graben basin filling models: new constraints on continental extensional basin development: Basin Research, v. 3, p. 123–141.

SCHLISCHE, R.W., AND OLSEN, P.E., 1990, Quantiative filling model for continental extensional basins with applications to early Mesozoic rifts of eastern North America: Journal of Geology, v. 98, p. 135–155.

SCHOLZ, C.A., AND ROSENDAHL, B.R., 1990, Coarse-clastic facies and stratigraphic sequence models from Lakes Malawi and Tanganyika, East Africa, *in* Katz, B.J., ed., Lacustrine Basin Exploration: American Association of Petroleum Geologists, Memoir 50, p. 151–167.

SCHOLZ, C.A., MOORE, T.C., HUTCHISON, D.R., GOLMSHTOK, A.J., KLITGORD, K.D., AND KUROTCHKIN, A.G., 1998, Comparative sequence stratigraphy of low-latitude versus high-latitude lacustrine rift basins: seismic data examples from the East African and Baikal rifts: Palaeogeography, Palaeoclimatology, Palaeoecology, v. 140, p. 401–420.

SHI, Y.F., KONG, Z.Z., WANG, S.M., TANG, L.Y., WANG, F.B., YAO, T.D., ZHAO, X.T., ZHANG, P.Y., AND SHI, S.H., 1993, Mid-Holocene climates and environments in China: Global and Planetary Change, v. 7, p. 219–233.

SPENCER, R.J., EUGSTER, H.P., AND JONES, B.F., 1985a, Geochemistry of Great Salt Lake, Utah II—Pleistocene–Holocene evolution: Geochimica et Cosmochimica Acta, v. 49, p. 739–747.

SPENCER, R.J., EUGSTER, H.P., JONES, B.F., AND RETTIG, S.L., 1985b, Geochemistry of Geat Salt Lake, Utah I—Hydrochemistry since 1850: Geochimica et Cosmochimica Acta, v. 49, p. 727–737.

SUMMERFIELD, M.A., AND HULTON, N.J., 1994, Natural controls of fluvial denudation rates in major world drainage basins: Journal of Geophysical Research, v. 99, p. 13,871–13,883.

VAIL, P.R., AUDEMARD, F., BOWMAN, S.A., EISNER, P.N., AND PEREZ-CRUZ, C., 1991, The stratigraphic signatures of tectonic, eustasy, and sedimentology—an overview, *in* Einsele, G., Ricken, W., and Seilacher, A., eds., Cycles and Events in Stratigraphy: Berlin, Springer-Verlag, p. 617–659.

WALLING, D.E., 1983, The sediment delivery problem: Journal of Hydrology, v. 65, p. 209–237.

WALLING, D.E., AND WEBB, B.W., 1983, The dissolved loads of rivers: a global overview, *in* Webb, B.W., ed., Dissolved Loads of Rivers and Surface Water Quantity/Quality Relationships: International Association of Hydrological Sciences, Publication 141, p. 3–20.

WALLING, D.E., AND WEBB, B.W., 1996, Erosion and sediment yield: A global overview, *in* Walling, D.E., ed., Erosion and Sediment Yield: Global and Regional Perspectives: International Association of Hydrological Sciences, Publication 236, p. 3–20.

WALTHAM, D., 1992, Mathematical modelling of sedimentary basin processes: Marine and Petroleum Geology, v. 9, p. 265–273.

WHITE, A.F., AND BLUM, A.E., 1995, Effects of climate on chemical weathering in watersheds: Geochimica et Cosmochimica Acta, v. 59, p. 1729–1747.

WILLIAMS, S.K., 1994, Late Cenozoic tephrostratigraphy of deep sediment cores from the Bonneville Basin, northwest Utah: Geological Society of America, Bulletin, v. 106, p. 1517–1530.

WOOD, W.W., AND SANFORD, W.E., 1990, Ground-water control of evaporite deposition: Economic Geology, v. 85, p. 1226–1235.

YAN, J.P., 1998, Numerical modelling of topographically-closed lakes: impact of climate on lake level, hydrochemistry and chemical sedimentation: Ph.D. thesis, University of Tübingen, Tübinger Geowissenschaftliche Arbeiten, v. C46, p. 1–145.

YAN, J.P., HINDERER, M., AND EINSELE, G., 2002, Geochemical evolution of closed lakes: general model and application to lakes Qinghai and Turkana: Sedimentary Geology, v. 148, p. 105–122.

YURETICH, R.F., 1979, Modern sediments and sedimentary processes in Lake Rudolf (Lake Turkana) eastern Rift Valley, Kenya: Sedimentology, v. 26, p. 313–331.

# OLORGESAILIE, KENYA: A MILLION YEARS IN THE LIFE OF A RIFT BASIN

ANNA K. BEHRENSMEYER
*Department of Paleobiology, MRC 121, National Museum of Natural History,*
*Smithsonian Institution, Washington, D.C. 20560, U.S.A.*
*e-mail: Behrensmeyer.Kay@NMNH.SI.edu*
RICHARD POTTS
*Department of Anthropology, MRC 122, National Museum of Natural History,*
*Smithsonian Institution, Washington, D.C. 20560, U.S.A.*
*e-mail: Potts.Rick@NMNH.SI.edu*
ALAN DEINO
*Berkeley Geochronology Center, 2455 Ridge Rd., Berkeley, California 94709, U.S.A.*
*e-mail: adeino@bgc.org*
AND
PETER DITCHFIELD
*Department of Geology, University of Bristol, Wills Memorial Building. Queens Road, Briston BS8 1RJ, Great Britain*
*e-mail: P.Ditchfield@bristol.ac.uk*

ABSTRACT: A sequence of lacustrine, volcaniclastic, and alluvial sedimentary deposits that record the past million years of the history in the Olorgesailie Basin, southern Kenya, provide an example of how tectonics, climate, and volcanism affect sedimentation in a rift valley. A series of radiometric dates on volcanic materials through this sequence permits relatively fine-scale calibration of the timing and duration of volcanic input to the depositional system, transgressive–regressive cycles of the lake, and intervals of valley cutting and filling. The Olorgesailie Formation, accumulated between 0.992 and 0.493 Ma, consists of relatively pure diatomites, reworked diatomites, primary volcanic and reworked volcaniclastic units, and alluvial deposits (clays, silts, and sands with several well-developed paleosols) that bear a rich archeological and paleontological record. After 0.493 Ma, increased tectonic activity initiated a series of valley cutting and filling cycles that continue into Recent times. A working hypothesis attributes the formation of the paleolake to a barrier on the southwest side of the basin, large-scale lacustrine versus alluvial phases of the Olorgesailie Formation to variations in subsidence rates operating on a time scale of $10^4$–$10^5$ yr, and transgressive–regressive cycles within these sedimentary packages to wet–dry climate cycles on time scales of $10^3$–$10^4$ yr. Episodes of volcanism were superimposed on these patterns but did not have significant long-term effects on the depositional system.

## INTRODUCTION

With the growth of understanding of global-scale climate cycles based on the marine record, there has been increased interest in finding evidence of these cycles in continental deposits. Although lake deposits can provide records of seasonal to orbital climate cycles if they represent sufficient time in a relatively stable or evenly subsiding basin, the expression of these cycles in shallow lacustrine to subaerial deposits, particularly in tectonically active basins, is much more problematic. The Olorgesailie Basin (36° 26' E, 1° 34' S), in the southern part of the East African Rift Valley in Kenya (Figures 1, 2), provides a well-calibrated succession of lacustrine, alluvial, and volcaniclastic sediments amenable to analysis of tectonic vs. climatic controls at various time scales over the past million years. Our goal in this paper is to characterize the broad patterns of sedimentation, erosion, and landscape stability at different scales and to present a preliminary assessment of the shifting processes responsible for filling this part of the Rift Valley of eastern Africa. We also discuss the potential for recognizing records of global-scale climate cycles in lake margin to subaerial deposits in African rift basins.

Olorgesailie has been the site of archeological and geological investigation since the 1940s (Shackleton, 1944, 1955, 1978; Leakey, 1952; Baker, 1958; Baker and Mitchell, 1976; Isaac, 1977, 1978; Owen and Renaut, 1981, 2000; Bye et al., 1987). The latest phase of research, which began in 1985 under the direction of R. Potts, has developed a multidisciplinary, landscape approach to the reconstruction of the geological, ecological, and paleoanthropological history of the basin (Potts, 1989, 1994; Deino and Potts, 1990; Potts et al., 1996; Potts et al., 1999; Sikes et al., 1999). Hominins have intermittently occupied the Olorgesailie area over the past million years, and although their skeletal remains have not been found, archeological materials and vertebrate, invertebrate, and plant fossils occur throughout the sequence. These remains provide a rich record of the biotic and cultural components of the basin's history, and combined with extensive work on the geology, geochronology, and paleogeography, constitute a unique data set for reconstructing paleoenvironmental history.

The Olorgesailie Basin is relatively small (present extent around 150 km²) and isolated within the context of the East African rift system, in that it does not currently have major drainage connections with other parts of this system. About 120 m of sediments, deposited in at least four major phases of basin filling, span the past million years and are calibrated by radiometric dating of volcanic ashes and pumices (Deino and Potts 1990, 1992) as well as magnetostratigraphy (Tauxe et al., 1992). Chemical fingerprinting of the volcanic ashes also permits detailed lateral correlations around the basin. The oldest of the four phases of basin filling is represented by the Olorgesailie Formation, which is about 80 m in maximum composite thickness and between 992 and 493 ka in age. This formation consists of 14 members (designated M1–M14) that are well exposed across the southern part of the basin (Figure 2) (Deino and Potts, 1990; Potts et al., 1996; Potts et al., 1999). Overlying this formation are more localized deposits representing at least three episodes of substantial valley cutting and filling that occurred between ~ 0.5 Ma and

Sedimentation in Continental Rifts
SEPM Special Publication No. 73, Copyright © 2002
SEPM (Society for Sedimentary Geology), ISBN 1-56576-082-4, p. 97–106.

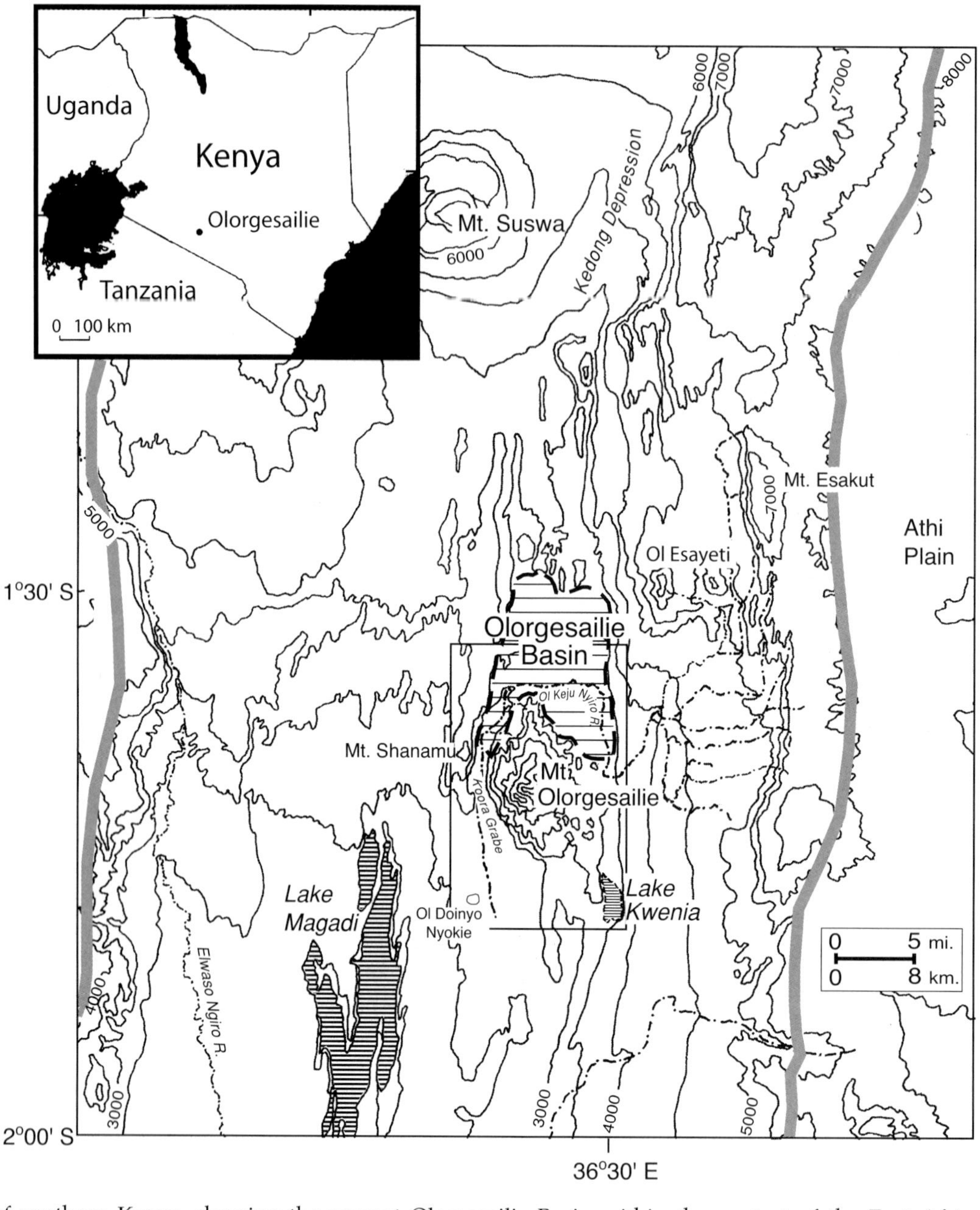


FIG. 1.—Map of southern Kenya, showing the present Olorgesailie Basin within the context of the East African Rift Valley (approximate boundaries delineated by heavy gray lines). Contour intervals in feet. Box shows area enlarged in Figure 2.

the end of the Pleistocene; the uppermost of these phases comprises the classic Oltepesi Beds (Isaac 1977), and the other two constitute a newly recognized, as yet undescribed formation between the Olorgesailie Formation and the Oltepesi Beds. Work on this complex story is ongoing; here we present a preliminary report on the past million years in the Olorgesailie rift basin as it is currently understood.

## OVERVIEW OF BASIN HISTORY

The Olorgesailie Basin today is bounded on the south by lavas associated with the extinct Pliocene volcano Mt. Olorgesailie, and on the north, east, and west by faulted ridges and outcrops of the widespread Magadi trachytes and older flood trachytes, phonolites, and basalts, which were extruded in the subsiding trough of the southern Kenya rift during the Plio-Pleistocene (Baker, 1958; Baker et al., 1972; Baker and Mitchell, 1976) (Figure 1). The rift floor in southern Kenya slopes gently from north to south, and this slope has exerted overall control on drainage into and out of the Olorgesailie Basin during the past million years. The present geomorphology of the basin is dominated by late Pleistocene faulting that has created a complex series of ridges, horsts, half-grabens, and grabens, complicating reconstructions of the paleobasin. Faults are generally north–south, with a secondary, east–west component. One of the north–south faults has uplifted the "lava peninsula" that divides the Legemunge and Oltepesi plains (Figure 2), with the downthrown side to the east. It is clear, however, that although some displacements began during sedi-

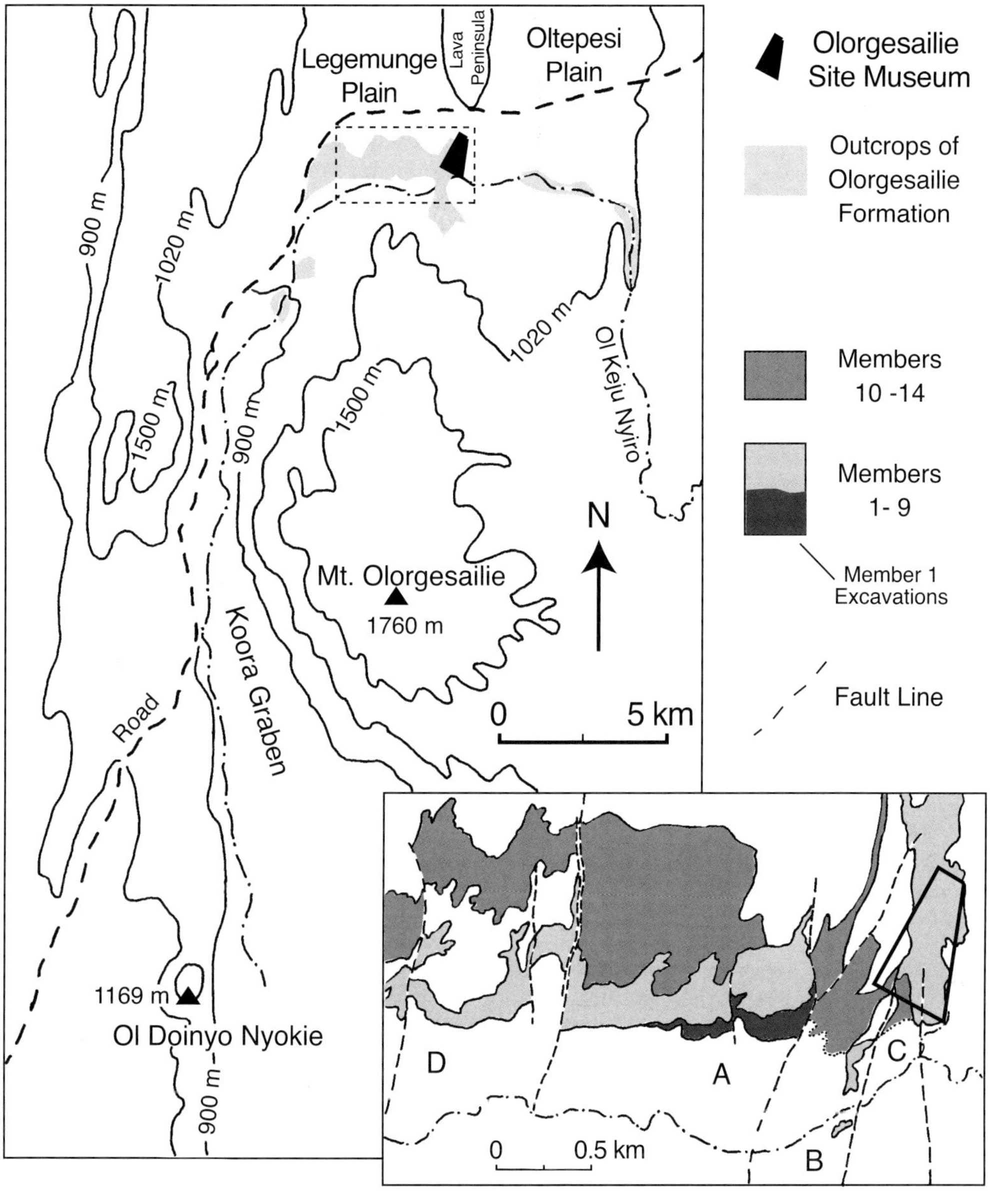


FIG. 2.—Detail of the vicinity of Mt. Olorgesailie showing principal areas of outcrop of the Olorgesailie Formation. Dashed-line box indicates area enlarged in inset map; letters on this map indicate Localities A–D, and the polygon shows the location of the Site Museum.

mentation, the basin was more laterally continuous from east to west between 1.0 and 0.5 Ma than after 0.5 Ma. The northern part of the present Olorgesailie basin is a flat, deflational plain, with few exposures to indicate the underlying extent of the Olorgesailie Formation. The Kedong depression to the north may have provided a conduit for volcanic debris derived from the volcanoes Suswa and Longonot (just off Figure 1, northeast of Suswa) to enter the Olorgesailie basin, but the geomorphology of this region has been altered by faulting and volcanic flows in the late Pleistocene (Baker and Mitchell, 1976), making reconstruction of its paleogeomorphology difficult. Drainage into the basin also was affected to the east and northeast by the faulted volcanic mountains Ol Esayeti and Esakut (Figure 2). To the southwest, the Koora Graben flanks Mt. Olorgesailie. This structure appears to have played an important role in controlling basin outflow versus ponding over the past million years. A series of north–south faulted ridges of Oltepesi basalts and Magadi trachytes lie to the east of Mt. Olorgesailie, and there is little direct evidence that the Pleistocene depositional basin extended around the mountain to the southeast. However, a low area that forms an ephemeral lake (Lake Kwenia) during unusually wet intervals occurs southeast of the mountain, and this may be a surviving analogue for some of the shallow lacustrine phases that are well expressed in Olorgesailie Formation sediments (Isaac, 1977).

The Olorgesailie Formation records the formation of a lake shortly before 992 ka, with subsequent alternations between lacustrine and subarial environments through the next ~ 500 kyr but no episodes of major downcutting (Figure 3). Between ~ 0.5

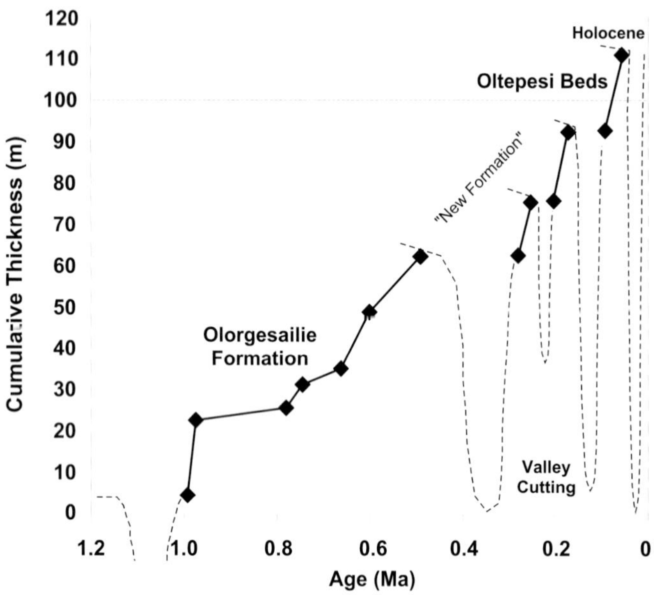


FIG. 3.—Proposed large-scale pattern of sediment accumulation through the past million years in the Olorgesailie Basin, showing alternating intervals of sedimentation and erosion that removed portions of the preexisting deposits but provided sites for subsequent episodes of valley fill. Radiometric dates are based on Deino and Potts (1990) and magnetostratigraphy on Tauxe et al. (1992).

Ma and the present, however, the basin experienced phases of valley cutting associated with base-level fluctuations of 50 m or more over periods of tens to hundreds of thousands of years. This is demonstrated by sharp erosional unconformities created by channels that cut into the preexisting deposits, plus the coarse-grained basal filling of these channels overlain by finer clastics and volcaniclastics that differ markedly from the lithologies of older deposits. These distinctive lithologies plus the erosional paleotopography at their base occur in discontinuous outcrops along the Ol Keju Nyiro River, from the Oltepesi area in the east to the western side of the Legemunge Plain (Figure 2), on the northeastern flank of Mt. Olorgesailie, and along the eastern margin of the Koora Graben (Figures 1, 2). Each of these depositional packages represents a distinct valley cut-and-fill cycle based on axial thicknesses of 10–30+ m, lateral continuity of the deposits up to hundreds of meters perpendicular to the drainage axis, and evidence for significant development of tributary channels. Downcutting was focused near the present course of the Ol Keju Nyiro River, along the southern margin of the basin next to the flank of Mt. Olorgesailie, and paleocurrent indicators show that drainage flowed west and southwest, as it does today. Remnants of a paleoescarpment cut by the first major phase of erosion are preserved in the southern part of Locality A (Figure 2), indicating that local drainage development was arrested by valley filling before it could remove significant portions of the Olorgesailie Formation (Figure 2, inset map). The volume of sediment that must have been removed and deposited in these relatively short-term cycles is impressive, and remnants of sediments from at least one of the filling phases occur on the plain north of the present escarpment, ~ 40 m above the Ol Keju Nyiro channel. Moreover, at least one of the valley fills includes massive rounded boulders estimated at 5 tons or more, which were noted by Baker and Mitchell (1976) and attributed by them to a catastrophic flood caused by the breaching of a volcanic dam north of the basin.

Deposits overlying the Olorgesailie Formation have been recognized for some time as the Oltepesi Beds (Baker and Mitchell, 1976; Shackleton, 1978; Deino and Potts, 1990). A number of radiometric dates on the volcanic components of these deposits provide the calibration points shown in Figure 3. Only in the past several years, however, have we recognized the complexity of the post-Olorgesailie deposits, with successive valley cut-and-fill cycles that lie stratigraphically between the Olorgesailie Formation proper and the classic Oltepesi Beds (Isaac, 1977). The latter consist of a distinctive suite of relatively fine-grained, buff-brown sands and silts, reworked tuffs, and pumice gravels and represent one of the latest valley-filling cycles. Prior to recognition of the erosional unconformities, the other valley-fill deposits were attributed either to the Oltepesi Beds or to the Olorgesailie Formation.

## THE OLORGESAILIE FORMATION

### *Lithologies and Structure*

The main depocenter for the Olorgesailie Formation was north of Mt. Olorgesailie; limited exposures of diatomites and volcaniclastic sediments west and southwest of Mt. Olorgesailie in the Koora Graben may represent extensions of these deposits but are not yet well understood. The composite section of the Olorgesailie Formation (Figure 4) is based on representative shorter section logs measured along the south-facing exposures in the southern Legemunge Plain (Figure 2), with a total composite thickness of about 65 m. The sediments are dominated by primary and reworked diatomites, pedogenically modified diatomites, fine-grained siliciclastics (silts, clays), and reworked volcanic materials (Table 1) (Isaac, 1977, 1978; Potts et al., 1999; Owen and Reneaut, 1981, 2000). Members 4 and 10 consist of fluvially reworked, pumice-rich sandy tuffs, and Member 12 is a pumice-rich, reworked diatomite; otherwise, volcanic units consist of primary ash falls or reworked tuffs or volcanic sands, with or without pumice. Specific sources for volcanics in the Olorgesailie Formation have yet to be determined; proposed correlations to the volcanic cone Ol Doinyo Nyokie, south of Mt. Olorgesailie (Bye et al., 1987), are unlikely given an age of 1.05 ± .014 Ma on one of the youngest ignimbrites from this cone (Deino and Potts, 1990). Possible sources to the north include the volcanoes Suswa and Longonot (Figure 1).

Previous work by Shackleton demonstrated considerable lateral variability in thickness caused by penecontemporaneous faulting and uneven subsidence across the basin (Isaac, 1977, 1978; Shackleton, 1978; Owen and Renaut, 1981, 2000). In spite of this structural deformation, many of the members as well as several distinctive paleosols and volcanic marker beds can be traced west to east from Legemunge to Oltepesi, and correlations for some of the tuffs have been confirmed using chemical signatures (Deino, unpublished data). This evidence plus the lack of major erosional breaks and angular unconformities within the stratigraphic sequence indicates net subsidence in the basin during much of Olorgesailie Formation time.

### *Paleoenvironmental Interpretation*

Enough work has been done within the Olorgesailie Formation to permit an assessment of transgressive–regressive cycles during the 0.5 Myr of net subsidence. These are indicated next to a composite stratigraphic column in Figure 4, which represents Localities A and C in the central part of the Legemunge

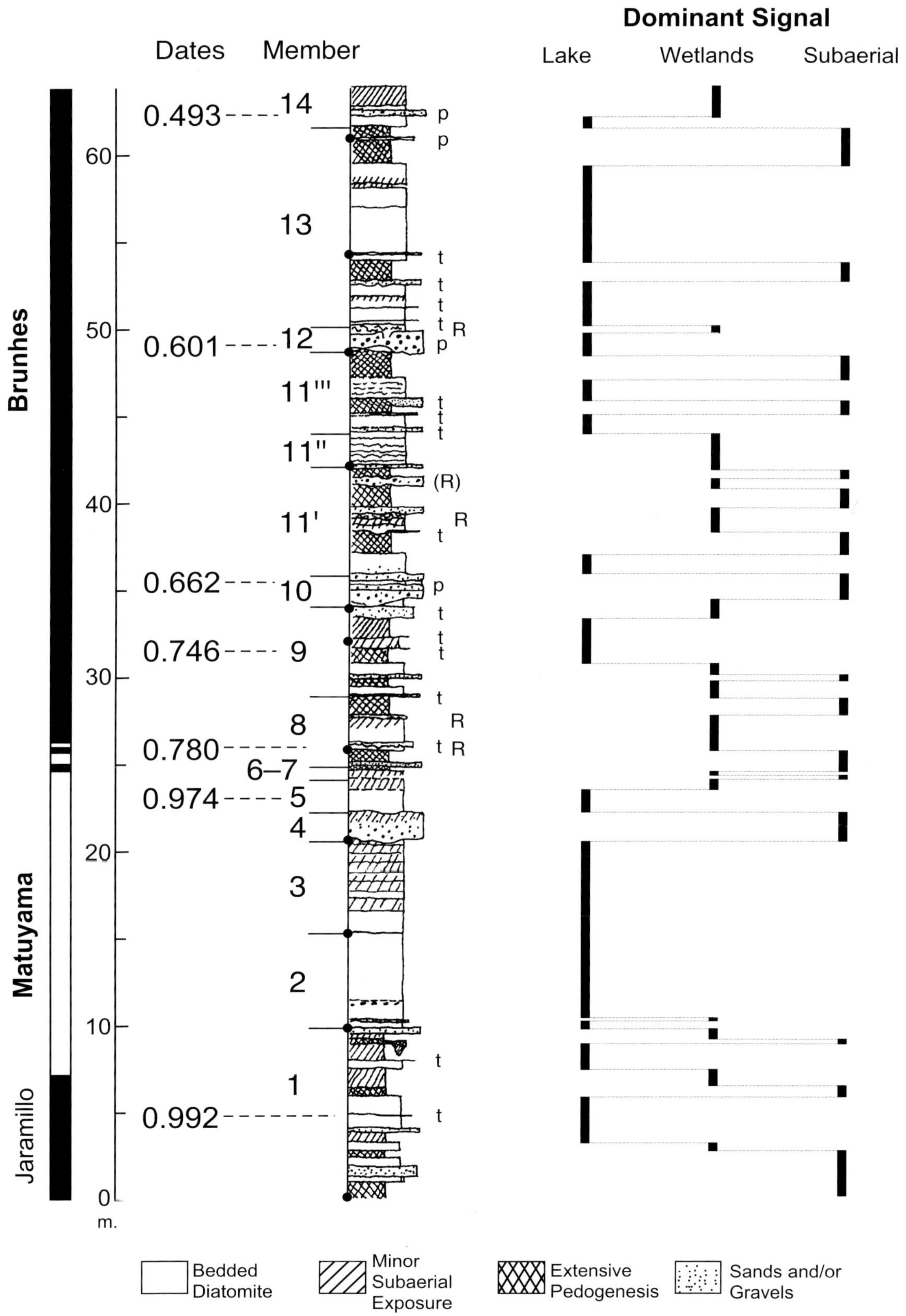


FIG. 4.—Composite stratigraphic section for the Olorgesailie Formation, based on measured section logs from the central area of exposures of the Legemunge Plain (Localities A and C; Potts et al., 1999). To the right of the section is an interpretation of the environments of deposition, showing both gradual and abrupt transitions from lacustrine to subaerial conditions. Black dots on left side of section indicate tops of individual section logs, "t" = tuff, "p" = pumice, "R" = red zone indicating plant combustion and baking of sediments, "(R)" = minor red zone (W. Melson, personal communication, 2001).

TABLE 1.—Lithofacies and inferred paleoenvironments of sediments that make up the Olorgesailie Formation.

| Lithofacies | Lithology | Internal Sedimentary Structures | Color | Fossils and Artifacts | Inferred Environment |
|---|---|---|---|---|---|
| 1. | Diatomite | Fine lamination, little evidence of bioturbation. $CaCO_3$ nodules | White to Yellow | Phytoliths, ostracods | Lake or lake margin |
| 2. | Silty diatomite | Massive or contorted bedding, relict lamination, root and stem traces, footprints and game trails | White, light green, yellow | Bones, artifacts, gastropods, ostracods | Bioturbated wetlands |
| 3. | Reworked diatomite | Massive or irregular bedding, root and stem traces | Dirty white, buff-green, yellow-green | Occasional bones and artifacts | Bioturbated lake margin or emergent lakebed. |
| 4. | Clayey diatomite | Massive, with brecciated diatomite, root traces | Green, yellow green | Occasional bones and artifacts | Pedogenically modified lakebed, lake margin. |
| 5. | Clay or silty clay | Ped structures, clay skins, root traces, $CaCO_3$ nodules | Dark green, greenish brown | Bones, artifacts locally abundant | Paleosol |
| 6. | Silt, sandy silt | Massive or irregular bedding, root traces, $CaCO_3$ nodules | Light brown, buff to greenish buff | Artifacts, bones, silicified macroplants (rare) | Fluvial overbank and paleosol, possible eolian components. |
| 7. | Sand, gravel | Poor sorting, cross-stratification | Light brown, greenish brown | Artifacts, bones locally abundant, reworked plant stems | Fluvial channels, sheetwash or proximal splays |
| 8. | Silt-grade ash | Well-sorted, laminated, laterally extensive; burrow and root traces | Dark to light gray | Silicified plant stems (uncommon) | Air-fall tuffs |
| 9. | Silty to sandy tuffs | Moderate to poor sorting, laminated to massive, laterally variable, burrow and root traces | Dark to light gray and gray-brown | Rare bones and artifacts, occasional plant impressions | Fluvial and aeolian reworked volcanic debris |
| 10. | Sand to gravel-grade volcaniclastics | Poorly sorted, bedded to massive, cross-stratification, contorted bedding, burrow structures | Dark gray, brown | Occasional bones and artifacts, plant casts | Fluvial channels and low-angle alluvial fans. |

Primary reddened (baked) zones occur mainly in the diatomite or silty diatomite facies, but reworked clasts of these reddened sediments may also occur in the coarser (fluvial) facies.

exposures (Figure 2). The section is generally representative of the vertical lithofacies changes for this part of the paleo-basin but does not incorporate details of lateral facies changes from other localities to the east (Oltepesi), south (Locality B), and west (Locality D). The dominant environmental signal was inferred on the basis of lithological and paleontological characteristics; numbers in the following list refer to lithofacies (LF) in Table 1. Lake = laminated diatomites with minimal bioturbation (LF 1 with occasional 8); Wetlands = bioturbated and / or physically reworked diatomites and primary volcanic tuffs or volcaniclastics with root traces, relict lamination, and occasional red baked zones (W. Melson, personal communication, 2001) as evidence of combustion of wetlands vegetation (LF 2, 3 with occasional 8, 9 components); Subaerial = clayey paleosols, siltstones, and diatomites with pedogenic fabric, fluvial overbank, and channel deposits (LF 4–10). Archeological materials and land vertebrate remains are most common in the subaerial environments; fish bones and gastropod molds are present in the lake and wetlands environments. Some of the paleosols were formed on sediments that accumulated in lake or wetlands environments, but their dominant environmental signal is recorded here as subaerial because pedogenesis thoroughly overprinted original depositional features. The paleoenvironmental assignments in Figure 4 thus somewhat underestimate the amount of primary diatomite and clay deposition.

Diatomite-depositing lakes and wetlands were the dominant paleoenvironments throughout the Olorgesailie Formation (Figure 5A), judging from thicknesses of different lithologies in the composite section (Figure 4). Aside from probable south-directed drainages entering from the north and northeast, the basin apparently was isolated from major clastic input throughout much of the deposition of the Olorgesailie Formation. There is a notable lack of coarse, cobble-grade material from the underlying and marginal basalts and trachytes, except within a few meters of outcrops of these rocks that formed areas of positive relief during Olorgesailie Formation deposition. Occasional large blocks of basalt and trachyte also occur associated with the paleosols, and some of these could be manuports related to stone-tool manufacture. Even in channel deposits of M7, where large handaxes abound, there are relatively few volcanic cobbles that are not artifacts or closely associated with artifacts (Potts et al., 1999), indicating a lack of fluvial energy sufficient for the transport of such materials into the basin from the surrounding terrain.

On the basis of environmental affinities of the diatom species and geochemical evidence, the lake was fresh for much of the time interval spanned by the Olorgesailie Formation, with a few periods of increased alkalinity (Owen and Renaut, 1981, 2000; Potts et al., 1999), and lacustrine phases alternated with subaerial periods when paleosols formed throughout the basin. Dry-land phases were accompanied by the formation of channels but not by large-scale erosive dissection of the landscape; transgressive surfaces generally are flat, and during periods of lake regression, even relatively coarse fluvial units do not cut more than a few meters into the preexisting strata. Hence, the basin appears to have maintained quasi-equilibrium with respect to base level, while subsidence continued between about 1.0 and 0.5 Ma. Most channel directions through the formation are oriented toward west or southwest, indicating outflow drainage through the Koora Graben (Figure 1).

Estimates of the time represented by the sedimentary units suggest that most of the Olorgesailie Formation was deposited in time intervals of less than $10^3$ years (Figure 5B). These estimates are based on radiometric calibration points, published information on average rates of diatomite accumulation, and judgments based on sedimentary features of alluvial and volcaniclastic

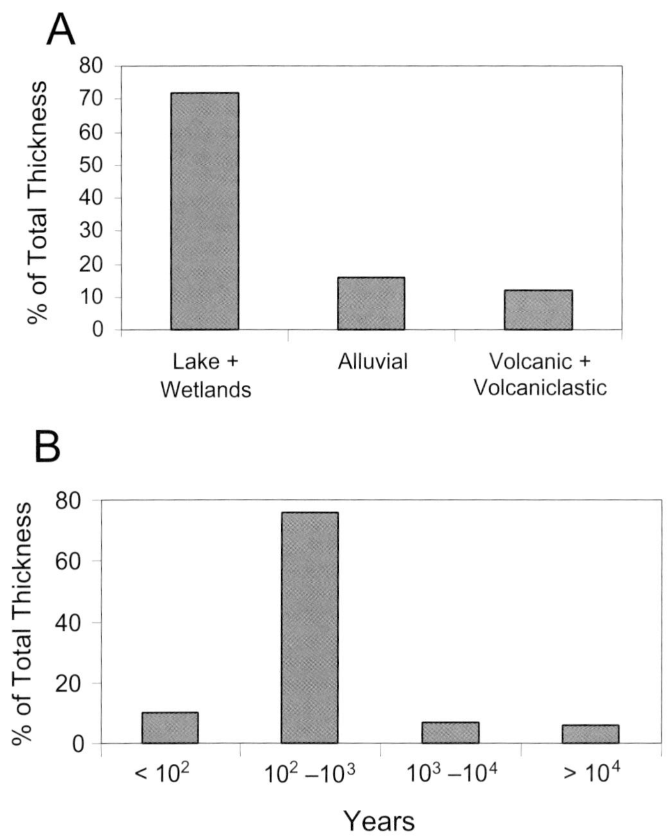


FIG. 5.—Summary characteristics of the Olorgesailie Formation. **A)** Percentage of different types of primary deposits; only fluvial sediments and well-developed paleosols are included in the "alluvial" category. **B)** Durations of time represented during intervals of active sedimentation, based on radiometric dates plus estimates of the temporal components of landscape development and depositional features of the strata. For further details, see Potts et al. (1999).

depositional events and the maturity of paleosols (Potts et al., 1999). Diatomites can accumulate quickly in silica-charged waters, filling available accommodation space at a rate of 0.8–2.0 m/kyr (Isaac, 1977; Potts et al., 1999). Much of the 0.5 Myr time span of the Olorgesailie Formation is compressed into the paleosols and minor erosional disconformities, as is typical for terrestrial deposits (Sadler, 1981).

## CONTROLS ON THE DEPOSITION OF THE OLORGESAILIE FORMATION

The fluctuations in depositional environment portrayed in Figure 4 represent the net result of basin subsidence, sediment influx, climate, and the tectonic processes that controlled the barrier that formed the lake. Our goal here is to develop a general working hypothesis concerning these controls at different time scales, using a combination of sedimentological evidence and chronostratigraphy provided by radiometric and magnetostratigraphic dates throughout the formation. What follows must be subject to further testing, but here we use it as an example of how this approach could contribute to understanding of rift-valley deposits. Tectonic, volcanic, or climatic processes could have controlled deposition as proposed in the following hypotheses:

1) *Tectonic control of a drainage barrier*. Tectonic processes raised and lowered a barrier to through-flow drainage from the Olorgesailie Basin into the Koora Graben and caused changes in lake level, recorded by transgressive and regressive cycles of lacustrine sedimentation.

2) *Tectonic control of subsidence*. Periods of subsidence in the basin provided increased accommodation space, causing lake expansion, while periods of arrested subsidence led to lake regression and paleosol formation when continuing sediment input filled the available accommodation space.

3) *Volcanic control*. Episodic influxes of volcaniclastic or pyroclastic materials from nearby eruptions blocked the through-flowing drainage of the basin, causing high lake stands, or filled the available accommodation space, causing a shift to subaerial deposition. Diatomite deposition also was controlled by the availability of dissolved silica in the lake, a function of overall volcanic influence on the chemistry of the lake and groundwater.

4) *Climate control*. Wet–dry climate fluctuations controlled by regional to global-scale climate cycles caused alternations between lacustrine and subaerial deposition through changes in the balance of rainfall and rates of evaporation from the lake.

The transgressive and regressive shifts portrayed in Figure 4 indicate four intervals when lacustrine sedimentation was dominant, alternating with three intervals with well-developed paleosols and alluvial deposits. These are summarized in Figure 6, which also shows that the initial lacustrine phase had the fastest net rate of accumulation. Within these intervals are shorter-term shifts between lake, wetlands, and dry land. A number of these shifts show a progressive change from lake to wetlands to land or vice versa, while in other cases wetlands are absent. The major volcaniclastic units, M4, M9–10, and M12, are spaced at relatively regular stratigraphic intervals throughout the formation, and the lower two are associated with lake regressions while M12 is coincident with transgression.

We propose that tectonic forces were the primary control on the major intervals of lacustrine vs. subaerial deposition on the basis of the following reasoning. The accumulation of relatively pure diatomites deposited in fresh to mildly alkaline water indicates high lake levels and through-flow drainage, although these conditions may not have lasted more than a few thousand years before accommodation space was filled, given the likely rates of diatomite deposition (0.8–2.0 m/yr; Isaac 1977; Potts et al., 1999). The absence of major periods of downcutting and deposits of clastic sediments, the lateral continuity of individual lithosomes throughout the basin, and the persistence of the fresh-water lake indicate stability of the depositional system on a time scale of $10^5$–$10^6$ yr (i.e., between 1.0 and 0.5 Ma.) We propose that these features of the Olorgesailie Formation are best explained by the early establishment and continued presence of a barrier between the Olorgesailie Basin and the Koora Graben, with variation in the balance of subsidence rates and sediment input on a time scale of $10^4$–$10^5$ yr superimposed upon this larger-scale stability. Periods of increased subsidence in the Olorgesailie Basin led to rapid lacustrine sediment accumulation (primarily diatomite), alternating with alluvial intervals when the available accommodation space was temporarily filled and the local drainage system over-

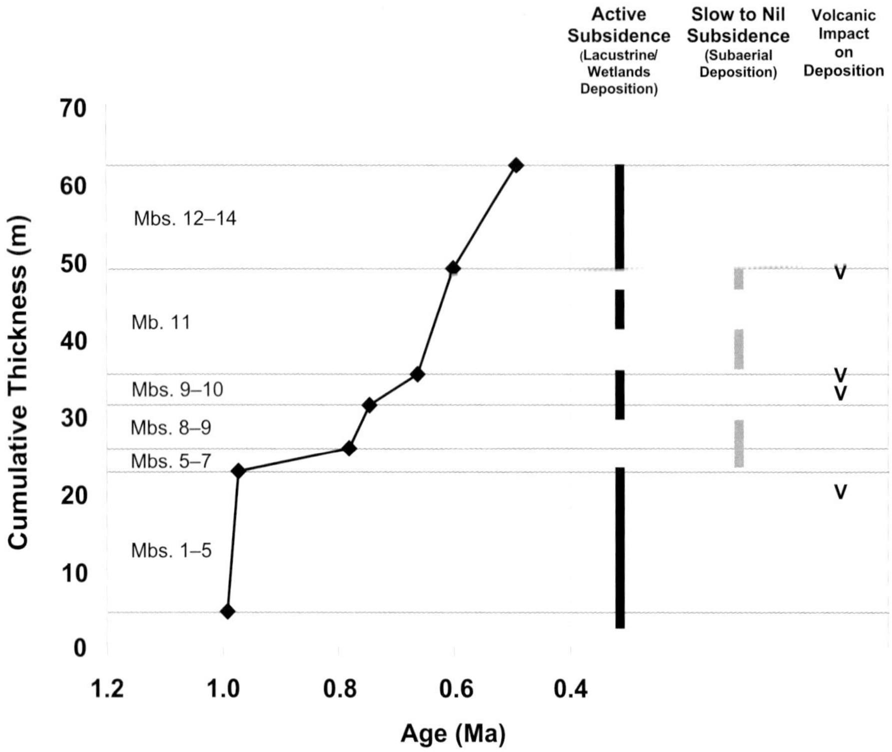


FIG. 6.—Plot of thickness versus age for the Olorgesailie Formation, showing the stratigraphic intervals of lacustrine or wetlands deposition that indicate significant intervals of active subsidence, alternating with subaerial deposits and paleosols that indicate slow to nil subsidence. Episodes of volcanic input (Vs in the right-hand column) are superimposed upon this overall pattern. Interpretations of depositional controls are based on the composite section in Figure 4 (also see text).

topped the barrier to the southwest, allowing outflow to the Koora Graben. This barrier was not significantly eroded during the alluvial intervals, and it may have consisted of a broad, relatively elevated area of lava flows and scoriaceous cones in the northern part of the Koora Graben rather than a more localized blockage of the outflow channel. We favor basin subsidence rather than barrier movement as a control on Olorgesailie Formation deposition at the $10^4$–$10^5$ yr time scale because: (1) the barrier would have had to be remarkably well behaved through time, moving up and down over small distances to generate the observed lake-to-land or land-to-lake shifts, and (2) during elevated phases, even if relatively broad, it would have been subject to erosion and changes in drainage dynamics that should be expressed progressively through the Olorgesailie deposits. Instead, the overall consistency of the lacustrine intervals suggests a simpler mechanism, i.e., variable basin subsidence. The formation of the barrier probably did influence the initial, relatively rapid filling of the lake basin, but subsequently it remained stable until the first major phase of post–Olorgesailie Formation valley cutting. Erosion on this scale indicates a significant lowering of base level and may well reflect faulting and subsidence of the Koora Graben and the barrier that separated it from the Olorgesailie Basin to the north.

The shorter-term transgressive–regressive cycles expressed in the Olorgesailie strata could represent fault-controlled pulses of up-and-down movement superimposed on overall basin subsidence. Given the rift-valley context of the basin and evidence for differential subsidence across faults, some degree of small-scale tectonic influence on sedimentation is likely. However, we regard climatic cycles as a more plausible hypothesis for these short-term patterns, particularly in the upper part of the formation. By counting the number of lake, wetlands, and subaerial environments between each of the dated horizons, we can examine the periodicity of the cycles in the Olorgesailie Formation (Figure 1, Table 2). Between the base of M11 (dated by the age of 0.662 Ma on M10) and the base of M12 (dated at 0.601 Ma) (Deino and Potts, 1990), there are five subaerial intervals alternating with seven lacustrine or wetland intervals, amounting to about one cycle every 10.2 kyr. This is not obviously related to established orbital periodicities for global climate cycles, although millennial-scale climate forcing could be indicated (Bond et al., 1992; Raymo et al., 1998). Between M12 and the base of M14, however, dated at 0.601 to 0.493 kyr, there are three lacustrine and two pedogenic units, or about one cycle every 43.2 kyr. This is close to the 41 kyr obliquity period and could indicate a link to global-scale climate cycles; however, we have yet to detect evidence of the 22 kyr precessional periodicities, which should be manifested more strongly at low latitudes (deMenocal et al., 1993). The lower part of the formation is even more problematic in terms of variable subsidence effects on rates of sediment accumulation, and cycles range from 5 to 78 kyr duration (Figure 1, Table 2).

We regard the volcanic influxes in M4, M10, and M12 as notable but temporary influences on basin sedimentation. M4 is a fluvial volcaniclastic unit with pedogenic modification that represents rapid deposition of several meters of poorly sorted pumice-rich sands followed by a period of subaerial exposure before the lake returned in basal M5. We interpret M4 as temporarily filling up the available accommodation space when the lake was relatively shallow. Similarly, M9–10 represent a coarsening-upward influx of volcaniclastics, initially mixed with diatomites

TABLE 2.—Numbers of transgression–regression cycles between dated ash beds in the Olorgesailie Formation, based on the composite section in Figure 4.

| Stratigraphic Interval | Radiometric Dates | # Years | # Cycles | Years per Cycle |
|---|---|---|---|---|
| M1–middle M5 | 0.992 ± .039–0.974 ± .007 | 18,000 | 3.5 | 5,100 |
| M7/8–middle M9 | 0.780 ± .006–0.746 ± .006 | 34,000 | 2.5 | 13,600 |
| M10–M12 | 0.662 ± .004–0.601 ± .003 | 61,000 | 6 | 10,200 |
| M12–lower M14 | 0.601 ± .003–0.493 ± .001 | 108,000 | 2.5 | 43,200 |

Dates are from Deino and Potts (1990) except the M7/8 date for the Brunhes–Matuyama magnetochron boundary (Cande and Kent, 1995; Tauxe et al., 1992).

in M9 followed by the fluvial deposits of M10. This volcanic influx also temporarily exceeded available accommodation space but soon was followed by a return to lacustrine deposition. M12 consists of an airfall ash overlain by a relatively pure diatomite with dispersed angular pumices that may represent direct airfall into a shallow lake. Deposition of M12 coincided with a transgressive phase of the lake, but the volume of volcanic material is minor compared with M4 and M9–10, and there is no evidence for a causal link between volcanism and the transgression. None of these units had any long-lasting effect on overall sedimentation. Volcanism provided high levels of dissolved silica to the lake waters and periodically swamped the basin with volcanic debris, but its effects were superimposed on the broader patterns of deposition controlled by subsidence.

This preliminary analysis of the Olorgesailie Formation is subject to further testing and refinement of the composite section (Figure 4) through lateral studies of facies variation throughout the basin. On the basis of the above discussion, however, we propose that a barrier to outflow into the Koora Graben caused a lake to form north of Mt. Olorgesailie, subsidence within this basin controlled deposition over intervals of $10^4$–$10^5$ yr, and within the intervals of more active subsidence, climatic cycles controlled lake transgressions and regressions on scales of $10^3$–$10^4$ yr. These cycles are recorded at different time scales because of variable rates of basin subsidence and sedimentation, and their relationship to orbital climate forcing is unknown. The major soils record periods when subsidence slowed and accommodation space was minimal; at these times drainage flowed through the basin but did not substantially incise preexisting deposits and the land surfaces were stable for $10^2$ up to $10^5$ yr. The barrier on the southwest side of the basin existed throughout the deposition of the formation, without significant erosion. The low level of input of sand and silt to the basin, combined with the scarcity of coarse-grade materials, also indicates that the basin margins to the north, east, and west were lower topographically than today and tectonically stable throughout the time of deposition of the Olorgesailie Formation.

## CONCLUSIONS

The above analysis demonstrates that characterizing rift-basin deposition at different scales can help to sort out tectonic versus climatic controls on basin history. We summarize our working hypothesis for the past million years of the Olorgesailie Basin as follows. At the scale of the past million years, Olorgesailie Basin deposition can be divided into two distinct intervals, the first half in a relatively stable, subsiding basin with a persistent barrier to outflow drainage to the southwest, and the second in an unstable, tectonically active basin with marked alternation of erosional and depositional conditions. At the scale of $10^4$–$10^5$ yr, lacustrine vs. subaerial deposition represented by the Olorgesailie Formation was controlled by subsidence and available accommodation space for sediment accumulation (primarily diatomite). At the scale of $10^3$–$10^4$ yr, transgressive–regressive cycles of the lake during the Olorgesailie Formation were controlled by climate and, in a few cases, by influxes of volcanic materials that overfilled the basin and caused temporary regressions. During the later phase (~ 0.5–0.0 Ma), increased faulting in the Koora Graben and along the Olorgesailie basin margins initiated a new depositional regimen, resulting in cycles of valley erosion and filling. This shift in depositional mode is consistent with the previously recognized increase in tectonic and volcanic activity in the Rift Valley of southern Kenya during the late Pleistocene (Baker, 1958; Baker and Mitchell, 1976).

This hypothesis provides a framework for testing based on our current understanding of the geology and may change with further field and laboratory research. It suggests, however, that there is potential for recovering climate records in transgressive–regressive cycles of rift-valley lakes as well as within continuous lacustrine deposits. The issues of climate vs. tectonics encountered in Olorgesailie are likely to be similar in other small-scale rift basins because we expect that: (1) variable subsidence will run the "climate recorder" at different speeds, even in relatively stable intervals of a basin's history, and (2) rapid filling of the available accommodation space during intervals of active subsidence means that most of the lacustrine deposits will represent time intervals too short to capture global-scale orbital cycles of $10^3$–$10^4$ yr. Transgressive–regressive cycles recorded when net deposition rates are in quasi-equilibrium with subsidence thus may be more useful for revealing climatic controls on lake fluctuations than continuous lacustrine deposits. Fine-scale calibration of rift-basin sequences using closely spaced, high-precision radiometric dates makes it possible to pin down the frequency of the transgression–regression cycles and is key to isolating possible climate signals.

## ACKNOWLEDGMENTS

We thank Gail Ashley and Robin Renaut for inviting us to contribute to this volume, and for their patience in seeing the manuscript through to completion. The Olorgesailie project is conducted in collaboration with the National Museums of Kenya (NMK), and we thank G. Abungu, M.G. Leakey, and the NMK Palaeontology Department for their support. We gratefully acknowledge our field crews, led by J.M. Nume, for their excellent geologic trenches and excavations; and also the contributions of J. Clark, D. Deocampo, W.F. Keyser, W.G. Melson, M. Noll, T.W. Plummer, N. Sikes, and other scientists, technical assistants, and students who have participated in the research. We appreciate Bill Melson's generosity in sharing the results of his study of the genesis of the Olorgesailie red beds. This project has been funded primarily by the Smithsonian's Human Origins Program, Scholarly Studies Program, and the National Museum of Natural History, Smithsonian Institution.

## REFERENCES

BAKER, B.H., 1958, Geology of the Magadi area: Geological Survey of Kenya, Report 42.

BAKER, B.H., AND MITCHELL, J.G., 1976, Volcanic stratigraphy and geochronology of the Kedong–Olorgesailie area and the evolution of the South Kenya rift valley: Geological Society of London, Journal, v. 132, p. 467–484.

BAKER, B.H., MOHR, P.A., AND WILLIAMS, L.A.J., 1972, Geology of the Eastern Rift System of Africa: Geological Society of America, Special Paper 136, 67 p.

BOND, G., HEINRICH, H., BROECKER, W., LABEYRIE, L., MCMANUS, J., ANDREWS, J., HUON, S., JANTSCHIK, R., CLASEN, S., SIMET, C., TEDESCO, K., KLAS, M., BONANI, G., AND IVY, S., 1992, Evidence for massive discharges of icebergs into the North Atlantic ocean during the last glacial period: Nature, v. 360, p. 245–249.

BYE, B., BROWN, F.H., CERLING, T.E., AND MCDOUGALL, I., 1987, Increased age estimate for the Lower Palaeolithic hominid site at Olorgesailie, Kenya: Nature, v. 329, p. 237–239.

CANDE, S.C., AND KENT, D.V., 1995, Revised calibration of the geomagnetic polarity timescale for the Late Cretaceous and Cenozoic: Journal of Geophysical Research, v. 100B, p. 6093-6095.

DEINO, A., AND POTTS, R., 1990, Single crystal $^{40}Ar/^{39}Ar$ dating of the Olorgesailie Formation, southern Kenya rift: Journal of Geophysical Research, v. 95, p. 8453–8470.

DEINO, A., AND POTTS, R., 1992, Age–probability spectra for examination of single-crystal $^{40}Ar/^{39}Ar$ dating results: Examples from Olorgesailie, southern Kenya rift valley: Quaternary International, v. 7/8, p. 81–89.

DEMENOCAL, P.B., RUDDIMAN, W.F., AND POKRAS, E.M., 1993, Influences of high- and low-latitude processes on African terrestrial climate: Pleistocene eolian records from equatorial Atlantic Ocean Drilling Program Site 663: Paleoceanography v. 8, p. 209–242.

ISAAC, G.Ll., 1977, Olorgesailie: Archeological Studies of a Middle Pleistocene Lake Basin in Kenya: Chicago, University of Chicago Press, 272 p.

ISAAC, G.Ll., 1978, The Olorgesailie Formation, *in* Bishop, W.W., ed., Geological Background to Fossil Man: Edinburgh, Scottish Academic Press, p. 173–206.

LEAKEY, L.S.B., 1952, the Olorgesailie Prehistoric Site, *in* Leakey, L.S.B., and Cole, S., eds., Proceedings of the Pan-African Congress on Prehistory, 1947, p. 209.

OWEN, R.B., AND RENAUT, R.W., 1981, Paleoenvironments and sedimentology of the Middle Pleistocene Olorgesailie Formation, southern Kenya Rift Valley: Paleoecology of Africa, v. 13, p. 1147–174.

OWEN, R.B., AND RENAUT, R.W., 2000, Spatial and temporal facies variations in the Pleistocene Olorgesailie Formation, southern Kenya Rift Valley, *in* Gierlowski-Kordesch, E., and Kelts, K., eds, Lake Basins Through Space and Time: American Association of Petroleum Geologists, Memoir 46, p. 553–560.

POTTS, R., 1989, Olorgesailie: new excavations and findings in Early and Middle Pleistocene contexts, southern Kenya rift valley: Journal of Human Evolution, v. 18, p. 477–484.

POTTS, R., 1994, Variables vs. models of early Pleistocene hominid land use: Journal of Human Evolution , v. 27, p. 7–24.

POTTS, R., JORSTAD, T., AND COLE, D., 1996, The role of GIS in interdisciplinary investigations at Olorgesailie, Kenya, a Pleistocene archeological locality, *in* Aldenderfer, M., and Maschner, H.D.G., eds., Anthropology, Space, and Geographic Information Systems: New York, Oxford University Press, p. 202–213.

POTTS, R., BEHRENSMEYER, A.K., AND DITCHFIELD, P., 1999, Paleolandscape variation and early Pleistocene hominid activities: Members 1 and 7, Olorgesailie Formation, Kenya: Journal of Human Evolution, v. 37, p. 747–788.

RAYMO, M.E., GANLEY, K., CARTGER, S., OPPO, D.W., AND MCMANUS, J., 1998, Millennial-scale climate instability during the early Pleistocene epoch: Nature, v. 392, p. 699–702.

SADLER, P., 1981, Sediment accumulation rates and completeness of sedimentary sections: Journal of Geology, v. 89, p. 569–584.

SHACKLETON, R.M., 1944, Preliminary report on the Olorgesailie Prehistoric Site: Unpublished Report, Mines and Geology Department, Nairobi.

SHACKLETON, R.M., 1955, Pleistocene movements in the Gregory Rift Valley: Geologische Rundschau, v. 43, p. 257–263.

SHACKLETON, R.M., 1978, A geological map of the Olorgesailie area, *in* Bishop, W.W., ed., Geological Background to Fossil Man: Edinburgh, Scottish Academic Press, map insert, p. 171.

SIKES, N.E., POTTS, R., AND BEHRENSMEYER, A.K., 1999, Early Pleistocene habitat reconstruction in Member 1 Olorgesailie, based on paleosol stable isotopes: Journal of Human Evolution, v. 37, p. 721–746.

TAUXE, L., DEINO, A.D., BEHRENSMEYER, A.K., AND POTTS, R., 1992, Pinning down the Brunhes/Matuyama and upper Jaramillo boundaries: A reconciliation of orbital and isotopic time scales: Earth and Planetary Science Letters, v. 109, p. 561–572.

# SEDIMENTATION PATTERNS IN A PLIO-PLEISTOCENE VOLCANICLASTIC RIFT-PLATFORM BASIN, OLDUVAI GORGE, TANZANIA


GAIL M. ASHLEY

*Department of Geological Sciences, Rutgers University, Piscataway, New Jersey 08854, U.S.A.*
*e-mail: gmashley@rci.rutgers.edu*

AND

RICHARD L. HAY

*Department of Geosciences, University of Arizona, Tucson, Arizona 85721 U.S.A.*
*e-mail: rhay@dakotacom.net*



ABSTRACT: The Plio-Pleistocene sedimentary deposits of upper Bed I and lowermost Bed II in the Olduvai basin, Tanzania, are analyzed in terms of sedimentary processes, paleoenvironments, and paleoclimate. The sedimentary basin was a shallow, closed depression (~ 50 km wide with strata ~ 100 m thick) located adjacent to the East African Rift between Precambrian basement rocks to the west and the large Plio-Pleistocene Ngorongoro volcanic complex to the east. Olduvai lies on the west flank of the main rift valley, and this rift platform area was subjected to crustal extension resulting in normal block faulting and block tilting. Through time the axial part of the sedimentary basin episodically migrated eastward toward the main rift valley. Syndepositional volcanism included basalt flows, pyroclastic air falls, ground surges, and ignimbrites from eruptive centers of the Ngorongoro volcanic complex. Hay (1976) interpreted the paleolandscape to have had a central lake basin that was filled with detrital sediment, carbonate rocks, and tephra. Detrital sediment was provided both by rivers sourced in the Ngorongoro complex and by Precambrian rocks of the Serengeti to the north and west. Recent high-resolution stratigraphic studies of paleoenvironments in a time slice (1.85–1.75 Ma) found that the sedimentary facies vary temporally and spatially because of the interplay among five independent variables affecting the sedimentary record: (1) sediment flux (pyroclastic air fall, intermittent fluvial runoff and mudflows, and eolian), (2) sediment composition (Precambrian source rocks vs. volcanic source rocks), (3) hydrology and chemistry (lake water, surface water, and groundwater), (4) soil formation, and (5) tectonics associated with rift extension. Lake cycles, ephemeral fluvial systems, soil formation on the interfluves and lake margin, well as a large groundwater-fed wetland (> 3 $km^2$) are all part of the Olduvai rift-margin basin record and provide insights into the Plio-Pleistocene paleoclimate history of East Africa.


## INTRODUCTION

Models for sedimentation in continental rifts are usually based on studies of depositional basins within the main rift valley, despite the potential for the development of many small basins within the rift platform (flank). These smaller, subsidiary basins are usually interiorly drained and shorter lived than the main valley. Because of their small size, however, they are likely to be highly responsive to climate change and thus could contain valuable paleoclimatic information. The sedimentary records of platform basins when combined with those of the main rift valley should produce a more complete understanding of the regional rifting history.

A small rift-platform basin (Olduvai) adjacent to the Gregory rift in northern Tanzania is being studied by an interdisciplinary team of scientists looking at hominid land-use patterns during a narrow window of time, ~ 1.85–1.75 Ma (Fig. 1). The study has utilized over 100 archaeological trenches along a 20 km basin transect and has yielded a high-resolution stratigraphic record that is ideal for interpreting rift-related tectonism, volcanism, and sedimentation in an arid setting.

The modern Gorge exposes a two-million-year sedimentary record in an incised river valley draining eastward from the Serengeti Plains (Fig. 2). The Gorge cuts across a 50-kilometer-wide rift-platform basin located between Precambrian basement to the west and the Plio-Pleistocene Ngorongoro Volcanic Highland to the east. The Plio-Pleistocene basin fill, now disrupted by rift-parallel faults, is surprisingly thin (100 m) and composed largely of reworked volcaniclastic sediment and air-fall tuffs deposited in a shallow semiarid fluvial–lacustrine basin (Hay, 1976). The sedimentary record contains a rich faunal and cultural record of early hominids and thus is also important to studies of human evolution (Leakey, 1971; Blumenschine and Masao, 1991; Peters and Blumenschine, 1995; Blumenschine and Peters, 1998; Feibel, 1999). The results presented here build on the sedimentation study of Hay (1976) but are viewed here in a new structural context of a "rift-platform basin".

The objectives of this paper are: (1) to describe the Plio-Pleistocene volcaniclastic sedimentary record in a shallow, rift-platform basin (Olduvai, Tanzania) and (2) to interpret the deposits in terms of sedimentary processes, paleoenvironmental setting, and paleoclimate.

## GEOLOGIC CONTEXT

Crustal extension in Africa was underway in the Suez region by Late Oligocene to Early Miocene (30–20 Ma) (Baker et al., 1972). By 12–15 million years ago, areas to the south (Ethiopia, Kenya, and Tanzania) began to experience broad regional uplift, local doming, increased heat flow, and the development of magma centers in the upper crust (Robson, 1971; Garfunkel and Bartov, 1977; Morley, 1999). Rift systems propagated southward along two major lineaments, Western (Albertine) and Eastern (Gregory), that commonly bifurcate and terminate (Baker and Wohlenberg, 1971; Baker et al., 1972; Chapman et al., 1978; Morley, 1999) (Fig. 1). Geophysical information reveals that structurally the rifts range from full to half grabens (up to 100 km wide, 500–800 km long, and 4 km deep), and that rift-parallel faulting may affect up to another 200 km on either side (rift platform or flank) of the main rift valley (Rosendahl, 1987; Morley, 1999).

Rifting generally begins with volcanism and gentle subsidence in the main valley trough, followed by uplift of the rift flanks, continued subsidence in the trough, deposition in rift-


Sedimentation in Continental Rifts
SEPM Special Publication No. 73, Copyright © 2002
SEPM (Society for Sedimentary Geology), ISBN 1-56576-082-4, p. 107–122.

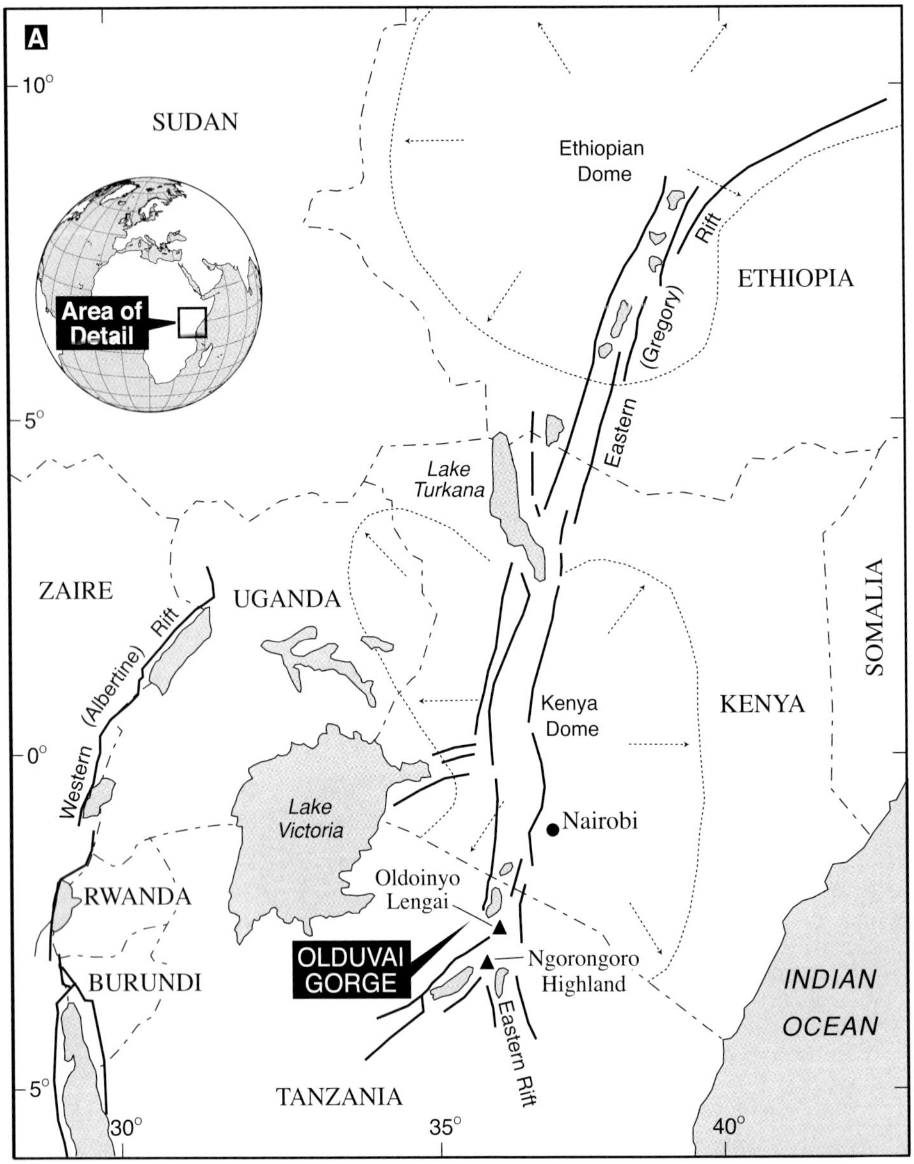


FIG. 1.—Map showing the Eastern (Gregory) and Western (Albertine) rifts of East Africa, the Ethiopian and Kenya Domes, and the location of Olduvai Gorge on the western margin of the Gregory rift in northern Tanzania (modified from Sutcliffe, 1985).

platform basins and in the main graben, and intense normal faulting (Withjack, 1979). An excellent example of a rift platform that through time might have produced many small, short-lived depositional centers is the Mesozoic rift system on the western margin of the Norwegian Sea (Fig. 3), and extensive geophysical data during oil exploration have provided exceptionally detailed records (Blystad et al., 1995). Beds thin and pinch out away from the main rift valley, and the sediment package is segmented by extension-related, mainly normal faults. The thickness and geometry of the infill of both the main rift valley and rift platform basins are controlled both by uplift and erosion of rift flank rocks and the magnitude of rift-related volcanism.

Ancient rifts, like this submerged example from the Norwegian Sea, have sedimentary records whose broad-scale nature is revealed by 3-D seismic data and cores. However, details of the sedimentary deposits, such as spatial variability of depositional environments, biological or cultural records, or even the paleoclimatic context, are difficult to interpret from widely spaced seismic lines and cores. Thus, rift-valley and rift-platform sediments in subaerial settings that have extensive laterally continuous outcrop exposures are desirable to augment the growing body of literature on rift sedimentation (Coward et al., 1987; Katz, 1990; Busby and Ingersoll, 1995; Frostick, 1997; Selley, 1997; Morley, 1999). Olduvai Gorge in northern Tanzania is one such example (Fig. 1).

## PHYSICAL SETTING

The Olduvai sedimentary basin was formed at ~ 2.0 Ma on the western flank of the Gregory rift in response to faulting and the growth of a large volcanic complex (Ngorongoro) within a

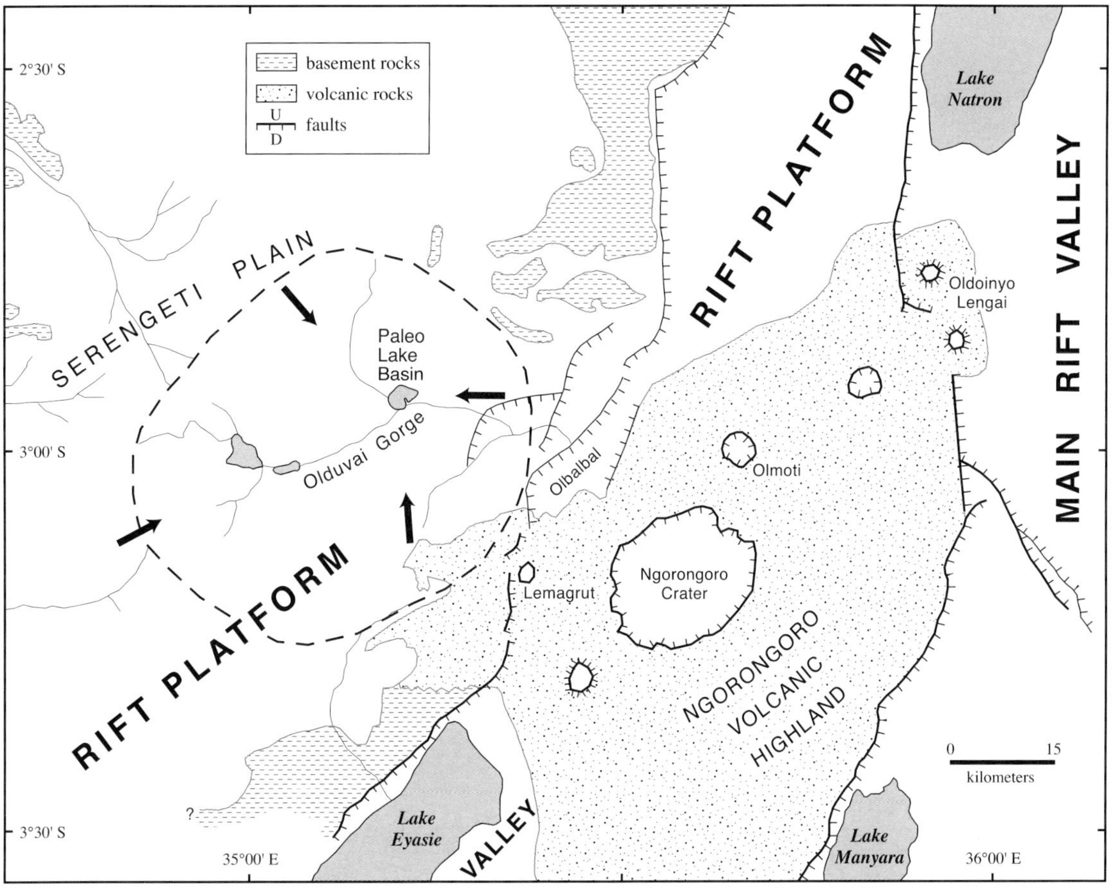


FIG. 2.—Map shows the general location of the main rift valley and the rift platform at 3 degrees south. The drainage pattern of the modern Olduvai Gorge and approximate position of the paleolake in the center of the basin is depicted. The approximate location of the paleobasin is indicated by a dashed line; maximum width was ~ 50 km. Arrows represent directions of sediment flux into the paleo basin (generalized from Hay, 1976). The main Gregory Rift valley is on the east and the Serengeti Plain lies to the west. The Ngorongoro Volcanic Highland is composed of several eruptive centers, of which only one, Oldoinyo Lengai, is still active.

prominent bifurcation in the main rift valley (Fig. 1). The Ngorongoro Volcanic Highland is an ~ 4,000 $km^2$ massif composed of 8–10 eruptive centers of alkali magma compositions ranging from olivine basalts, trachyandesites, and trachytes to nephelinites and carbonatites (Dawson, 1962, 1964; Hay, 1976, 1989). Except for the Laetolil Beds (~ 3.5 Ma) to the south and exposed in the Side Gorge (Hay, 1976), the earliest Olduvai deposits were laid down on an eastward-sloping undulating surface of metamorphic basement rocks of Precambrian age (Mozambique Belt), suggesting that nearly all rocks younger then Precambrian were removed prior to basin infilling (Hay, unpublished data).

During the Plio-Pleistocene, the basin was an estimated 3,500 $km^2$ in area, and it was bordered on the south and east by the volcanic highlands and on the west and north by metamorphic terrane. The catchment area including Lakes Masek and Ndutu on the western margin and the Ngorongoro Highlands on the eastern margin was 50 km in diameter (Fig. 2). The sedimentary record preserved in the basin center is 30 km in diameter, with the lake at maximum expansion up to 15 km in diameter. The area was semiarid, and annual evaporation vastly exceeded precipitation. The Ngorongoro Highland (> 3,000 m high) produced an effective rain shadow from the easterly trade winds blowing off the Indian Ocean, particularly during the monsoons.

Climate appears to have fluctuated, probably as a consequence of Milankovitch astronomical cycles, but with an overall trend toward aridity during the 2-million-year history (Cerling and Hay, 1986; Kappelman, 1986; deMenocal, 1995; Sikes, 1996; Ashley and Driese, 2000). Precipitation is estimated at ~ 800–900 mm/yr for the early part of the record, compared to modern rainfall of ~ 560 mm/yr (Bonnefille and Riollet, 1980, Sikes and Ashley, in press).

## STRATIGRAPHY

Establishing the stratigraphic framework of Olduvai began with the early efforts of Reck (1914, 1951) and Pickering (1960), culminating in the synthesis of the Gorge geology by Hay (1976). The stratigraphic record above the basal volcanics is divided into six units (Fig. 4), although lateral variation in lithofacies and rift-related faulting complicate correlation. Nevertheless, a basin-wide stratigraphic framework has been established using detailed field mapping, mineral chemistry, argon/argon dating, and paleomagnetic studies (Gromme and Hay, 1971; Hay, 1976; Walter et al., 1991).

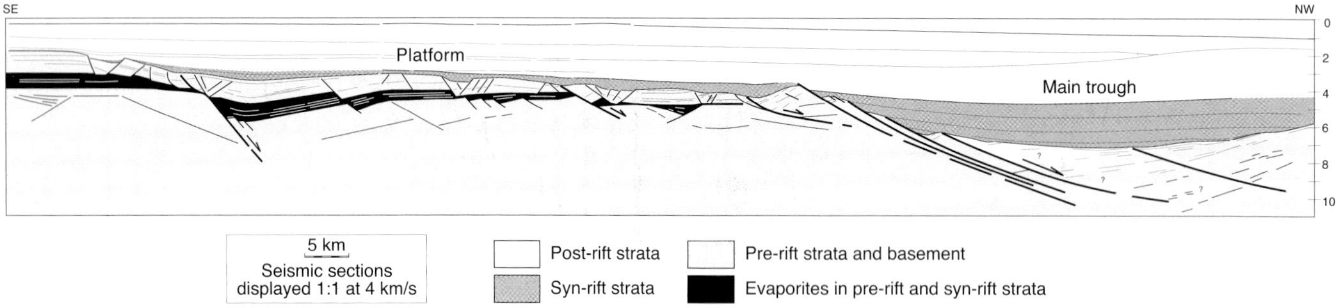


FIG. 3.—Line drawing of segment of seismic line GMNR94-310 through the Mesozoic Voring rift basin of offshore Norway. The basin and associated platform-margin east–west cross section shows the typically thin platform margin deposits compared to the thick accumulations in the main trough (rift valley). Rift-related faulting extends into the shallow basement rocks. Small isolated basins may develop on the platform during evolution of the rift (modified from Blystad et al., 1995; Withjack et al., this volume).

Much of the interdisciplinary research on paleoenvironments and hominid land use during the Plio-Pleistocene (Blumenschine and Masao, 1991; Blumenschine et al., 1997; Blumenschine et al., 1999; Ashley, 2000) has involved extensive excavations. Fieldwork began in 1989, and since then members of the Olduvai Landscape Paleoanthropology Project (OLAPP) have excavated more than 100 trenches throughout the paleobasin. These trenches are confined to a narrow "time slice" defined by mappable tuffs within upper Bed I and lowermost Bed II (Fig. 4). The high-resolution sedimentary record present in the archaeological trenches revealed a significant variability of lithofacies in time and space. The sedimentary record tracks a slowly changing landscape affected by volcanism, climate change, and tectonics. The lithofacies and their inferred paleoenvironments within the 100,000 year time slice (~ 1.85–1.75 Ma) are summarized below.

## PALEOENVIRONMENTS AND SEDIMENTARY RECORD

The early Pleistocene Olduvai paleobasin was a roughly circular depression adjacent to the Ngorongoro Volcanic Highland with relief ~ 1500 m (Figs. 2, 5, and 6). A lake occupied the central lowland and was surrounded by an extensive margin that was inundated during wetter periods. The lake may have dried up completely during the most arid periods. Some material was transported into the basin as pyroclastic flow deposits (hot ash flows and lahars) and as fallout deposits distributed by the prevailing easterly trade winds. Sediments were also transported by surface runoff, mass movement, and wind. These various systems built a fluvial plain and delta on the western and northern basin margins and a pyroclastic–alluvial fan complex on the eastern and southern basin margins (Fig. 5). A small freshwater wetland, fed mainly by groundwater, developed on the southeastern margin of the lake (Ashley, 1996; Deocampo, 1997; Deocampo and Ashley, 1999; Liutkus, 2000).

The lithologic variability within the sedimentary package of the rift-platform basin from west to east can be represented by four stratigraphic sections from key environments of deposition: fluvial plain, lake, lake margin–distal fan, and pyroclastic–alluvial fan (Fig. 7). They are composed of primary fallout deposits (tuffs), reworked tephras (tuffaceous claystones), siliciclastic and volcaniclastic sandstones and siltstones with minor conglomerate, and carbonates (dolostones and varied forms of limestones). The most common lithofacies are described and interpreted with respect to paleoenvironment below.

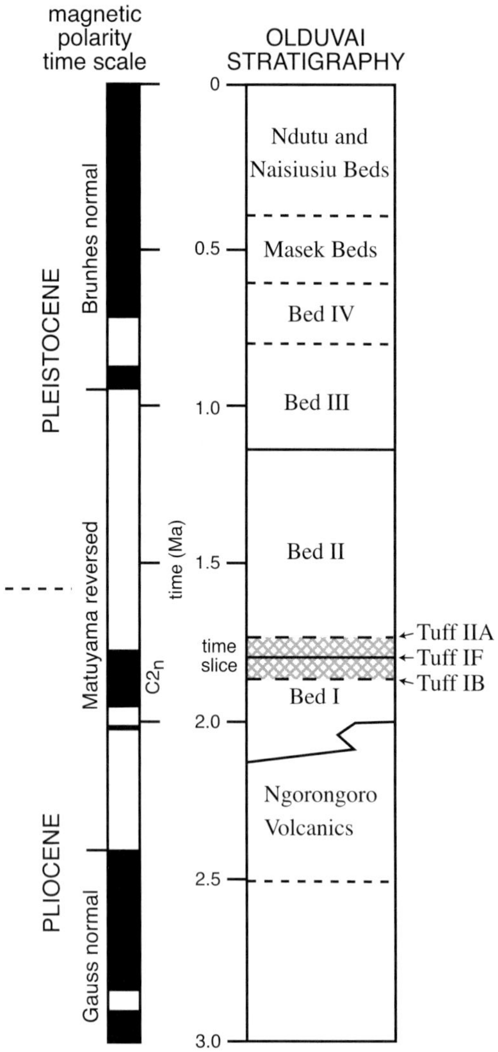


FIG. 4.—Late Pliocene–Pleistocene magnetic time scale (normal polarity in black, reversed in white) (modified from Hay, 1976). The Olduvai subchron (C2n) is a period of normal polarity (1.94–1.785 Ma) during the Matuyama reversed chron. The Olduvai volcaniclastic record began about 2.00 Ma. The sedimentary package (time slice) studied in detail is highlighted in a hachured pattern.

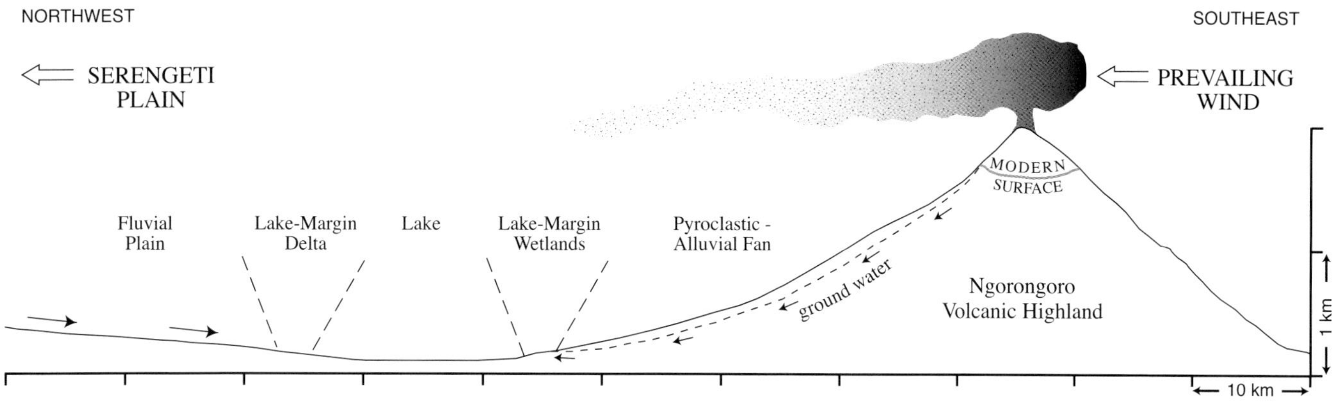


FIG. 5.—Interpreted early Pleistocene topography (oriented northwest to southeast) shows a 60-km-wide paleobasin and equally wide Ngorongoro Volcanic Highland. Ngorongoro provided an orographic barrier to the prevailing easterly trade winds, creating a rain shadow at paleo Olduvai. These winds carried pyroclastic material from erupting volcanoes into the basin. Intermittent streams drained from the Serengeti and Ngorongoro Volcanic Highland. Groundwater emerged at the base of the slope, forming a lake-margin wetland. Topographic relief is > 1500 m.

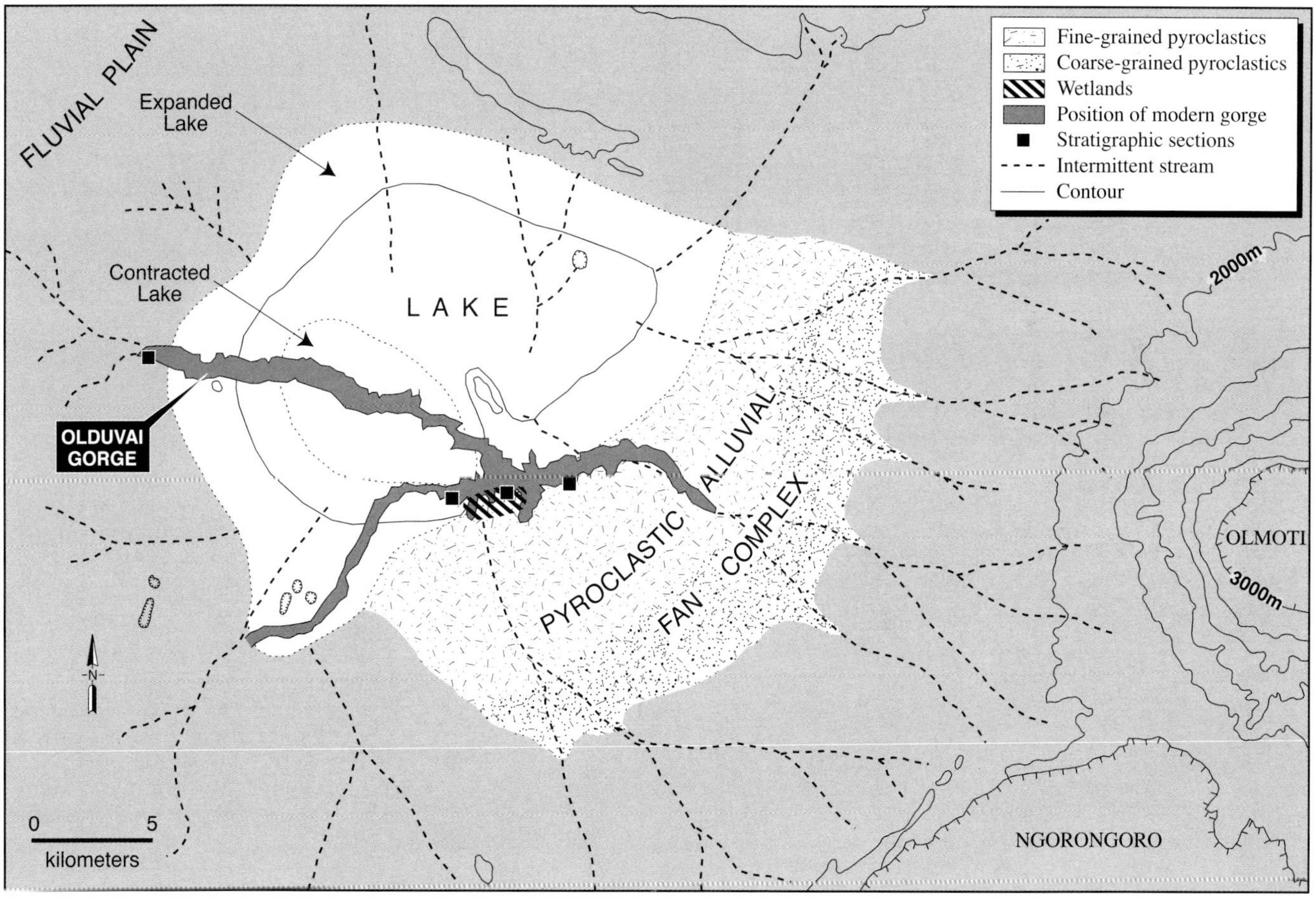


FIG. 6.—Paleogeographic reconstruction of the Olduvai basin during early Pleistocene time (modified from Peters and Blumenschine, 1995; Hay, 1976). The outline of the modern Olduvai Gorge is shown for reference. This map shows the area covered during lake expansions, the location of the wetland, and the fine-grained and coarse-grained facies of the pyroclastic–alluvial fan. Localities of stratigraphic sections in Figure 7 are indicated by squares.

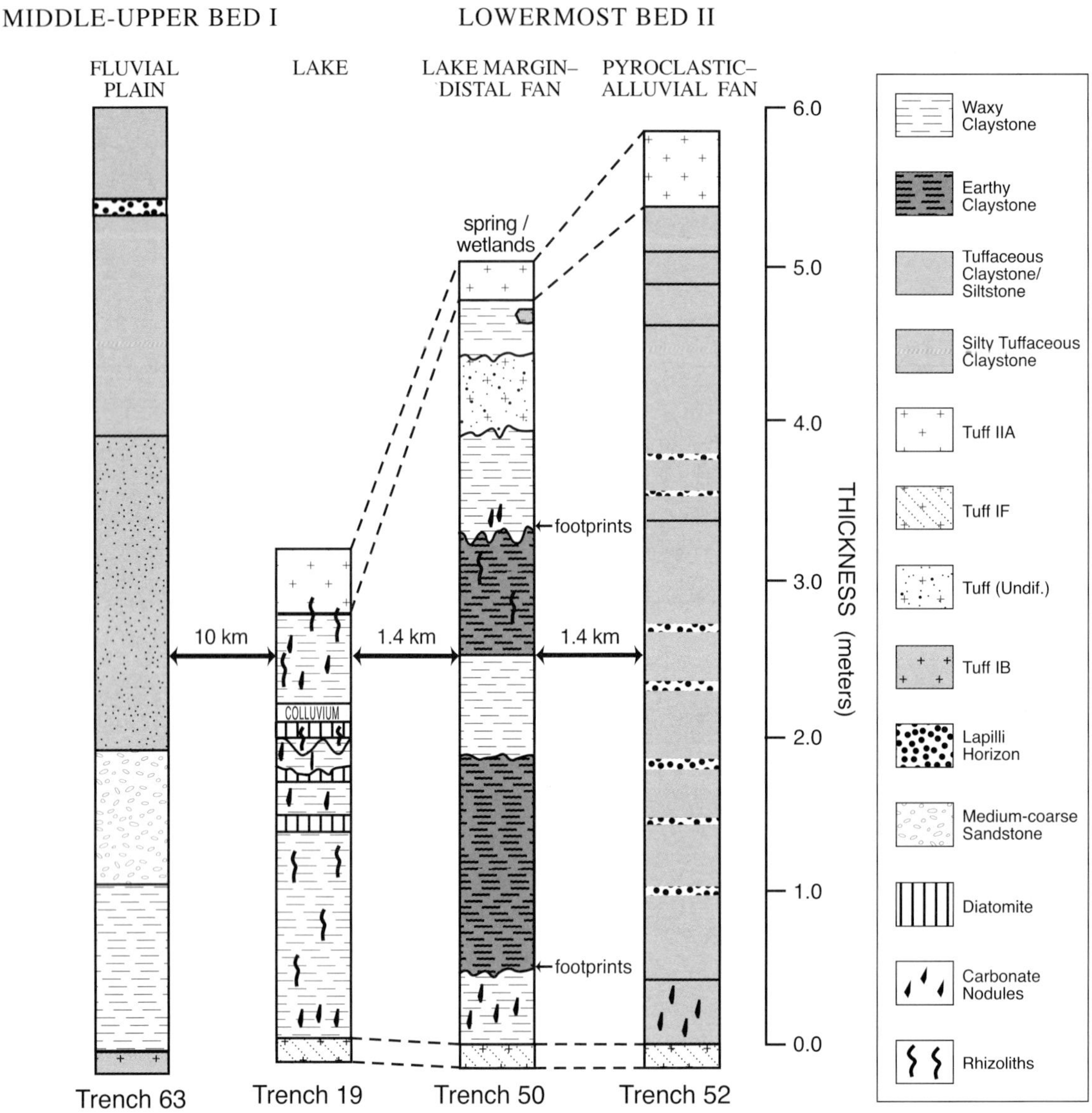


FIG. 7.—Representative lithologic sections (1 m wide and up to 6 m thick) from four distinctive depositional environments: fluvial plain, lake, lake-margin and distal fan (spring and wetlands), and proximal fan. Localities are shown on Figure 6. The fluvial plain is from upper Bed I, lake, and lake-margin, and pyroclastic–alluvial fan sections are from lowermost Bed II.

## *Pyroclastic Sediments*

### *Description.—*

Pyroclastic sediments are of two compositional types, dominantly trachytic or mafic. Trachytic tephra is characterized by pumice lapilli tuffs, and the trachytic deposits are crudely stratified and may exhibit a diamictite texture, with pumice lapilli in a clayey fine-grained matrix (Fisher and Schmincke, 1984). Most tephras have been reworked by wind or water (Fig. 8A). Individual beds have sharp bases, are 10–20 cm thick, and commonly exhibit manganese-stained root or animal traces throughout. Trachytic tuffs increase in abundance eastward and upslope in the pyroclastic–alluvial fan deposits, and most beds pinch out westward near the eastern edge of the lake-margin deposits.

The mafic tuffs are of medium sand size and contain abundant augite (Hay, 1976; 1996), and they are generally reworked to form augite-rich sandstones, many of which contain root markings or root casts. Mafic tephra also occurs as dispersed grains in claystones in the eastern lake-margin zone. Mafic ejecta are most common in deposits of the eastern lake-margin zone, and only small amounts were noted in the pyroclastic–alluvial fan deposits.

### *Interpretation.—*

Two eruptive sources are indicated by the different tephra types and their distribution. The trachytic deposits contributed to the growth of the pyroclastic–alluvial fan, and their distribution, texture, and sedimentary structures favor deposition by streams, slurries, and/or lahars (Fisher and Schmincke, 1984; Busby-

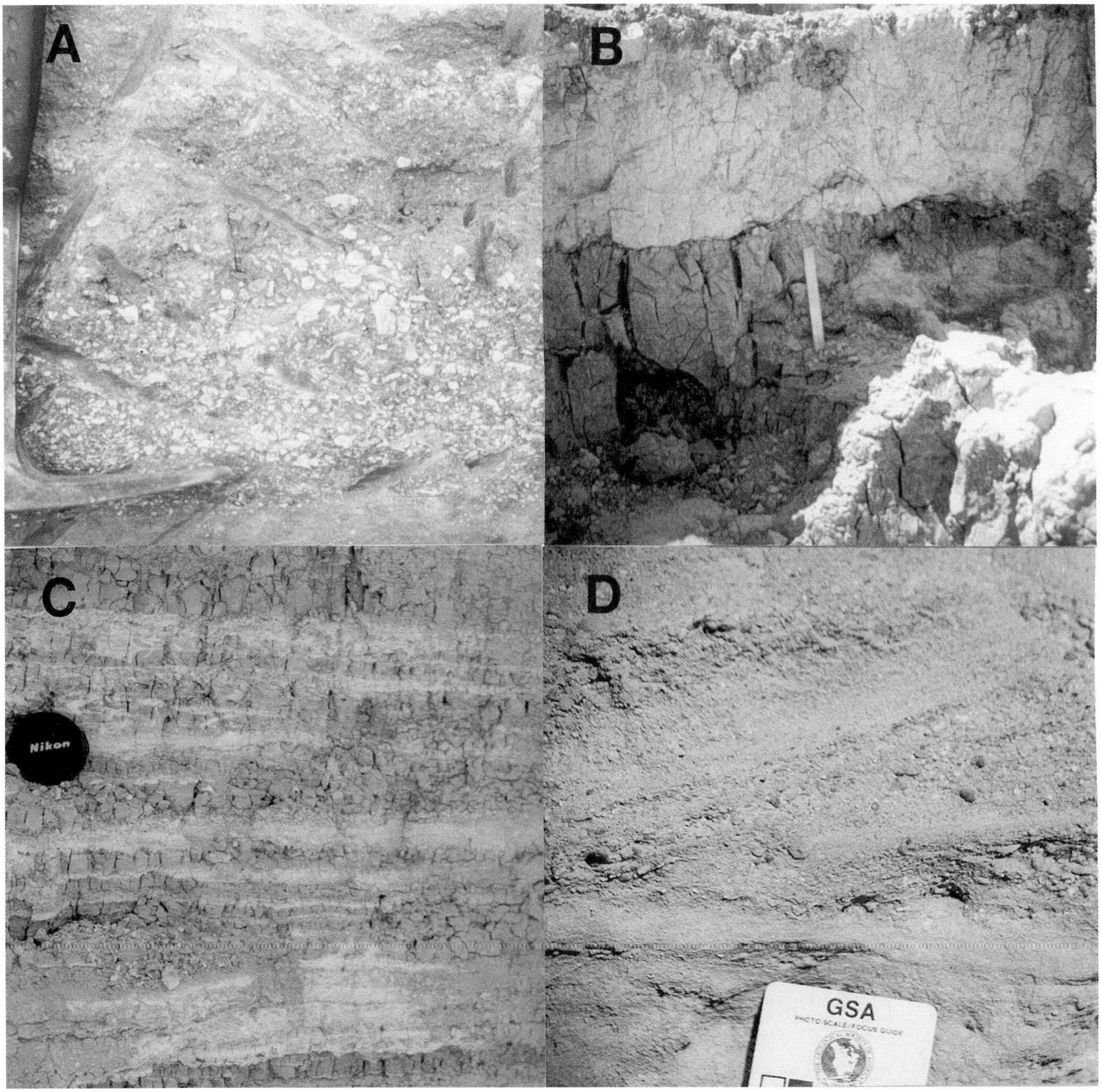


FIG. 8.—Photographs of lithofacies. **A)** Pyroclastic lapilli-rich bed that may have been deposited as a lahar. Note the poor sorting, matrix-supported fabric, and ill-defined internal organization. **B)** Lake-margin sediments; light-colored earthy claystone overlies the darker waxy claystone. The lower contact is sharp. The upper contact of the earthy claystone is convoluted and infilled with waxy claystone. The uneven contact is the vertical view of a highly irregular bedding plane interpreted to be a foot-printed surface bioturbated by large hoofed animals. A 30 cm ruler for scale. **C)** Interlaminated waxy claystones and calcite crystal sand from the center of the lake. **D)** Fluvial cross bedding in gently dipping, pebbly, quartzose–feldspathic sandstone. Flow from right to left. Card is 6 cm wide.

Spera, 1986). Seasonal runoff incised widely separated channels in the tuffs and transported poorly sorted material down slope (Riggs and Busby-Spera, 1990). Biogenic structures (root traces) suggest that vegetation (mainly grasses) stabilized interfluves on the fan. Mafic tephra contributed to the growth of the lake-margin zone, which was vegetated at times of low lake level; mafic ash was reworked by streams, sheetwash, and probably wind. Diagenetic zeolite formation (analcime) in a paleosol developed on the distal end of the pyroclastic alluvial fan points to an arid, alkaline subaerial setting (Renaut, 1993; Ashley and Driese, 2000)

### *Tuffaceous Claystone*

*Description.—*

Tuffaceous claystones, typically light brown to reddish-brown, are a mixture of ash-size pyroclasts (crystal, glass, and lithic), clay, and minor silt. They are commonly root marked and contain claystone–pellet aggregates (Hay, 1976). An overall prismatic to blocky structure is characteristic. Slickensides, deep cracks, redoximorphic mottles, iron glaebules, clay cutans, and hypocoatings are abundant throughout. Some deposits have a greater clastic component (10–20%) and are best described as silty and / or sandy tuffaceous claystones, depending on grain size.

*Interpretation.—*

The widespread evidence for pedogenesis (prismatic structure, mottles, slickensides, and clay cutans) suggests intense weathering and fluctuating groundwater table in an alkaline environment (Ashley and Driese, 2000). The depositional setting was likely a low-gradient surface at the distal end of the pyroclastic fan (Talbot et al., 1994). Clasts of ejecta were probably added to the surface periodically by air fall and subsequently weathered and reworked by wind, water, and mass movement.

### *Earthy Claystone*

*Description.—*

A distinctive earthy luster characterizes these beige to white porous deposits (Fig. 8C). They are interbedded with waxy claystones and have sharp upper and lower contacts. Bed thickness ranges from 20 to 60 cm; average is 30 cm. They are composed of Mg-smectitic clays and weathered and unweathered volcanic minerals and are highly siliceous (phytoliths, diatoms, chrysophyte algae fragments, and siliceous plant remains) (Hay, 1976; Liutkus, 2000). Carbonate rhizoliths are common. Earthy claystone beds are massive and commonly reveal vertebrate footprints on upper contacts that are infilled with overlying sediment (Fig. 8C) ( Ashley et al., 2000).

*Interpretation.—*

The earthy claystones are a component of the lake-margin record (Hay, 1976); more specifically they are a record of a wetlands (Ashley, 1996; Liutkus, 2000; Ashley, 2001). The wetlands shift laterally or are drowned during periods of high lake level. Wetland vegetation, ranging from aquatic species anchored in shallow water to terrestrial plants tapping into groundwater with a root system, provided a baffle and acted as a sediment trap. Mineral matter transported to the site by wind or surface runoff supplemented the biogenic sediment generated by plants and animals. Massive bedding is likely due to extensive bioturbation by vertebrates (Laporte and Behrensmeyer, 1980; Ashley et al., 2000).

### *Waxy Claystone*

*Description.—*

Waxy claystones from the center of the playa lake are gray-olive; those deposited at the margins of the lake are olive to brown to yellow. Claystones from the center exhibit repetitious laminae of calcite crystal sand, tuffs, rare dolomite, and magadi-type chert (Hay, 1976) (Fig. 8D). Rootmarkings and / or rhizoliths are present in about half of the lake-margin claystones, and zeolites are present in most. Calcite nodules and concretions are also common in lake-margin claystones. The dominant clay mineral is Mg-rich smectite, with or without interstratified illite. Sand and silt are chiefly volcanic minerals, and they rarely exceed 10%. Dispersed in claystones at most localities are small pebbles, generally 0.5–1.0 cm in diameter and consisting largely of quartz and pink K-feldspar (Hay, 1996).

*Interpretation.—*

The waxy claystones suggest that variation occurred in lake level, detrital source, and depositional processes. The clay minerals were formed by the interaction of detrital pedogenic clay (composed of smectite, kaolinite–halloysite, etc.) with saline, highly alkaline lake water (Singer, 1984; Hay et al., 1991). The Mg-rich clay consists of ~ 65–70% of lacustrine solutes (Mg, $SiO_2$, and alkalies), and the rest is the detrital fraction. Minor amounts of zeolite (generally analcime) also formed from the detrital clay (Hay, 1970). Most root markings and rhizoliths found in lake-margin claystones were probably formed at times of subaerial exposure. The Ngorongoro Volcanic Highland was the principal source of sand and silt. Reworked volcaniclastic sediments were transported by ephemeral streams and wind. The pebbles of quartz and feldspar are Precambrian. They were probably derived from local inselbergs and transported to the lake at least partly by sheetwash (Hay, 1996).

### *Carbonate*

*Description.—*

Carbonate occurs as isolated nodules (5 mm–10 cm in diameter), root casts (1–2 cm in diameter; 5–15 cm long), and limestone beds and masses (Klappa, 1980). The limestones beds include (1) coarsely crystalline concretions cemented together; (2) dense fine-grained nodular micritic beds (Hay, 1976), and (3) irregular fenestral limestone masses (~ 1–2 $m^2$) that have an openwork structure with waxy claystone in the interstices. It is these latter masses that are associated with footprinted surfaces (Ashley et al., 2000).

*Interpretation.—*

The rhizoliths and isolated nodules are likely phreatic carbonate precipitated from groundwater near the water table (Mount and Cohen, 1984; Wright and Tucker, 1991; Renaut, 1993). The localized limestone masses could have formed when $CO_2$ escaped as groundwater seeped both into the sediments at the lake margin, as well as directly into shallow lake waters (Hay, et al., 1986; Deocampo, 1997; Liutkus, 2000; Ashley, 2001). Extensive bioturbation by vertebrates in association with the carbonate accumulation suggests that groundwater discharged at the surface and provided drinking water (Ashley, 1996; Ashley et al., 2000).

### *Sandstone: Arkosic and Lithic Wacke*

*Description.—*

Arkosic and lithic wacke occurs only on the eastern and southern part of the basin. They consist of 50% VRFs (volcanic rock fragments), 5–25% plagioclase + anorthoclase, > 5% augite,

TABLE 1.—Composition of clasts in fluvial sediments from east and west sides of basin.

**EAST SIDE OF BASIN**

| Sample | [1]VRF % | Pumice % | Mafic glass % | Augite % | Plagioclase % Anorthoclase | [2]SRF % | Microcline % | Quartz % | Cement | [3]Olduvai Localities |
|---|---|---|---|---|---|---|---|---|---|---|
| GA-39-99 | 65 | 5–10 | rare | 5 | 25 | — | — | 1 | zeolite | THC |
| GA-108-95 | 60 | 20 | few | 5 | 15 | — | — | — | zeolite | LOC 9 |
| GA-48-99 | 50 | 0 | few | 30–35 | 3–5 | 15 | — | — | — | HWK-E |
| CML-17-99 | 40 | 0 | few | 05 | 3 | 40 | — | — | carbonate | HWK-E |
| GA-47-99 | 25 | ~ 10 | few | 50 | 10 | 10 | — | — | — | HWK-E |

**WEST SIDE OF BASIN**

| Sample | [1]VRF % | Pumice % | Mafic glass % | Augite % | Plagioclase % Anorthoclase | [2]SRF % | Microcline % | Quartz % | Cement | [3]Olduvai Localities |
|---|---|---|---|---|---|---|---|---|---|---|
| GA-07-96 | 3–5 | 40–45 | —— | 1–2 | 20–30 | 10 | 3–5 | 10 | clay cement | Loc 64 |
| GA-45-96 | 1–2 | 3–5 | —— | 2 | 3–5 | 5–10 | 10 | 65 | clay cement | 63a, 62 |
| GA-99-96 | 1 | —— | —— | trace | 5–10 | 50 | 3–5 | 40–50 | clay matrix clay cement | Loc 64 |
| GA-82-96 | 0 | 20 | —— | trace | 5 | 50 | 1–2 | 20 | clay matrix clay cement | Loc 66a |
| GA-57-96 | 0 | 30–40 | —— | 1 | 5–10 | 10–15 | 3–5 | 35–40 | clay cement | Loc 66 |

[1]VRF = Volcanic rock fragments [2]SRF = Sedimentary rock fragments [3]Localities in Hay (1976).

and no quartz; sedimentary rock fragments (SRFs) are common on the distal pyroclastic–alluvial fan (Table 1; Fig. 9A). The sediments are both texturally and mineralogically immature. They are generally poorly sorted (clay-rich matrix), and the maximum clast size fines downstream from occasional boulders and cobbles–pebbles on the upper parts of the fan to pebbly sand at the distal end. Stratification ranges from massive to vague bedding to distinct cross bedding (10–20 cm sets) containing abundant scour-and-fill structures. Zeolites cement occurs in the upper parts of the fan; carbonate cement is more common in the distal end of the fan (near the lake margin).

*Interpretation.—*

The sandstones consist of pyroclastic material that was reworked by intermittent streams flowing from the upper slopes of the Ngorongoro Volcanic Highlands during episodes of seasonal runoff (Fig. 3). Deposits with massive bedding and poor sorting are likely mud flows; those with better defined internal stratification are probably fluvial. Both flow types carried mineralogically and texturally immature volcaniclastic sediment loads but rarely reached the lake margin. Instead they percolated into the subsurface, leaving poorly sorted deposits. No well developed deltaic deposits occur on the eastern side of the lake, suggesting only ephemeral surface runoff. The high total percent of friable pumice and VRFs also suggest rapid, short-lived events typical of semiarid environments affected by monsoonal rainfall. Hay (1976) invoked sheet flow to explain the occurrence of small pebbles in an otherwise fine-grained deposits (Talbot et al., 1994). Overall, these sandstone deposits resemble fluvial sandstones described by Riggs and Busby-Spera (1990) in a Lower Jurassic multi-vent volcanic complex. We envision the depositional environment to have been similar to that presented in their model, a system of ephemeral, shallow, multichannel flows (Fig. 10).

*Sandstone: Quartzose and Feldspathic Wacke*

*Description.—*

Quartzose wacke and feldspathic wacke are exposed only on the western margin of the Olduvai basin. They are coarse to medium grained, moderately sorted sandstone ranging from 0.15 to 1.5 m thick and exhibit crude bedding and occasional cross-bedded sets (Fig. 8D). Most are rich in quartz (~ 20–65%), plagioclase–anorthoclase (5–10%), and microcline (~ 5%); they have variable pumice (3–45%) and contain minor augite (1–2%) and VRFs (< 3%) (Fig. 9B). They have clay matrix and / or are cemented with clay (Table 1). Mapping revealed a single river channel ~ 50 m wide and a few meters deep with a meander wavelength of ~ 1 km (Sikes and Ashley, in press). Mean paleocurrent direction was directly east, judging from a few cross-bed measurements and the mapping of the channel sand body.

*Interpretation.—*

The sediments were transported from Precambrian basement rocks by streams draining the Serengeti from west and northwest (Fig. 3). However, the highly variable pumice content (Table 1) indicates that pyroclastic air falls intermittently reached the western side of the basin. Dust devils capable of reworking loess-size sediments are common today in East Africa and were likely important in the early Pleistocene. The prevailing easterly winds would have transported dust to the fluvial plain, where it was likely trapped in vegetation or reworked by seasonal surface runoff (Figs. 5, 6, 10).

## THE TIME-SLICE SYNTHESIS

The landscape paleoanthropology approach to data collection enables us to reconstruct a pattern of lithologic facies during a relatively narrow window of time across the rift-platform basin (upper Bed I to lowermost Bed II)(Figs 4 and 10). This highly variable sedimentary time slice is well bracketed between Tuff IB at the base and Tuff IIA (and the laterally equivalent Lemuta Member) at the top (Fig. 10). The average thickness of the sediment section decreases with distance from the volcanic source (Ngorongoro) to the basin center and then thickens again in the western fluvial plain (Fig. 7). Lithostratigraphy varies both laterally and vertically, reflecting changes in depositional environment in both space and in time.

Coarse-grained, thick-bedded and crudely stratified pyroclastic sediments are proximal to the stratovolcano and grade laterally into thin-bedded tuffaceous claystones at the distal end of the fan. Waxy claystones are interbedded in the east with earthy claystone (freshwater wetlands) and extend westward across the lake-margin zone into the lake. The fluvial plain in the west, mainly on the basis of data from upper Bed I (Fig. 7), consists of tuffaceous claystones with varying amounts of detritus, and a thin (1–2 m) ribbon of quartzose–feldspathic sandstone and conglomerate (> 200 m wide) was deposited in a river channel that fed a delta at the shore of the lake (Sikes and Ashley, in press).

The lateral variability in lithology is highlighted by contrasting the sediments from representative trenches: Trench 53 (pyroclastic–alluvial fan), Trench 50 (spring–wetlands), Trench 19 (saline lake), and Trench 63 (fluvial plain)(Fig. 7). The pyroclastic–alluvial fan is composed of a series of stacked tephras and lapilli-rich tuffs. Much if not most of the volcanic sediment is reworked tuff. In addition, most beds have been altered somewhat by pedogenesis, with the best developed paleosols in the finer-grained, lower slope areas. Localized drainage channels are incised into the tuffs and contain mineralogically and texturally immature sediments (Fig. 9A, B; Table 1). The composition of the fluvial deposits is distinct and reflects different sources, volcanics in the east and south and metamorphic quartzose–feldspathic basement in the west and north (Fig. 2).

The spring (groundwater seeps) and wetland area is characterized by interbedded waxy claystones and earthy claystones (Fig. 8C). The contacts between these beds are sharp except where there

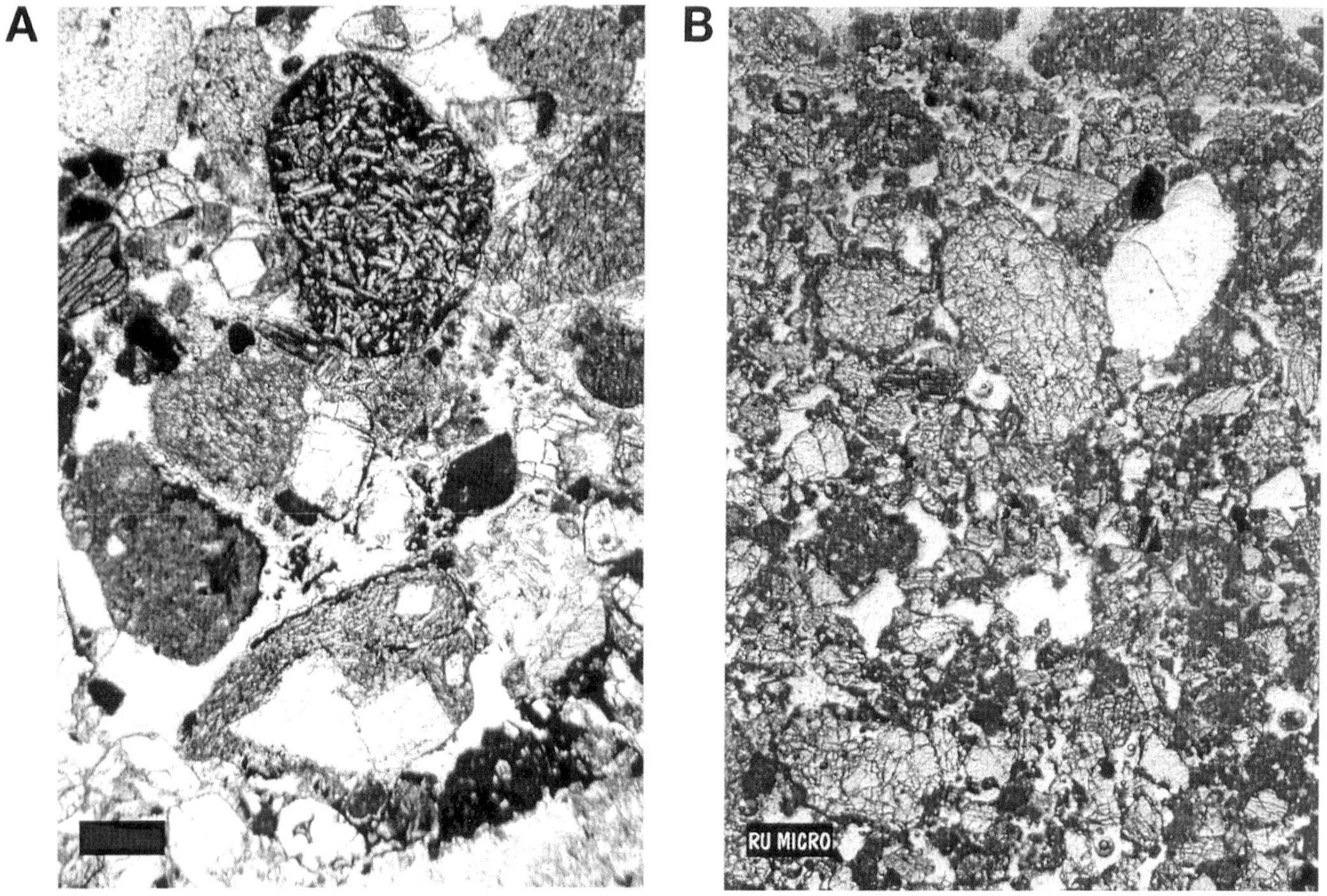


FIG. 9.—Photomicrographs of petrographic thin sections; **A)** Arkosic wacke (GA-108-95, Loc 9) from lowermost Bed II. 60% VRF, 20% pumice, 5% augite, and 15% plagioclase documents the volcanic provenance of the fluvial sediments on the east side of the basin (Table 1). Field of view, 7 mm. **B)** quartzose–feldspathic sandstone from middle Bed I (GA-99-96, Loc 64). High proportion of sedimentary clasts (~ 50%), quartz (40%), and microcline (5%) suggests that the main sediment source was metamorphic basement rocks of the north and west side of the basin. LOC localities from Hay (1976). Field of view, 7 mm.

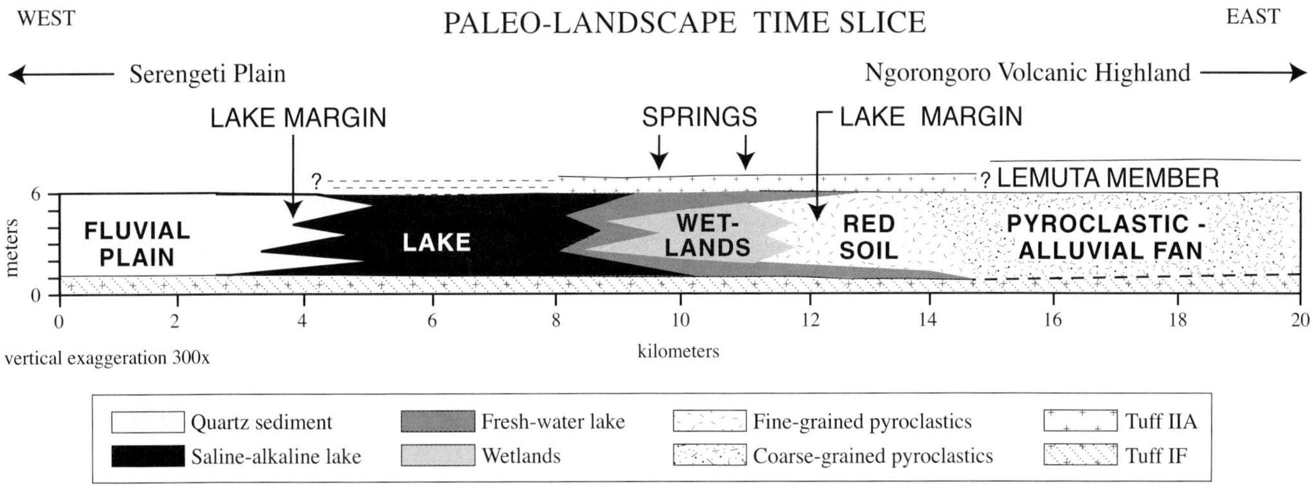


FIG. 10.—A high-resolution paleoenvironmental reconstruction of the paleo-landscape during a narrow window of time (~ 100 kyr) beginning at ~ 1.85 Ma (Walter et al., 1991). The sedimentary record with a scale of 6 m (vertical) by 20 km (long) reveals lateral facies changes that reflect the paleogeography and depositional setting (Fig. 6) and vertical facies changes that suggest a shift in depositional setting with time. A volcaniclastic fan built into the basin from the east as soils developed near the lake. The lake periodically expanded and contracted over the toe of the fan (lake-margin area). Lake waters varied from saline–alkaline water in the center of the basin to slightly fresher water at the southeastern lake shore near the site of a groundwater-fed wetland (> 3 km$^2$). The western 2.2-kilometer-wide record of the basin is from upper Bed I; the other 17.8 kilometers is from lowermost Bed II.

has been intense footprinting by large vertebrates. Wetland deposits contain large amounts of opaline silica from phytoliths and diatom frustules, as well as carbonate rhizoliths (root casts), nodules, and diatomite beds (Liutkus, 2000; Ashley, 2001). Lacustrine deposits interfinger with fluvial sediments at the basin margin. Periodically the lake expanded, covering large areas of the playa lake basin (Fig. 6) and occasionally even drowning the wetlands.

## DISCUSSION

### *Tectonics and Volcanism*

Extension tectonics in this rift-platform setting had several effects. (1) Prior to the beginning of sedimentation in the basin, uplift and erosion of nearly all rocks younger than the Precam-

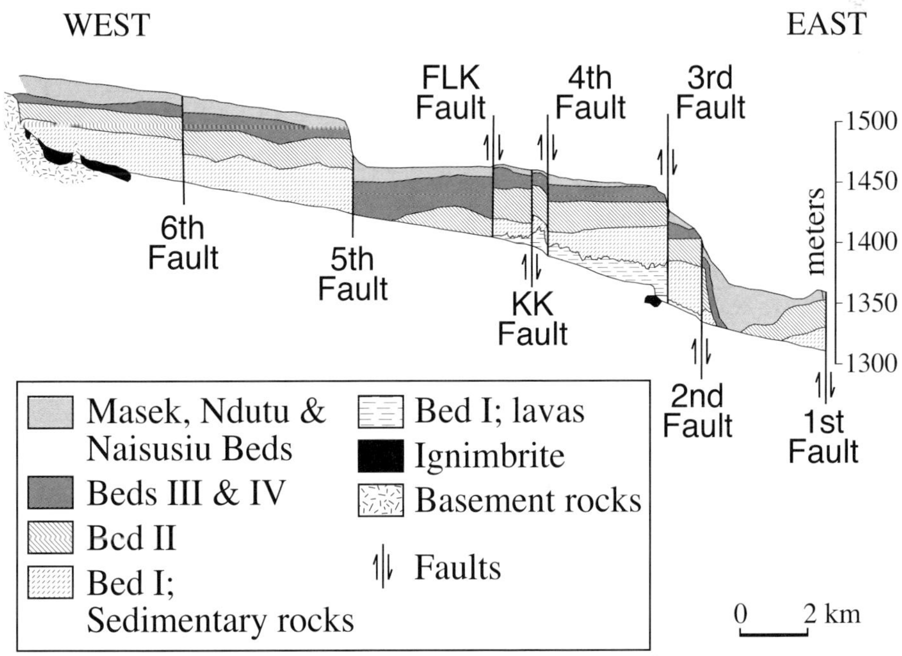


FIG. 11.—West-to-east cross section of the middle part of the Olduvai Basin (20 km wide) shows rift-related (mainly normal) faulting that includes Precambrian crystalline basement rocks.

brian basement along the western margin of the rift occurred, leaving an undulating surface with moderate relief. The land sloped eastward toward the lake and, judging by the present drainage divide (Fig. 2), the drainage divide might have been located up to 75 km to the west of the main rift valley. The outpouring of lava and pyroclastic eruptions from the several eruptive centers in the Ngorongoro area created an imposing landform that formed the eastern and southern margin of the basin (Fig. 5). The highlands were a source of water (surface water and groundwater) as well as volcanic sediments. (2) Continued extension created rift-parallel normal faults. The faulting segmented the sedimentary record into blocks which dropped down toward the main rift valley to the east (Fig. 11). This syndepositional faulting affected base level and stream gradients, initiating river incision that locally eroded and reworked sediments and created hiatuses in the basin record. (3) The center of the enclosed basin shifted eastward through time as a consequence of the continued subsidence in the main rift graben.

Sedimentation was intermittently dominated by explosive eruptions of ash and lapilli that periodically blanketed the landscape with tephra. Modern studies (Mt. St. Helens, U.S.A.; Mt. Pinatubo, Philippines; and Raupehu, New Zealand) have shown that volcanic eruptions can occur frequently over a one or two years, decades, or a few hundred years, after a period of quiescence. Flora and fauna generally recover within a few years to a few tens of years after a volcanic eruption (Harris and Van Couvering, 1995). There may have been four to six eruptive periods during lowermost Bed II time, with tephra redistributed by wind, intermittent surface runoff (sheet flows), and channelized flow. The sedimentary record closest to a volcano is typically obscure (Fig. 12). Proximal deposits consist of undifferentiated stacked lapilli horizons, and individual eruptions cannot be distinguished. Farther from the volcanic center, in the medial and distal facies of the pyroclastic fan and lake floodplain, there is a suggestion of pulses of active volcanism alternating with quiescent periods of weathering and soil formation (Ashley and Dreise, 2000).

## *Paleoclimate and Paleohydrology*

The hydrologic budget, especially in semiarid environments, is a key factor controlling the nature of the biological, physical, and geochemical environments. Near the equator, seasonal temperature and evaporation rates typically do not vary substantially, but precipitation does. Episodic changes in climate (i.e., especially variations in precipitation) can be inferred from expansion and contraction of the basin lake (Hay, 1976, 1990). Extensive, laterally continuous deposits of waxy claystone indicate higher lake levels, which in a closed basin coincide with prolonged wetter periods (Street-Perrott and Roberts, 1983) or stronger monsoonal rains (Pokras and Mix, 1985, 1987; Gasse et al., 1989; Sirocko, 1996). During such episodes, the waxy claystones (representing the lake) extend farther and with increasing frequency onto the distal fan and wetland environments. Thus, in the lake-margin environment, waxy claystones (lacustrine) are typically intercalated with tuffaceous claystones (distal-fan or fluvial-plain sediments) and earthy claystones (freshwater wetlands) (Fig. 10).

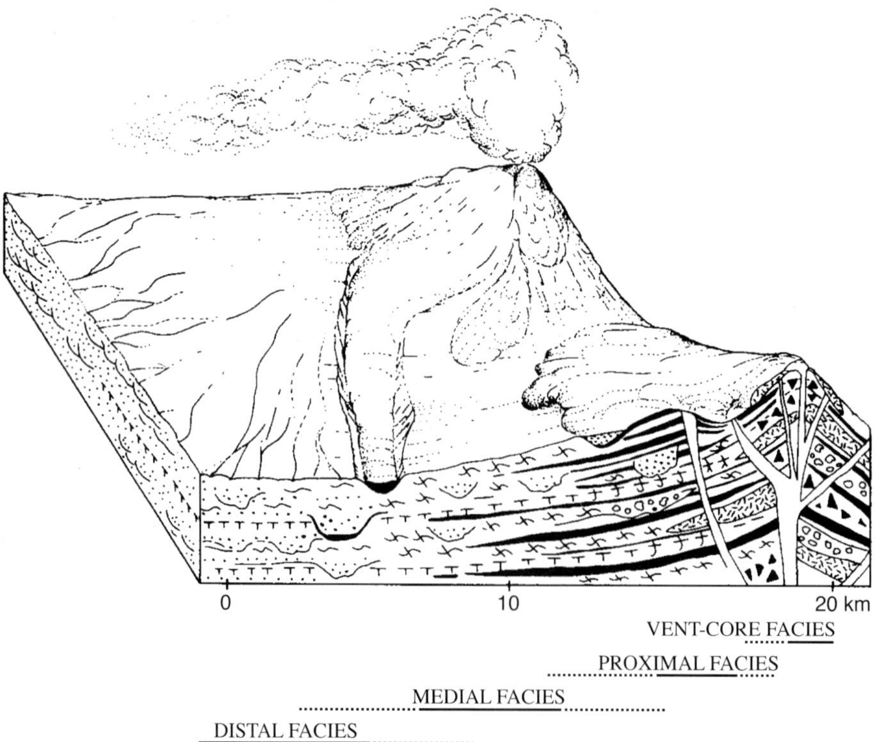


FIG. 12.—A depositional model of volcaniclastic sediments erupted from a stratovolcano and redistributed by wind, gravity-driven mass transport, and fluvial processes (from Riggs and Busby-Spera, 1990). Vent-core facies has coarse ejecta, lava flows, and mass-flow deposits. Proximal facies have lava flows, welded ignimbrites and debris flows. Medial facies have nonwelded ignimbrites, debris-flow deposits, and coarse epiclastic units. Distal facies have epiclastic deposits, minor lava flows, pyroclastic deposits, reworked pyroclastics, and some fluvial deposits.

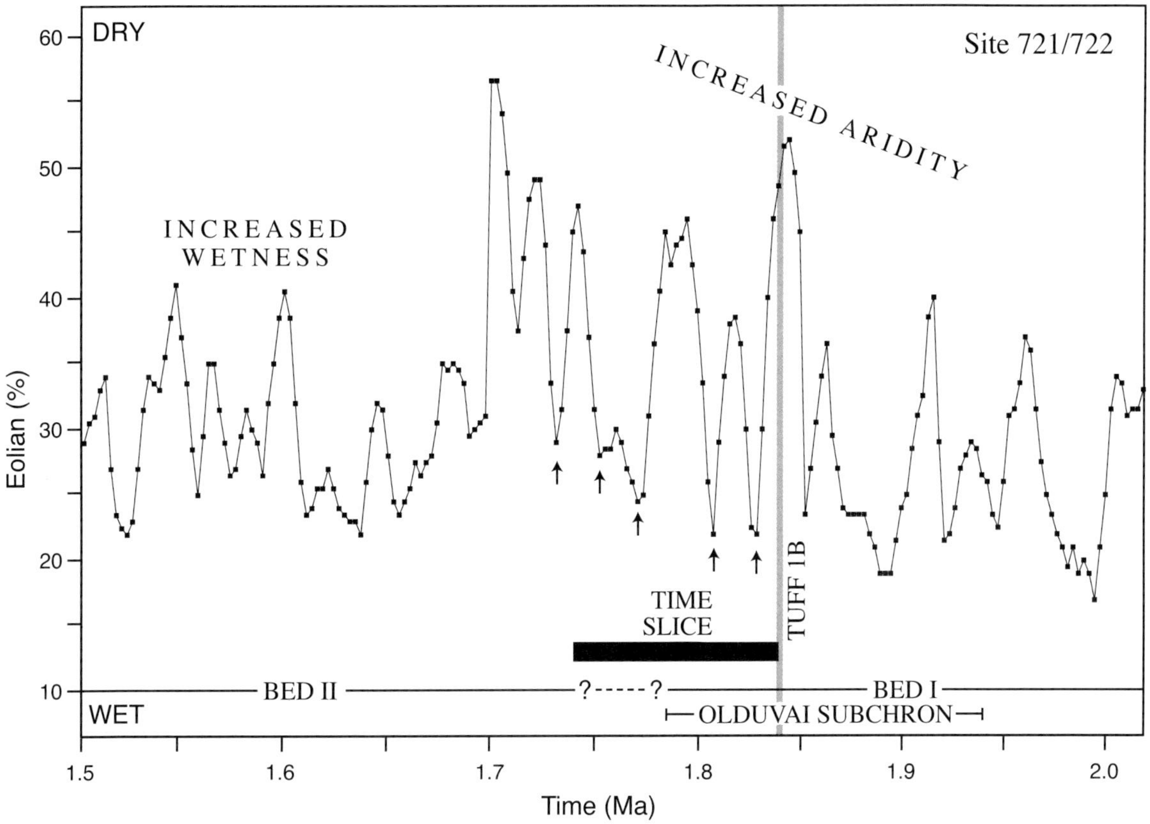


Figure 13.—Records of eolian dust from Plio-Pleistocene marine cores off Africa collected by the Ocean Drilling Program (ODP) 721-722 suggest fluctuations in climate that have been linked to Milankovitch astronomical cycles (obliquity, 41 kyr and precession, 10–21 kyr) by deMenocal (1995). Twenty-thousand to forty-thousand year fluctuations in moisture were interpreted from the dust records between 1.5 Ma and 2.0 Ma, which brackets the time period under study here. Encompassing these shorter fluctuations are longer-term trends (100–400 kyr in length) of aridity and increased wetness. Using dates from Walter et al. (1991), the duration of the Olduvai time slice is depicted. Five lake expansions are recorded in the stratigraphic record (Ashley and Driese, 2000; Sikes and Ashley, in press), and five wet periods are indicated by the marine dust record within the same 100,000 yr period. We speculate that the sedimentary lake record in the shallow platform Basin at Olduvai has faithfully recorded the regional wet and dry cycles.

Using dust records from Plio-Pleistocene marine cores off Africa, deMenocal and Bloemendal (1995) produced a curve that is interpreted as a proxy for climate (wet vs. dry) (Fig. 13). deMenocal (1995) suggested that the 1.85–1.65 million year period (which includes our time slice) was an unusually dry period within a long-term trend of increasing aridity during the Plio-Pleistocene. However, definite wet periods occur within this "drier" period (Fig. 13). These wet phases may be linked to the Milankovitch-driven global climate fluctuations of ~ 21,000 yrs precession cycles or obliquity (41,000 yr cycles).

Five wet periods (lake expansions) were noted in the time-slice stratigraphy. At least one discrete lake expansion (possibly two) were recognized in upper Bed I deposits (1.84–1.79 Ma) (Sikes and Ashley, in press) and two major and possibly one minor lake expansion occurred during lowermost Bed II (1.79–1.74 Ma) (Ashley and Driese, 2000). The same number of wet periods (5) occur on the dust curve during the 100,000 year period representing the time slice (Fig. 13), suggesting a precession-cycle effect (~ 21,000 yrs). Although intriguing, this correlation is speculative, and only high-resolution dating of Beds I and II can confirm whether or not the lake fluctuations match wet periods of the dust periods and are related to orbital rhythms.

Most of the fluvial sediments are texturally immature wackes (Table 1), suggesting transportation by short-lived events and/or nearby sources. During the monsoons, streams drained the uplands and moved sediments into the basin. This is consistent with a playa-basin setting characterized by seasonal rains (Hooke, 1967).

Groundwater, probably from meteoric water sourced in the volcanic highlands, moved south and east year round (Fig. 5). The groundwater system served as a reservoir of fresh water for both the lake-margin wetlands and the lake, as well as springs and seeps (Rosen, 1994; Ashley, 1996). Groundwater seeping into the lake freshened its southeast corner during wet periods (i.e., high-lake-level periods). During times of higher lake levels, groundwater probably flowed onto the surface in small (~ 1–2 meter wide) springs. During dry periods of the Plio-Pleistocene, the groundwater discharge zone probably supported extensive marshes and wet meadows (1–3 square kilometers) that may have contained some trees (Bonnefille and Riollet, 1980; Sikes, 1996; Liutkus, 2000) and could have been an important source of potable water for animals and hominids in what otherwise would have been a very harsh environment.

## CONCLUSIONS

The sedimentary fill of Olduvai, a Plio-Pleistocene rift-platform basin, reflects the interplay between five independent variables affecting the sedimentary record: (1) tectonics associated with rift extension, (2) sediment flux (pyroclastic air fall, fluvial runoff, and eolian), (3) sediment composition (Precambrian source rocks vs. volcanic source rocks), (4) hydrology and geochemistry (lake water, surface water, and groundwater), and (5) soil formation. The synthesis of high-resolution stratigraphic studies of the paleoenvironments of upper Bed I and lowermost Bed II (~ 1.85–1.75 Ma) determined that the sedimentary facies vary temporally (paleoclimate fluctuations) and spatially (changing depositional environment).

The paleoenvironmental reconstruction reveals a slowly changing landscape affected by changes in hydrology (precipitation vs. evaporation), by variations in explosive volcanism on a long time scale (thousands of years), and by short-term changes over decades to hundreds of years. Lake cycles, ephemeral fluvial systems, soil formation on the interfluves and lake margin, as well as a large groundwater-fed wetland (> 3 km$^2$) are all part of the Olduvai rift-margin basin record and provide insights into Plio-Pleistocene paleoclimate history of East Africa.

## ACKNOWLEDGMENTS

The research was supported by National Science Foundation grants (Archaeology BNS 9000099, SBR 9601065, SBR 9602478 to R.J. Blumenschine and F.T. Masao and GEO 9903258 to G.M. Ashley) and grants to Blumenschine and Masao from the L.S.B. Leakey Foundation, the Wenner-Gren Foundation, the National Geographic Society, the Sokkia Corporation, and Rutgers University. We thank the hard-working crew of the Olduvai Landscape Paleoanthropology Project and the hospitality of the Olduvai Museum staff, O. Kileo, J. Paresso, and Godfrey Olle Moita. We have benefited greatly over the years from discussions with Kay Behrensmeyer, Rob Blumenschine, Thure Cerling, Al Deino, Dan Deocampo, Steve Driese, Craig Feibel, Cindy Luitkus, Lindsay McHenry, Fidelis Masao, Godwin Mollel, Charles Peters, Rick Potts, Robin Renaut, Nancy Sikes, and Carl Swisher. We are grateful to Paul Buchheim, B. Charlotte Schreiber, and Frederick Schwab for very helpful reviews.

## REFERENCES

ASHLEY, G.M., 1996, Springs, pools and adjacent wetlands, a newly recognized habitat, lowermost Bed II, Olduvai Gorge (abstract): Geological Society of America, Abstracts with Programs, v. 28, p. 28.

ASHLEY, G.M., 2000, Geologists probe hominid environments: GSA Today, Geological Society of America, Boulder, Colorado, v. 10, no. 2, p. 24–29.

ASHLEY, G.M., 2001, Archaeological sediments in springs and wetlands, *in* Stein, J.K., and Farrand, W.R., eds., Sediments in Archaeological Contexts: Salt Lake City, University of Utah Press, p. 183–210.

ASHLEY, G.M., AND DRIESE, S.G., 2000, Paleopedology and paleohydrology of a volcanic paleosol interval: implications for early Pleistocene stratigraphy and paleoclimate record, Olduvai Gorge: Journal of Sedimentary Research, v. 70, p. 1065–1080.

ASHLEY, G.M., DEOCAMPO, D.M., AND LIUTKUS, C.M., 2000, Track records of large mammals in lake-margin wetlands, LMBedII, Olduvai Gorge, Tanzania (abstract): International Association of Sedimentologists, Regional Meeting, Dublin, Ireland.

BAKER, B.H., AND WOHLENBERG, J., 1971, Structure and evolution of the Kenya rift valley: Nature, v. 229, p. 538–542.

BAKER, B.H., MOHAIR, P.A., AND WILLIAMS, L.A.J., 1972, Geology of Eastern Rift System of Africa: Geological Society of America, Special Paper 136, 67 p.

BLUMENSCHINE, R.J., AND MASAO, F.T., 1991, Living sites at Olduvai Gorge, Tanzania? Preliminary landscape archaeology results in the basal Bed II lake-margin zone: Journal of Human Evolution, v. 21, p. 451–462.

BLUMENSCHINE, R.J., AND PETERS, C.R., 1998, Archaeological predictions for hominid land use in the paleo–Olduvai Basin, Tanzania, during lowermost Bed II times: Journal of Human Evolution, v. 34, p. 565–607.

BLUMENSCHINE, R.J., MASAO, F.T., ASHLEY, G.M., CLARKE, R.J., EBERT, J.I., PETERS, C.R., SIKES, N.E., AND VAN DER MERWE, N.J., 1997, New paleoanthropologic discoveries and tests of hominid land use models in the western lacustrine plain of the lowermost Bed II Olduvai basin (abstract): Journal of Human Evolution, v. 32, p. A5.

BLUMENSCHINE, R.J., PETERS, C.R., MASAO, F.T., ASHLEY, G.M., AND EBERT, J.I., 1999, Preliminary tests of paleoanthropological predictions for hominid land use in the east-central portion of the lowermost Bed II Olduvai Basin, Tanzania (abstract): Journal Human Evolution, v. 36, p. A2–A3.

BLYSTAD, P., BREKKE, H., FAERSETH, R.B., LARSEN, B.T., SKOGSEID, J., AND TORUDBAKKEN, B., 1995, Structural elements of the Norwegian continental shelf, Part II—The Norwegian Sea Region: Norwegian Petroleum Directorate, Bulletin 8, 45 p.

BONNEFILLE, R., AND RIOLLET, G., 1980, Palynologie, végétation et climates de Bed I et Bed II à Olduvai, Tanzania: 8th Pan-African Congress of Prehistory and Quaternary Studies, Nairobi, 1977, Proceedings, p. 123–127.

BUSBY-SPERA, C.J., 1986, Depositional features of rhyolitic and andesitic volcaniclastic rocks of the Mineral King submarine caldera complex, Sierra Nevada, California: Journal of Volcanology and Geothermal Research, v. 27, p. 43–76.

BUSBY, C.J., AND INGERSOLL, R.V., EDS., 1995, Tectonics of Sedimentary Basins: Cambridge, Massachusetts, Blackwell Science, 579 p.

CERLING, T.E., AND HAY, R.L., 1986, An isotopic study of paleosol carbonates from Olduvai Gorge: Quaternary Research, v. 25, p. 63–78.

CHAPMAN, G.R., LIPPARD, S.J., AND MARTYN, J.E., 1978, The stratigraphy and structure of the Kamasia range, Kenya rift valley: Geological Society of London, Journal, v. 135, p. 265–281.

COWARD, M.P., DEWEY, J.F., AND HANCOCK, P.L., EDS., 1987, Continental Extensional Tectonics: Geological Society of London, Special Publication 28.

DAWSON, J.B., 1962, The geology of Oldoinyo Lengai: Bulletin Volcanologique, v. 24, p. 349–387.

DAWSON, J.B., 1964, Carbonatite volcanic ashes in northern Tanganyika: Bulletin Volcanologique, v. 27, p. 1–11.

DEMENOCAL, P.B., 1995, Plio-Pleistocene African climate: Science, v. 270, p.53–59.

DEMENOCAL, P.B., AND BLOEMENDAL, J., 1995, Plio-Pleistocene climatic variability in subtropical Africa and the paleoenvironment of hominid evolution: a combined data–model approach: *in* Vrba, E.S., Denton, G.H., Partridge, T.C., and Burckle, L.H., eds., Paleoclimate and Evolution with Emphasis on Human Origins: New Haven, Connecticut, Yale University Press, p. 262–288.

DEOCAMPO, D.L., 1997, Modern sedimentation and geochemistry of freshwater springs: Ngorongoro Crater, Tanzania: Unpublished M.S. thesis, Rutgers University, New Brunswick, New Jersey, 140 p.

DEOCAMPO, D.M., AND ASHLEY, G.M. 1999, Siliceous island in a carbonate sea: Modern and Pleistocene records of spring-fed wetlands in Ngorongoro Crater and Oldupai Gorge, Tanzania: Journal of Sedimentary Research, v. 69, p. 974–979.

FEIBEL, C.S., 1999, Basin evolution, sedimentary dynamics, and hominid habitats in East Africa: an ecosystem approach: *in* Bromage, T., and Schrenk, F., eds., African Biogeography, Climate Change, and Human Evolution: Oxford, U.K., Oxford University Press, p. 276–281.

FISHER, R.V., AND SCHMINCKE, H.-U., 1984, Pyroclastic Rocks: Berlin, Springer-Verlag, 472 p.

FROSTICK, L.E., 1997, The East African rift basins, *in* Selley, R.C., ed., African Basins: Amsterdam, Elsevier Science, Sedimentary Basins of the World: p. 187–209.

GARFUNKEL, Z., AND BARTOV, Y., 1977, The tectonics of the Suez rift: Geological Survey of Israel, Bulletin 71, p. 1–44.

GASSE, F., LEDEE, V., MASSAULT, M., AND FONTES, F.C., 1989, Water level fluctuations of Lake Tanganyika in phase with oceanic changes during the last glaciation and deglaciation: Nature, v. 342, p. 57–59.

GROMME, C.S., AND HAY, R.L., 1971, Geomagnetic polarity epochs: age and duration of the Olduvai normal polarity event: Earth and Planetary Science Letters, v. 10, p. 179–185.

HARRIS, J., AND VAN COUVERING, J., 1995, Mock aridity and the paleoecology of volcanically influenced ecosystems: Geology, v. 23, p. 253–596.

HAY, R.L., 1970, Silicate reactions in three lithofacies of a semiarid basin, *in* Morgan, B.A., ed., Fiftieth Anniversary Symposium: Mineralogy and Petrology of the Upper Mantle, Sulfides, and Mineralogy, and Geochemistry of Non-Marine Evaporites: Mineralogical Society of America, Special Paper 3, p. 237–255.

HAY, R.L., 1976, Geology of the Olduvai Gorge; A Study of Sedimentation in a Semiarid Basin: Berkeley, California, University of California Press, 203 p.

HAY, R.L., 1989, Holocene carbonatite–nephelinite tephra deposits of Oldoinyo Lengai, Tanzania: Journal of Volcanology and Geothermal Research, v. 37, p. 77–91.

HAY, R.L., 1990, Olduvai Gorge: A case history in the interpretation of hominid paleoenvironments in East Africa, *in* Laporte, L.F., ed., Establishment of a Geologic Framework for Paleoanthropology: Geological Society of America, Special Paper 242, p. 23–37.

HAY, R.L., 1996, Stratigraphy and lake-margin paleoenvironments of lowermost Bed II in Olduvai Gorge, *in* Magori, C.C. Saanane, C.B., and Schrenck, T., eds., Four Million Years of Evolution in Africa: Papers in Honour of Mary Douglas Leakey's Outstanding Contributions to Paleoanthropology: Kaupia-Darmstadter Beiträge zur Naturgeschichte, v. 6, p. 223–230.

HAY, R.L., SEXTON, R.E., TEAGUE, T.T., AND KYSER, T.K., 1986, Spring-related carbonate rocks, Mg clays, and associated minerals in Pliocene deposits of the Amargosa Desert, Nevada and California: Geological Society of America Bulletin, v. 97, p. 1488–1503.

HAY, R.L., GULDMAN, S.G., MATTHEWS, J.C., LANDER, R.H., DUFFIN, M.E., AND KYSER, T.K., 1991, Clay mineral diagenesis in core KM-3 of Searles Lake, California: Clays and Clay Minerals, v. 39, p. 84–96.

HOOKE, R.LEB, 1967, Processes on arid-region alluvial fans: Journal of Geology, v. 75, p. 438–460.

KAPPELMAN, J., 1986, Plio-Pleistocene marine–continental correlation using habitat indicators from Olduvai Gorge, Tanzania: Quaternary Research, v. 25, p. 141–149.

KATZ, B.J., ED., 1990, Lacustrine Basin Exploration—Case Studies and Modern Analogs: American Association of Petroleum Geologists, Memoir 50, 340 p.

KLAPPA, C.F., 1980, Rhizoliths in terrestrial carbonates: classification, recognition, genesis, and significance: Sedimentology, v. 27, p. 613–629.

LAPORTE, L.F., AND BEHRENSMEYER, A.K., 1980, Tracks and substrate reworking by terrestrial vertebrates in Quaternary sediments of Kenya: Journal of Sedimentary Petrology, v. 50, p. 1337–1346.

LEAKEY, M.D., 1971, Olduvai Gorge: Excavations in Beds I and II; 1960–1963: Cambridge, U.K., Cambridge University Press. 309 p.

LIUTKUS, C.L., 2000, Sedimentology, stratigraphy and stable isotopes of a wetlands in a volcaniclastic setting, Olduvai Gorge, Tanzania: Unpublished M.S. thesis, Rutgers University, New Brunswick, New Jersey, 159 p.

MOUNT, J.F., AND COHEN, A.S., 1984, Petrology and geochemistry of rhizoliths from Plio-Pleistocene fluvial and marginal lacustrine deposits, East Lake Turkana, Kenya: Journal of Sedimentary Petrology, v. 54, p. 263–275.

MORLEY, C.K., 1999, Geoscience of Rift Systems—Evolution of East Africa: American Association of Petroleum Geologists, Studies in Geology 44, 250 p.

PETERS, C.R., AND BLUMENSCHINE, R.J., 1995, Landscape perspectives on possible land use patterns for early hominids in the Olduvai Basin: Journal of Human Evolution, v. 29, p. 321–362.

PICKERING, R., 1960, A preliminary note on the Quaternary geology of Tanganyika, *in* Joint Meeting, Leopoldville, 1958, of east-central, west-central, and southern committees for geology: Commission for Technical Cooperation in Africa South of the Sahara, Publication 44, p. 77–89.

POKRAS, E.M., AND MIX, A.C., 1985, Eolian evidence for spatial variability of Late Quaternary climates in tropical Africa: Quaternary Research, v. 24, p.137–149.

POKRAS, E.M., AND MIX, A.C., 1987, Earth's precession cycle and Quaternary climatic change in tropical Africa: Nature, v. 326, p. 486–487.

RECK, H., 1914, Erste vorläufige Mitteilung über den Fund eines fossilen Menschenskelets aus Zentralafrika: Gesellschaft Naturforschender Freunde zu Berlin, Sitzungsberichte, no. 3, p. 83–95.

RECK, H., 1951, A preliminary survey of the tectonics and stratigraphy of Olduvai, *in* Leakey, L.S.B., Olduvai Gorge: Cambridge, U.K., Cambridge University Press, p. 5–19.

RENAUT, R.W., 1993, Zeolite diagenesis of late Quaternary fluviolacustrine sediments and associated calcrete formation in the Lake Bogoria Basin, Kenya Rift Valley: Sedimentology, v. 40, p. 271–301.

RIGGS, N.R., AND BUSBY-SPERA, C.J., 1990, Evolution of a multi-vent volcanic complex within a subsiding arc graben depression: Mount Wrightson Formation, Arizona: Geological Society of America, Bulletin, v. 102, p. 1114–1135.

ROBSON, D.A., 1971, The structure of the Gulf of Suez (Clysmic) rift with special reference to the eastern side: Geological Society of London, Journal, v. 127, p. 247–276.

ROSEN, M.R., 1994, The importance of groundwater in playas: a review of playa classification and sedimentology and hydrology of playas *in* Rosen, M.R., ed., Paleoclimate and Basin Evolution of Playa Systems: Geological Society of America, Special Paper 289, p. 1–18.

ROSENDAHL, B.R., 1987, Architecture of continental rifts with special references to East Africa: Annual Review of Earth and Planetary Sciences, v. 15, p. 445–503.

SELLEY, R.C., ED., 1997, African Basins, *in* Amsterdam, Elsevier Science, Sedimentary Basins of the World, v. 3, 394 p.

SIKES, N.E., 1996, Hominid habitat preferences in lower Bed II: stable isotope evidence from paleosols: Kaupia, v. 6, p. 231–238.

SIKES, N.E., AND ASHLEY, G.M., in press, Stable isotope signatures of pedogenic carbonates as indicators of paleoecology in the Plio-Pleistocene (middle–upper Bed I, western margin of Olduvai Basin, Tanzania: Journal of Human Evolution.

SINGER, A., 1984, The paleoclimatic interpretation of clay minerals in sediments—A Review: Earth-Science Reviews, v. 21, p. 251–293.

SIROCKO, F., 1996, Past and present subtropical summer monsoons: Science, v. 274, p. 937–938.

STREET-PERROTT, F.A., AND ROBERTS, N., 1983, Fluctuations in closed-basin lakes as an indicator of past atmospheric circulation patterns, *in* Street-Perrott, F.A., Beran, M., and Ratcliffe, R.A.S., eds., Variations in the Global Water Budget: Dordrecht, The Netherlands, Reidel, p. 331–345.

SUTCLIFFE, A.J., 1985, On the Track of Ice Age Mammals: Cambridge, Massachusetts, Harvard University Press, 224 p.

TALBOT, M.R., HOLM, K., AND WILLIAMS, M.A.J., 1994, Sedimentation in low-gradient desert margin systems: a comparison of the Late Triassic of northwest Somerset (England) and the late Quaternary of east-central Australia, *in* Rosen, M.R., ed., Paleoclimate and Basin

Evolution of Playa Systems: Geological Society of America, Special Paper 289, p. 97–117.

WALTER, R.C., MANEGA, P.C., HAY, R.L., DRAKE, R.E., AND CURTIS, G.H., 1991, $^{40}Ar/^{39}Ar$ dating of Bed I, Olduvai Gorge, Tanzania: Nature, v. 354, p.145–149.

WITHJACK, M.O., 1979, Continental rifts: their tectonic evolution and deformation: Tulsa, Oklahoma, Cities Service Company, Research Report #63, G79-15, 60 p.

WRIGHT, V.P., AND TUCKER, M.E., 1991, Calcrete: Introduction: International Association of Sedimentologists, Reprint Series 2, p. 1–22.

# FOSSIL HOT-SPRING TRAVERTINE IN THE TURKANA BASIN, NORTHERN KENYA: STRUCTURE, FACIES, AND GENESIS

ROBIN W. RENAUT
*Department of Geological Sciences, University of Saskatchewan, Saskatoon, Saskatchewan S7N 5E2, Canada*
CHRIS K. MORLEY
*Department of Geology and Petroleum Geology, Kings College, Aberdeen University, Meston Building, Aberdeen, AB24 3UE, Scotland, U.K.*
*present address: Department of Petroleum Geoscience, University of Brunei, Darussalam, Bandar, Seri Begawan, Brunei Darussalam*
AND
BRIAN JONES
*Department of Earth and Atmospheric Sciences, University of Alberta, Edmonton, Alberta T6G 2E3, Canada*

ABSTRACT: The Ngakoringora Ridge is a large (300 m long), linear mound of limestone, dolomitized limestone, and chert that rises abruptly from the desert floor on the southwestern edge of the Lothidok Hills, west of Lake Turkana, Kenya. The origin of the ridge has been controversial. It has previously been considered either a hot-spring deposit or an uplifted inlier of pre-Cenozoic marine carbonates. Interpretation has been hampered by the extensive diagenetic alteration of the rocks and the lack of identifiable fossils.

A preliminary examination of the ridge morphology and the facies and fabrics of the rocks confirms a hydrothermal origin. The carbonates contain radial calcite fans, micrite–microsparite laminae, stromatolites, coated grains, peloids, intraclasts, Mn-shrubs, and other fabrics that characterize hot-spring travertines. Many of the carbonates have been partly silicified and dolomitized.

The ridge is interpreted to be a fissure-ridge travertine, precipitated from thermal waters that discharged along a medial fissure. Fluids flowed laterally from five main mounds that were probably active at different times. Crystal fabrics along the ridge crest are compatible with abiotic precipitation from alkaline spring waters undergoing rapid degassing of $CO_2$. Microbially influenced fabrics, including stromatolites, become more common distally. Silicified plants, filamentous microbial mats, and thin chert beds are locally present in distal slope settings. Plant silicification and chert formation may have taken place in shallow terrace pools from spring fluids undergoing cooling and evaporation.

The Ngakoringora Ridge formed after the faulting and tilting that formed the Lothidok Hills, but its age is difficult to constrain. Silicification and dolomitization of the carbonates resulted from contact with hydrothermal fluids, and possibly from circulating ground water or lake water after deposition.

Hydrothermal activity in rifts migrates with the evolving structural configuration. In the Kenya Rift, this is evident as a migration of hydrothermal activity toward the rift axis. Fossil spring deposits can provide much useful paleoenvironmental information even though they are of small lateral extent.

## INTRODUCTION

Hydrothermal activity is very common in continental rifts, particularly those with abundant volcanism. In the Kenya Rift, most modern thermal springs discharge waters at sites on fault lines in and along the margins of the axial rift depression (Renaut and Owen, 1988; Tole, 1988, 1997; Allen et al., 1989; Clarke et al., 1990; Allen and Darling, 1992; Dunkley et al., 1993). Although more than five hundred thermal features are active in Kenya, few fossil thermal spring deposits have yet been recognized in the geological record of the Kenya Rift. Dunkley et al. (1993) described fossil sinters and travertines from several late Quaternary volcanic centers in the northern Kenya Rift, and Pleistocene travertine and tufa deposits are known from several places (e.g., Casanova, 1986, 1994). Behr and Röhricht (2000) and Behr (this volume) have interpreted some of the cherts at Lake Magadi as precipitates of hydrothermal waters. Watkins (1983) discussed Miocene hydrothermal deposits from northeast of Lake Turkana. Records of past geothermal activity, however, are sparse when compared to the modern distribution and abundance of thermal springs, fumaroles, and areas of hydrothermal alteration.

Fuchs (1934, 1939) briefly described a large fossil "calc-tufa" deposit in the Lothidok Hills, southwest of Lake Turkana (Fig. 1) and suggested that it precipitated from hot spring waters. This deposit, which is called the Ngakoringora Ridge, is a linear mound of limestone, dolostone, and chert, about 300 m long, that rises abruptly from the desert floor (Figs. 1, 2). The origin of this ridge has been controversial because the rocks are highly altered and most of the original fabrics and textures are replaced or poorly preserved. Walsh and Dodson (1969), who undertook reconnaissance geological mapping of the region, supported a hydrothermal origin for the ridge. Boschetto (1988), however, implied that the carbonate rocks might be of marine origin and of pre-Cenozoic age, partly on the basis of plant-like fossils that are found locally in the deposits. The latter interpretation, if correct, would carry many implications for the regional geological history of the Turkana Basin. Amoco geologists, exploring for petroleum in the region during the 1980s, did not reach firm conclusions about the origin of the enigmatic carbonates.

The morphology, structure, and sedimentary fabrics of the Ngakoringora Ridge were studied to resolve whether the rocks are hydrothermal or marine in origin. In this paper, we describe the main features of the ridge and confirm that it is a fossil hot-spring deposit. The morphology and petrographic evidence show that the travertine ridge formed from several vents that discharged thermal waters along a linear fissure. Although its age is uncertain, the Ngakoringora Ridge provides evidence for long-lived hydrothermal activity in the northern Kenya Rift. After describing the ridge, we consider some of the broader implications of hydrothermal activity in continental rift sedimentation.

## GEOLOGICAL AND ENVIRONMENTAL SETTING

The Ngakoringora Ridge lies at the southwest margin of the Lothidok Hills, and is adjacent to the north side of the track

Sedimentation in Continental Rifts
SEPM Special Publication No. 73, Copyright © 2002
SEPM (Society for Sedimentary Geology), ISBN 1-56576-082-4, p. 123–141.

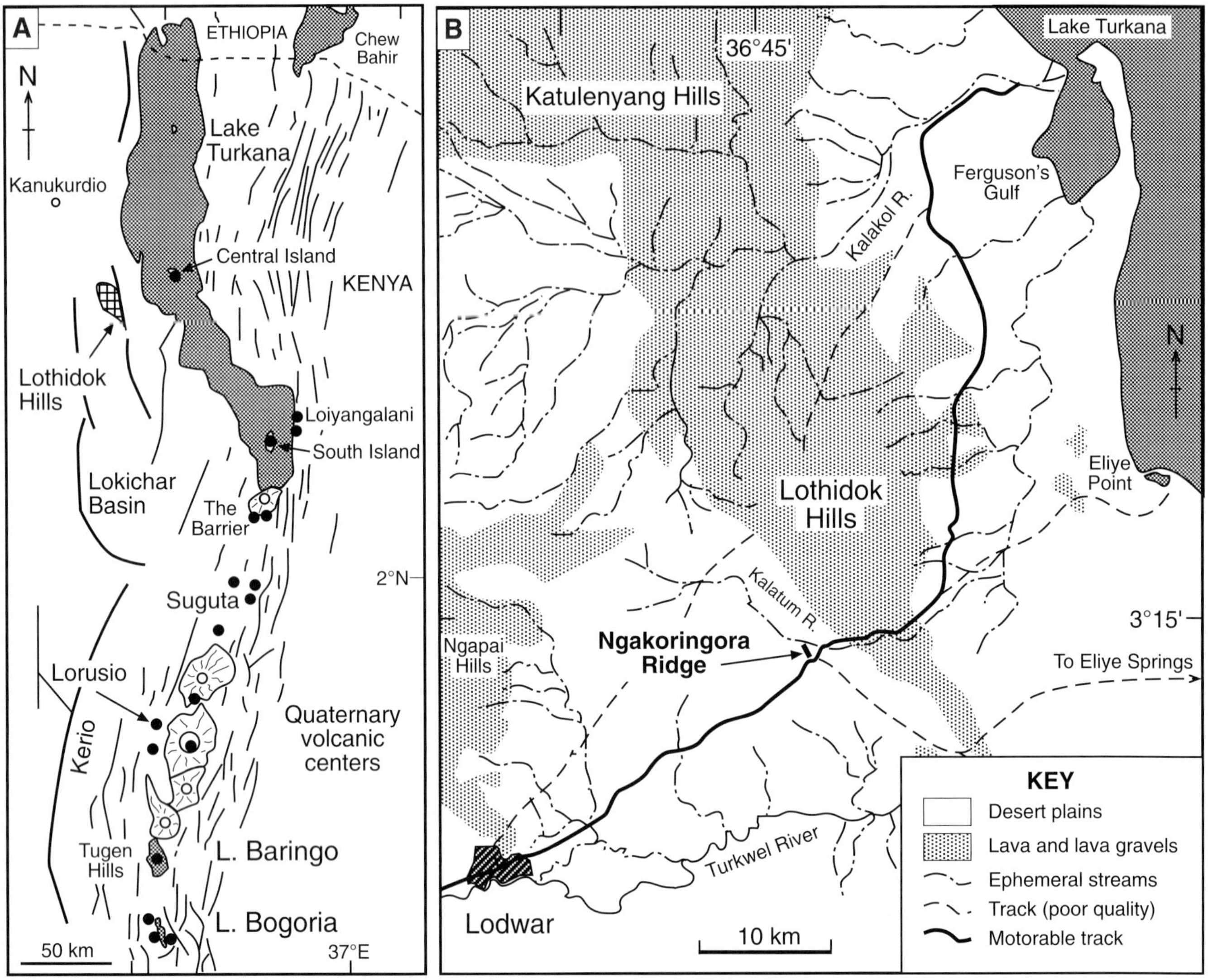


FIG. 1.—Location map of the Ngakoringora Ridge. **A)** The northern Kenya Rift Valley, showing localities mentioned in the text. Black dots: Hot spring groups. Only major faults are shown. **B)** Geographical setting of the Ngakoringora Ridge.

between Lodwar and Ferguson's Gulf (Figs. 1, 2A). The Lothidok Hills lie in the Turkana Depression about 30 km west of Eliye Springs, a village on the midwestern shore of Lake Turkana (Fig. 1B). The elevation of the ridge is poorly constrained, but published topographic maps place its base between about 460 and 480 m above sea level. The Lothidok Hills comprise a tilted and faulted series of Lower and Middle Miocene lava flows, intercalated with volcaniclastic and siliciclastic sedimentary rocks (Fig. 3A). These rocks lie disconformably upon Precambrian gneisses and amphibolites (Fuchs, 1939; Arambourg, 1943; Walsh and Dodson, 1969).

The Kalakol Basalts (27–18 Ma) form the lower part of the exposed Tertiary sequence. These strata are overlain by the lavas, volcaniclastic deposits, and sedimentary rocks of the Lothidok Formation (18–13.2 Ma). This formation is overlain unconformably by the Loperi Basalts (~ 12 Ma). These lavas, in turn, are unconformably overlain by the conglomerates and arkosic sandstones informally termed the Turkwel beds (< 11.9 Ma: Boschetto,

→

FIG. 2 (opposite page).—The Ngakoringora Ridge. **A)** View of the ridge from the eastern margin, looking northward from the Lodwar–Ferguson's Gulf track. Mound 1 (left) lies at the southern edge. **B)** The western side of the ridge crest, showing mounds 2 (right) to 5 (left). **C)** View south from ridge crest, showing proximal slopes of the western side of the ridge. Rubble litters the upper slopes. **D)** Ridge crest, showing coarsely jointed carbonates. **E)** Rubble-strewn eastern marginal slopes. Small terraces, coated by desert varnish, are visible in foreground. **F)** Detail of terraces shown in Part E. **G)** Small mound at the northern end of the ridge. The Kalatum River, with riparian vegetation, lies north of the ridge; Lake Turkana lies on distant horizon. Arrow points to the fossil pool-rim dam at the end of the mound, illustrated in Fig. 8B.

A
B
C
D
E
F
G

1988; Boschetto et al., 1992). The Neogene strata in the southern Lothidok Hills dip at 10–20° to the southwest because of repeated normal faulting.

A seismic line (TVK-3) acquired by Amoco passes through the southern tip of the Ngakoringora Ridge (Morley, 1999, fig. A5). The interpretation (Fig. 3B) shows that the Lothidok Hills lie in the uplifted footwall block to a large, basin-bounding fault that is aligned north–south along their eastern margin. The Oligocene–Middle Miocene units of the Lothidok Hills comprise an eastward-thickening sequence that expands toward a west-dipping boundary fault. A younger east-dipping boundary fault formed on the eastern margin of the Lothidok Hills during the Late Miocene and Pliocene, and controlled the development of the Turkana Basin. Footwall uplift caused exposure and erosion of the hanging wall of the Oligocene–Middle Miocene (Lothidok) basin. West of the Lothidok Hills, another westward-thickening half graben opened in the Late Miocene–Pliocene, which was controlled by the northern segment of the Lokichar Fault. Today, therefore, the Lothidok Hills lie on the flexural margin of the North Lokichar Basin, and the uplifted footwall block of the Turkana Basin.

The region is semiarid desert with very little surface water. The Kalatum River, which lies ~ 100 m to the north and east of the ridge, is a sandy ephemeral stream. The plains surrounding the ridge have low scrubby, xerophytic vegetation and are discontinuously mantled by stone pavements or loose eolian sands (Champion, 1937; Hemming and Trapnell, 1957). No thermal springs have been reported near the ridge. The nearest thermal springs are probably those on Central Island in Lake Turkana (Worthington, 1932; Dunkley et al., 1993), 40 km to the east (Fig. 1A).

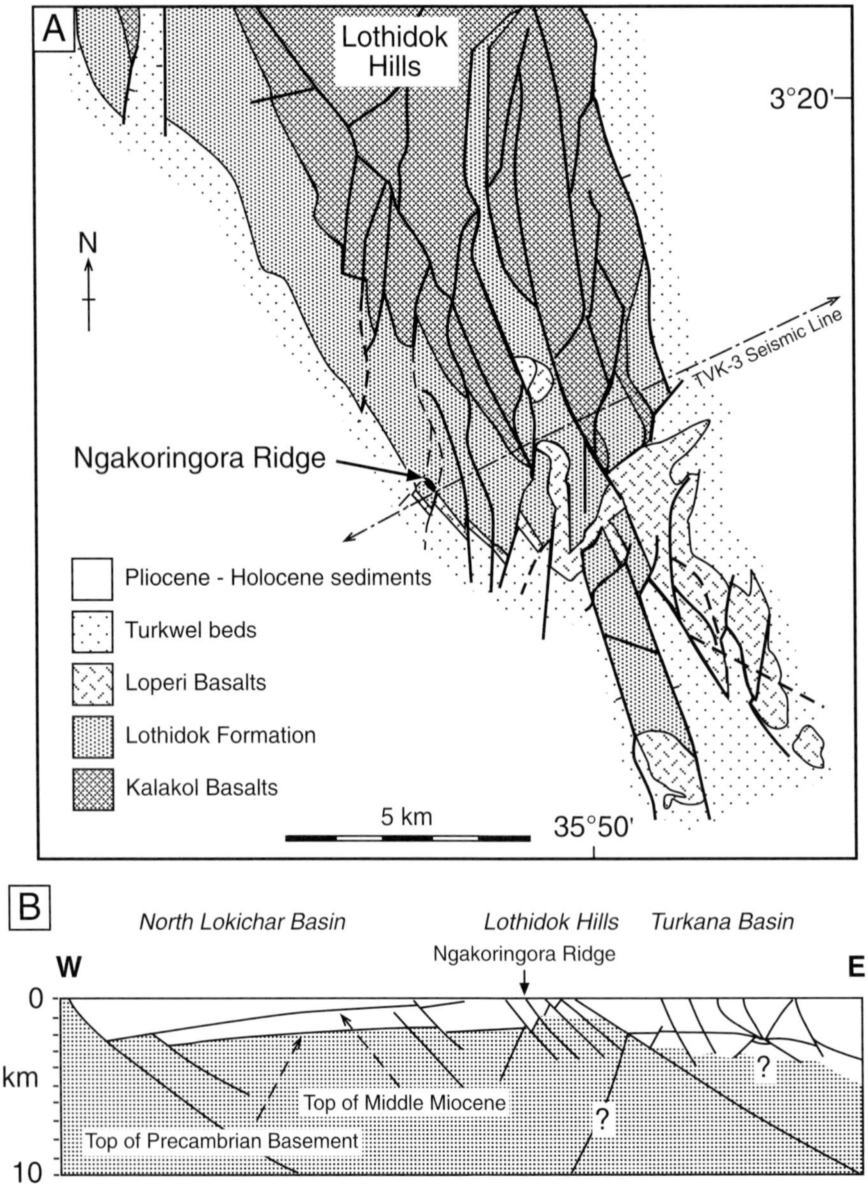


FIG. 3.—Geological setting of Ngakoringora Ridge. **A)** Geological map of southern Lothidok Hills (simplified after Boschetto et al., 1992); **B)** Structural cross section (after Morley et al., 1999), based on seismic line TVK-3, part of which is shown in Part A.

## THE NGAKORINGORA RIDGE

The Ngakoringora Ridge is ~ 300 m long, up to 15 m high by ~ 6–8 m wide at its crest, and has flanking carbonate deposits up to 100 m in width (Figs. 2, 3A). The ridge axis is oriented NNW–SSE at 300°, parallel to the overall strike of the Lothidok Hills. The ridge lies along a poorly exposed outcrop belt of the upper Lothidok Formation, a series of volcaniclastic rocks with interbedded lava flows that dip southwestward at 12–15° near the ridge (Boschetto, 1988; Boschetto et al., 1992). Boschetto (1988) mapped the northern and southern limits of the ridge as NE–SW trending faults (Fig. 3A).

The ridge is formed by five main mounds (numbered 1–5 from south to north) that are connected along the ridge axis (Fig. 2A). The mounds, four of which have similar heights, are elongated parallel to the ridge axis, and most have flat linear tops instead of well-defined summit cones. Mound 1, which is only half as high as the other four mounds (Fig. 2A), is formed by coalescence of two smaller mounds. The ridge has a steep upper section near the ridge crest, which passes downslope into more gently sloping lateral terraces (Fig. 2B, C). Distally, these terraces either pass gradually through a carbonate rubble zone into siliciclastic desert sands and volcanic gravels or terminate locally in a small (< 1 m high) cliff. The original terrace morphology is best preserved along the western margins of the ridge.

The upper parts of the ridge are badly eroded in places, and the morphology of the mounds is partly disguised by coarse boulders and rubble that cover their upper slopes below the ridge crest (Fig. 2C). The coarse rubble that litters the upper slopes reflects the strong jointing in the crestal carbonates (Fig. 2D). Joints broadly parallel and normal to the ridge axis commonly form conjugate sets with a spacing of 0.5–1 m. Weathering and release of these blocks has hampered detailed reconstruction of the original mound morphology. Inspection of the ridge crest, however, shows that its former margins were nearly vertical in places but elsewhere had primary depositional slopes of 60–70°.

The proximal marginal slopes below the ridge crest are poorly exposed. The slopes are broken in places by low (< 50 cm) steps that may mark the edges of small sloping (~ 2–10°) terraces. Distally, these subhorizontal terraces become more evident, particularly on lower slopes of the western margins of the ridge (Fig. 2E, F). These terraces range from narrow, gently sloping stepped surfaces about 1–2 m wide to broad subhorizontal platforms up to 10 m across that are locally overstepped by higher terraces. The lowest terraces typically have only a few decimeters of vertical relief, with smooth convex-downward edges at their distal margins (Fig. 2F).

Two much smaller mounds lie at the northern end of the ridge. North of mound 5 (Fig. 2G), a low (~ 4 m high), broadly conical mound lies along the same axis as the ridge. About 10 m north of that mound (arrow in Fig. 2G), a poorly preserved, circular wall of carbonate < 1 m high and < 2 m wide encloses some small (50–100 cm) weathered mounds of carbonate. The circular wall has a radius of ~ 15 m. Four small, eroded mounds < 2 m high lie at the southern margin of the ridge near the Lodwar–Ferguson's Gulf track.

## METHODS

Rock samples were obtained from different parts of the ridge, including a crest-to-distal-slope transect along the midwestern side at Mounds 4 and 5. Large thin sections were stained for dolomite, ferroan calcite, and ferroan dolomite, using standard methods (Miller, 1988). The mineralogy was determined by standard petrographic methods and by X-ray diffraction (XRD) of powdered cavity mounts on a Rigaku RU-200 diffractometer. The percentage Mg in the dolomite was estimated using the method of Goldsmith et al. (1961).

## LITHOLOGY AND STRUCTURE OF THE RIDGE

The rocks that form the ridge are mainly carbonates but include at least 10% quartz, which fills fractures on several scales and locally forms thin beds and lenses of chert. The rocks are highly altered, and the primary depositional fabrics are poorly preserved (Boschetto, 1988). Locally, however, weathered surfaces highlight subtle differences in fabric and mineralogy that provide clues to the original nature of the carbonates and cherts.

### *Ridge-Crest Carbonates*

The ridge crest comprises the highest, medial parts of the ridge, which pass downward laterally into the upper marginal slopes (Fig. 2A–D). The carbonates forming the ridge crest are highly indurated limestones, silicified limestones, and dolomitic limestones. The carbonates are mottled gray, brownish gray, or dark brown on weathered surfaces, and are pale yellowish gray or white on fresh surfaces. A thin (< 3 mm) opaque dark bluish brown to black patina covers many exposed surfaces. Texturally, the carbonates range from cryptocrystalline to coarsely crystalline, but recrystallization and patchy dolomitization have obscured most of the original variations in crystal morphology and size. Dense, finely crystalline, white carbonates are the most common rocks. Where extensively silicified, the carbonates are flinty or porcellaneous, and fracture conchoidally. They have knobby surfaces where the silicification is patchy.

Bedding structures, lamination, and other sedimentary structures are difficult to recognize on most fresh exposures, but coarse bedding features are visible on many weathered surfaces. Some bedding and coarse (~ 0.5–1 cm) laminae are highlighted by slight color changes or differential weathering. Where the carbonates are well jointed, beds range from a few centimeters to a few decimeters thick. The beds are horizontal or subhorizontal near the mound crests, but they gradually increase in dip laterally (up to ~ 30°) within a few meters of the mound axis.

An important variation in the bedding orientation is seen in mounds 4 and 5. Exposures at the apices of these mounds show an abrupt change from subhorizontal beds on the eastern and western margins of the ridge crest to a central axial zone of vertically oriented banding (Fig. 4). This contact is sharp and continuous over an exposed vertical range of up to 3 m. In places, the subhorizontal beds curve upward within 50 cm of their contact with the adjacent vertically banded carbonates. The vertically banded zone is about 70–100 cm wide and is composed of carbonates that in places show hints of crystal growth normal to the vertical bands, based on color changes and differential weathering. Some of the vertically banded carbonates are partly silicified and flinty. Locally, the axis of the vertically banded zone deviates several degrees from vertical and then returns to a vertical orientation.

Despite their high induration, many exposed carbonates at the ridge crest have some porosity, as noted by Fuchs (1934). Most of that porosity (up to ~ 20%) appears to be secondary, and related to carious weathering and opening of preexisting joints. Fenestral porosity parallel to bedding surfaces is common locally, together with irregular vugs up to 10 cm across. Many joints at the ridge crest are open, although some open joints are lined by sparry calcite. Older generations of fractures, joints, and veins are common in the lithified carbonates. Some of these are filled by brown-stained quartz and form positive relief on weathered surfaces; others are filled by calcite.

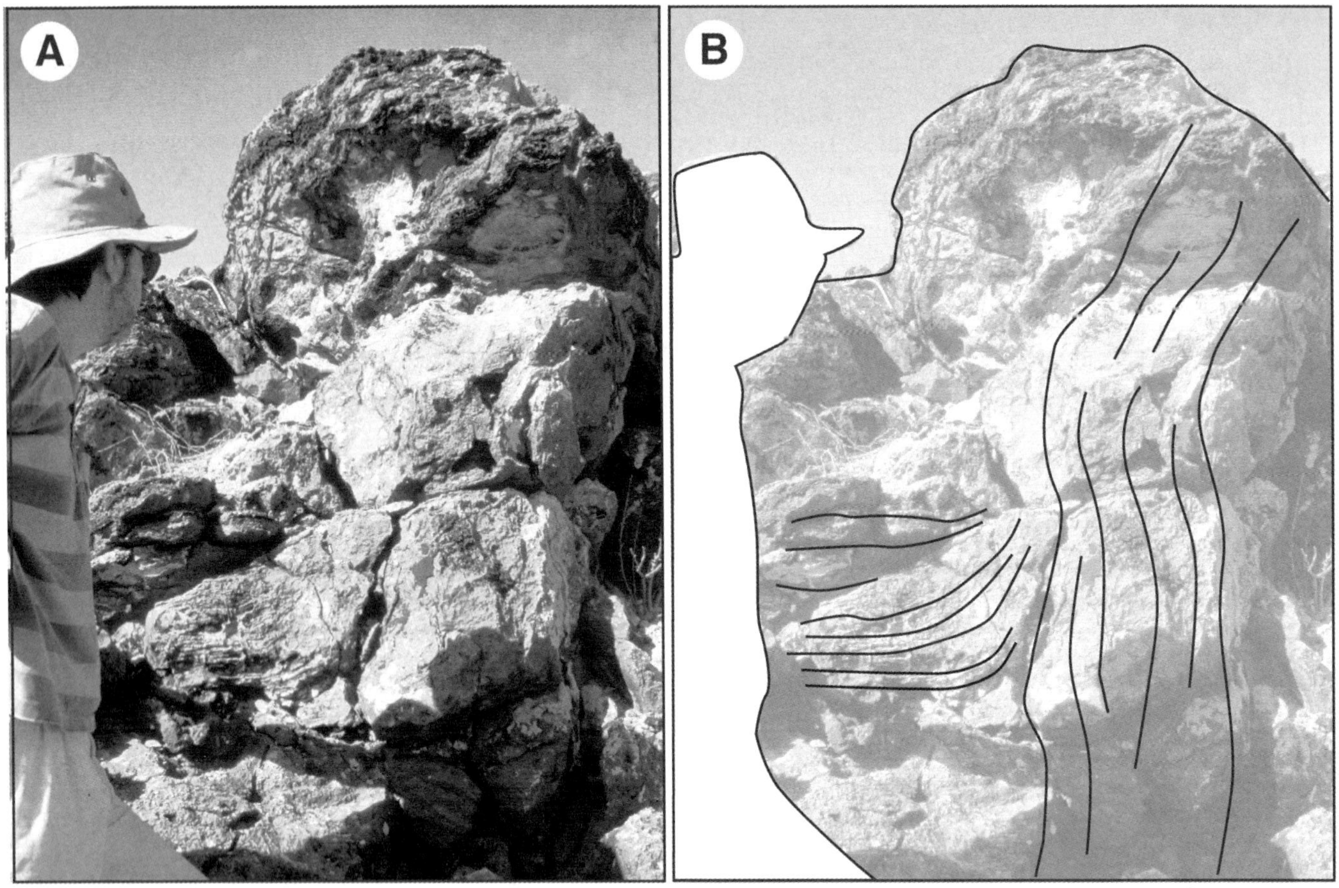


FIG. 4.—**A)** Bedding relationships and **B)** their interpretation at the ridge crest. Cross section across medial axis of Mound 5. The former medial fissure has been completely filled by vertically banded carbonates and silica.

### *Marginal Slope Carbonates and Chert*

The *in situ* carbonates on the upper slopes are similar in lithology to those at the ridge crest, but bedding and its orientation are difficult to recognize because of the poor exposure. On the mid- to lower slopes (< 15°), however, the *in situ* bedrock is locally exposed as low terraces and gently sloping surfaces below a discontinuous gravel cover (Fig. 2E, F).

Both carbonates and cherts are common on the marginal slopes. The carbonates include limestone, dolomitic limestone, and dolostone. Unlike the ridge-crest rocks, bedding and lamination can commonly be seen (Fig. 5A), but primary crystal textures are still difficult to identify. The terraces consist of horizontal and gently sloping (< 10°) beds of carbonate, each bed typically being 1–5 cm thick. Many terraces end in convex beds or laminae that dip downward onto the next lower terrace at 40–60° (Fig. 2F). Locally, the beds show alternating lighter (coarser) and darker (finer) laminae, but in many cases the crystal size in these different-colored laminae is similar. Discontinuities (minor disconformities) in the bedding and laminae are common. Most carbonate beds are composed of finely crystalline calcite, or dolomite that has replaced $CaCO_3$ and lacks recognizable fabrics. However, stromatolitic laminae, intraclasts (0.5–3 cm), and coated grains (0.2–2 cm diameter) were found in mid-slope to lower-slope sites. Angular and subangular intraclasts are locally abundant in some horizons, and form discontinuous lenses of breccia of variable thickness (Fig. 5A). Many carbonate beds are broken by subvertical linear joints, irregular fractures, and veins (5–30 mm wide), some of which are filled by coarse calcite and (or) quartz. Downward-tapering desiccation cracks (1–2 cm deep) are present locally on horizontal surfaces of the midslope and distal terraces.

The most prominent terrace deposits are resistant chert beds, which are easily recognized by their dark brown, lustrous patina of desert varnish (Figs. 2F, 5B). These chert beds, which individually are up to 5 cm thick, are commonly smooth or have a rippled surface morphology. Internally, they lack any clear features except lamination. Clumps of silicified microbial filaments ("streamers") are present on a few surfaces. The chert beds cap, or are interbedded with, limestone and dolomitic limestone laminae or beds. The dark desert varnish passes downward transitionally into lighter unaltered chert or partly silicified carbonates (Fig. 5B).

In several sites near the lower middle and distal parts of the marginal slopes, dark brown chert beds are rich in tubelike fossils that are interpreted to be silicified plant remains (Figs. 5C, D). These are the only recognizable macrofossils in the Ngakoringora Ridge deposits apart from silicified microbial filaments. The fossilized plant remains are confined to the (sub)horizontal strata of the middle to lower terraces. They are present mainly as dense clusters of silicified "stems" that have several orientations and fabrics. Individual stems are 1–4 mm in diameter and up to at least 7 cm long; most are 0.5–4 cm long. The stems commonly are

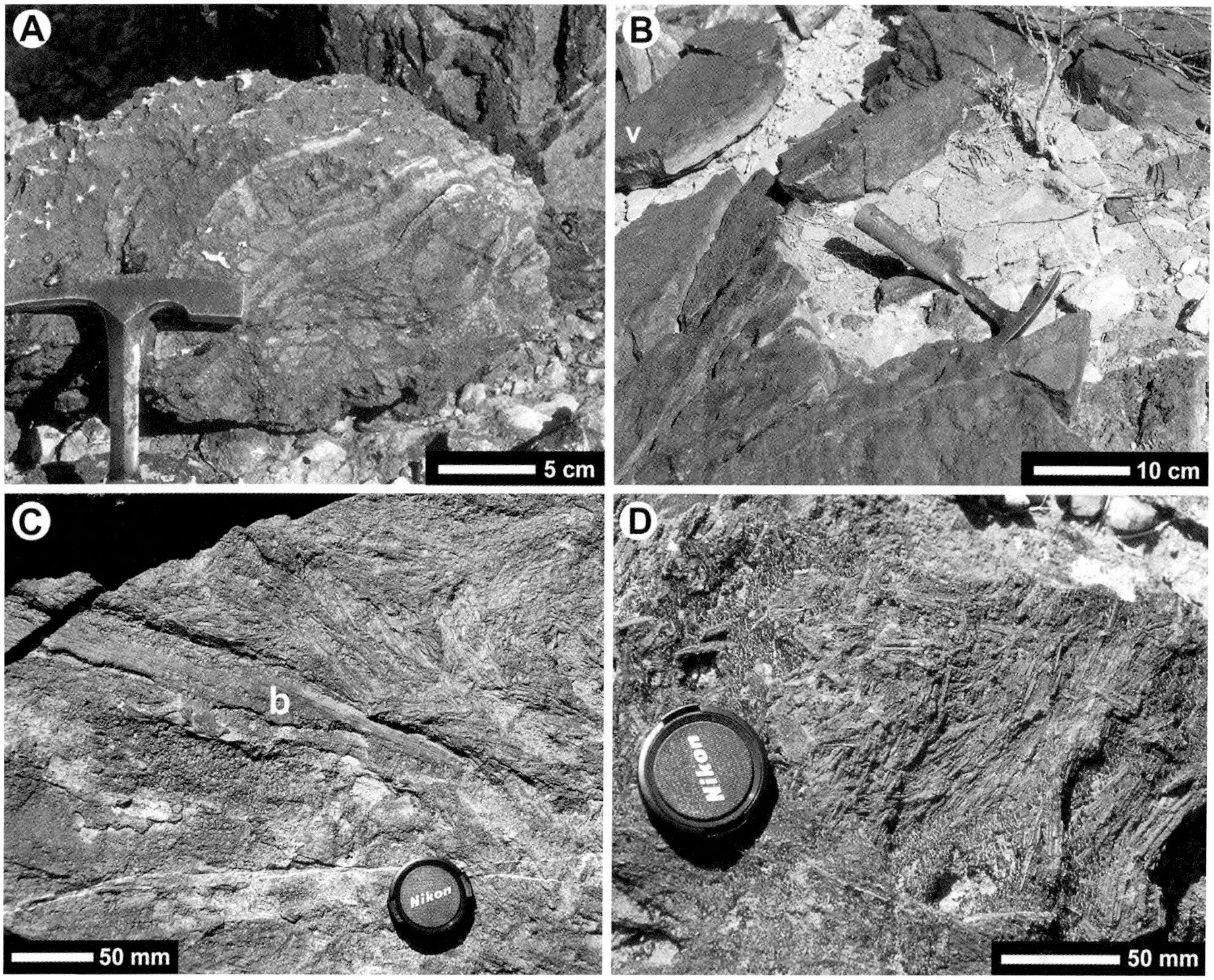


FIG. 5.— Detail of features seen in outcrop, lower mid-slope, midwestern side of ridge. **A)** Curved bedding and lamination on edge of small terrace. Intraclasts in breccia horizon are visible right of hammer. Knobby surface characteristic of patchy silicification is visible upper left. **B)** Desert varnish on silicified dolomitic limestones. The transition from varnished to unvarnished chert is seen in the block at top left (v). Hammer lies in a former terrace pool; sloping edge of terrace is to its left. **C)** Silicified plant remains on bedding planes. A piece of silicified wood bark (b) extends from top left to center. Although many stems are random, most are oriented subparallel to the bark. **D)** Dense clusters of silicified stems, several layers thick, with variable orientation.

grouped in dense clusters, 0.5–1 cm thick. In places, the stems show a preferential alignment (Fig. 5C), whereas elsewhere they are randomly oriented, but most have their long axes parallel to bedding planes or laminae (Fig. 5D). Many stems have a sharp termination at one or both ends. Some subvertical stems are *in situ*, embedded in partly silicified, poorly laminated carbonates. Associated with the stems are silicified impressions of large plants and silicified wood bark (Fig. 5C) that have their long axes aligned parallel or subparallel to the clusters of silicified stems.

## PETROLOGY AND MINERALOGY

The controversy on the origin of the Ngakoringora Ridge carbonates arose mainly because of their degree of diagenetic alteration and unfamiliar textures. Although many primary fabrics have been overprinted or destroyed, some surviving features offer clues to the depositional environment of the ridge-forming carbonates. Only ~ 30% of the 42 thin sections used in this study showed recognizable features. A full petrographic description of all the textures and microfabrics present is beyond the scope of this paper; only the most important fabrics and those that are considered diagnostic are summarized.

X-ray diffraction analyses were made of 14 carbonate samples from the ridge and marginal slopes. They confirm the presence of low-Mg calcite, dolomite, and quartz. The dolomite is Ca-enriched, with a compositional range $Ca_{51-56}Mg_{44-49}(CO_3)_2$. Accessory (< 5% in total) minerals include K-feldspar (mainly sanidine), Fe-oxides, and analcime.

## Ridge-Crest Facies

The subhorizontal carbonates near the ridge crest are massive, weakly laminated, or thick bedded in outcrop. In thin section, most of these carbonates are composed of alternating dense gray micrite or microsparite laminae (Fig. 6A, B), both of which may be variably replaced by anhedral or subhedral dolomite crystals. Many of the carbonates are massive and lack lamination. Most microsparite laminae are thicker than the micrite laminae, but in places micrite replaces the underlying microsparite with an irregular diffuse contact (Fig. 6B). Some microsparite or sparite laminae show hints of branching, with dendritic, feather-like or shrub-like upward growth (Fig. 6B, H). Crystal fans up to 1.5 cm in radius are preserved in subhorizontal beds near the ridge crest. These fans consist of alternating lighter microsparite laminae and darker, thin micritic laminae (Fig. 6C).

Many ridge-crest carbonates consist of massive or poorly defined mosaics of micrite and microsparite that lack an ordered fabric. Clotted fabrics, with irregular patches of dark micrite surrounded by clear microsparite, are common, but the contacts between the micrite and microsparite are normally sharp (Fig. 6D). Many of the ridge-crest carbonates show vague subhorizontal and curvilinear laminae, wisps of dark brown organic matter, and vuggy and fenestral porosity, some of which is filled by sparry calcite and (or) quartz. The quartz is cryptocrystalline or forms coarser subhedral mosaics. Vague spherulitic structures, which have a darker micritic nucleus that is surrounded by a lighter zone of clear radial or partly radial microsparite (Fig. 6E), are present in some laminae. In rare examples, Mn-rich "shrubs" are present (cf. Chafetz et al., 1998), either as clusters along individual laminae or as scattered clumps throughout the micrite–microsparite groundmass (Fig. 6F).

The fabrics of the vertically banded carbonates that form the axial zone of the ridge crest are similar to those forming the adjacent subhorizontal carbonates. Oriented samples, however, show that the traces of large (centimeter scale) crystals seen in outcrop are represented by calcite or dolomite pseudosparite in thin section; some of the calcite, in turn, is locally replaced by fine anhedral silica mosaics. Relatively coarse (0.5–2.0 mm) quartz crystals are commonly the latest cements, partly filling fractures and cracks in the vertically banded rocks.

Fine (< 3 mm wide) desiccation cracks, which are locally present on many (sub)horizontal surfaces, mark breaks in carbonate precipitation and periods of subaerial exposure. Many ridge-crest carbonates are penetrated by larger cracks, some up to several centimeters wide and at least several decimeters deep, that have been filled by calcite, quartz, or calcite that has been replaced by dolomite. Some of these crack-filling cements are symmetrically banded inward from the outer margins (Fig. 6G), implying inward growth from the open crack walls. The quartz in such cracks shows anhedral to subhedral mosaics, but euhedral growth is seen where crystals grew in an open fissure. Some crack-filling quartz replaces older (sub)vertical micrite–microsparite bands, now left as residual vestiges in the quartz mosaic (Fig. 6H).

## Marginal-Slope Carbonates and Cherts

The marginal-slope carbonates, like those at the crest, are dominated by massive, weakly bedded and laminated micrite and microsparite, which may be formed of calcite and (or) dolomite. These rocks have textures and fabrics that are similar to those found at the ridge crest. Some of the components, however, are different. These include stromatolitic fabrics, coated grains, peloids, intraclasts, silicified plant debris, and lenses, laminae, and beds of chert.

Samples from the floors of the terraces of the mid- to lower slopes commonly display laminated and stromatolitic fabrics composed of irregularly alternating light and dark laminae on a scale of < 100 μm to > 1 mm thick (Fig. 7A). The stromatolitic laminae are composed of small laterally linked hemispheroids 1–20 mm in diameter. The laminae are variably composed of calcite, or $CaCO_3$ variably replaced by dolomite and (or) quartz (Fig. 7B). Dark brown, prostrate microbial filaments and clots are present in a few samples. Small Mn-shrubs, similar to those near the ridge crest, are present locally in terrace pool samples. The carbonate laminae may be micrite, microsparite, or composed of acicular and fibrous calcite crystals that are oriented normal to the laminae. Many sparite and microsparite crystals have been partly replaced by micrite. The quartz crystals, which are pale yellow in thin sections, commonly fill former fenestral pores that lie parallel to the lamination.

Beds and laminae that contain small (< 1 cm) coated grains and peloids are locally associated with the stromatolites. Most coated grains have a nucleus that consists of micritic carbonate of indeterminate origin, surrounded by thinly (100–500 μm) laminated microsparite or alternating microsparite–micrite. Lamellae can be discontinuous, implying that the grains were sometimes stationary during accretion. Some of these grains (Fig. 7C) resemble pseudohexagonal calcite crystals in travertines from Oklahoma illustrated by Love and Chafetz (1990, fig. 24B). The "coatings" in the latter crystals represent inclusion-rich zones linked to interrupted growth rather than microbial laminae. The peloidal fabrics have subrounded clots of dark gray micrite encased in microsparite (Fig. 7D). Gray and pale brown micritic intraclasts are locally common across the marginal slopes. These range from < 100 μm to more than 5 cm in maximum dimension. Most intraclasts are cemented by microsparite and sparite. Some clasts are distinctly elongate, 0.1–5 mm thick, and have an isopachous rim of microsparite.

The cherts on the distal parts of the western marginal slopes contain elongate tube-like structures 1–3 mm in diameter that are filled by coarse pale yellow quartz crystals, a fine mosaic of microcrystalline silica, or, more rarely, fibrous chalcedony. Some

→

FIG. 6 (opposite page).—Thin-section photomicrographs of travertines from the ridge crest (Mounds 4 and 5). All plane-polarized light except Parts D and H, which are taken using cross-polarized light. **A)** Laminated bands of microsparite (light gray) and thinner micrite (dark gray). Arrows denote a level with possible sparmicritization. **B)** Subhorizontal laminae of microsparite–sparite (light) and micrite (dark gray). The upper part of the sparite shows hints of branching crystals and possible sparmicritization (arrows). **C)** Radial fibrous calcite fan, partly replaced by dolomite. Arrow indicates upward direction. **D)** Clotted micrite and pseudosparite? Both are calcite. Black pores are open. **E)** Radial spherulitic structures. Dark gray is partly dolomitized micrite and microsparite. Lighter crystals are calcite sparite. **F)** Mn shrubs embedded in calcite microsparite. **G)** Quartz-filled vein (Q) showing symmetrical inward growth from fracture walls. The irregular dark gray matrix in center of the vein is calcite. The quartz appears to have selectively replaced older carbonates in the vein. **H)** Quartz mosaic (Q) replacing fibrous calcite laminae (C). Vestiges of the calcite survive in the quartz mosaic (arrows). Possible calcite dendrites are present at bottom right (d).

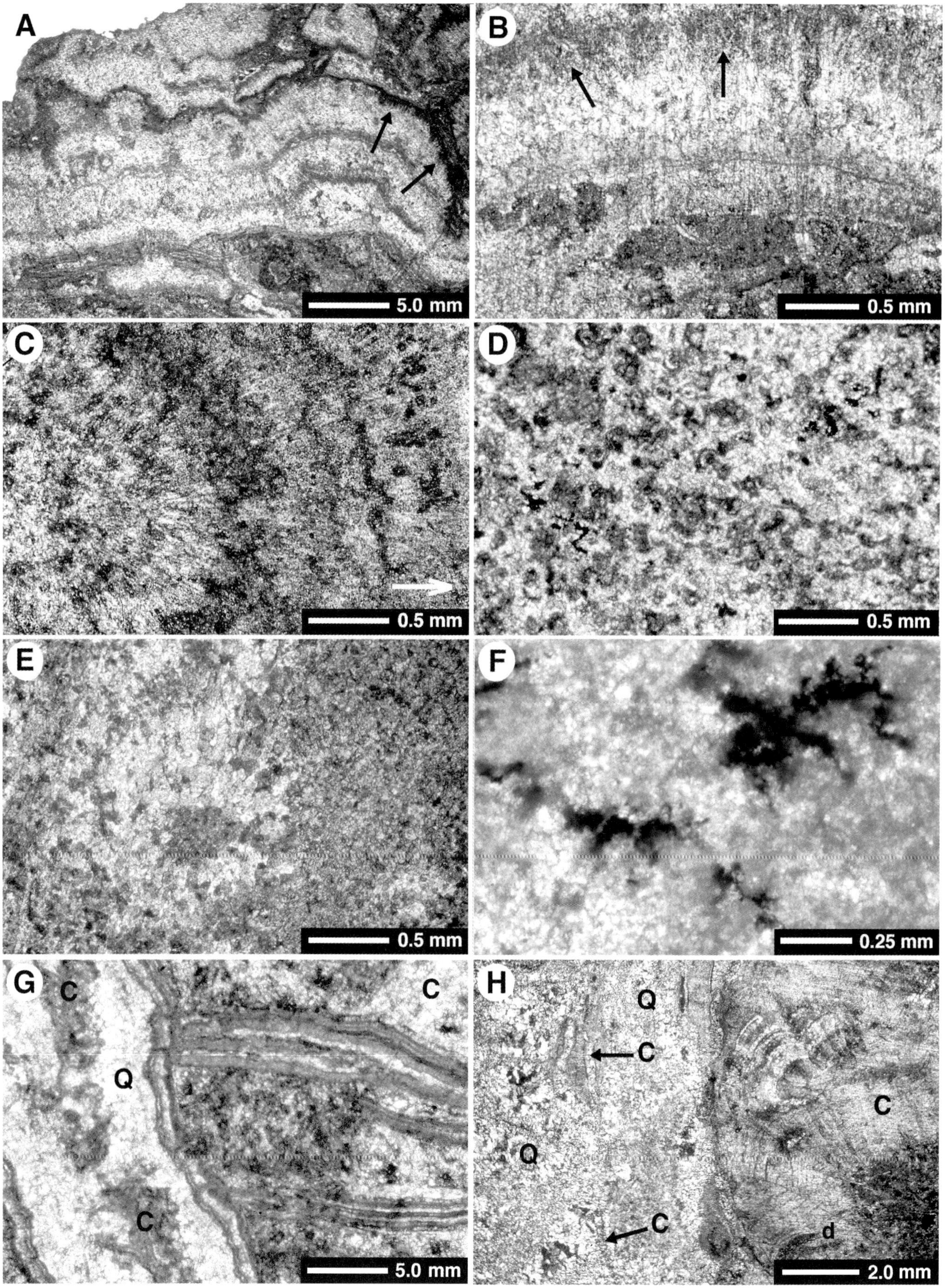
A
5.0 mm
B
0.5 mm
C
0.5 mm
D
0.5 mm
E
0.5 mm
F
0.25 mm
G
C
C
Q
C
5.0 mm
H
Q
C
C
Q
C
d
2.0 mm

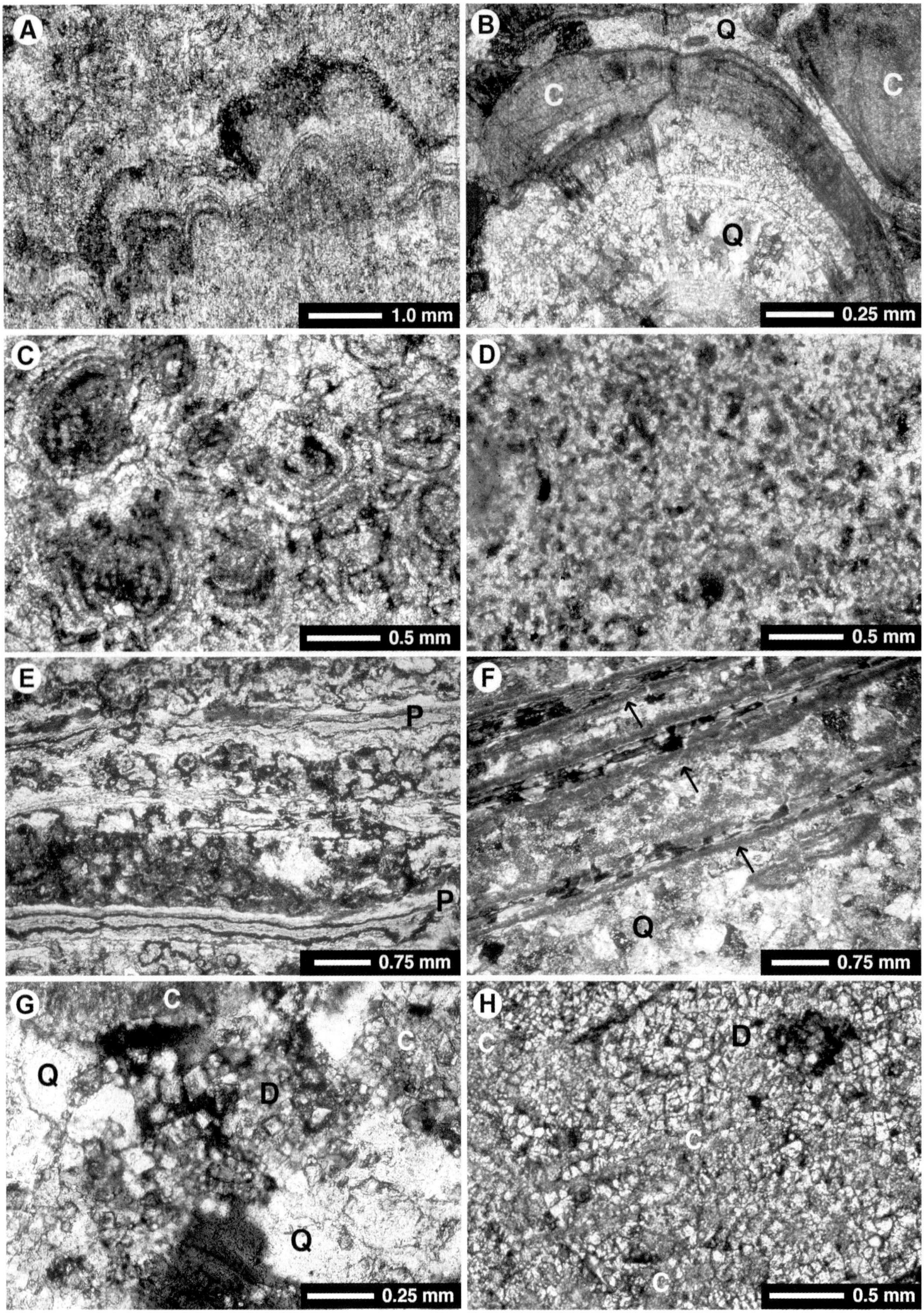
A
1.0 mm
B
Q
C
C
Q
0.25 mm
C
0.5 mm
D
0.5 mm
E
P
P
0.75 mm
F
Q
0.75 mm
G
C
Q
C
D
Q
0.25 mm
H
C
D
C
C
0.5 mm

tubes are lined internally by dark brown organic films and are enclosed in a matrix variably composed of gray micritic peloids, spherulites, and intraclasts that are cemented by clear calcite microsparite, or carbonate partly replaced by quartz (Fig. 7F). Some quartz crystals are equant (salt pseudomorphs?), up to 1 mm across, and have been partly replaced by dolomite. This dolomite is present as small euhedral rhombs in both the quartz and the enclosing calcite matrix (Fig. 7G). Some of the matrix dolomite shows clear rims surrounding a gray, dusty, inclusion-rich interior. The more distal carbonates contain up to 90% dolomite, which forms idiotopic or xenotopic mosaics (Fig. 7H).

## INTERPRETATION OF THE RIDGE

Although the rocks are highly altered, the new evidence supports the original interpretation of Fuchs (1934, 1939) that the Ngakoringora Ridge is a thermal spring deposit. The ridge is a large fissure-ridge travertine, similar to examples described from the western United States (Bargar, 1978; Chafetz and Folk, 1984; Chesterman and Kleinhampl, 1991; Wohletz and Heiken, 1992), Pamukkale in Turkey (Altunel and Hancock, 1993a, 1993b; Ford and Pedley, 1996), and several sites in Italy (Chafetz and Folk, 1984; Guo and Riding, 1999). The Ngakoringora Ridge is large compared to those described from the western United States and Italy, but it is much smaller than those at Pamukkale, which are up to 1.5 km long.

Fissure ridges form where hydrothermal fluids rise to the land surface along open fractures or faults and flow laterally from the crest of the medial fissure. Calcium carbonate may be precipitated along the inner walls of the medial fissure, and it forms dense vertical bands of carbonate crystals, most of which have their long axes oriented normal to the vent walls (Bargar, 1978, fig. 6; Altunel and Hancock, 1993a, fig. 7; Guo and Riding, 1999, fig. 5). In contrast, the waters that spill over the subaerial edges of the fissure ridge precipitate more porous carbonate crystal beds, which may be steeply sloping near the vent axis but attain lower angles nearer the distal base of the mound and lie nearly parallel to the surface topography (Chafetz and Folk, 1984). This juxtaposition of bedding structures is exposed near the crests of at least two of the mounds along the fissure ridge (Fig. 4). This relationship is difficult to explain by any other plausible depositional mechanism. The fissure is not preserved, having been completely filled by inward growth of carbonates and quartz to produce the vertically banded axial zone. The mounds are unlikely to have been precipitated subaqueously because of the presence of well-defined terraces along the distal margins, and they lack the chimney-like morphology found in many sublacustrine thermal-spring carbonates (cf. Fontes and Puchan, 1975; Tiercelin et al., 1993; Pentecost and Viles, 1994).

The individual mounds that formed along the ridge crest represent point sources of fluid flow that were active for some time. Fissure ridges rarely discharge water along their whole length at the same time, but flow may occur at one or more places simultaneously. When precipitation takes place in the medial vent and along the rim of the ridge, the precipitated carbonates may cut off the flow, causing diversion to another part of the ridge (Bargar, 1978). Lateral flow from the ridge may happen on both sides simultaneously or on one side at a time. Flow takes place along one side of the ridge crest until carbonate precipitation raises the level of the substrate to the point where the opposite margin is slightly lower, at which point flow may change direction. Sometimes, however, sheets of water may pour down both sides of the ridge, building up the ridge crest symmetrically. A second reason for migration of the fluid point source relates to the elevation of the hydrostatic head or artesian pressure. Once the ridge attains a certain height, upward flow cannot carry water over the crest of the mound and the water migrates to a lower site of discharge. Four of the mounds grew to almost the same height (Fig. 2A), which implies that their elevation was controlled by the contemporary hydrostatic head. Today the regional base level for flow is Lake Turkana, which is a closed-basin lake 30 km to the east, with a surface elevation (~ 375 m) that lies ~ 80–100 m below the base of the ridge.

The relative ages of the individual mounds cannot be determined. Some or all might have grown simultaneously, but each one represents a successive locus of precipitation, with perhaps only one or two mounds actively discharging water at any time. Some fissure ridges form by the coalescence of several point-sourced mounds, but they lack a continuous medial fissure. Other fissure ridges form along a continuous medial fissure, as at Orange Spring Mound at Mammoth Hot Springs, Yellowstone (Bargar, 1978, p. 24–25). Although a continuous medial fissure may have existed, the individual mounds show that the Ngakoringora Ridge grew from several point-sourced springs located along the fissure axis.

The low, ring-like wall of carbonate at the northern limit of the ridge is a fossil pool-rim dam that, when actively forming, completely surrounded a hot-spring pool about 1.5 m deep. Similar structures are present at Loburu hot springs, Lake Bogoria (Fig. 8). Similar pools have been recognized at other fissure ridges, including Hot Tub Ridge, California (Chesterman and Kleinhampl, 1991, photo 1) and active fissure ridges at Çukurbag, Turkey (Altunel and Hancock, 1993b, photo 5). The fossil pool-rim dam may represent the site of an incipient mound that never fully formed because the thermal activity or fluid discharge declined. This site may be analogous to a "fault-tip" hot spring of Curewitz and Karson (1997), implying that hydrothermal activity might have moved progressively northward as the axial fissure propagated in that direction.

←

FIG. 7 (opposite page).—Thin-section photomicrographs of travertines and cherts from the marginal slopes below mounds 3–4. All cross-polarized light except Parts A and E, which are taken using plane-polarized light. **A)** Alternating laminae of microsparite (light) and micrite (dark gray). Black layer is rich in Mn-oxides. Stromatolitic layer is sandwiched between subhorizontal laminae of partly silicified carbonates. **B)** Stromatolitic laminae from small terrace pool. Quartz mosaic (Q) has replaced calcite (C). Abrupt contact between quartz and silica may reflect differences in original porosity or mineral composition (i.e., calcite vs. aragonite). **C)** Coated grains or zoned pseudohexagonal calcite crystals cemented in calcite sparite from terrace pool floor. **D)** Micritic peloids cemented by calcite sparite, terrace pool floor. **E)** Chert with plant stem (?) remains (P). The rock is entirely silica, except vestiges of gray calcite micrite and minute scattered dolomite rhombs. A brown ragged organic film lines interior margin of each stem wall. **F)** Chert with replaced plant stem (?) remains indicated by arrows. Each stem is filled by clear quartz crystals. The matrix consists of subhedral quartz (Q) crystals, some of which enclose patches of micritic calcite. **G)** Euhedral dolomite rhombs (D) that have replaced quartz (Q) and earlier calcite (C). The calcite was silicified before formation of the dolomite. H) Xenotopic dolomite (D) replacing calcite micrite (C).

FIG. 8.—Travertine pool-rim dams. **A)** Subfossil pool-rim dam of calcite travertine at Loburu, Lake Bogoria. The dam is not actively forming; the modern spring at this site (foreground) contains < 2 mg $l^{-1}$ of calcium and is precipitating only thin peripheral crusts of opaline silica. **B)** Heavily eroded wall of fossil pool-rim dam at the northern edge of Ngakoringora Ridge. Photo taken from center of former pool.

Many fabrics and structures preserved in the ridge can be related to the facies reported from recent fissure-ridge travertines (Fig. 9). Fissure-ridge travertines typically show rapid facies changes (Guo and Riding, 1999). This is also evident in the Ngakoringora Ridge, although commonly obscured by diagenesis. The proximal fabrics are commonly associated with abiotic precipitates that result from $CO_2$ degassing. Guo and Riding (1999) noted that the precipitation along the crestal region of a fissure ridge at Rapolano Terme is characterized by "crystalline crust, paper-thin raft and shrub lithotypes." The laminated fabrics, comprising alternating (micro)sparite and micrite couplets in the Ngakoringora ridge carbonates (Fig. 6A, B, H), are similar to the crystalline crusts at Rapolano Terme, although the laminae are generally thinner in the Ngakoringora ridge. The microsparite and sparite laminae consist mainly of vertically oriented crystals, but they are commonly micritized (Fig. 6B). Similar spar-micritization has been described from other travertines (compare fig. 3E in Chafetz et al., 1994, with Fig. 6B). Radial crystal fan fabrics (Fig. 6C) and possible feather crystals (Fig. 6B) are also common in thermal-spring travertines (e.g., Folk et al., 1985; Guo and Riding 1992). Shrub fabrics are rare along the Ngakoringora Ridge crest, although scattered lenses are present. Shrubs are scarce in other Kenyan travertines, possibly because most alkaline springs in the Kenya Rift have low concentrations of $H_2S$ (cf. Chafetz and Folk, 1984, p. 305). Crystal rafts, which form in shallow pools along ridge crests (Guo and Riding, 1999), have not been recognized at Ngakoringora Ridge, but some intraclasts may have originally been rafts.

The distal slopes and terraces show a higher proportion of biotic remains. By analogy with Rapolano Terme and other sites, those carbonates formed mainly in local pools and on gently sloping surfaces where waters were cooler than at the ridge crest. The terrace pool deposits comprise laminated crusts similar to those on the upper ridge slopes (Fig. 7A). Coated grains formed around nuclei in the low-relief pools, peloids by micritization of coarse particles, or as bacterial precipitates (cf. Chafetz, 1986; Folk and Chafetz, 2000). The plant remains, which are interpreted to be mainly the silicified stems of reeds (see below), lie in a setting analogous to the "reed travertines" found in distal parts of Rapolano Terme (cf. Guo and Riding, 1999, figs. 14, 15).

## FLUID GEOCHEMISTRY AND MINERAL PRECIPITATION

Modern thermal springs in the northern Kenya Rift fall into two main groups. High-enthalpy springs, which discharge at or near local boiling point (80–100°C), are mainly of $Na–HCO_3–Cl$ composition with salinities of 1–5 g $l^{-1}$ total dissolved solids (TDS), pH of 7.5–9.5, and aqueous silica concentrations of ~ 40–270 mg $l^{-1}$ $SiO_2$ (Allen and Darling, 1992; Dunkley et al., 1993; Renaut and Tiercelin, 1994; Renaut et al., 2002). Ca concentrations are typically < 3 mg $l^{-1}$, with Mg < 1 mg $l^{-1}$. Low-enthalpy (warm) springs in the rift have a $Na–Ca–HCO_3$ composition, with salinities of < 2 g $l^{-1}$ TDS, pH of 7.0–9.0, and aqueous silica concentrations of 20–80 mg $l^{-1}$ $SiO_2$ (Renaut and Tiercelin, 1987; Tole, 1988; Dunkley et al., 1993). Concentrations of dissolved Ca range from 0.5 to 15 mg $l^{-1}$, with 0.1–5 mg $l^{-1}$ Mg (Renaut and Tiercelin, 1987; Tole, 1988; Allen and Darling, 1992; Dunkley et al., 1993). Most low-enthalpy springs discharge at temperatures between 30 and 45°C.

The precipitates from modern high-enthalpy springs can be $CaCO_3$ (calcite or aragonite) or silica. Both carbonate and silica may precipitate from the same fluid. Calcite, rather than aragonite, is precipitating from fluids with temperatures of 80–100°C when the waters meet the atmosphere and rapidly exsolve $CO_2$ at the hot springs at Lake Bogoria (Jones and Renaut, 1995, 1998; Renaut and Jones, 1997) and Lorusio (Dunkley et al., 1993; Renaut et al., 1999). At Chemurkeu (Lake Bogoria), both calcite and aragonite have formed from boiling fluids (Jones and Renaut, 1996; Renaut and Jones, 1997). Silica, as opal-A, precipitates as thin crusts on microbial substrates at many sites, where waters have cooled and undergone evaporative concentration (Renaut et al., 1998; Renaut et al., 2002). Fossil siliceous sinter (paleosinter) formed from high-enthalpy fluids at several late Quaternary central volcanoes between Lake Baringo and Lake Turkana (Dunkley et al., 1993) but is not actively forming at the land surface because present water tables are lower than during the late Pleistocene (Darling et al., 1993; Sturchio et al., 1993).

In contrast, low-enthalpy thermal springs generally precipitate only carbonates because their dissolved silica concentrations are too low to precipitate silica upon cooling alone. Most low-temperature travertine and tufa deposits of late Quaternary age

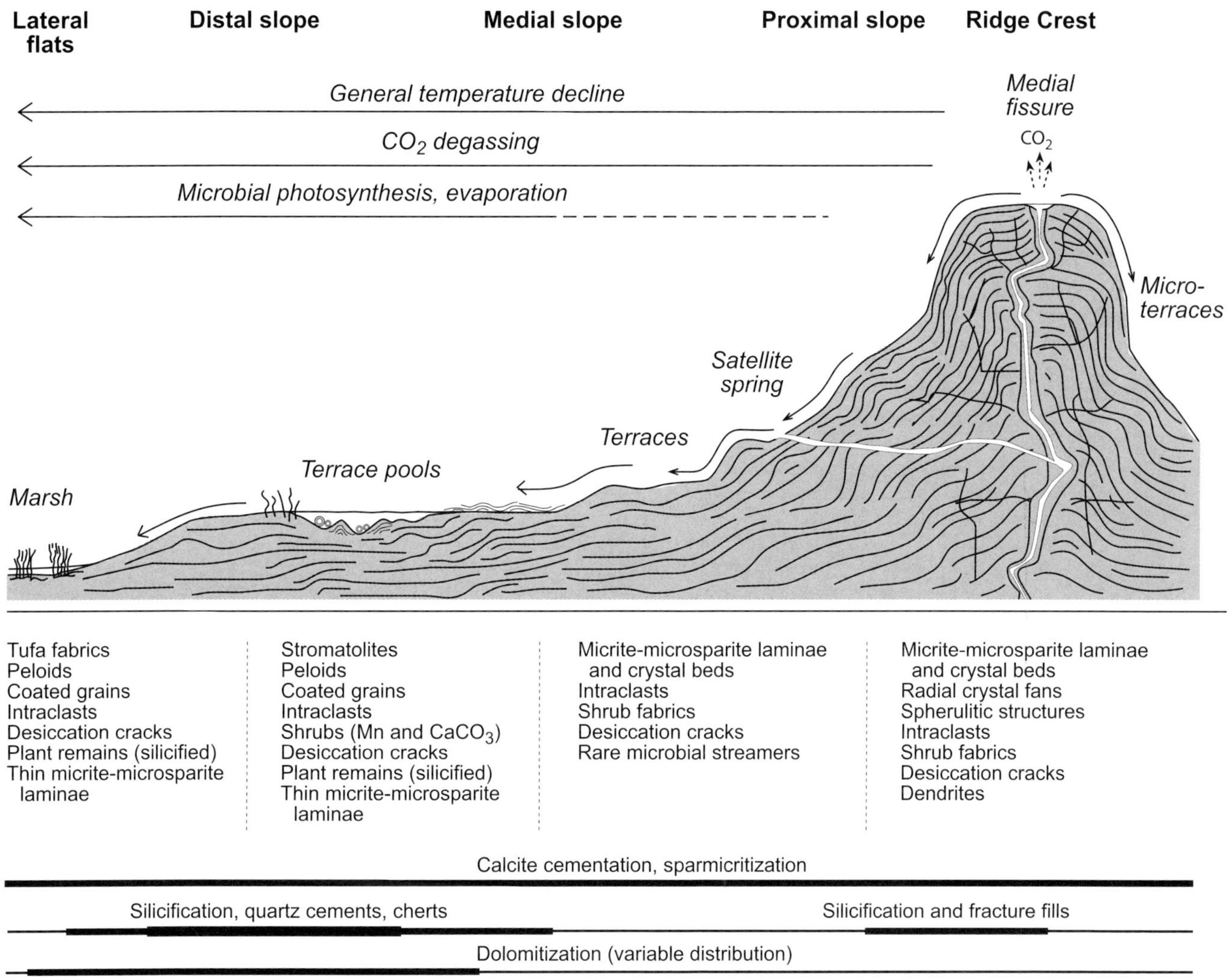


FIG. 9.—Summary diagram of facies, features, and relative abundance of diagenetic minerals in the Ngakoringora Ridge.

in the Kenya Rift are calcite (Casanova, 1986, 1994), but few have been studied in detail.

The composition and temperature of the fluids that formed the Ngakoringora Ridge are uncertain, but clues are preserved in the rocks. The dominance of crystal bed fabrics near the crest can be interpreted in several ways—the waters were too hot for extensive microbial colonization (i.e. > 70°C); the high rate of degassing was such that abiotic crystal growth masked biotic controls of fabric development (cf. Chafetz and Guidry, 1999); or harsh fluids (e.g., sulfide rich) inhibited most organisms. The last option is unlikely because most thermal springs with travertine deposits in Kenya are poor in sulfides. It is not possible to determine the likely maximum temperatures of carbonate precipitation from the preserved fabrics, but the lack of vegetal remains in the proximal parts of the ridge implies that the fluids were hot (e.g., bryophyte tufa is absent). At temperatures near boiling point (> 80°C), high disequilibrium fabrics characterize other travertines in the Kenya Rift (Jones and Renaut, 1995, 1996, 1998), and include dendrites and skeletal crystals. Possible dendritic fabrics were recognized in two travertine samples from the ridge crest (Fig. 6B, H), but their high porosity would make them prone to diagenetic alteration or masking by cementation. The original mineralogy may have been calcite or aragonite, both of which can form fine crystal beds and radial fans (e.g., Chafetz and Folk, 1984; Guo and Riding, 1992, 1994, 1998; Koban and Schweigert, 1993; Chafetz and Guidry, 1999). The extensive recrystallization and selective dolomitization may be evidence that some of the carbonate may have been aragonite. Although stable-isotope analysis may help to constrain the temperature of the carbonate precipitation, its reliability in thermal-spring travertines is uncertain (cf. Chafetz and Lawrence, 1994). Diagenetic alteration, possible exchange, and the uncertainty of the isotopic composition of the original fluid preclude reliable isotopic estimation of paleotemperature.

The maximum range of spring-water temperature cannot be confirmed from lithological evidence. The fabrics, however, are compatible with waters that discharged at between 40 and 80°C, on the basis of modern analogs. Cooling took place as fluids moved downslope and locally became ponded in distal terraces. Plants, locally rooted in the ponds, imply local water temperatures of < 40°C. The stromatolites and coated grains are likely to have formed in waters < 45°C because they are locally interlayered with plant debris.

### *Chert Formation and Silicification*

Silica replaced carbonates, mainly in the mid-slope and distal parts of the ridge, and fills fractures, joints, and other open porosity throughout the ridge, including the crest. Where replacing or cementing carbonates, the silica is clearly younger than the travertine. It is difficult from exposed field relationships and petrography, however, to determine if the silica was precipitated from thermal waters while the carbonates were still precipitating elsewhere along the ridge, or whether the silica entirely postdates ridge formation. It is also unclear whether the silica formed in one or several phases. There are several possible ways to explain the origin of the silica.

In many modern travertine-precipitating thermal springs in Kenya, silica precipitates downstream from the vent where fluids, undersaturated upon discharge, attain silica saturation by rapid cooling and evaporation (Renaut et al., 1998; Renaut et al., 1999; Renaut et al., 2002). The chert beds and silicified plant debris at the Ngakoringora Ridge formed in terrace pools about 50–70 m from the vents. The plant debris is interpreted to be the silicified lower stems of reeds and grasses that grew in spring-fed pools. Silicification might have occurred while the reeds or grasses were alive or shortly after their death. Silica-rich fluids may have been drawn up the lower hollow stems by evapotranspiration (Fig. 10) and capillary draw, replacing and (or) encrusting the organic tissue of the plant wall. Similar processes have taken place at modern hot spring–fed pools at Reporoa in New Zealand (Jones et al., 1998). The upward capillary draw was insufficient to transmit fluids up the whole stem, so only the basal few centimeters were silicified. The unsilicified part of the plant may have decayed or broken off and fallen into the pool, perhaps also becoming silicified. As pools periodically dried up or stagnated, the silica concentration would increase by evaporative concentration, cementing and eventually replacing and cementing earlier carbonate beds in the pool. The decay of vegetation in the pool may have decreased the pH, locally enhancing silicification by promoting carbonate dissolution (cf. Deocampo and Ashley, 1999). Some clumps of silicified stems may have been reworked and current deposited, but most of the stem accumulations could be mats of reed remains that accumulated on pool floors or low-angle slopes (cf. Fig. 5C). Although this hypothesis is tentative, it explains the local development of the chert beds and plant detritus, and it implies that at least some of the silicification was contemporary with carbonate ridge development.

The model presented for the chert beds cannot explain the silica that fills joints and locally replaces and cements carbonates elsewhere in the ridge. If the fluids that formed the chert beds resulted from cooling and evapotranspiration, then it is likely that these waters originally were high in silica even if undersaturated when discharged at the vents. Such fluids, therefore, could potentially precipitate silica in existing porous travertine upon cooling or evaporation anywhere in the mound where they attained saturation. At Chemurkeu (Lake Bogoria), opaline silica forms thin veins and fills pores in travertine; it does not precipitate subaqueously at the spring vents (Renaut et al., 1986). Most of that silica will transform to quartz with time. Quartz is present as late cements in the vertically banded zone at the crest of the ridge (Fig. 6G). Thermal waters, therefore, probably supplied most of the silica.

An alternative possibility would be that the fluids changed composition and (or) increased temperature through time, becoming silica enriched. Major temporal changes in fluid composition are known from other geothermal areas (e.g., Waiotapu,

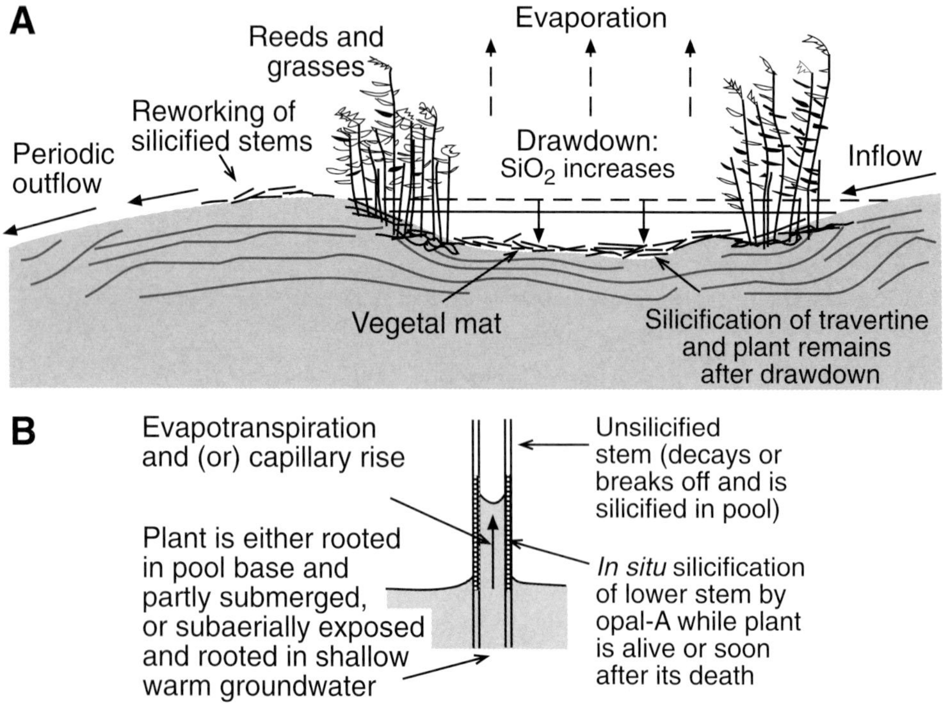


FIG. 10.—Schematic model for the origins of the silicified plant remains and associated cherts on the western marginal slopes of Ngakoringora Ridge. **A)** Section through terrace pool showing inferred distribution of plants and their remains. **B)** Silicification (replacement) of a reed or grass stem by evapotranspiration and capillary rise of silica-rich spring fluids.

New Zealand; Lloyd, 1959), and at Lake Bogoria carbonate precipitation has declined while silica continues to precipitate (Fig. 8A; Renaut and Jones, 1997). A change of fluid composition could explain precipitation of late silica in joint fills and interstitial pores, but subaerial sinters are not present, making this explanation less likely.

A third possibility is that some silica precipitation postdates ridge formation and the decline of hydrothermal activity. This scenario would have been difficult to attain unless the carbonate ridge was completely buried and exposed to silica-bearing groundwaters then exhumed by erosion, or was submerged by lake water. To mobilize silica, that groundwater would need to be alkaline, warm, or contain sufficiently high concentrations of silica by local dissolution of silicate minerals or volcanic glass. Although burial and exhumation of the ridge are possible, much of the silica is facies controlled, which implies an association with contemporary fluids.

### *Origin of the Dolomite*

The origin of the dolomite is enigmatic and merits further investigation. Dolomite has only rarely been reported from recent travertines (e.g., Folk, 1994; Renaut and Jones, 1997). Minute (< 1 μm) rhombs were found in high-temperature travertines at Chemurkeu, Lake Bogoria, but their origin was not resolved (Renaut and Jones, 1997).

As with silica, the fluids that led to dolomitization could have been thermal spring waters or later unrelated fluids if, for example, the ridge was buried or submerged. Early (primary or early diagenetic) dolomite formation in travertines may be limited mainly by the low Mg/Ca ratio in most thermal waters. The molar Mg/Ca ratio at the vent is typically < 0.5 in most Kenyan hot springs (Renaut and Tiercelin, 1987; Cioni et al., 1992; Dunkley et al., 1993). Although the Mg/Ca ratio of spring waters may theoretically increase downflow after precipitation of calcite or aragonite near the vent, dolomite is generally not present even in the distal parts of spring deposits. Indeed, in some Kenyan hot springs, Mg-bearing silicates, such as sepiolite and smectite, precipitate on carbonate crystal surfaces in travertine, and may deplete the pore fluids in $Mg^{2+}$ (Renaut and Jones, 2000, fig. 3J). Consequently, no significant increase in Mg/Ca ratio is recorded downflow.

Although fluids with Mg > Ca are rare in the Kenya Rift, such waters *are* present in the Turkana Basin today. Hot springs (T: 70.5°C) on Central Island, and at Loiyangalani (T: 39.2°C) (Fig. 1A) are unusual because they have an Mg/Ca ratio of ~ 2. Ambient groundwaters (T: 33°C) discharging at Kanukurdio, 70 km northwest of Ngakoringora Ridge (Fig. 1A), also have an Mg/Ca ratio of > 1. The Central Island spring fluids may derive from deep geothermal fluid that has mixed with lake water (Dunkley et al., 1993, p. 158).

Although the presence of Mg-rich fluids in this region is unusual in the context of the rift as a whole, Fuchs (1939, p. 270–272) described lacustrine sediments from the region (Loperot) that are composed mainly of dolomite. There is, however, no direct evidence to link these lake dolostones with those present in the ridge. They show, however, that Mg-enriched fluids have evolved within the southwest Turkana region several times.

Without detailed geochemical investigation, it is difficult to connect the dolomitization directly to hydrothermal fluids. The petrographic evidence shows that dolomitization postdates calcite (or aragonite) precipitation and at least some of the quartz precipitation and silicification (Fig. 7G, H). The $Mg^{2+}$ for dolomitization most likely, therefore, comes from thermal waters that have a high Mg/Ca ratio similar to those on Central Island possibly during later stages of ridge formation, from groundwater following burial of the ridge or from lake water during a period of submergence.

## AGE AND PALEOENVIRONMENTAL SETTING OF THE RIDGE DEPOSITS

Gravity and seismic reflection data obtained in the region during the 1980s showed several extensive rift basins in this region (Morley et al., 1992; Morley et al., 1999); the age of the deeper basinal section, however, could not be established by direct correlation with reflections passing into surface outcrops (Morley et al., 1992; Hendrie et al., 1994). The Anza Graben, of Mesozoic–Tertiary age, which lies to the east of Lake Turkana, represents the known southeastern extent of the Central African rift system (Bosworth and Morley, 1994). During the Amoco exploration project, it was unclear whether basins linked to this system also existed on the western side of the lake (e.g., Williamson and Savage, 1986; Hendrie et al., 1994). Consequently, there remained the possibility of a pre-Cenozoic carbonate section being present, which would have had many paleogeographic implications, and added interesting hydrocarbon exploration possibilities.

With the interpretation of the Ngakoringora Ridge as a fossil hot-spring deposit, there is no reason to invoke Paleozoic or Mesozoic marine strata southwest of Lake Turkana. Nevertheless, the age of the rocks that form the Ngakoringora Ridge is uncertain. The carbonates clearly postdate the phase of faulting that created the fissure for hydrothermal fluid flow, but their stratigraphic relationship to the Lothidok Formation is difficult to establish in the field. The ridge is surrounded by alluvial sediments, and there are no well-exposed contacts between ridge rocks and the Neogene formations. There are two main possibilities: either the ridge is contemporary with the mid-Miocene sequence that forms Lothidok Hills, or the ridge postdates deposition and deformation of the Miocene rocks.

The ridge is clearly a subaerial deposit, and it became a positive topographic feature on the landscape during its formation. The fluids that flowed to the surface to precipitate the carbonates moved up an open fissure that followed the same structural trend as the Lothidok Formation, and, according to Boschetto (1988), the ridge is terminated north and south by faults. The carbonates of the former medial fissure are aligned vertically (Fig. 4), whereas the sedimentary strata of the Lothidok Formation in which the ridge is found are tilted at 10–15° to the southwest. This implies that the ridge formation postdates the faulting and tilting of the Miocene Lothidok strata. It could be argued that the ridge formed on the eastern margin of the North Lokichar Basin while the latter was accumulating sediments. The lack of tilting of the ridge strata and their inferred termination by faults that are younger than the Lothidok Formation make this scenario possible, but unlikely.

An alternative model is that hydrothermal activity began during the Pliocene or Pleistocene after the faulting and tilting that produced the Lothidok Hills. During the period of ridge formation, the local (and probably regional) hydrostatic head probably was much higher than today unless major uplift has occurred west of Lake Turkana since ridge formation. A higher hydrostatic head and locally higher geothermal gradient were needed to produce the thermal springs. Such conditions might have existed following uplift of the Lothidok Hills but before the phase of subsidence that produced the Turkana Depression. A Pliocene or Pleistocene age is considered possible for the age of the ridge, but present evidence precludes a more exact age determination.

Following its subaerial formation, the ridge might have been buried by alluvial sediments or submerged by rising lake water. Although the detailed regional paleogeography is poorly known, mainly because of erosion, the precursors of modern Lake Turkana are known to have existed west of its present location at several stages during the Pliocene, Pleistocene, and Holocene (e.g., Owen et al., 1982; Feibel, 1988; Feibel et al., 1991; Cerling, 1994; Ward et al., 1999). A late Pleistocene–Holocene age is improbable, however, because the ridge probably lies above the inferred maximum level of Lake Turkana (~ 445 m: Owen et al., 1982) during the early and mid-Holocene, and the degree of weathering and patination preclude a late Quaternary age. Furthermore, the locus of volcanism and hydrothermal activity had then already shifted to its present location by the late Pleistocene. From stratigraphic and structural evidence, the ridge must be younger than the deformation that produced the Lothidok Hills and older than the last major Pleistocene rise of Lake Turkana.

## DISCUSSION

Although of small size and extent, fossil travertines such as Ngakoringora Ridge are common in continental rifts and can provide useful paleoenvironmental information of both local and regional significance. Most modern hydrothermal activity is found near sites of active or relatively young faults and (or) volcanism. Most high-enthalpy thermal springs in the Kenya Rift are associated either with faults of late Pleistocene or Holocene age or with older faults that have been rejuvenated periodically and have remained open conduits for fluid flow. The hottest thermal springs tend to lie near sites of recent volcanism and high geothermal gradients, but not necessarily at or near central volcanoes. Chapman et al. (1978) showed that the rift depocenter at the latitude of Lake Baringo (Fig. 1A) migrated eastward progressively from the Miocene to Holocene. As the locus of major faulting and extension moved toward its present position, volcanism gradually ended in the west, and the rift floor became lower to the east. One result of this lateral migration was erosional downcutting and regional depression of water tables as the rift floor subsided. Similar processes have happened elsewhere in the rift with the progressive abandonment of depositional basins and volcanism and their migration to the present axial rift depression (e.g., at Lake Magadi in southern Kenya).

It is highly likely that hydrothermal activity would also migrate from former centers of discharge to new locations, in synchronism with rift faulting, volcanism, and falling groundwater levels. At Lake Baringo, for example, hydrothermal quartz veins dated at 198–345 ka (Le Gall et al., 2000) cut trachytes 7 km west of the present spring activity. In the Tugen Hills west of Lake Baringo, quartz veins line fault planes in the Pliocene Kaparaina Basalts near Kibingor and trachytes near Poi. Although associated surface spring deposits have not been reported, it is probable that these veins record hydrothermal activity that was contemporaneous with that volcanic activity. The same temporal migration of hydrothermal activity is recorded by the travertines of Ngakoringora Ridge. Spring discharge probably ended following regional lowering of the water table and migration of volcanic activity eastward, where hot springs at Central Island, South Island, Loiyangalani and the Barrier (Fig. 1) are the current sites of surface hydrothermal activity and recent volcanism at this latitude. Hydrothermal activity in rifts normally migrates laterally toward the axis of contemporary extension.

In the Kenya Rift, most thermal fluids have acquired a Na–Ca–$HCO_3$–Cl composition, similar to those that evolve from weathering of volcanic rocks by silicate hydrolysis. Those springs with shallow circulation that are fed by groundwater with short residence time tend to discharge fluids containing high $Ca^{2+}$ and relatively low $Cl^-$ concentrations. These fluids commonly precipitate carbonates at the surface, but no sinter. In contrast, fluids that have deeper circulation, a longer residence time, and show less evidence of mixing with near-surface groundwaters commonly discharge waters poor in $Ca^{2+}$ but with high $Cl^-$ and $SiO_2$ concentrations. The Cl-rich springs tend to be associated with border faults or transverse structures that are deeply rooted in basement rocks, such as the hot springs on Ol Kokwe Island in Lake Baringo (Dunkley et al., 1993; Renaut et al., 2002). The abundance of carbonates at Ngakoringora Ridge is more compatible with shallow rather than very deep circulation, or it may reflect deeper Cl-rich fluids that have mixed with shallower Ca-bearing fluids before surface discharge.

A close relationship exists between climate and the presence and composition of spring deposits in continental rifts, such as those at the Ngakoringora Ridge (cf. Sturchio et al., 1993; Darling et al., 1993; Renaut and Jones, 1997). During humid phases, fumaroles and steam vents may become hot springs and geysers as the groundwater recharge increases and water tables rise. The converse is also true: thermal-spring activity may decline or stop during dry climatic phases, or the composition of the fluids may change as the volume of shallow recharge and mixing declines (Renaut and Jones, 1997). Spring deposits do not form constantly, and, depending on the depth and source of recharge, may be good indicators of paleoclimate. During dry climatic phases, deeply sourced springs that tap "purer" geothermal fluids may continue to flow for a prolonged period, while springs with short residence time that are fed by shallow aquifers may dry up.

Faulting can potentially confuse any paleoclimatic signature preserved in spring deposits, by changing the locations of groundwater discharge or modifying the hydrogeology of the geothermal reservoir. Although the effects of faulting have been recognized in other geothermal localities, their impact in the northern Kenya Rift is unclear. Faulting may either open new permeable conduits for fluid flow or destroy or seal existing pathways.

The main imprint of thermal springs on rift sedimentation is as a source of recharge at times when other sources of surface water are absent or in decline (Renaut and Tiercelin, 1994). For the Ngakoringora Ridge, it is difficult to assess that role because the basinal context and age are poorly constrained. In large lakes, and where springs discharge into river channels or extensive wetlands, the thermal springs may contribute only a small proportion of the recharge and be undetectable in the geological record. In small lakes, playas, and semiarid wetlands, however, springs may contribute a high proportion of the annual recharge (cf. Deocampo and Ashley, 1999). It can be important, therefore, to detect the hydrothermal inflow when trying to interpret ancient rift strata, particularly from a paleoclimatic perspective (Renaut and Tiercelin, 1994; Renaut et al., 2002). Large spring deposits, like the Ngakoringora Ridge, provide unambiguous proof of hydrothermal inflow into the contemporary basin, but this is not the case for some thermal springs. Several hot springs that flow into lakes Bogoria (at Loburu) and Magadi, for example, contain < 1 mg $l^{-1}$ $Ca^{2+}$ and < 100 mg $l^{-1}$ $SiO_2$. Although these springs discharge many liters per second of Na–$HCO_3$–Cl water, neither travertine nor sinter is forming around their subaerial vents. Despite the lack of sedimentary record, thermal springs could supply > 50% of the annual recharge during dry years in some basins. The only sedimentary record of the hydrothermal activity at these sites may be chaotically disturbed clastic sediments produced by slumping into the inactive vent, or a sedimentary fill in the submerged vent depression if drowned by rising lake water.

Another impact on rift sedimentation is the potential role of the hydrothermal fluids in diagenesis, not only on the spring deposits themselves but also on other sediments. Sublacustrine thermal fluids circulating through lake sediments, for example, may accelerate some diagenetic reactions, produce atypical mineral assemblages, or promote early maturation of organic matter. At Lake Bogoria, for example, thermonatrite ($Na_2CO_3 \cdot H_2O$), which is stable at high temperatures (> 35°C: Eugster, 1966), is found in lake sediments near sublacustrine hot-spring seepages. Elevated $PCO_2$ may develop in lake sediments where $CO_2$ is exsolved from deeper thermal fluids, resulting in carbonate dissolution or formation of minerals such as nahcolite ($NaHCO_3$).

Large spring deposits like the Ngakoringora Ridge, therefore, can yield potentially valuable palaeoenvironmental information, but like many other facets of rift sedimentology, separating climatic from tectonic roles in the formation of the deposits can be a challenge. Further detailed studies of individual spring deposits, including geochemical analyses and microstratigraphic studies, may eventually yield precise information on seasonal and long-term paleoclimatic variations in continental rifts.

## CONCLUSIONS

The Ngakoringora Ridge is a fissure-ridge travertine mound precipitated by thermal waters that discharged along an open medial fissure. The ridge was originally precipitated as calcite (and) or aragonite. Carbonate precipitation near the vent resulted mainly from $CO_2$ degassing, producing laminated crystal beds. Distally, stromatolites, coated grains, and other fabrics formed in cooler waters of marginal terraces. During and after ridge formation, the carbonates underwent diagenetic alteration that obscured many textures, including local replacement of carbonates by silica, silicification of plants and formation of thin chert beds, and patchy to pervasive replacement of calcite and aragonite by dolomite. At least some diagenetic alteration resulted from thermal waters.

The age of the ridge is poorly constrained. It postdates the faulting that affected the mid-Miocene Lothidok Formation, but it could be as young as Pleistocene. Following ridge formation, hydrothermal activity migrated eastward toward modern Lake Turkana. Hydrothermal sediments can potentially provide useful paleoenvironmental information in rift basins. When interpreting ancient continental rift-basin fills, it is important to recognize the role that hydrothermal processes might have played in sedimentation and diagenesis.

## ACKNOWLEDGMENTS

We are grateful to the Natural Sciences and Engineering Research Council of Canada, which funded this research (grants GP0000629 to RWR; A6090 to BJ). We thank the Office of the President, Government of Kenya, for permission to undertake this research, and the National Oil Corporation of Kenya, especially Mr. F. Mbatau, for logistical support. We also thank Christophe Coussement and Vincent Kipiego for help in the field; Hank Chafetz, Li Guo, and Robert Riding, for their very helpful reviews; and Gail Ashley, who offered advice and handled the manuscript.

## REFERENCES

ALLEN, D.J., AND DARLING, W.G., 1992, Geothermics and hydrogeology of the Kenya Rift Valley between Lake Baringo and Lake Turkana: British Geological Survey, Research Report SD/92/1, 39 p.

ALLEN, D.J., DARLING, W.G., AND BURGESS, W.G., 1989, Geothermics and hydrogeology of the southern part of the Kenya Rift Valley with emphasis on the Magadi–Nakuru area: British Geological Survey, Research Report, Hydrogeology Series, SD/89/1, 68 p.

ALTUNEL, E., AND HANCOCK, P.L., 1993a, Morphology and structural setting of Quaternary travertines at Pamukkale, Turkey: Geological Journal, v. 28, p. 335–346.

ALTUNEL, E., AND HANCOCK, P.L., 1993b, Active fissuring and faulting in Quaternary travertines at Pamukkale, western Turkey: Zeitschrift für Geomorphologie, Neue Folge, Supplementband 94, p. 285–302.

ARAMBOURG, C., 1943, Contribution á l'étude géologique et paléontologique du bassin du Lac Rudolphe et de la Basse Vallé de l'Omo. Mission Scientifique de l'Omo, 1932–1933: Géologie-Anthropologie, v. 1, p. 1–74.

BARGAR, K.E., 1978, Geology and thermal history of Mammoth Hot Springs, Yellowstone National Park, Wyoming: U.S. Geological Survey, Bulletin 1444, 55 p.

BEHR, H.-J., AND RÖHRICHT, C., 2000, Record of seismotectonic events in siliceous cyanobacteria sediments (Magadi cherts), Lake Magadi, Kenya: International Journal of Earth Sciences, v. 89, p. 268–283.

BOSCHETTO, H.B., 1988, Geology of the Lothidok Range: M.Sc. Thesis, University of Utah, 203 p.

BOSCHETTO, H.B., BROWN, F.H., AND MCDOUGALL, I., 1992, Stratigraphy of the Lothidok Range, northern Kenya, and K/Ar ages of its Miocene primates: Journal of Human Evolution, v. 22, p. 47–71.

BOSWORTH, W., AND MORLEY, C.K., 1994, Structural and stratigraphic evolution of the Anza Rift, Kenya: Tectonophysics, v. 236, p. 93–115.

CASANOVA, J., 1986, East African Rift stromatolites, *in* Frostick, L.E., Renaut, R.W., Reid, I., and Tiercelin, J.-J., eds., Sedimentation in the African Rifts: Geological Society of London, Special Publication 25, p. 201–210.

CASANOVA, J., 1994, Stromatolites from the East African Rift: a synopsis, *in* Bertrand-Safarti, J., and Monty, C., eds., Phanerozoic Stromatolites, II: Dordrecht, The Netherlands, Kluwer, p. 193–226.

CERLING, T.E., 1994, Lake Turkana and its precursors in the Turkana Basin, East Africa (Kenya and Ethiopia), *in* Gierlowski-Kordesch, E., and Kelts, K.K., eds., Global Geological Record of Lake Basins, Volume 1: Cambridge, U.K., Cambridge University Press, p. 341–343.

CHAFETZ, H.S., 1986, Marine peloids: a product of bacterially induced precipitation of calcite: Journal of Sedimentary Petrology, v. 56, p. 812–817.

CHAFETZ, H.S., AND FOLK, R.L., 1984, Travertines: depositional morphology and the bacterially constructed constituents: Journal of Sedimentary Petrology, v. 54, p. 289–316.

CHAFETZ, H.S., AND GUIDRY, S., 1999, Bacterial shrubs, crystal shrubs, and ray-crystal shrubs: bacterial vs. abiotic precipitation: Sedimentary Geology, v. 126, p. 57–74.

CHAFETZ, H.S., AKDIM, B., JULIA, R., AND REID, A., 1998, Mn- and Fe-rich black travertine shrubs: bacterially (and nanobacterially) induced precipitates: Journal of Sedimentary Research, v 68, p. 404–412.

CHAFETZ, H.S., AND LAWRENCE, J.R., 1994, Stable isotopic variability within modern travertines: Géographie Physique et Quaternaire, v. 48, p. 257–273.

CHAFETZ, H.S., SRDOC, D., AND HORVATINCIC, N., 1994, Early diagenesis of Plitvice Lakes waterfall and barrier travertine deposits: Géographie Physique et Quaternaire, v. 48, p. 247–255.

CHAMPION, A.M., 1937, Physiography of the region to the west and southwest of Lake Rudolf: Geographical Journal, v. 89, p. 97–118.

CHAPMAN, G.R., LIPPARD, S.J., AND MARTYN, J.E., 1978, The structure of the Kamasia Range, Kenya Rift Valley: Geological Society of London, Journal, v. 135, p. 265–281.

CHESTERMAN, C.W., AND KLEINHAMPL, F.J., 1991, Travertine Hot Springs, Mono County, California: California Geology, v. 44, p. 171–182.

CIONI, R., FANELLI, G., GUIDI, M., KINYARIRO, J.K., AND MARINI, L., 1992, Lake Bogoria hot springs (Kenya): geochemical features and geothermal implications: Journal of Volcanology and Geothermal Research, v. 50, p. 231–246.

Clarke, M.C.G., Woodhall, D.G., Allen, D., and Darling, G., 1990, Geological, volcanological and hydrogeological controls on the occurrence of geothermal activity in the area surrounding Lake Naivasha, Kenya: Report of the Ministry of Energy, Kenya. Nairobi, 138 p.

Curewitz, D., and Karson, J.A., 1997, Structural settings of hydrothermal outflow: Fracture permeability maintained by fault propagation and interaction: Journal of Volcanology and Geothermal Research, v. 79, p. 149–168.

Darling, W.G, Allan, D.J., and Spiro, B., 1993, Oxygen isotopes in relict sinters as a palaeoindicator of Rift Valley climate, *in* Isotope Techniques in the Study of Past and Current Environmental Changes in the Hydrosphere and the Atmosphere: International Atomic Energy Agency, Vienna, Proceedings Series, STI/PUB/908, p. 555–559.

Deocampo, D., and Ashley, G.M., 1999, Siliceous islands in a carbonate sea: modern and Pleistocene spring-fed wetlands in Ngorongoro Crater and Oldupai Gorge, Tanzania: Journal of Sedimentary Research, v. 69, p. 974–979.

Dunkley, P.N., Smith, M., Allen, D.J., and Darling, W.G., 1993, The geothermal activity and geology of the northern sector of the Kenya Rift Valley: British Geological Survey, Research Report SC/93/1, 185 p.

Eugster, H.P., 1966, Sodium carbonate–bicarbonate minerals as indicators of $PCO_2$: Journal of Geophysical Research, v. 71, p. 3369–3377.

Feibel, C.S., 1988, Paleoenvironments of the Koobi Fora Formation, Turkana Basin, Northern Kenya: Ph.D. Dissertation, University of Utah, 330 p.

Feibel, C.S., Harris, J.M., and Brown, F.H., 1991, Neogene paleoenvironments of the Turkana Basin, *in* Harris, J.M., ed., Koobi For a Research Project, Volume 3, Stratigraphy, Artiodactyls and Paleoenvironments: Oxford, U.K., Clarendon Press, p. 321–346.

Folk, R.L., 1994, Interaction between bacteria, nannobacteria, and mineral precipitation in hot springs of central Italy: Géographie Physique et Quaternaire, v. 48, p. 233–246.

Folk, R.L., and Chafetz, H.S., 2000, Bacterially induced microscale and nanoscale carbonate precipitates, *in* Riding, R.E., and Awramik, S.M., eds., Microbial Sediments: Berlin, Springer, p. 40–49.

Folk, R.L., Chafetz, H.S., and Tiezzi, P.A., 1985, Bizarre forms of depositional and diagenetic calcite in hot spring travertines, central Italy, *in* Schneidermann, N., and Harris, P., eds., Carbonate Cements: SEPM, Special Publication 36, p. 349–369.

Fontes, J.C., and Puchan, D., 1975, Les cheminées du lac Abbé (TFAI): stations hydroclimatiques de l'Holocène: Academie des Sciences [Paris], Comptes Rendus, v. 280, p. 383–386.

Ford, T.D., and Pedley, H.M., 1996, A review of tufa and travertine deposits of the world: Earth-Science Reviews, v. 41, p. 117–175.

Fuchs, V.E., 1934, The geological work of the Cambridge Expedition to the East African Lakes, 1930–31: Geological Magazine, v. 71, p. 97–116.

Fuchs, V.E., 1939, The geological history of the Lake Rudolf basin, Kenya Colony: Royal Society [London], Philosophical Transactions, Section B, v. 560, p. 219–274.

Goldsmith, J.R., Graf, D., and Heard, H.C., 1961, Lattice constants of the calcium–magnesium carbonates: American Mineralogist, v. 46, p. 453–457.

Guo, L., and Riding, R.E., 1992, Micritic aragonite laminae in hot water travertine crusts, Rapolano Terme, Italy: Sedimentology, v. 39, p. 1067–1079.

Guo, L., and Riding, R.E., 1994, Origin and diagenesis of Quaternary shrub fabrics, Rapolano Terme, Italy: Sedimentology, v. 41, p. 499–520.

Guo, L., and Riding, R.E., 1998, Hot-spring travertine facies and sequences, Late Pleistocene, Rapolano Terme, Italy: Sedimentology, v. 45, p. 163–180.

Guo, L., and Riding, R.E., 1999, Rapid facies changes in Holocene fissure ridge hot spring travertines, Rapolano Terme, Italy: Sedimentology, v. 46, p. 1145–1158.

Hemming, C.F., and Trapnell, C.G., 1957, A reconnaissance classification of the soils of the South Turkana Desert: Journal of Soil Science, v. 8, p. 167–183.

Hendrie, D.B., Kusznir, N.J., Morley, C.K., and Ebinger, C.J., 1994, Cenozoic extension in northern Kenya: a quantitative model of rift basin development in the Turkana area: Tectonophysics, v. 236, p. 409–438.

Jones, B., and Renaut, R.W., 1995, Noncrystallographic calcite dendrites from hot-spring deposits at Lake Bogoria, Kenya: Journal of Sedimentary Research, v. 65, p. 154–169.

Jones, B., and Renaut, R.W., 1996, Morphology and growth of aragonite crystals in hot spring travertines at Lake Bogoria, Kenya Rift Valley: Sedimentology, v. 43, p. 323–340.

Jones, B., and Renaut, R.W., 1998, Origin of platy calcite crystals in hot-spring deposits of the Kenya Rift Valley: Journal of Sedimentary Research, v. 68, p. 913–926.

Jones, B., Renaut, R.W., Rosen, M.R. and Klyen, L. 1998. Primary siliceous rhizoliths from Loop Road hot springs, North Island, New Zealand: Journal of Sedimentary Research, v. 68, p. 115–123.

Koban, C.G., and Schweigert, G., 1993, Microbial origin of travertine fabrics—two examples from southern Germany (Pleistocene Stuttgart Travertines and Miocene Riedöschingen Travertine): Facies, v. 29, p. 251–264.

Le Gall, B., Tiercelin, J.-J., Richert, J.-P., Gente, P., Sturchio, N.C., Stead, D., and Le Turdu, C., 2000, A morphotectonic study of an extensional fault zone in a magma-rich rift: the Baringo Trachyte Fault System, central Kenya Rift: Tectonophysics, v. 320, p. 87–106.

Lloyd, E.F., 1959, The hot springs and hydrothermal eruptions of Waiotapu: New Zealand Journal of Geology and Geophysics, v. 2, p. 141–176.

Love, K.M., and Chafetz, H.S., 1990, Petrology of Quaternary travertine deposits, Arbuckle Mountains, Oklahoma, *in* Herman, J.S., and Hubbard, D.A., eds., Travertine–Marl: Stream Deposits in Virginia: Virginia Division of Mineral Resources, Publication 101, p. 65–77.

Miller, J., 1988, Microscopical techniques: I. Slices, slides, stains and peels, *in* Tucker, M.E., ed., Techniques in Sedimentology: Oxford, U.K., Blackwell, p. 86–107.

Morley, C.K., ed., 1999, Geoscience of Rift Systems—Evolution of East Africa: American Association of Petroleum Geologists, Studies in Geology no. 44, 250 p.

Morley, C.K., Wescott, W.A., Stone, D.M., Harper, S.T., Wigger, S.T., and Karanja, F.M., 1992, Tectonic evolution of the northern Kenya Rift: Geological Society of London, Journal, v. 149, p. 333–348.

Morley, C.K., Wescott, W.A., Stone, D.M., Harper, S.T., Wigger, S.T., Day, R.A., and Karanja, F. M., 1999, Geology and geophysics of the Western Turkana Basins, Kenya, *in* Morley, C.K., ed., Geoscience of Rift Systems—Evolution of East Africa: American Association of Petroleum Geologists, Studies in Geology no. 44, p. 19–54.

Owen, R.B., Barthelme, J., Renaut, R.W., and Vincens, A., 1982. Palaeolimnology and archaeology of Holocene deposits north-east of Lake Turkana, Kenya: Nature, v. 298, p. 523–529.

Pentecost, A., and Viles, H., 1994, A review and reassessment of travertine classification: Géographie Physique et Quaternaire, v. 48, p. 305–314.

Renaut, R.W., and Jones, B., 1997, Controls on aragonite and calcite precipitation in hot spring travertines at Chemurkeu, Lake Bogoria, Kenya: Canadian Journal of Earth Sciences, v. 34, p. 801–818.

Renaut, R.W., and Jones, B., 2000. Microbial precipitates around continental hot springs and geysers, *in* Riding, R.E., and Awramik, S.M., eds., Microbial Sediments: Berlin, Springer, p. 187–195.

Renaut, R.W., and Owen, R.B., 1988, Opaline cherts associated with sublacustrine hydrothermal springs at Lake Bogoria, Kenya Rift Valley: Geology, v. 16, p. 699–702.

Renaut, R.W., Jones, B., and Le Turdu, C., 1999, Calcite lilypads and ledges at Lorusio Hot Springs, Kenya Rift Valley: travertine precipitation at the air–water interface: Canadian Journal of Earth Sciences, v. 46, p. 649–666.

Renaut, R.W., Jones, B., and Tiercelin, J.J., 1998, Rapid *in situ* silicification of microbes at Loburu hot springs, Lake Bogoria, Kenya Rift Valley: Sedimentology, v. 45, p. 1083–1103.

Renaut, R.W., Jones. B., Tiercelin, J.-J., and Tarits, C., 2002, Sublacustrine precipitation of hydrothermal silica in rift lakes: evidence from Lake Baringo, central Kenya Rift Valley: Sedimentary Geology, v. 148, p. 235–257.

Renaut, R.W., and Tiercelin, J.J., 1987, Alimentation, hydrologie, *in* Tiercelin, J.J., and Vincens, A., eds., Le demi-graben de Baringo–Bogoria, Rift Gregory, Kenya: Centres Recherches Exploration–Production Elf–Aquitaine, Bulletin, v. 11, p. 284–309.

Renaut, R.W., and Tiercelin, J.J., 1994, Lake Bogoria, Kenya Rift Valley—a sedimentological overview, *in* Renaut, R.W., and Last, W.M., eds., Sedimentology and Geochemistry of Modern and Ancient Saline Lakes: SEPM, Special Publication 50, p. 101–123.

Renaut, R.W., Tiercelin, J.-J., and Owen, R.B., 1986, Mineral precipitation and diagenesis in the sediments of the Lake Bogoria basin, Kenya Rift Valley, *in* Frostick, L.E., Renaut, R.W., Reid, I., and Tiercelin, J.-J., eds., Sedimentation in the African Rifts: Geological Society of London, Special Publication 25, p. 159–176.

Sturchio, N.C., Dunkley, N., and Smith, M., 1993. Climate-driven variations in geothermal activity in the northern Kenya rift valley: Nature, v. 362, p. 233–234.

Tiercelin, J.-J., Pflumio, C., Castec, M., Boulègue, J., Gente, P., Rolet, J., Coussement, C., Stetter, K.O., Huber, R., Buku, S., Mifundu, W., 1993, Hydrothermal vents in Lake Tanganyika, East African Rift System: Geology, v. 21, p. 499–502.

Tole, M.P., 1988, Low enthalpy geothermal systems in Kenya: Geothermics, v. 17, p. 777–783.

Tole, M.P., 1997, Geothermal energy research in Kenya: a review: Journal of African Earth Sciences, v. 23, p. 565–575.

Walsh, J., and Dodson, R.G., 1969, Geology of Northern Turkana: Geological Survey of Kenya, Report 82, 42 p.

Ward, V.V., Leakey, M.G., Brown, B., Brown, F., Harris, J., and Walker, A., 1999, South Turkwel: A new Pliocene hominid site in Kenya: Journal of Human Evolution, v. 36, p. 69–95.

Watkins, R.T., 1983, The geology of the Suregei–Asille District and the Upper Bakate Valley, northern Kenya: Ph.D. thesis, University of London, 390 p.

Williamson, P.G., and Savage, R.J.G., 1986, Early rift sedimentation in the Turkana basin, northern Kenya, , *in* Frostick, L.E., Renaut, R.W., Reid, I., and Tiercelin, J.-J., eds., Sedimentation in the African Rifts: Geological Society of London, Special Publication 25, p. 267–283.

Worthington, E.B., 1932, Scientific results of the Cambridge Expedition to the East African lakes, 1930–31: Journal of the Linnean Society, Zoology, v. 38, p. 99–119.

Wohletz, K., and Heiken, G., 1992, Volcanology and Geothermal Energy: Berkeley, California, University of California Press, 432 p.

# PALEOHYDROLOGICAL SIGNATURES AND RIFT TECTONICS IN THE INTERIOR OF GONDWANA DOCUMENTED FROM UPPER PERMIAN LAKE DEPOSITS, THE MID-ZAMBEZI RIFT BASIN, ZAMBIA

KEDDY YEMANE

*Earth and Environmental Science Department, University of Pennsylvania, 240 South 33rd Street, Philadelphia, Pennsylvania 19104-6316, U.S.A.*

DJORDJE GRUJIC

*Department of Earth Sciences, Dalhousie University, Halifax, Nova Scotia B3H 3J5, Canada*

AND

IMASIKU NYAMBE

*Geology Department, School of Mines, University of Zambia, Lusaka, Zambia*

ABSTRACT: Paleoenvironmental signatures deduced from the sedimentary record of the Upper Permian Madumabisa Formation in the Mid-Zambezi Basin indicate moderate climate and positive hydrologic balance in continental interiors of Gondwana (50–60° S) during the Late Permian. The inferred climate conditions are congruent with the prolific Late Permian terrestrial ecosystems that characterized south-central Africa and equivalent latitudes in Gondwana. Our results do not support the severe aridity and extreme seasonal temperature variations simulated by numerical climate models. The Madumabisa Formation is a lacustrine sequence that represents the maximum extensional phase of the Mid-Zambezi Rift basin. The basin preserves > 5 km of continental sedimentary rocks, of which the Upper Permian Madumabisa Formation comprises freshwater lacustrine deposits up to 700 m thick. Deposition occurred in a fault-controlled asymmetric half graben formed by passive rifting driven by plate boundary stresses that originated from the southern edge of Gondwana. The sedimentary sequence is dominated by black and dark greenish gray mudstones and limestones interbedded with minor gray siltstones and sandstones. The mudstones are predominantly internally massive or bioturbated, but thin-bedded and thinly laminated dark shales with moderate quantities of organic matter and carbonate occur in the lower part of the sedimentary sequence. Laminated limestones are interbedded with the mudstones and comprise fossil-rich beds and inorganic micrite beds precipitated from lake waters, most likely during periods of low clastic supply. Minor current-rippled siltstones and sandstones indicate that silt-laden flows reached deep parts of the basin. Judging from the distribution of the remains of the Madumabisa Formation, the ancient "Madumabisa Lake" covered 140,000 km$^2$, and was larger than most modern lakes. The mainly dark color of the rocks and thinly laminated shales and carbonate content suggest that the lake was deep and experienced periodic, perhaps seasonal, thermal stratification. Abundant biogenic remains including conchostracods, bivalves, fish, and algae indicate a highly productive freshwater lake. Lack of features indicative of emergence and desiccation, such as evaporites, soil features, mudcracks, or cavities, suggests that the lake was perennial. Preserved plant debris including leaves and wood logs indicate that uplands in the watershed were vegetated. For most of the history of the lake system, the perennial lake remained hydrologically closed because of a steady subsidence and moderate axial sediment supply. At the end of the Permian, pulses of basinal shortening caused shallowing of the lake. Basin uplift associated with postulated basin inversion led to changes in sediment supply patterns from axial to rift flanks, as shown by paleocurrent directions. The lake sediments were eroded and buried rapidly, as indicated by the lack of paleosol features, and the lake was filled by thick, red conglomeratic sandstones of the Lower Triassic Escarpment Grit Formation. Contrary to predictions of numerical climate models, abundant moisture and cool (to moderately warm) temperatures would have been essential to sustain a large, perennial, freshwater lake such as "Madumabisa Lake" was, deep within the continental interior of south Gondwana.

## INTRODUCTION

The unique paleogeography of vast continental interiors (e.g., Scotese et al., 1999) and the greatest mass extinction event at the end of the Permian (e.g., Erwin, 1999) makes the study of Late Permian climate changes in Gondwana crucial to the understanding of climate dynamics in the interiors of supercontinents (Fig. 1). Early geological studies have established that climate in Gondwana changed progressively from frigid to warm between the Carboniferous and the Jurassic as Gondwana drifted northward from the South Pole to lower latitudes (e.g., Du Toit, 1954). The distribution of climate-sensitive rocks (Robinson, 1973), pollen assemblages (Kremp, 1977), and phytogeographic patterns (Ziegler, 1990) show that the Permian southern cool temperate humid belt in Gondwana lay between 45° and 60°. During this time, central and southern Africa lay within these paleolatitudes deep in the continental interior of southern Gondwana. Floral and faunal remains of the Late Permian terrestrial ecosystem, representative of this cool temperate biome, are well preserved in fluviolacustrine deposits throughout southern Gondwana, including central to southern Africa. The preponderance of Upper Permian fluvio-lacustrine deposits throughout the southern African subcontinent indicates the existence of large perennial freshwater lakes within major fluvial frameworks (e.g., Smith et al., 1993; Yemane, 1993; Yemane and Kelts, 1990), the magnitude of which is unparalleled in the interiors of modern continents.

Gondwana-age deposits in central and southern Africa comprise continental sedimentary successions of Late Carboniferous (≈ 286 Ma) to Early Jurassic (180 Ma) age, regionally known as the Karoo Supergroup (Table 1). The sedimentary record of Upper Permian deposits in south-central Africa indicates that a cool to moderate climate characterized the continental interior of Gondwana (e.g., Blenkinsop and Oesterlen, 1996; Drysdall and Weller, 1966; Kreuser et al., 1990; Oesterlen and Millsteed, 1994; Wopfner, 1989; Yemane, 1993; Yemane et al., 1996; Yemane

Sedimentation in Continental Rifts
SEPM Special Publication No. 73, Copyright © 2002
SEPM (Society for Sedimentary Geology), ISBN 1-56576-082-4, p. 143–162.

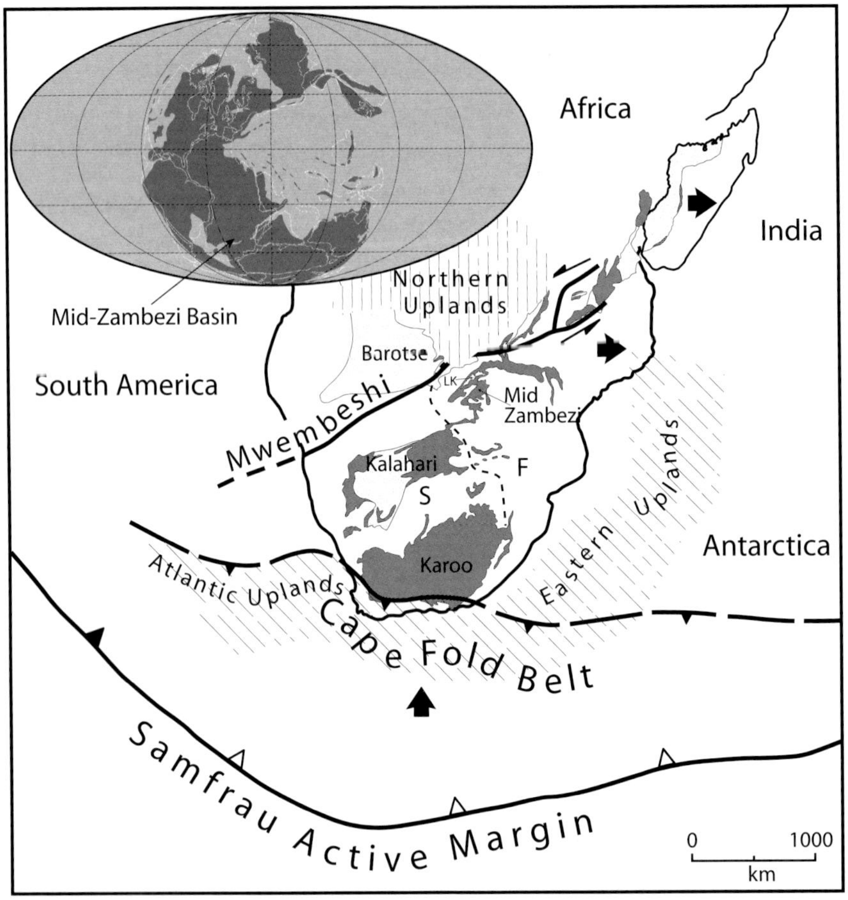


FIG. 1.—Tectonic model for the structure of Karoo basins. Karoo basins in southern Africa developed as foreland basins or sag basins in front of the Cape fold and thrust belt (which was part of the Samfrau Orogen along the southern Panthalassan margin of Gondwana). Karoo rift basins in Central Africa formed by strike-slip and transtensional movement along major fault zones that are mostly reactivated pan-African structures. Outcrop (gray) and subsurface (stippled) extent for all basins shown is from Veevers et al. (1994a). An approximately north–south line divides Karoo rift basins in the east from the open basins (sags) in the southwest. **S**, Karoo terrane dominated by open basin and swell; **F**, "Zambezian" terrane dominated by faults (after Veevers et al., 1994b; Rust, 1975); **K**, Lake Kariba. Uplands after Veevers et al. (1994a). Plate reconstruction and orientations of the major structures are from Daly et al. (1989). Inset: paleogeographic map of Gondwana for the Late Permian (Kazanian age) after Ziegler et al. (1997).

and Kelts, 1990, 1996; Yemane et al., 1998; Yemane et al., 1989). However, paleoclimate simulations based on numerical models (e.g., Crowley, 1993; Fawcett and Barron, 1993; Kutzbach and Gallimore, 1989; Kutzbach and Ziegler, 1993; Moore et al., 1992; Moore and Peoples, 1990; Moore et al., 1990), have consistently produced seasonally extreme climates for the Late Permian at the high latitudes in the interior of Gondwana. Model simulations for the paleolatitudes of central-southern Africa in particular show summer temperatures in the range of 40 to 60°C and winter temperatures below -40°C. Such climate extremes would have rendered southern Africa inhospitable to both plant and animal life, when in fact the primary Late Permian fossil sites for the most diverse and abundant mammal-like reptiles, amphibians, and aquatic fauna are located in central to southern Africa (South Africa, Zambia, Zimbabwe, Malawi, and Tanzania). The paleobotanical data also show that luxurious *Glossopteris* woodland forests covered uplands and riverbanks (cf. Anderson, 1977), providing ample foliage, stems, and undergrowth for the thriving herds that roamed the mudflats and floodplains (cf. Hotton et al., 1986).

Despite the abundance of geologic and biologic data, discord remains regarding comparisons between model results and empirical data for the interiors of Gondwana at high latitudes. Discrepancies between numerically simulated climates and those deduced from the rock record for Gondwana have been attributed to selective preservation of local climate signatures in the geologic record. Crowley (1993) and Crowley et al. (1989) suggest that lake-margin and interpaludal depositional environments may have been selectively represented from Late Permian Gondwana times, whereas Rees et al. (1999) postulate that riparian flora is preferentially preserved in the flora of Early Permian Gondwana. Yemane (1993) and Ziegler (1993) argued that the presence of vast perennial, inland freshwater lake systems and interconnecting rivers would have strongly damped the continentality of the climate in interiors of large continents. Subsequent simulations incorporating the lakes and paleotopography show amelioration of seasonal surface temperatures, suggesting regional damping due to dynamical and hydrological effects (Kutzbach and Ziegler, 1993).

In this paper, we present climate signatures deduced from the sedimentary record of the Mid-Zambezi Basin, a high-latitude rift lake basin in central Gondwana (Fig. 1). The sedimentary record of the Mid-Zambezi Basin is suitable for documenting climate and paleohydrology conditions, for the following reasons: (1) The Basin was located at the heart of Gondwana (cf. Scotese and Langford, 1995), far removed from coastal meteoric influences, where the variance between model simulations and the rock record is greatest (Crowley, 1993; Kutzbach and Gallimore, 1989; Moore and Peoples, 1990; Moore et al., 1990). (2) Biostratigraphy within the Basin is well constrained (Bond, 1974; Falcon, 1973; Nyambe and Utting, 1997). (3) The thick sedimentary sequence is well preserved, continuous, and accessible (Nyambe, 1993; Tavener-Smith, 1958). In evaluating the attributes of climate signatures archived in the sedimentary record, we present preliminary structural analyses of both the rift sediments and the basement rocks in order to estimate the influence of basin tectonics on sedimentation. We compare our results to those from well-studied equivalent neighboring Karoo rift basins within the tectonic framework of southern Gondwana. We conclude that climate signatures preserved in this high-latitude rift basin are not artifacts of selective representation but represent the paleoclimate and paleohydrologic conditions that existed in the continental interiors of Gondwana during the time of deposition.

## TECTONIC SETTING OF THE MID-ZAMBEZI BASIN

Southwestern Gondwana consisted of a series of stable cratonic nuclei (basement) separated by the pan-African system of mobile belts representing zones of structural weaknesses (Daly et al., 1989) (Fig. 2). During the late Paleozoic the tectonic activity in south Gondwana was mainly controlled by subduction of the paleo-Pacific (also known as Panthalassa) plate under the Gondwana plate (Visser, 1992; Visser and Praekelt, 1996; Pysklywec and Mitrovica, 1999), which led to formation of the Gondwanides or Samfrau Orogen (e.g., Du Toit, 1937). Transpressive strike-slip motion along the active paleo-Pacific

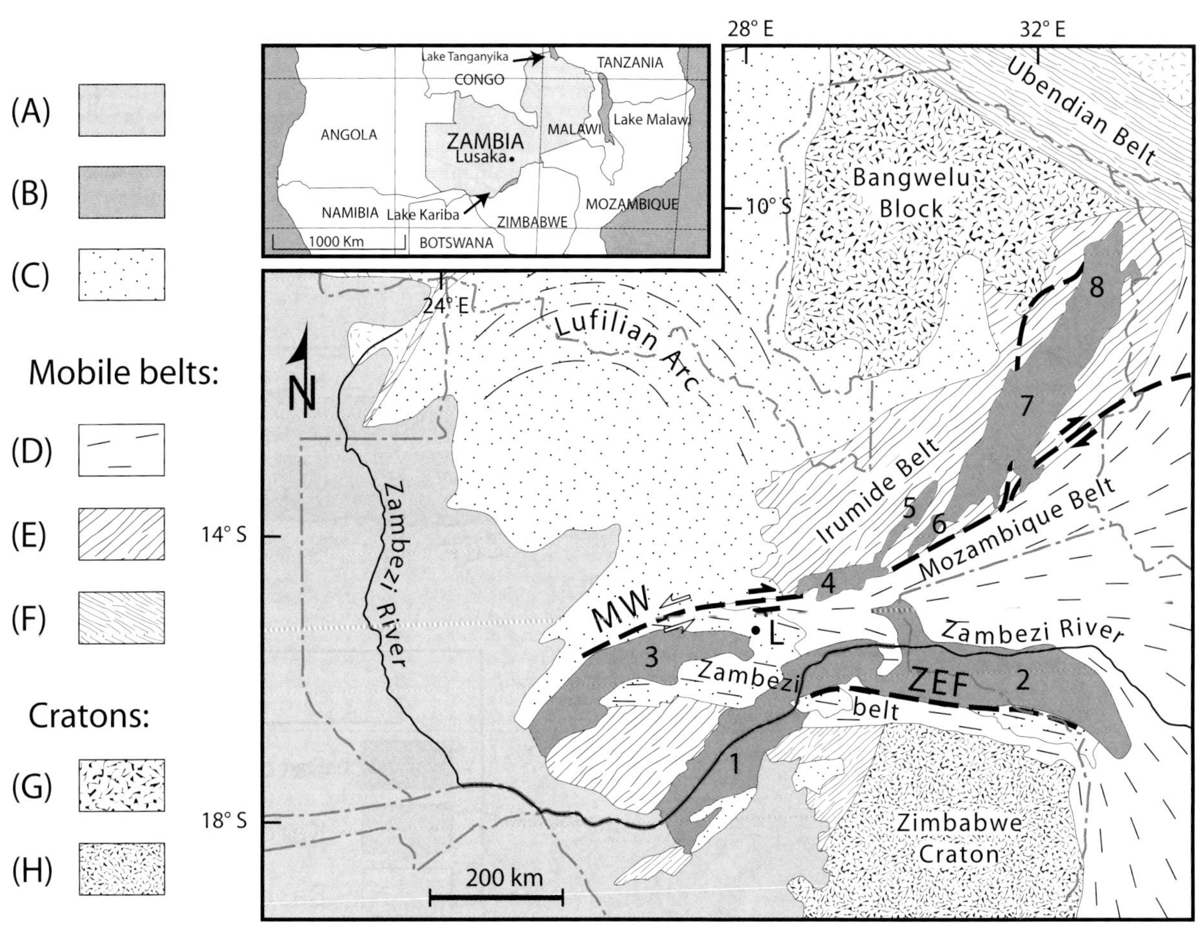


FIG. 2.—Regional setting of Karoo sedimentary basins within the tectonic framework of southern African mobile belts and cratons (from Daly et al., 1989; Hanson et al., 1988). A) Phanerozoic cover sediments; B) main Karoo rift basins (1, Mid Zambezi; 2, Lower Zambezi; 3, Kafue; 4, Luano; 5, Lukusashi; 6, South Luangwa; 7, Mid Luangwa; 8, North Luangwa); C) Katangan (1310–870 Ma) and possible equivalents; D) Pan-African (656–503 Ma); E) Middle Proterozoic; F) Lower Proterozoic; G) Lower Proterozoic; H) Archean gneisses. **L**, Lusaka; **MW**, Mwembeshi shear zone; empty half-arrows indicate pan-African sinistral shear, and black half-arrows indicate Karoo dextral reactivation of the zone; **ZEF**, Zambezi Escarpment fault (after Oesterlen and Blenkinsop, 1994). International borders of Zambia and adjacent countries are shown by gray broken lines. Inset: location map for Zambia and the Mid-Zambezi basin.

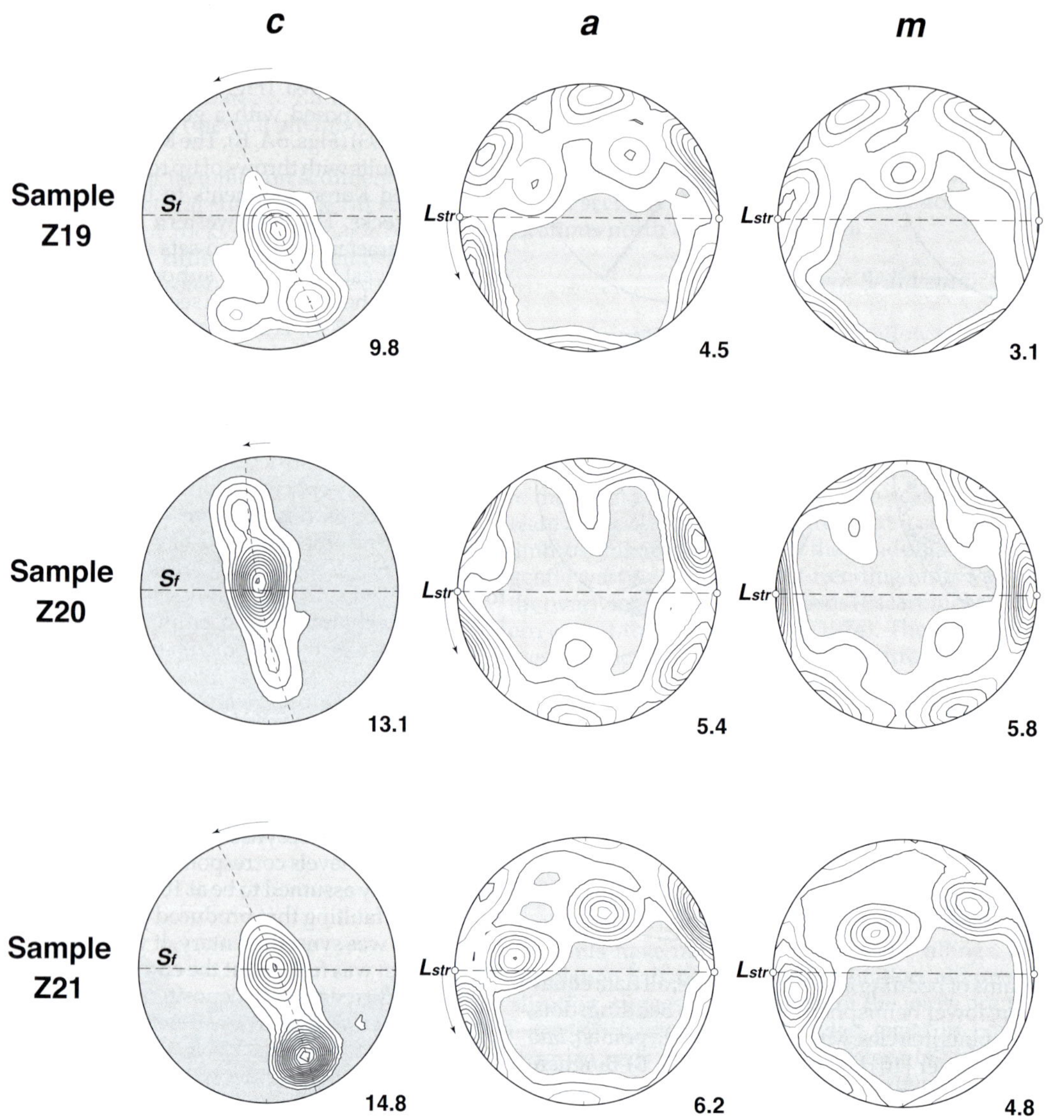


FIG. 7.—Quartz textures from the Mwembeshi shear zone from locations north of Lusaka (Fig. 3). Crystallographic axes *c*, *a*, and *m* are plotted. The crystallographic axes are plotted in the sample reference frame such that the mylonitic foliation is vertical striking east–west and the stretching lineation is horizontal also striking east–west. All data equal-area projection, lower hemisphere. Contours are isolines of "times uniform" distribution. The maximum densities are indicated next to each diagram, gray below density contour 1.

partial girdle of *c* axes against the mylonitic foliation and the obliquity of the largest maximum of *a* axes relative to the macroscopic lineation is consistent and indicates simple shear strain (Fig. 7). Because the foliation dips moderately to the southwest and the lineation plunges to the west-southwest, all three studied outcrops show characteristics of oblique-slip fault, in this case dextral-normal fault (Fig. 3). The different sense of shear (e.g., sinistral strike-slip) than is commonly assumed for the pan-African event (Hanson et al., 1993) and the lower metamorphic conditions of deformation suggest that the Mwembeshi shear zone was reactivated. The deformation mechanisms in the quartz mylonites are indicative of greenschist facies conditions, as shown by the topologies of the crystallographic axes (Fig. 7), in which the dislocation glide in the quartz crystals was mainly on the rhomb glide planes in the <*a*> direction. This is consistent with the observed characteristic structures, such as dynamic recrystallization by subgrain rotation and "sweeping" undulose extinction. The peak metamorphic conditions of the mylonites and thus of the activity of the Mwembeshi shear zone are, however, of the eclogite facies (Vrána et al., 1975), which occurred during the Late Proterozoic. Thus, the oblique normal fault shearing that produced the measured quartz textures at greenschist facies conditions occurred after this part of the Mwembeshi shear zone had been exhumed to the corresponding crustal level, either at the end of pan-African orogeny or at the initial rifting stages. Later sinistral strike-slip brittle movement is not recorded in the investigated outcrops, but it can be inferred from regional observations (e.g., Daly et al., 1989).

There are indications of both dextral and sinistral strike-slip movement along the Mwembeshi shear zone during the Paleozoic. Although both the Mid-Zambezi and the Luangwa basins are oriented subparallel to the Mwembeshi shear zone (Fig. 1), the sense of strike-slip faulting in the Mid-Zambezi basin is not conclusive. The kinematics of the rifting of Karoo basins is, therefore, not yet elucidated. On the basis of internal fault geometry Daly et al. (1989) postulated an east-southeast extension regime for the Mid-Zambezi Basin (Fig. 1). The northwest margin of the Mid-Zambezi Basin (Fig. 3) is defined by two sets of faults. Longer segments dip to the southeast and are connected by shorter faults that dip to south to south-southeast (Oesterlen and Blenkinsop, 1994). The southeast-dipping faults constitute a northeast-trending en echelon array of faults, and the left-hand stepover of these fault segments indicate a dextral shear component along the array. A dextral strike-slip component along a border fault has also been observed in outcrop (Fig. 5A). By contrast, the Lower Zambezi and Luangwa basins appear to have formed as pull-apart basins, in which the southern margin of the Lower Zambezi Basin is defined by a dextral strike-slip fault (the Zambezi Escarpment fault) reactivated along a pan-African sinistral shear zone (Orpen et al., 1989). However, more detailed fault analysis (Oesterlen and Blenkinsop, 1994) has shown that the Lower Zambezi Basin is an extensional basin formed by maximum extension oriented perpendicular to the rift zone. This observation is supported by the interpretation of the seismic data that show that the the Zambezi Escarpment fault (Figs. 2, 3) is a north-dipping normal fault (Hiller and Buttkus, 1996). This, however, does not exclude a potential strike slip (out of plane of the section) displacement along the fault. In addition, the likely dextral shear component along the northwestern margin of the Mid-Zambezi Basin is consistent with postulated later dextral movement along the Mwembeshi shear zone that caused the phases of basin inversion in the similarly oriented Luangwa Basin (Banks et al., 1995). The orientation of the folds in Karoo deposits is compatible with dextral strike-slip movements and supports arguments of basin inversion. Detailed analyses of fault and fold kinematics from all the rift basins, with age control on the deformation events, are needed to distinguish between the existing models (or to establish a new one).

## DEPOSITION IN THE MID-ZAMBEZI BASIN

### *The Madumabisa Formation*

The Mid-Zambezi Basin subsided continuously from the Carboniferous through the Jurassic and preserved > 5 km of synrift sedimentary rocks (Table 1). The rocks unconformably overlie the Precambrian basement or rest locally on the Sinakumbe Group (Ordovician–Devonian), which directly overlies the Precambrian Basement Complex. The Upper Permian succession of the Karoo deposits, the Madumabisa Formation, comprises 500 to in excess of 700 m of lacustrine deposits. Early geologic and sedimentologic work on the Zambian portion of the mid-Zambezi Basin were conducted with a focus on energy resources, primarily on coal-bearing formations (Bond, 1981; Gair, 1956; Gair , 1959; Tavener-Smith, 1953, 1956; Tavener-Smith, 1957, 1958, 1960, 1962). Only a few studies are available on the Upper Permian Madumabisa Mudstone Formation (e.g., Bond, 1955; Nyambe, 1993; Nyambe and Dixon, 2000; Tavener-Smith, 1958).

Natural outcrop sections of the Madumabisa Formation along the banks of the Mulungwa and Lusitu rivers (Fig. 4A) show conformable and gradational relationships with the underlying fluvial sandstones of the Gwembe Coal-Bearing Formation (Nyambe, 1999a). The contact with the overlying Escarpment Grit Formation is abrupt and disconformable, and the base of that formation contains bright red, cross-bedded, tabular feldspathic sandstone with flute casts, elongated green rip-up clasts derived from the mudstone below, and fist-size spheroidal concretions. At the contact with the Triassic Escarpment Grit, the uppermost 50 cm of the Madumabisa Formation is streaked with reddish stains, probably from erosion. The similarity of abrupt lithological breaks in equivalent sequences throughout south-central Africa has led many geologists to infer a hiatus representing a major Permo-Triassic unconformity (Drysdall and Kitching, 1963; Lister, 1979; Tavener-Smith, 1958; Watson, 1962). However, absence of soil features and presence of flute casts and elongated rip-up clasts, which would have formed before the lake sediments consolidated, suggest erosion. Dixey (1930) earlier suggested that the break in deposition did not result in significant erosion in the Mid-Zambezi Basin, unlike elsewhere in southern Africa. Recent seismic studies in the Luangwa Basin show that no angular unconformity exists between the Upper Permian and Lower Triassic rocks (Banks et al., 1995).

### *Description and Interpretation of Facies*

The primary lithologies of the Madumabisa Formation consist of mudstones and carbonates with minor siltstones and rare sandstones and conglomerates. Outcrop sections at both the Lusitu and Mulungwa riverbanks (Fig. 4A) were described, measured, and sampled in detail. Lithology and sedimentary structures observed in the field, and mineral assemblages from petrographic and XRD studies, allow identification of seven lithofacies. Facies descriptions and interpretations of depositional environments are discussed below. Representative lithostratigraphy of the Madumabisa Formation is given in Fig. 8.

*Massive Mudstone and Calcareous Mudstone Facies.—*

This lithofacies makes up ca. 40% of the studied section and consists of massive, mainly calcareous, mainly dark gray to dark green mudstones, with common mauve and reddish brown mudstones in the upper half of the outcrops. The massive mudstones are interbedded with shales and limestones, and locally with siltstone beds; faint lamination is visible in the mudstones at the contact. The mudstones are thin-bedded (10 cm) to thick-bedded (up to 20 m thick), and highly friable, and they commonly lack internal structures (Fig. 9A). Mudstones with higher silt content display conchoidal fractures, which present resistant horizons in the outcrops. Mineral composition includes abundant clay minerals and quartz, with disseminated micrite and minor feldspars. In some beds, micrite carbonate makes up 15% of the mudstones. The clay-mineral assemblage is made up of predominantly illite with traces of chlorite and illite–smectite (Nyambe, 1993). Amorphous organic matter, fine carbonaceous fragments, and subrounded to rounded, small argillaceous and micritic intraformational clasts occur throughout. The mudstones host random isolated, greenish gray to dark gray calcite concretions, which are spheroidal to elliptical in shape and 5 to 20 cm in diameter (Fig. 9B). The majority of the concretions contain fibrous calcite-filled septarian cracks that radiate out from the center. Some of the concretions preserve remains of conchostracans, bivalves, gyrogonites, and relict fine lamination.

This lithofacies was deposited in a quiet lacustrine environment characterized by a steady influx of sediments into basinal waters. The biogenic remains indicate an open freshwater lake. Mud clasts in the mudstones and relict lamination in the concretions indicate that the absence of internal structures resulted from a combination of slow deposition and thorough bioturbation enhanced by oxygenated bottom waters. Uninterrupted sedimentation and homogenization have been identified as the cause for mixing of siliciclastics and carbonates (e.g., Missimer and Ginsburg, 1998). The occurrence of reddish hues in the upper part

TABLE 1.—Stratigraphy of Karoo deposits in the Mid-Zambezi Basin, Zambia and Zimbabwe, and in the Luangwa Basin, northern Zambia (modified from Nyambe, 1993).

| Period (Ma) | Super-group | Group | | Mid-Zambezi Zambia | Mid-Zambezi Zimbabwe | Luangwa Zambia |
|---|---|---|---|---|---|---|
| (144) Early Jurassic | KAROO | UPPER KAROO | Drakens-berg | Batoka Basalt Fm | Batoka Basalt Fm | Dolerite Dykes |
| (213) Late Triassic | KAROO | UPPER KAROO | Forme- "Stormberg" | Red Sandstone Fm | Forest Sandstone Pebbly Arkose | Upper Grit |
| Early Triassic | KAROO | UPPER KAROO | Forme- "Stormberg" | Interbeded Sandstone and Mudstone Fm | Red Marly Sandstone Ripple marked Flagstone | Red Marl Ntawere |
| Early Triassic | KAROO | UPPER KAROO | Beaufort | Escarpment Grit Fm | Escarpment Grit Fm | Escarpment Grit Fm |
| (248) Late Permian | KAROO | LOWER KAROO | Beaufort | MADUMABISA MUDSTONE FM | Madumabisa Mudstone | Madumabisa Mudstone |
| (258) Early Permian | KAROO | LOWER KAROO | Ecca | Gwembe Coal Fm | Upper Wankie Sandstone<br>Black Shale & Coal Group<br>Lower Wankie Sandstone | Luwumbu Coal<br>Fimba & Mpwashi members |
| (286) Carboniferous | KAROO | LOWER KAROO | Dwyka | Siankondobo Sandstone Fm | Dwyka Glacial Beds | Basal Beds |
| Devonian (360)<br>Silurian | CAPE | SINAKUMBE | Upper | Zongwe Sandstone Fm | | |
| (440) Ordovician | CAPE | SINAKUMBE | Lower | Sikalamba Conglomerate Fm | | |

Data for absolute ages are from Harland et al. (1982); regional stratigraphy nomenclature is the South African Committee on Stratigraphy (SACS, 1980); stratigraphic succession in the Mid-Zambezi Basin, Zambia, from Nyambe (1997a, 1997b), Nyambe and Utting (1997), Denman and Money (1968), Denman and Money (1971), Denman et al. (1968), Gair (1959), Nyambe (1993), and Tavener-Smith (1960); succession in Mid-Zambezi Basin, Zimbabwe, from Bond (1967) and Stagman et al. (1978); and succession in Luangwa Basin from Drysdall and Kitching (1962, 1963) and Utting (1970).

of the section is explained by oxidation of iron present in the sediments, inasmuch as the mineral assemblage is similar to those of the dark mudstones. More arid periods would have allowed syndepositional alteration of iron-rich minerals (cf. Walker, 1967). The inorganic muddy limestones within the mudstones formed during periods of diminished clastic influx into the basin, when carbonates precipitated from saturated lake waters.

*Shale and Calcareous Shale Facies.—*

This thinly laminated shale facies is closely associated with the massive mudstones and constitutes ca. 42% of the studied section. The shales are horizontally layered, laterally continuous, thin- to medium-bedded, rhythmic, greenish gray to black, and commonly calcareous with up to 20% of micrite (Fig. 9C). Biogenic remains comprise conchostracans and bivalves. The shales contain submillimeter- to centimeter-scale laminae, which are highly fragile and frequently organic-rich. Lamination is produced by the alternation of light and dark layers that differ in composition (clay, silt, and micrite), grain size (clay to silt), and organic-matter content (organic-rich, organic-lean). The laminae occur in sets of organic matter–micrite–clay laminae and clay–silt laminae, the latter of which is thicker (up to a centimeter) and light-colored because of greater silt content. The rhythmic lamination is comparable to those observed in Upper Permian shales in Malawi (Yemane, 1994). Shale horizons with microlaminated couplets have a highly fissile texture (paper shale) and contain up to 45% micrite, up to 10% organic matter, and traces of framboidal pyrite. Upon weathering, these shales disintegrate into whitish powder, similar to those observed in the organic-rich Upper

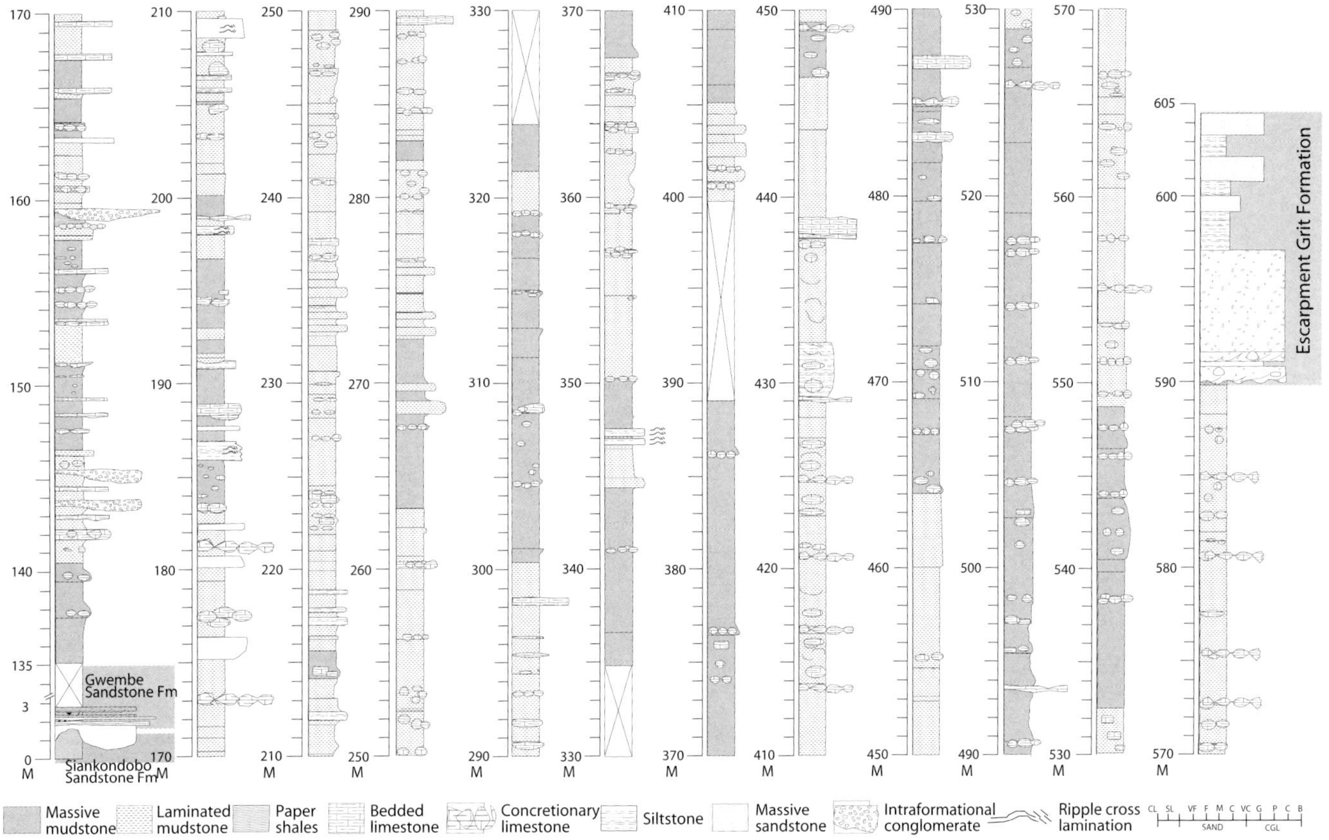


FIG. 8.—Lithostratigraphy of the Madumabisa Formation, compiled from outcrops along the Mulungwa and Lusitu Rivers.

Permian paper shales in Malawi (Yemane et al., 1989). The clay-mineral assemblage is dominated by illite with traces of chlorite and mixed-layer clays.

The horizontal lamination, fine grain size of the detritus, and absence of current- or wave-induced structures, slumping, or other deformational features indicate a quiet depositional environment. Absence of desiccation features in the laminae suggests that deposition occurred from suspension fallout below wave base in relatively deep water. Abundance of micrite and absence of particulate plant organic debris indicate high organic-matter productivity in the lake. Pyrite framboids and lack of bioturbation indicate anoxic bottom waters. The organic-lean laminae sets may have formed from episodic detrital pulses into basinal waters (Sturm, 1979), whereas the organic-rich laminae set may represent seasonal productivity (Kelts and Hsü, 1978) and were preserved in a lake with stratified water column.

*Laminated to Massive Limestone Facies.—*

This facies makes up ca. 6% of the studied section. It comprises dense decimeter-thick, massive to laminated, dark gray, tabular limestones and is interbedded with the shale and calcareous shale facies (Fig. 10A). The laminae are mostly a few millimeters to a centimeter thick, but subordinate submillimeter laminae are present. Isolated, small spheroidal calcite concretions are more common in the massive limestones than in the laminated limestones. In the latter case, the laminae are contorted around the concretion (Fig. 10B). These concretions contain septarian cracks filled with fibrous sparite and organic remains, like those found in the massive mudstones (Fig. 10C). The limestones are composed of up to 85% micritic calcite, with traces of microsparite filling veinlets and replacing shell remains. Detrital siliciclastic components include clays, angular to subangular fine-grained quartz silt, and traces of fine-grained feldspars. Chlorite is the dominant clay mineral in this facies, with minor illite. Biogenic constituents include conchostracans, bivalves, and fish scales, and are most abundant in the massive limestones.

The absence of desiccation features and current or wave structures, and the scarcity of intraclasts, indicate carbonate deposition from fallout in deep lake waters. Abundance of fossils indicates significant biogenic calcite production, whereas the fossil-free micritic beds may have been derived from precipitation in lake water supersaturated with carbonate. Preservation of thin laminae within the bedded limestones may be explained by periodic anoxia owing to lake-water stratification. Calcite-filled septarian cracks and contorted lamination around concretions suggest synsedimentary growth.

*Stratabound Concretionary Limestone Facies.—*

This facies constitutes ca. 11% of the studied section and is commonly found overlying shales and, locally, mudstones (Fig. 10D). The contact with the shale varies from diffuse to abrupt and is rarely erosional. The beds are compact and light to dark gray (with orange-weathering streaks) and vary in thickness from 0.10 m to 1.2 m, with thicker beds being more common in the lower part of the section. The beds have irregular bases and tops, and most are continuous over distances of hundreds of meters, but

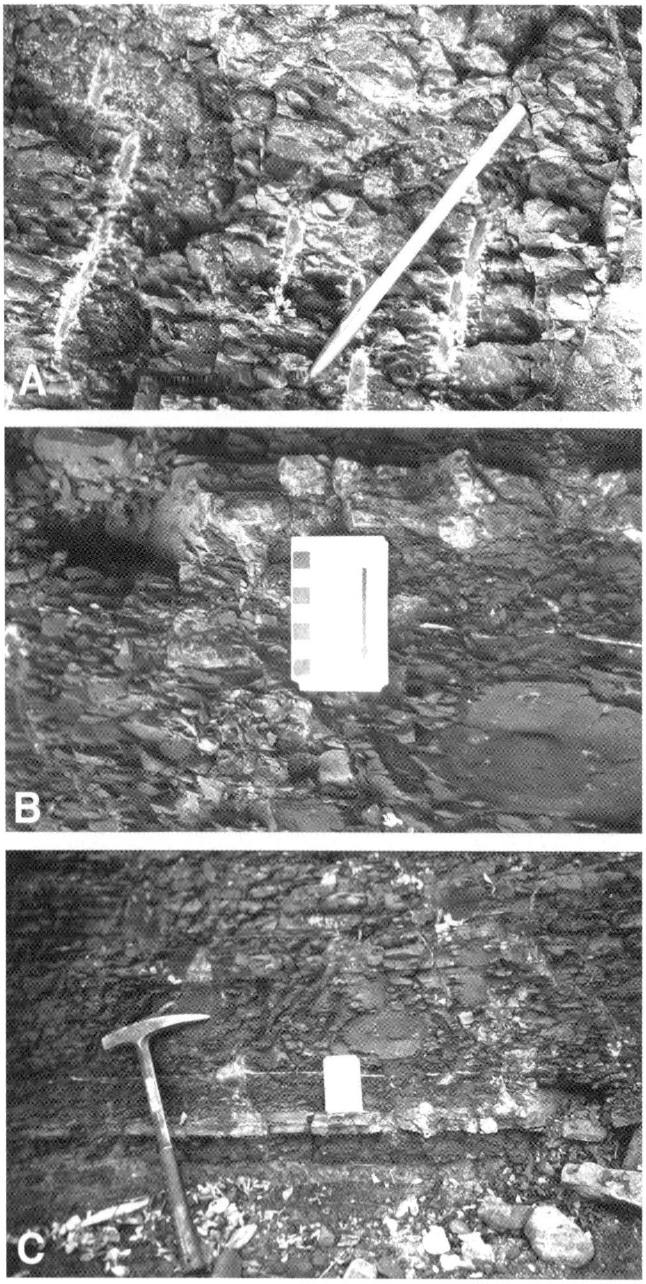

FIG. 9.—Representative photographs of mudstones and shales. **A)** Massive mudstones (vertical lines are pick marks). Pen is 16 cm long. **B)** Conchoidally fracturing massive mudstones with elongated concretions (scale in centimeters). **C)** Thin-bedded shales (hammer 54 cm long).

some are locally discontinuous. Fractures and stylolites are common. Some beds appear to have resulted from the coalescence of individual concretions. The beds are massive at outcrop scale, but irregular and discontinuous thin laminae with disrupted boundaries, and algal mat-like structures, are observed on polished slabs. Abundant conchostracans, bivalves, fish remains, bone (phosphatic) fragments, amorphous and globular organic matter, and subangular to subrounded elongated micrite and mudstone intraclasts (up to 2 cm across) occur throughout. The limestones are composed mainly of micrite and minor clay minerals, with angular, fine-grained quartz silt. The clay-mineral assemblage comprises illite and chlorite. Sparite replaces bivalve shells as well as fills septarian veins.

Nodular lacustrine carbonates are commonly interpreted to indicate palustrine environments or pedogenesis in littoral settings (Platt, 1989). The characteristic features associated with palustrine carbonates, e.g., desiccation cracks, rhizoliths and rhizoconcretions, microkarst, and color mottling (Wright and Platt, 1995), are lacking in this facies. The abundance of bioclasts, the high organic-matter content, and the presence of microlaminae indicate that biogenic calcite precipitation occurred in the photic zone. The intensity of bioturbation, abundant intraclasts, and irregularity of bed surfaces indicate that deposition occurred in a lacustrine setting with decreased suspended load, perhaps at lower lake levels, or during lake turnover. The absence of mud cracks and root traces strongly suggests, however, that the lake floor was not exposed. Concretion growth is attributed to precipitation from supersaturated pore waters generated by *in situ* organic-matter decay (Raiswell and Fisher, 2000). The abundant organic remains in this facies, the commonly oblate shape of the concretions, and the calcite-filled septarian veins suggest early diagenetic growth in unconsolidated sediments.

*Laminated Siltstone and Muddy Siltstone Facies.—*

This facies is minor (ca. 1%) in the outcrop examined and comprises laterally consistent, horizontally stratified, thin fine-grained siltstones and muddy siltstones, commonly associated with the massive mudstones (Fig. 11A). Except for locally erosional contacts, beds have sharp planar bases. The framework mineralogy of the siltstones is composed mainly of subangular quartz (up to 65%), feldspars (up to 15%), and minor micrite. Nyambe (1993) reported heavy minerals including epidote, sphene, tourmaline, garnet, and ilmenite. Observed internal structures include thin horizontal lamination, small-scale cross-lamination (including climbing ripples), with rare flaser and lenticular laminae in the muddy siltstones. The current ripples show paleocurrent directions to the southeast (Nyambe, 1993) (Fig. 11B). The light gray siltstones in general have little organic matter (< 0.3%), but a few beds contain fish scales, skeletal fragments, conchostracans, and carbonaceous fragments. Bioturbation is common in beds with abundant vertical and oblique burrows, disrupting laminae and filled with materials of slightly different color or grain size. Intraclasts and fist-size septarian concretions are scattered randomly throughout. Micrite occurs as cement in some siltstone beds (up to 25%) and as matrix in the muddy siltstones. The dark gray muddy siltstones contain up to 40% clay, with illite as the dominant clay mineral, and have higher content of organic debris. Comminuted leaves of *Glossopteris* and *Annularia* occur on bedding planes of a few beds. Two mineralized (silicified) tree trunks, ca.30 cm in diameter and 65 cm in length, are preserved parallel to the bedding plane in the section at the Lusitu riverbank. Similar tree trunks were found elsewhere in the basin, e.g., the Siankondobo area (Fig. 4B).

Horizontal lamination, lack of oscillatory-flow sedimentary structures, absence of structures indicating subaerial exposure, and the lateral persistence of horizontal laminae and beds are attributed to deposition below wave base. The presence of ripple cross-lamination in the siltstones indicates deposition from lower-flow-regime unidirectional currents. Climbing ripples form out of suspension at rapid rates of deposition, as in decelerating flows such as river floods or turbidity currents (Allen, 1972). Parallel-laminated siltstones with ripple cross-laminated tops may represent turbulent flows with decreased

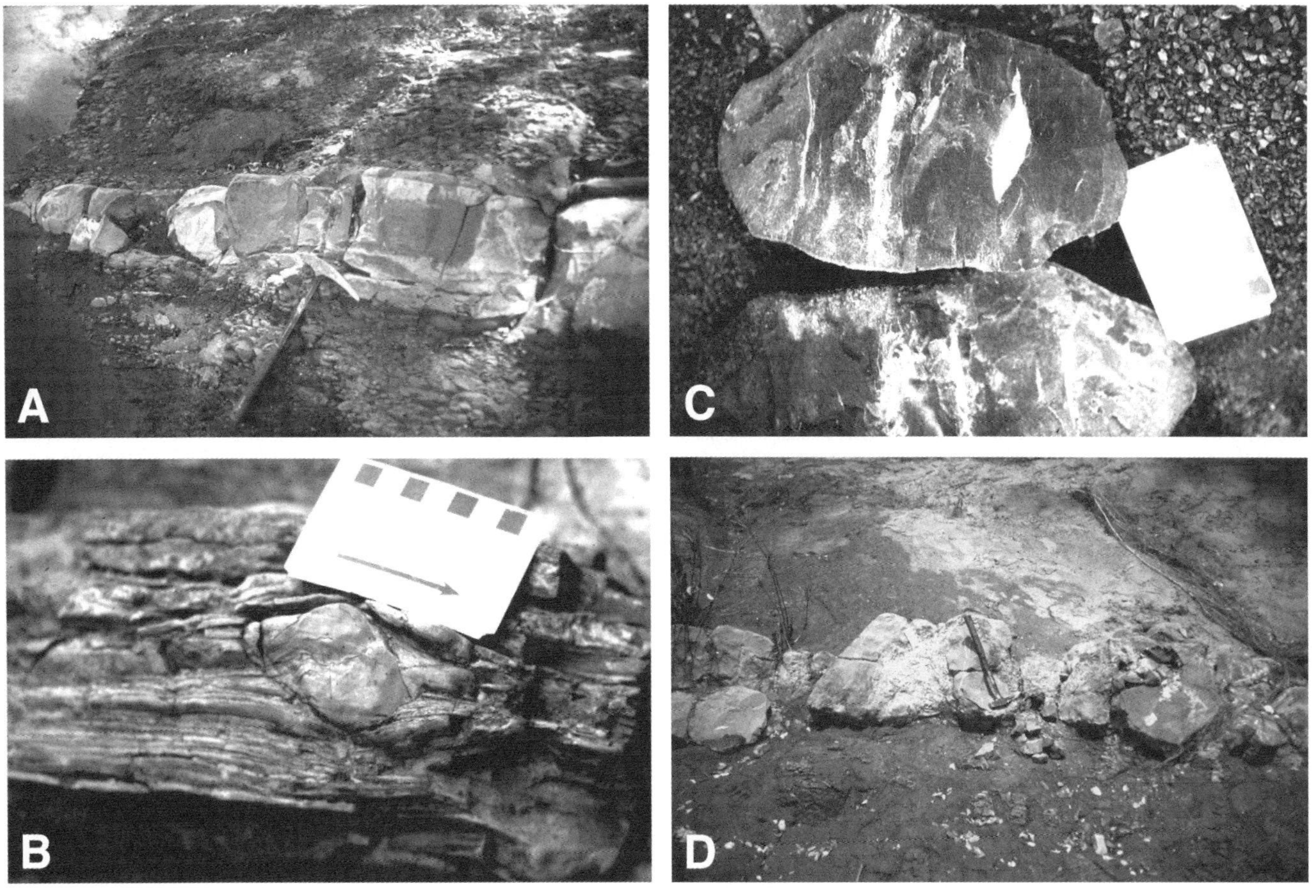


FIG. 10.—Representative photographs of limestones. **A)** Bedded limestone within massive mudstones (hammer 54 cm long). **B)** Laminated limestone with isolated concretion (scale in centimeters). **C)** Individual concretions with sparite-filled septarian cracks (scale in centimeters). **D)** Concretionary limestone bed weathered out of massive mudstones (hammer 54 cm long).

velocity. Flaser and lenticular laminae form in environments in which sediment supply is rhythmic or periodic, as in delta fronts or river floodplains (Leeder, 1982). The assemblage of these sedimentary structures suggests that this lithofacies was deposited from sediments that were derived from a delta prograding into the lake. The tree trunks were swept into the lake by rain-swollen streams from vegetation surrounding the lake and riverbanks.

*Massive Sandstone Facies.—*

This facies is minor in the studied Mulungwa section (< 1%) and occurs in only three beds. The sandstone beds are associated with shales. Bed thickness varies between 0.25 m and 1.5 m. Very fine- to coarse-grained, tabular-bedded sandstones have sharp planar contacts with the underlying shales. Two of the sandstone beds are massive, but one bed with intercalated shale shows ripples at the base. The sandstone is composed of abundant angular quartz (up to 80%), feldspars (up to 10%), and mudclasts within a clayey matrix. Primarily vertical burrows occur throughout.

The lack of structures indicative of subaerial exposure and the massive nature of the sandstones suggests subaqueous deposition. The non-erosional contact with the underlying thinly laminated shales and presence of ripples at the base of one of the sandstone beds point to rare turbulent flows that reached the deep part of the lake floor.

*Intraformational Conglomerate Facies.—*

This facies is the least represented (< 1%) in the studied section and occurs in only three beds, each no thicker than 10 cm. The conglomerates consist of matrix-supported, poorly sorted angular to subangular intraclasts (Fig. 11C). The intraclasts are 1 to 5 cm long and are made of mud and micrite, and the cement is micrite (ca. 30%). The beds have no internal structures. Two of the beds have planar contacts with the underlying black shale, whereas the third erodes into the underlying limestone bed.

The angular texture, intraclast composition, and erosional contacts indicate that the conglomerates were deposited from subaqueous erosion of carbonate facies areas by high-energy currents after periodic erosional events. The non-erosional contacts between the conglomerate beds and shales indicate that deposition may have occurred during high lake stands in deeper lacustrine areas.

## DISCUSSION AND CONCLUSION

*Depositional Environment*

The various facies of the Madumabisa Formation described above can be grouped into a single perennial lacustrine assemblage. Karoo rocks in the Mid-Zambezi rift basin cover approximately 140,000 km$^2$ (estimated from fig. 2 in Drysdall, 1962). From

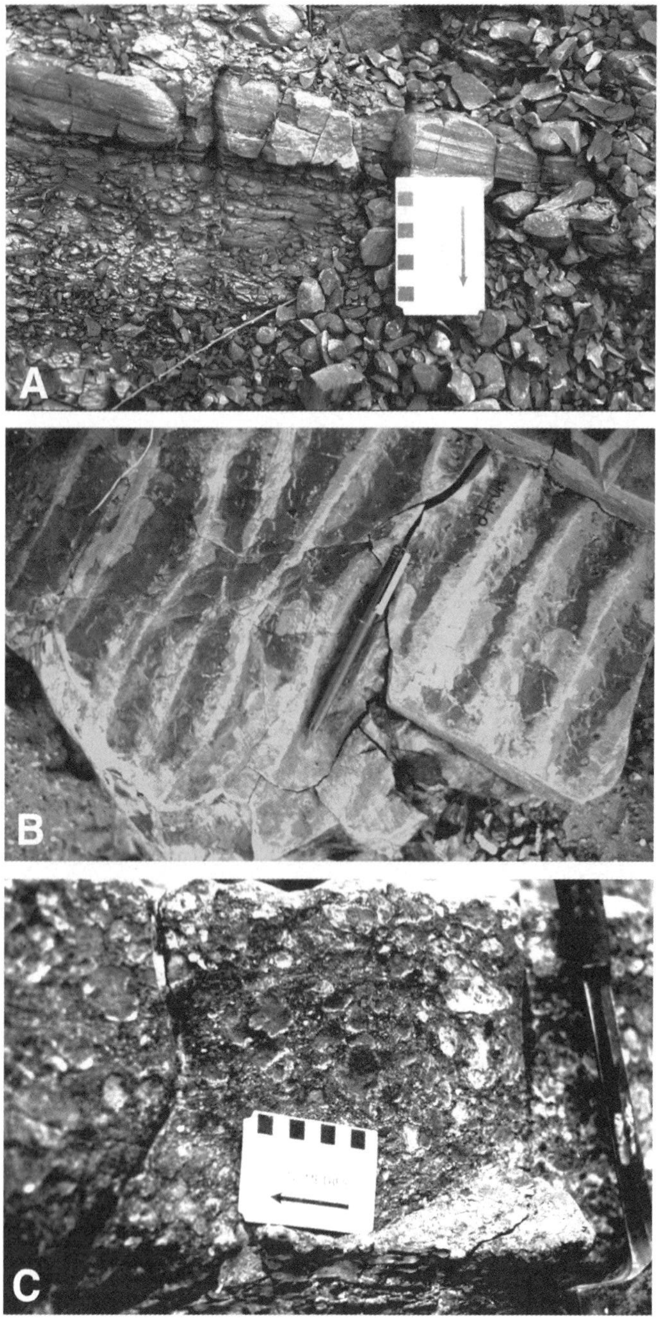

FIG. 11.—Representative photographs of siltstones and conglomerates. **A)** Thin-bedded laminated siltstone within massive mudstones (scale in centimeters). **B)** Planar view of a rippled siltstone bed, with bifurcating ripples draped with dark clay (pen 14 cm long). **C)** Intraformational conglomerate with subangular mud and micrite clasts (scale in centimeters).

the basinwide distribution of the Madumabisa Formation, it can be inferred that the ancient "Madumabisa Lake" had a larger surface area than most modern lakes.

The lack of sedimentary structures and lithologic features indicative of lake-floor emergence and the diverse freshwater faunal assemblage and aquatic flora (Bond, 1954, 1964, 1965) indicate that the ancient freshwater lake was perennial. Deposition in Madumabisa Lake was characterized mainly by suspension fallout in deep-water, low-energy environments. The dominance of intensely bioturbated massive mudstones indicates slow and continuous sedimentation in an open lake with oxygenated bottom waters. The lake experienced seasonal thermal stratification, as indicated by the accumulation of rhythmic, thinly laminated, thin-bedded shales and carbonates, in reducing bottom-water conditions. Precambrian basement rocks, including schists, gneisses, granulites, and granites, were major sources for the detritus, and are also reflected in the illite-dominated clay mineral assemblage.

Detrital carbonate grains are absent from the lacustrine beds; therefore, the micrite in the lake beds is interpreted to be authigenic. Remains of conchostracans, bivalves, and gastropods make up the biogenic components of the carbonates. Algal-driven seasonal carbonate precipitation, analogous to modern carbonate production in freshwater temperate lakes, is evident from remains of gyrogonites. Unlike the laminated carbonate beds, the nodular carbonates accumulated during periods of lower lake levels in marginal lacustrine environments. Outcrops of Middle and Upper Proterozoic Katangan Supergroup limestone and dolomite marbles to the north of the basin could have provided dissolved calcium (Munyanyiwa and Hanson, 1988).

The low-energy lacustrine milieu was interrupted by periodic influx of coarser clastics, resulting in the emplacement of sand, gravel, and tree trunks in deep parts of the lake. Lack of scouring indicates, however, that turbidity currents were not an important depositional mechanism. Rare rock fragments suggest that mass wasting was not active on the slopes of the surrounding hills, perhaps owing to thick vegetation cover.

### *Basin Tectonics*

The primary control on facies distribution in continental rift lakes has been attributed to climate (e.g., Frostick and Reid, 1989; Olsen, 1986), or tectonics (e.g., Lambiase, 1990; Lambiase and Bosworth, 1995), or both (e.g., Frostick and Reid, 1990; Leeder and Gawthorpe, 1987). Both climate and tectonics exert coeval small-scale and large-scale influences on sedimentation (Bohacs et al., 2000; Carroll and Bohacs, 1999), and it may prove difficult to distinguish contributions from each (Ruffell and Shelton, 1999). The influence of tectonics on the Madumabisa Lake is reflected in the overall deep-water facies. Lakes can exist only during that phase of a basin's history when rift shoulders and accommodation zones form a closed, sediment-starved basin (Lambiase, 1990). Maximum extension and subsidence causes rift lakes to become deep and restricted, during which time mainly fine-grained clastic sediments accumulate (Frostick and Reid, 1990; Hamblin and Rust, 1989). The uniformity of sedimentary conditions during the Late Permian time interval indicates that sedimentation was compensated by subsidence.

On the basis of the gentle folding in Lower Permian deposits underlying the Madumabisa Formation, we propose that a structural inversion in the Mid-Zambezi Basin occurred at the end of the Permian. The fold axes are in general perpendicular to the maximum shortening direction and parallel to the axis of the Mid-Zambezi Basin, indicating shortening perpendicular to the basin. Although the folds have low amplitudes resulting from only few percent shortening, this might have been sufficient to shallow the lake (by halting of subsidence and / or by basin-floor uplift) and change depositional conditions. Lister (1979) proposed that folding and faulting at the close of the Permian increased the relief of the region. Consequently, the quiescent lacustrine regime was terminated abruptly during the Early Triassic by the deposition of the Escarpment Grit Formation (Prasad and Lehtonen, 1977; Tavener-Smith, 1958). The thick (500

m) reddish, coarse-grained, stratified conglomeratic sandstones and intercalated mudstones are characteristic of high-energy braided fluvial environments (Money et al., 1979; Nyambe, 1993). Decrease in basin subsidence lowers tectonic drainage divides, which allows the development of throughgoing river systems (Carroll and Bohacs, 1999). Thus, the abrupt termination of lacustrine fill by the fluvial sediments of the Escarpment Grit may have been more the result of basin tectonics than of climate changes. Similar disconformable contacts between Upper Permian and Lower Triassic deposits, observed in other Karoo basins of south-central Africa (e.g., Bond, 1955; Lister, 1979; Watson, 1962). This attests to the regional nature of basin inversion and may be related to changes in strike-slip movements along the major shear zones, such as the Mwembeshi shear zone (Banks et al., 1995). When the movement on the Mwembeshi shear zone was sinistral, there was concomitant extension in the Karoo rift basins; when the movement became dextral, the suitably oriented basins were compressed and inverted.

Banks et al. (1995) postulated that structural inversions resulted from changes in the stress pattern in the African continent. Pysklywec and Mitrovica (1999) state that the subduction of the paleo–Pacific plate below southern Gondwana and the ensuing lithospheric deflection caused the entire observed long-wavelength subsidence of the Karoo foreland basin. They conclude that lithospheric response to the detachment of the subducted plate, thought to have occurred at approximately at the Permian–Triassic boundary, caused rapid uplift 0 to 20 Myr later. Information on the stress distribution and magnitude of this uplift is yet unknown, but one can speculate that the rapid uplift in the south may have caused changes in stress inland. These changes include renewed rifting movements along the major faults, erosion and material influx into the rifts far to the north, changes in subsidence rate, inversion tectonics and even changes in climate. The rapid uplift event appears to be in good agreement with the onset of deposition of the Escarpment Grit Formation and its equivalent throughout south–central Africa. Banks et al. (1995) suggested that the Escarpment Grit sedimentation in neighboring Luangwa Basin (Fig. 3) resulted from regional tectonic uplift and an increasingly hot and / or arid climate. Structural and sedimentologic results for the Mid-Zambezi Basin indicate that the change from lacustrine to fluvial sedimentation at the end of the Permian was caused by slowing subsidence rate or by uplift of the basin (or basin floor). It is unlikely that tectonic uplift alone could have resulted in the thick, coarse Early Triassic major clastic deposition in an arid climate setting, because it would have required high-energy fluvial transport (cf. Griffin, 1999). Because all Karoo rift basins in central Africa appear to have been similarly affected, the Early Triassic fluvial deposition may be linked more to the reorganization of stresses at the Gondwana plate boundaries than to a major climate change.

### *Paleoclimate and Paleohydrologic Conditions*

The relative balance between tectonics and climate dictates the occurrence, distribution, and general character of lakes (Bohacs et al., 2000; Carroll and Bohacs, 1999). The rock types in the area influence the type of sediment input and ground-water chemistry (Gierlowski-Kordesch, 1998; Platt and Wright, 1991). The highland drainage to lake area, elevated burial rates, and frequent water-level fluctuations serve to enhance and amplify depositional signatures in lakes (Scholz, 1997; Sly, 1978). These qualities render lakes reliable archives of records of paleoclimatic and paleohydrologic conditions in continental environments (Kelts, 1988; Wohl and Enzel, 1995). The lacustrine Madumabisa Formation in the Mid-Zambezi Basin provides a unique opportunity to obtain a snapshot of the hydrologic and climate conditions that existed deep within the interior of southern Gondwana during the Late Permian.

The facies assemblage indicates that temperate, humid climate conditions prevailed and that basin subsidence, sediment supply, and positive water budget sustained a perennial lake. This is consistent with the "balanced-fill" lake basin type (Carroll and Bohacs, 1999) in which sediment supply is closely linked to lake level (Bohacs et al., 2000). Similar modern large high-latitude rift lake systems tend to be overfilled and have stable water levels except in extremely arid conditions (Scholz et al., 1998). The vertical facies association in the Madumabisa Formation shows two depositional cycles represented by the bioturbated–laminated mudstone and mudstone–carbonate successions. Detrital–chemical cycles represent depositional cycles characterized by throughflow during a relatively humid period and an arid period of internal drainage, respectively (Collinson, 1979). The two cycles in the Madumabisa Formation are attributed to changes in water energy levels, basinal water chemistry, and sediment input to the ancient lake. The bioturbated–laminated mudstons cycles reflect periodic changes in bottom-water circulations and in amount of organic-matter input in the basin. Coarser grain size and lean organic content in these mudstones indicate clastics-laden freshwater input. On the other hand, the mudstone–carbonate cycle reflects stronger climate variations, which may have affected both humidity and temperature regimes. Increased temperature and / or reduced siliciclastic input promote carbonate precipitation (e.g., Digerfeldt et al., 2000; Dodson and Ramrath, 2001). The close association of the carbonates with laminated mudstones suggests that curtailed sediment input favored carbonate precipitation.

The stratigraphic placement of the Madumabisa Formation between the Lower Permian Gwembe Coal Formation (Utting, 1970, 1978a, 1978b) and the Lower–Middle Triassic Escarpment Grit (Nyambe and Utting, 1997) shows that the lake existed for approximately 10 Myr. The great thickness of the lacustrine sequence and lack of evidence for lake-floor emergence suggest that the lake existed continuously throughout the Late Permian. Modern rift lakes in active extensional phases are long-lived and contain thick synrift sedimentary fills. Lakes Malawi and Tanganyika have been operating for 5 and 9–12 Myr, respectively, whereas the high-latitude Lake Baikal is ca. 35 Ma old (Scholz et al., 1998). The Mid-Zambezi Basin preserves over 5 km of Karoo sediments (Upper Carboniferous to Lower Jurassic), and the duration occupied by the lacustrine phase of sedimentation is comparable to modern large extensional rift basins at high latitudes (e.g., Scholz et al., 1998). The illite-dominated clay-mineral assemblage reflects not only the mineralogy of the source area (mica) but also the moderate weathering path (mica to illite) characterizing climates at temperate high latitudes.

The sedimentary record of the Madumabisa Formation is consistent with lacustrine deposition under a cool to moderate, seasonal, humid climate regime and is consistent with the prolific Late Permian terrestrial ecosystem that occupied the paleolatitudes of southern Africa. The severe aridity and extreme temperature ranges simulated by numerical climate models for the high latitudes of southern Gondwana during this time interval could not have supported a long-lived, deep, perennial freshwater lake such as "Lake Madumabisa", deep within the interior of the supercontinent.

## ACKNOWLEDGMENTS

We thank the following personnel of the Geology Department of the University of Zambia for aid during field work: Dr. B.

Mapani, Acting Head of the Geology Department, for facilitating field logistics; Drs. F. Tembo and B. Mapani for discussion and for leading field trip to the Mwembishi shear zone, and W. Nundwe for organizing field vehicles. Sylvester and B. Shaka Kai, Headmasters, provided lodging in Siamungo and Sulungondwe elementary schools, respectively. Mischeck, Oliver, Paulo, and Arisona assisted in the field. We thank A. Chipanzi, Managing Director of the Mamba Collier Ltd, for access the Mamba Coal Mine; E. Ngwata, Technical Superintendent, for discussion on the geology of the mine; and the clinic staff and Izuma Lodge at Maamba for their kind hospitality. Measurements of quartz crystallographic orientation were done by x-ray texture goniometer at ETH Zurich, Switzerland. Comments on earlier drafts by E. Gierlowski-Kordesch, B. Lepage, and H. Pffeferkorn as well as constructive criticism from J. Hutton and an anonymous reviewer improved the quality of the text. We thank G. Ashley for thoughtful and practical suggestions. This project was funded by National Science Foundation grant GER 95-50414 (VPW) to K. Yemane with field support to D. Grujic. K.Y. is grateful for Margrete Klein's continued support.

## REFERENCES


Allen, J.R.L., 1972, A theoretical and experimental study of climbing-ripple cross-lamination, with a field application to the Uppsala esker: Geografiska Annaler, Series A, Physical Geography, v. 53, p. 157–187.

Anderson, J.M., 1977, The biostratigraphy of the Permian and Triassic, Part 3. A review of Gondwana Permian palynology with particular reference to the northern Karoo basin, South Africa: Botanical Survey of South Africa, Memoirs, v. 41, 133 p.

Banks, N.L., Bardwell, K.A., and Musiwa, S., 1995, Karoo rift basins of the Luangwa Valley, Zambia, *in* Lambiase, J.J., ed., Hydrocarbon Habitat in Rift Basins: Geological Society of London, Special Publication 80, p. 285–295.

Blenkinsop, T.G., and Oesterlen, P.M., 1996, Extension directions and strain near the failed triple junction of the Zambezi and Luangwa rift zones, Southern Africa; discussion and reply: Journal of African Earth Sciences, v. 22, p. 621–622.

Bohacs, K.M., Carroll, A.R., Neal, J.E., and Mankiewicz, P.J., 2000, Lake-basin type, source potential, and hydrocarbon character: an integrated sequence-stratigraphic–geochemical framework, *in* Gierlowski-Kordesch, E.H., and Kelts, K.R., eds., Lake Basins through Space and Time: American Association of Petroleum Geologists, Studies in Geology 46, p. 3–34.

Bond, G., 1954, Lamellibranchia and plants from the lower Karroo beds of Northern Rhodesia: Geological Magazine, v. 91, p. 189–192.

Bond, G., 1955, The Madumabisa (Karroo) shales in the middle Zambesi region: Geological Society of South Africa, Transactions, v. 58, p. 71–99.

Bond, G., 1964, A new phyllopod from the Karroo of Northern Rhodesia: Arnoldia (National Museums of Rhodesia, Series of Miscellaneous Publications), v. 1, p.4.

Bond, G., 1965, Some new fossil localities in the Karroo system of Rhodesia: Arnoldia (National Museums of Rhodesia, Series of Miscellaneous Publications), v. 2, p.4.

Bond, G., 1967, A review of the karoo sedimentation and lithology in southern Rhodesia: First Symposium on Gondwana Stratigraphy, Papers and Proceedings, Mar del Plata, Argentina, p. 173–195.

Bond, G., 1974, The stratigraphic distribution of plant and animal macrofossils in the Karoo System of the mid-Zambezi region: Arnoldia (National Museums of Rhodesia, Series of Miscellaneous Publications), v. 7, p. 6.

Bond, G., 1981, Late Palaeozoic (Dwyka) glaciation in the Middle Zambezi region, *in* Hambrey, M.J., and Harland, W.B., eds., Earth's Pre-Pleistocene Glacial Record: Cambridge, U.K., Cambridge University Press, p. 55–57.

Carroll, A.R., and Bohacs, K.M., 1999, Stratigraphic classification of ancient lakes: Balancing tectonic and climatic controls: Geology, v. 27, p. 99–102.

Clemson, J., Cartwright, J.A., and Swart, R., 1999, The Namib rift: a rift system of possible Karoo age, offshore Namibia, *in* Cameron, N.R., Bate, R.H., and Clure, V.S., eds., The Oil and Gas Habitats of the South Atlantic: Geological Society of London, Special Publication 153, p. 381–402.

Collinson, J.D., 1979, Lakes, *in* Reading, H.G., ed., Sedimentary Environments and Facies: New York, Elsevier, p. 61–79.

Crowley, T.J., 1993, Seasonal cycle variations on Pangea (abstract): Canadian Society of Petroleum Geologists and Global Sedimentary Geology Program, Annual Convention, Program and Abstracts, Carboniferous to Jurassic Pangea, p. 64.

Crowley, T.J., Hyde, W.T., and Short, D.A., 1989, Seasonal cycle variations on the supercontinent of Pangaea: Geology, v. 17, p. 457–460.

Daly, M.C., 1986, The intracratonic Irumide Belt of Zambia and its bearing on collision orogeny during the Proterozoic of Africa, *in* Coward, M.P., and Ries, A.C., eds., Collision Tectonics: Geological Society of London, Special Publication 19, p. 321–328.

Daly, M.C., Chakraborty, S.K., Kasolo, P., Musiwa, M., Mumba, P., Naidu, B., Namateba, C., Ngambi, O., and Coward, M.P., 1984, The Lufilian Arc and Irumide Belt of Zambia: results of a geotraverse across their intersection: Journal of African Earth Sciences, v. 2, p. 311–318.

Daly, M.C., Chorowicz, J., and Fairhead, J.D., 1989, Rift basin evolution in Africa: the influence of reactivated steep basement shear zones, *in* Cooper, M.A., and Williams, G.D., eds., Inversion Tectonics: Geological Society of London, Special Publication 44, p. 309–334.

Daly, M.C., Lawrence, S.R., Kimun'a, D., and Binga, M., 1991, Late Paleozoic deformation in central Africa: a result of distant collision?: Nature, v. 350, p. 605–607.

de Swardt, A.M.J., and Drysdall, A.R., 1964, Precambrian geology and structure in central Northern Rhodesia: Northern Rhodesia Geological Survey, Memoir 2, 82 p.

de Swardt, A.M.J., Garrard, P., and Simpson, J.G., 1965, Major zones of transcurrent dislocation and superposition of orogenic belts in parts of central Africa: Geological Society of America, Bulletin, v. 76, p. 89–102.

Denman, P.D., and Money, N.J., 1968, The coal resources of the Zambezi valley; V, Siankondobo, the north-eastern area, preliminary report: Zambia, Geological Survey Department, Report 23, 14 p.

Denman, P.D., and Money, N.J., 1971, Stratigraphy and sedimentology of the Sinakumbe Formation and Siankondobo Sandstone Formation in the Gwembe region, Mid-Zambezi Valley: International Union of Geological Societies, Commission on Stratigraphy, Subcommittee on Gondwana Stratigraphy, Palaeontology, Gondwana Symposium Proceedings Paper, v. 2, p. 409–420.

Denman, P.D., Money, N.J., and Radosevic, B., 1968, Stratigraphy of the lower Karroo sediments in the Siankondobo and Mulungwa areas of the mid-Zambezi valley: Republic of Zambia, Geological Survey, Records, v. 11, p. 1–8.

Digerfeldt, G., Olsson, S., and Sandgren, P., 2000, Reconstruction of lake-level changes in Lake Xinias, central Greece, during the last 40 000 years: Palaeogeography, Palaeoclimatology, Palaeoecology, v. 158, p. 65–82.

Dixey, F., 1930, A provisional correlation of the Karoo north of the Zambezi: International Geological Congress, 15th Session, South Africa, Comptes Rendus, p. 146–160.

Dodson, J.R., and Ramrath, A., 2001, An upper Pliocene lacustrine environmental record from south-western Australia; preliminary results: Palaeogeography, Palaeoclimatology, Palaeoecology, v. 167, p. 309–320.

DRYSDALL, A.R., 1962, The Karoo system in Northern Rhodesia: Correlation and sedimentation: Salisbury, Geological Survey of North Rhodesia, Occasional Papers, series 32, p. 113–138.

DRYSDALL, A.R., AND KITCHING, J.W., 1962, The Karoo succession of the upper Luangwa valley, Northern Rhodesia: Geological Society of South Africa, Transactions, v. 65, p. 75–90.

DRYSDALL, A.R., AND KITCHING, J.W., 1963, A re-examination of the Karoo succession and fossil localities of part of the upper Luangwa valley, Northern Rhodesia: Geological Survey of Northern Rhodesia, Memoir 1, 62 p.

DRYSDALL, A.R., AND WELLER, R.K., 1966, Karoo sedimentation in Northern Rhodesia: Geological Society of South Africa, Transactions, v. 69, p. 39–69.

DU TOIT, A.L., 1937, Our Wandering Continents; An Hypothesis of Continental Drifting: Edinburgh, Oliver & Boyd, 366 p.

DU TOIT, A.L., 1954, The Geology of South Africa: London, Oliver & Boyd, 611 p.

ERWIN, D.H., 1999, Biospheric perturbations during Gondwanan times; from the Neoproterozoic–Cambrian radiation to the end-Permian crisis: Journal of African Earth Sciences, p. 115–127.

FALCON, R., 1973, Palynology of the Middle Zambezi Basin: Rhodesia Geological Survey, Bulletin 70, p. 43–71.

FAWCETT, P.J., AND BARRON, E.J., 1993, A comparison of the paleoclimates of Late Permian Pangea and Early Cretaceous post-Pangea with the GENESIS climate model (abstract): Canadian Society of Petroleum Geologists and Global Sedimentary Geology Program, Annual Convention, Program and Abstracts, Carboniferous to Jurassic Pangea, p. 95.

FROSTICK, L.E., AND REID, I., 1989, Is structure the main control of river drainage and sedimentation in rifts?: Journal of African Earth Sciences, v. 8, p. 165–182.

FROSTICK, L.E., AND REID, I., 1990, Structural control of sedimentation patterns and implication for the economic potential of the East African Rift basins: Journal of African Earth Sciences, v. 10, p. 307–318.

GAIR, H.S., 1956, A summary of the structure and tectonic history of the mid-Zambezi valley: Scientific Council for Africa South of the Sahara, United Kingdom, Publication, series no. 1956, p. 123–127.

GAIR, H.S., 1959, The Karroo system and coal resources of the Gwembe District, northeast section: Salisbury, Northern Rhodesia, Department of Geological Survey, B, series 1, Report, 88 p.

GIERLOWSKI-KORDESCH, E.H., 1998, Carbonate deposition in an ephemeral siliciclastic alluvial system: Jurassic Shuttle Meadow Formation, Newark Supergroup, Hartford Basin, USA: Palaeogeography, Palaeoclimatology, Palaeoecology, v. 140, p. 61–184.

GRIFFIN, D.L., 1999, The Late Miocene climate of northeastern Africa: unravelling the signals in the sedimentary succession: Geological Society of London, Journal, v. 156, p. 817–826.

GRILL, H., 1996, The Permo-Carboniferous glacial to marine Karoo record in southern Namibia: sedimentary facies and sequence stratigraphy [unpublished Ph.D. dissertation]: Universtät Würzburg, Würzburg, 129 p.

HAMBLIN, A.P., AND RUST, B.R., 1989, Tectono-sedimentary analysis of alternate-polarity half-graben basin-fill successions; Late Devonian–Early Carboniferous Horton Group, Cape Breton Island, Nova Scotia: Basin Research, v. 2, p. 239–255.

HANSON, R.E., WILSON, T.J., BRUECKNER, H.K., ONSOTT, T.C., WARDLAW, M.S., JOHNS, C.C., AND HARDCASTLE, K.C., 1988, Reconnaissance geochronology, tectonothermal evolution, and regional significance of the middle Proterozic Choma–Kalomo block, southern Zambia: Precambrian Research, v. 42, p. 39–61.

HANSON, R.E., WARDLAW, M.S., WILSON, T.J., AND MWALE, G., 1993, U–Pb zircon ages from the Hook granite massif and Mwembeshi dislocation: constraints on Pan-African deformation, plutonism, and transcurrent shearing in central Zambia: Precambrian Research, v. 63, p. 89–209.

HARLAND, W.B., COX, A., LLEWELLYN, P.G., PICKTON, C.A.G., SMITH, A.G., WALTERS, R., AND FANCETT, K.E., 1982, A Geologic Time Scale: Cambridge, U.K., Cambridge University Press, 131p.

HILLER, K., AND BUTTKUS, B., 1996, Structural and sedimentary thicknesses in the Zambezi Rift valley, Zimbabwe—Investigation of the potential for hydrocarbons: Zeitschrift für Angewandte Geologie, v. 42, p. 132–137.

HOTTON, N., III, MACLEAN, P.D., ROTH, J.J., AND ROTH, E.C., 1986, The Ecology and Biology of Mammal-Like Reptiles: Washington, D.C., Smithsonian Institution Press, 326 p.

KELTS, K., 1988, Environments of deposition of lacustrine petroleum source rocks: an introduction, *in* Fleet, A.J., Kelts, K., and Talbot, M.R., eds., Lacustrine Petroleum Source Rocks: Geological Society of London, Special Publication 40, p. 3–26.

KELTS, K., AND HSÜ, K.J., 1978, Freshwater carbonate sedimentation, *in* Lerman, A., ed., Lakes: Chemistry, Geology, Physics: New York, Springer-Verlag, p. 295–323.

KING, B.C., 1978, A comparison between the older (Karroo) rifts and the younger (Cenozoic) rifts of eastern Africa, *in* Ramberg, I.B., and Neumann, E.R., eds., Tectonics and Geophysics of Continental Rifts: Dordrecht, The Netherlands, D. Reidel, p. 347–350.

KREMP, O.W., 1977, The positions and climatic changes of Pangaea and five southeast Asian plates during Permian and Triassic times: Paleo Data Banks, v. 7, p. 1–21.

KREUSER, T., WOPFNER, H., KAAYA, C.Z., MARKWORT, S., SEMKIWA, P.M., AND ASLANIDIS, P., 1990, Depositional evolution of Permo-Triassic Karoo basins in Tanzania with reference to their economic potential: Journal of African Earth Sciences, v. 10, p. 151–167.

KUTZBACH, J.E., AND GALLIMORE, R.G., 1989, Pangaean climates, megamonsoons of the megacontinent: Journal of Geophysical Research, D, Atmospheres, v. 94, p. 3341–3357.

KUTZBACH, J.E., AND ZIEGLER, A.M., 1993, Simulation of Late Permian climate and biomes with an atmosphere ocean model-comparisons.: Royal Society [London], Philosophical Transactions, Series B, Biological Sciences, v. 311, p. 327–340.

LAMBIASE, J.J., 1990, A model for tectonic control of lacustrine stratigraphic sequences in continental rift basins, *in* Katz, B.J., ed., Lacustrine Basin Exploration; Case Studies and Modern Analogues: American Association of Petroleum Geologists, Memoir 50, p. 265–276.

LAMBIASE, J.J., AND BOSWORTH, W., 1995, Structural controls on sedimentation in continental rifts, *in* Lambiase, J.J., ed., Hydrocarbon Habitat in Rift Basins: Geological Society of London, Special Publication 80, p. 117–144.

LEEDER, M.R., 1982, Sedimentology; Process and Product: London, George Allen & Unwin, 360 p.

LEEDER, M.R., AND GAWTHORPE, R.L., 1987, Sedimentary models for extensional tilt-block/half-graben basins, *in* Coward, M.P., Dewey, J.F., and Hancock, P.L., eds., Continental Extensional Tectonics: Geological Society of London, Special Publication 28, p. 139–152.

LISTER, L.A., 1979, The geomorphic evolution of Zimbabwe Rhodesia: Geological Society of South Africa, Transactions, v. 82, p. 363–370.

MCKENZIE, D., AND BRUNE, J.N., 1972, Melting on fault planes during large earthquakes: Geophysical Journal of the Royal Astronomical Society, v. 29, p. 65–78.

MISSIMER, T.M., AND GINSBURG, R.N., 1998, Homogenized carbonates and siliciclastics in the Tertiary of Southwest Florida, (abstract): American Association of Petroleum Geologists, Bulletin, v. 82, p. 1787–1788.

MONEY, N.J., DENMAN, P.D., AND DRYSDALL, A.R., 1974, Geology of the coalfields of the Mid-Zambezi valley, Zambia: Geological Survey of Zambia, unpublished Bulletin, 6 p.

MONEY, N.J., PRASAD, R.S., AND THEISME, J.G., 1979, Geology of Escarpment Grit Formation of Kariba area and its economic potential (Zambia): International Gondwana Symposium, series no. 4, p. 886–888.

MOORE, G.T., AND PEOPLES, C.J., 1990, The paleoclimate of the Kazanian (early Late Permian) world: AAPG annual convention, technical

YEMANE, K., NYAMBE, I., AND GRUJIC, D., 1998, Temperate climate during the Late Permian in the continental interiors of southern Gondwana: Sedimentary record from a rift lake (abstract): American Geophysical Union, Abstracts, v. 79, p. F478.

ZIEGLER, A.M., 1990, Phytogeographic patterns and continental configurations during the Permian period, *in* McKerrow, W.S., and Scotese, C.R., eds., Palaeozoic Palaeogeography and Biogeography: Geological Society of London, Memoir 12, p. 363–379.

ZIEGLER, A.M., 1993, Gondwanaland: models come in from the cold: Nature, v. 361, p. 16–17.

ZIEGLER, A.M., HULVER, M.L., AND ROWLEY, D.B., 1997, Permian world topography and climate, *in* Martini, I.P., ed., Late Glacial and Postglacial Environmental Changes—Quaternary, Carboniferous–Permian and Proterozoic: New York, Oxford University Press, p. 111–146.

# Fluvial and Alluvial Depositional Systems in Rift Basins

# TEMPORAL AND SPATIAL VARIABILITY OF ALLUVIAL-FAN AND AXIAL-FLUVIAL SEDIMENTATION IN THE PLIO-PLEISTOCENE PALOMAS HALF GRABEN, SOUTHERN RIO GRANDE RIFT, NEW MEXICO, U.S.A.

GREG H. MACK
*Department of Geological Sciences, New Mexico State University, Las Cruces, New Mexico 88003, U.S.A.*
MIKE LEEDER
*School of Environmental Sciences, University of East Anglia, Norwich, Norfolk NR4 7TJ, United Kingdom*
AND
STEPHEN L. SALYARDS
*Department of Earth and Space Sciences, University of California at Los Angeles, Los Angeles, CA 90024, U.S.A.*

**ABSTRACT**: The Palomas basin in the Rio Grande rift of southern New Mexico is a Neogene half graben, whose Pliocene–lower Pleistocene fill (Palomas Formation) has been partly exhumed, providing a nearly three-dimensional view of basin-fill architecture. The distribution of Palomas Formation depositional systems and their lithofacies is strongly asymmetrical. Footwall-derived sediment was deposited on small (radial length < 2 km), steep (slopes 1–4°) fans and consists primarily of turbulent-flow, channel conglomerates and hyperconcentrated-flow, pebbly sand, whereas hanging wall–derived sediment was deposited on broader (radial length 15 km), gentler (slopes 0.7–1°) fans and displays a greater proportion of channels with lateral accretion sets and overbank mudstones, suggesting that medial and distal hanging wall–derived fan channels behaved like gravel-bed streams. Axial-fluvial sediment, deposited by the ancestral Rio Grande, occupies a narrow (< 5 km) belt near the footwall block and consists of multistory pebbly medium–coarse channel sand, mottled overbank mudstone, and fine sand deposited as crevasse splays or in small channels. One to three excursions of the axial-fluvial system toward the footwall, each lasting $10^5$–$10^6$ years, were accomplished by toe cutting of the footwall-derived fans and resulted in deposition of multistory channel sands, while the western edge of the floodplain was dominated by deposition of overbank mudstone and crevasse-splay / small-channel fine sands. Tectonic tilting of the half graben is the probable cause of the fluvial incursions toward the footwall, because toe cutting of the fans is largely restricted to the footwall side of the basin and the number of incursions correlates to distinct segments of the border fault system.

## INTRODUCTION

The Palomas basin of south-central New Mexico provides an excellent opportunity to study sedimentation in a still active but partially exhumed continental half graben (Fig. 1). Developed during the most recent stage of crustal extension in the southern Rio Grande rift, the Palomas basin experienced continuous sedimentation on alluvial fans and by an axial river (the ancestral Rio Grande) for approximately 4.2 million years, from early Pliocene to early Pleistocene time. Beginning approximately 0.78 Ma, the ancestral Rio Grande and its tributaries began to alternately incise and partially backfill the basin, ultimately exposing from 50 to 100 m of previously deposited Pliocene–lower Pleistocene sediment (Mack et al., 1993; Mack et al., 1998). Also locally preserved as gently sloping geomorphic surfaces are the constructional tops of the paleofans that formed just prior to the onset of incision. Thus, the Palomas basin provides a nearly three-dimensional view of Plio-Pleistocene basin fill.

Previous work on the Palomas basin by our research group, which is summarized here, has concentrated on footwall-derived, alluvial-fan and axial-fluvial sediment and the distribution of authigenic carbonate (Mack and Leeder, 1999; Mack et al., 2000). Also previously presented is an interpretation of small-scale paleoclimatic cycles in footwall-derived, alluvial-fan sediment, which will not be repeated here (Mack and Leeder, 1999). This paper provides previously unpublished data on hanging wall-derived, alluvial-fan sediment and its relationship to the axial-fluvial system, as well as new magnetostratigraphic correlations throughout the basin. This study builds upon previously published work by (1) relating differences between footwall-derived and hanging wall–derived alluvial-fan strata to fan gradients and size of the catchments, (2) using the basin-wide distribution of alluvial-fan and axial-fluvial sediment as a test of models of sedimentation in half grabens, (3) illustrating differences in fluvial lithofacies and interaction with alluvial fans on the eastern and western sides of the floodplain, and (4) providing a more detailed chronologic framework for understanding lateral shifts in the axial-fluvial system across the basin.

## GEOLOGIC SETTING

Crustal extension in the southern Rio Grande rift began in early Oligocene time and has continued in several pulses to the present day (Seager et al., 1984; Mack et al., 1994a; Mack et al., 1994b). The Palomas basin first developed in latest Oligocene and early Miocene time with initial uplift of the Caballo Mountains and Black Range (Fig. 1; Mack et al., 1994b). The basin narrowed and took on its modern form during the most recent pulse of block faulting in latest Miocene and early Pliocene time, when faulting stepped into the former basin and uplifted the Red Hills, Animas Hills, and Salado Hills fault blocks (Fig. 1; Mack et al., 1994b).

Approximately 40 km long and 25 km wide, the Pliocene–Holocene Palomas basin is an east-tilted half graben (Fig. 1). Exposed on the footwall scarp of the Caballo Mountains and Red Hills are Proterozoic granitic and metamorphic rocks overlain by about one kilometer of Paleozoic sedimentary rocks, mostly carbonates. In addition, Permian redbeds and Cenozoic volcanic and volcaniclastic rocks are exposed in Apache Valley, the structurally and topographically low area between the Caballo Mountains and Red Hills fault blocks (Fig. 1). The hanging-wall source region contains primarily Cenozoic volcanic and volcaniclastic rocks with lesser amounts of Precambrian crystalline basement

Sedimentation in Continental Rifts
SEPM Special Publication No. 73, Copyright © 2002
SEPM (Society for Sedimentary Geology), ISBN 1-56576-082-4, p. 165–177.

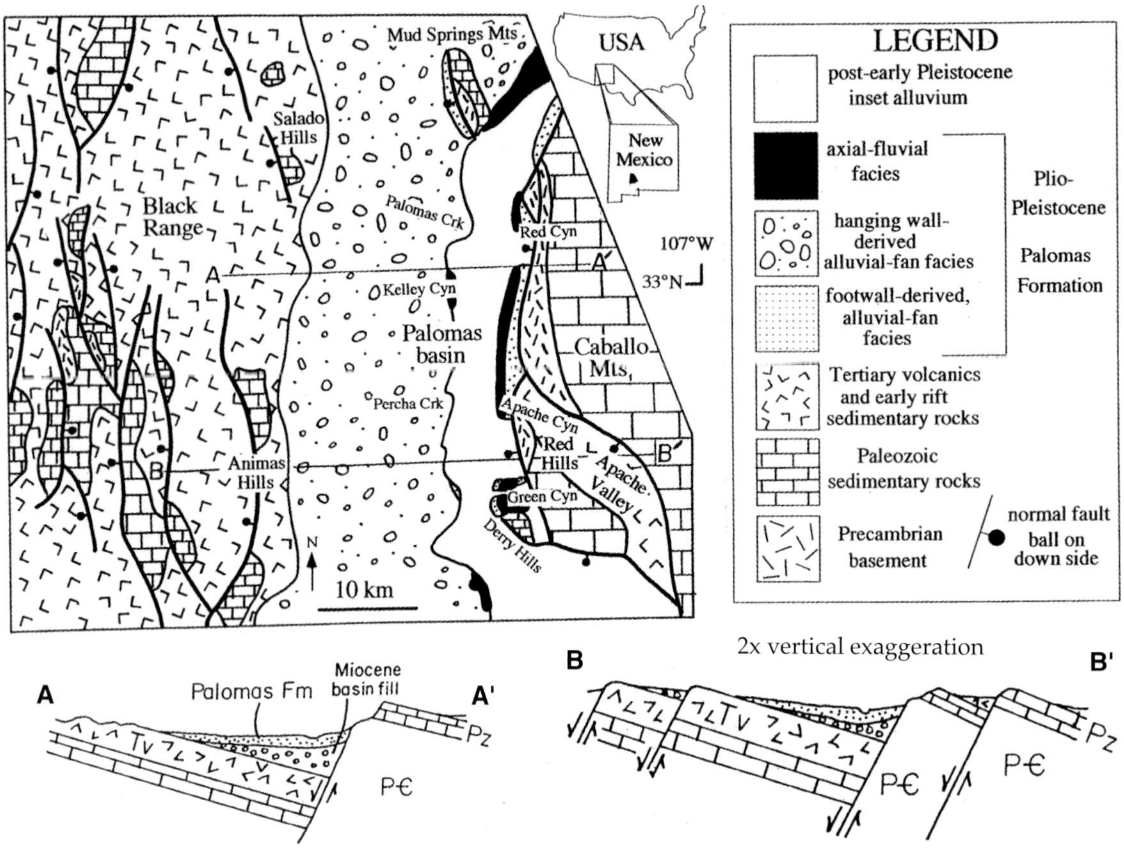


FIG. 1.—Generalized geologic map and cross sections of the Palomas basin and adjacent mountains in the southern Rio Grande rift, adapted from Seager et al. (1982).

and Paleozoic sedimentary rocks (Fig. 1). The border faults of the Caballo Mountains have experienced about 30 m of post–early Pleistocene movement, as well as several meters of Holocene offset (Machette, 1987; Foley et al., 1988).

The young basin fill of the Palomas basin is the Palomas Formation (Lozinsky and Hawley, 1986a, 1986b), which consists of about 100 m of gravel/conglomerate, sand/sandstone, and mudstone deposited on alluvial fans and by the ancestral Rio Grande. The Palomas Formation and the coeval Camp Rice Formation in basins to the south have been dated as Pliocene to early Pleistocene by a combination of vertebrate biostratigraphy (Hawley et al., 1969; Tedford, 1981; Repenning and May, 1986; Lucas and Oakes, 1986; Morgan et al., 1998), radioisotopic dates of underlying and interbedded volcanic rocks (Bachman and Mehnert, 1978; Seager et al., 1984; Mack et al., 1996; Mack et al., 1998), and reversal magnetostratigraphy (Repenning and May, 1986; Vanderhill, 1986; Mack et al., 1993; Mack et al., 1998; Leeder et al., 1996). Approximately 0.78 Ma ago, at or near the boundary between the Matuyama and Brunhes geopolarity chrons, the ancestral Rio Grande and its tributaries began to alternately incise and partially backfill the basin, lowering the level of the modern river by about 100 m and exposing all or part of the Palomas Formation (Fig. 1; Mack et al., 1998). Locally present is the gently sloping Cuchillo geomorphic surface, which represents the constructional top of the paleofans of the Palomas Formation. The Cuchillo surface preserves the slope of the paleofans that existed just prior to basin entrenchment.

The nature of the outcrops of the Palomas Formation varies within the basin (Fig. 2). Sediments on the eastern side of the basin are highly dissected by the principal incised drainages and their distributaries, as well as by drainages that head within the sediment, providing closely spaced outcrops both parallel to and perpendicular to the basin axis. Although of equal depth, incised drainages on the western side of the basin are fewer in number and more widely spaced. However, locally near the mouths of the western drainages there are badlands several kilometers square that provide excellent three-dimensional exposures. The Palomas Formation has largely been removed from the center of the basin, with the exception of isolated exposures at the southern end.

## REVERSAL MAGNETOSTRATIGRAPHY

Reversal magnetostratigraphic data from the Palomas basin are part of a larger set of data collected throughout the southern Rio Grande rift of south-central New Mexico (Vanderhill, 1986; Repenning and May, 1986; Mack et al., 1993; Mack et al., 1998; Leeder et al., 1996). Eighteen stratigraphic sections of the Palomas Formation and coeval Camp Rice Formation have been analyzed for polarity reversals by us, one by Repenning and May (1986), and three by Vanderhill (1986). The sections vary from 10 to 100 m thick and were sampled on the scale of 0.5 to 2 m. Seven sections have dated volcanic ash, pumice conglomerates, or basalt above, below, or within the Palomas or Camp Rice Formations, which provide accurate tie lines to the polarity reversal time scale. It has been consistently possible to identify the Gilbert–Gauss and Gauss–Matuyama chron boundaries, and in five cases all of the subchrons within

FIG. 2.—Photographs of incised strata of the Pliocene–lower Pleistocene Palomas Formation. **A)** View to southeast of incised footwall-derived, alluvial-fan strata with Caballo Mountains footwall in background. Depth of incision is approximately 50 m. **B)** View to southeast of approximately 50 m of incised proximal, footwall-derived, alluvial-fan sediment in Apache Canyon with Red Hills fault block in the background. Gently sloping surface in right-center of photo is the Cuchillo surface. **C)** View to southeast of incised hanging wall–derived, alluvial-fan strata. The gently sloping surface in the center of the photograph is the Cuchillo surface. **D)** Badlands exposures of distal, hanging wall–derived, alluvial-fan strata near the mouth of Palomas Creek.

a chron have been encountered. The constructional top of the formations has been shown in 13 sections to be at or just below the Matuyama–Brunhes boundary. Because magnetostratigraphic patterns are consistent between sections and with other stratigraphic and chronologic constraints, we have a high level of confidence in magnetostratigraphic correlations within the Palomas basin.

The primary magnetization in Palomas and Camp Rice sediments is carried by magnetite, as indicated by the demagnetization behavior (Mack et al., 1993). The typical demagnetization path shows a magnetic component removed by heating from 100°C to about 250°C with a strong linear component removed from about 250°C to 500°C. While the low-temperature component is probably carried by secondary hydrous iron oxides formed by weathering, the component interpreted as the primary has stable thermal behavior consistent with single-domain magnetite. This magnetite would have been deposited in the sediments and aligned with the earth's magnetic field within minutes of deposition (Collinson, 1965). While there are numerous postdepositional processes that can influence the coherency of the magnetic direction of the sediments (Tucker, 1983), little disruptive effects are seen in our samples. While some sections show a noticeable inclination flattening, most pass the paleomagnetic reversal test. Furthermore, magnetostratigraphic patterns are consistent between sections and with other stratigraphic and chronologic constraints on the age of the sediments.

## LITHOFACIES OF THE PALOMAS FORMATION

### *Alluvial-Fan Lithofacies*

*Turbulent-flow conglomerate/gravel.—*

This lithofacies is present in both footwall–derived and hanging wall-derived alluvial-fan strata, and consists of calcite-cemented, boulder to granule conglomerate and interbedded sandstone in the former and mostly unconsolidated gravel and sand in the latter. The difference in lithification between footwall-derived and hanging wall–derived sediment may be related to the abundance of carbonate clasts in the former and their absence in the latter. The carbonate clasts may have acted as a source of and / or as nucleation sites for calcite cement (Mack et al., 2000). The conglomerate–gravel occupies the floors of channels that range from 1 to 6 m deep and tens to hundreds of meters wide. The channels extend from fan head to fan toe and are best exposed near the mouths of, and on the north-trending cliffs that connect the mouths of, modern incised drainages. Deposition by turbulent water flow is indicated by the presence of trough and planar crossbeds and horizontal beds with well developed imbrication. For example, illustrated in Figure 3 is an outcrop of distal, footwall-derived, alluvial-fan strata oriented perpendicular to the westward paleoflow of the fan channels. The shallow fan channels, shown in longitudinal view, contain thin (0.3 m), imbricated conglomerates, which probably were deposited as longitudinal bars, and downstream-accreting foresets that probably represent bars with slip faces. Also present in the upper right side of the panel is a crossbed whose foresets dip at nearly right angles (north) to the channel paleoslope (west) and may have developed as a cross-channel bar at an anabranch confluence (Ashmore, 1993; Mack and Leeder, 1999).

Some channels, particularly distal, hanging wall–derived channels, have evidence of lateral accretion sets indicative of gravelly point bars. For example, illustrated in Figure 4 is a multistory gravel and sand channel oriented perpendicular to regional paleoflow and to flow indicators in the outcrop. The lower story has a 2-m-thick lateral accretion set dipping to the south toward the cut bank. The upper story also has evidence in the southern part of the outcrop of a lateral accretion set dipping gently to the south. Also well developed in this ourcrop are reddish-brown, overbank mudstones.

*Hyperconcentrated-flow pebbly sand.—*

Interbedded with turbulent-flow conglomerate / gravel in both footwall-derived and hanging wall–derived, alluvial-fan strata are beds 0.5 to 6 m thick of argillaceous fine to medium sand with primarily pebble-size and secondarily cobble- and granule-size clasts (Fig. 3). In some beds the gravel-size clasts are matrix supported, suggesting deposition as debris flows. Other beds display thin (< 30 cm) lenses of weakly grain-supported granules and pebbles with indistinct upper and lower contacts. These gravel lenses generally lack internal fabric, although in a few cases there is indistinct normal grading or imbrication. The weakly grain-supported variety of pebbly sand is interpreted to have been deposited as hyperconcentrated flood flows in which both turbulence and internal dispersive pressure acted on the sediment (cf. Smith, 1986). Both the debris flows and the hyperconcentrated flood flows resulted from higher sediment-to-water ratios than existed during deposition of turbulent-flow conglomerates / gravels.

Virtually all pebbly sands have superimposed calcic paleosols, which impart a distinctive light color to the beds. Described in detail by Mack et al. (2000), the calcic paleosols contain root traces of various sizes and consist of a brownish argillic B horizon overlying a calcic B horizon. The argillic B horizons experienced downward movement (translocation) of particulate clay into the horizon, as evidenced by clay coats (argillans) around detrital grains. Most calcic horizons consist of scattered nodules and tubules of pedogenic microcrystalline calcite, defining stage II morphology of Gile et al. (1966). Less commonly, calcic horizons are massive, well indurated micrite up to 1 m thick and are designated K or Bkm horizons. These paleosols indicate relatively long periods ($10^3$–$10^4$ yr) of stability and pedogenesis following deposition by hyperconcentrated flows.

*Reddish-brown mudstone.—*

Largely restricted to hanging wall–derived, alluvial-fan sediment is reddish-brown mudstone in beds 0.3 to 1 m thick. Many of the mudstones can be traced laterally for several hundred meters, but others are truncated after only a few tens of meters by gravel channels. Most reddish-brown mudstones contain pedogenic nodules and tubules of microcrystalline calcite, as well as vertic features, including slickensides, desiccation cracks, and wedge-shaped peds produced by shrinking and swelling of expandable clays (Mack et al., 2000).

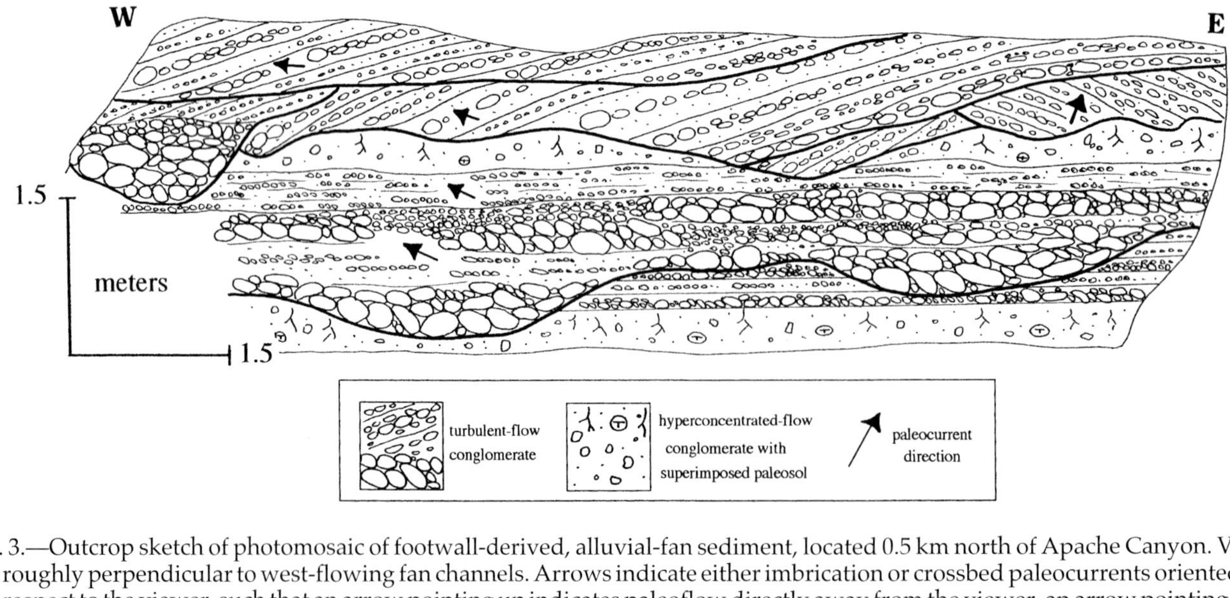


FIG. 3.—Outcrop sketch of photomosaic of footwall-derived, alluvial-fan sediment, located 0.5 km north of Apache Canyon. View is roughly perpendicular to west-flowing fan channels. Arrows indicate either imbrication or crossbed paleocurrents oriented with respect to the viewer, such that an arrow pointing up indicates paleoflow directly away from the viewer, an arrow pointing down represents paleoflow directly toward the viewer, an arrow pointed to the right represents paleoflow to the viewer's right, and an arrow pointed left indicates paleoflow to the viewer's left. Bold lines represent boundaries between channel stories.

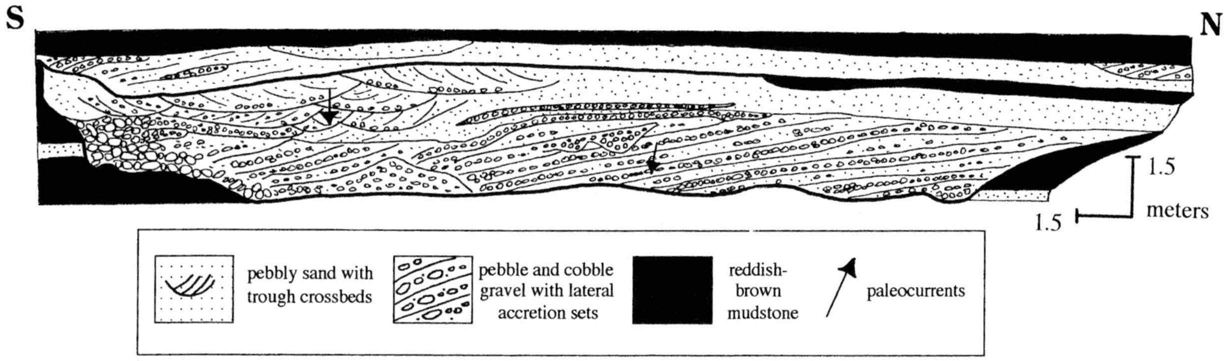


FIG. 4.—Outcrop sketch of photomosacic of hanging wall–derived, alluvial-fan sediment, located 6 km north of Percha Creek. Panel is oriented perpendicular to channel with paleoflow toward the viewer. Bold lines represent boundaries between channel stories. Paleocurrent arrows as in Figure 3.

*Ground-water carbonate.—*

Although authigenic rather than a primary depositional lithofacies, a prominent rock type interbedded with alluvial-fan sediment is massive, microcrystalline calcite in beds 1.5 to 3 m thick. The carbonate beds are restricted to hanging wall–derived fan strata and are traceable for kilometers parallel to the basin axis, but they pinch out within a few kilometers in an upfan direction. Sharp upper and lower contacts, internal horizontal layering, vesicular texture, and rhizoliths near the bed tops suggest an origin by precipitation by laterally flowing, shallow groundwater, perhaps locally associated with springs (Mack et al., 2000). These features are in sharp contrast to calcic paleosols, which display vertical arrangement of peds, calcic tubules, and root traces, and generally have a clay-rich B horizon gradationally overlying the calcic B horizon, which itself has a gradational basal contact (Mack et al., 2000).

*Comparison of depositional processes on footwall-derived and hanging wall–derived fans.—*

Both footwall-derived and hanging wall–derived fan sediment contain turbulent-flow conglomerates/gravels and hyperconcentrated-flow pebbly sands deposited in channels that extended from fan head to toe (Figs. 5, 6). However, distal hanging wall–derived strata contain a larger number of channels with lateral accretion sets, indicating greater channel sinuosity near the toes of the hanging wall–derived fans. Moreover, hanging wall–derived strata are characterized by a greater proportion of reddish-brown mudstone, expecially in midfan and distal fan settings, but also present in proximal-fan settings (Figs. 5, 6). Lateral tracing of these mudstone beds reveals that most represent overbank deposition between fan channels. However, some relatively thick intervals of mudstone and ground-water carbonate are not interbedded with channel gravels, nor can they be traced laterally over distances of hundreds of meters into channel gravels. These fine-grained intervals may have been deposited on very low-gradient alluvial flats beyond the toes of the hanging wall–derived fans.

These observations suggest significantly different alluvial-fan processes on either side of the Palomas basin. Footwall-derived fans behaved more like modern alluvial fans in the semiarid western U.S.A. (cf. Bull, 1972), in which deposition took place by turbulent flow in fan channels and by hyperconcentrated flows within channels and perhaps spilling onto interchannel areas. The channels were closely spaced and/or actively spread across the fan surface, inhibiting deposition of interchannel fines or effectively eroding fine sediment following its deposition. In contrast, greater channel sinuosity and preservation of overbank muds suggest that hanging wall–derived fan channels behaved more like gravel-bed rivers, particularly on the distal fan and where the fan channels periodically flowed onto the alluvial flat or floodplain of the ancestral Rio Grande (cf. Boothroyd and Ashley, 1975; Boothroyd and Nummedal, 1978).

## *Axial-Fluvial Lithofacies*

*Channel sand.—*

Axial-fluvial sediment is primarily represented by pebbly and cobbly, medium to coarse sand. Gravel is scattered throughout the sand and exists as lags up to 1 m thick above story boundaries. Mack and Leeder (1999) recognized two distinct types of channel lags. One consists of well rounded pebbles dominated by volcanic rock types from upstream sources, and the other is a bimodal mixture of rounded volcanic pebbles and more angular cobbles of granitic, metamorphic, and sedimentary rocks. The cobble fraction of the bimodal lags was derived from the Caballo and Red Hills footwall uplifts and was too coarse to be transported out of the basin by the ancestral Rio Grande. Channel sands with bimodal lags are commonly interbedded with footwall-derived, alluvial-fan strata, whereas channel sands with volcanic pebble lags are most common in a more basinward position. Also present in basal lags are large (up to 1 m long) rip-up clasts of reddish-brown and green mudstone.

Axial-fluvial sand exists as single-story or multistory sheets (Fig. 7). Story boundaries are defined by gravel/rip-up clast lags, and individual stories range from 1 to 5 m thick. The most common sedimentary structures within the channel sands are trough crossbeds and horizontal laminae with parting lineation. Planar crossbeds and asymmetrical ripple cross-laminae are less common, with the latter restricted to fine sand near the tops of channel-fill sequences (Fig. 7). Because the axial-fluvial sands are generally unlithified and there are few laterally extensive outcrops, a detailed architectural analysis of the channels in the Palomas basin is not possible. However, such a study was performed in outcrops 65 km southeast of the Palomas basin by Perez-Arlucea et al. (2000), who interpreted the ancestral Rio Grande sediment at this location to have been deposited by broad, low-sinuosity channels up to 3 m deep, floored by three-

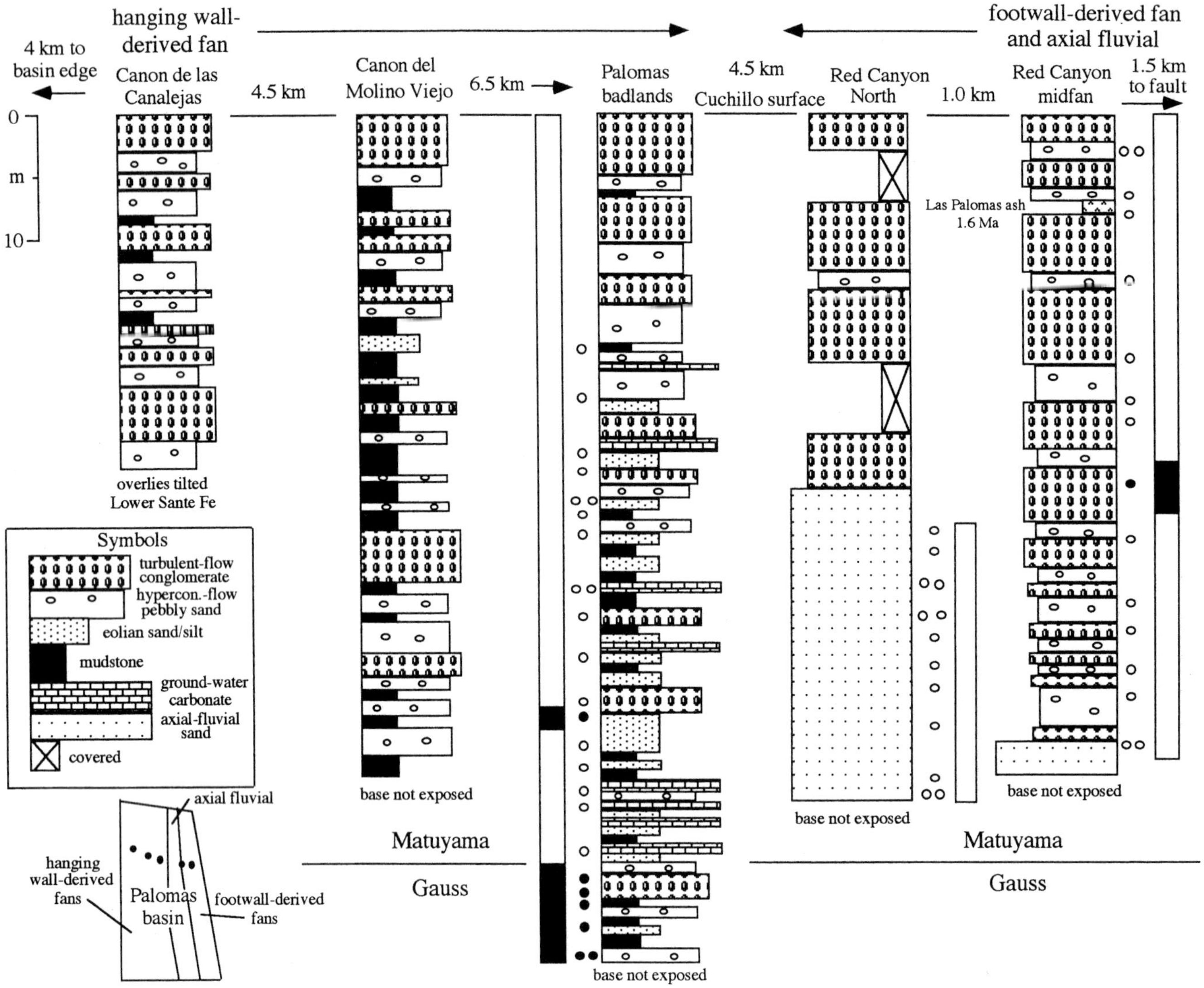


FIG. 5.—Measured sections of axial-fluvial and footwall-derived, alluvial-fan sediment at Red Canyon, and hanging wall–derived, alluvial-fan sediment in Palomas Creek and its tributaries. Paleomagnetic samples are shown as dark circles (normal polarity) and open circles (reversed polarity); bar scale is interpretation of the reversal magnetostratigraphic record. See Figure 9 for magnetic polarity reversal time scale. Inset of the Palomas basin shows the location of the measured sections within the basin and with respect to facies belts.

dimensional dunes, low-angle dunes, and downstream-accreting bars.

*Mottled mudstone.—*

Constituting a small percentage of axial-fluvial strata is reddish-brown and green mudstone locally interbedded on the scale of a few centimeters with light gray siltstone. Like mudstone in hanging wall–derived, alluvial-fan strata, the fluvial mudstones commonly contain root traces, desiccation cracks, wedge-shaped peds, and slickensides. Unique to the fluvial mudstones, however, is red and green mottling (gley) and horizontal beds and stringers of calcite, both of which are interpreted to have formed in response to a high and fluctuating water table (Mack et al., 2000). One thin (0.7 m), green mudstone contains gastropods and may represent a small pond. Along the eastern edge of the floodplain, where axial-fluvial and footwall-derived fan sediment interfinger, mudstone–siltstone exists as beds up to 2 m thick that occupy the tops of fluvial channels. These discontinuous beds probably were deposited following abandonment of the channels. In contrast, on the western side of the floodplain, more laterally continuous beds up to 1 m thick of mottled mudstone were probably deposited as overbank fines on the floodplain.

*Fine sand.—*

Along the western edge of the floodplain are beds 0.5 to 2.5 m thick of fine to very fine sand interbedded with mottled mudstones. Some beds of fine sand contain granules and pebbles, either scattered throughout or as thin (< 20 cm) lags at the bases of beds. Many of the fine sands exhibit horizontal laminae or

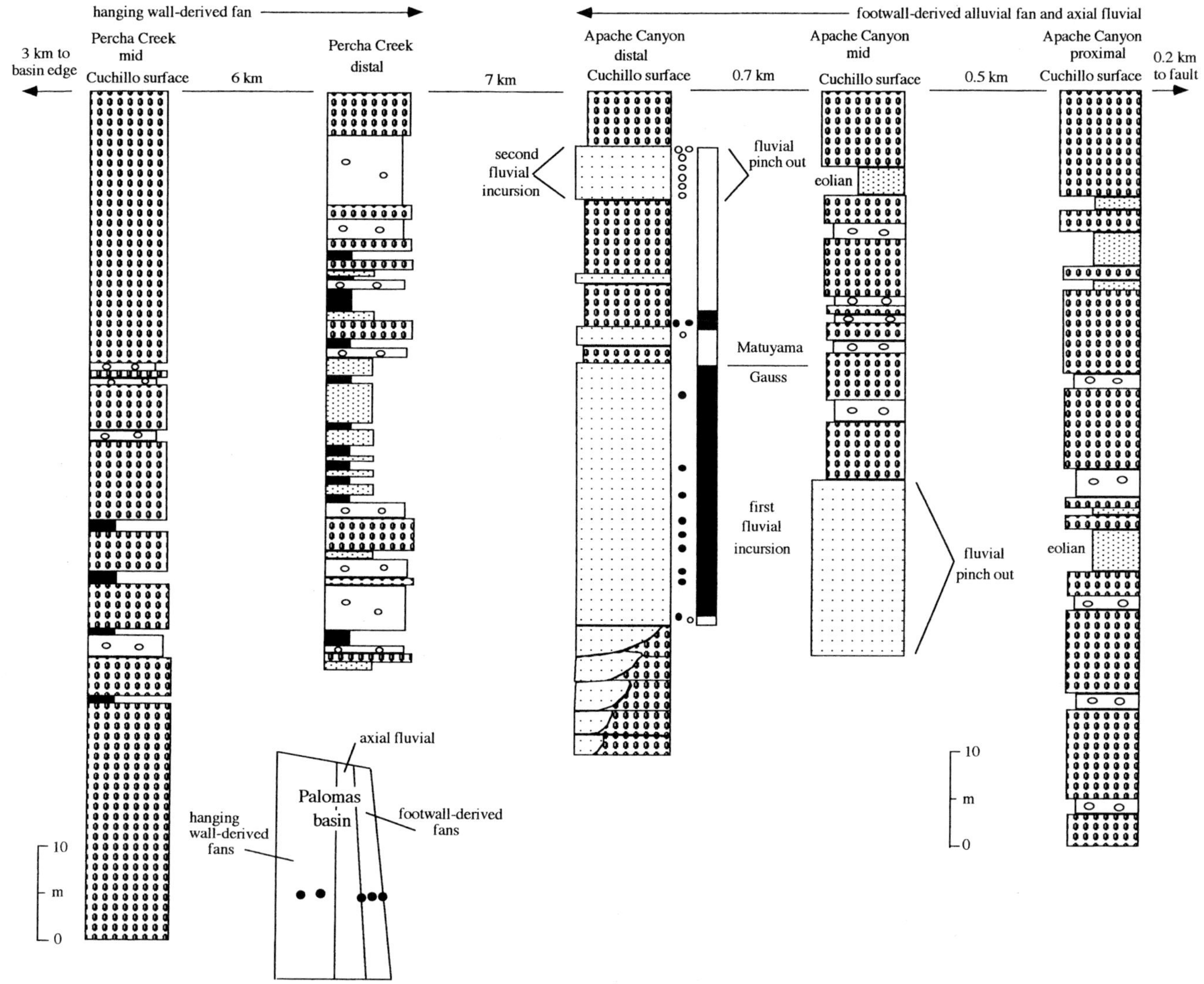


FIG. 6.—Measured sections of axial-fluvial and footwall-derived, alluvial-fan sediment at Apache Canyon and hanging wall–derived, alluvial-fan sediment at Percha Creek. See Figure 5 for description of symbols and paleomagnetic data base. Inset of the Palomas basin shows the location of the measured sections within the basin and with repsect to the facies belts.

ripple cross-laminae, although a few beds lack current structures. Few exposures and limited lateral extent of outcrops make this lithofacies difficult to interpret. Because of its fine grain size and association with overbank mudstones, the fine sand may represent crevasse splays and / or small, secondary channels that occupied the edge of the floodplain.

## Eolian Lithofacies

A relatively rare lithofacies consists of very fine sand and silt in beds 0.3 to 2 m thick. The sand and silt is generally structureless, although rarely small-amplitude climbing ripple cross-laminae or planar crossbeds are present. An eolian origin for this facies is based on fine grain size, excellent sorting, and ripple cross-laminae that are identical to those produced by climbing wind ripples (Hunter, 1977; Fryberger and Schenk, 1981). However, the bulk of the eolian sand and silt is structureless and resembles loessites described by Johnson (1989) in Permo-Pennsylvanian rocks of Colorado. Scattered pedogenic calcic nodules and tubules are common in the eolian sand and silt (Mack et al., 2000).

Although not abundant volumetrically, eolian sediment is widely distributed throughout the basin. It is interbedded with hanging wall–derived, alluvial-fan sediment in midfan and distal fan locations (Figs. 5, 6). An abundance of volcanic rock fragments and feldspar suggests that this eolian sand / silt represents reworking of abandoned hanging wall–derived fan channels or that the sand / silt blew out of fan channels into interchannel areas, where it became interbedded with reddish-brown mudstones. Less commonly, eolian sand / silt is interbedded with footwall-derived, alluvial-fan strata, particularly in Apache Canyon (Fig. 6, Mack and Leeder, 1999). Far richer in quartz, this eolian sand / silt probably blew out of the axial river channel onto the footwall-derived fans. Indeed, in Apache Canyon, eolian sand beds are located immediately upfan from thick axial-fluvial sand bodies (Fig. 6; Mack and Leeder, 1999). Eolian sand / silt is also interbedded with fluvial sediment along the western margin of the floodplain.

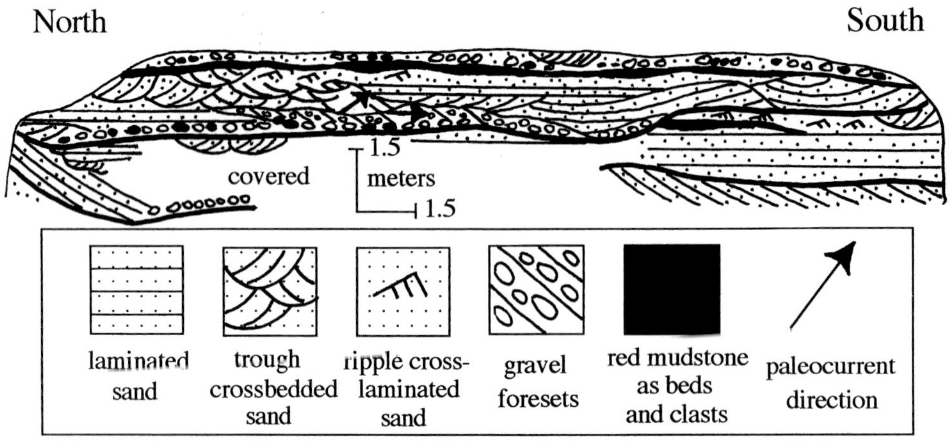


FIG. 7.—Outcrop sketch of photomosaic of multistory channels in axial-fluvial sediment, located 1.5 km north of Apache Canyon. View is perpendicular to paleoflow, which was to the south (toward right side of panel). Bold lines represent boundaries between channel stories. Paleocurrent arrows as in Figure 3.

## BASIN-WIDE DISTRIBUTION OF DEPOSITIONAL SYSTEMS

The basin-wide distribution of alluvial-fan and axial-fluvial strata of the Palomas Formation has been mapped at the scales of 1:24,000 (Seager and Mack, 1991) and 1:125,000 (Seager et al., 1982). The asymmetrical distribution of these depositional systems is consistent with theoretical and empirical models describing tectonic controls on sedimentation in continental half grabens (Fig. 8; Leeder and Gawthorpe, 1987; Blair and Bilodeau, 1988; Mack and Seager, 1990; Leeder and Jackson, 1993; Peakall, 1998). Most footwall-derived paleofans were small in area, because they had small catchments on the footwall scarps of the Caballo Mountains and Red Hills (Fig. 8; Leeder and Gawthorpe, 1987; Leeder and Jackson, 1993). Exceptions include the paleofans fed by the large Apache, Green, Red, Mescal, and Ash catchments. The Apache and Green catchments are large, because their drainages head in the Caballo Mountains and cut across the young Red Hills fault block, suggesting an antecedent origin (cf. Figs. 1, 8). The large Mescal and Ash catchments are situated on a north-tilted relay ramp (Mack and Seager, 1995), whereas the origin of the large size of the Red catchment is not known, but may be due to active headward erosion. The slopes of the footwall-derived paleofans, preserved as the Cuchillo surface, range from 1.1° to 4.7° and vary inversely with the area of the catchments, a relationship recognized on modern alluvial fans (Bull, 1972).

Hanging wall–derived paleofans were much larger (15 km radial length) and gentler (slopes 0.7–1°) than footwall-derived paleofans, presumably as a result of the large area of the hanging-wall catchments and eastward tilt of the basin (Fig. 8; Leeder and Gawthorpe, 1987; Leeder and Jackson, 1993). Low slopes of the paleofans may explain why the hanging wall–derived fan channels behaved like gravel-bed rivers, particularly the meandering character near the paleofan toes and the preservation of interchannel overbank mudstone. The hanging wall–derived fan channels also may have had greater discharge than those adjacent to the footwall because of greater precipitation in the higher-elevation hanging-wall catchments. This is certainly the case today, in that most of the Black Range is covered in tall pine forests and the incised drainages that head in the Black Range experience perennial flow to within a kilometer or less of the modern floodplain of the Rio Grande. In contrast, the lower-elevation Caballo Mountains receive less annual precipitation than the Black Range, resulting in vegetative cover of small junipers and pinyon pines and ephemeral fan drainages. Like Apache and Green drainages, the major hanging-wall drainages may be antecedent in origin, originally heading in the Black Range and subsequently cutting across the younger Animas and Salado fault blocks as they rose.

The narrow (< 5 km) floodplain of the ancestral Rio Grande and its location near the footwall blocks is a response to eastward tilt of the basin and the confining effect of the large hanging wall–derived paleofans (Leeder and Gawthorpe, 1987; Mack and Seager, 1990). The distribution of axial-fluvial channel and floodplain lithofacies is also asymmetrical. On the eastern side of the floodplain, axial-fluvial sediment consists almost exclusively of multistory channel sands, whereas mudstone–siltstone is rare and restricted to the tops of abandoned channels. On the western side of the floodplain, however, channel sands are more likely to be single story or consist of fewer stories than on the eastern side of the basin, and there is greater preservation of overbank mudstone and fine sand of crevasse-splay and/or small-channel origin. Thus, the main axial river appears to have spent less time on the western side of the floodplain than on the eastern side.

## FLUVIAL–FAN INTERACTION

Axial-fluvial sediment is well exposed along the eastern edge of the basin, where it is interbedded with footwall-derived, alluvial-fan conglomerates (Mack and Leeder, 1999). In the northern part of the basin there is one relatively thick (up to 80 m) fluvial lithosome that can be mapped to within 0.5 to 1 km of the border fault (Fig. 5). At Apache Canyon (Fig. 6), where the Red Hills fault turns northeastward to intersect the Caballo fault, there are two fluvial lithosomes, the lower one extending to within 400 m of the border fault and the upper one to within 1.2 km. In the southern part of the basin (Fig. 9), three mappable fluvial lithosomes are present, each successively younger one extending closer to the border fault, with the uppermost one mapped to within 100 m of the fault (Seager and Mack, 1991). Throughout the eastern side of the basin the final depositional event was progradation of the footwall-derived fans several kilometers into the basin (Figs. 5, 6, 9).

A

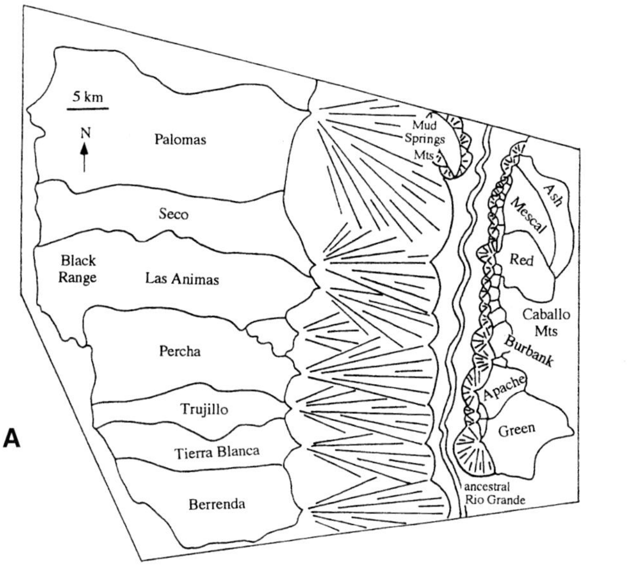


B

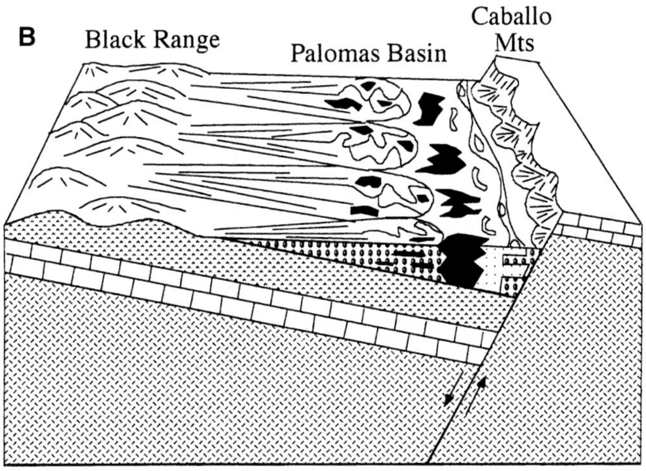


FIG. 8.—**A)** Simplified paleogeographic map of the Palomas Basin in Pliocene and early Pleistocene time. Labeled catchments (Green, Percha, etc.) are based on their modern extent. Location and size of the footwall-derived alluvial fans is based on a combination of provenance of clasts, slope direction of the Cuchillo geomorphic surface, and paleocurrents (Mack and Leeder, 1999). Hanging wall–derived alluvial fans are drawn symmetrically about those modern incised drainages that head in the Black Range and incise the Palomas Formation. Location of axial-fluvial system is based on geologic maps by Seager et al. (1982) and Seager and Mack (1991). **B)** Schematic block diagram of the distribution of alluvial-fan and axial-fluvial facies in the Palomas basin.

Reversal magnetostratigraphic data allow general estimates of the duration of fluvial incursions toward the footwall. At Red Canyon (Fig. 5), maximum incursion took place early in the Matuyama Chron (~ 2.6–2.1 Ma), followed by fan progradation into the basin prior to eruption of the Las Palomas ash (1.6 Ma; Mack et al., 1998). In contrast, at Apache Canyon (Fig. 6), the first fluvial incursion began during the Gauss Chron, and perhaps earlier, and the fan had subsequently prograded at least 1.2 km into the basin prior to the Gauss–Matuyama boundary (~ 2.6 Ma). The later fluvial incursion at Apache Canyon can be dated only as entirely within the Matuyama Chron. In the southern part of the basin (Fig. 9), the intitial fluvial incursion toward the footwall began in the Gilbert Chron, reaching its maximum extent near the end of the Cochiti subchron (~ 4.3 Ma). Following a period of footwall fan progradation lasting through the early Gauss (~ 800 kyr), the second fluvial incursion began near the beginning of the Mammoth subchron (~ 3.3 Ma) and lasted at least into the lower part of the Matuyama (~ 2.1 Ma). The final fluvial incursion is entirely within the Matuyama Chron, but it cannot be dated more accurately.

Suitably oriented outcrops show that movement of the axial river toward the footwall was accomplished by toe cutting of the footwall-derived fans (Fig. 10). During net movement of the river toward the footwall, the toes of the footwall fans must have exhibited prominent cliffs equal in height to the depth of incision by the axial river, up to 9 m in the case of the Palomas basin (Fig. 10). Following river avulsion away from the toe-cut fans, footwall-derived fan gravels prograded and downlapped the abandoned axial channel and floodplain sediments ("healing wedges" of Fig. 10). The downlapping fan wedges emanated from the primary fan channels and possibly also from "blind channels" that incised the toe of the fan in response to a reduction in base level brought on by toe cutting. Such "blind channels" are present on modern toe-cut fans in the Big Lost River basin, Idaho (M. Leeder, unpublished data).

Although interbedded axial-fluvial and footwall-derived, alluvial-fan sediment is well exposed along the eastern margin of the basin, there are few areas where axial-fluvial and hanging wall–derived, alluvial-fan sediment are interbedded (Fig. 1). The best of these exposures is in Kelly Canyon (Fig. 11), where subdivision of the strata into several distinct time intervals is possible by walking out key marker beds. Simplified paleogeographic maps for these seven time intervals are shown in Figure 12.

In the Kelly Canyon area there is evidence of avulsion of the main axial-fluvial channel toward (Tk2, Tk4, Fig. 12) and away from (Tk3, Tk5, Fig. 12) the western margin of the floodplain of the ancestral Rio Grande. During periods when the main channel avulsed eastward, away from the hanging wall–derived fans, overbank mud was deposited along the western edge of the floodplain, with occasional incursions by crevasse channels and/or small axial-fluvial channels containing a fine sand bedload. During those periods when the main fluvial channel avulsed toward the western edge of the floodplain it was separated from the hanging wall–derived fan channels by mudstone, eolian sand, and spring carbonate deposited near the toe of the fans or on a very gently sloping alluvial flat. Unlike on the eastern side of the basin, there is no evidence at Kelly Canyon and elsewhere in the western half of the basin for toe cutting of the hanging wall–derived fans by the axial river. Moreover, hanging wall–derived fan progradation appears to have occurred when the main channel was located along the eastern edge of or entirely east of the Kelly Canyon transect (Tk1, Tk3, Tk5, Tk6, Fig. 12). Indeed, the final depositional event in the Kelly Canyon area was eastward progradation of the hanging wall–derived fans beyond the easternmost measured section (Tk7, Fig. 12).

Several lines of evidence suggest that periods of fault activity and consequent eastward tilting of the basin were responsible for lateral shifting of the axial river toward the footwall blocks. Firstly, toe cutting of the fans by the river took place only on the footwall side of the basin. If the process of toe cutting was driven by climate or was autocyclic, then it should have occurred equally on both sides of the basin. Secondly, the number of fluvial

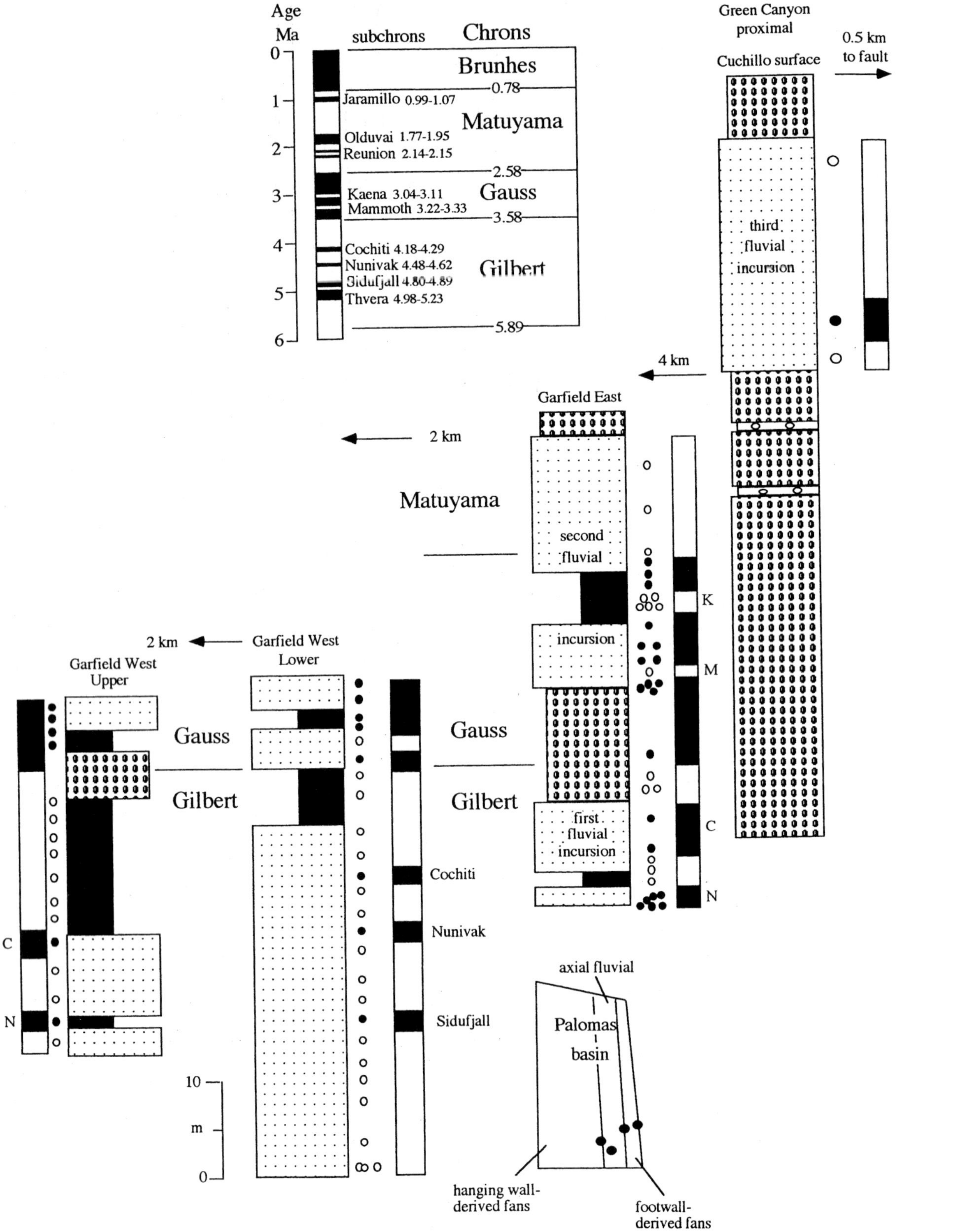


FIG. 9.—Measured sections of footwall-derived, alluvial-fan and axial-fluvial sediment at Green Canyon and areas to the west of the mouth of Green Canyon. Conglomerate at "Garfield West Upper" section was deposited on a hanging wall–derived alluvial fan. See Figure 5 for description of symbols and paleomagnetic data base. K = Kaena subchron; M = Mammoth subchron; C = Cochiti subchron; N = Nunivak subchron. Magnetic polarity reversal time scale is from Berggren et al. (1995). Inset of Palomas basin shows the location of the measured sections within the basin and with respect to the facies belts.

incursions toward the footwall varies within the basin. If climate was driving the process, then the number of incursions should be the same throughout the basin. In contrast, independent histories of fault movement along the length of and between the Caballo and Red Hills faults is to be expected and would have resulted in the different number and magnitude of fluvial incursions within the basin. Finally, the geographic scale of the fluvial incursions and their protracted nature argue for a tectonic origin. In some cases the river moved over a kilometer toward the footwall block over periods of $10^5$ to $10^6$ years, virtually obliterating locally the footwall-derived fans (Seager and Mack, 1991; Mack and Leeder, 1999). There is no evidence in size of the bedload, sedimentary structures, or channel geometries to suggest that the lateral shifts of the river were related to marked changes in discharge, nor is there evidence of paleoclimate changes of the magnitude required to produce such large-scale lateral shifts of the river.

## CONCLUSIONS

Post-Matuyama (< 0.78 Ma) incision of the Palomas half graben in the southern Rio Grande rift provided an opportunity to document basin-wide architecture of Pliocene–lower Pleistocene alluvial-fan and axial-fluvial sediment. As predicted in theoretical and empirical models, the Palomas half graben has an asymmetrical distribution of facies belts, characterized by narrow belts of footwall-derived, alluvial-fan and axial-fluvial sediment and a broader belt of hanging wall–derived, alluvial-fan sediment. Contrary to the suggestion by Blair and McPherson (1994) that deposition on alluvial fans is primarily by either sheetfloods or debris flows, deposition on the Palomas paleofans took place primarily in channels. Moreover, distal, hanging wall–derived, alluvial-fan sediment closely resembles sediment deposited by gravel-bed rivers, because of the presence of lateral accretion sets and common preservation of overbank mudstones. An asymmetrical distribution of lithofacies is also apparent in axial-fluvial sediment, such that strata on the hanging-wall side of the floodplain is dominated by overbank mudstones and fine-grained crevasse-splay / small-channel sands, while kilometer-scale incursions of the main river channel toward the footwall was accomplished by toe cutting the footwall-derived fans and resulted in deposition of multistory channels sands.

## ACKNOWLEDGMENTS

This research was sponsored by National Science Foundation Grant EAR-9627468 awarded to G. Mack. M. Leeder was supported by a grant from the Royal Society of London. Andrew Miall, Torbjörn Törnqvist, and Gail Ashley read an earlier version of the manuscript and made many helpful suggestions for its improvement.

## REFERENCES

ASHMORE, P., 1993, Anabranch confluence kinetics and sedimentation processes in gravel-bed streams, *in* Best, J.L., and Bristow, C.S., eds., Braided Rivers: Geological Society of London, Special Publication 75, p. 129–146.

BACHMAN, G.O., AND MEHNERT, H.H., 1978, K–Ar dates and late Pliocene and Holocene geomorphic history of the Rio Grande region, New Mexico: Geological Society of America, Bulletin, v. 89, p. 283–292.

BERGGREN, W.A., HILGEN, F.J., LANGEREIS, C.G., KENT, D.V., OBRADOVICH, J.D., RAFFI, I., RAYMO, M.E., AND SHACKLETON, N.J., 1995, Late Neogene chronology: New perspectives in high-resolution stratigraphy: Geological Society of America, Bulletin, v. 107, p. 1272–1287.

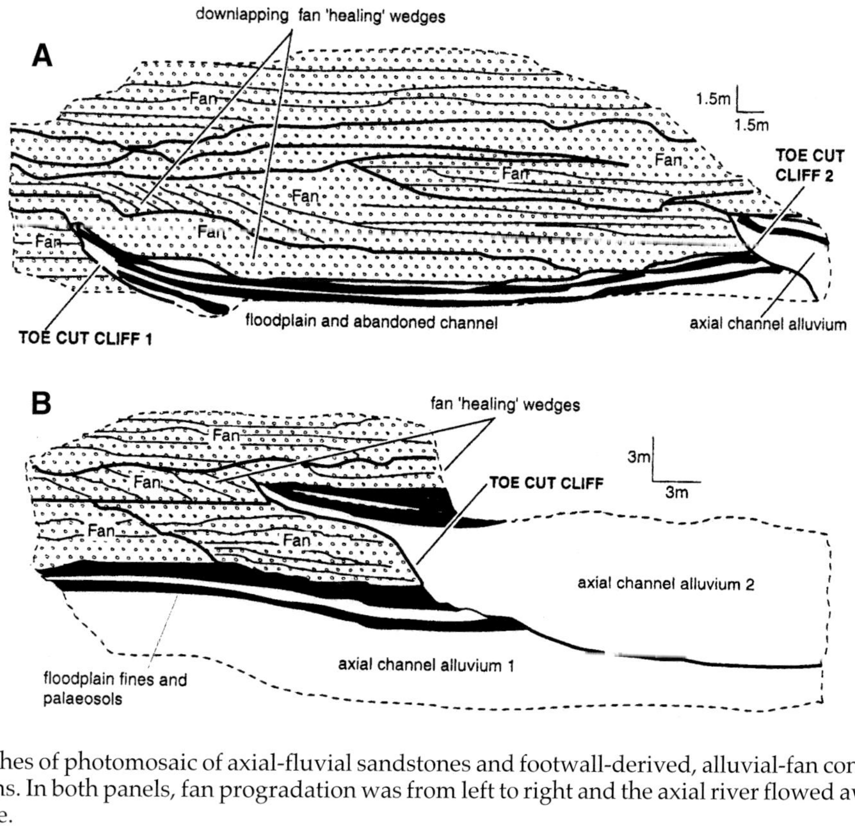


FIG. 10.—Outcrop sketches of photomosaic of axial-fluvial sandstones and footwall-derived, alluvial-fan conglomerates illustrating toe cutting of the fans. In both panels, fan progradation was from left to right and the axial river flowed away from viewer. Black = mottled mudstone.

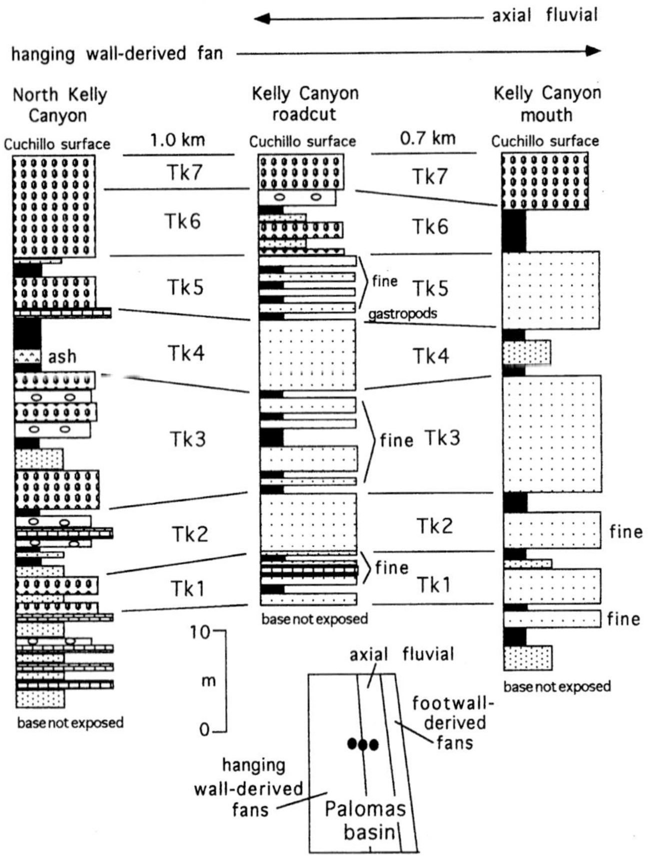


FIG. 11.—Measured sections of axial-fluvial and hanging wall–derived, alluvial-fan sediment at Kelly Canyon. Time intervals (Tk1 etc.) are based on physical correlation of key marker beds. See Figure 5 for symbols. Inset of the Palomas basin shows the location of the measured sections within the basin and with respect to the facies belts.

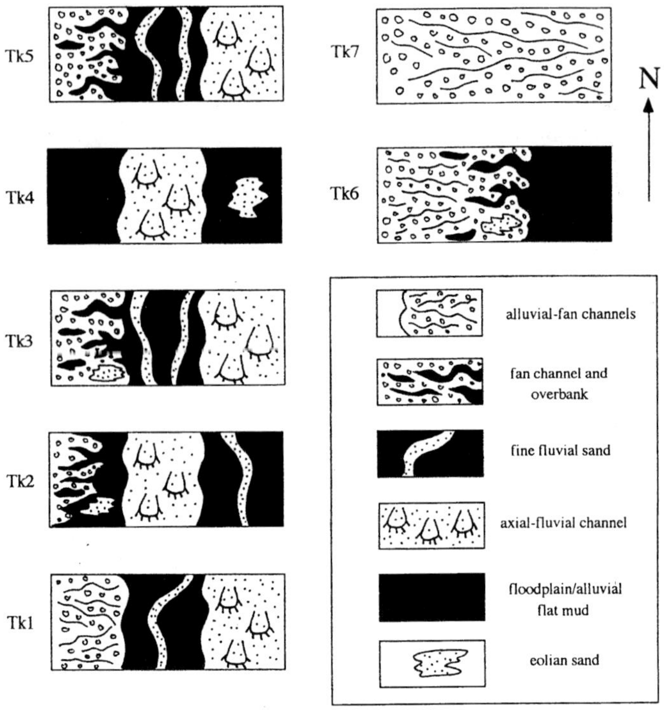


FIG. 12.—Simplified paleogeography in the Kelly Canyon area corresponding to the time intervals shown in Figure 11.

BLAIR, T.C., AND BILODEAU, W.L., 1988, Development of tectonic cyclothems in rift, pull-apart, and foreland basins: Sedimentary response to episodic tectonism: Geology, v. 16, p. 517–520.

BLAIR, T.C., AND MCPHERSON, J.G., 1994, Alluvial fans and their natural distinction from rivers based on morphology, hydraulic processes, sedimentary processes, and facies assemblages: Journal of Sedimentary Research, v. A64, p. 450–489.

BOOTHROYD, J.C., AND ASHLEY, G.M., 1975, Processes, bar morphology, and sedimentation structures on braided outwash fans, northeastern Gulf of Alaska, *in* Jopling, A.V., and McDonald, B.C., eds., Glaciofluvial and Glaciolacustrine Sedimentation: Society of Economic Paleontologists and Mineralogists, Special Publication 23, p. 193–222.

BOOTHROYD, J.C., AND NUMMEDAL, D., 1978, Proglacial braided outwash: a model for humid alluvial-fan deposits, *in* Miall, A.D., ed., Fluvial Sedimentology: Canadian Society of Petroleum Geologists, Memoir 5, p. 641–668.

BULL, W.B., 1972, Recognition of alluvial fan deposits in the stratigraphic record, *in* Rigby, J.K., and Hamblin, W.K., eds., Recognition of Ancient Sedimentary Environments: Society of Economic Paleontologists and Mineralogists, Special Publication 16, p. 63–83.

COLLINSON, D.W., 1965, Depositional remanent magnetization in sediments: Journal of Geophysical Research, v. 70, p. 4663–4668.

FOLEY, L.L., LAFORGE, R.C., AND PIETY, L.A., 1988, Seismotectonic study for Elephant Butte and Caballo dams, Rio Grande Project, New Mexico: U.S. Bureau of Reclamation, Seismotectonic Report 88-9, 60 p.

FRYBERGER, S.G., AND SCHENK, C., 1981, Wind sedimentation tunnel studies on the origins of eolian strata: Sedimentology, v. 28, p. 805–821.

GILE, L.H., PETERSON, F.F., AND GROSSMAN, R.B., 1966, Morphological and genetic sequences of carbonate accumulation in desert soils: Soil Science, v. 101, p. 347–360.

HAWLEY, J.W., KOTTLOWSKI, F.E., SEAGER, W.R., KING, W.E., STRAIN, W.S., AND LEMONE, D.V., 1969, The Santa Fe Group in the south-central New Mexico border region: New Mexico Bureau of Mines and Mineral Resources, Circular 104, p. 52–76.

HUNTER, R.E., 1977, Basic types of stratification in small eolian dunes: Sedimentology, v. 24, p. 361–387.

JOHNSON, S.Y., 1989, Significance of loessite in the Maroon Formation (middle Pennsylvanian to Lower Permian), Eagle basin, northwest Colorado: Journal of Sedimentary Petrology, v. 59, p. 782–791.

LEEDER, M.R., AND GAWTHORPE, R.L., 1987, Sedimentary models for extensional tilt-block / half-graben basins, *in* Coward, M.P., Dewey, J.F., and Hancock, P.L., eds., Continental Extensional Tectonics: Geological Society of London, Special Publication 28, p. 139–152.

LEEDER, M.R., AND JACKSON, J.A., 1993, The interaction between normal faulting and drainage in active extensional basins with examples from the western United States and central Greece: Basin Research, v. 5, p. 79–102.

LEEDER, M.R., MACK, G.H., AND SALYARDS, S.L., 1996, Axial-transverse fluvial interactions in half-graben: Plio-Pleistocene Palomas basin, southern Rio Grande rift, New Mexico, USA: Basin Research, v. 12, p. 225–241.

LOZINSKY, R.P., AND HAWLEY, J.W., 1986a, The Palomas Formation of south-central New Mexico—a formal definition: New Mexico Geology, v. 8, p. 73–78, 82.

LOZINSKY, R.P., AND HAWLEY, J.W., 1986b, Upper Cenozoic Palomas Formation of south-central New Mexico: New Mexico Geological Society, Guidebook 37, p. 239–247.

LUCAS, S.G., AND OAKES, W., 1986, Pliocene (Blancan) vertebrates from the Palomas Formation, south-central New Mexico: New Mexico Geological Society, Guidebook 37, p. 249–255.

MACHETTE, M.N., 1987, Preliminary assessment of Quaternary faulting near Truth or Consequences, New Mexico: U.S. Geological Survey, Open File Report 87-652, 40 p.

MACK, G.H., AND LEEDER, M.R., 1999, Climatic and tectonic controls on alluvial-fan and axial-fluvial sedimentation in the Plio-Pleistocene Palomas half graben, southern Rio Grande rift: Journal of Sedimentary Research, v. 69, p. 635–652.

MACK, G.H., AND SEAGER, W.R., 1990, Tectonic control on facies distribution of the Camp Rice and Palomas Formations (Plio-Pleistocene) in the southern Rio Grande rift: Geological Society of America, Bulletin, v. 102, p. 45–53.

MACK, G.H., AND SEAGER, W.R., 1995, Transfer zones in the southern Rio Grade rift: Geological Society of London, Journal, v. 152, p. 551–560.

MACK, G.H., SALYARDS, S.L., AND JAMES, W.C., 1993, Magnetostratigraphy of the Plio-Pleistocene Camp Rice and Palomas Formations in the Rio Grande rift of southern New Mexico: American Journal of Science, v. 293, p. 49–77.

MACK, G.H., NIGHTENGALE, A.L., SEAGER, W.R., AND CLEMONS, R.E., 1994a, The Oligocene Goodsight–Cedar Hills half graben near Las Cruces and its implications to the evolution of the Mogollon–Datil volcanic field and to the southern Rio Grande rift: New Mexico Geological Society, Guidebook 45, p. 135–142.

MACK, G.H., SEAGER, W.R., AND KIELING, J., 1994b, Late Oligocene and Miocene faulting and sedimentation, and evolution of the southern Rio Grande rift, New Mexico, USA: Sedimentary Geology, v. 92, p. 79–96.

MACK, G.H., MCINTOSH, W.C., LEEDER, M.R., AND MONGER, H.C., 1996, Plio-Pleistocene pumice floods in the ancestral Rio Grande, southern Rio Grande rift, USA: Sedimentary Geology, v.103, p. 1–8.

MACK, G.H., SALYARDS, S.L., MCINTOSH, W.C., AND LEEDER, M.R., 1998, Reversal magnetostratigraphy and radioisotopic geochronology of the Plio-Pleistocene Camp Rice and Palomas Formations, southern Rio Grande rift: New Mexico Geological Society, Guidebook 49, p. 229–236.

MACK, G.H., COLE, D.R., AND TREVINO, L., 2000, The distribution and discrimination of shallow, authigenic carbonate in the Plio-Pleistocene Palomas Basin, southern Rio Grande rift: Geological Society of America, Bulletin, v. 112, p. 643–656.

MORGAN, G.S., LUCAS, S.G., AND ESTEP, J.W., 1998, Pliocene (Blancan) vertebrate fossils from the Camp Rice Formation near Tonuco Mountain, Dona Ana County, southern New Mexico: New Mexico Geological Society, Guidebook 49, p. 237–249.

PEAKALL, J., 1998, Axial river evolution in response to half-graben faulting: Carson River, Nevada, USA: Journal of Sedimentary Research, v. 68, p. 788–799.

PEREZ-ARLUCEA, M., MACK, G., AND LEEDER, M., 2000, Reconstructing the ancestral (Plio-Pleistocene) Rio Grande in its active tectonic setting, southern Rio Grande rift, New Mexico, USA: Sedimentology, v. 47, p. 701–720.

REPENNING, C.A., AND MAY, S.R., 1986, New evidence for the age of lower part of the Palomas Formation, Truth or Consequences, New Mexico: New Mexico Geological Society, Guidebook 37, p. 257–259.

SEAGER, W.R., AND MACK, G.H., 1991, Geology of Garfield quadrangle, Sierra and Dona Ana Counties, New Mexico: New Mexico Bureau of Mines and Mineral Resources, Bulletin 128, 24 p.

SEAGER, W.R., CLEMONS, R.E., HAWLEY, J.W., AND KELLEY, R.E., 1982, Geology of northwest part of Las Cruces 1° x 2° sheet: New Mexico Bureau of Mines and Mineral Resources, Geologic Map 53, scale 1:125,000.

SEAGER, W.R., SHAFIQULLAH, M., HAWLEY, J.W., AND MARVIN, R.F., 1984, New K–Ar dates from basalts and the evolution of the southern Rio Grande rift: Geological Society of America, Bulletin, v. 95, p. 87–99.

SMITH, G.A., 1986, Coarse-grained nonmarine volcaniclastic sediment: terminology and depositional process: Geological Society of America, Bulletin, v. 97, p. 1–10.

TEDFORD, R.H., 1981, Mammalian biochronology of late Cenozoic basins of New Mexico: Geological Society of America, Bulletin, v. 92, p. 1008–1022.

TUCKER, P., 1983, Magnetization of unconsolidated sediments and theories of DRM, *in* Creer, K.M., Tucholka, P., and Barton, C.E., eds., Geomagnetism of Baked Clays and Recent Sediments: Amsterdam, Elsevier, p. 9–19.

VANDERHILL, J.B., 1986, Lithostratigraphy, vertebrate paleontology, and magnetostratigraphy of Plio-Pleistocene sediments in the Mesilla basin, New Mexico: Unpublished Ph.D. dissertation, University of Texas, Austin, 305 p.

# THE OKAVANGO DELTA—SEMIARID ALLUVIAL-FAN SEDIMENTATION RELATED TO INCIPIENT RIFTING

T. S. MCCARTHY
*Department of Geology, School of Geosciences, University of the Witwatersrand, P.O. Box 3, WITS 2050, Johannesburg, South Africa,*
*e-mail: mccarthyt@geosciences.wits.ac.za*
N. D. SMITH
*Department of Geosciences, University of Nebraska, Lincoln, Nebraska 68588-0340, U.S.A.*
*e-mail: nsmith3@unl.edu*
W. N. ELLERY
*Department of Geographical and Environmental Sciences, University of Natal, 4001 Durban, South Africa,*
*e-mail: ellery@nu.ac.za*
AND
T. GUMBRICHT
*Department of Geology, School of Geosciences, University of the Witwatersrand, P.O. Box 3, WITS 2050, Johannesburg, South Africa,*
*e-mail: thomas_gumbricht@hotmail.com*

ABSTRACT: The Okavango "Delta" is a large (25,000 $km^2$), low-gradient (1:3400) alluvial fan which has developed in a depression between two basement arch–horst systems which mark the tips of incipient rifts. Base flow in the Okavango River sustains about 4000 $km^2$ of permanent wetland in the Delta, while the seasonal flood can expand the area of inundation to more than 12,000 $km^2$. The climate over the Delta is semiarid, and only 2% of combined inflow plus rainfall (ca. 15 $km^3$/yr) leaves as surface flow, the remainder being lost to the atmosphere by evapotranspiration. Sedimentologically, the system is virtually closed, and about 600,000 tonne of sediment is deposited annually in the Delta, resulting in a very low average sedimentation rate. Two distinct types of sedimentation occur on the fan: clastic on the upper fan and chemical on the lower fan. In both, biotas play a key role. Although the region is tectonically active, there is little active faulting within the Delta itself, and apart form the rising arches associated with the propagating rifts, the major tectonic process in the Delta appears to be crustal sagging, induced by sediment accumulation. This has caused localized depression of the southeastern arch and associated faulting. Seismicity is localized to the area of the fan, and is also attributed to gravitational loading. Continued propagation of the rift is likely to result in diversion of the Okavango River into the Zambezi River.

## INTRODUCTION

Depositional conditions in rifts vary widely, depending particularly on the stage of rifting and local climate. Many examples of rift sedimentation have been documented, particularly along the East African Rift (e.g., Frostick et al., 1986). The majority of studies have been carried out in well defined rift systems, while few have examined sedimentation associated with the very early stages of continental rifting. The southern tip of the southwest ward propagating East African Rift affords an opportunity to examine this early stage of sedimentation. Here, the courses of the Kwando and Okavango Rivers have been disrupted by incipient rifting, creating local depositional basins, in which the Linyanti and Okavango wetlands have developed (Fig. 1).

The Okavango "Delta" wetland is a large (25,000 $km^2$), low-gradient (1:3400) alluvial fan occupying a local depocenter within the larger, intracontinental Kalahari Basin. The Delta is situated in a semiarid environment, but base flow in the Okavango River at the head of the fan is sufficient to sustain about 4000 $km^2$ of permanent swamp. This is flanked by seasonal swamp, which may attain up to 12,000 $km^2$ in area. The system is essentially closed, inasmuch as virtually no sediment leaves, and 98% of the inflow is lost to the atmosphere by evapotranspiration. Moreover, anthropogenic impact on the Delta is minimal, and both the catchment and the Delta remain undeveloped. Little is known of the subsurface stratigraphy, but depositional processes presently active on the fan have been studied in considerable detail. Structural and seismic studies have provided insight into the local tectonic setting of the Delta.

The Okavango Delta therefore provides an opportunity to examine fluvial style and sedimentary processes in the very early phases of rifting, albeit under relatively unusual conditions. Moreover, the Okavango Delta provides insight into fluvial behavior in bedload-dominated systems under quasi-steady flow conditions as well as sedimentary processes under conditions of extreme evapotranspirational loss. In addition, it provides excellent examples of the myriad roles played by biotas as they bear on geomorphic and sedimentologic processes. This paper emphasizes geomorphologic aspects of the Delta, because of the lack of subsurface information.

## REGIONAL SETTING OF THE OKAVANGO DELTA

### *The Kalahari Basin*

The Delta is situated within, and is an integral component of, the intracontinental Kalahari Basin, and it is necessary to examine the regional settings and interrelationships of both in order to differentiate between intracontinental and rift-related sedimentation.

The gross geomorphology of southern and eastern Africa is dominated by the African Superswell (Nyblade and Robinson, 1994), a region of relatively elevated terrain which manifests itself both on continental Africa and on the ocean floor to the south-

Sedimentation in Continental Rifts
SEPM Special Publication No. 73, Copyright © 2002
SEPM (Society for Sedimentary Geology), ISBN 1-56576-082-4, p. 179–193.

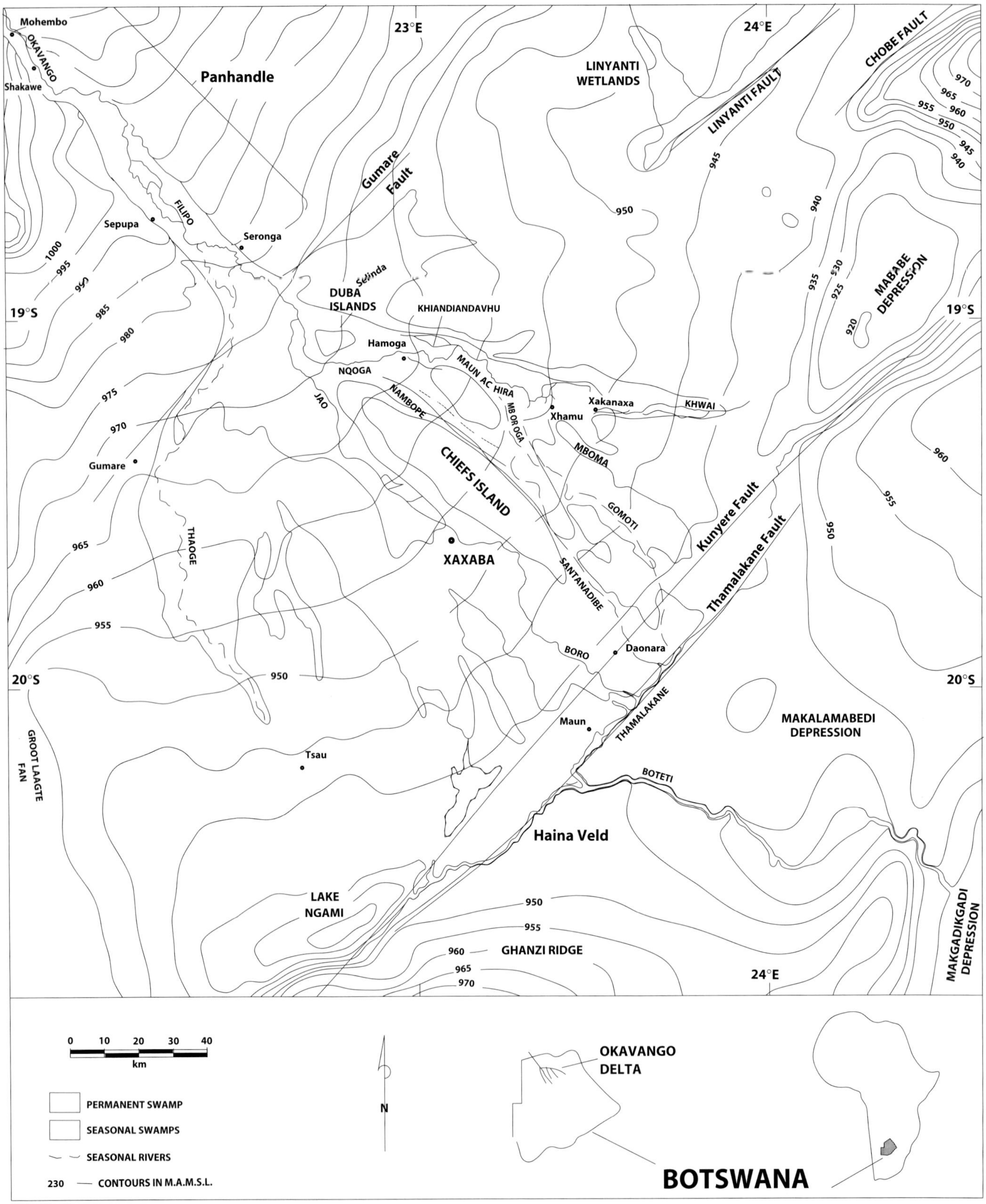


FIG. 1.—The Okavango Delta. Topographic contours are based on a digital elevation model. Details of model construction have been published elsewhere (T. Gumbricht, T. S. McCarthy, and C. L. Merry, 2001). The area can be divided into three regions, separated by northeast-striking faults: the Panhandle region, to the northwest of the Gumare fault; the Ghanzi Ridge, southeast of the Thamalakane fault; and the central region, which includes the Delta.

west. It includes almost the entire subcontinent, where it is clearly delineated by the 1000 m topographic contour (Fig. 2A). The terrain within the Superswell lies at an average elevation at least 500 m higher than equivalent crustal rocks elsewhere on the world's continents (Nyblade and Robinson, 1994). It appears to have had a multiphase uplift history, which also involved tilting to the west. During the Miocene the eastern escarpment was raised by about 300 m and the western escarpment by about 150 m, and during the Pliocene they were raised again by 900 m and 100 m, respectively (Partridge, 1998).

The African plate is also characterized by large basin-like depressions bounded by elongated regions of local relative uplift which Burke (1996) termed "sub-swells" and which he believes reflect zones of local mantle upwelling, some of which accommodate Cenozoic sedimentary material. These sub-swells also play an important role in delimiting drainage basins within southern Africa (du Toit, 1933).

One of these depressions forms the Kalahari Basin, which is situated within the African Superswell (Fig. 2B; Haddon, 1999). It is displaced to the western side of the subcontinent by the higher terrain of the eastern margin of the Superswell. The stratigraphy of the Kalahari Group as a whole is poorly known, and it evidently displays considerable lateral variation. A widely distributed basal conglomerate is developed, overlain by fluvial sand, marl, eolian sand, and calcrete beneath a veneer of mobile eolian sand (the Gordonia Formation, or "Kalahari sands") (Thomas and Shaw, 1991). Partridge (1998) has suggested that the lowermost units of the Kalahari Group are probably upper Cretaceous inasmuch as these overlie Cretaceous kimberlites in which crater facies are still preserved.

### *Seismicity on the Subcontinent*

The investigation by Scholz et al. (1976) into the seismicity of the Okavango region gave rise to the notion that the Delta has developed in an extension of the East African Rift system, and subsequent workers have largely accepted this view (Hutchins et al., 1976; Shaw and Thomas, 1992; McCarthy et al., 1993a; Modisi et al., 2000). Seismic activity on the subcontinent is concentrated mainly along the Western Rift, its southern extension, and in the higher terrain along the eastern side of the Superswell (Fig. 2C; Graham and Brandt, 2000). Apart from the Okavango region, the Kalahari Basin is virtually devoid of seismic activity.

A diffuse zone of seismicity is developed to the west of the sigmoidal arc defined by the Western Rift and its southern extension. Within this diffuse zone, several southwest-trending belts of increased seismicity can be discerned, two of which are particularly prominent and represent incipient rift zones which accommodate lakes and constrain river courses. In their northeastern reaches, these rift zones are characterized by central depressions with marginal uplifts, and Lake Bangweulu and the Lukanga swamps lie in the intervening depression (Fig. 2D). The Okavango region is also seismically active, but this activity is located northwest of, and is detached from, the Luangwa–Kariba seismic axis (Fig. 2C). Focal-mechanism studies by Scholz et al. (1976) indicate predominantly dip-slip motion on the faults in the Okavango area.

Although Burke (1996) suggested that the interior of the southern part of the African Plate may be under compression because of ridge-push effects, Coblentz and Sandiford (1994) and Andreoli et al. (1996) suggest that the Superswell is experiencing tension, possibly related to the excess topography of the region (Coblentz and Sandiford, 1994). The location of active and incipient rifts is closely related to preexisting structure, and rift zones usually follow old, suitably oriented continental sutures (Fairhead and Stuart, 1982; McCarthy et al., 1993a; Sebagenzi, 1997; Modisi et al., 2000).

### *Southern African Drainages*

The drainage patterns of southern Africa are strongly influenced by the Superswell and its associated rifting. The Eastern and Western Rifts are characterized by numerous lakes and highly disrupted, commonly internal drainages (Fig. 2D). The northern and western slopes of the Superswell delimit the Congo Basin to the northwest. Uplift of the Superswell has accentuated the marginal escarpment of southern Africa, which is characterized by short, steep rivers, particularly along the southeastern coastline. The marginal escarpment has been breached by three major rivers: the Zambezi, the Limpopo, and the Orange, and these rivers drain much of the interior. Within the Superswell in central southern Africa, two internal drainage basins are developed: the Etosha ari and the Okavango–Makgadikgadi Basin.

The evolution of the major rivers of central southern Africa has been partly established, although certain aspects remain controversial. It is believed that the Kafue, upper Zambezi, Kwando, and Okavango systems probably once formed a single drainage system (Thomas and Shaw, 1988) which possibly flowed southward into the ancestral Orange River system (McCarthy, 1982). Uplift along the sub-swell now defining the southern margin of the Kalahari Basin disrupted this link, and an internal drainage basin developed which later was diverted into the Limpopo River. Uplift to the northeast along the Kalahari–Zimbabwe axis isolated the Zambezi–Okavango system from the Limpopo and reestablished the internal drainage system, giving rise to Lake Makgadikgadi, possibly during the Pliocene (Thomas and Shaw, 1988). Meanwhile, headward erosion by the lower Zambezi, as well as local tectonism (see below), led to progressive capture and eastward diversion of the Kafue, upper Zambezi, and Kwando systems by the lower Zambezi during the late Pliocene and mid-Pleistocene (Thomas and Shaw, 1991; Shaw and Thomas, 1992). Local tectonism also disrupted the Okavango River. Lake Makgadikgadi was thus deprived of most of its inflow, and degenerated to an ephemeral pan. The formation of the Okavango Delta therefore most likely began during the late Pliocene or early Pleistocene and postdates formation of the Kalahari Basin.

### *Regional Climate*

Climatic determinants in southern Africa are generally well understood (Tyson and Preston-Whyte, 2000) and have recently been reviewed by McCarthy et al. (2000) insofar as they affect the Okavango region. Overall, southern Africa is characterized by pronounced declining rainfall gradients in north-to-south and east-to-west directions. This climatic pattern developed over a protracted period.

The Late Cretaceous was evidently characterized by mesic climates which produced deep weathering and kaolinization. The end of the Cretaceous saw a major desiccation of southern Africa (Partridge, 1998). Prior to the epeirogenic uplift of the Superswell, the western part of southern Africa was sufficiently humid to support woodland communities, but conditions became more arid during the Miocene, partly bcause of uplift but especially influenced by the initiation of upwelling associated with the Benguela current system, which followed the formation of the East Antarctic ice sheet at about 14 Ma. Massive uplift of the Superswell during the Pliocene greatly increased aridity in the interior because of the orographic effects of the eastern escarpment. Aridification was further accentuated by the onset of

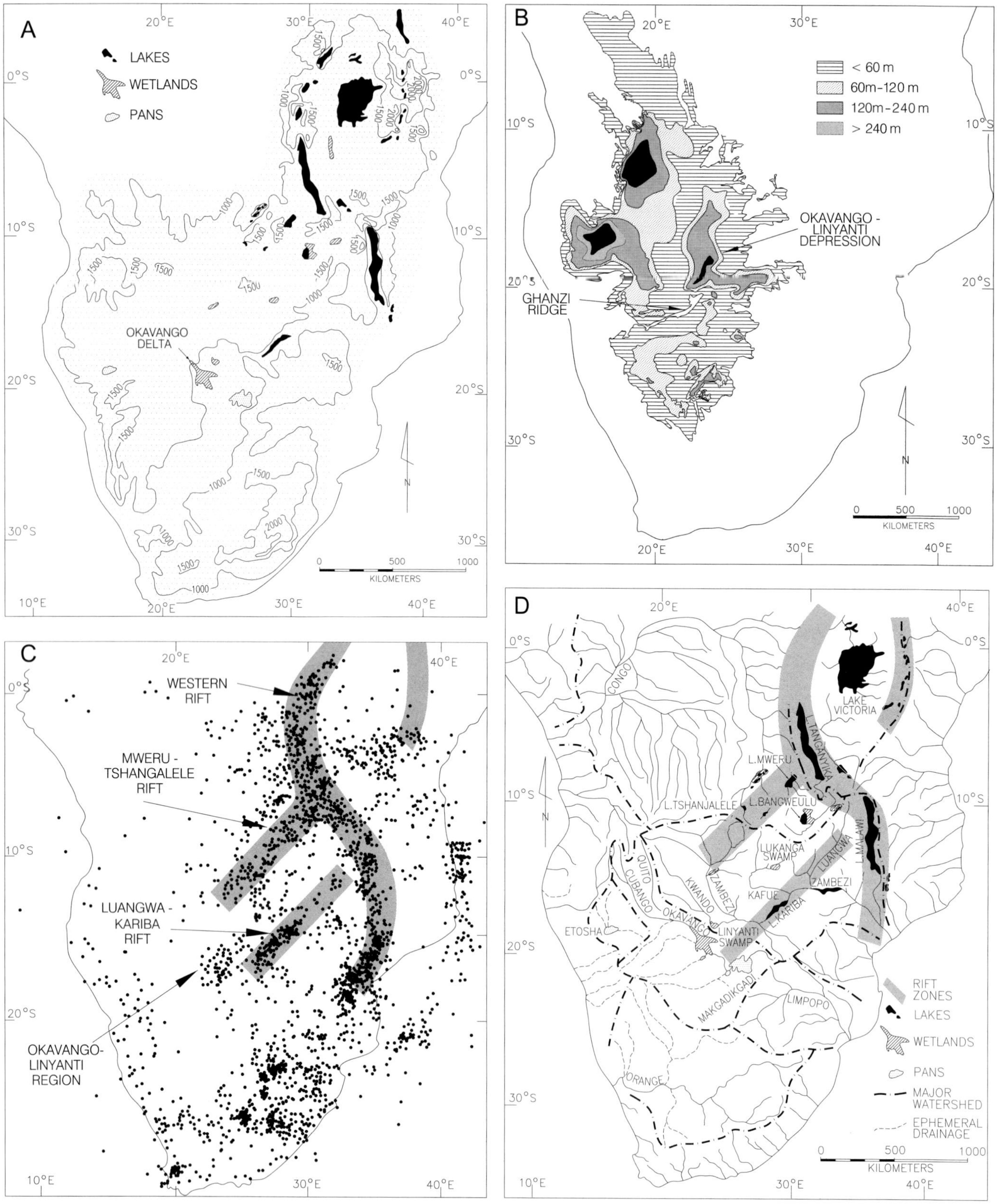


FIG 2.—**A)** The topography of the African Superswell, a region of elevated topography marked by the 1000 meter contour, which includes most of southern Africa. **B)** Thickness of the Kalahari Group (modified from Haddon, 1999), a Cenozoic intracontinental sedimentary sequence which fills the Kalahari Basin. The Ghanzi Ridge divides the Kalahari Basin into two sub-basins. **C)** Seismicity of southern and east Africa (Graham and Brandt, 2000), which clearly delineates active and incipient rifts on the subcontinent. **D)** The rivers, major lakes, and wetlands of southern and east Africa.

northern hemisphere glaciation, giving rise to desert conditions in the Kalahari Basin.

Climatic changes during the late Quaternary have been well documented from geomorphological evidence. The presence of linear dunes north of the Okavango Delta implies an annual precipitation of less than 150 mm, which obtained at various times between 57,000 to 21,000 yr BP (Thomas et al., 2000). These were interspersed with wetter periods during which local rainfall may have approached 1000 mm / yr. Since 12,000 yr BP, generally drier conditions have prevailed (Shaw and Thomas, 1996). The aridity of the late Quaternary has resulted in extensive eolian sand cover over the entire Kalahari Basin, extending from the Orange River in the south to the Congo ari  in the north.

## THE GEOMORPHOLOGY OF THE OKAVANGO DELTA

The Okavango Delta lies within a fault-bounded trough (Fig. 3), and the four major faults (Kunyere, Thamalakane, Linyanti, and Chobe) downthrow to the northwest. The Thamalakane and Kunyere faults dip to the northwest (Scholz et al., 1976). In contrast, the Gumare fault downthrows to the southeast. Maximum thickness of the sediment fill, indicated by drilling, is about 300 m, which includes underlying continental sediments of the Kalahari Group that predate formation of the Delta. The graben also hosts remnants of the Kwando Delta, now beheaded by the Linyanti and Chobe faults, and the inactive Groot Laagte Delta in the west (Shaw and Thomas, 1992).

The topography of the Okavango Delta is illustrated by the contour map in Figure 1. The region can be divided into three structural domains: the northwest, which includes the Panhandle; the central depression, which accommodates the Delta; and the Ghanzi Ridge to the southeast.

The northwestern region slopes gently to the southeast (1:2000), and the offset of contours across the Panhandle indicates northwest-striking faulting within the Panhandle, with downthrow to the southwest. Faulting in this orientation is regionally widespread (Sinding-Larsen et al., 1991), and influences not only the orientation of the Panhandle but also certain reaches of the Cubango, Quito, Kwando, and Zambezi Rivers (Fig. 2D). The Panhandle may occupy a graben structure, but it is also incised into the surrounding terrain. Valley gradient in the Panhandle is 1:5600, substantially less than the regional slope (1:2000). Outside the Panhandle, the terrain is characterized by the presence of large, west-northwest striking linear dunes, now heavily degraded (Jacobberger and Hooper, 1991), which reflect particularly arid climates in the past (Thomas et al., 2000)

The Ghanzi Ridge to the southeast is characterized by the central Makalamabedi depression, which is separated from the Delta domain by the Thamalakane fault. Southeast of this ridge lies the Makgadikgadi depression. The Boteti River, the only outlet of the Delta, is superimposed across the ridge. This river is underfit and once carried greater discharges (Shaw, 1988). The Kalahari cover is limited to thin, discontinuous sand in the southwest but thickens into the Makalamabedi depression. The

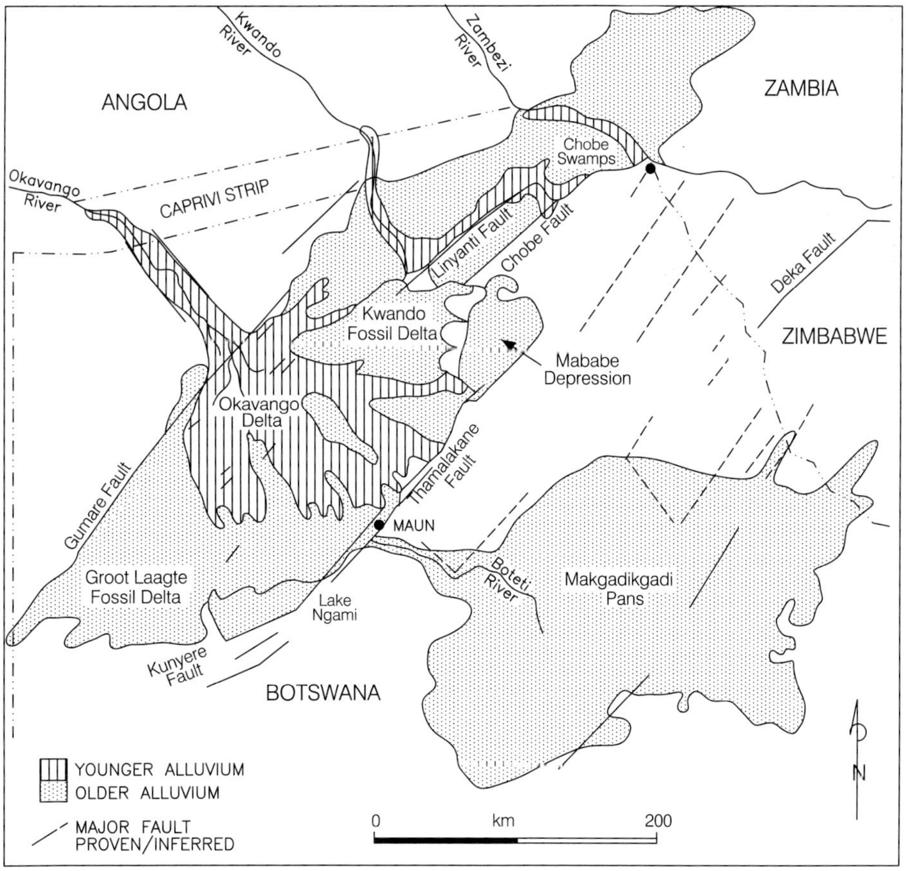


FIG. 3.—Distribution of alluvial sediments in the Okavango–Linyanti depression and their relationship to major faults in the region (modified after Shaw and Thomas, 1992).

area is characterized by relatively short west-northwest striking dunes, currently inactive.

The central Delta depression is dominated by the Okavango alluvial fan. The gradient on the fan surface is 1:3400 and is radially very uniform. However, a distinct bulge occurs on the fan surface southwest of the Linyanti swamps, which is particularly evident in the 950 to 960 m contours, with a local depression to the north in which the Selinda spillway is developed. The Maunachira channel flows along the southern edge of this bulge. In the southwest, contours swing to the south-southeast, marking the terminus of the Groot Laagte fan. Lake Ngami (currently dry) lies in the triangular depression bounded by the Kunyere fault and the Okavango and Groot Laagte fans. In the northeast, the arcuate contours of the Okavango fan swing to the northeast, subparallel to the Linyanti and Chobe faults. The Mababe Depression, a former lake, is confined by the Kunyere fault in the southeast, the Okavango fan in the west, and the horst associated with the Chobe fault in the northeast. The two former lakes show a number of distinct paleo-shorelines indicating previous highstand episodes (Shaw, 1988). The highest of the shorelines is particularly well developed on the western (leeward) shores of Lake Ngami and the Mababe Depression, and is also very well developed on the western shore of paleo–Lake Makgadikgadi. The age of this shoreline is not known. It is believed that all of the lakes were connected at the time this shoreline formed (Shaw, 1988). The shoreline lies at an elevation between 935 m and 940 m around paleo–Lake Makgadikgadi and at a similar elevation at Lake Ngami. Around the Mababe Depression, however, this shoreline rises from 930 m in the south to 945 m in the north, indicating substantial tilting of the lake bed, which has been lowered at its southern end by 5 m and raised at the northern end by 10 m relative to paleo–Lake Makgadikgadi.

The area occupied by the Okavango fan is presently devoid of eolian features. On a local scale, the topography is gently undulating, with relief of generally less than two meters. Higher ground forms islands, which become more common towards the distal margins of the fan. Alluvial features such as scroll bars and topographically inverted channel fills are common in the upper and mid fan, where they form long, curved or sinuous islands. Three prominent large islands occur on the fan: Chief's Island, Duba Island cluster, and Mboma Island. The digital terrain model indicates that none of these are unusually elevated and most show evidence of previous inundation. Detailed analysis has indicated that the Delta surface is transected by numerous lineaments, which locally appear to influence water distribution and may be responsible for the larger islands (Hutchins et al., 1976; McCarthy et al., 1993a; McCarthy et al., 1997; Smith et al., 1997). These lineaments are very subtle features, however, and show variable strike orientations.

## HYDROLOGY OF THE OKAVANGO SYSTEM

Local rainfall over the Okavango Delta falls mainly in the summer between November and March. Average annual rainfall is about 500 mm/yr (McCarthy et al., 2000). Mean annual evaporation is about 2200 mm/yr (McCarthy et al., 1998a); much of the rainfall is therefore lost to evaporation, but some contributes to replenishment of groundwater. Although rainfall is relatively low, the area of the Delta is large, and the annual contribution of rain to the water budget is about $6.1 \times 10^9$ $m^3/yr$ (McCarthy et al., 1998a).

The Okavango River receives most of its water from rainfall on the highlands of central Angola via its two main tributaries, the Quito and Cubango rivers (Fig. 2D). These contribute 43% and 57% respectively, to the $9.2 \times 10^9$ $m^3/yr$ (292 $m^3/s$) discharge of the Okavango River (McCarthy et al., 1998a). Rainfall is strongly seasonal, and peak discharge in the Okavango River at the head of the Panhandle occurs in April. The flood wave passes into the Panhandle, and water leaks from the channel to flood the surrounding backswamps. The extent of lateral flooding is limited by the edges of the Panhandle (Fig. 1), however, and seasonal stage variations of nearly two meters occur (Wilson and Dincer, 1976). At the lower end of the Panhandle, lateral confinement ceases, and floodwaters spread laterally across the gently sloping fan surface, greatly extending the area of inundation. Seasonal stage variations in this region are small, typically in the order of 20 cm (Wilson and Dincer, 1976). Base flow in the Okavango River sustains about 4000 $km^2$ of permanent wetlands, but the area of inundation can seasonally expand to over 12,000 $km^2$. The area of seasonal inundation is highly variable, subject to variations in local rainfall as well as rainfall in the catchment (McCarthy et al., 1998a; McCarthy et al., 2000).

The seasonal flood wave takes about four months to traverse the 250 km distance from the apex of the Panhandle to the Thamalakane fault scarp, which defines the distal limit of the fan, and peak flooding in this distal region occurs in August. The floodwaters flow along the scarp in the Thamalakane River, where they split, some flowing towards Lake Ngami and the remainder flowing via the Boteti River towards the Makgadikgadi depression (Fig. 1). Only 2% of combined inflow and rainfall leaves the Delta as surface flow, and it is believed that little if any leaves via groundwater flow (Gieske, 1995; McCarthy et al., 1998a). Most of the approximately 15 $km^3$ of water that enters the Delta each year is lost to the atmosphere by evapotranspiration.

## HYDROLOGICAL AND SEDIMENT BUDGET OF THE OKAVANGO DELTA

The overall sediment budget of the Okavango system is relatively simple. There is only one source of alluvial sediment, and virtually none of this escapes the fan system. Moreover, the overall hydrological budget is well constrained, so it is possible to quantify the present hydrological and sediment budget for the entire system (Table 1; McCarthy and Ellery, 1998a).

Clastic sediment is derived from two sources, aerosol fallout and fluvial transport by the Okavango River. Aerosol fallout, which is related to the regional anticyclonic circulation, contributes significantly to the overall sediment budget (Table 1). This contribution is not specific to the Okavango but is spread widely over the entire central Kalahari region, and is therefore not considered further, although it may play an important role in the nutrient budget of the Delta ecosystem (Garstang et al., 1998). More important is fluvially introduced sediment, which is derived almost entirely from erosion of the eolian deposits of the Kalahari Group (Figs. 2B, D) and consists primarily of well-sorted, quartz-rich, fine to medium sand (0.2 mm average grain size) carried as bedload. Suspended-sediment concentrations are relatively low because of the eolian sand source. This source material also results in low solute concentration in the river water (~ 40 ppm total dissolved solids) as well as extremely low nutrient levels. However, water loss from the quasi-endoreic Okavango system is such that most of the solute load is precipitated and deposited, resulting in a large chemical contribution of sediment to the overall sediment budget (Table 1). The sedimentation rate, when averaged over the whole Delta, is very low, and amounts to about 24 $tonnes/km^2/yr$, or about 0.05 mm/yr (excluding aerosol fallout).

TABLE 1.—Sediment and water budget for the Okavango fan.

| | Inflow (per year) | Surface outflow (per year) |
|---|---|---|
| Rainfall | $6.14 \times 10^9$ m$^3$ | |
| Okavango River | $9.2 \times 10^9$ m$^3$ | $0.24 \times 10^9$ m$^3$ |
| Bedload | 170,000 t | nil |
| Suspended load | 39,000 t | nil |
| **Total clastic load** | **209,000 t** | **nil** |
| Solute load | | |
| $CaCO_3$ | 114,900 t | 5310 t |
| $MgCO_3$ | 19,100 t | 1640 t |
| $SiO_2$ | 147,000 t | 8300 t |
| $NaHCO_3$ | 67,100 t | 5600 t |
| $KHCO_3$ | 33,000 t | 2600 t |
| **Total solute load** | **381,100 t** | **23,450 t** |
| **Aerosol fall-out** | **250,000 t** | |

t = tonnes

## SEDIMENTATION IN THE OKAVANGO DELTA

### *The Panhandle*

At the apex of the Panhandle, the Okavango River is about 90 m wide, but width decreases downstream to about 40 m at its southern end (McCarthy et al., 1991a). The channel displays well-developed meanders with reach-averaged sinuosities ranging from 1.80 to 2.36 and meander wavelengths ranging from 160 to 460 m. These pattern variations appear to be responses to tectonically induced changes in valley slope (Smith et al., 1997), where local steepening increases sinuosity. Comparison of aerial photographs taken at different times indicates that meander-loop migration is relatively slow (~ 0.5 m/yr). Point-bar tops and some channel margins are colonized by the giant sedge *Cyperus papyrus*, which forms a dense, submerged mat of thick, interlocking rhizomes and roots from which the vertical stalks rise. This mat is buoyant, rising and falling with the seasonal flood. The root mat traps suspended sediment, but because of the low suspended load, near-channel floodplain deposits are volumetrically dominated by plant material (McCarthy et al., 1991a). On straight reaches, the dominant vegetation species is the reed *Phragmites australis*, which forms dense stands. The reeds trap fine sand and suspended sediment from seasonal floodwater, resulting in the formation of low levees, the composition of which is also dominated volumetrically by organic matter. Channel migration and lateral point-bar deposition produce upward-fining successions from medium to fine sand overlain by clay-rich peat. The heights of the point bars are determined by river stage, and past periods of sustained high discharge are reflected as scroll bars, which form islands within the meander loops. Occasional neck cutoffs occur, resulting in oxbow lakes.

The water surface in the channels is elevated above that in the backswamp areas, so that water constantly leaks from channels through the flanking vegetation, although this is laterally limited. The rate of leakage increases during rising stages associated with seasonal floods. Water is also lost via hippopotamus trails, which commonly breach levees and connect the main channel to backswamp areas. As a consequence, discharge decreases steadily downstream (McCarthy et al., 1991a), falling by about 20% down the length of the Panhandle during low stages and probably considerably more at high stages. Channel width decreases as a result. Bedload, however, is not lost from the channel, and much of this probably is accumulating presently in the Panhandle as channel and point-bar deposits.

Some of the water lost from the channel collects in backswamp streams along former channels and hippo trails, and may return to the main channel (Smith et al., 1997). By headward propagation, these channels may ultimately connect back to the main channel (usually along a hippo trail), leading to anastomosis. Once a connection has been established, the former hippo trail becomes widened by erosion. Channel anastomosis, which is best developed in one 26-km reach of the lower central Panhandle, is driven by elevation differences resulting from neotectonism and probably also from aggradation within the active meander belt (Smith et al., 1997). Ultimately, complete avulsion may occur, and the abandoned channel becomes overgrown with *papyrus* and ultimately filled by peat. Although the alluvial landforms of the Panhandle are dominated by both active and abandoned meander-related features, infrequent and episodic channel avulsion is a more important process in dispersing sandy bedload throughout the entire Panhandle. Meandering reaches are continually being abandoned and reoccupied through avulsion, and the associated meander-related landforms have probably accumulated over many reoccupation cycles. These processes maintain a valley gradient of about 1:5600 (McCarthy et al., 1997).

### *The Permanent Swamps on the Upper and Mid-Fan*

The Panhandle is terminated by the Gumare fault, where the Okavango fan commences. Gradient increases abruptly to 1:3400, coincident with the loss of lateral confinement. In this region, the Okavango River divides into three distributary channels: the Thaoge, flowing to the southwest; the Jao/Boro system, to the southeast; and the Nqoga, to the east (Fig. 1). In the period since 1850, discharge on the fan has switched from the Thaoge to the Nqoga–Mboroga–Santantadibe system, and currently is moving to the Nqoga–Jao–Boro system (Shaw, 1985; McCarthy et al., 1991a; Stigand, 1923; McCarthy et al., 1992; McCarthy and Ellery, 1995a; Wilson, 1973), which now diverts about 25% of the discharge from the Nqoga channel. Nevertheless, the Nqoga and its distributary channels remain the principal distributary system on the fan today (McCarthy et al., 1991a).

The Nqoga channel is essentially a continuation of the Okavango River from the Panhandle. Although sinuous (average sinuosity = 1.52; Stanistreet et al., 1993), it does not actively meander, and channel plan has not changed appreciably since the earliest aerial photography in 1937. Channel width decreases from about 50 m at the Thaoge junction to about 20 m at the lower end of the Nqoga at Hamoga Island, although depth remains constant at about 3–4 m. Much of the associated decrease in discharge is due to outflow to the Thaoge and Jao distributary channels. Between the Jao junction and Hamoga Island, however, discharge decreased from 64 $m^3/s$ to 35 $m^3/s$ during one period of measurement because of leakage through the channel margins (Ellery et al., 1990)

The Nqoga channel has a sandy bed, but channel margins consist of peat, stabilized by roots, and support luxuriant stands of *papyrus* (McCarthy et al., 1988a; Ellery et al., 1995). Water surface in the channel is elevated relative to that in the backswamp by as much as 50 cm (McCarthy et al., 1991a), and there is continuous leakage from the channel. Rise in stage during the seasonal flood is limited to about 20 cm because increasing stage is accompanied by increased loss through the *papyrus* fringe. Discharge leaked from the channel sustains vast, permanently flooded backswamp areas dominated by an aquatic grass (*Miscanthus junceus*) and various open-water species (Ellery et al., 1990).

Channel leakage causes suspended sediment to be deposited in the porous channel margins, but because the suspended load is so low, channel margins are nevertheless dominated by peat. Bedload dominates the clastic sediment load, and as water is lost from the channel, its ability to transport this sediment declines, initiating aggradation of the channel bed. An important feedback loop appears to operate in these aggrading channels. As the bed aggrades, water level in the channel rises and leakage increases. The increased leakage through the channel margins promotes *papyrus* growth and hence peat production, so that the channel margins aggrade in sympathy with the aggrading bed, allowing channel depth to remain essentially constant. Hydro gauges in such reaches record a steady rise in water level with time, and a rate of 6 cm/yr has been recorded for the gauge at Hamoga Island since 1977 (McCarthy et al., 1992). Such aggradation results in a decline of channel gradient in and upstream from the channel reach where deposition is occurring, and hence flow velocity declines, promoting further aggradation. Thus, between the Jao junction and Hamoga Island, average flow velocity was observed to decrease from 0.62 m/s to 0.48 m/s (authors' unpublished data). The feedback loop leads to a runaway effect which culminates in channel blockage by vegetation, and ultimately abandonment. This process has been well documented along the lower Nqoga (McCarthy et al., 1992; Ellery et al., 1995; McCarthy and Ellery, 1995a).

Channel blockage and abandonment initiates a dramatic transformation of the wetland landscape. Deprived of the sustaining water leaked from the channel, the flanking floodplain begins to desiccate, and the peat and other dead vegetal material eventually catches fire and burns off in swaths over several years (Ellery et al., 1989). A 4–5 m thick peat accumulated during the aggradational phase of the channel is reduced to a layer of compacted ash 20 to 30 cm thick (McCarthy et al., 1988b; Stanistreet et al., 1993) and the aggraded beds of the former channel and its minor distributaries (usually hippo trails) stand in positive relief above this ash layer (Fig. 4) (Cairncross et al., 1988; Stanistreet et al., 1993). The area then reverts to dry land or seasonal wetland. Mineralogical analysis of ash from the mid-fan regions indicates that phytoliths (silica bodies formed in plants) and diatoms are an important component, constituting between 30 and 70% by weight. The remainder consists of fine quartz and kaolinite (McCarthy et al., 1989).

Channel failure is accompanied by increased loss of water and steeper water-surface gradients away from the channel, and flanking backswamp areas therefore experience greater throughflow (McCarthy et al., 1992). These areas are typically transected by networks of hippo trails (McCarthy and Ellery, 1998b), which offer more efficient pathways for water dispersal. Those trails oriented close to the zones of steepest hydraulic gradient experience increased discharge, and they eventually become widened and deepened by erosion to form a secondary water distributary system (McCarthy et al., 1992; McCarthy and Ellery, 1997). These secondary channels propagate up-gradient by headward erosion as the primary channel fails and the two channels overlap. Direct connection is not easily established between the channels, however, because of the prolific growth of *papyrus* along the failing reach (McCarthy et al., 1992), and the failing primary channel and its associated secondary channel remain separated by a "filter" of aquatic vegetation. However, avulsion into such secondary channels may occur where the channel water surface is particularly elevated relative to the backswamp (McCarthy et al., 1991a). Such avulsions can lead to radical shifts in water distribution on the fan and are probably the major mechanism for maintaining the uniform fan gradient. Once avulsion has occurred, externally derived bedload enters the formerly erosive secondary channel, and aggradation of the channel bed commences. Shifts in water distribution which have occurred in historic time suggest that the active life of a primary channel is on the order of 200 yr.

Because the secondary channels receive no sediment from external sources, they tend to be erosive in their upper reaches but aggradational in downstream reaches as the eroded sediment (mainly sand) is redeposited (McCarthy and Ellery, 1997). These distal channels are not, however, invaded by aggressive aquatic vegetation, possibly because of reduced nutrient levels in the water. Channel margins are dominated by the non-invasive grass *Miscanthus junceus*. In their distal reaches, the channel bed becomes vegetated by a variety of submerged aquatic plants; however, the channels remain open to water flow, facilitating the supply of water to the fringing seasonal wetlands.

Although the channels of the mid-fan are not actively meandering today, this has not always been the case. The region of permanent swamps where the secondary Maunachira channel currently flows (Fig. 1) is characterized by a number of meander-related geomorphological features including oxbow lakes, scroll bars, point bars, and sinuous islands which represent inverted channel beds (McCarthy et al., 1993b). The scale of these features implies a channel system about 200 m wide, substantially greater than the present 25 m width of the Maunachira channel, which was active about 2000 yr BP during a particularly wet period. These features suggest that during periods of higher discharge, the meander belt of the Panhandle prograded out onto the fan.

### *The Seasonal Swamps*

The permanent swamps gradually give way down-fan to the seasonal swamps. The transition occurs over a diffuse zone characterized by an increasing abundance of islands separated by sinuous channels and shallow, sedge- and grass-covered wetlands during peak discharge periods. At present, the seasonal swamps are not uniformly distributed across the fan, but are more extensively developed west of Chief's Island. The area east of Chief's Island receives more uniform flow throughout the year,

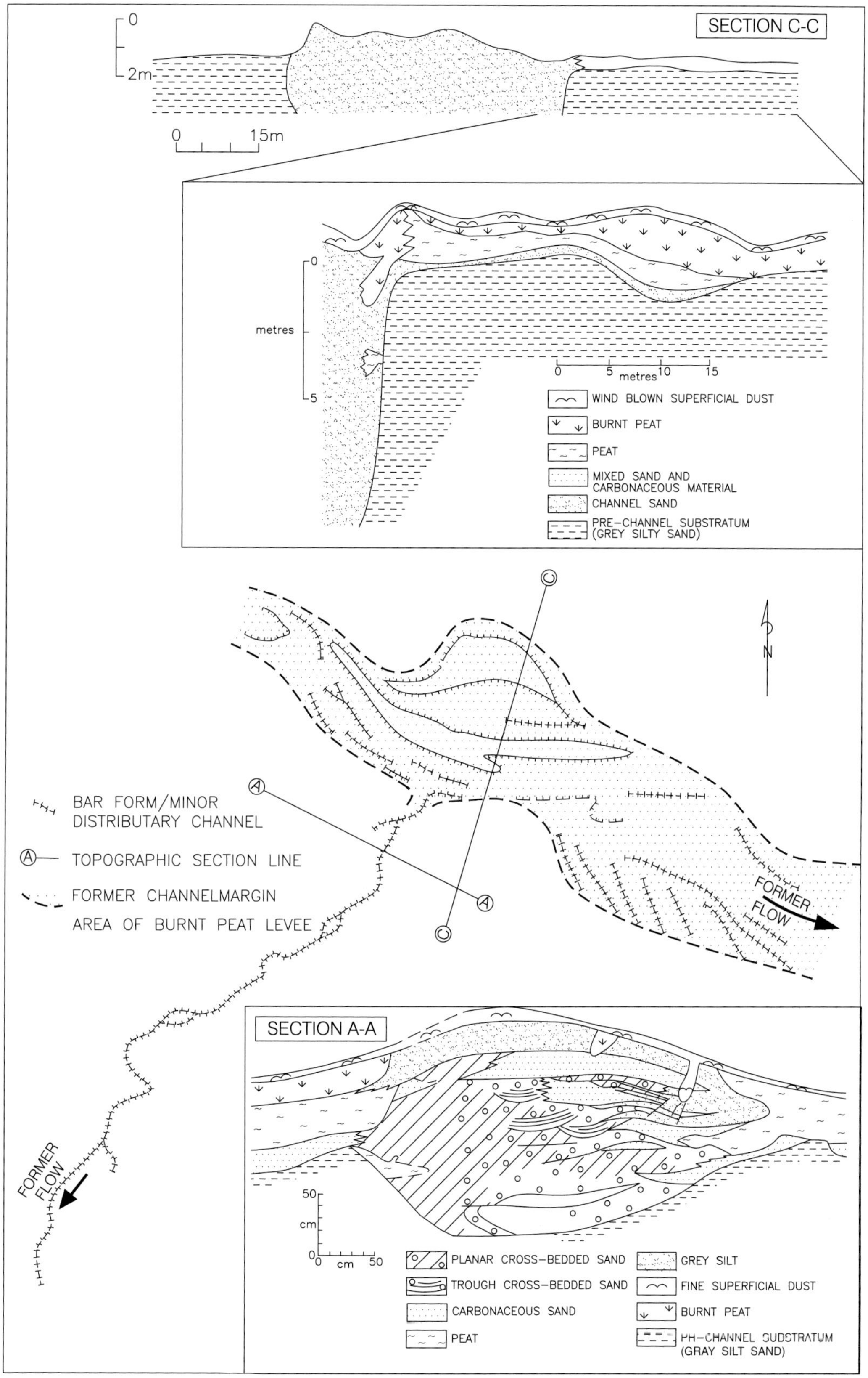


FIG. 4.—The topography and internal architecture of a section of abandoned channel on the Nqoga channel system (Stanistreet et al., 1993). The section of channel is shown in plan view in the central diagram, with cross sections along section lines A–A and C–C as insets.

and hence only a small area of seasonal swamp is developed in this region (McCarthy et al., 1998a; Gumbricht et al., 2000).

Local relief in the seasonal swamps is in the order of 1–2 m except for occasional termite mounds which may rise to heights of several meters. Some islands are sinuous and appear to represent inverted channel fills, whereas many are irregular in plan, varying in size from a few square meters to tens of hectares (Fig. 5). The larger islands commonly have a slightly elevated rim surrounding a central depression, which may contain an ephemeral saline pan.

The duration of flooding of the seasonal swamps is highly variable. On the fringes of the permanent swamps, inundation may last for several months, and in wetter periods flooding may persist throughout the year. Towards the distal areas of the fan, the duration of flooding decreases, and in drier periods they may not be inundated for several successive years.

Trees in the delta are confined almost exclusively to islands; very few species inhabit either permanently or seasonally flooded areas. Tree species commonly show a distinct zonation on islands. The margins are dominated by broad-leaf evergreen trees; these give way to deciduous trees and then to the ivory palm *(Hyphaenae petersiana)*. Toward the island interiors, trees disappear, and the shrub *Pechuel-loeschea leubnitziae* and the couch grass *Cynodon dactylon* dominate. This community gives way to the grass *Sporobolus spicatus* and eventually to bare soil (Ellery et al., 2000), which is often covered by a crust of trona and thermonatrite (McCarthy et al., 1986). This zonation, caused by increasing salinity of the groundwater beneath islands (see below), reflects the varying salt tolerance of different species (Ellery et al., 1993).

Sedimentation in the seasonal swamps is predominantly chemical. The seasonal swamps receive little clastic sediment from the Okavango River, because this is filtered out during its passage through the permanent swamps. Okavango River water has very low concentrations of dissolved solids, typically 40 ppm, dominated by silica (~ 20 ppm) and calcium and magnesium bicarbonate, with lesser amounts of sodium bicarbonate (Hutton and Dincer, 1976; Sawula and Martins, 1991; Cronberg et al., 1995). The extremely large evapotranspirational loss ensures that most of this solute load is deposited in the Delta (Table 1). The processes involved in chemical sedimentation have been extensively studied (McCarthy et al., 1991b; McCarthy et al., 1993c; McCarthy and Ellery, 1994; McCarthy and Ellery, 1995b).

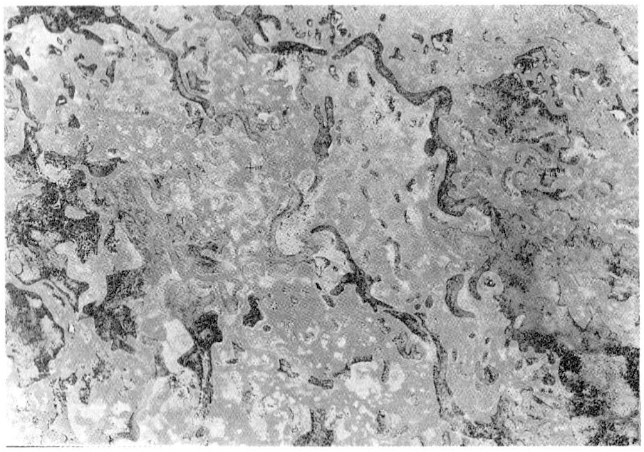

FIG. 5.—Sinuous and irregular islands in the seasonal swamps of the Okavango Delta (extracted from the 1:50,000 scale map series of the Okavango Delta). The area covered is 10.5 km x 15 km.

As the seasonal flood advances, infiltration occurs through the bed of the floodplain, recharging the groundwater. Grasses, sedges, and lilies growing in the sediment shoot, and the water surface is quickly covered by emergent vegetation. Some evaporation of surface water occurs, but transpiration dominates over evaporation (McCarthy and Ellery, 1995b). As a consequence, the concentration of dissolved solutes rises in the pore water in the root zone of the floodplain soil. Transpired water is replaced by infiltration of overlying surface water, and ultimately nearly all surface water is lost to the atmosphere by evapotranspiration. The rise in groundwater salinity induces silica saturation (McCarthy and Ellery, 1995b), resulting in precipitation of amorphous silica in the soil, forming a weak cement. Transpiration by aquatic plants continues, however, and the water table falls below the soil surface. Trees are far more effective transpirers of water per unit surface area than are sedges and grasses, and moreover, their root systems penetrate to greater depths. Islands therefore are very effective transpirational pumps in the seasonal swamps. Water lost by transpiration from trees is replaced by inflow from the groundwater reservoir beneath the floodplains (McCarthy and Ellery, 1994), causing further increases in groundwater salinity beneath the islands. Saturation and eventual precipitation of (magnesian) calcite occurs within the soil of the island fringe, causing vertical expansion and island growth (Fig. 6) (McCarthy et al., 1993c), and an inward gradient of increasing salinity. Zonation of plant species (see above) develops in response to this gradient. Numerical modeling indicates that the interior of an island permanently surrounded by wetlands will become completely salinized in about 200 years (McCarthy et al., 1993c).

Islands appear to be initiated primarily in two ways: through topographic inversion following channel abandonment (described above), which gives rise to long, sinuous islands (Fig. 5); and through the action of termite colonies *(Macrotermes michaelseni)*, which nucleate the smaller, irregular islands (McCarthy et al., 1998b). Areas of floodplain that are abandoned for long periods because of drought or shifts in water distribution are colonized by the termite, which builds large termitaria that may reach several meters in height. Colonization of mounds by terrestrial plants sets in motion groundwater processing as described above, which leads to calcite precipitation. This, together with further termite activity, results in island growth. *M. michaelseni*, like the hippopotamus, thus occupies a disproportionately important role in sedimentary processes in the Okavango Delta (Dangerfield et al., 1998).

The accumulation of sodium salts beneath island surfaces is temporary. When an area of wetland reverts to dry land because of shifts in water distribution, the water table falls, and rainwater both dilutes the groundwater and flushes the salts from the soil. In contrast, accumulated calcite and silica remain. After reflooding, vegetation on the islands will be unzoned, but over time, salinity gradients will again develop, leading to vegetation zonation. The presence of sodium salts therefore relates only to the current flooding episode, whereas silica and calcite within the island represent accumulation over many wet episodes. Unpublished results indicate that once islands are colonized by trees, they act as traps for dust generated by thermals on the surrounding flood plains during the dry periods, which further contributes to island growth (M. Krah, personal communication, 2001).

Although the dominant form of sedimentation in the seasonal swamps at present is chemical, eolian sand is nevertheless widespread in these areas, and probably reflects the incursion of eolian sand during more arid periods in the past. Areas surrounding the

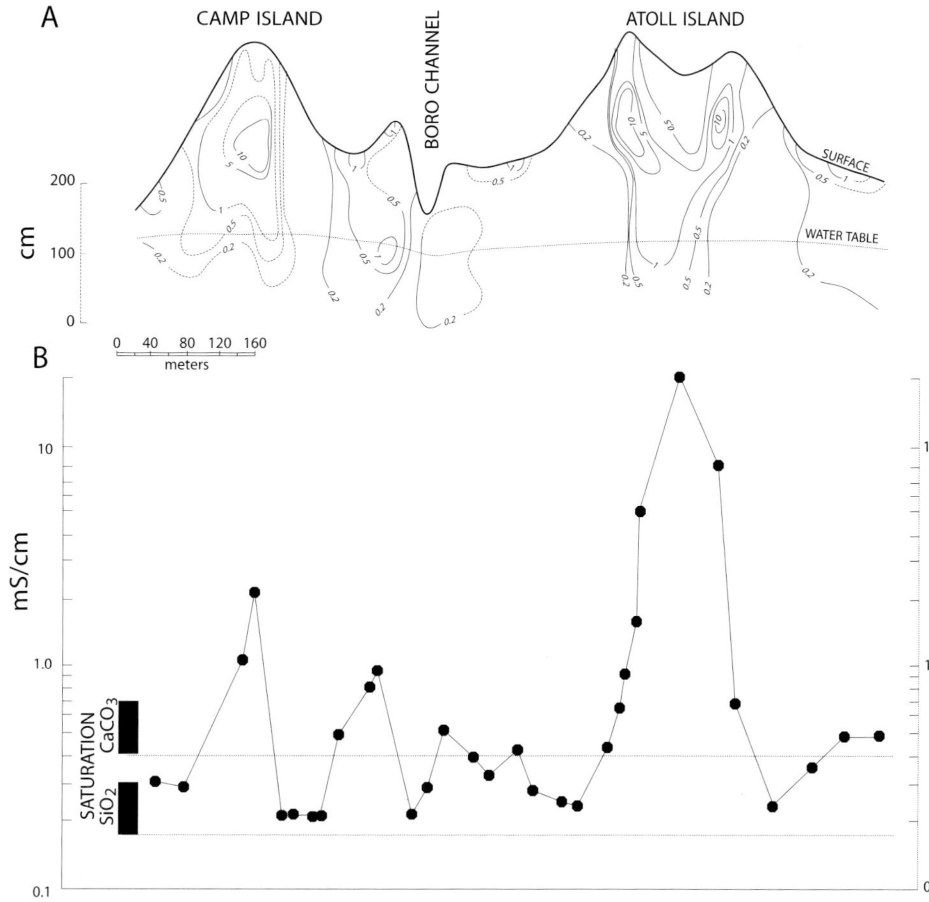


FIG. 6.—The upper diagram shows the surface topography, subsurface distribution of calcite (indicated by CaO), and the water table along a section in the seasonal swamps. The lower diagram shows the electrical conductivity (in mS / cm) of the underlying ground water. The conductivites at which silica and calcite saturate are shown (McCarthy and Ellery, 1995b) on the lower diagram. Elevated ground is enriched in calcite and is underlain by more saline groundwater. The groundwater is everywhere saturated in silica, but calcite saturation occurs only beneath elevated ground (islands).

Delta provide abundant geomorphic evidence for such dry periods. However, eolian landforms are extremely rare in the seasonal swamps, and appear to have been obliterated by processes associated with flooding. The nature of these processes is unknown.

### *The Lakes*

Two former lakes, the Mababe Depression and Lake Ngami, lie astride the Okavango Delta (Fig. 1). These lakes were filled at various times in the past during wetter periods or at times when they were fed directly by primary distributary channels, as in the case of Lake Ngami during the 19th century (Shaw, 1985).

Sedimentation in these two lakes has not been studied in detail. Cursory examination indicates that the lake-bed sediment consists of fine-grained material resembling ash from peat fires described above and is probably mineralogically similar, although the proportion of autochthonous diatoms and phytoliths appears to be higher (Thomas and Shaw, 1991), possibly because of the distal position of the lakes. Lake bed sediments must have originally contained a considerable amount of organic matter, inasmuch as peat fires were widespread in the region of Lake Ngami during its desiccation in the early part of the 20th century (Stigand, 1923).

During wetter periods in the past, the two lakes were connected when a single water body developed along the Thamalakane and Kunyere fault scarps (Shaw, 1988). Lake sediments therefore probably occur across the entire lower fan, although they are presently buried beneath eolian sand. Fine-grained sediment has been intersected by boreholes in the lower fan and probably represents such lake sediments. These are interlayered with more sandy units which represent eolian and possibly seasonal swamp deposition during drier periods.

## DISCUSSION

Although an integral part of the Kalahari Basin, the Okavango Delta is genetically unrelated to this larger feature. The Kalahari Basin appears to have been initiated during the Late Cretaceous by regional warping ("sub-swells" of Burke, 1996) within the African Plate, which created a large, intracratonic basin in which fluvial and lacustrine sediments of the lower Kalahari Group accumulated under the prevailing mesic climate. Uplift of the African Superswell during the Miocene and Pliocene resulted in increasing aridification in the interior and accumulation of calcretes and eolianites in the basin. The excess topography resulting from uplift appears to have initiated rifting, which is propagating south and southwestwards across the subcontinent, following suture zones defined by Precambrian metamorphic belts. During the late Pliocene or early Pleistocene, one of these incipient rifts (Luangwa–Kariba) began to transect the Kalahari Basin, and it has given rise to the Okavango Delta. Fluvial and other sediments of the Delta therefore overlie the Kalahari Group.

While previous workers have emphasized graben development in the Okavango depression (McCarthy et al., 1993a; Shaw and Thomas, 1992; Modisi et al., 2000), examination of isopachs of the Kalahari Group (Fig. 2B) indicate that the major tectonic style is really one of uparching and successive upthrow along the northeast-striking faults, resulting in the Ghanzi Ridge, which penetrates and divides the Kalahari Basin into two sub-basins. To the northeast, the ridge broadens and is cut by a progressively widening axial rift which accommodates the Kariba gorge and the Luangwa valley (Fig. 2D). The depression northwest of this ridge hosts the Bangweulu and Lukanga swamps. In traversing this depression, the Kwando and Zambezi Rivers develop extensive alluvial plains (Thomas and Shaw, 1991). The Linyanti and Okavango wetlands lie on the southwestern extension of this depression (Fig. 2D). These relationships suggest that the Okavango graben is not an incipient rift graben, but rather a depression on the flank of a basement arch that marks the tip of the southwest-propagating Luangwa–Kariba rift. Such inter-rift depressions normally have a lower level of seismic activity than the adjacent rifts, but this is not the case for the Okavango–Linyanti depression (Fig. 2C).

The Okavango and Kwando Rivers earlier flowed across the Kalahari Basin, discharging into the Limpopo, but uplift of the sub-swell marking the western boundary of the basin terminated this connection, and gave rise to palaeo–Lake Makgadigadi. Uplift associated with the developing rift dammed the Kwando, producing the Linyanti swamps and deflecting flow to the east into the Zambezi River. It also created the Ghanzi Ridge, although discharge of the Okavango River was sufficient to maintain flow across this feature. Discharge has evidently declined over time, due to diversion of the Kwando (and upper Zambezi) and possibly to changing climate, resulting in the present underfit condition of the Boteti River and its seasonal flow. The rise of the Ghanzi Ridge has created a trap for sediments carried by the Okavango River.

The origin of the Panhandle is more uncertain. It occupies a minor fault zone that may represent a graben structure, but it appears to be incised and to have been widened by lateral erosion, and basement rocks are exposed at its apex at Popa Falls, a short distance upstream of Mohembo. The edges of the Panhandle now confine the flood plain, which is progressively more incised towards the north. Within the Panhandle, the Okavango River has established a valley gradient of 1:5600, while the regional slope is 1:2000. Small-displacement faults crossing the flood plain cause local steepening, to which the river has responded by change in meander wavelength. These relationships suggest that the area north of the Panhandle is experiencing uplift, possibly related to the Mweru–Tshanjalele rift zone.

The southern end of the Panhandle is defined by the Gumare fault. The origin of this fault is unclear, and it may be antithetic to the Kunyere fault, but it is more likely related to a rising arch to the north of the Panhandle. The flood plain of the Okavango River loses confinement as it crosses this fault, expanding into an alluvial fan, and a new equilibrium gradient of 1:3400 is established. Channel meandering ceases, but this appears to be more related to discharge than to gradient, because there is geomorphological evidence for the previous existence of meandering channels below the Panhandle, but with far greater discharge than the present channels.

Two types of deposit dominate the fan: clastic sediments in the upper to mid-fan region, and chemical sediments on the distal fan. In both, biota play a key role. The low suspended load in the Okavango River accentuates the role of aquatic plants, and they assume a key function normally fulfilled by fine clastic sediment: they confine channels. However, the vegetal banks are permeable and thus allow water to escape, which sustains large backswamp areas, and supplies the seasonal swamps during the annual flood. Furthermore, their vigorous growth sustains hydrologically unstable conditions for protracted periods (decades), and forces in-channel sedimentation. Hippopotami also play an important role in this environment, maintaining flow in failing channels, and more importantly, creating nuclei for new channel formation. Clastic sedimentation is largely confined to the vicinity of distributary channels. Although extensive areas of peat accumulate, the preservation potential of this material is very low, because of the arid climate. Chemical sedimentation dominates the distal fan and is a consequence of the high evapotranspirational losses. Here too, biota play an important role. Termites construct mounds above the level of flooding and thus create nuclei for island formation. Plants ensure that transpiration exceeds evaporation, inducing subsurface formation of silcrete and calcrete and preventing the formation of saline surface water. Clastic and chemical sedimentation evidently maintain a balance, sustaining the uniform gradient on the fan.

The present sedimentation rate, when averaged over the entire Delta, is very low. Under present conditions, however, less than half of the Delta is inundated by the seasonal flood, which increases average sedimentation rate to about 0.1 mm/yr. Sedimentation is not uniformly distributed, however, and fluvial sedimentation may locally exceed 6 cm/yr in failing channels. During wetter periods, the clastic input is likely to increase, although the chemical component decreases correspondingly because of reduced evapotranspiration. During more arid periods, eolian sand incursions probably occur in the distal parts of the Delta, and indeed, the wide distribution of eolian sand throughout the seasonal swamps attests to the importance of such incursions. The present conditions may well be atypical, because such low sedimentation rates are unlikely to have constructed the present form of the fan. Nevertheless, it is likely that the long-term average sedimentation rate over the entire fan is low, probably less than 1 mm/yr. The land surface of the fan is, therefore, generally relatively old.

The fundamental regional structural control on the Delta is upfaulting associated with the Ghanzi Ridge, possibly accompanied by uplift in the Panhandle area. However, it seems that this uplift is being accentuated by subsidence centered on the Delta itself, which appears to have induced buckling of the Ghanzi Ridge to create the Makalamabedi depression, tilted the Mababe Depression, and may also have resulted in the formation of the Thamalakane fault, by detaching a sliver of the rising Ghanzi Ridge along the edge of the Makalamabedi depression (Fig. 1). It is suggested that this subsidence is also responsible for the localized seismicity in the Delta. Enhanced local subsidence is centered on the Delta, and may be the result of gravitational loading by sedimentation and flooding. Many examples of localized crustal subsidence, and seismicity, induced by point loading are known: for example, Lake Bonneville, which was 350 m deep, induced crustal subsidence of 50 m (May et al., 1991), and it is well known that seismicity is often associated with the filling of dams (Simpson et al., 1988).

Apart from the bulge in the east, the alluvial fan has a remarkably uniform surface, and given the relatively slow sedimentation rates, implies conditions of considerable stability within the fan. Locally, minor lineaments appear to influence water dispersal on the fan (McCarthy et al., 1993a), but their orientations bear little relationship to basement structures in the region, and therefore are not expressions of displacements within the basement. It is suggested that many of these may be local accommodation structures formed in response to differential subsidence across the fan. An exception is the bulge on the fan surface to the southwest of the Linyanti swamps. This has a marked depression on its northern side, which now accommodates the Selinda spillway. The Delta sediments are clearly being deformed in this region. The position of the bulge suggests it represents an extension of the Linyanti fault, while the Selinda depression is probably a graben between the Linyanti and Gumare faults. This has important implications for the long-term fate of the Delta: it seems probable that as the Linyanti fault develops, an increasing proportion of Okavango discharge will enter the Selinda spillway, and ultimately the entire flow will be captured by the Linyanti system and the Zambezi River.

## CONCLUSIONS

The alluvial sediment supplied to the Okavango Delta is predominantly sandy bedload. Bedload-dominated rivers are often braided, but the Okavango River is meandering. In the channels of the Okavango Delta, roles normally fulfilled by fine-grained sediment are assumed by hydrophytic vegetation which constructs levees, stabilizes channel margins, and forms extensive overbank accumulations of organic material. Where channel margins are composed mainly of active vegetation, their high permeability limits the life of channels and ultimately regulates avulsion processes. Hippos fulfill ancillary but important roles in maintaining the dynamic variability of Okavango channel systems and water distribution patterns.

The huge evapotranspirational loss of water over the Delta results in extensive chemical sedimentation over much of the distal fan regions. Here too, plants play a dominant role, inasmuch as they ensure that transpiration exceeds evaporation, thus minimizing the presence of saline surface water. Silica and bicarbonates of calcium and magnesium, which precipitate from the groundwater as benign minerals and have no deleterious influence on the vegetation, form the bulk of the solute load. The only potentially toxic salt is sodium bicarbonate, which inevitably becomes concentrated to toxic abundance levels. The growth patterns of plants, however, ensures that the resulting toxicity is localized to the centers of islands. In this environment, termites play an important role because they create nuclei for island growth and actively transport nutrients to these nuclei, promoting island development. Biota clearly play an important role in sedimentary processes in the Okavango Delta. Notwithstanding the extensive chemical sedimentation in the seasonal swamps, eolianite forms a major component of the sediment, reflecting more arid periods in the past. However, eolian landforms are not developed.

Sedimentation rate in the Delta is evidently extremely slow, and hence the land surface must be relatively old. The fan surface shows very limited evidence of tectonic disruption, implying relatively stable conditions of sedimentation, in spite of its location in a seismically active area. In contrast, the surrounding areas, particularly to the southeast and east, show abundant evidence of active tectonism. Analysis of the regional geomorphological setting appears to have resolved this paradox. The Okavango is evidently not a rift graben, as was previously assumed, but appears to have developed in the depression between two propagating rift systems. As such, the Okavango offers insight into the response of a river system of moderate size to the onset of rifting. A depression has been created as the bulges develop at the leading edges of the rifts. Accumulation of sediment in this depression induces further sagging and may also accentuate local seismicity. Lakes often form in such settings, but in the case of the Okavango, a large, low-gradient alluvial fan has formed instead. The reasons for this are probably complex, and most likely include: the slow rate of subsidence; the accentuation of faults, particularly the Gumare fault, by gravitational loading; the semiarid environment; and the relatively slow rate of deformation within the depression. These factors, together with the low suspended load, have resulted in an environment of slow sedimentation, in which biota have assumed a preeminent role.

In the long term, it seems likely that continued southeastward propagation of the Ghanzi Ridge and its associated faults will result in diversion of the Okavango River to the east via the Selinda spillway. The Okavango may thus be destined to become a tributary of the Zambezi River, culminating a process of successive river capture in central southern Africa which began in the Pliocene.

## ACKNOWLEDGMENTS

We gratefully acknowledge the many contributions made by our co-workers over nearly two decades of research in the Okavango, as well as the logistical support provided by safari companies operating in the Delta. We thank the funding agencies which have made this research possible, particularly the University of the Witwatersrand, the Claude Harris Leon Foundation, the Skye Foundation, the Foundation for Research Development, the Jim and Gladys Taylor Trust, the University of Nebraska, the National Geographic Society, the National Science Foundation (EAR-9304104), the Royal Swedish Academy of Sciences, and De Beers Consolidated Mines Limited.

## REFERENCES

Andreoli, M.A.C., Doucoure, M., van Bever Donler, J., Brand,t, D., and Anderson, N.J.B, 1996, Neotectonics of southern Africa—a review: African Geoscience Review, v. 3, p. 1–16.

Burke, K., 1996, The African Plate: South African Journal of Geology, v. 99, p. 341–409.

Cairncross, B., Stanistreet, I.G., McCarthy, T.S., Ellery, W.N., and Grobicki, T.S.A., 1988, Palaeochannels (stone-rolls) in coal seams: modern analogues from fluvial deposits of the Okavango Delta, southern Africa: Sedimentary Geology, v. 57, p. 107–118.

Coblentz, D.D., and Sandiford, M., 1994, Tectonic stresses in the African plate: constraints on the ambient lithospheric stress state: Geology, v. 22, p. 831–834.

Cronberg, G., Gieske, A., Martin, E., Nengu, J., and Stenstrom, I.M., 1995, Hydrobiological studies of the Okavango Delta and Kwanda/Linyanti/Chobe river, Botswana. I. Surface water quality analysis: Botswana Notes and Records, v. 27, p. 151–225.

Dangerfield, J.M., McCarthy, T.S., and Ellery, W.N., 1998, The mound-building termite *Macrotermes michaelsensi* as an ecosystem engineer: Journal of Tropical Ecology, v. 14, p. 507–520.

Du Toit, A.L., 1933, Crustal movement as a factor in the geographical evolution of South Africa: South African Geography Journal, v. 16, p. 3–20.

Ellery, W.N., Ellery, K., McCarthy, T.S., Cairncross, B., and Oelofse, R., 1989, A peat fire in the Okavango Delta, Botswana, and its importance as an ecosystem process: African Journal of Ecology, v. 27, p. 7–21.

Ellery, K., Ellery, W.N., Rogers, K.H., and Walker, B.H., 1990, Formation, colonization and fate of foating sudds in the Maunachira river systems of the Okavango Delta, Botswana: Aquatic Botany, v. 38, p. 315–329.

# PEDOGENIC MUD AGGREGATES IN RIFT SEDIMENTATION

ELIZABETH H. GIERLOWSKI-KORDESCH
*Department of Geological Sciences, 316 Clippinger Labs, Ohio University, Athens, Ohio 45701-2979, U.S.A.*
*e-mail: gierlows@ohio.edu*
AND
MARTIN R. GIBLING
*Department of Earth Sciences, Dalhousie University, Halifax, Nova Scotia B3H 3J5, Canada*
*e-mail: mgibling@is.dal.ca*

**Abstract:** Pedogenic mud aggregates are sand-size particles composed of clay flakes and silt grains, produced in vertic soils with expandable clays during seasonal cycles of wetting and drying. The aggregates are commonly reworked from floodplain soils and transported as bedload, but they tend to be destroyed by compaction during burial. Their identification is important for interpreting floodplain dynamics, i.e., bedload transport rather than suspension settling of clay. Expandable clays, principally smectite, are necessary for aggregate formation, and are commonly produced during weathering of mafic volcanic rocks and ash, which are prominent constituents of many rift-basin fills. To date, pedogenic mud aggregates have been documented from some ancient rift fills.

A new occurrence of mud aggregates is documented from the upper part of the Triassic–Jurassic New Haven Arkose of the Hartford rift basin, Connecticut. The sand- and silt-size aggregates are present in laminated and cross-laminated sandstones and siltstones, where the larger framework grains have protected the aggregates from destruction during compaction. Blue-light microscopy was especially helpful in aggregate identification. The host strata represent shallow bedload channels and floodplain sandstone sheets interbedded with thin red mudstones, interpreted as immature paleosols, and the aggregates are inferred to have originated principally from pedogenic activity in the mudstones and / or other adjacent paleosols. The high-energy floodplains were frequently reworked by overbank floods, and some mudstones may have been deposited by active flows capable of bedload transport of aggregates. Mobilization of soil aggregates is probably promoted in rift settings similar to that of the Hartford basin, where floodplains are small and frequently reworked by high-discharge events.

## INTRODUCTION

Pedogenic mud aggregates are predominantly sand-size particles composed of differing amounts of clay flakes and silt grains, produced in soils containing expandable clays during cycles of wetting and drying. Contrary to the common view that all mud is deposited from suspension settling under aqueous conditions, pedogenic mud aggregates are sufficiently stable to be transported as bedload in high-energy fluvial environments. Recognition of these aggregates in the rock record is difficult because they are easily destroyed by compaction. With the possibility that mud can be transported as bedload, the interpretation of continental depositional systems and their paleoclimatic conditions can be significantly altered (e.g., Gierlowski-Kordesch and Rust, 1994).

Continental rifts may be ideal sites for the formation of pedogenic mud aggregates. Production of aggregates requires soils that contain expandable clays, principally smectite, and experience wetting–drying cycles; both requirements are commonly met in rift settings. Pedogenic mud aggregates have been identified within ancient rift sediments. Recognizing aggregates in rock samples, as well as understanding how and why they form, has important implications for interpreting rift depositional systems. This paper outlines processes that form soil aggregates, the conditions and materials necessary to produce them, and examples of modern and ancient pedogenic mud aggregates in rift and other settings. Methodology for identification of pedogenic mud aggregates in the rock record is discussed, and we document a newly discovered occurrence of aggregates in a Mesozoic rift basin.

## PEDOGENIC MUD AGGREGATES

### *Soil Structure and Processes*

Formation of aggregates is a common process associated with soils containing more than 15% clay (Buol et al., 1997; Horn and Baumgartl, 2000) and is an important soil characteristic for agricultural purposes. The degree and type of aggregation in a soil profile control permeability and aeration, which are important factors for plant growth. Classification of soil structure is based on the shape and arrangement of these aggregates, which are commonly termed peds (Buol et al., 1997). Peds can be discrete particles or poorly formed masses in soils; soil structure is termed complete when the peds are separate particles within the soil matrix. Ped shapes range from columnar or blocky to prismatic or wedge-like to granular or crumbly, and ped size can vary from less than one millimeter to greater than 10 cm in diameter. For granular soils, the term "aggregate" can be used to describe peds that are approximately 1–10 mm in diameter. The terms ped and aggregate are not equivalent for soil-structure classes larger than granular, especially when mechanical alteration (tillage) occurs (Kay and Angers, 2000).

Ped type and stability depends on many variables (FitzPatrick, 1983; Kay and Angers, 2000). These include clay type, percentage of clay and other minerals, amount of organic material, presence of cementing substances (calcite, iron oxides, and aluminum oxide), exchangeable ions (e.g., sodium and calcium), extent of biologic activity (both plant and animal), climate (temperature gradient, rate of wetting), and amount of tillage.

Sedimentation in Continental Rifts
SEPM Special Publication No. 73, Copyright © 2002
SEPM (Society for Sedimentary Geology), ISBN 1-56576-082-4, p. 195–206.

Of special relevance to the geologic record is the formation of mud aggregates as robust discrete particles in granular soils, capable of transport as bedload and incorporation into traction deposits. These granular aggregates or peds have been recognized in modern and ancient sedimentary deposits and have been termed pedogenic mud aggregates (Rust and Nanson, 1989).

### *General Features of Pedogenic Mud Aggregates*

Pedogenic mud aggregates are predominantly of sand size and are produced during wetting-and-drying cycles of swelling clays in soils. The soil processes needed to produce such robust granular aggregates are termed "self mulching" and include shrink–swell phenomena (Wilding and Tessier, 1988) in Vertisols and other vertic soils containing expandable clays. Wet clays increase in volume, and pressures within the soil are augmented, shifting soil material upward and outward. With drying, deep cracks split the soil, with the formation of granular peds at the surface and angular peds at depth. Displacement of granular surface aggregates to subsurface levels occurs through top soil falling into cracks, as well as through illuviation. Horizonation processes are limited because of the repeated churning of the vertic soil during wet–dry cycles. Although the presence of smectite is strongly implicated in aggregate formation, the exact amount of smectite needed to form stable pedogenic mud aggregates is still unclear. Aggregates can be clay-rich (>60%) (Maroulis and Nanson, 1996), and Vertisols can crack with a minimum of at least 15% smectite (Dudal and Eswaran, 1988).

The granular aggregates formed at the surface of these vertic soils can be mobilized during floods and transported as bed load and suspended load. Reworked aggregates have been described in particular detail from channel and floodplain deposits of the Channel Country of central Australia (Rust and Nanson, 1991; Maroulis and Nanson, 1996; Gibling et al., 1998). In the Channel Country soils, granular peds range from less than 1 mm to about 2 mm in diameter, and reworked aggregates range in mean diameter from 0.13 mm (dry) to 0.70–0.75 mm (wet) (Maroulis and Nanson, 1996). The aggregates are composed of densely packed clay flakes with minor silt-size quartz and feldspar; their form ranges from simple, rounded, and relatively equant particles to composite particles composed of silt-size grains and clay flakes (Gibling et al., 1998).

The aggregates are extremely durable during transport in water (Singer et al., 1992) and, in some experimental trials, disaggregration occurred only after 24 hours of continuous flume operation. Disaggregated clay can also reaggregate after several wet–dry cycles in a flume (Norrish and Tiller, 1976; Singer et al., 1992; Maroulis and Nanson, 1996). Aggregate stability is attributed to the dominance of the expandable clay, smectite, in which a large electrostatic interaction exists between clay sheets (Norrish and Tiller, 1976). Microbial binding of clay was not considered important in the semiarid climate of the Channel Country, although it may be of minor importance in more humid climates (Rust and Nanson, 1989). The presence of calcium ions may assist aggregate formation, but salts in general were not found to be a ubiquitous factor in aggregate formation in fluvial environments (Rust and Nanson, 1989). In contrast, salts are important in the formation of eolian clay pellets that accumulate in dunes that border playa-lake muds in arid areas (e.g. Butler, 1956; Bowler, 1973).

Fragments of semiconsolidated mud (intraclasts) are commonly produced by natural erosive processes in rivers but are probably not a major source of mud aggregates. Such intraclasts are produced through episodic erosion over limited areas, and are unlikely to be as uniform in size as mud aggregates produced by pedogenic processes. Soil processes, on the other hand, can produce large volumes of aggregates of similar size over very large areas.

### *Factors for Production of Aggregates*

Vertic soils rich in smectite cover a wide climatic range; for example, Vertisols have a wide global distribution in humid and arid zones with seasonal rainfall, and they have been reported from over 80 countries (Dudal and Eswaran, 1988). Key factors in forming pedogenic mud aggregates are the existence of a source rock with the potential for smectite formation during weathering and pronounced wet–dry cycles. Primary smectite in soils commonly originates through weathering of source rocks in the drainage area (e.g., Potter et al., 1975). The weathering and neoformation of minerals such as olivine, pyroxene, amphibole, chlorite, biotite, and muscovite can lead to smectitic clays as byproducts (Nahon, 1991; McBride, 1994), and some of these minerals are especially abundant in mafic volcanic rocks. The weathering of basalt in particular has been shown to yield smectites (Macaire et al., 1994; Buol et al., 1997). This is illustrated by the distribution of clay minerals in the Columbia River area (Knebel et al., 1968): upper reaches with plutonic and metamorphic rocks produce mostly illite, and lower reaches with basaltic rocks produce mostly montmorillonite (smectite). Smectite is also produced through the weathering of volcanic ash (Zielinski, 1982; Van der Gaast et al., 1986). Thus, the formation of pedogenic mud aggregates is especially likely if source areas and/or basins contain mafic rocks and ash. Neoformation of smectites also occurs in association with Na-rich saline waters, and smectites are prominent in closed lake systems under alkaline, saline conditions (Deike and Jones, 1980; Jones, 1986; McBride, 1994; Hay et al., 1995).

Many continental rift basins are characterized by the presence of mafic volcanics and ash (Surdam and Sheppard, 1978), for example, the East African Rift Valley (Hay, 1986; Van der Gaast et al., 1986). Smectite-rich clays are present in significant quantities in the alluvial and lacustrine sediments of the modern Lake Turkana basin, whose source area is dominated by mafic volcanic rocks (Yuretich, 1986). Smectites and vertic soils can also develop in rift settings where sediments are derived from nearby basement source rocks such as gneiss, schist, and phonolite (Stephen et al., 1956). Closed lake systems are characteristic of many continental rift basins (e.g., Lambiase and Bosworth, 1995). For example, Lake Abert, an alkaline saline lake in Oregon, is surrounded by mostly basalt with some andesites and tuffs and is rich in smectite (Deike and Jones, 1980; Jones and Weir, 1983).

Marked seasonal contrasts may occur within rifts, especially because of the placement and elevation of rift shoulders (Hay et al., 1982; Lambiase and Bosworth, 1995). Thus, rift topography may enhance the wet–dry cycles, which influence soil formation.

### *Aggregate Occurrences*

Rift settings, ancient and modern, may be important sites for the formation of pedogenic mud aggregates, for the reasons noted above. Many circumstances promote the production of smectitic clay, including local climate conditions that produce strong seasonality for wet–dry cycles. Unfortunately there are no published studies of these aggregates in modern rifts. The most thoroughly studied modern example of pedogenic mud aggregates transported as bedload—the Channel Country of central Australia—lies within a cratonic basin (Rust and Nanson, 1989; Maroulis and Nanson, 1996; Gibling et al., 1998). Other documented modern localities are from foreland basins, including the badlands of the San Juan basin in New Mexico (Gutierrez, 1983) and the Biskra region of the Algerian Sahara (Williams, 1970).

Table 1 lists documented occurrences of pedogenic mud aggregates in the geologic record. Examples are found in the East Berlin and Shuttle Meadow Formations of the Newark Supergroup in the Hartford rift basin, of special relevance to the present study.

## METHODS FOR DETECTION

Burial diagenesis precludes using smectite clay content as a primary test in the search for pedogenic mud aggregates in the geologic record. Smectite in soil and playa sediments can be altered to illite during burial (Jones and Weir, 1983; Robinson and Wright, 1987; Retallack, 1991; Caudill et al., 1996; Mora et al., 1998; Driese et al., 2000). In addition, postdepositional compaction reduces the preservation potential of mud aggregates. The discrete aggregates are vague to unrecognizable within a few decimeters below the surface at Cooper Creek in the Channel Country of Australia because of compaction and addition of clay from suspension (Rust and Nanson, 1989).Thus, a mudstone body composed of mud aggregates tends to lose most traces of structures indicating traction deposition during burial and compaction. Aggregate particles have a better chance of preservation and detection when isolated among quartz and other framework mineral grains (Figs. 1A, B) in sandstones (Rust and Nanson, 1989; Gierlowski-Kordesch and Rust, 1994; Gierlowski-Kordesch, 1998) or preserved through early cementation.

Clay in sandstones may also originate from authigenic formation in pores, from the *in situ* alteration of labile grains (such as feldspars and glass) (Cummins, 1963), and from the settling of matrix mud. The key in detection is to identify possible mud aggregates within sandstone bedforms where the aggregated mud was clearly deposited as part of the bedload, for example, as cross-laminae in ripple cross-sets, or within horizontal laminae attributed to upper-regime flow (Figs. 1C, D). Such occurrences of sand-size mud grains in sandy traction deposits contrast with the more ubiquitous distribution of mud as disseminated matrix, authigenic flakes, and alteration products of framework grains (Gierlowski-Kordesch and Rust, 1994).

Mud aggregates are not commonly identified in petrographic studies of sandstones because they are difficult to resolve where partially compacted and an original matrix exists. A method for detection of pedogenic mud aggregates is the use of blue (even ultraviolet) light microscopy in thin section (Anadón et al., 1992; Talbot et al., 1994). Mud particles stand out well against quartz and feldspar (Figs. 1D, E) because high-energy light is blocked effectively by the clay and not by the crystal grains. This leads to a high contrast between crystal grains and clay so that the patterns of clay distribution and shape can be recognized more easily. Unfortunately, crushed aggregrates cannot be differentiated from suspension clays using this method.

## NEWLY IDENTIFIED DEPOSIT OF PEDOGENIC MUD AGGREGATES

### *Geologic Context*

Using the petrographic techniques outlined above, we have identified an additional occurrence of rift-related pedogenic mud aggregates in the New Haven Arkose of the Hartford basin of Connecticut. The New Haven Arkose is the basal formation of the Hartford basin (Fig. 2) with a probable age of ?lower Norian (Upper Triassic) to perhaps basal Hettangian (Lower Jurassic); its thickness is unknown but postulated to be 2400 m or more (Huber and Gierlowski-Kordesch, 1995). The Hartford basin (Fig. 2) is part of a series of rift basins along the eastern coast of North America, the fills of which are collectively named the Newark Supergroup. The rift basins developed in reponse to crustal extension and incipient rifting of the supercontinent Pangea during the early Mesozoic (Manspeizer, 1988; Schlische, 1993). The Hartford basin is presently 140 km by 30 km, although

TABLE 1.—Occurrences of pedogenic mud aggregates in the geologic record (Siluro-Devonian to Miocene) as recorded in recent literature.

| OCCURRENCE | AGE | BASIN TYPE | REFERENCES |
|---|---|---|---|
| Red Marls (Anglo-Welsh basin) England (Wales) | Siluro-Devonian | Foreland | Marriott and Wright, 1996 |
| Ridgeway Conglomerate Formation (Anglo–Welsh basin) England (Wales) | Lower Devonian | Foreland | Ékes, 1993 |
| Maringouin Formation (Fundy basin) eastern Canada | mid-Carboniferous | Strike-Slip | Rust and Nanson, 1989 |
| Hawkesbury Sandstone (Sydney basin) eastern Australia | Triassic (Anisian) | Rift? | Rust and Nanson, 1989 |
| Mercia Mudstone Group (Bristol Channel basin) southern England | Triassic (Norian) | Rift | Talbot et al., 1994 |
| New Haven Arkose (Hartford basin) eastern U.S. | Triassic (Norian?)–Jurassic (Hettangian) | Rift | This paper |
| East Berlin Formation (Hartford basin) eastern U.S. | Jurassic (Hettangian) | Rift | Rust and Nanson, 1989; Gierlowski-Kordesch and Rust, 1991 |
| Shuttle Meadow Formation (Hartford basin) eastern U.S. | Jurassic (Hettangian) | Rift | Gierlowski-Kordesch, 1998 |
| Miocene Units (Madrid basin) central Spain | Miocene | Foreland | J.P. Calvo (pers. comm.); Gierlowski-Kordesch, 1998 |

FIG. 3.—Outcrop photo of part of Section B in the New Haven Arkose near Meriden, Connecticut (see Fig. 2 for location). **A)** Original photo and **B)** tracing. ss = sandstone, md = mudstone. Channel body in upper part of photo is 3.2 m thick, with a nearly structureless central body of medium–coarse sandstone and a wing that thins to the left. The base of the central body includes a lens of small-pebble conglomerate composed of extrabasinal clasts. The central body (C) truncates an underlying channel wing that thickens to the left (beyond photo) into a central body. Sheet complexes above and below the channel body comprise interstratified fine–medium sandstone and silty mudstone sheets.

laterally offset with deeply incised bases, resulting in complex, amalgamated sediment bodies. Fills are predominantly coarse sand at the base, with local layers of extrabasinal gravel, fining to medium sand at body tops. Fine to very fine sand predominates in channel wings. Mudstone intraclasts a few centimeters in diameter are present locally in Sections A–C but are not abundant.

Sets of inclined heterolithic strata are common components of the channel fills, and are up to 2.9 m thick with apparent lateral extents of 20–60 m. The inclined surfaces strike nearly parallel to bidirectional paleoflow indicators, indicating that the sets represent lateral accretion of bank-attached bars. In contrast, the thickest and coarsest channel bodies have nearly structureless fills. Trough cross-sets, plane beds, and isolated scours are present in some channel fills, but stratification is diffuse and many beds appear virtually structureless, as noted in outcrops elsewhere in the region by Krynine (1950, p. 103). In coarser fills, discrete thin layers of sandstone and conglomerate define a crude horizontal stratification. Drab, near-vertical pipes and patches, up to 1 m long, that taper and branch downwards from channel-body tops are interpreted as root traces.

Sandstone and mudstone sheet complexes range from a few meters to more than 15 m thick. Channel wings thin away from the central body to become part of the adjacent sheet complexes, indicating a close dynamic relationship between sheet complexes and channel bodies. The sandstone sheets are mainly very fine sand to silt in grade, red and micaceous, and some fine upward. Most sheets are a few decimeters thick, but a few exceed 1 m. They form continuous, nearly planar sheets with abrupt to diffuse contacts that are traceable across the outcrop (~ 100 m) (Fig. 3). Rare sedimentary structures visible in the sheets include trough cross-sets, low-angle cross-stratification, ripple cross-lamination, and scour fills. The top surfaces of many planar-based sandstone sheets show mounds with a maximum relief of 50–100 cm (above the sheet base), an apparent width of ~ 5 m, and a quasi-regular spacing along a continuous bed. Form-concordant stratification is faintly visible within the mounds. Mounded sheets are also components of channel wings or cap central bodies. Although the three-dimensional form of the mounds could not be ascertained, the orientation of their margins suggests that they are linear, flow-parallel features.

Bioturbation is prominent in many sandstone sheets, and some beds have been pervasively penetrated by sinuous tubes, interpreted as *Scoyenia* burrows by Hubert et al. (1978), and root casts. Clay surfaces with a slickensided, onion-skin weathering appearance may be traces of peds or partially sealed desiccation cracks.

Red mudstone sheets in the sheet complexes are typically a few decimeters thick, predominantly silt in grade and weakly stratified, with a platy to conchoidally fractured appearance. Sinuous tubes are locally very abundant and, where separately identifiable, are mainly flat-lying, which imparts a fissile appearance to the mudstone. Vertical drab zones that penetrate both sandstone and mudstone are probably root traces. Orange carbonate nodules, up to 2 cm in diameter, are sparsely distributed in some mudstone sheets but do not form petrocalcic horizons. Rare decimeter-thick layers of mudstone contain abundant millimeter-size concretions composed of hematite and silica.

*Interpretation.—*

During deposition of the upper New Haven Arkose, the alluvial plains were traversed by sets of relatively broad, shallow channels (probably ~ 2 m in depth, based on story thickness), with incised to low-angle margins. Floods commonly exceeded bankfull capacity, resulting in overbank flows that deposited proximal sheets of sandy sediment (wings). Episodic high-energy flow is suggested by the gravel and coarse sand channel fills, upper-regime plane beds, and prominent scours at channel bases and within fills. A complex erosional and depositional history, probably linked to flood events, is indicated by the superimposed and amalgamated stories. Bank-attached bars accreted laterally in some channels, but vertical aggradation is also indicated by horizontally layered sand and gravel and nearly structureless fills in some bodies. Channel activity frequently resulted in erosion of muddy banks, inasmuch as mudstone blocks and small intraclasts are present in most channel bodies.

Sheet sandstone and mudstone complexes indicate active overbank (sheetflood) deposition, with an abundance of sand deposited as extensive layers adjacent to channels. There is little evidence for elevated sediment belts adjacent to the channels, and overbank flows may have spilled over gentle channel banks rather than feeding onto the floodplains through levee crevasses. The sets of stream-parallel mounds on the tops of sandstone sheets suggest periodic energetic flow in shallow water over channel tops and margins.

Sandstone and mudstone sheets, as well as upper channel fills, have been substantially modified through a combination of intense organic activity (roots), desiccation, and soil development. Soils were probably cumulative and, apart from intense bioturbation and sparse carbonate nodules, there is little indication of mature soil development; soil horizons are not apparent in the outcrops, and slickensided surfaces are rare. Millimeter-scale hematite-rich concretions may be concentrated into layers by erosion and indicate the beginnings of soil ferritization, which is the accumulation of iron oxides in a highly weathered soil (FitzPatrick, 1983; Buol et al., 1997; Beinroth et al., 2000).

No detailed analysis of these outcrops has been published to date. Hubert (1977, 1978), Hubert et al. (1978), and Weddle and Hubert (1983) inferred that braided rivers were generally predominant in the New Haven Arkose, with meandering rivers at some stratigraphic levels, on the basis of the presence of lateral accretion surfaces (McInerney, 1993). The presence of sand- and gravel-bed rivers with aggradational style, as well as channels with a component of lateral accretion of bars, is in general accord with these previous studies.

The geomorphic setting of the studied New Haven Arkose sections is unclear. Judging from paleoflow trends, bidirectional indicators suggest NE–SW drainage courses—nearly parallel to the axis of the Hartford basin. This may imply that the drainage formed part of a trunk system. The presence of both trunk and transverse flow systems within the formation was inferred from facies and paleoflow analysis by Weddle and Hubert (1983) and Hubert et al. (1992).

*Petrographic data.—*

Pedogenic mud aggregates were identified through preliminary petrographic analysis of thin sections of sandstones of the New Haven Arkose from a channel body in Section A (Fig. 4) and a rippled muddy siltstone sheet from Section G (Fig. 5). The channel sandstone from Section A has bimodal grain size with the pedogenic mud aggregates preserved between the smaller grains closer to their size (Fig. 4). The rippled muddy siltstone (Fig. 5) contains clearly segregated mud-dominated laminae with the aggregates preserved in part through calcite cementation of the framework grains. Although some grains were partially crushed, discrete silt-size mud bodies can be recognized.

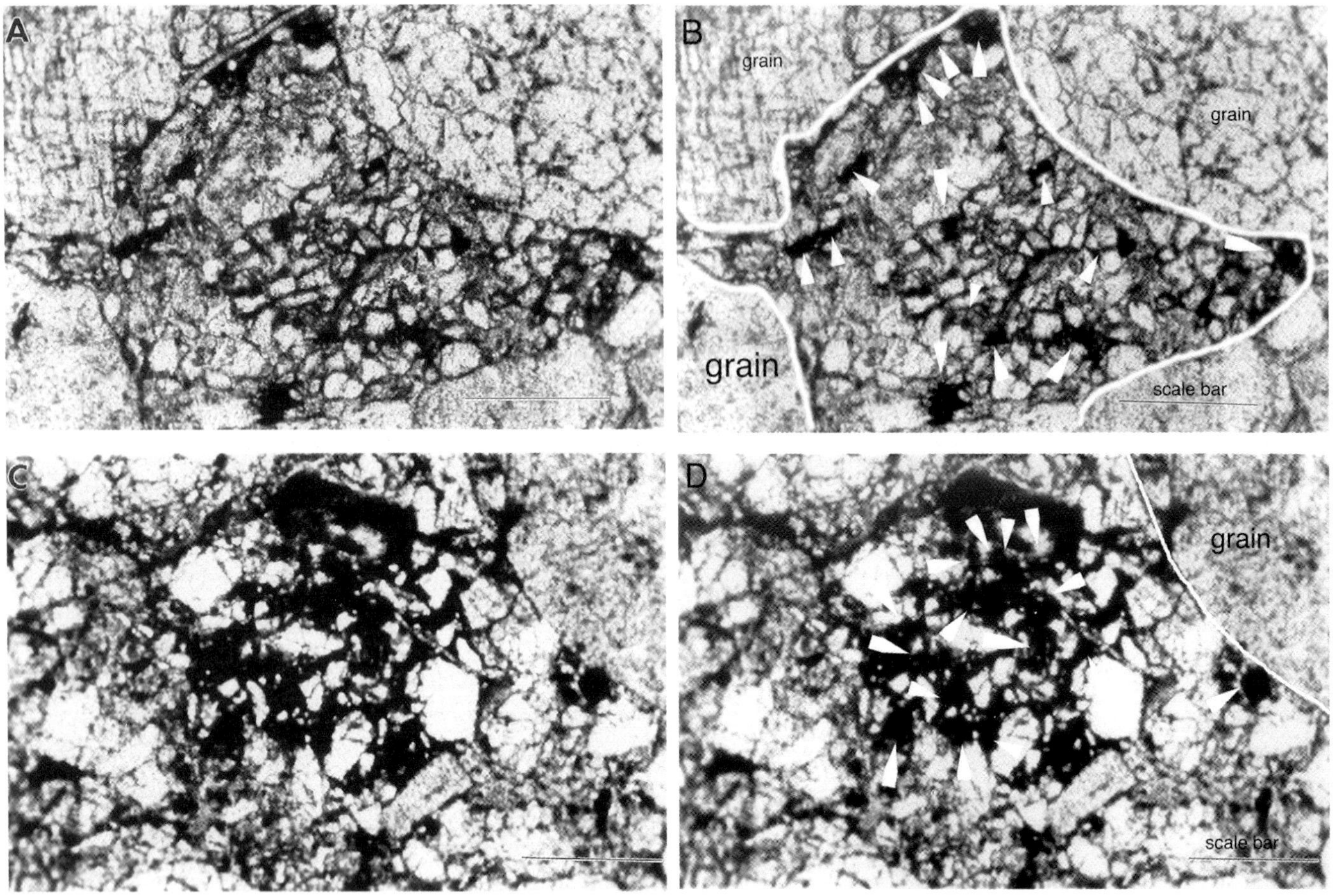


FIG. 4.—Pedogenic mud aggregates from Section A of the New Haven Arkose, Hartford basin, Meriden, Connecticut (see Fig. 2 for location). **A)** Thin-section photo. Aggregates associated with bimodal-grain-size sandstone, Bed 17. The aggregates are well isolated by the smaller mineral grains. Field of view: 4.8 mm. Scale bar: 1 mm. **B)** Scan of Part A. White line defines pore space surrounded by larger mineral grains. White arrows point to some pedogenic mud aggregates isolated by smaller mineral grains. **C)** Thin-section photo. Aggregates clustered together in a bimodal-grain-size sandstone, Bed 17. Preservation of pedogenic mud aggregates may be enhanced by hematite cementation. Field of view: 4.8 mm. Scale bar: 1 mm. **D)** Scan of Part C. White line defines large mineral grain at right, and white arrows point to pedogenic mud aggregates found in a cluster mixed with smaller mineral grains.

The pedogenic mud aggregates found in this limited sampling are silt- to sand-size particles, rounded to subangular in shape. The discrete form of the aggregates was preserved because larger grains in the bimodal sand population protected them from compaction and because early calcite cementation lithified the muddy siltstone. Other structureless, rooted sandstones from Sections A and C exhibit diffuse mud bodies in thin section, but none could be clearly identified as mud aggregates. The mud bodies are more randomly distributed and mostly crushed between grains; they could have originated as aggregates, zones of true mud matrix, or altered mineral grains. Preservation of the pedogenic mud aggregates in the New Haven Arkose in general is not as good as in the East Berlin and Shuttle Meadow Formations.

## DISCUSSION

### *Depositional Processes*

This study and previous studies (Rust and Nanson, 1989; Gierlowski-Kordesch and Rust, 1994) of channel and sheet sandstones and siltstones in the Hartford rift basin show that mud aggregates are widely distributed in the upper New Haven Arkose and the East Berlin and Shuttle Meadow Formations. The aggregates documented in these studies have a limited size range, and they are prominent in both channel and floodplain settings where they are components of bedforms. We infer that the bulk of the aggregates in the upper New Haven Arkose originated as pedogenic aggregates in the associated floodplain soils, which show intense bioturbation, local desiccation, sparse carbonate nodules, and hematite concretions. Although vertic paleosol features are prominent in the East Berlin and Shuttle Meadow strata, floodplain mudstones in the studied New Haven Arkose sections are immature and appear to lack vertic features: it remains uncertain whether soil development at these locations included aggregate formation. Alternatively, the aggregates could have originated in more mature soils at nearby localities or during weathering of more distant bedrock, although Maroulis and Nanson (1996) noted that aggregates break down during prolonged transport.

Mudstone intraclasts are present, but not abundant, in channel sandstones in the studied sections. This suggests that semiconsolidated channel banks and, possibly, proximal regions

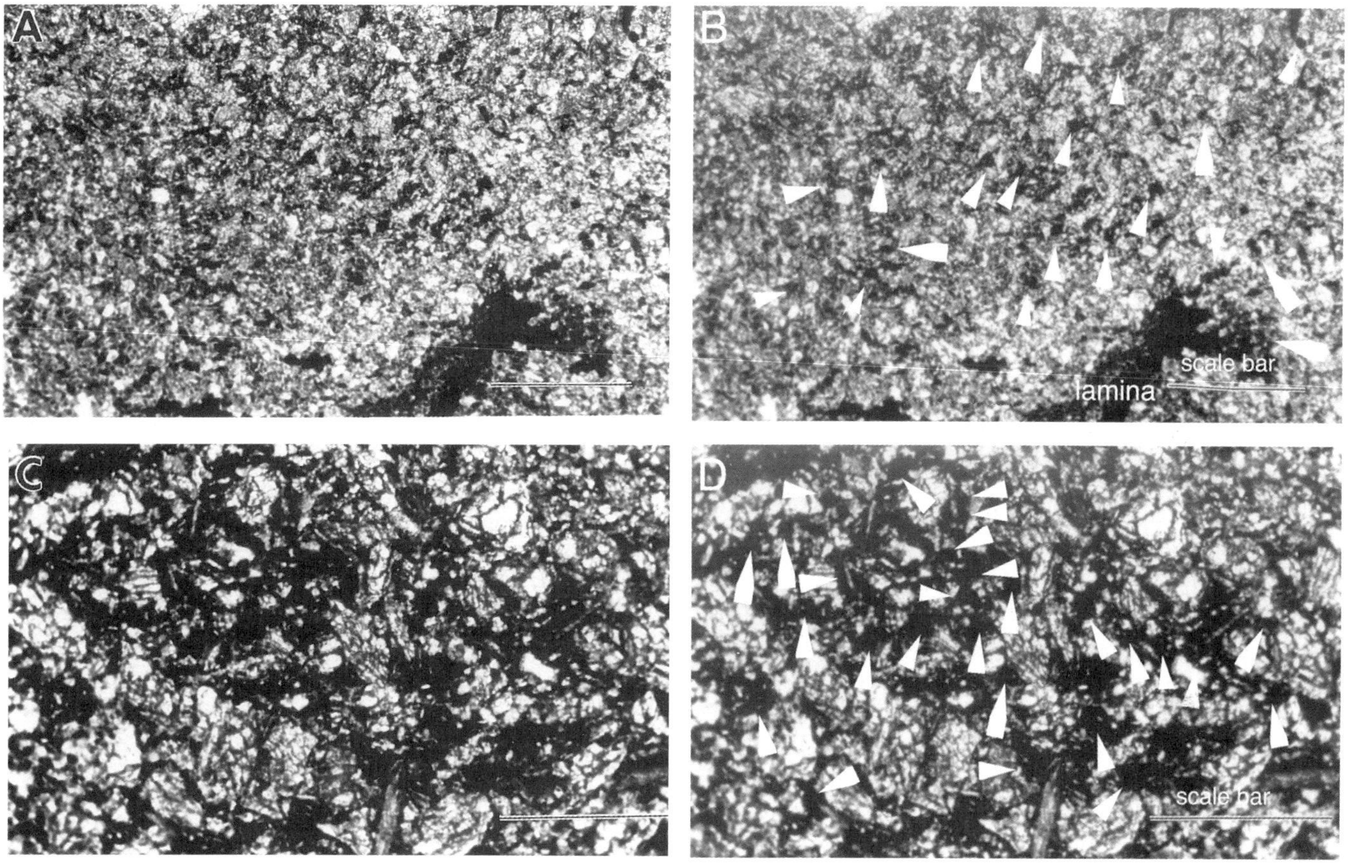


FIG. 5.—Pedogenic mud aggregates from Section G of the New Haven Arkose, Hartford basin, Southington, Connecticut (see Fig. 2 for location). **A)** Thin-section photo. Ripple cross-laminated muddy siltstone. Note calcitic groundmass interspersed with mud bodies. Broken and contorted lamina at bottom of photo. Field of view: 4.8 mm. Scale bar: 1 mm. **B)** Scan of Part A. White arrows point to some clusters of the better preserved pedogenic mud aggregates. **C)** Close-up of Part A showing calcitic groundmass containing mud bodies interpreted to be crushed pedogenic mud aggregates. Field of view: 0.9 mm. Scale bar: 0.25 mm. **D)** Scan of Part C. White arrows point to some pedogenic mud aggregates.

of floodplains were episodically eroded to yield mud fragments. No petrographic criteria currently allow us to make the distinction between eroded fragments and pedogenic aggregates, and some of the sand- and silt-size mud aggregates that we observed may have originated through erosion of mud layers. However, the large number of aggregates of relatively uniform size in the sample suite tends to support a predominantly pedogenic, rather than erosional, origin for the aggregates—inasmuch as eroded mud fragments are not especially common in the studied sections of the New Haven Arkose.

The abundance of aggregates in floodplain sandstone and siltstone sheets raises the possibility that some red mudstone sheets, which are composed of mixed clay and silt, were deposited largely as aggregates. If so, flows capable of moving silt- to sand-size materials in traction are implicated, rather than deposition of fine, suspended clay in floodbasins, as might otherwise have been inferred. The mudstones in the studied sections have been intensely bioturbated, and there is little indication of cryptic sedimentary structures that might indicate deposition from active flows. However, in a lower stratigraphic interval exposed near Southington (McInerney, 1993), diffuse traces of scours and cross-beds are common in mudstones similar to those in the studied sections.

The abundance of mud aggregates is in accord with sedimentologic evidence that the floodplains of the New Haven Arkose experienced frequent, catastrophic high-energy events. Sandstones form a high proportion of the strata, in the form of channel bodies and floodplain sheets, and networks of small channels probably switched location frequently to cut new flow lines and reoccupy older channels. The presence of sandy channel wings shows that floods frequently exceeded bankfull discharge.

Sandstone sheets with locally erosional bases and stream-parallel mounds in their upper parts suggest that overbank flows were commonly energetic. Under these conditions, incipient channel-bank and floodplain soils would have been reworked frequently and would have yielded aggregates, if they were present in the soils. There is no evidence in the studied sections of thick continuous mudstones that might indicate persistent floodbasins, isolated from destructive floods, where fines could accumulate from slow-moving or quiescent waters. However, the road cuts may preferentially expose resistant, sandstone-dominated intervals of the New Haven Arkose.

Floodplains of the upper New Haven Arkose accord broadly with the medium- to high-energy floodplains described from modern settings by Nanson and Croke (1992). The closest analogs are probably high-energy floodplains that these authors termed unconfined vertical-accretion sandy floodplains (group A3) and cut-and-fill floodplains (group A4). There are also some similarities with braided-river floodplains (group B1) of the medium-energy floodplain suite.

Floodplains in many rift basins are likely to be high-energy sites, prone to frequent floods and reworking of surficial muds. The relatively small size and narrow form of most rifts, including the Hartford basin, means that floodplains will be areally restricted, especially where trunk and tributary drainage systems "compete" for space. Additionally, many rifts are fed by short, high-gradient fluvial systems where strong discharge events transport detritus from bordering uplands of high relief. Depositional systems in such settings include alluvial fans with networks of shallow channels that experience frequent high-energy flows, promoting rapid aggradation and channel switching. The literature contains numerous examples of such depositional systems in ancient fault-bounded basins (e.g. Hill, 1989; Nakayama, 1999; Arche and López-Gómez, 1999), as well as in other basin settings with bordering uplands (Allen et al., 1983; Rust, 1984; Bentham et al., 1993). Thus, floodplain area limitation due to small rift size and the presence of multiple alluvial systems and dense channel networks, coupled with episodic high-discharge conditions, suggest that reworking of mud aggregates is an important process in rifts, where soil-forming conditions promote aggregate generation.

### *Paleoclimatic Implications*

Krynine (1950) inferred a tropical humid climate for the Hartford basin fill in general (with particular emphasis on the New Haven Arkose). His interpretation was based on criteria that included the prevailing red color and the abundance of arkose, which he took to indicate rapid downcutting into fresh source rock during high-precipitation events, accompanied by wash of feldspar and red soil material into the sedimentary basin, where it was rapidly buried. Other evidence presented by Krynine for a humid climate included the presence of ferruginous concretions, typical of tropical soils, and fossilized wood. The identification of "caliche" horizons by Hubert (1977, 1978) led him to infer instead a semiarid climate during the deposition of the New Haven Arkose.

Climate modeling (Wilson et al., 1994) as well as general global patterns of the Late Triassic (Sims and Ruffell, 1990; Parrish, 1993; Manspeizer, 1994; Olsen and Kent, 1996) point toward a tropical monsoonal climate for the New Haven Arkose depositional period, with the Hartford basin drifting across approximately 7° to 11° N latitude (Olsen, 1997; Kent and Olsen, 2000), normally a tropical humid zone. The interpreted drier climatic conditions in the Hartford basin at Late Triassic tropical latitudes was attributed to topographic rain shadows from high rift shoulders (Hay et al., 1982; Manspeizer, 1994). However, the Hartford basin drainage was not closed during the deposition of the New Haven Arkose (Smoot, 1991), implying that subsidence may have not been high enough to produce such extreme topographic differences to affect local climate until later in the Hettangian.

Preliminary sedimentologic data on the New Haven Arkose sediments suggest depositional conditions requiring somewhat wetter conditions than previously interpreted. This may agree better with the Late Triassic general circulation model of monsoonal circulation and decrease the unnaturally "steep latitudinal gradient" between evaporation and precipitation regimes in the tropical zone of the Late Triassic as modeled by Kent and Olsen (2000) for the Newark Supergroup basins. Continued sedimentologic and petrographic analysis of the New Haven Arkose may shed light on this dichotomy of wet (intense rooting, ferrigenous concretions, intense flooding, poor preservation of paleosol horizons and pedogenic mud aggregates) and dry (caliche) sedimentary features in the formation. Research is ongoing.

## CONCLUSIONS

Pedogenic mud aggregates are predominantly sand-size particles produced during seasonal wetting and drying in soils with expandable clays, principally smectite. To date, aggregates have been identified in only a few ancient deposits. Pedogenic aggregates are identified in the upper part of the New Haven Arkose (Upper Triassic to Lower Jurassic) of the Newark Supergroup in the Hartford basin, Connecticut. The sand- to silt-size aggregates are present in laminated and cross-laminated channel sandstones and floodplain sandstone and siltstone sheets. The presence of larger framework grains as well as early calcite cementation protected the aggregrates from destruction during compaction, and blue-light microscopy was valuable in identifying the aggregates. Sedimentologic evidence indicates that the New Haven floodplains were traversed by shallow bedload channels that aggraded and switched rapidly, with minor lateral accretion of bank-attached bars. These high-energy, sand-dominated floodplains were frequently reworked by overbank floods. Thin red mudstones interstratified with the sandstone sheets are immature paleosols that underwent intense bioturbation.

We infer that the bulk of the aggregates in the New Haven sandstones were reworked from these paleosols or from nearby more mature floodplain or upland soils. Although evidence for bedload transport of aggregates is lacking in the red mudstones, diffuse bedforms are visible in some red mudstones, and part of the clay component may have been deposited by active flows rather than from suspension settling in floodbasins.

Although pedogenic mud aggregates have not been documented in modern rift systems, the weathering of mafic volcanics and ash, which are prominent components of many rifts, could produce soils rich in expandable clays and suitable for aggregate development. Additionally, rain-shadow effects linked to rift topography may enhance climatic seasonality in many rifts.

Mobilization of soil aggregates is probably promoted in many rift basins with a sedimentologic setting similar to that of the Hartford basin. In these settings, limitations to floodplain size are imposed by rift dimensions, multiple alluvial systems, and shallow channel networks, allowing the floodplains to be swept by high-energy flood events, which are characteristic of basins with topographic relief. Reworking of mud aggregates is likely to be an important process in rifts, where soil-forming conditions promote aggregate generation.

## ACKNOWLEDGMENTS

Many thanks to W. Crawford Elliott for information on clay formation. M.E. Kordesch supported field work in Connecticut as well as microscopic analysis. David Kidder kindly lent his microscope for photography. M.R.G. is grateful for financial support from the Natural Sciences and Engineering Research Council of Canada. Peter M. LeTourneau, Margaret J. Guccione, and the editors contributed useful review comments.

## REFERENCES

Allen, P.A., Cabrera, Ll., Colombo, F., and Matter, A., 1983, Variations in fluvial style on the Eocene–Oligocene alluvial fan of the Scala Dei Group, SE Ebro Basin, Spain: Geological Society of London, Journal, v. 140, p. 133–146.

Anadón, P., Rosell, L., and Talbot, M.R., 1992, Carbonate replacement of lacustrine gypsum deposits in two Neogene continental basins, eastern Spain: Sedimentary Geology, v. 78, p. 201–216.

April, R.H., 1981a, Clay petrology of the Upper Triassic/Lower Jurassic terrestrial strata of the Newark Supergroup: Sedimentary Geology, v. 29, p. 283–307.

April, R.H., 1981b, Trioctahedral smectite and interstratified chlorite/smectite in Jurassic strata of the Connecticut Valley: Clays and Clay Minerals, v. 29, p. 31–39.

Arche, A., and López-Gómez, J., 1999, Subsidence rates and fluvial architecture of rift-related Permian and Triassic alluvial sediments of the southeast Iberian Range, eastern Spain, *in* Smith, N.D., and Rogers, J., eds., Fluvial Sedimentology VI: International Association of Sedimentologists, Special Publication 28, p. 283–304.

Beinroth, F.H., Eswaran, H., Uehara, G., and Reich, P.F., 2000, Oxisols, *in* Sumner, M.E., ed., Handbook of Soil Science: Boca Raton, Florida, CRC Press, p. E373–E392.

Bentham, P.A., Talling, P.J., and Burbank, D.W., 1993, Braided stream and floodplain deposition in a rapidly aggrading basin: the Escanilla Formation, Spanish Pyrenees, *in* Best, J.L., and Bristow, C.S, eds., Braided Rivers: Geological Society of London, Special Publication 75, p. 177–194.

Bersier, A., 1958, Séquences détritiques and divagations fluviales: Eclogae Geologicae Helvetiae, v. 51, p. 854–893.

Blakey, R.C., and Gubitosa, R., 1984, Controls of sandstone body geometry and architecture in the Chinle Formation (Upper Triassic), Colorado Plateau: Sedimentary Geology, v. 38, p. 51–86.

Bowler, J.M., 1973, Clay dunes: their occurrence, formation, and environmental significance: Earth-Science Reviews, v. 9, p. 315–338.

Buol, S.W., Hole, F.D., McCracken, R.J., and Southard, R.J., 1997, Soil Genesis and Classification: Ames, Iowa, Iowa State University Press, 527 p.

Butler, B.E., 1956, Parna—an eolian clay: Australian Journal of Science, v. 18, p. 145–151.

Caudill, M.R., Driese, S.G., and Mora, C.I., 1996, Preservation of a paleo-Vertisol and an estimate of Late Mississippian paleo-precipitation: Journal of Sedimentary Research, v. 66, p. 58–70.

Cummins, W.A., 1963, The greywacke problem: Liverpool and Manchester Geological Journal, v. 3, p. 51–72.

Deike, R.G., and Jones, B.F., 1980, Provenance, distribution, and alteration of volcanic sediments in a saline alkaline lake, *in* Nissenbaum, A., ed., Hypersaline Brines and Evaporitic Environments: Amsterdam, Elsevier, p. 167–193.

Driese, S.G., Joeckel, R.M., Mora, C.I., Nordt, L.C., and Stiles, C.A., 2000, Mass balance reconstruction of a modern Vertisol: implications for interpreting the geochemistry and burial alteration of paleo-Vertisols: Geoderma, v. 95, p. 179–204.

Dudal, R., and Eswaran, H., 1988, Distribution, properties and classification of Vertisols, *in* Wilding, L.P., and Puentes, R., eds., Vertisols: Their Distribution, Properties, Classification, and Management: College Station, Texas, Texas A&M University Printing Center, p. 1–22.

Ékes, C., 1993, Bedload transported pedogenic mud aggregates in the Lower Old Red Sandstone in southwest Wales: Geological Society of London, Journal, v. 150, p. 469–471.

FitzPatrick, E.A., 1983, Soils; Their Formation, Classification, and Distribution: Hong Kong, Longman Scientific and Technical, 353 p.

Gibling, M.R., Nanson, G.C., and Maroulis, J.C., 1998, Anastomosing river sedimentation in the Channel Country of central Australia: Sedimentology, v. 45, p. 595–619.

Gierlowski-Kordesch, E.H., 1998, Carbonate deposition in an ephemeral siliciclastic alluvial system: Jurassic Shuttle Meadow Formation, Newark Supergroup, Hartford Basin, USA: Palaeogeography, Palaeoclimatology, Palaeoecology, v. 140, p. 161–184.

Gierlowski-Kordesch, E., and Rust, B.R., 1994, The Jurassic East Berlin Formation, Hartford Basin, Newark Supergroup (Connecticut and Massachusetts): A saline lake–playa–alluvial plain system, *in* Renaut, R.W., and Last, W.M., eds., Sedimentology and Geochemistry of Modern and Ancient Saline Lakes: SEPM, Special Publication 50, p. 249–265.

Gottfried, R.M., and Kotra, R.K., 1988, Comparative mineralogy of clay-rich strata in selected early Mesozoic basins of the eastern United States: U.S. Geological Survey, Bulletin 1776, p. 99–102.

Gutierrez, A.A., 1983, Geomorphic processes and sediment transport in badland watersheds, San Juan County, New Mexico, *in* Wells, S.G., Love, D.W., and Gardner, T.W., Chaco Canyon Country. A Field Guide to the Geomorphology, Quaternary Geology, Paleoecology, and Environmental Geology of Northwestern New Mexico: American Geomorphology Field Group, Guidebook, p. 113–120.

Hay, R.L., 1986, Role of tephra in the preservation of fossils in Cenozoic deposits of East Africa, *in* Frostick, L.E., Renaut, R.W., Reid, I., and Tiercelin, J.-J., eds., Sedimentation in the African Rifts: Geological Society of London, Special Publication 25, p. 339–344.

Hay, R.L., Hughes, R.E., Kyser, T.K., Glass, H.D., and Liu, J., 1995, Magnesium-rich clays of the Meerschaum mines in the Amboseli basin, Tanzania and Kenya: Clays and Clay Minerals, v. 43, p. 455–466.

Hay, W.W., Behensky, J. F., Jr., Barron, E.J., and Sloan, J.L., II, 1982, Late Triassic–Liassic paleoclimatology of the protocentral North Atlantic rift system: Palaeogeography, Palaeoclimatology, Palaeoecology, v. 40, p. 13–30.

Hill, G., 1989, Distal alluvial fan sediments from the Upper Jurassic of Portugal: controls on their cyclicity and channel formation: Geological Society of London, Journal, v. 146, p. 539–555.

Horn, R., and Baumgartl, T., 2000, Dynamic properties of soils, *in* Sumner, M.E., ed., Handbook of Soil Science: Boca Raton, Florida, CRC Press, p. A19–A51.

Huber, P., and Gierlowski-Kordesch, E., 1995, Lake sequences of the central Hartford Basin, Newark Supergroup: State Geological and Natural History Survey of Connecticut, Guidebook no. 7, p. B1–39.

Hubert, J.F., 1977, Paleosol caliche in the New Haven Arkose, Connecticut: record of semiaridity in Late Triassic–Early Jurassic time: Geology, v. 5, p. 302–304.

Hubert, J.F., 1978, Paleosol caliche in the New Haven Arkose, Newark Supergroup, Connecticut: Palaeogeography, Palaeoclimatology, Palaeoecology, v. 24, p. 151–168.

Hubert, J.F., Reed, A.A., Dowdall, W.L., and Gilchrist, J.M., 1978, Guide to the Mesozoic redbeds of central Connecticut: State Geological and Natural History Survey of Connecticut, Guidebook no. 4, 129 p.

Hubert, J.F., Feshbach Meriney, P.E., and Smith, M.A., 1992, The Triassic–Jurassic Hartford rift basin, Connecticut and Massachusetts: evolution, sandstone diagenesis, and hydrocarbon history: American Association of Petroleum Geologists, Bulletin, v. 76, p. 1710–1734.

Jones, B.F., 1986, Clay mineral diagenesis in lacustrine sediments: U.S. Geological Survey, Bulletin 1578, p. 291–300.

Jones, B.F., and Weir, A.H., 1983, Clay minerals of Lake Abert, an alkaline, saline lake: Clays and Clay Minerals, v. 31, p. 161–172.

Kay, B.D., and Angers, D.A., 2000, Soil structure, *in* Sumner, M.E., ed., Handbook of Soil Science: Boca Raton, Florida, CRC Press, p. A229–276.

Kent, D.V., and Olsen, P.E., 2000, Magnetic polarity stratigraphy and paleolatitude of the Triassic–Jurassic Blomidon Formation in the Fundy basin (Canada): implications for early Mesozoic tropical climate gradients: Earth and Planetary Science Letters, v. 179, p. 311–324.

Knebel, H.J., Kelly, J.C., and Whetten, J.T., 1968, Clay minerals of the Columbia River: a qualitative, quantitative, and statistical evaluation: Journal of Sedimentary Petrology, v. 38, p. 600–611.

Krynine, P.D., 1950, Petrology, stratigraphy, and origin of the Triassic rocks of Connecticut: Connecticut State Geology and Natural History Survey, Bulletin 73, 247 p.

Lambiase, J.J., and Bosworth, W., 1995, Structural controls on sedimentation in continental rifts, *in* Lambiase, J.J., ed., Hydrocarbon Habitat in

Rift Basins: Geological Society of London, Special Publication 80, p. 117–144.

LORENZ, J.C., 1988, Triassic–Jurassic Rift Basin Sedimentology—History and Methods: New York, Van Nostrand Rheinhold, 315 p.

MACAIRE, J.J., COCIRTA, C., KARRA, L., AND PERRUCHOT, A., 1994, Basalt weathering and fluvial sedimentary particles: comparison of two watersheds in the Middle Atlas Mountains, Morocco: Journal of Sedimentary Research, v. A64, p. 490–499.

MANSPEIZER, W., 1988, Triassic–Jurassic rifting and opening of the Atlantic: an overview, *in* Manspeizer, M., ed., Triassic–Jurassic Rifting: Continental Breakup and the Origin of the Atlantic Ocean and Passive Margins, Part A: Amsterdam, Elsevier, p. 41–79.

MANSPEIZER, W., 1994, The breakup of Pangea and its impact on climate. consequences of Variscan–Alleghanide orogenic collapse, *in* Klein, G.D., ed., Pangea: Paleoclimate, Tectonics, and Sedimentation During Accretion, Zenith, and Breakup of a Supercontinent: Geological Society of America, Special Paper 288, p. 169–185.

MARRIOTT, S.B., AND WRIGHT, V.P., 1996, Sediment recycling on Siluro-Devonian floodplains: Geological Society of London, Journal, v. 153, p. 661–663.

MAROULIS, J.C., AND NANSON, G.C., 1996, Bedload transport of aggregated muddy alluvium from Cooper Creek, central Australia: a flume study: Sedimentology, v. 43, p. 771–790.

MCBRIDE, M.B., 1994, Environmental Chemistry of Soils: New York, Oxford University Press, 406 p.

MCINERNEY, D., 1993, Fluvial architecture and contrasting fluvial styles of the lower New Haven Arkose and the mid-upper Portland Formations, Early Mesozoic Hartford basin, central Connecticut: Master's Thesis, University of Massachusetts, Amherst, 271 p.

MORA, C.I., SHELDON, B.T., ELLIOTT, W.C., AND DRIESE, S.G., 1998, An oxygen isotope study of illite and calcite in three Appalachian Paleozoic vertic paleosols: Journal of Sedimentary Research, v. 68, p. 456–464.

NAHON, D.B., 1991, Introduction to the Petrology of Soils and Chemical Weathering: New York, John Wiley & Sons, Inc., 313 p.

NAKAYAMA, K., 1999, Sand- and mud-dominated alluvial-fan deposits of the Miocene Seto Porcelain Clay Formation, Japan, *in* Smith, N.D., and Rogers, J., eds., Fluvial Sedimentology VI: International Association of Sedimentologists, Special Publication 28, p. 393–407.

NANSON, G.C., AND CROKE, J.C., 1992, A genetic classification of floodplains: Geomorphology, v. 4, p. 459–486.

NORRISH, K., AND TILLER, K.G., 1976, Subplasticity in Australian soils. V. Factors involved and techniques of dispersion: Australian Journal of Soil Research, v. 14, p. 273–289.

OLSEN, P.E., 1997, Stratigraphic record of the early Mesozoic breakup of Pangea in the Laurasia–Gondwana rift system: Annual Review of Earth and Planetary Sciences, v. 25, p. 337–401.

OLSEN, P.E., AND KENT, D.V., 1996, Milankovitch climate forcing in the tropics of Pangaea during the Late Triassic: Palaeogeography, Palaeoclimatology, Palaeoecology, v. 122, p. 1–26.

PARRISH, J.T., 1993, Climate of the supercontinent Pangea: Journal of Geology, v. 101, p. 215–233.

POTTER, P.E., HELING, D., SHIMP, N.F., AND VAN WIE, W., 1975, Clay mineralogy of modern alluvial muds of the Mississippi River basin: Centre de Recherches de Pau (Société National de Petroles d'Aquitaine) Bulletin, v. 9, p. 353–389.

PRATT, L.M., SHAW, C.A., AND BURRUSS, R.C., 1988, Thermal histories of the Hartford and Newark Basins inferred from maturation indices of organic matter: U.S. Geological Survey, Bulletin 1776, p. 58–62.

RETALLACK, G.J., 1991, Untangling the effect of burial alteration and ancient soil formation: Annual Review of Earth and Planetary Sciences, v. 19, p. 183–206.

ROBINSON, D., AND WRIGHT, V.P., 1987, Ordered illite–smectite and kaolinite–smectite: pedogenic minerals in a Lower Carboniferous paleosol sequence, South Wales?: Clay Minerals, v. 22, p. 109–118.

RUST, B.R., 1984, Proximal braidplain deposits in the Middle Devonian Malbaie Formation of eastern Gaspé, Quebec, Canada: Sedimentology, v. 31, p. 675–695.

RUST, B.R., AND NANSON, G.C., 1989, Bedload transport of mud as pedogenic aggregates in modern and ancient rivers: Sedimentology, v. 36, p. 291–306.

RUST, B.R., AND NANSON, G.C., 1991, Reply to Discussion: Bedload transport of mud as pedogenic aggregates in modern and ancient rivers: Sedimentology, v. 38, p. 158–160.

SCHLISCHE, R.W., 1993, Anatomy and evolution of the Triassic–Jurassic continental rift system, eastern North America: Tectonics, v. 12, p. 1026–1042.

SIMMS, M.J., AND RUFFELL, A.H., 1990, Climatic and biotic change in the late Triassic: Geological Society of London, Journal, v. 147, p. 321–327.

SINGER, M.J., SOUTHARD, R.J., WARRINGTON, D.N., AND JANITZKY, P., 1992, Stability of synthetic sand–clay aggregates after wetting and drying cycles: Soil Science Society of America, Journal, v. 56, p. 1843–1848.

SMOOT, J.P., 1991, Sedimentary facies and depositional environments of early Mesozoic Newark Supergroup basins, eastern North America: Palaeogeography, Palaeoclimatology, Palaeogeography, v. 84. p. 369–423.

STEPHEN, I., BELLIS, E., AND MUIR, A., 1956, Gilgai phenomena in tropical black clays of Kenya: Journal of Soil Science, v. 7, p. 1–9.

SURDAM, R.C., AND SHEPPARD, R.A., 1978, Zeolites in saline, alkaline-lake deposits, *in* Sand, L.B., and Mumpton, F.A., eds., Natural Zeolites: Occurrence, Properties, Use: Oxford, U.K., Pergamon Press, p. 145–174.

TALBOT, M.R., HOLM, K., AND WILLIAMS, M.A.J., 1994, Sedimentation in low-gradient desert margin systems: A comparison of the Triassic of northwest Somerset (England) and the late Quaternary of east-central Australia, *in* Rosen, M.R., ed., Paleoclimate and Basin Evolution of Playa Systems: Geological Society of America, Special Paper 289, p. 97–117.

VAN DER GAAST, S.J., MIZOTA, C., AND JANSEN, J.H.F., 1986, Curved smectite in soils from volcanic ash in Kenya and Tanzania: a low-angle x-ray powder diffraction study: Clays and Clay Minerals, v. 34, p. 665–671.

WEDDLE, T.K., AND HUBERT, J.F., 1983, Petrology of Upper Triassic sandstones of the Newark Supergroup in the northern Newark, Pomperaug, Hartford, and Deerfield basins: Northeastern Geology, v. 5, p. 8–22.

WILDING, L.P., AND TESSIER, D., 1988, Genesis of Vertisols: shrink–swell phenomena, *in* Wilding, L.P., and Puentes, R., eds., Vertisols: Their Distribution, Properties, Classification, and Management: College Station, Texas, Soil Managment Support Services, Technical Monograph 18, p. 55–81.

WILLIAMS, G.E., 1970, Piedmont sedimentation and late Quaternary chronology in the Biskra region of the northern Sahara: Zeitschrift für Geomorphologie, v. 10 (supplement), p. 40–63.

WILSON, K.M., POLLARD, D., HAY, W.W., THOMPSON, S.L., AND WOLD, C.N., 1994, General circulation model simulations of Triassic climates: preliminary results, *in* Klein, G.D., ed., Pangea: Paleoclimate, Tectonics, and Sedimentation During Accretion, Zenith, and Breakup of a Supercontinent: Geological Society of America, Special Paper 288, p. 91–116.

YURETICH, R.F., 1986, Controls on the composition of modern sediments, Lake Turkana, Kenya, *in* Frostick, L.E., Renaut, R.W., Reid, I., and Tiercelin, J.-J., eds., Sedimentation in the African Rifts: Geological Society of London, Special Publication 25, p. 141–152.

ZIELINSKI, R.A., 1982, The mobility of uranium and other elements during alteration of rhyolite ash to montmorillonite: a case study in the Troublesome Formation, Colorado, U.S.A.: Chemical Geology, v. 35, p. 185–204.

# PALEOPEDOLOGY AND STABLE-ISOTOPE GEOCHEMISTRY OF LATE TRIASSIC (CARNIAN–NORIAN) PALEOSOLS, DURHAM SUB-BASIN, NORTH CAROLINA, U.S.A.: IMPLICATIONS FOR PALEOCLIMATE AND PALEOATMOSPHERIC $PCO_2$

STEVEN G. DRIESE AND CLAUDIA I. MORA

*Department of Geological Sciences, University of Tennessee-Knoxville, Knoxville, Tennessee 37996-1410, U.S.A.*
*e-mail: sdriese@utk.edu; cmora@utk.edu*

ABSTRACT: Paleosols in rift-related Upper Triassic (Carnian–Norian) terrestrial deposits of the Durham sub-basin of North Carolina record evidence of a shift from wetter to drier conditions, and slightly elevated atmospheric levels of carbon dioxide. Paleosols in siltstone interbeds of Lithofacies Association IIa (upper Carnian to lower Norian) are characterized by abundant micrite-cemented crayfish burrows, *Scoyenia* trace fossils, and numerous reduction spots and haloes indicating wet, but not hydric, soil conditions. Paleosols in fluvial–lacustrine sandstone–shale facies of Lithofacies Association IIb (lower Norian) record deposition in semiarid to subhumid climate characterized by seasonal soil-moisture deficits and high evapotranspiration. Pedogenic carbonate in both paleosols includes micritic rhizolith, nodule and burrow-fill morphologies. The carbon-isotope compositions of pedogenic carbonate differ significantly between paleosols within the two lithofacies, indicating different controls imposed by the pedogenic environments. The $\delta^{13}C$ values of pedogenic, micritic rhizoliths in the paleosols suggest somewhat elevated $pCO_2$ during the late Carnian–early Norian, two to three times modern levels, and support general estimates of $pCO_2$ based on similar carbonates in time-equivalent paleosols.

## INTRODUCTION

Rift-related terrestrial deposits offer a tremendous potential for paleoclimate reconstructions, with much past and current attention focused on interpreting Milankovitch-scale climate cyclicity recorded by lacustrine facies (Olsen, 1984, 1986, 1997; Hu and Textoris, 1994; Olsen et al., 1996). Paleosols also have great potential for paleoclimate studies of terrestrial deposits (e.g., Retallack, 1990; Retallack and Alonzo-Zarza, 1998; Joeckel, 1999; Kraus, 1999; Mora and Driese, 1999; Ashley and Driese, 2000). In addition to paleoenvironmental conditions that may be inferred from paleosol morphology and mineralogy, carbon isotope compositions of pedogenic carbonate provide a proxy for paleoatmospheric carbon dioxide levels (Cerling, 1991). Soil-carbonate carbon isotope $CO_2$ paleobarometry has been successfully applied to paleosols of Late Silurian and younger age (e.g., Mora et al., 1996; Ekart et al., 1999 and references therein). Recent studies of Triassic paleosols (Retallack and Alonso-Zarza, 1998; Krull and Retallack, 2000) and soil carbonate paleobarometry (Ekart et al., 1999) suggest very different histories of Triassic $pCO_2$ and its direction of change through the Triassic. Paleosols in the Upper Triassic (Carnian–Norian) terrestrial deposits of the Deep River Basin of North Carolina have not been previously examined in detail. These continental rift-related deposits may hold important clues concerning Late Triassic paleoenvironments and atmospheric $pCO_2$. We consider the impact of pedogenic and early post-pedogenic environments on the isotopic compositions of micritic carbonate in the paleosols, and, consequently, on estimates of atmospheric $pCO_2$ thus demonstrating the complications that may arise in this kind of study. Estimated Late Triassic atmospheric $CO_2$ levels are similar to, or slightly lower than, those suggested by other Late Triassic paleosol studies (Ekart et al., 1999; Jones, 1999) or the carbon mass-balance model of Berner (1994).

## GEOLOGIC SETTING

### *Structure*

The Upper Triassic rocks of the Deep River Basin crop out across central North Carolina, near Raleigh and Durham (Fig. 1). The Deep River Basin is divided into three sub-basins, which are, from north to south, the Durham, the Sanford, and the Wadesboro sub-basins (Fig. 1) (Brown et al., 1985). The Durham and Sanford sub-basins are separated from each other by the Colon cross-structure (Gore, 1986a). The Jonesboro fault bounds the Durham sub-basin on the east, with corresponding development of coarse fanglomerate and alluvial conglomerate facies proximal to the fault, and floodplain to floodbasin and lacustrine siltstone, claystone, and shale facies grading laterally into the basin center (Bain and Harvey, 1977; Gore, 1986a) (Fig. 1).

### *Stratigraphy*

The Upper Triassic rocks of the Durham sub-basin are assigned to the Chatham Group of the Newark Supergroup, and are further subdivided into three major lithostratigraphic mapping units, designated as "Lithofacies Associations" by Hoffman and Gallagher (1989) and assigned ages based on biostratigraphic (vertebrate) correlations by Huber et al. (1993). The western association (Lithofacies Association I) consists of arkosic sandstone with interbedded siltstone deposited in a western belt of anastomosing braided streams, and is mainly upper Carnian (Fig. 1). The central association (Lithofacies Association II) is composed of two lithofacies, arkosic sandstone with interbedded siltstone (IIa), and siltstone–claystone with interbedded arkosic sandstone (IIb), deposited in meandering-stream, floodplain, and lacustrine basins (Fig. 1). Lithofacies Association II is mainly lower Norian. The eastern association (Lithofacies Association III) is composed of four arkosic and lithic arenite lithofacies, namely, sandstone, pebbly sandstone, sandstone with interbedded conglomerate, and conglomerate deposited in alluvial fans in proximity to the Jonesboro border fault (Fig. 1). Lithofacies Association III is mainly lower Norian. The lacustrine and distal floodbasin claystone and shale of Lithofacies Association II are important sources of mineable brick clay in the Durham sub-basin, and low-grade coal deposits were once mined extensively in the Cumnock Formation of the Sanford sub-basin (Robbins and Textoris, 1986; Textoris and Robbins, 1988).

The Upper Triassic rocks of the Durham sub-basin have been considered as potential repository media for low-level nuclear waste (Bain and Harvey, 1981; Wooten et al., 1996). A proposed

Sedimentation in Continental Rifts
SEPM Special Publication No. 73, Copyright © 2002
SEPM (Society for Sedimentary Geology), ISBN 1-56576-082-4, p. 207–218.

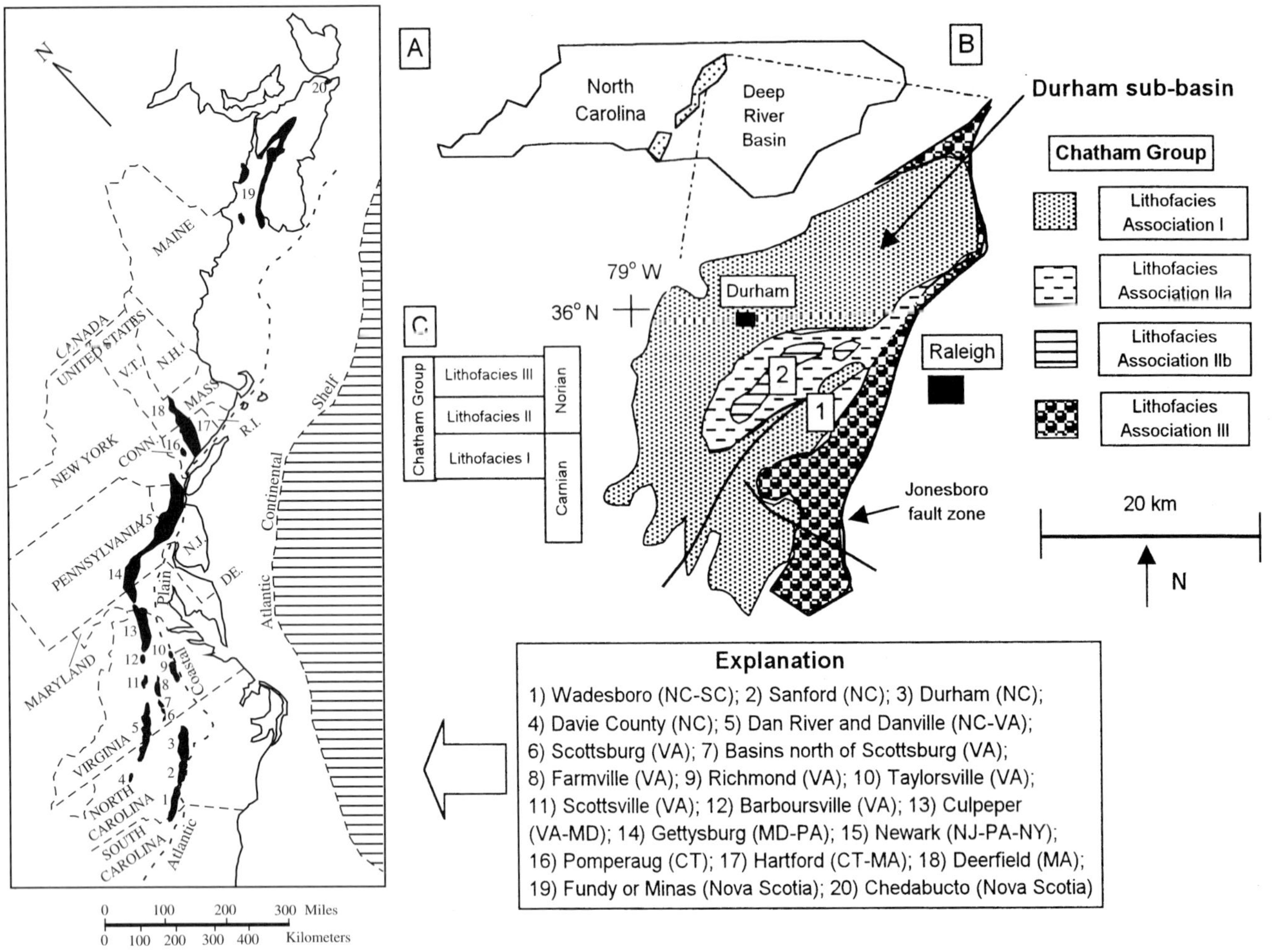


FIG. 1.—Sampling location and stratigraphy. **A)** Locality map showing distribution of Triassic–Jurassic extensional basins in eastern North America (adapted from Froelich and Olsen, 1984). **B)** Locality map showing Deep River Basin and Durham sub-basin in central North Carolina. Inset map shows locations of Spillway outcrop section (#1), proposed Wake/Chatham waste-disposal site (#1), and Triangle Brick Quarry outcrop section (#2). **C)** Generalized Upper Triassic stratigraphy of Durham sub-basin, following terminology proposed by Hoffman and Gallagher (1989).

waste site in Wake and Chatham Counties (Fig. 1), situated within anastomosing braid-plain and alluvial-fan-fringe deposits of Lithofacies Association IIa, was extensively cored as part of the site assessment. We visually inspected the cores in 1996 to confirm that they contained paleosols and pedogenic carbonate, but for proprietary reasons the cores could not be sampled. Although currently the subject of litigation, these cores may provide a valuable source of stratigraphic and sedimentologic information on paleosols from this part of the basin, when eventually released to the general public. Logs of the cores are available from the North Carolina State Geological Survey in Raleigh (Scott, 1993).

Macroflora in Deep River Basin strata include both seed plants (gymnosperms, cycads, and cycadeoids) as well as pteridophytes (ferns and fern allies), which are consistent with paleoclimate conditions that were warm and moist, but with seasonal rainfall (Gensel, 1986). Palynological studies of these rocks, in addition to constraining age relationships, indicate increasing abundance of more xeric palynomorphs up-section (Traverse, 1986).

### *Sedimentology*

Paleosols were examined at two outcrops of the Chatham Group (Fig. 1). Locality 1 included outcrops located 100 m south of the spillway beneath the Shearon Harris Reservoir dam, located in Chatham County, North Carolina, which were described by Cavaroc (1977), Gore (1986b), and Wooten et al. (1996). The deposits are mapped as the arkosic sandstone and siltstone facies of Lithofacies Association IIa and include gray, brown, and maroon conglomerate, sandstone, siltstone, claystone, and shale. Generalized (map-scale) stratigraphy is presented in Wooten et al. (1996), but they did not describe the paleosols present in the section. Locality 2 includes exposures in the Triangle Brick Quarry near Genlee, in Durham County, North Carolina, and is described in Olsen (1977) and Gore (1986c). These deposits are mapped as the red sandstone and siltstone–mudstone facies of Lithofacies Association IIb and include gray, brown, and maroon sandstone, siltstone, claystone, and shale.

## METHODS

Paleosol and pedogenic carbonate macrostructural features were described in the field, and detailed stratigraphic sections were constructed using modern soil-horizon designations (Birkeland, 1984), Munsell Soil Color Charts (1975) soil colors, and descriptive modifiers proposed by Retallack (1990) and Wright et al. (1995). Paleosol microfabrics (micromorphology) were characterized in thin section. Carbonate microfabrics were characterized using the terminology of Wright (1990, 1991) and FitzPatrick (1993), and cathodoluminescence (CL) petrography was utilized to interpret the history of calcite precipitation in pedogenic carbonate deposits and to detect recrystallization, if present (Machel and Burton, 1991).

Criteria for differentiating groundwater carbonate from pedogenic carbonate (Wright, 1991; Spötl and Wright, 1992; Mora et al., 1993) guided both sampling of carbonate material for isotopic analysis and its interpretation. Carbonate was sampled from thin-section billets and polished slabs using a small dental drill. Distinctive phases micro-sampled included: (1) micritic rhizoliths, (2) micritic concretions formed around, and within, crayfish burrows, (3) micritic nodules, (4) calcite spar cement filling septarian or geodal voids in micritic nodules, and (5) calcite spar cement filling tectonic fractures or lining pedogenic slickensides. Carbonate samples were roasted at 375°C for 1 hr to remove volatile organic material and reacted with phosphoric acid, following standard technique. Oxygen and carbon isotope ratios of evolved $CO_2$ were measured on the Finnigan MAT Delta Plus isotope ratio mass spectrometer housed at the University of Tennessee and are reported in standard δ-permil notation relative to PDB, with a precision of ± 0.1‰.

## DESCRIPTION

### *Paleosols*

*Spillway site.*—

About 5–6 m of section are exposed at this locality, consisting of very coarse-grained feldspatholithic sandstone and poorly sorted, polymictic, clast-supported to matrix-supported conglomerate toward the base of the section, grading upward into red to maroon siltstone and claystone (Fig. 2A). The rocks are arranged into crude fining-upward sequences consisting of about 1–2 m of conglomerate, overlying scour surfaces with up to 2 m of relief, and overlain by as much as 1–3 m of siltstone and mudstone (Wooten et al., 1996). Although initially interpreted as alluvial fan in origin (Cavaroc, 1977; Gore, 1986b), K. Farrell (personal communication, 1994) has suggested that these rocks are more correctly interpreted as anastomosed channel and floodplain deposits. Gibling et al. (1998) described sediments, with modern soils, associated with anastomosed river systems in the Channel Country, central Australia, that are comparable to the Lithofacies Association IIa strata exposed at the Spillway section.

Paleosols are present locally in the sandstone and conglomerate but are more common in interbedded siltstone strata toward the base of the section (Fig. 2B). We studied two siltstone paleosols near the top of the outcrop. These paleosols are grayish-red (Munsell color 5R 4/2) to dark reddish-brown (10R 3/4) in color, are very thin (< 1 m thick), and exhibit: (1) angular to subangular blocky ped structures, (2) greenish-gray (5GY 6/1) reduction spots, (3) downward-branching and tapering, 0.5–1 cm diameter, micritic root traces (rhizoliths) with light olive gray (5Y 6/1) outer walls, olive gray (5Y 4/1) centers, and drab greenish-gray reduction haloes, (4) complex three-dimensional galleries 2–5 cm in diameter, knobby-walled micritic concretions having light brownish-gray (5YR 6/1) outer walls and light olive gray centers, and (5) rare pedogenic slickenside surfaces. *Scoyenia* ichnofacies trace fossils are very abundant (Fig. 2C).

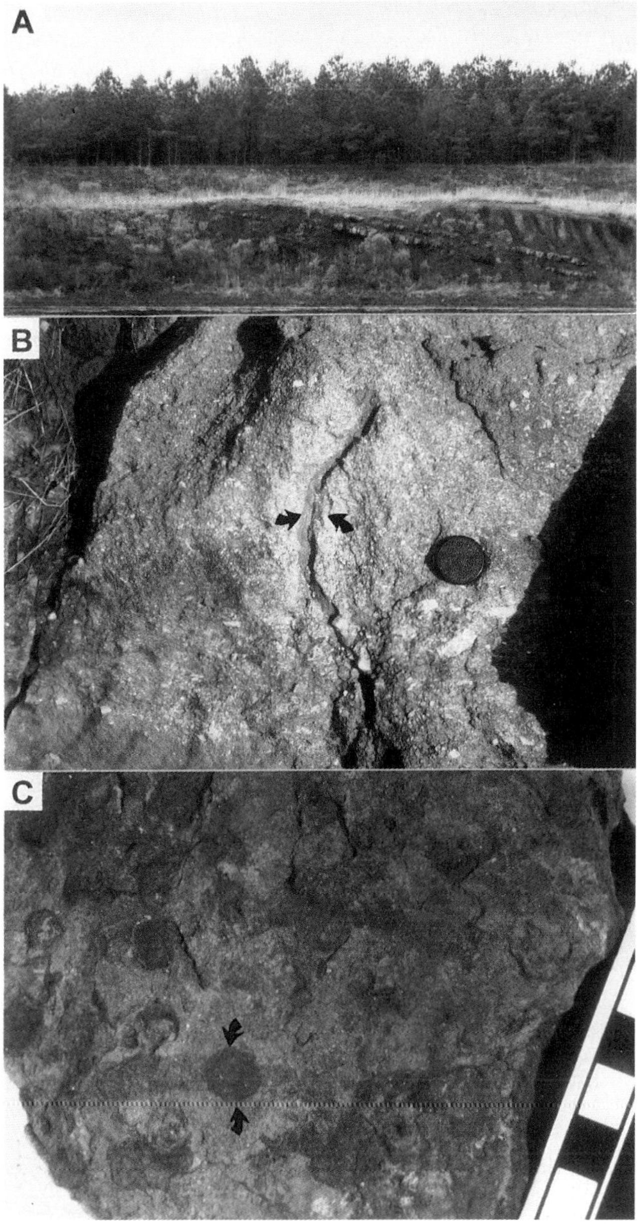


FIG. 2.—Field photographs from Spillway section (Locality 1 in Fig. 1). **A)** Exposures of fine-grained sandstone and siltstone deposits containing paleosols in the upper part of the stratigraphic section. Outcrop face is about 3 m high and 20 m long. **B)** Example of drab-haloed root trace in sandstone paleosol. Lens cap is 5.5 cm in diameter. **C)** *Scoyenia* continental trace fossils common in outcrop section. Scale bar in centimeters.

*Triangle brick quarry.—*

About 3–4 m of section are exposed at this locality, consisting of 20–50 cm thick, sheet-like beds of medium-grained feldspatholithic sandstone intercalated with 0.5–2 m thick beds of grayish-red siltstone and claystone (Figs. 3, 4A). These Lithofacies Association IIb strata are arranged into crude, 0.5–1.25 m thick, fining-upward sequences consisting of trough cross-stratified to ripple cross-laminated sandstone, which grade upward into hackly weathering, rooted, clayey siltstone (Olsen, 1977; Smoot, 1985; Gore, 1986c). Previous workers interpreted these rocks as intertonguing fluvial and lacustrine deposits containing a diverse lacustrine limestone and shale biota that included ostracodes, conchostracans, unionid and corbiculid? pelecypods, crayfish, insects, fish scales, bones, and teeth, phosphatized coprolites, and *Scoyenia* trace fossils. An articulated, poposaurid rauisuchian archosaur and a nearly complete, articulated sphenosuchian crocodylomorph were collected from the same quarry (Peyer et al., 2001; Sues et al., 2001).

The paleosols are grayish-red (5R 4/2) to dark reddish-brown (10R 3/4) clayey siltstone to silty claystone, which is characterized by abundant arcuate slickensides that dip from 10 to 60° and form bowl-shaped surfaces lined with black manganese oxide stains and locally with coarse calcite spar (Figs. 3, 4B). The paleosols commonly have medium to coarse, subangular to angular blocky ped structures, and downward-branching and tapering root traces with drab olive gray (5Y 4/1) reduction haloes present locally (Fig. 4C). Layers rich in 2–10 cm diameter, light olive gray (5Y 6/1) to olive gray, tuberose, micritic nodules, and 1–5 cm diameter, vertically stacked, micritic nodules are abundant in the lower parts of paleosols, where the slickensides diminish in abundance and are at shallower dip angles (Fig. 4D).

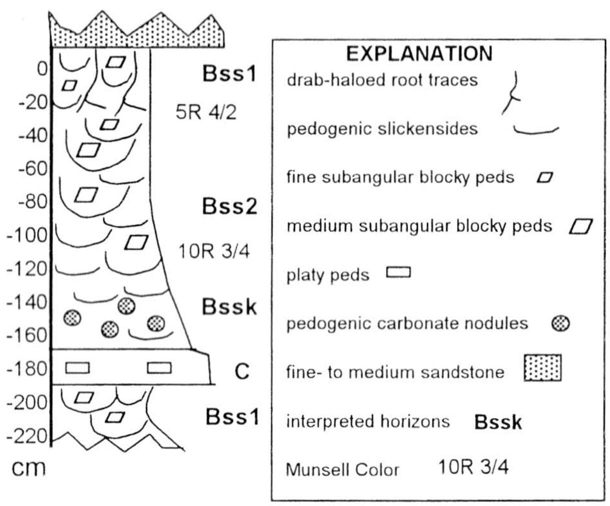


FIG. 3.—Stratigraphic column of paleoVertisol developed in distal floodplain and lacustrine deposits at Triangle Brick Quarry (Locality 2 in Fig. 1). The high clay content, abundance of pedogenic slickensides, and dominance of angular to subangular, fine- to medium-blocky ped structures support interpretation of this as a paleoVertisol, analogous to modern Vertisols. Note also the drab-haloed root traces and pedogenic carbonate nodules and rhizoliths. Horizons defined include Bss1 and Bss2 (slickensided B horizons), Bssk (slickensided B horizon with pedogenic carbonate accumulations), and C (weathered parent material). Munsell colors are also shown.

## Pedogenic Carbonate

The siltstone–claystone paleosols at both the Spillway section and the Triangle Brick Quarry section thus contain pedogenic carbonate of three types:

(1) *Micritic rhizoliths,* < 1 cm diameter and up to 30 cm long, consist of a dense, light olive gray micrite with sparry calcite internal pore-filling cements, and stand out in sharp contrast to the surrounding grayish-red claystone matrix. The rhizoliths branch and bifurcate downward, and are roughly circular in cross section.

(2) *Knobby-walled, stacked concretions,* up to 5 cm in diameter and over 1 m long, consist of light olive gray micrite with abundant calcite spar cement–filled cracks up to 0.5 cm in width. The concretions have knobby or bumpy walls composed of coarse pellet-like features, and form anastomosing three-dimensional galleries that do not bifurcate or branch downward like rhizoliths. Meniscate structures are locally present in some knobby concretions.

(3) *Micritic nodules* up to 10 cm in diameter, but more commonly 0.5–5 cm diameter, consist of light olive gray to greenish-gray micrite and abundant calcite spar cement–filled cracks up to 0.5 cm in width. These ubiquitous nodules commonly have slickensides on their outer surfaces as well as adhering grayish-red to reddish-brown claystone matrix, and are composed of micronodules welded by micrite.

## Micromorphology

Micritic rhizoliths are composed of a dense micrite that contains some occluded dark (organic?) material. Alveolar-septal structures are common in rhizoliths, as are calcite spar cement–filled shrinkage pores (Fig. 5A). Many of the pores have geopetal floors defined by vadose silt, although overall there is very little detrital clay and silt occluded within the micrite (Fig. 5B). Iron reduction is localized around the margins of many rhizoliths. Rhizolith micrite exhibits a dull red-brown luminescence, or no luminescence at all, whereas sparry calcite pore-filling cements show initial nonluminescent cements nucleated on pore walls, followed by later pore-central cements that have bright orange-red luminescence (Figs. 6A, B).

Knobby-walled concretions are composed of large numbers of 1–5 mm diameter micritic micronodules, welded together by a micritic matrix that is locally pelletal (Fig. 5C). Septarian shrinkage cracks and circumgranular cracks form complex networks in the centers of concretions; these crosscut the micrite and are filled with coarse, sparry calcite cements that are generally prismatic to bladed on pore walls and equant in pore centers. Meniscate structure is locally well developed within some of the concretions and crosscuts the welded micritic nodules and pelleted micrite. The cathodoluminescence of micrite and spar in the knobby concretions are similar to the same phases in rhizoliths. Bands of spar cement are disrupted by micro-unconformities marking numerous (up to 5–10) growth bands, followed by later pore-central cements that have bright orange-red luminescence (Figs. 6C, D). Rare, coarse, botryoidal to fan-like cements are present as encrustations on the periphery of the knobby concretions or lining pedogenic slickensides. Late, fine fractures filled with calcite spar cement that is very bright orange luminescent crosscut all petrographic constituents within knobby concretions.

The micromorpology of micritic nodules is similar to that of the knobby concretions, i.e., masses of micritic micronodules that are engulfed and welded together by micrite; however, this micrite includes a significant amount of paleosol matrix material

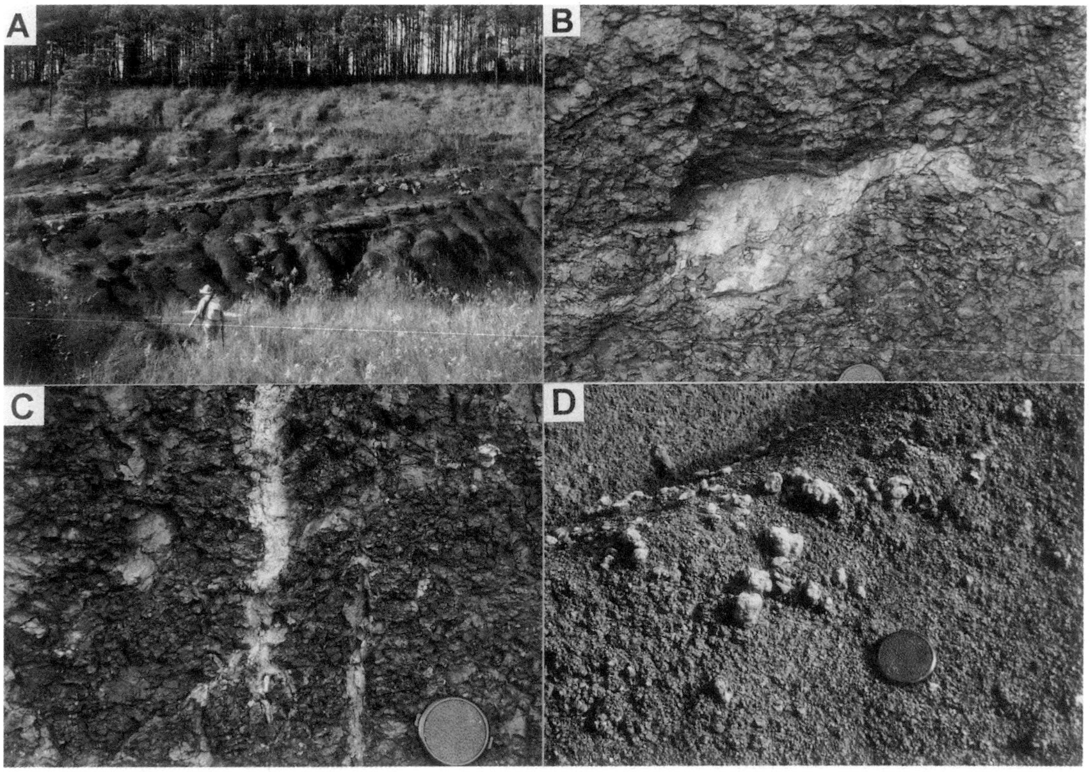


FIG. 4.—Field photographs from Triangle Brick Quarry (Locality 2 in Fig. 1). **A)** Exposures of silty claystone consisting of paleosols, interbedded with subordinate fine-grained sandstone and siltstone. Person is approximately 2 m tall for scale. **B)** Paleosol claystone with angular to subangular, medium blocky ped structure and pedogenic slickenside surfaces. **C)** Paleosol claystone with downward-branching, drab-haloed root traces. **D)** Pedogenic carbonate rhizocretions consisting of vertically stacked micritic nodules. Lens cap is 5.5 cm in diameter in parts B, C, and D.

(reddish clay and quartz silt). In addition, some nodules contain banded Fe concretions or have Fe concretions that serve as nuclei for micritic micronodules. Sparry calcite cements commonly fill septarian and circumgranular cracks, as described previously (Fig. 5D). Spherulitic masses of prismatic crystals (spherules < 1 mm in diameter) appear to replace some areas of sparry calcite and micrite.

### *Stable Isotopes*

Carbon and oxygen isotope compositions for the principal carbonate constituents are shown in Figure 7. Carbon isotope compositions of all constituents in the Spillway paleosols are distinct from their counterparts in the Triangle Brick Quarry paleosols. In the Spillway paleosols, the average $\delta^{13}C$ (PDB) value of all micritic pedogenic phases is –9.45‰, with slight differences in the compositions of micritic rhizoliths ($\delta^{13}C$ = –10.58 to –9.24‰, average –9.92‰), micritic nodules ($\delta^{13}C$= –8.91 to –9.52, average –9.21‰), and knobby concretions ($\delta^{13}C$ = –8.72 to –9.02‰, average –8.77‰). Compositions of septarian and rhizolith spar-filling cements are similar to the micritic phases. Pedogenic micritic phases in the Triangle Brick Quarry paleosols are more variable, with nodules having $^{13}C$-enriched compositions ($\delta^{13}C$= –6.44 to –7.36‰) compared to rhizoliths (–8.45‰) or micritic knobby concretions (–8.78‰). Septarian spar cement had $\delta^{13}C$ = –7.27‰. Values of $\delta^{18}O$ (PDB) range from –4.95 to –7.46‰ for micritic nodule phases in both paleosols, with average $\delta^{18}O$ (PDB) = –6.40 ± 0.62‰. Coarse, botryoidal to fan-like calcite occurring on the exteriors of nodules and lining pedogenic slickenside surfaces were significantly different ($\delta^{13}C$ = –13.3‰, $\delta^{18}O$ = –3.9‰).

## INTERPRETATION

### *Paleosols*

Carbonate-bearing Upper Triassic redbed deposits at both sample localities are interpreted as paleosols on the basis of their constituent downward-branching and tapering plant root traces, *Scoyenia* ichnofacies continental invertebrate traces, and other soil structures described previously. The clay-rich mineralogy and presence of angular blocky ped structures, arcuate (bowl-shaped) slickensides, and infilled deep vertical cracks in the paleosols at the Triangle Brick Quarry are all characteristic of the Vertisol soil order (Dudal and Eswaran, 1988; Coulombe et al., 1996). Modern Vertisols (also vertic features in other soil types or vertic intergrades) form in clay-dominated soils having a high proportion of expandable clay minerals. Soil structures reflect formation in seasonally wet climates, or climates characterized by a seasonal soil-moisture deficit (Coulombe et al., 1996). Mora and Driese (1999) recently summarized Late Ordovician to Early Permian occurrences of paleoVertisols in the Paleozoic redbed formations of the Appalachian basin in the U.S. and Canada, which share many similarities with these Late Triassic paleosols formed in the Durham sub-basin. Vertic features are particularly well developed in the Triangle Quarry claystone paleosols, in comparison with the Spillway paleosols, where vertic features

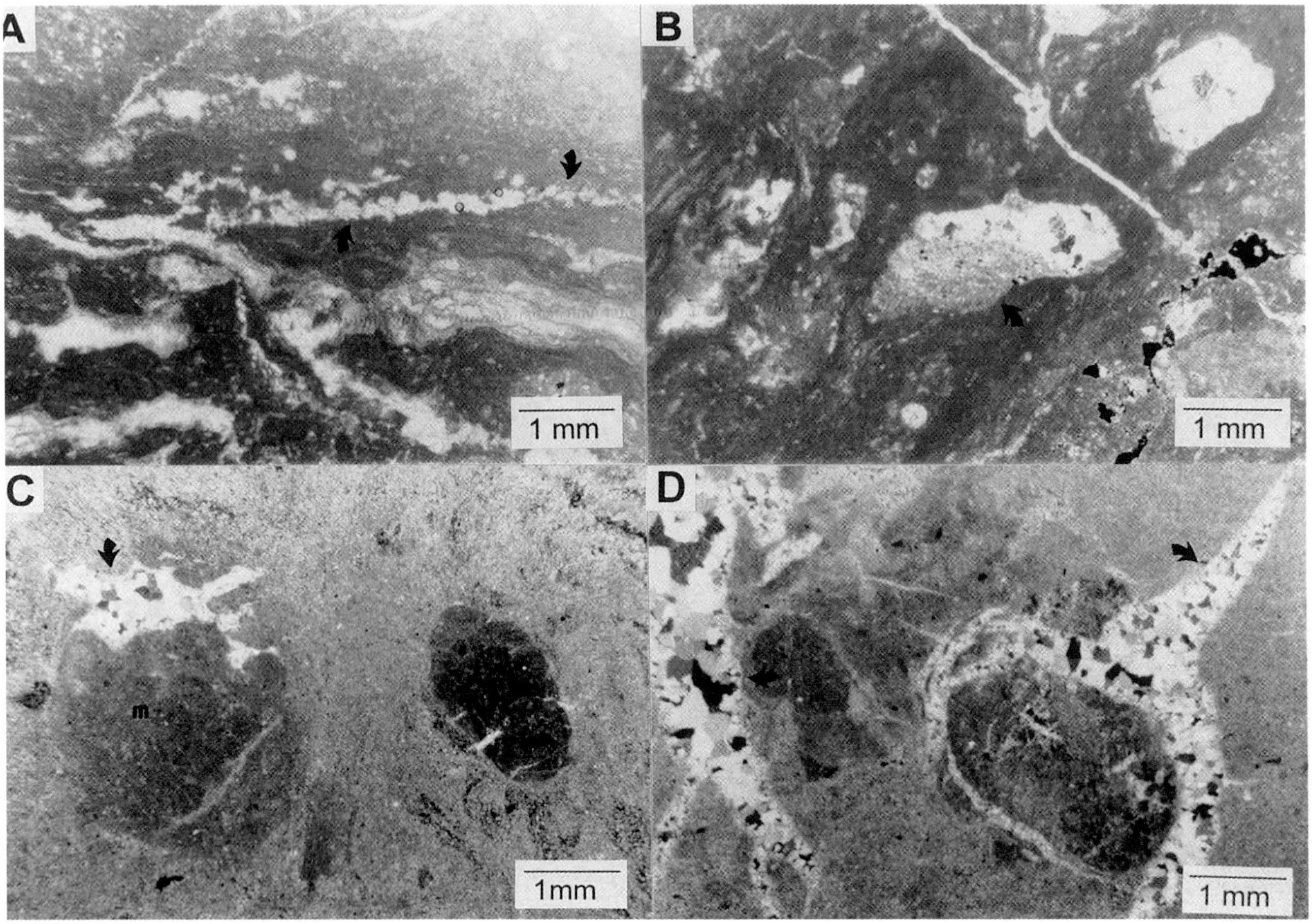


FIG. 5.—Micromorphology of pedogenic carbonate in Upper Triassic paleosols, Spillway locality, Durham sub-basin, North Carolina. **A)** Longitudinal section through micritic rhizolith showing well-developed alveolar septal structures (arrows) that project into calcite spar cement–filled pores. Plane-polarized light. **B)** Geopetal vadose carbonate and siliciclastic silt (arrow) filling pore space within micritic rhizolith. **C)** Micritic crayfish concretion composed of welded micritic micronodules (m), one of which has prominent calcite spar cement–filled septarian crack (arrow). **D)** Micritic nodule exhibiting well-developed calcite spar cement–filled septarian and circumgranular cracks (arrows). Parts B, C, and D are photographed with crossed nicols.

either did not form due to pedogenesis under different climatic conditions or were suppressed by paleosol development in much siltier parent sediments.

The paleosols at the Spillway section are interpreted to have formed in a paleoclimate that was wetter than that of the Triangle Brick Quarry paleosols because of the presence of abundant drab-colored reduction mottles and spots, more abundant soil animals, as evidenced by crayfish burrows (krotovina) cemented by micrite (see discussion below) and by meniscate *Scoyenia* trace fossils, and more residual organic C preserved as inclusions in micrite. Micritic concretions more than 1 m long, with knobby, pelleted walls and anastomosing three-dimensional galleries, likely represent decapod crustacean (i.e., crayfish) burrows, judging by their distinctive morphologies (e.g., Hasiotis and Mitchell, 1993; Hasiotis and Honey, 2000). Despite abundant crayfish burrows and redoximorphic features (*sensu* Vepraskas et al., 1994), we do not interpret periodically saturated soil conditions for the Spillway paleosols. The depth of crayfish burrows, in this case extending almost through the entire paleosol profile, is helpful to define the depth of the paleo–water table (Hasiotis and Honey, 2000). Hydric soil indicators (hydric soils experience seasonal periods of saturation and development of anaerobic conditions; Hurt et al., 1998) such as organic C-rich layers, gleyed (low-chroma) layers, and zones of extensive redoximorphic features are commonly reported in very wet, hydromorphic, seasonally saturated paleosols (Kraus and Aslan, 1993; Driese et al., 1995; Wright et al., 2000) but are not seen in the Spillway paleosols.

The Spillway section soils formed on a landscape characterized by anastomosing sand-filled channels and adjacent silt–clay levees and floodplains.The modern anastomosed river belts and associated floodplains of central Australia (Gibling et al., 1998) serve as a potential modern analog for the Spillway section paleosols and laterally coeval strata, although the Australian site is much drier (MAP 40–50 cm / yr) than what we infer for the Spillway paleosols (see discussion below). The climatic significance of the Upper Triassic paleosols can be further evaluated by direct analogy with carbonate morphologies seen in modern Vertisols formed from red alluvium of the Upper Pleistocene (< 35,000 years B.P.) Beaumont Formation, Texas

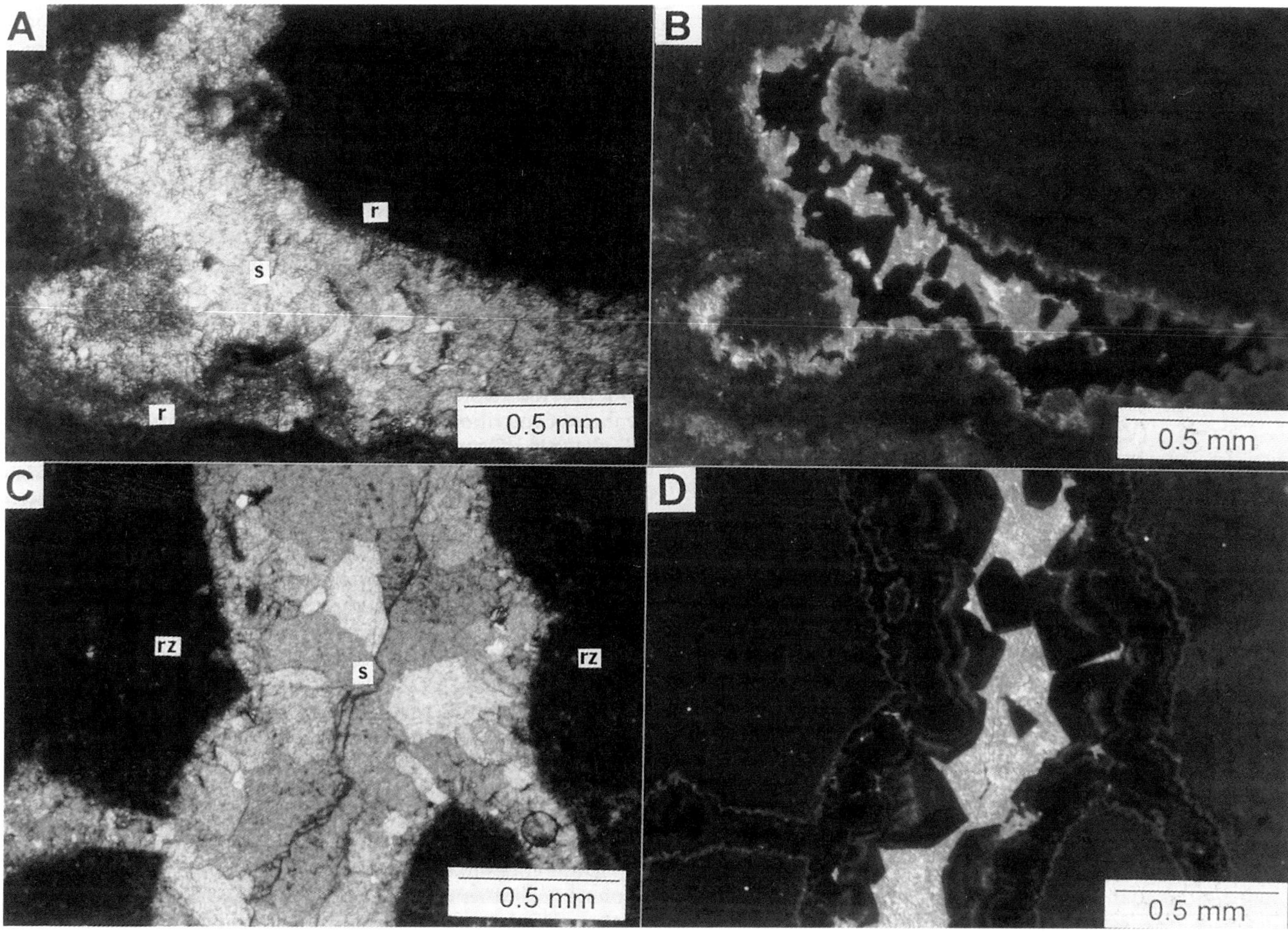


FIG. 6.—Cathodoluminescence (CL) micromorphology of pedogenic carbonate in Triassic paleosols, Spillway locality, Durham subbasin, North Carolina. **A)** Enlargement of area of micritic rhizolith depicted in Figure 5A, showing dense micritic sheath (r) and sparry calcite central pore-filling cement (s). Plane-polarized light. **B)** Same field as in Part A, but viewed under CL. Note that micritic sheath is dull-brown luminescent, whereas sparry calcite pore-filling consists of moderate to dull orange-red luminescent and nonluminescent cement microbands that alternate in tripartite sequence. **C)** Calcite spar cement–filled septarian cracks within micritic crayfish concretion (rz). Plane-polarized light. **D)** Same field as in Part C, but viewed under CL. Note that micrite nodule is dull-brown luminescent, whereas sparry calcite pore-filling consists of earlier nonluminescent to weakly luminescent cement microbands, followed by later bright orange-red luminescent center.

Coast Prairie, in moisture regimes ranging from 60 to 140 cm/yr MAP but also characterized by high seasonal soil-moisture deficits due to high evapotranspiration (Driese et al., 2000; Stiles et al., 1999; Stiles et al., 2001; Mora et al., 2000). Pedogenic carbonate (micrite) nodules are abundant in the modern Vertisols along the entire climosequence transect, although the depth of maximum pedogenic carbonate development is systematically deeper in wetter areas. Pedogenic rhizoliths are generally rare in the modern Vertisols but are more common in the drier part (< 100 cm/yr MAP) of the climosequence transect. Thus, the mere presence of pedogenic carbonate nodules in the Triassic paleosols, by analogy with the modern Vertisols, is not indicative of a particular precipitation regime, although depth to pedogenic carbonate horizon is a rough proxy for (paleo)precipitation in some modern and ancient soils (Retallack, 1994; Royer, 1999; Stiles et al., 2001). Crayfish krotovina, some lined with or cemented by carbonate nodules, are present in the modern Vertisols only where there is at least 100 cm/yr MAP, and are most abundant in the wettest areas (100–140 cm/yr). In the case of the Triassic paleosols reported here, the co-occurrences of several morphological types of pedogenic carbonate that have climatic sensitivity, namely, rhizoliths and crayfish concretions, help constrain the Late Triassic paleoprecipitation estimate in North Carolina to possibly within the range of 60–140 cm/yr, with the slightly older Spillway paleosols representing the wetter paleoclimate endmember and the slightly younger Triangle Brick Quarry the drier endmember. Our paleoclimate interpretations based on the paleosols are thus consistent with existing palynological and macrofloral evidence (Cornet, 1977; Gensel, 1986; Traverse, 1986).

## *Carbonate Micromorphology*

The micromorphology of carbonate in the Late Triassic paleosols is consistent with its interpretation as pedogenic. Carbonate is especially abundant within and below parts of the paleosols

semiarid climate (Agnew and Stubbs, 1972; Berry, 1972). Lake Malawi is presently an open basin, with an outflow via the Shire River at the south end. The lake water has a total dissolved solids (TDS) of 200 mg/l, a pH of 7.3 to 8.5, and an alkalinity between 2.31 and 2.51 meq/l (Beadle, 1981; Gonfiantini et al., 1979). The lake is stably stratified, with a chemocline at a depth near 250 m (Halfman, 1996).

The Quaternary sediments of Lake Malawi have already yielded insight into the paloenvironmental changes of this large lake. Finney and Johnson (1991) studied several piston cores collected from this lake, which span an interval of ~ 10 kyr. They concluded that the lake has undergone changes in its water level and composition during this period. Specifically, an increased abundance of $CaCO_3$ in core intervals dated from 6 to 10 ka points to the existence of a more alkaline lake. Carbon-to-nitrogen (C/N) ratios in the sediments also increase during this time, suggesting an increased input of terrestrial organic matter, which may have resulted from a lower lake level. Finney and Johnson (1991) estimate this level to be ~ 150 m below the present lake surface. The chemocline was also probably depressed a similar amount. For the past 6 ka the lake level has remained around its present level (i.e., overflowing), but the concentration of iron found in oxidized sediments at certain locations, and variations in diatom abundance and speciation, indicate that the depth of the chemocline did not remain as constant. Variations in the type and abundance of clay minerals can help interpret the environmental conditions that may have existed during these episodes.

*Lake Turkana.—*

Lake Turkana, in northern Kenya, is the largest lake in the eastern branch of the African Rift Valley. It has a maximum depth of 125 m in a small basin in the southern extremity of the lake, but it is otherwise quite shallow, with an average depth of 35 m. The basin is hydrologically closed, and the water is moderately saline (TDS = 2,500 mg/l) and alkaline (pH = 9.2; alkalinity = 20 meq/l) (Yuretich, 1979; Yuretich and Cerling, 1983). The climate in the immediate vicinity of the lake is very arid, with an average rainfall around 250 mm per year. Over 90% of the water entering the lake comes from the Omo River, which drains the Ethiopian highlands to the north (Ferguson and Harbott, 1982). Cenozoic volcanic rocks dominate the drainage basin of the river, and the annual rainfall in this area is 1,500 to 2,000 mm (Butzer, 1971). Weathering and erosion of these rocks is the primary source of lake sediments, which are very fine-grained and accumulate at a high rate, as much as 3 mm/year (Halfman et al., 1994; Yuretich, 1979).

Clay minerals dominate the sediments in Lake Turkana. Yuretich and Cerling (1983) postulated a prominent role for them in maintaining the chemical equilibrium of the lake water, especially with respect to Mg. Specifically, they hypothesized that a Mg-rich trioctahedral smectite, such as stevensite, was being produced in the sediments and served as a sink for Mg. Bailey (1986) studied the detailed structure and composition of several samples from the uppermost sediments in the lake. She found that the smectite in the North Basin is an Fe-beidellite, with an average composition of 0.24 Mg per unit cell. The Mg of smectite in the South Basin is considerably greater at 0.55 per unit cell, with a layer charge mostly on the octahedral layer, which makes the mineral montmorillonite. Cerling (1996) studied the interstitial water chemistry of two long cores from the northern and southern parts of the lake. Pore-water profiles show little change in the concentration of dissolved species with depth for the core from the North Basin, but Na, Cl, alkalinity, and Mg all have progressive increases with depth in the southern core. These trends are consistent with the existence of a more saline lake in the deeper South Basin during a very arid climatic interval prior to the time represented by the core. The South Basin core was dated by AMS radiocarbon methods, and spanned the time interval from 1.3 to 5.2 ka (Halfman et al., 1994). Similar trends in chloride profiles for the North Basin and the South Basin were found in cores collected on a previous expedition (Cerling, 1986). The Mg increase with depth is proportionally very large, and Cerling (1996) concluded that stevensite or a similar Mg-rich smectite was dissolving postdepositionally and caused the observed Mg flux. However, a more detailed analysis of the smectite composition in Lake Turkana is needed to confirm this process.

## METHODS

We obtained samples of cores from Lake Malawi, which had been collected in 1986 and 1987 under the auspices of Project PROBE (Johnson et al., 1988). In order to examine the distribution of sedimentary mineralogy in the modern sediments of the lake, we sampled the tops (upper 2 cm) of all available cores. Thirty-six cores from all parts of the lake were sampled in this way (Kalindekafe et al., 1996). From this initial group, we selected twelve cores that were representative of the different bathymetric and sedimentary provinces of the lake (Fig. 1B). Among these were five cores previously studied for carbonates, diatoms, and $^{14}C$ ages by Finney and Johnson (1991). The cores were sampled at intervals of approximately 1 m, with additional samples from layers or laminae exhibiting distinctive colors or features. A total of 101 samples were collected.

Samples of Lake Turkana sediments were taken from two preexisting cores. Core TK is a 12-m-long core collected in 1980 by C. Barton from the North Basin of the lake, and core LT84-2P was collected by Duke University in 1984 from the South Basin (Fig. 1C). Halfman et al. (1994) examined the chronology of the latter core by AMS $^{14}C$ dating, and Cerling (1996) analyzed the pore waters. Cerling (1986) examined the chloride profile in the interstitial water of core TK. Subsamples of each core were taken at approximately 50 cm intervals down the lengths of the cores. In addition, we examined two samples of surface sediments collected by the first author from the central part of the lake (LR73-10T, 12T). These two samples are comparable to those used in a detailed study of the clay minerals of Lake Turkana by Bailey (1986).

Routine processing of all samples proceeded along similar lines. Preliminary drying of the sediment was done in an oven at 60°C. The dried sediment was crushed gently, weighed, and put into centrifuge bottles with distilled water. These were allowed to stand overnight, then they were shaken to break up loose clumps and dispersed using an ultrasonic dismembrator. The preparations were centrifuged to settle the > 2 μm fraction to the bottom of the flask. The cloudy supernatant was transferred to a beaker and allowed to evaporate (at 60°C) until the clays had formed a paste. The paste was smeared on glass slides to produce oriented mounts for X-ray diffraction analysis.

The > 2 μm size fraction in the bottom of the centrifuge bottles was passed through a 63 μm sieve to separate the sand and silt size fractions. These were dried and weighed to calculate the approximate grain-size distribution of the samples. Clay minerals were identified by X-ray diffraction using a Siemens® Type F® goniometer and a Databox® digital upgrade. Diffraction patterns were obtained using Cu Kα radiation, scanning from 2° 2θ to 30° 2θ with a step size of 0.05° and a counting time of 1 s. Samples were run after drying in air, and then after solvating with ethylene glycol. Selected samples were also heated to 550°C or Mg-

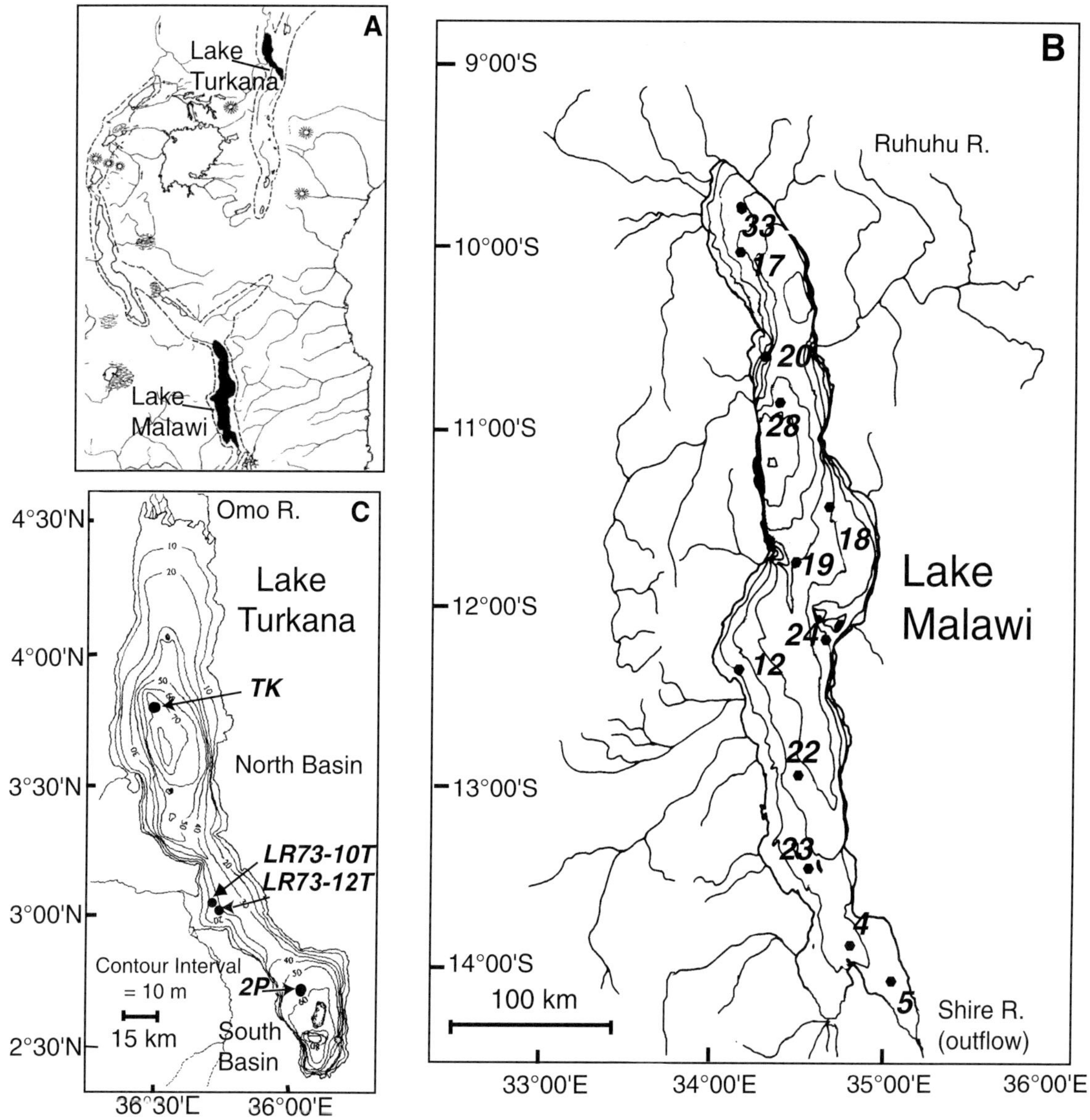


FIG. 1.—**A)** The East African Rift Valleys showing the locations of Lakes Malawi and Turkana. **B)** Lake Malawi, with major rivers, bathymetry, and core locations discussed in text. All cores numbers have the prefix "M86-" and the suffix "P", used in the text. **C)** Lake Turkana, with bathymetry, sample sites, and place names mentioned in text.

saturated and solvated with glycerol to further assist in the identification of the clay minerals (Brown and Brindley, 1980; Hardy and Tucker, 1988). Relative abundance of clay minerals was determined by ratios of diffractogram peak height of the 001 diffraction line of each clay on glycolated samples (Griffin, 1971). Although this does not provide a true estimate of the actual proportion of each clay mineral, it is very reliable for comparing relative changes among the samples. Further details concerning the methods of sample preparation and mineral identification are given in Yuretich et al. (1999).

The < 2 μm preparations of the Lake Malawi long cores were further analyzed by scanning electron microscopy. The smear slides were sputter-coated with carbon, and these were examined with a Zeiss® DSM 960 SEM. Chemical analyses were determined using a Link® eXL® energy-dispersive spectrometer. The counts were normalized to 100% using a feldspar standard, which was powdered and smeared onto a glass slide to duplicate the matrix effects of the clay preparation.

For the sediments from Lake Turkana, we further separated the < 0.1 μm grain size by means of high-speed centrifugation. This fraction was concentrated into a gel, which was spread onto glass slides as a thin film. These slides were used both for mineralogical and chemical characterization by X-ray diffraction and scanning electron microscopy as discussed previously. Several of these very fine-grained Lake Turkana samples were analyzed chemically on a Cameca® SX-50® Electron Microprobe after sputter coating with carbon. The electron beam was defocused to an area of 10 μm to avoid damaging the clay structures, but operating conditions were otherwise standard (15 kV, 15 nA). A range of natural and synthetic silicate standards was used for instrument calibration, and the chemical composition of each sample was based on the average of a transect of 30 measurements across the slide.

## RESULTS

### *Lake Malawi*

Details of the modern clay minerals in Lake Malawi have been summarized previously by Kalindekafe et al. (1996). The clay-size fraction forms between 5 and 50% of the sediments in Lake Malawi. X-ray diffractograms indicate the presence of three major clay minerals in Lake Malawi sediments: smectite, kaolinite, and illite (Fig. 2). The proportions of these clays vary depending upon the location in the lake, but on average, smectite is the most abundant clay mineral, followed by kaolinite and then illite.

The highest concentration of kaolinite in the uppermost sediments occurs in the northern sections of the lake, especially in the region extending from the Ruhuhu Delta across the lake and along the eastern shoreline. Illite abundance is much more equally distributed throughout the lake. Smectite shows a distinctive pattern of abundance, with a maximum in the southern part of the lake and a secondary maximum in the north (Kalindekafe et al., 1996). In the extreme south of the lake (cores 4P and 5P), the X-ray diffractograms show a poor crystallinity of the minerals. Diatoms are very abundant in the sediments here, and the presence of amorphous silica is confirmed by the very high $SiO_2$ in the < 2 μm size fraction (Table 1). In a few localities in the south, however, the sediments consist almost exclusively of very crystalline smectite, which has a high iron content, e.g., sample M86-23P (Fig. 2; Table 1). This is most likely nontronite, which has been reported previously from the sediments of Lake Malawi (Müller and Förstner, 1973; Owen et al., 1996).

*Clay-mineral variations with depth in sediment cores.—*

The local differences in the composition of the uppermost sediments are reflected in the profiles from the cores, but other persistent patterns are also recognized. Most generally, as seen in Fig. 2, the relative abundance of smectite, and often its crystallinity, increases with depth. We focus here on trends in smectite, because this mineral is often the most sensitive to small environmental fluctuations. In the cores from the northern part of the lake (33P, 17P, and 20P), smectite constitutes between 20% and 30% of the clay minerals near the sediment surface (Fig. 3). With depth in all cores, the proportion of smectite increases to 60% to 70%. Cores 17P and 20P exhibit very similar profiles with depth, with the maximum smectite occurring at approximately 200 cm depth. In core 33P, the smectite maximum occurs near 400 cm, perhaps owing its proximity to multiple river deltas in the extreme north end and a correspondingly higher sedimentation rate (Fig. 1). The chemical analyses of core 33P show little vertical variability, with the exception of Si/Al, which exhibits a pronounced maximum at a depth of 300 cm, coincident with a rise in the abundance of kaolinite (Fig.3). The lack of variability in the Fe/Al or Mg/Al ratios suggests that the smectite is of the montmorillonite or beidellite variety. Finney and Johnson (1991) also found that the Fe/Al ratio in core 17P does not vary appreciably with depth.

The central parts of the lake basin, as shown by cores 18P, 12P, 22P, and 24P, exhibit few obvious trends in clay-mineral abundance except for a general increase in smectite with depth (Fig. 4A). All of these cores have radiocarbon dates for at least one horizon (Finney and Johnson, 1991; Finney et al, 1996), and when the smectite data are plotted against $^{14}C$ age of the cores, the covariance among the various sites improves (Fig. 4B). The abundant smectite in core 18P at a depth of 240 cm is different from the "run-of the mill" smectite in the rest of the core (Fig. 5).

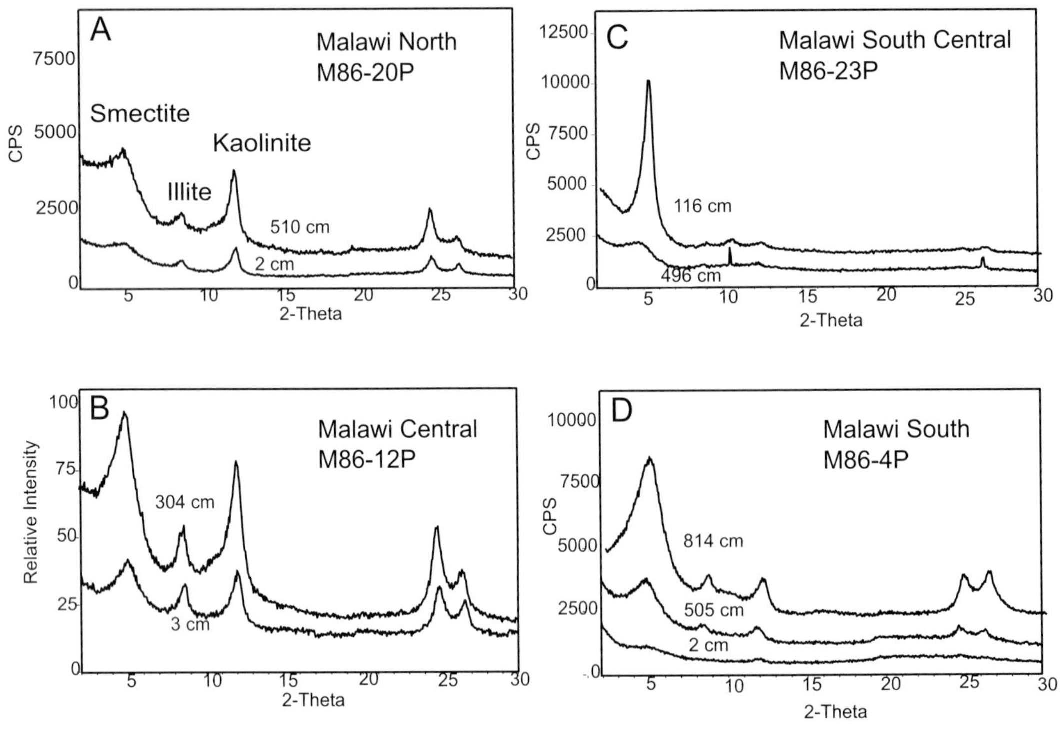


FIG. 2.—Representative X-ray diffractograms of the < 2 μm size fraction of Lake Malawi sediments. All samples have been treated with ethylene glycol. Numbers refer to depth in core. 001 diffraction peaks of smectite, illite, and kaolinite are identified.

TABLE 1.—Chemical composition of < 2 μm size fraction of surface sediments from Lake Malawi core sites as determined from SEM–EDS analyses.

| | 33P | | 18P | | 12P | | 23P | | 5P | | 4P | |
|---|---|---|---|---|---|---|---|---|---|---|---|---|
| $SiO_2$ | 61.82 | ± 0.27 | 67.22 | ± 5.89 | 60.27 | ± 1.98 | 56.98 | ± 1.00 | ***85.38*** | ± 1.43 | ***86.45*** | ± 0.44 |
| $Al_2O_3$ | 27.07 | ± 0.14 | 15.31 | ± 8.71 | 23.23 | ± 2.76 | 13.67 | ± 0.80 | 9.80 | ± 0.85 | 8.94 | ± 0.20 |
| $TiO_2$ | 1.26 | ± 0.08 | 0.80 | ± 0.43 | 1.42 | ± 0.37 | 0.74 | ± 0.13 | 0.55 | ± 0.11 | 0.50 | ± 0.14 |
| MgO | 1.21 | ± 0.07 | 2.54 | ± 0.86 | 2.36 | ± 0.45 | 3.55 | ± 0.22 | 0.38 | ± 0.00 | 0.51 | ± 0.12 |
| FeO | 6.79 | ± 0.21 | 4.53 | ± 3.53 | 9.36 | ± 1.63 | ***24.56*** | ± 1.27 | 2.93 | ± 0.12 | 3.21 | ± 0.14 |
| MnO | 0.20 | ± 0.01 | 0.18 | ± 0.02 | 0.25 | ± 0.04 | 0.26 | ± 0.06 | 0.26 | ± 0.01 | 0.21 | ± 0.04 |
| CaO | 0.86 | ± 0.04 | 3.53 | ± 2.04 | 1.47 | ± 0.79 | 0.17 | ± 0.03 | 1.08 | ± 1.26 | 0.28 | ± 0.03 |
| $Na_2O$ | 0.97 | ± 0.01 | 5.13 | ± 4.13 | 1.22 | ± 0.43 | 0.92 | ± 0.05 | 1.16 | ± 0.01 | 1.04 | ± 0.14 |
| $K_2O$ | 1.33 | ± 0.03 | 1.55 | ± 0.47 | 1.59 | ± 0.11 | 0.46 | ± 0.04 | 0.56 | ± 0.08 | 0.44 | ± 0.07 |
| Total | 101.52 | | 100.81 | | 101.17 | | 101.30 | | 102.10 | | 101.57 | |

It has XRD similarities to the nontronite (Fe-smectite) in the surface sediments of core 23P (Fig. 2). The color of this layer in the core is dusky green (Munsell 5G 3/2), in contrast to the rest of the sediment, which ranges from olive gray (5Y 3/2) to dark yellowish brown (10YR 4/2). EDS analyses of this layer confirm a high iron content (FeO = 19.43%) comparable to that found on the sediment surface in core 23P (Table 1).

Clays in cores from the extreme southern parts of the lake have the most distinctive characteristics. The crystallinity of smectite in cores 4P and 5P is very low, as shown by the characteristically broad and low-intensity X-ray diffraction peaks (Fig. 6). Exceptions are the two lowermost samples from below a distinct unconformity in core 4P, which exhibit excellent intensities for clay minerals. Core 4P and 5P have unusually

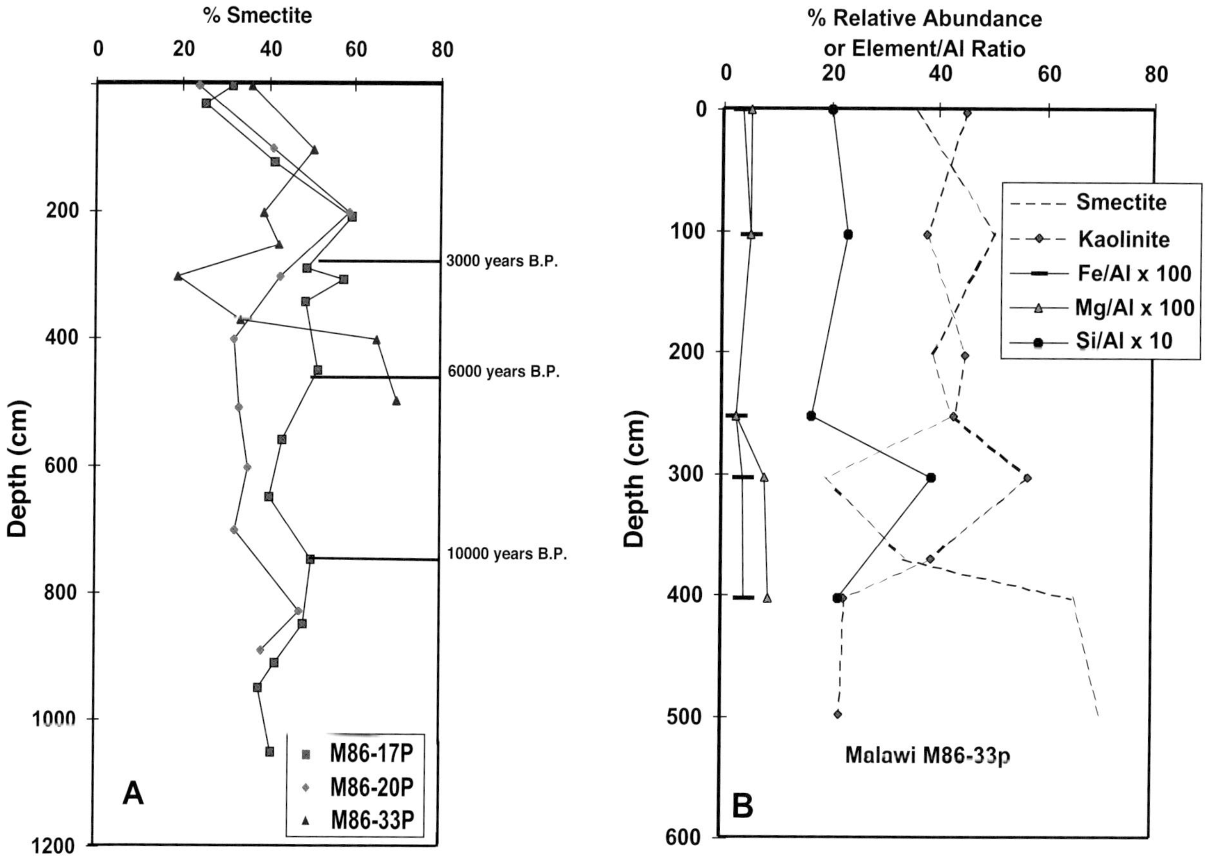


FIG. 3.—Clay mineralogy and chemistry of cores from the northern part of Lake Malawi. **A)** Depth profiles of smectite abundance as determined from X-ray diffraction. Chronology of 17P is based upon $^{14}C$ dates of Finney and Johnson (1991). **B)** Depth profiles of smectite, kaolinite, Fe/Al, Mg/Al, and Si/Al in core 33P.

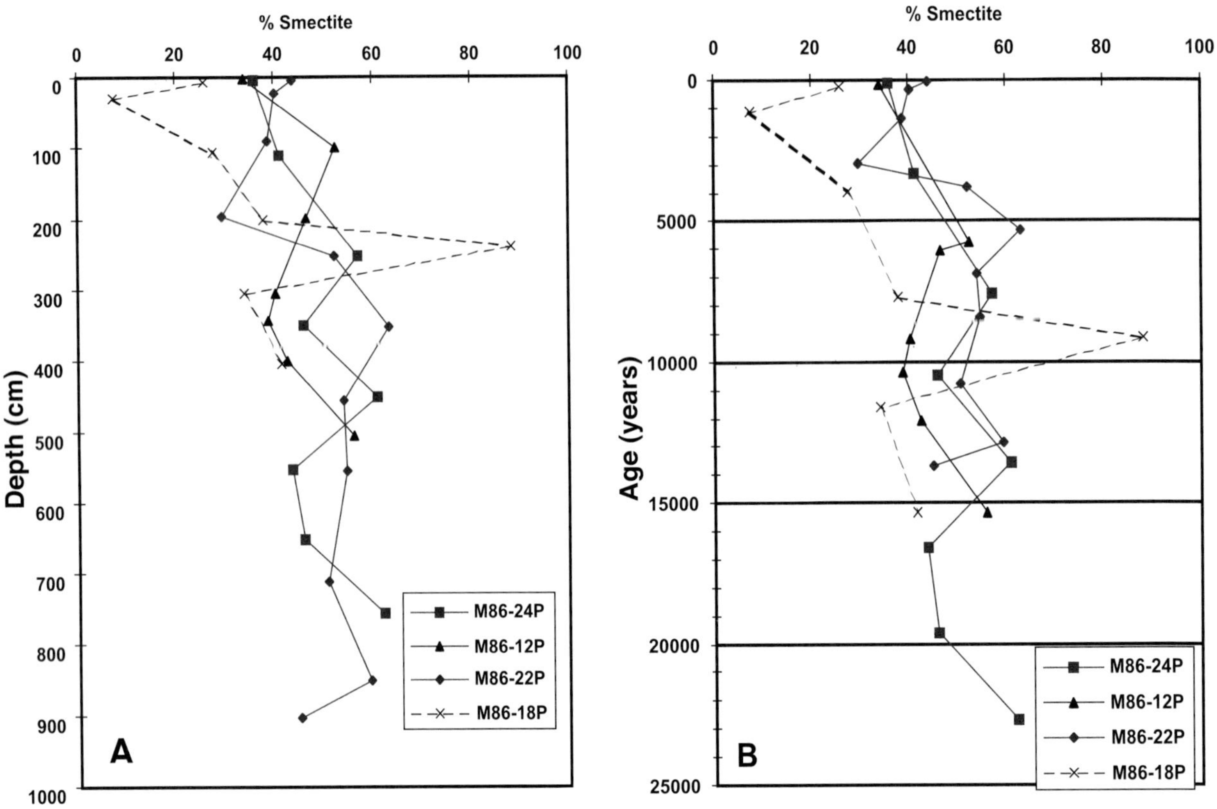


FIG. 4.—Smectite abundance in cores from the central part of Lake Malawi. **A)** Smectite abundance vs. depth. **B)** Smectite abundance vs. age of sediment based upon the chronology of Finney and Johnson (1991).

high amounts of silica, owing to the prevalence of diatoms in sediments at this site. In addition, three samples in the middle of core 5P have a sharp peak at 7.6 Å, which corresponds to the 020 diffraction line for gypsum (or brushite, $CaHPO_4 \cdot 2H_2O$, which is isostructural, but unusual). As discussed previously, core 23P contains an extraordinarily well crystallized smectite, presumably nontronite, on the basis of its XRD pattern and distinctive chemical composition (Fig. 6; Table 1). In this core, the relative abundance of smectite decreases with depth, most notably below 360 cm.

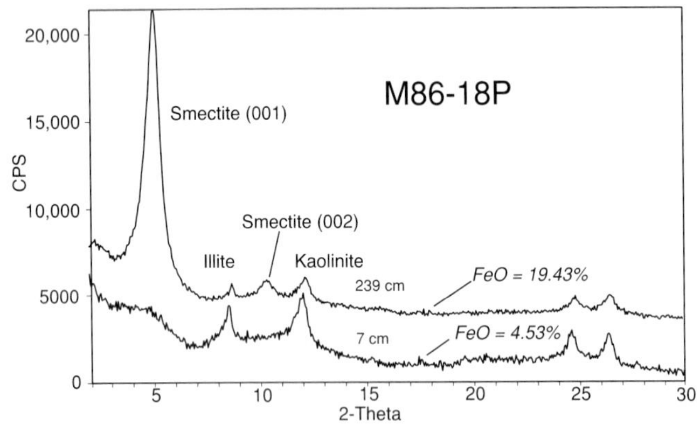


FIG. 5.—Representative X-ray diffractograms of two samples from 18P in central Lake Malawi. The FeO content of the sample from 239 cm (approximately 9.1 ka) is much higher than that of the sample from 7 cm (0.25 ka).

### *Lake Turkana*

Smectite is the most important clay mineral at each site, as it is throughout much of the lake (cf. Yuretich, 1979). This is in stark contrast to the very heterogeneous distribution of clay minerals observed in Lake Malawi. The < 0.1 μm fraction of cores TK from the North Basin and LT84-2P from the South Basin contains 95–100% smectite, as estimated from the X-ray diffractograms (Fig. 7). Illite, kaolinite, and other coarser-grained detrital species are removed during the process of high-speed centrifugation. Although the X-ray diffraction patterns at the two sites are very similar, there are differences in the *d* spacing of the glycolated 001 smectite peak. In the North Basin, this value remains reasonably constant throughout the core profile, whereas in the southern core there is an apparent decrease in the *d* spacing from 19.2 Å to 18 Å with depth (Fig. 8).

Because the < 0.1 μm size fraction is nearly pure smectite, the results of the electron microprobe analyses of this size fraction from core LT84-2P were recast into structural formulae for this mineral (Table 2). Several interesting generalities emerge from this analysis. First, the smectite is dioctahedral, with about half of the octahedral cations consisting of Al. Secondly, Fe is an important octahedral cation. Although all iron in this case has been cast as Fe(III), the same pattern occurs if the calculations are done assuming all ferrous iron. The samples from 550 to 116 cm depth contain a higher Fe content than those from shallower depths. Mg is also an important octahedral cation. Thirdly, layer charge is on the high end for smectite, although there is a decrease in the two deepest samples (the unusually high layer charge for the sample from 215-cm depth raises questions about the validity of this data point). Lastly, the distribution of the layer charge also shows differences with depth. The samples from 85 cm to 215 cm are montmorillonite, with the layer charge dominantly in the octahedral sheet. The layer charge at 550 cm, 873 cm, and 1165 cm is almost evenly divided between the octahedral and tetrahedral layers, such that these minerals should be classified as beidellite. Even so, they contrast with the structural formulae for smectite from the northern parts of the lake (LR73-10T, 12T), which are more aluminous, the layer charge is mostly on the tetrahedral layer, and there is much lower Mg substitution in the octahedral layer (Table 2). The trioctahedral smectite stevensite is not found in any of the sediments analyzed. The chemical analyses also reveal interesting trends within core LT84-2P, most noticeably the shift in Mg and Fe concentration with depth (Fig. 9).

## DISCUSSION

### *Lake Malawi*

*Geographic variations in sedimentary clay minerals.—*

The most important factors controlling the distribution of clay minerals in Lake Malawi are the geology of the catchment area and climate (Kalindekafe et al.,1996). The greater rainfall in the northern part of the lake catchment produces red, thoroughly leached soils and high runoff into the lake (Agnew and Stubbs,

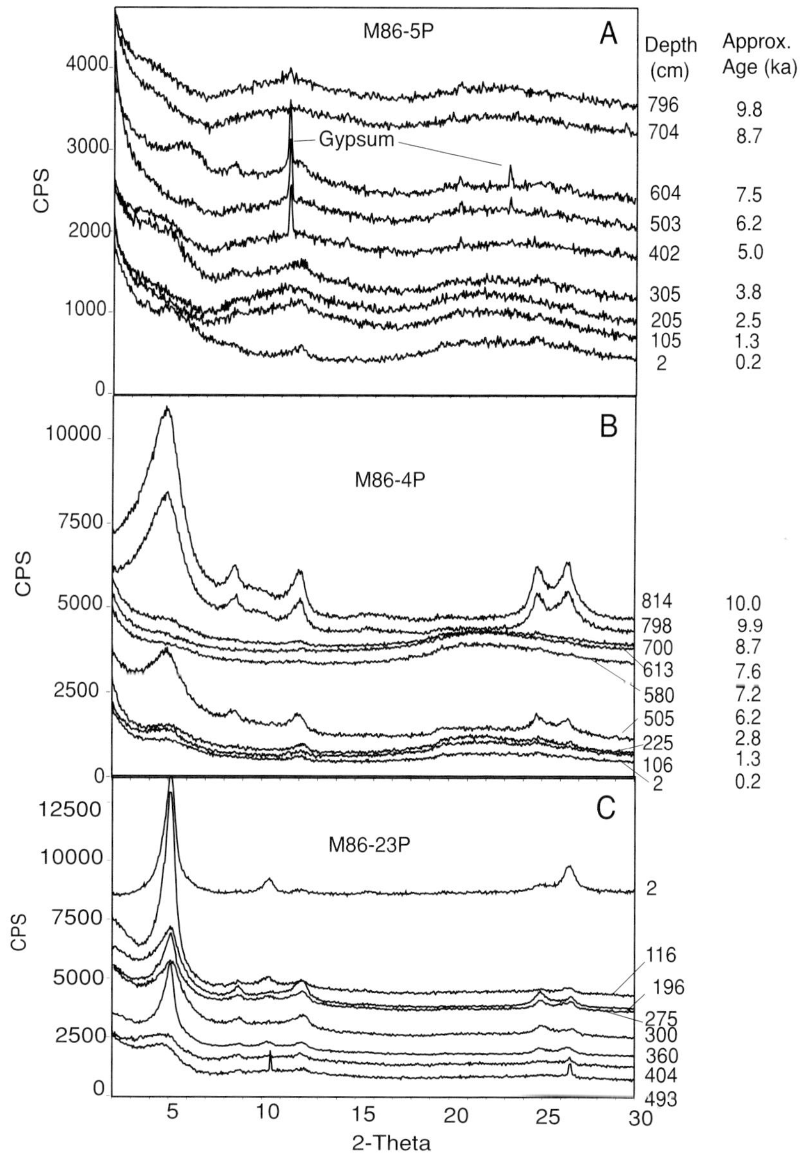


FIG. 6.—X-ray diffractograms of the < 2 µm size fraction of cores from southern Lake Malawi. All samples have been treated with ethylene glycol. Poor crystallinity of clays in M86-4P and M86-5P prevent semiquantitative estimates of clay-mineral abundance. Depth of sample in core is given at right of figure, along with approximate age calculated from data of Finney and Johnson (1991).

1972; Berry, 1972). Kaolinite is an expected weathering product under these conditions. The feldspathic sandstones of the Karroo and Cretaceous sedimentary rocks are also likely to decompose to kaolinitic clays. The climate at the southern end of the lake is more arid, which favors the formation of smectite-rich soils (Chamley, 1989), and a higher proportion of soil smectite will survive transport to the lake. Poorly crystalline varieties of smectite in the southernmost part of the lake may have been transported the longest distance, thereby degrading the crystal structure of the minerals. The cores from the extreme southern part of the lake (cores 4P and 5P) consist largely of diatom ooze (Johnson et al., 1988). Some of the poorly crystalline smectite in these cores may be diagenetic, formed by the reaction of warm pore-water solutions with the diatoms, because hydrothermal activity may be a source of silica to the lake (Johnson and N'ganga, 1990).

*Variations in clay mineralogy with depth in sediments.—*

Most of the long cores preserve the source-area signature found in the uppermost sediments throughout the length of the core. That is, cores in the southern part of the basin generally have a greater relative abundance of smectite than cores from the more northern parts of the lake, which have proportionally greater kaolinite. Nevertheless, there are vertical variations in the clay mineral assemblages in several of the cores, which may correspond to changes in weathering rates and the distribution of clays within the lake related to climatic fluctuations. In their studies of several cores from Lake Malawi, Finney and Johnson (1991) and Finney et al. (1996) used the presence of carbonate layers and the relative abundance of different diatom genera to reconstruct the history of lake level. On the basis of these criteria and $^{14}C$ dates, Finney and Johnson (1991) proposed that Lake Malawi was a closed basin with a much lower lake level during the period 10 to 6 ka, when extensive carbonate sediment is found in the lake cores. This is accompanied by a high ratio of the number of diatoms belonging to the genus *Stephanodiscus* to those belonging to *Melosira*. A high *Stephanodiscus* to *Melosira* ratio, but without concomitant carbonate precipitation, also distinguishes the period from 3 to 2.5 ka. Finney and Johnson (1991) interpreted this period as a time of high productivity related to upwelling water that tapped a nutrient-rich and relatively shallow chemocline. The clay minerals of Lake Malawi respond to these episodes of change in lake level and climate, but in different ways in different regions of the lake. In the central part of the lake, smectite abundance is higher during the presumed closed-basin episode, between 40 and 60% compared with 25 to 40% in modern sediments, and relatively stable during this interval (Fig. 4). This is consistent with our current understanding of the conditions favorable to the formation and preservation of smectite, namely alkaline conditions in a semiarid environment.

However, the clays in the northern cores have more kaolinite than smectite during the closed-basin episode from 10 to 6 ka,

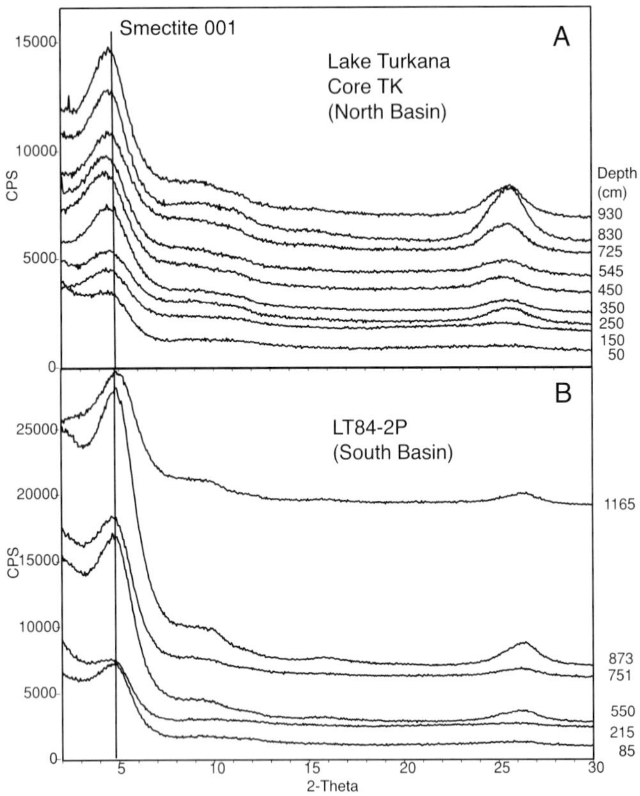


FIG. 7.—X-ray diffractograms of < 0.1 μm size fraction in cores from Lake Turkana. **A)** TK (North Basin). B**)** LT84-2P (South Basin). All samples have been treated with ethylene glycol. Size fraction consists of > 95% smectite at all depths in cores.

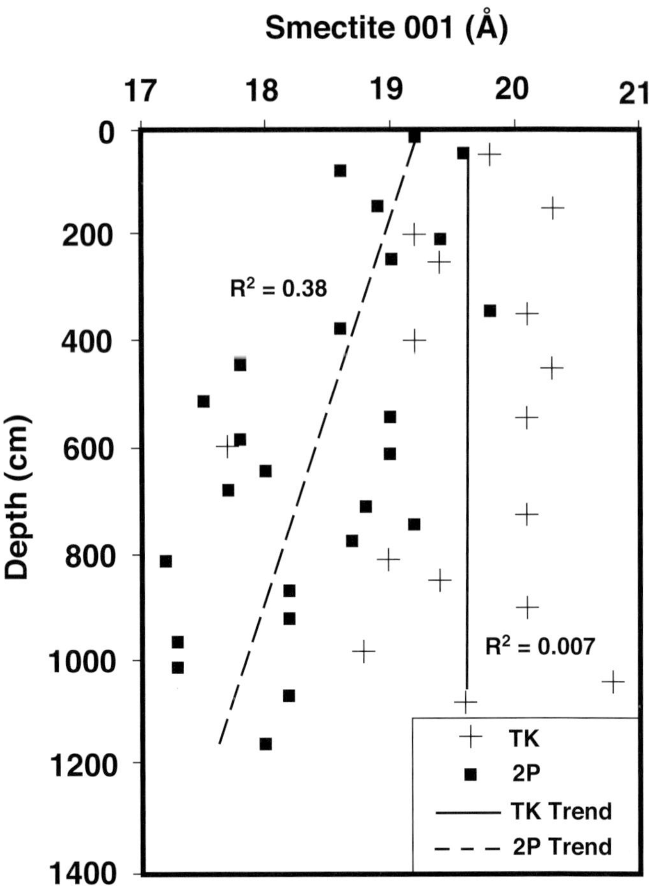


FIG. 8.—Vertical trends in the expandability of smectite clay minerals in the Lake Turkana cores TK and LT84-2P. Horizontal axis gives the measured *d* value of smectite 001 peak on glycolated samples.

TABLE 2.—Structural formulae of smectite in the < 0.1 μm size fraction of sediments from Lake Turkana.

| | Core LT84– 2P | | | | | | | LR 73 | |
|---|---|---|---|---|---|---|---|---|---|
| **Depth (cm)** | **85** | **150** | **215** | **550** | **751** | **873** | **1165** | **10T** | **12T** |
| **Si** | 3.92 | 3.71 | 4.00 | 3.73 | 3.72 | 3.70 | 3.75 | 3.39 | 3.45 |
| **Al (IV)** | 0.08 | 0.29 | 0.00 | 0.27 | 0.28 | 0.30 | 0.25 | 0.61 | 0.55 |
| ***Total Tetrahedral*** | *4.00* | *4.00* | *4.00* | *4.00* | *4.00* | *4.00* | *4.00* | *4.00* | *4.00* |
| **Al (VI)** | 1.05 | 1.00 | 0.91 | 1.00 | 0.99 | 1.01 | 1.07 | 1.21 | 1.28 |
| **Fe (as $Fe^{3+}$)** | 0.35 | 0.44 | 0.32 | 0.51 | 0.50 | 0.56 | 0.53 | 0.63 | 0.62 |
| **Mn** | 0.01 | 0.01 | 0.00 | 0.01 | 0.01 | 0.01 | 0.00 | 0.01 | 0.01 |
| **Mg** | 0.57 | 0.62 | 0.49 | 0.58 | 0.53 | 0.49 | 0.45 | 0.23 | 0.22 |
| ***Total Octahedral*** | *1.98* | *2.07* | *1.72* | *2.10* | *2.03* | *2.07* | *2.05* | *2.08* | *2.13* |
| **Ca** | 0.21 | 0.15 | 0.25 | 0.09 | 0.12 | 0.10 | 0.12 | 0.07 | 0.04 |
| **Na** | 0.43 | 0.33 | 0.69 | 0.24 | 0.39 | 0.27 | 0.15 | 0.21 | 0.20 |
| **K** | 0.11 | 0.11 | 0.10 | 0.13 | 0.12 | 0.13 | 0.14 | 0.12 | 0.12 |
| ***Total Exchangeable*** | *0.76* | *0.59* | *1.04* | *0.46* | *0.63* | *0.50* | *0.41* | *0.40* | *0.36* |
| ***Tetrahedral Charge*** | *– 0.08* | *– 0.29* | *0.00* | *– 0.27* | *– 0.28* | *– 0.30* | *– 0.25* | *– 0.61* | *– 0.55* |
| ***Octahedral Charge*** | *– 0.64* | *– 0.42* | *– 1.33* | *– 0.29* | *– 0.45* | *– 0.29* | *– 0.24* | *0.00* | *– 0.07* |
| ***Total Layer Charge*** | *– 0.72* | *– 0.71* | *– 1.33* | *– 0.56* | *– 0.73* | *– 0.59* | *– 0.49* | *– 0.61* | *– 0.62* |

although this based solely on the chronology available from core 17P (Fig. 3). Finney and Johnson (1991) postulate a lowering of the lake level by 100 m to 150 m during this time. As previously mentioned, the northern part of the lake currently receives more kaolinite-rich sediments from the surrounding source areas. Kaolinite is coarser grained than smectite and is normally deposited in proximal parts of deltas (Gibbs, 1977; Kalindekafe et al., 1996). During the lowstand, these northern sites (cores 33P, 17P, and 20P) were in shallower water and closer to the sediment source. The finer-grained smectite clays were more easily transported into deeper parts of the basin, giving rise to an apparent enrichment in kaolinite at the location of the core. Consequently, the clay-mineral assemblage here is influenced as much or more by the water depth and proximity to sediment influx as by changes in the climate of the surrounding watershed.

In the southernmost part of the lake, the poor crystallinity of the clays in cores 4P and 5P does not seem to respond to the postulated changes in lake level. However, closer inspection reveals that there are some subtle but significant changes in these sediments. In M86-5P, the three samples at 402, 503, and 604 cm correspond to an age range of ca. 5 to 7.5 ka, if the sedimentation rates are comparable to those calculated for M86-4P by Finney and Johnson (1991). The intensity of the smectite peaks at these depths is noticeably more pronounced than in other samples (Fig. 6). Moreover, the gypsum peaks at 7.56 and 4.27 Å may indicate an evaporative episode in the lake. Judging from available data for the concentration of $Ca^{2+}$ (0.484 x $10^{-3}$ m/l) and $SO_4^{2-}$ (0.065 x $10^{-3}$ m/l) in modern Lake Malawi water (Wüest et al., 1996), the ion activity product (IAP) of gypsum is $10^{-7.56}$. Because the Ksp for gypsum @ 25°C is $10^{-4.56}$, Lake Malawi would have to be concentrated by about 3 orders of magnitude during the lowstand to achieve gypsum saturation. It is unlikely that the lake water became so highly concentrated; nevertheless, the gypsum may have been produced diagenetically from elevated $Ca^{2+}$ and $SO_4^{2-}$ in the pore fluids. In core M86-4P, the smectite abundance increases noticeably in the sample at 505 cm, which has an approximate age of 6.2 ka. This could have resulted from more smectite production in the sediment source area or greater preservation in a more alkaline lake environment. In either case, these mineralogical data support a lowstand during the 6 to 10 ka period.

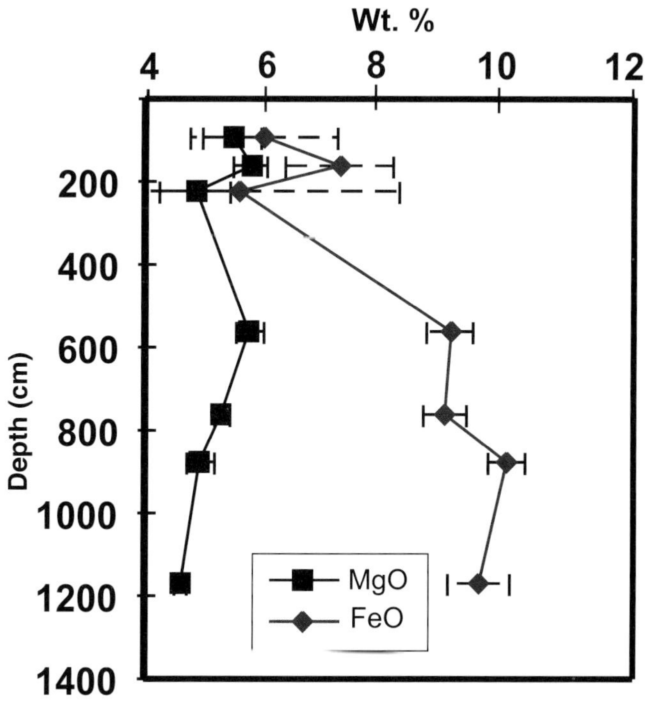


FIG. 9.—Vertical changes in the MgO and FeO of the < 0.1 μm size fraction in core LT84-2P as determined by the electron microprobe.

Finney and Johnson (1991) also propose a period of high productivity during the period from 3 to 2.5 ka. This interval is characterized by an abundance of *Stephanodiscus* (over *Melosira*), but without any noticeable carbonate precipitation. If increased rainfall and elevated weathering rates are part of the reason for this productivity increase, then a decrease in smectite (or an increase in kaolinite) is an expected result. The cores from the central part of the lake indicate that smectite decreased between ca. 3 and 1 ka (Fig. 4B). The data are better from cores 22P and18P in this regard. Very subdued smectite peaks are found in the southern cores (4P, 5P) during this time (Fig. 6). In particular, samples from core 5P at depths of 105 cm (~ 1.3 ka) and 205 cm (~ 2.5 ka) show almost no distinguishable smectite. High diatom productivity during this period would also diminish the importance of clays in the sediment. This pattern does not hold for the cores from the northern part of the lake (Fig. 3A). Smectite is at a maximum in core 17P near 3 ka, perhaps caused by complications in sediment dispersal during the rising lake level as kaolinite was retained closer to shore in newly flooded areas. Moreover, because a strong climatic gradient exists along the length of the lake today, it is also reasonable to postulate that runoff may have been greater from streams entering the central lake during this earlier period, bringing with it more smectite from these geologically distinct sources. However, because the dating of these northern cores is not well constrained, we will need to wait for a more robust chronology before reaching a conclusion.

The geochemistry of iron in Lake Malawi, and its interplay with the clay mineralogy, shows itself in the Fe-smectite or nontronite identified in two of the cores, 23P and 18P. In core 23P, the nontronite occurs mostly near the top of the core, at depths of 2 cm and 116 cm (Fig. 6C). This site is presently located at a depth of 156 m below the present lake level, and falls within the depth range of 80 to 160 m, which Owen et al. (1996) found to be the preferred accumulation zone of ferromanganese nodules. These authors proposed that the Fe–Mn nodules form just above the anoxic–oxic boundary, or chemocline, where pore fluids enriched in $Fe^{2+}$ and $Mn^{2+}$ mix with oxidizing waters and precipitate the oxide phases. If sufficient silica or a detrital smectite precursor is also present, the oxidized $Fe^{3+}$ may substitute for $Al^{3+}$ in a recrystallizing smectite, giving rise to a mineral of nontronite composition. This is also compatible with the mechanism of nontronite formation in Lake Malawi originally proposed by Müller and Förstner (1973). Although no age data are available for core 23P, the XRD patterns indicate that nontronite abundance was reduced in the past, perhaps as a lower lake level removed $Fe^{2+}$-charged pore waters from this site.

A tantalizing affirmation of this idea comes from the mineralogical data of core 18P. The very crystalline smectite at a depth of 240 cm is enriched in Fe (Fig. 5) and the sediment is dark green in color, all of which point toward a nontronite mineralogy. The extrapolated age of this horizon is ~ 9.1 ka, placing it at a time when the lake level was lower by ~ 150 m (Finney and Johnson, 1991). This means that the core site at that time was in a water depth of ~ 138 m, or just above the paleochemocline proposed by Finney and Johnson (1991). Accordingly, the presence of these Fe-smectite minerals indicates that the chemocline was 50 to 100 m deeper than the core site at the time the Fe-smectite minerals were formed.

### *Lake Turkana*

Cerling (1996) noted differing trends in the pore water between cores from the North and South Basins in Lake Turkana. In the North Basin, there is little change in ion concentrations with depth, except for a gradual increase in sodium by about 10% at the bottom of the core (12 m depth). This may be related to ion exchange processes removing the Na from the clay-mineral surfaces. In contrast, all the ions in the southern core (LT84-2P) show dramatic increases, with a proportionally large increase in the Mg ion. The rate of increase of Mg in the pore waters with depth rises noticeably below 5.5 m in core 2P. Some of this increase is presumably related to diffusion of ions from more concentrated waters at depth, a function of the near-desiccation of Lake Turkana during a time period earlier than the 5 ka range of this core. However, the Mg change could also be coupled to changes in the smectite composition as indicated by the XRD data and microprobe analyses. Expansion of glycolated smectite is related to layer charge of the clay-mineral structure: those with low layer charge generally show a greater expansion with glycol than high-layer-charge varieties (Moore and Reynolds, 1997). Therefore, the shift in expandability down core 2P may result from progressive substitution in the clay lattice that alters the layer charge. Chemical analysis of the $< 0.1$ μm fraction using the electron microprobe confirms a decrease of Mg in the clays with depth in the core, but the layer charge decreases with depth, which is the opposite of the expected direction (Fig. 9; Table 2). Perhaps the degree of expansion of the smectite is being influenced by something other than total layer charge. Two other trends emerge from the data: $Fe^{3+}$ replaces $Mg^{2+}$ in the octahedral sites with depth, thereby reducing the octahedral layer charge (Table 2). At the same time, $Al^{3+}$ substitutes for $Si^{4+}$ and raises the charge on the tetrahedral layers. Compositionally and mineralogically, the clays from the deeper parts of core 2P approach the Fe-rich, low-Mg beidellite found commonly in the northern parts of the lake (Bailey, 1996).

Such trends could represent postdepositional modifications of originally Mg-rich clay. However, evidence from terraces around Lake Turkana shows that the lake level was very high (and water less alkaline) from 10 to 4 ka. The basin became hydrologically closed subsequently, and, over the long term, the level has been dropping (Johnson, 1996). As the lake water became more saline and alkaline over time, the uptake of Mg into the smectite structure became increasingly favored. If this is true, the Mg uptake occurs dominantly in the smectite of the southern end of Lake Turkana. Perhaps the lower sedimentation rate in this part of the lake, coupled with a longer settling time for suspended fine-grained material, accounts for the localization of the process.

The process of Mg removal into clay minerals in Lake Turkana is much more complicated than originally envisioned by Yuretich and Cerling (1983) and Cerling (1996). Despite numerous analytical forays, we have found no traces of a trioctahedral Mg-smectite, such as stevensite. It is possible that such a mineral formed prior to 10 ka when the lake was smaller and more alkaline than at present. These deposits would be restricted to a small area in the deeper South Basin, which remained submerged during an arid episode of much lower lake level (Fig. 10A). During the subsequent open-basin period from 10 to 4 ka, these possible stevensite clays were buried under a blanket of detrital Fe-smectite, most likely with a source from the volcanic terrain to the north of the lake (Fig. 10B). After the lake basin closed again at ca. 4 ka and the alkalinity of the water mass increased, $Mg^{2+}$ removal from the water commenced anew (Fig. 10C). The process occurs more vigorously in the southern part of the lake away from the dominant fluvial input and where the sedimentation rate is slower, permitting more contact time between smectite particles and lake water. As a result, the smectite here is more Mg-rich with a layer-charge distribution of montmorillonite.

If this scenario is true, then the increasing ion concentrations in pore waters with depth in core LT84-2P found by Cerling (1996) results from the upward diffusion of more saline waters contained within sediments below the bottom of the core (Fig. 10C). The lower Mg content of the smectite in the bottom of the core reflects the composition of the clays at the time of deposition during the fresh-water phase of Lake Turkana. Longer cores that penetrate deposits older than ~ 10 ka are needed to test this idea definitively.

## CONCLUSIONS

The clay minerals in the sediments of Lakes Malawi and Turkana can help establish paleoenvironmental conditions, especially when used in conjunction with other sedimentological and geochemical data. During a closed-basin episode ca. 6 to10 ka, the sediments in the central and southern parts of Lake Malawi contained elevated smectite, which is consistent with the onset of more arid climatic conditions. The crystallinity of the smectite, as determined from X-ray diffractograms, is also greater during this time. In the northern parts of the lake, kaolinite was more abundant during this time period. Kaolinite is often indicative of a warm and wet climate, but kaolinite is also abundant in the source rocks of the northern lake drainage. Because this mineral also accumulates closer to shore, the lowered lake level during the closed-basin episode resulted in relative kaolinite enrichment in the northern sediments. During a high-productivity episode in the lake approximately 3 to 2.5 ka, kaolinite content increases in the central region and decreases in the northern lake, presumably in response to more kaolinite production in the source areas during a wetter climate. In the north, this kaolinite increase is not observed in the studied cores, perhaps because the relatively coarse-grained kaolinite is being retained near shore as the lake expands.

Nontronite, dioctahedral Fe-smectite of likely diagenetic origin, is found in the recent sediments in some parts of Lake Malawi ~ 50 to 100 m above the modern chemocline. It forms from the oxidation of $Fe^{2+}$ and reaction with silica in anoxic pore waters coming into contact with oxygenated lake water. Nontronite occurrence in older sediment layers of Lake Malawi cores provides a good indicator that the paleodepth of the chemocline was ~ 50 to 100 m deeper than the core site depth at the time of nontronite formation.

The smectite-dominated clays in Lake Turkana are involved in early diagenetic reactions with the lake water. The Mg content of smectite increases toward the southern end of the lake, away from the primary detrital source of the Omo River. This Mg is apparently being incorporated into the octahedral sites of the smectite during transport and sedimentation. Such reactions can account successfully for the rapid removal of Mg from the lake water that has been documented previously. However, the clay minerals in a 12 m core of ca. ~ 5 kyr duration from the South Basin of Lake Turkana reflect the original hydrological and chemical conditions of the lake, with only minimal postdepositional alteration. Fe-smectite characterizes the clays deposited during a time of a higher lake level and a hydrologically open basin. Increasing substitution of Mg for Fe in smectite towards the top of the core documents the growing salinity and alkalinity of the lake during the past 4 kyr. High $Mg^{2+}$ and other ions in the pore waters from this core most likely originate from the upward diffusion of more saline waters of an earlier closed-basin episode (> 10 ka). In both Lake Turkana and Lake Malawi the climate and lake-water chemistry at the time of deposition are the principal influences on clay mineralogy. Clay minerals, therefore, are important tools for reconstructing past environments in rift-valley lakes.

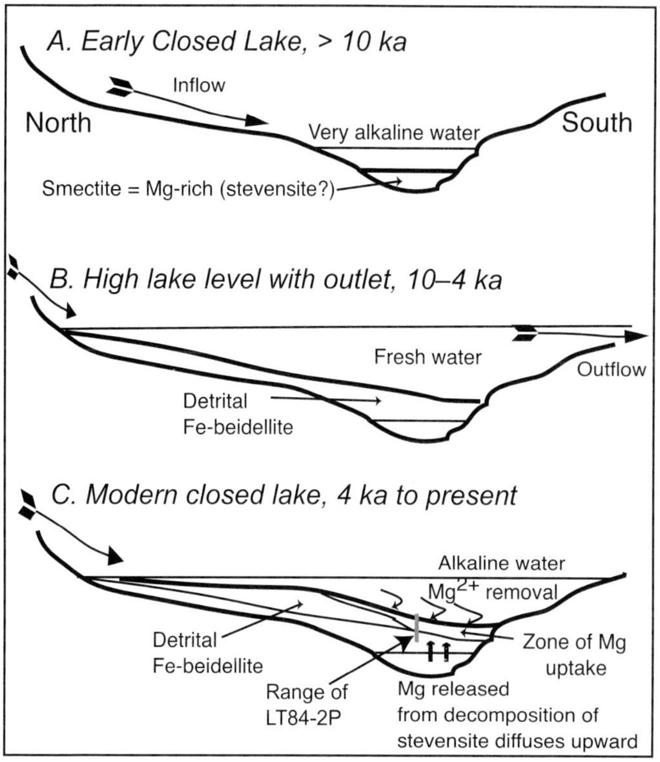


FIG. 10.—Cartoon summarizing the proposed relationships among lake level, hydrology, water chemistry, and clay composition in the sediments of Lake Turkana during the past 10 kyr.

## ACKNOWLEDGMENTS

Several people helped with various aspects of the investigation. Tom Johnson gave us access to the cores from Lake Malawi and provided samples from the Lake Turkana core. Lori Barg assisted with the X-ray diffraction analyses, and Christopher Dorey carried out the chemical analyses of the Lake Turkana sediments. Mike Jercinovich guided us through the operation of the electron microprobe, and Peter Crowley let us use the SEM facilities at Amherst College. Discussions with many colleagues helped refine our ideas. Most importantly, the thoughtful and constructive comments of an anonymous SEPM reviewer were essential for correcting errors and rectifying inconsistencies in the original draft. She deserves a medal for her efforts! The research presented here was supported by the National Science Foundation, through award EAR-901752.

## REFERENCES

AGNEW, S., AND STUBBS, M., 1972, Malawi in Maps: London, University of London Press, 143 p.

BAILEY, J.A., 1986, Structural Chemistry and Possible Origin of Illite and Smectite in Lake Turkana, Kenya: Unpublished M.S. Thesis, University of Massachusetts, Amherst, 105 p.

BEADLE, L.C., 1981, The Inland Waters of Tropical Africa, 2nd Edition: New York, Longman Inc., 475 p.

BERRY, L., 1972, Tanzania in Maps: Graphic Perspectives of a Developing Country: New York, Africana Publishing Corp., 172 p.

BROWN, G., AND BRINDLEY, G.W., 1980, X-RAY diffraction procedures for clay mineral identification, *in* Brindley, G.W., and Brown, G., eds., Crystal Structures of Clay Minerals and Their X-ray Identification: London, The Mineralogical Society, p. 305–359.

BUTZER, K.W., 1971, Recent History of an Ethiopian Delta: University of Chicago, Department of Geography, Research Paper 136, 184 p.

CERLING, T.E., 1986, A mass-balance approach to basin sedimentation: constraints on the recent history of the Turkana basin: Palaeogeography, Palaeoclimatology, Palaeoecology, v. 64, p. 63–86.

CERLING, T.E., 1996, Pore water chemistry an alkaline lake: Lake Turkana, Kenya, *in* Johnson, T.C., and Odada, E., eds., The Limnology, Climatology and Paleoclimatology of the East African Lakes: Amsterdam, Gordon & Breach, p. 225–240.

CHAMLEY, H., 1989, Clay Sedimentology: Berlin, Springer-Verlag, 623 p.

FERGUSON, A.J., AND HARBOTT, B.J., 1982, Geographical, physical and chemical aspects of Lake Turkana, *in* Hopson, A.J., ed., Lake Turkana: A Report of the Findings of the Lake Turkana Project, 1972–1975: London, Overseas Development Administration, p. 1–107.

FINNEY, B.P., AND JOHNSON, T.C., 1991, Sedimentation in Lake Malawi (East Africa) during the past 10,000 years: a continuous paleoclimatic record from the southern tropics: Palaeogeography, Palaeoclimatology, Palaeoecology, v. 85, p. 351–366.

FINNEY, B.P., SCHOLZ, C.A., JOHNSON, T.C., AND TRUMBORE, S., 1996, Late Quaternary lake-level changes of Lake Malawi, *in* Johnson, T.C., and Odada, E., eds., The Limnology, Climatology and Paleoclimatology of the East African Lakes: Amsterdam, Gordon & Breach, p. 495–508.

GIBBS, R.J., 1977, Clay mineral segregation in the marine environment: Journal of Sedimentary Petrology, v. 47, p. 237–243.

GONFIANTINI, R., ZUPPI, G.M., ECCLES, D.H., AND FERRO, W., 1979, Isotope investigations of Lake Malawi, *in* Proceedings on the Application of Nuclear Techniques to the Study of Lake Dynamics: Vienna, International Atomic Energy Agency Symposium, p. 195–205.

GRIFFIN, G.M., 1971, Interpretation of X-ray diffraction data, *in* Carver, R.J., ed., Procedures in Sedimentary Petrology: New York, Wiley, p. 541–570

HALFMAN, J.D., 1996, CTD-Transmissometer profiles from Lakes Malawi and Turkana, *in* Johnson, T.C., and Odada, E., eds., The Limnology, Climatology and Paleoclimatology of the East African Lakes: Amsterdam, Gordon & Breach, p. 169–182.

HALFMAN, J.D., JOHNSON, T.C., AND FINNEY, B.P., 1994, New AMS dates, stratigraphic correlations and decadeal climatic cycles for the past 4 ka at Lake Turkana, Kenya: Palaeogeography, Palaeoclimatology, Palaeoecology, v. 111, p. 83–98.

HARDY, R., AND TUCKER, M., 1988, X-ray powder diffraction of sediments, *in* Tucker, M., ed., Techniques in Sedimentology: London, Blackwell, p. 191–228.

KALINDEKAFE, L.S.N., DOLOZI, M.B., AND YURETICH, R., 1996, Distribution and origin of clay minerals in the sediments of Lake Malawi, *in* Johnson, T.C., and Odada, E., eds., The Limnology, Climatology and Paleoclimatology of the East African Lakes: Amsterdam, Gordon & Breach, p. 443–460.

JOHNSON, T.C., 1996, Sedimentary processes and signals of past climatic change in the large lakes of the East African Rift Valley, *in* Johnson, T.C., and Odada, E., eds., The Limnology, Climatology and Paleoclimatology of the East African Lakes: Amsterdam, Gordon & Breach, p. 367–412.

JOHNSON, T.C., AND N'GANGA, P., 1990, Reflections on a rift lake, *in* Katz, B.J., ed., Lacustrine Basin Exploration—Case Studies and Modern Analogs: American Association of Petroleum Geologists, Memoir 50, p. 113–135.

JOHNSON, T.C., DAVIS, T.W., HALFMAN, B.M., AND VAUGHAN, N.D., 1988, Sediment core descriptions: MALAWI 86, Lake Malawi, East Africa: Project PROBE Technical Report, Beaufort, North Carolina, Duke University Marine Laboratory.

MENKING, K.M., 1997, Climatic signals in clay mineralogy and grain-size variations in Owens Lake core OL-92, southeast California, *in* Smith, G.I., and Bischoff, J.L., eds., An 800,000-Year Paleoclimatic Record from Core OL-92, Owens Lake, Southeast California: Geological Society of America, Special Paper 317, p. 25–36.

MOORE, D.M., AND REYNOLDS, R.C., JR., 1997, X-Ray Diffraction and the Identification and Analysis of Clay Minerals, 2nd Edition: Oxford, U.K., Oxford University Press, 378 p.

MÜLLER, G., AND FÖRSTNER, U., 1973, Recent iron ore formation in Lake Malawi, Africa: Mineralium Deposita, v. 8, p. 278–290

OWEN, R.B., RENAUT, R.W., AND WILLIAMS, T.M., 1996, Characteristics of laminated ferromanganese nodules from Lake Malawi, Central Africa, *in* Johnson, T.C., and Odada, E., eds., The Limnology, Climatology and Paleoclimatology of the East African Lakes: Amsterdam, Gordon & Breach, p. 461–474.

WÜEST, A., PIEPKE, G., AND HALFMAN, J.D., 1996, Combined effects of dissolved solids and temperature on the density stratification of Lake Malawi, *in* Johnson, T.C., and Odada, E., eds., The Limnology, Climatology and Paleoclimatology of the East African Lakes: Amsterdam, Gordon & Breach, p. 183–202.

YURETICH, R.F., 1979, Modern sediments and sedimentary processes in Lake Rudolf (Lake Turkana), Eastern Rift Valley, Kenya: Sedimentology, v. 26, p. 313–331.

YURETICH, R.F., AND CERLING, T.E., 1983, Hydrogeochemistry of Lake Turkana, Kenya: Mass balance and mineral reactions in an alkaline lake: Geochimica et Cosmochimica Acta, v. 47, p. 1099–1109.

YURETICH, R., MELLES, M., SARATA, B., AND GROBE, H., 1999, Clay minerals in the sediments of Lake Baikal: a useful climate proxy: Journal of Sedimentary Research, v. 69, p. 588–596.

# SEDIMENTOLOGICAL CHARACTERISTICS AND ORIGINS OF DIATOMACEOUS DEPOSITS IN THE EAST AFRICAN RIFT SYSTEM

RICHARD B. OWEN

*Department of Geography, Hong Kong Baptist University, Kowloon Tong, Hong Kong*

ABSTRACT: Lacustrine diatomaceous deposits are widespread in the East African Rift System, occurring in Miocene to Holocene sedimentary units. Problems in their classification can be clarified with a common set of well-defined terms, including: diatomite, siliciclastic diatomite, subdiatomite, diatomitic, and diatom-poor. Additional descriptors refer to lamination, fragmentation, and dissolution. Diatom taphonomy and assemblage alteration reflect sedimentation, interface, and analysis controls, including: dissolution, settling rates, fragmentation, and floral mixing. The spatial and temporal distribution of East African diatomaceous deposits, especially in the Kenya and Malawi Rifts, suggest specific models for the development of a variety of diatomaceous facies. The models are based on factors that control the availability of diatom nutrients, including: fluvial inputs, water depth, seasonality of rainfall, wind regimes and their influence on upwelling and overturn, and turbidity currents. In turn, these factors reflect tectonic and geomorphic setting, drainage basin geology and resultant water chemistry, and climatic setting.

## INTRODUCTION

Diatomaceous deposits have been variously referred to as siliceous or diatomaceous earths, bog deposits, kieselguhr, tripolite, or moler (Dolley, 1999). Where diatoms dominate, the sediment is referred to as diatomite, though definition of this term varies widely. Diatomaceous sediments may be composed of monospecific floras or include several hundred different taxa. Preservation states vary widely, reflecting the degree of fragmentation and dissolution. Pure diatomite is gray to white, chalk-like, and very fine grained, with a very low density (floating on water until saturated). The high porosity and grain size of diatomite reflects the form and size of individual diatom cell walls, which consist of two interlocking, radially or bilaterally symmetrical, box-like frustules perforated by numerous pores. Diatom size varies considerably, ranging from about 2 to 2000 μm, though most individuals are between 5 and 200 μm.

Diatomaceous sediments accumulate in fresh, brackish, and marine environments, and are characterized by the presence of the opaline silica ($SiO_2 \cdot nH_2O$) found in diatom frustules. These may comprise nearly an entire deposit, or form only a small fraction, variously mixed with clastic particles, authigenic minerals, or evaporites. Diatomaceous sediments are frequently interbedded with siliciclastics, biochemical sediments, or organic materials. The largest deposits of pure diatomite are marine (Dolley, 1999), with lacustrine examples occurring over smaller areas (1–50 $km^2$). Most lacustrine diatomites are found in volcanic terranes associated with fault basins, such as those of the Basin and Range Province of the western United States (Krebs et al., 1987), the strike-slip grabens of northern Thailand (Owen and Utha-Aroon, 1999), or the rift valleys of East Africa (Owen, 1981).

Marine diatomaceous sediments date back to the Early Cretaceous (Harwood and Gersonde, 1990), with the oldest lacustrine examples being reported from lower Eocene rocks (Dolley, 1999). However, diatomaceous deposits are common only from the Miocene onwards. Krebs et al. (1987) report Miocene (ca. 16 Ma) examples from western North America. Fourtanier and Gasse (1988) note the appearance of diatoms in Miocene (ca. 11 Ma) lacustrine sediments from northern Africa, with Owen and Renaut (2000) describing their presence in the Miocene Ngorora Formation (ca. 12 Ma) of the central Kenya Rift.

Diatoms have been used for biostratigraphic research, and they have been particularly valuable as ecological indicators in East African paleoenvironmental studies. This paper addresses the nature and classification of diatomaceous deposits and attempts to explain some of the major controls involved in diatom deposition and preservation. In doing so, emphasis is placed on ancient (Kenya Rift) and modern (Malawi Rift) examples from East Africa.

## SEDIMENT CLASSIFICATION

The terminology used in describing diatomaceous sediments is often confused and misleading. Both volcanic ash and white-colored clay have been mistakenly called "diatomite", and sediments have been described as diatomite, or diatomaceous, or diatom-rich with little or no definition of the actual diatom content. Commercial approaches to classification rely on geochemical and/or physical properties and are concerned only with the definition and identification of relatively pure diatomite. These are sediments with a high silica content (> 85%), though deposits with lower silica are also referred to as diatomite. Other elements are present in small percentages, with aluminum and iron tending to be the next most common after silicon. In general, diatomite has a low specific gravity and high porosity (up to 80–90%). A few selected examples of geochemical and physical analyses of diatomite and siliciclastic diatomite are given in Table 1.

Geochemical classifications are important commercially, but they have not been widely applied to rift sediments, even though such studies can indicate changes in diatom abundance. For example, Finney and Johnson (1991) made use of Si/Al ratios as proxies for diatom content in Lake Malawi sediments. Most diatom investigations are based on changes in the percentage of individual species in an assemblage. In some cases, research has attempted to combine these data with quantitative estimates of diatom abundance, expressed as diatoms per gram of sediment (e.g., Gasse, 1974). However, because diatom sizes may vary a hundredfold, similar counts can represent quite different sediment types.

Figure 1 shows a simple classification, based on point-counting percentages, as observed in smear slides (Owen and Utha-Aroon, 1999). Three components (diatoms, clay, and silt/sand size siliciclastics) that are common in rift basins are used as end members in Figure 1A, and a fourth (volcanic ash) is added to the classification in Figure 1B. The scheme is based principally on the content of diatoms. The term diatomaceous is currently used to describe any sediment that contains diatoms. Given its long and

Sedimentation in Continental Rifts
SEPM Special Publication No. 73, Copyright © 2002
SEPM (Society for Sedimentary Geology), ISBN 1-56576-082-4, p. 233–246.

TABLE 1.—Major properties of deposits referred to as diatomite.

| Property | 1 | 2 | 3 | 4 |
|---|---|---|---|---|
| $SiO_2$ (%) | 75–80 | 84.5–91.2 | 89.70–91.20 | 82.94–82.97 |
| $Al_2O_3$ (%) | 10–12 | 0.62–3.72 | 0.62–3.72 | 2.89–3.59 |
| $Fe_2O_3$ (%) | < 5 | 0.20–1.86 | 0.20–1.09 | 0.97–1.13 |
| CaO (%) | — | 0.19–3.00 | 0.19–3.00 | 0.48–1.17 |
| Specific gravity | 2.2–2.25 | 2.0 | 2.0 | — |
| density ($g/cm^3$) | 0.5–0.65 | 0.32–0.64 | — | — |

Data from: 1. Inglethorpe et al. (1997); 2. Breese (1994); 3. Harben and Kuzvart (1997); 4. Wang et al. (1997)

often confused history, it should be retained with only a very loose meaning for sediment containing > 10% diatoms by volume. In contrast, other terms should be defined more rigorously. It is recommended that diatomite be used only for sediments composed almost entirely (> 90% by volume) of diatoms. As diatom content declines the sediment can be described as: siliciclastic–tuffaceous diatomite (70–90% diatoms); subdiatomite (50–70%); diatomitic (10–50%); and diatom-poor (< 10%). The relative abundance of clay, silt, sand, and ash determines their order of usage, as shown in Figure 1. For simplicity, mixed clay and silt can also be described as siliciclastic where the clays are detrital. For example, a sediment comprising 12% volcanic ash, 28% silicates (13% silt, 15% clay), and 60% diatoms would plot in area (2) of Figure 1B and be referred to as a tuffaceous, silty clayey (or siliciclastic), subdiatomite. Where clays are authigenic, the occurrence of clay should be noted. Other components (microfossils, carbonates, organics etc.) can be included as qualifiers if they form a significant part of the sediment.

The above classification takes no account of diatom fragmentation or dissolution, which are important aspects of diatomaceous sediments that offer clues as to the genesis of a deposit. It is suggested that suitable qualifiers indicating these aspects be added to descriptions. For fragmentation, terms might include: intact (> 90% unbroken), mostly intact (50–90% unbroken), mostly fragmented (10–50% unbroken), and fragmented (< 10% unbroken). Silica loss is often harder to quantify because weakly silicified diatoms might be completely dissolved, leaving no trace of their presence in a sediment. However, where researchers are confident about the degree of dissolution descriptive phrases such as uncorroded (> 90% undissolved), mostly uncorroded (50–90% undissolved), mostly corroded (10–50% undissolved), and strongly corroded (< 10% undissolved) may be applicable.

Diatomaceous deposits can also be divided into homogeneous and laminated varieties. The percentage of diatoms in the former are very variable, ranging from diatom-poor examples to diatomite. Laminated sediments often consist of diatomite or siliciclastic diatomite laminae that alternate with diatomitic mud. The diatomite laminae tend to be monospecific or dominated by just two or three taxa. In contrast, the diatomitic mud frequently

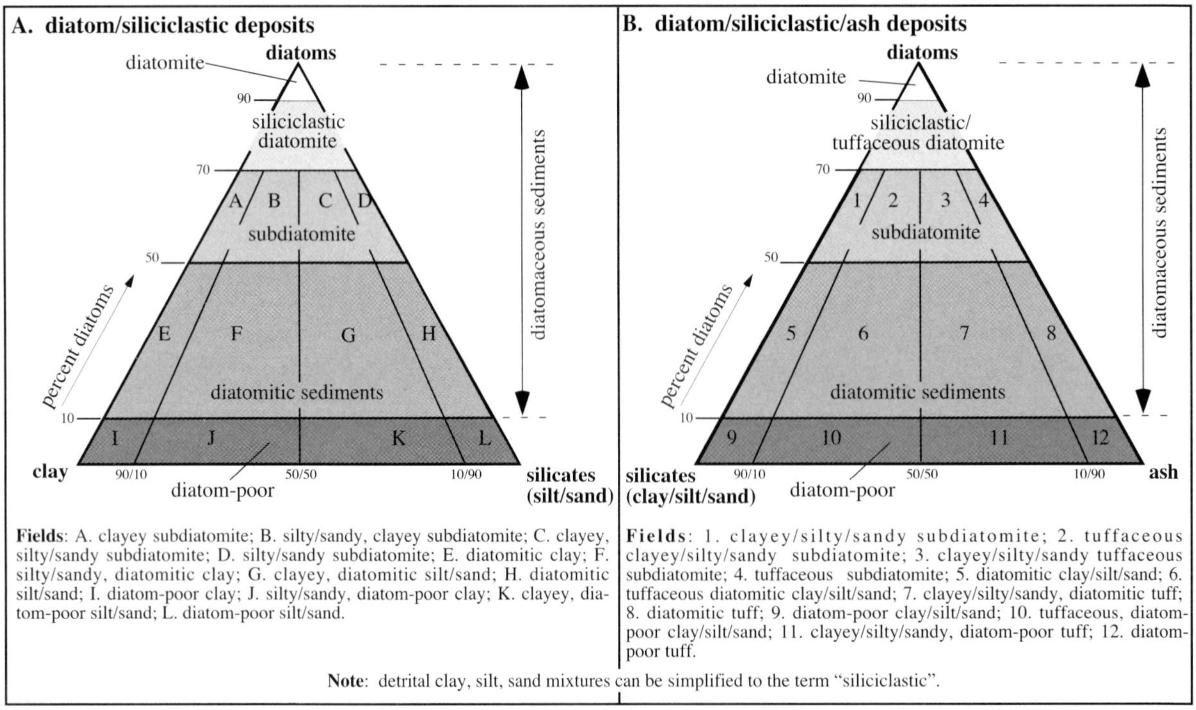


FIG. 1.— Classification of diatomaceous sediment. See text for discussion.

contains more varied assemblages. Where the two lamina types are of a similar thickness, and on a millimeter scale, the deposit should be referred to as "highly laminated". If the diatomitic mud is > 3x thicker than the diatomite laminae, the deposit is better referred to as "laminated". In some cases, isolated thin diatomite layers are present in otherwise homogeneous deposits, in which case, the term "intermittently laminated" can be applied.

## DIATOM TAPHONOMY

In East Africa, diatom paleoenvironmental research (Gasse, 1974; Haberyan, 1987; Owen and Crossley, 1992) has been based on comparisons with modern assemblages (Gasse, 1986, 1987; Gasse, et al., 1995; Haberyan, 1990, 1992; Holdship, 1976; Kilham and Kilham, 1975). Problems arise where no modern analogue is available or where taphonomic processes have altered the original floral composition (Gasse et al., 1997). The final recorded diatom assemblage develops in response to three major sets of controls, in addition to the habitat controls that determine the original flora (Fig. 2). These include: sedimentation controls, sediment–water interface controls, and analysis controls.

### *Sedimentation Controls and Alteration of Diatom Assemblages*

*Dissolution.—*

Diatoms are composed of opaline silica (opal-A), an x-ray amorphous substance with a high water content and porosity (Tribble et al., 1995). Susceptibility to dissolution varies between diatom taxa, because of differences in size, thickness, and ornamentation, which determines the surface area. Silica dissolution is first apparent on weakly silicified parts of diatoms (Barker, 1992; Flower, 1993). The first areas to disappear are fine silica membranes that project into the pores of many taxa (Fig. 3D, E, I). Subsequently, the pores expand, due to further dissolution (Fig. 3D, F). Figure 3A–C shows three examples of *Stephanodiscus* that illustrate successive stages in the expansion of pores and the relatively slower loss of heavily silicified ribs. In some cases, diatoms show both dissolution and precipitation of silica (Fig. 3K, L). At fossil hot springs near Lake Bogoria, Kenya (Renaut and Owen, 1988), diatoms may have acted as nucleation sites, fusing with surrounding opaline silica (Fig. 3M, N).

Differential dissolution can directly result in significant changes in species composition as a living assemblage becomes a death assemblage (Schrader, 1971). This is a particular problem in highly saline lakes (Barker et al., 1990), where floras may be significantly altered or totally lost. The rate of dissolution of diatom silica is related to several factors. Walther and Helgeson (1977) note that the solubility of all silica polymorphs increases with temperature and pressure. Volosov et al. (1972) report that dissolution increases with pH above 9, an observation largely confirmed by preservation states in saline–alkaline lakes in East Africa (Barker et al., 1994). Flower (1993) suggests that alkali metal carbonates (indirectly due to hydrolysis and pH effects) correlate with the diatom dissolution sequence, in the order $Na_2CO_3 > MgCO_3 > CaCO_3$, and that this is reflected in the alteration of diatoms into minerals such as magadiite and zeolites in Lake Magadi (Barker et al., 1990).

In shallow water bodies and marshes the fate of organic matter may also be important for diatom preservation. In perennial marshes organic acids keep pH low. In shallow lakes high microfloral productivity and respiration may elevate dissolved $CO_2$ levels (Liutkus, 2000), which in turn depress pH and enhance diatom preservation. However, pure diatomite will rarely develop due to dilution by moderate to high clastic inputs and/or the accumulation of organic matter.

Hecky et al. (1996) indicate that the dissolution of diatom silica is controlled by a hydration reaction, and that when silica in solution is low, diatom frustules will dissolve. This has major implications for patterns of diatom loss in rift lakes. For example, in the deep stratified rift lakes of East Africa, diatoms often bloom seasonally and extract silica from the epilimnion. As the bloom develops, silica concentrations drop, and upon death the diatoms tend to be dissolved from the water column. As diatom skeletons sink into deeper layers with higher silica concentrations, dissolution rates slow down, but it is only after

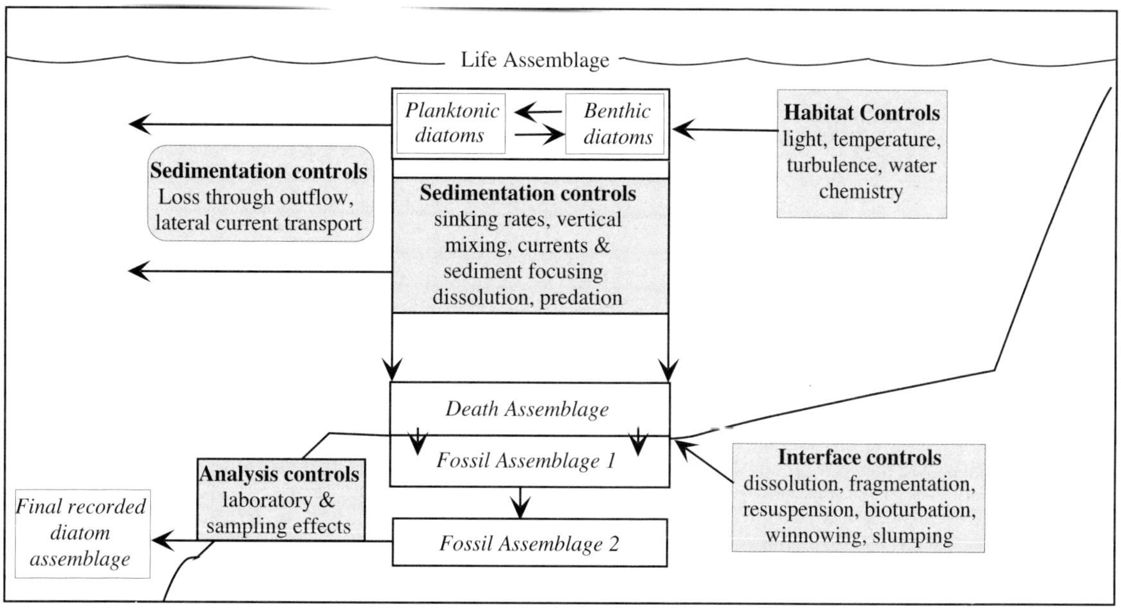


FIG. 2.—Schematic diagram showing the major controls that result in a final recorded diatom assemblage.

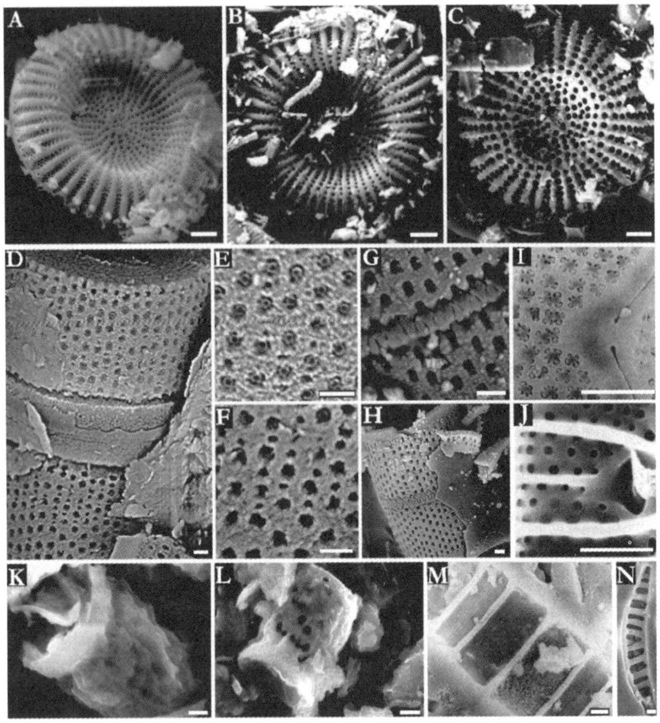


FIG. 3.—Diatom dissolution effects. **A–C)** *Stephanodiscus* showing gradual loss of silica. **D)** *Aulacoseira* showing loss of silica in the lower part and the first stages of fragmentation. **E)** detail of the upper part of Part D showing preserved silica membranes within the pore spaces. **F)** detail of the lower part of Part D showing loss of silica membranes. **G–H)** *Aulacoseira* showing membrane loss, but otherwise good preservation. **I)** *Epithemia* showing excellent preservation and intact membranes. **J)** Internal view showing good preservation of heavily silicified ribs. **K–L)** *Aulacoseira* showing both silica dissolution and reprecipitation. Note the irregular surface and enlarged pore spaces. **M)** *Aulacoseira* showing fusion with opaline silica spring deposits. **N)** *Epithemia* showing similar fusion of diatom and opaline silica. Scale bars are 2 μm long.

burial that silica concentrations, in pore water, are likely to become fully saturated and dissolution stop. The rate and manner of sinking is therefore important for diatom preservation or loss.

### *Sinking.—*

The vertical movement of planktonic diatoms downwards to a lake floor depends on the balance between sinking rates and vertical water movements. Experimentally determined sinking rates for marine diatoms (Table 2) in still water suggest that larger diatoms sink faster. Furthermore, the development of colonies (such as those of the common freshwater genus *Aulacoseira*) also tends to increase sinking rates (Smayda, 1970). On the other hand, fragmentation and/or dissolution can be expected to reduce the velocity of sinking. Natural water bodies are rarely totally calm, with wind-induced turbulence being common. Smayda (1971) notes, for example, that a minimum wind speed of 1 m $s^{-1}$ can produce downwelling and upwelling speeds in Langmuir convection cells of 1 cm $s^{-1}$, well in excess of diatom sinking rates. Diatoms should therefore remain in suspension, unless there are prolonged calm conditions, or they pass into deeper, less turbulent water. Loss of diatoms from the wind-mixed zone of lakes therefore necessitates accelerated sinking.

Several possible mechanisms for accelerated sinking have been suggested (Smayda, 1970, 1971), including: density inversion, aggregation, downwelling, and fecal pellet descent. Density inversion occurs when the density of a micro-layer of water and particulate organisms exceeds that of the water below (Bradley, 1965, 1969). As a result, a vertical density current may develop.

Aggregates of diatoms may form through entanglement, or because of embedding in mucilage (Grimm et al., 1997). Alldredge and Gotschalk (1989) observed marine diatoms aggregating within 24 hours of blooming, forming centimeter-size flocs of *Chaetoceros* that sank at 117 m ± 56 m $day^{-1}$, two orders of magnitude faster than individuals.

Rates of downwelling are highly variable but clearly have the potential to transport diatoms to lower water layers or to a lake floor. However, downwelling may result only in sedimentation locally, because diatoms carried down in a circulation cell are just as likely to be transported up again (Smayda, 1971). In contrast, the downward movement of density currents from lake margins is a one-way transport mechanism with the potential to deliver littoral floras to deeper water.

Encasement in fecal pellets is probably an important means of accelerating the downward sinking of diatoms. Schrader (1971) reports that the 50 μm skeleton of the marine diatom *Thalassiosira baltica* sinks at 10 m $day^{-1}$, but that a fecal pellet of 100 μm size will sink at a rate of 100 m $day^{-1}$. He also notes that membranes encasing pellets tend to preserve diatom silica from dissolution. Evidence for the importance of pellets in lacustrine deposits comes from Haberyan (1985), who reports that copepod fecal pellets account for over 40% of diatom deposition in Lake Tanganyika.

### *Assemblage Mixing and Fragmentation.—*

Mixing of diatom floras can occur both temporally and spatially, resulting in distortions of living assemblages. Vertical mixing (Fig. 4A) occurs when diatoms growing at different levels in stratified water sink and form combined floras on the lake bottom. Extraneous diatoms may also be introduced directly to a lake floor by hyperpycnal density currents, or via hypopycnal overflows and intraflows followed by pelagic settling (Fig. 4B). In addition, diatoms may be introduced by mixing of waters of similar density near river mouths (Fig. 4C). In some cases, different taxa may bloom in a seasonal succession (Fig. 4D). For example, *Aulacoseira* blooms often deplete silica in water and are replaced by diatoms, such as *Stephanodiscus*, that require less silica for their growth. After settling to the lake floor even slight bioturbation or reworking result in a mixed assemblage.

Fragmentation is common where diatoms are laid down in relatively high-energy settings (Round, 1964), and preferentially affects long slender diatoms, such as *Synedra* and *Nitzschia*, bringing about selective changes in diatom assemblages. Grazing is another cause of fragmentation, often followed by sedimentation within fecal pellets. Haberyan (1985) notes that this occurs in Lake Tanganyika and that selective grazing may be causing partial changes in diatom floras reaching the lake floor. Fragmentation may also result from current-induced transportation, perhaps related to sediment focusing (Hilton, 1985). A further breakage mechanism involves erosion and redeposition of preexisting diatomaceous sediments, particularly during stages of low lake level. Good examples of eroding sediments occur around Lake Turkana (Kenya), where early Holocene diatomaceous deposits

TABLE 2.—Sinking rates for diatoms of varying size.

| Marine centric taxa | diameter (μm) | sinking rate m $day^{-1}$ |
|---|---|---|
| *Cyclotella nana* | 6.7 | 0.16 |
| *Coscinodiscus* sp. | 43 | 1.9 |
| *Thalassiosira baltica* | 50 | 10 |
| *Coscinodiscus wailesii* | 140 | 8.0 |
| *Coscinodiscus concinnus* | 300 | 60 |
| *Ethmodiscus rex* | 1500 | 510 |
| Marine Pennate diatoms | | |
| *Nitzschia seriata* | 43–48 | 0.35–0.50 |
| *Thalassionema nitzschioides* | 47.6–50.0 | 0.35–0.78 |

are being rapidly eroded and transported by ephemeral river systems to the modern lake.

### *Interface and Diagenetic Controls on Death Assemblages*

Bioturbation at the sediment–water interface can result in the mixing of floras deposited at different times and under different conditions. The degree of mixing depends partly on sedimentation rates. Estimates suggest these range from 1.8 mm $yr^{-1}$ for Holocene deposits below Lake Manyara, Tanzania (Holdship, 1976) to 4 mm $yr^{-1}$ for Holocene diatomitic silts at Lake Turkana, Kenya (Owen, 1981). Laminated diatomite and diatomitic silts in northern Lake Malawi are accumulating at about 1 mm $yr^{-1}$ (Johnson and Ng'gang'a, 1990), with homogeneous diatomite forming in southern Lake Malawi at about 0.8–1.3 mm $yr^{-1}$, depending on the dating assumptions used (Owen and Crossley, 1992).

The rate of deposition is also important for diatom preservation because diatoms may be repeatedly ingested and broken by epifauna and burrowing infauna. Where oxygen is available, surficial sediment may be strongly bioturbated, to depths of 20 cm or more. Rapid sedimentation tends to remove diatoms from this zone quickly and favor preservation. In contrast, slow deposition may result in greater exposure to resuspension and fragmentation (Anderson and Battarbee, 1994).

Flower (1993) reports a further breakage mechanism. He notes that 50% of the large diatom *Campylodiscus clypeus*, from a shallow saline Moroccan lake, were well preserved in wet sediment samples, but that after laboratory drying fragmentation exceeded 99%. This suggests that breakage can be caused by volume changes related to drying. The same diatom also occurs in shallow-water sediments in parts of the Pleistocene Olorgesailie Formation (southern Kenya Rift), where it is nearly always broken. In this case, the fragmentation may be due to alternate wetting and drying during deposition, or perhaps result from postdepositional dewatering of the sediment.

Dissolution near the sediment–water interface continues as long as pore water is undersaturated with regard to silica (Hecky et al., 1996). Where sedimentation rates are slow, and diatoms remain close to the sediment surface for prolonged periods, exchange of pore and lake water results in greater dissolution. In contrast, if sedimentation rates are rapid and diatoms are quickly buried, pore water is likely to become saturated more quickly, thus inhibiting further dissolution.

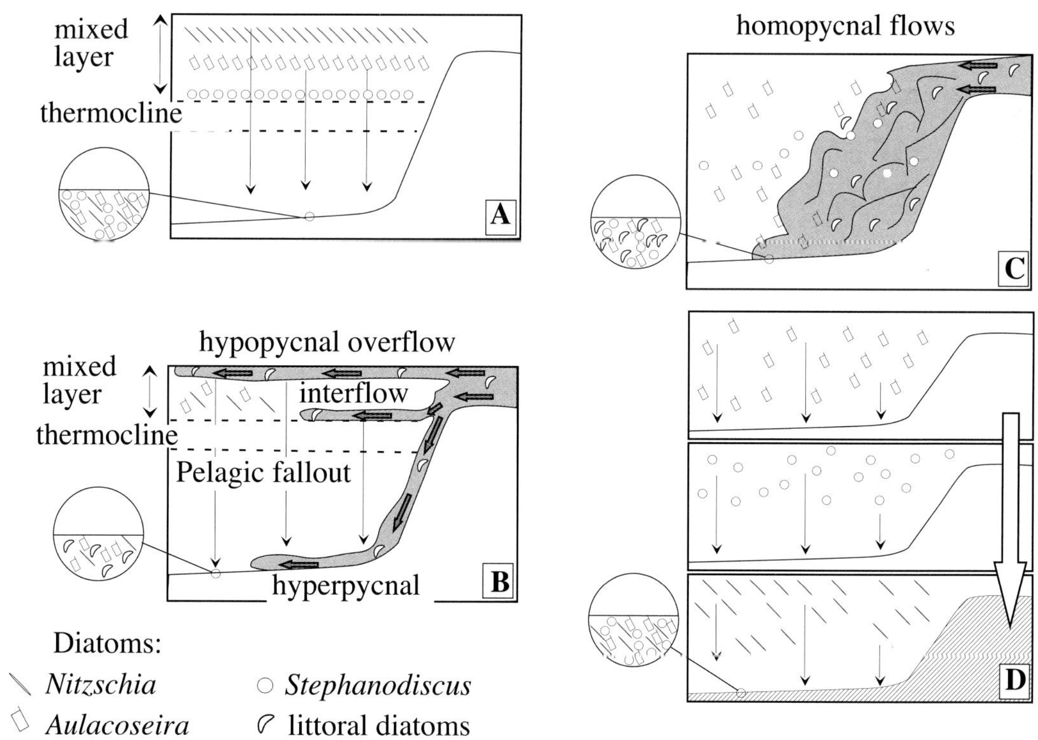


FIG. 4.—Floral mixing. Living diatom assemblages may be altered by mixing, which can occur **A)** as diatoms sink vertically; **B)** via horizontally introduced hyperpycnal and hypopycnal flows; **C)** through homopycnal mixing near river mouths; or **D)** temporally, when floras blooming at different times become mixed on a lake floor by bioturbation or resuspension.

### *Analysis Controls and Assemblage Alteration*

Though not central to a discussion of diatom sedimentology, it is important to note that alteration of diatom floras may continue during sample collection and laboratory preparation. Loose sediments collected from an outcrop are subject to mixing of sediment layers during sampling and transport, unless they are solidified. Core materials and rock samples suffer less from such problems, but a decision is usually made to "mix" a certain thickness of sediment for study.

Allowing samples to dry, for measurements of density or water content, can cause substantial diatom breakage, particularly of larger and more elongate taxa (Flower, 1993). Vigorous cleaning procedures such as the removal of organics with hot $H_2O_2$ can also cause damage, as will sediment grinding. Smear slides of untreated materials approximate the death assemblage most closely. The numbers of diatoms counted has a strong influence on repeatability of analyses, and decisions on which diatom fragments should be included in counts also have an impact on data reliability.

## DISTRIBUTION AND ORIGINS OF DIATOMACEOUS FACIES IN EAST AFRICA

Diatomaceous facies are widespread in the East African Rift System. The western branch (inset, Fig. 5) contains many large fresh lakes with sediments that date back up to several million years. However, sedimentological studies there are generally confined to the later Pleistocene and Holocene (Gasse et al., 1989; Haberyan and Hecky, 1987; Harvey, 1976; Johnson, 1996; Owen and Crossley, 1992; Stoffers and Hecky, 1978). In contrast, the eastern branch contains many well-exposed outcrops of both Tertiary and Quaternary diatomaceous sediments that were laid down in a series of distinct volcano-tectonic basins. These extend from the Afar region and the rifted highlands of Ethiopia (Gasse, 1974; Gasse and Fontes, 1989; Gasse and Street, 1978) through the Kenya Rift Valley (Owen, 1981; Owen and Renaut 1981, 1986, 2000; Richardson and Richardson, 1972) to the faulted grabens of northern Tanzania (Liutkus, 2000). The following sections consider examples of diatom deposition in the Kenya and Malawi Rifts.

### *Tertiary and Quaternary Diatomaceous Sedimentary Units of the Kenya Rift*

Figure 5 shows the major diatomaceous sedimentary units in the Kenya Rift, their stratigraphic relationships, and the occurrence of selected common diatoms. Although varied in the details of their formation, these sediments can be grouped together under three major types: (1) homogeneous diatomite and siliciclastic diatomite; (2) homogeneous subdiatomite and diatomitic sediments; and (3) laminated diatomaceous sediments.

#### *Homogeneous Diatomite and Siliciclastic Diatomite.—*

Homogeneous, intact and fragmented, diatomite and siliciclastic diatomite (Fig. 5) occur in parts of: the Pliocene Chemeron Formation, Baringo District (Fig. 6D, E); the middle Pleistocene high lake beds at Naivasha; the middle Pleistocene Olorgesailie Formation, southern Kenya Rift; the late Pleistocene to early Holocene Suguta Beds, Suguta Valley; and the late Pleistocene to early Holocene Kokwob Formation, Baringo District (Fig. 6C). The floras vary but are generally dominated by *Aulacoseira*, of which *A. granulata* is particularly common. Siliciclastic diatomite tends to be more variable in species composition, with *Synedra* and *Stephanodiscus* being abundant in several units (Fig. 5).

Both diatomite and siliciclastic diatomite require high diatom productivity and relatively low inputs of clastic sediment in order to develop. The frequent dominance of *Aulacoseira* in such deposits suggests high levels of dissolved silica and Si:P ratios (Kilham et al., 1986). An absence of clastic inputs and the introduction of dissolved silica may be brought about by: (1) rivers where sediments are trapped at a basin margin (e.g., by macrophytes); (2) rivers that carry a high solution load, but with low suspended and bed loads; (3) the release of silica held in the pore water of bottom sediments, combined with a thorough mixing of the water column; and (4) intrusions of deep lake water into shallower regions (e.g., by seiches or overturn events). In some cases direct ash fall into a lake may also trigger diatom blooms and / or floristic change (Haberyan, 1998).

#### *Homogeneous Subdiatomite and Diatomitic Sediments.—*

In the Baringo District, homogeneous subdiatomite and diatomitic sediments occur in parts of the Miocene Ngorora Formation (Fig. 6I) and in pond deposits in the Miocene Kaparaina Basalts Formation (Fig. 6H). Diatomitic silts are also common in the early–middle Holocene Galana Boi Formation of northern Kenya (Fig. 6A, B) and many of the modern lakes in the Kenya Rift. The development of these facies reflects lower diatom productivity and / or higher inputs of clastic particles. Their homogeneity may be achieved in several ways. For example, diatoms may be mixed in the water column with clay, silt, and littoral diatoms introduced by hypopycnal flows in deep areas, or by homopycnal mixing in littoral regions (Fig. 4). Alternatively, diatoms may bloom seasonally and at a time quite different fromthe introduction of clastic particles, but if the lake is fully mixed, and there is a thriving bottom benthos, bioturbation may cause mixing on the lake floor. In shallower lakes, wind-induced mixing may reach to the lake floor, causing destruction of any lamination.

#### *Laminated Diatomaceous Sediments.—*

Laminated deposits dominated by *Aulacoseira* and *Synedra* occur in the upper Miocene Lukeino Formation (Fig. 6F, G), and in the late Pleistocene to early Holocene Suguta Beds (Fig. 5). *Nitzschia*-dominated diatomitic silts alternate with layers of magadiite around Lake Magadi (Damnati et al., 1992; Roberts et al., 1993). Where diatomaceous deposits form couplets with clay, lamination is usually < 1–2 mm thick, but with coarser silt-size siliciclastics, layer thickness often increases. For example, the Holocene Galana Boi Formation of northern Kenya (Fig. 5), contains centimeter-scale diatomitic silt layers with variable diatom / clastic contents that may reflect seasonal changes in sediment supply.

Laminated diatomaceous sediments require both the production and the preservation of laminae. Chang et al. (1998) note that laminated marine sediments develop below a well-defined oxygen minimum zone that excludes bioturbating organisms. Anoxic bottom layers in lakes also exclude fauna, as would high salinity, with both conditions favoring lamina development. However, Cohen (1984) cautions that lamination can develop in tropical water bodies such as Lake Turkana, even where oxic conditions prevail on the lake floor.

### *Spatial Distribution of Diatomaceous Sediments in the Malawi Rift*

The Malawi Rift is occupied by the second-largest water body in Africa, by volume, and contains a variety of sedimen-

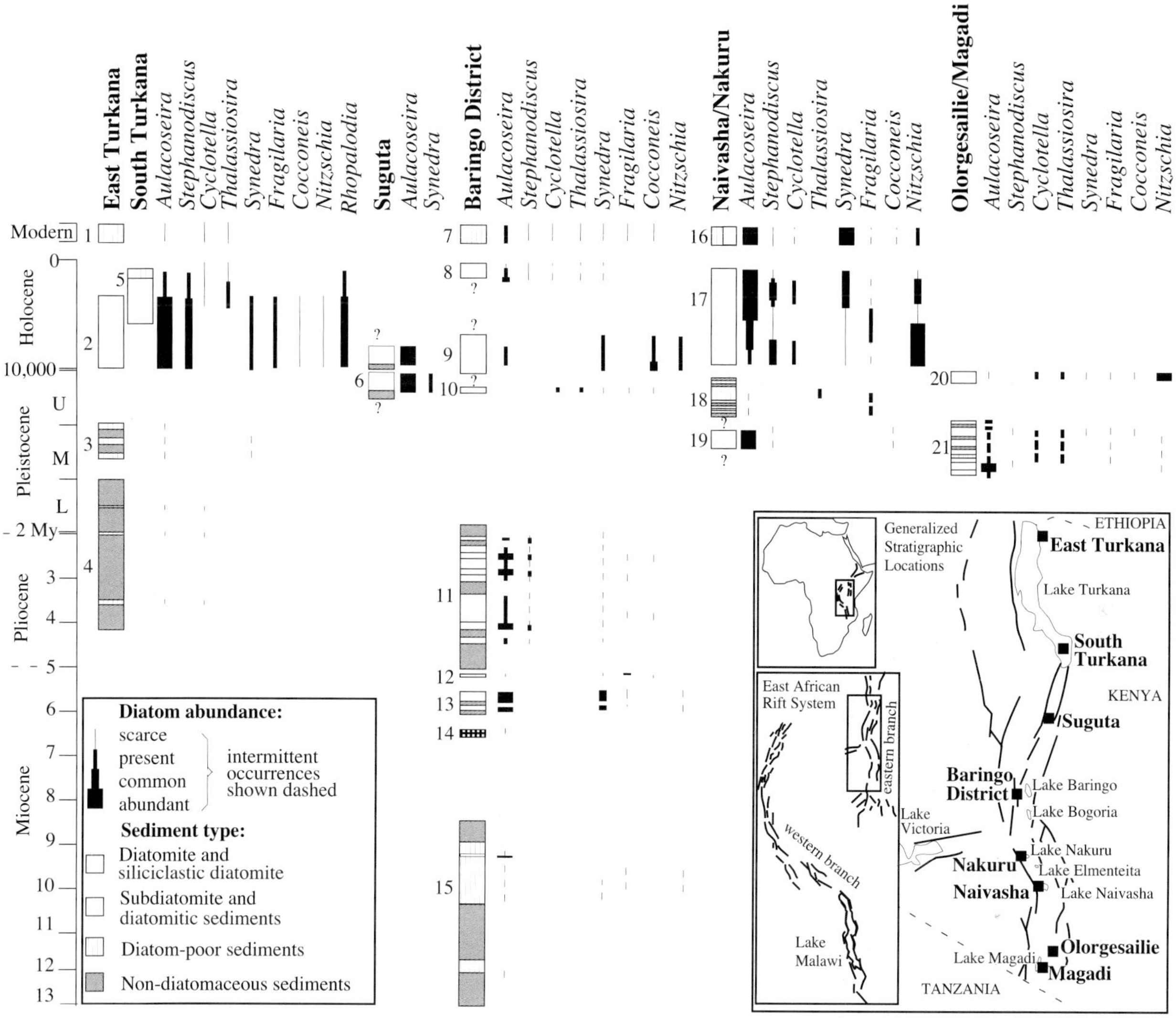


FIG. 5.—Stratigraphic and spatial distribution of diatomaceous sedimentary units and selected diatom genera in the Kenya Rift Valley. Note that *Fragilaria* also includes taxa now placed into *Staurosira, Staurosirella,* and *Pseudostaurosira.* Insets show the two main branches of the rift and the location of areas indicated in the stratigraphic columns. Sedimentary units are numbered as follows: 1 - Modern Lake Turkana sediment near Koobi Fora; 2 - Galana Boi Formation; 3 - Guomde Formation; 4 - Koobi Fora Formation; 5 - southern Lake Turkana deposits (Halfman et al., 1992); 7 - Modern Lake Baringo sediments; 8 - late Holocene Lake Baringo mud; 9 - Kokwob Formation; 10 - Ilosowuani Beds; 11 - Chemeron Formation; 12 - Paleo–pond sediments within Kaparaina Basalts Formation; 13 - Lukeino Formation; 14 - Mpesida Beds; 15 - Ngorora Formation; 16 - Modern Lake Naivasha deposits; 17 - Holocene lake sediment from Naivasha (Richardson and Richardson, 1972); 18 - Ol Njorowa Gorge deposits (Trauth and Strecker, 1996); 19 - Kariandusi Lake Beds; 20 - Lake Magadi High Lake Beds (Barker et al., 1990); 21 - Olorgesailie Formation.

tary facies (Crossley and Owen, 1988; Johnson and Ng'ang'a, 1990; Owen and Crossley, 1992; Scholz et al., 1993). Figure 7 shows the distribution of diatomaceous sedimentary facies in Lake Malawi, together with selected core logs for the upper 50–100 cm of lake-floor sediment. Seven diatomaceous facies occur: (1) highly laminated diatomite and diatomitic mud; (2) laminated diatomite and diatomitic mud; (3) homogeneous diatomitic mud and intermittent diatomite laminae; (4) homogeneous subdiatomite; (5) diatomite and siliciclastic diatomite; (6) proximal turbidite, diatomite laminae, and diatomitic mud; and (7) distal turbidite, diatomite laminae, and diatomitic mud.

*Laminated Facies.—*

Highly laminated (Facies 1) and laminated (Facies 2) diatomite and diatomitic mud occur in the northern parts of Lake Malawi (Fig. 7) (Pilskaln and Johnson, 1991; Francois et al., 1996). Facies 1 (subequal 1-mm-scale laminae) and Facies 2 (diatomite laminae 1 mm thick, with diatomitic mud laminae of 1–10 mm thickness) are separated spatially, with the former occurring in the extreme north (map, Fig. 7). Facies 3 is very variable and consists of isolated diatomite laminae in homogeneous diatomitic mud, and dominates in the south-

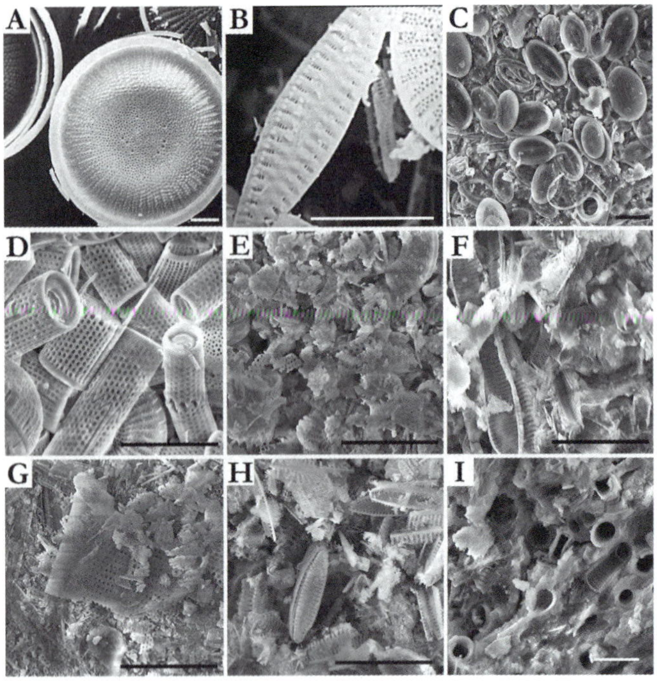


FIG. 6.—Diatomaceous sediments from East Africa. A–B are Holocene sediments from Lake Turkana. C–I are Miocene to Holocene sediments from the Baringo District, Kenya. **A)** *Stephanodiscus* in a siliciclastic diatomite, Galana Boi Formation. **B)** *Pseudostaurosira* in a siliciclastic diatomite, Galana Boi Formation. **C)** Siliciclastic diatomite, dominated by well-preserved *Cocconeis*, Holocene Kokwob Formation. **D)** Well-preserved *Aulacoseira* / *Stephanodiscus* forming a diatomite, Pliocene Chemeron Formation. **E)** Fragmented diatomite with *Aulacoseira* / *Stephanodiscus*. Note the well sorted nature of the fragments, Pliocene Chemeron Formation. **F)** *Synedra*-dominated siliciclastic diatomite lamina, upper Miocene Lukeino Formation. **G)** *Aulacoseira* / *Synedra* forming a siliciclastic diatomite lamina (alternates with "F" above), upper Miocene Lukeino Formation. **H)** *Fragilaria*-dominated subdiatomite, occurring in pond deposits within the upper Miocene Kaparaina Basalts Formation. **I)** *Aulacoseira*-dominated diatomitic clay. Note the individual *Aulacoseira* embedded in a clay matrix, Miocene, Ngorora Formation. Scale bars are 10 μm long.

ern half of Lake Malawi (Fig. 7). Locally, there may be no laminae present.

Lamination tends to become more common northwards and shows a strong depth control. It occurs in parts of cores recovered from water > 100 m deep (Fig. 8) but is dominant only where the lake is > 250 m deep (below the anoxic–oxic boundary). The diatomite laminae in Facies 1 and 2 are dominated by *Aulacoseira nyassensis* (Fig. 9C), with *Stephanodiscus rotula* and *Cyclostephanos damasii* being common (Owen and Crossley, 1992). Laminae in Facies 3 are similar, but with *Nitzschia nyassensis* (Fig. 9A) dominating close to submerged fault scarps. This particular species is long and sharp and may be avoided by grazers such as copepods (Haberyan, 1985). It forms entangled mats similar to those produced by the marine diatom *Thalassiothrix longissima*. The latter mats develop as small aggregates in the photic zone prior to sinking (Chang et al., 1998). Boden and Blackman (1996) report that *Thalassiothrix* mats (and perhaps also *Nitzschia* mats) are avoided by bottom organisms because of their high tensile strength and impenetrability, giving them an ability to survive even in oxic bottom conditions. The diatomitic mud in Facies 3 is generally composed of < 20% diatoms, mostly *Aulacoseira* (Fig. 9D) and *Stephanodiscus*.

### *Homogeneous Facies.—*

Facies 3 (noted above) locally grades into homogeneous diatomitic mud (Fig. 7) with no laminae, or occasionally into diatom-poor deposits. Facies 4 (homogeneous subdiatomite) occurs close to the faulted margins of northwestern Lake Malawi (map, Fig. 7) and is dominated by *Aulacoseira nyassensis*. Facies 5, diatomite and siliciclastic diatomite (Fig. 9B), occurs in the extreme south. The purest deposits contain < 0.1% siliciclastics, and diatom assemblages are much more variable than those noted above. Several species of *Aulacoseira* dominate, with other taxa also being common (Owen and Crossley, 1992).

Homogeneous facies show variable depth relationships (Fig. 8), probably reflecting a variety of modes of deposition, including distal turbidites, pelagic fallout in deep areas, and the effects of bioturbation in shallow regions.

### *Diatomaceous–Turbiditic Facies.—*

Facies 6 (proximal turbidite, diatomite laminae and diatomitic mud) gradually changes to Facies 7 (distal turbidite, diatomite laminae, and diatomitic mud) with distance from the shoreline (Fig. 7). The turbidites consist of upward-fining sand and silt that grade into diatomitic mud (Crossley and Owen, 1988). These sandy units are common in Facies 6 but become scarce, thinner, and finer-grained in Facies 7. Littoral diatoms are present locally. Diatomite laminae are varied and dominated by *Aulacoseira* or *Nitzschia*, with *Stephanodiscus* dominance being less common. In Lake Malawi, these facies generally occur in areas of deep water (> 300 m) near major border faults.

### *Models of Diatomaceous Sedimentation for Lake Malawi.—*

Figure 10A–E shows several possible models for the development of laminated diatomaceous facies in deep stratified parts of Lake Malawi. Figure 10A represents "background sedimentation" through the year, with siliciclastic particles being introduced by either hypopycnal or hyperpycnal flows (Fig. 4). Much of this material enters the lake during the rainy season (December to April). If no diatom blooms develop, a largely homogeneous sediment results. Dissolution takes place as diatoms settle. Fragmentation occurs in areas (or during seasons) with abundant pelagic grazers.

Where diatom blooms do develop (seasonally or periodically), and anoxic bottom conditions prevail, laminated sediments may form, composed of homogeneous diatomitic mud (background sedimentation) that alternates with diatomite or siliciclastic diatomite on a variety of scales. There are several potential mechanisms that may generate diatom blooms (Fig. 10). Model 1 (Fig. 10B) involves supply of nutrients from deep water to stimulate diatom growth. This may be caused by the strong seasonal southerly winds (April to August) that depress the thermocline in northern Lake Malawi (Patterson and Kachinjika, 1993) and / or result from an increased depth of wind-induced mixing. Model 2 (Fig. 10C) involves the introduction of nutrients from lateral sources, principally rivers. This may be significant close to fluvial inputs, but it is unlikely to have a major influence lakewide.

Model 3 (Fig. 10D) involves the introduction of density currents to produce the turbidites typical of Facies 6 and 7. Back-

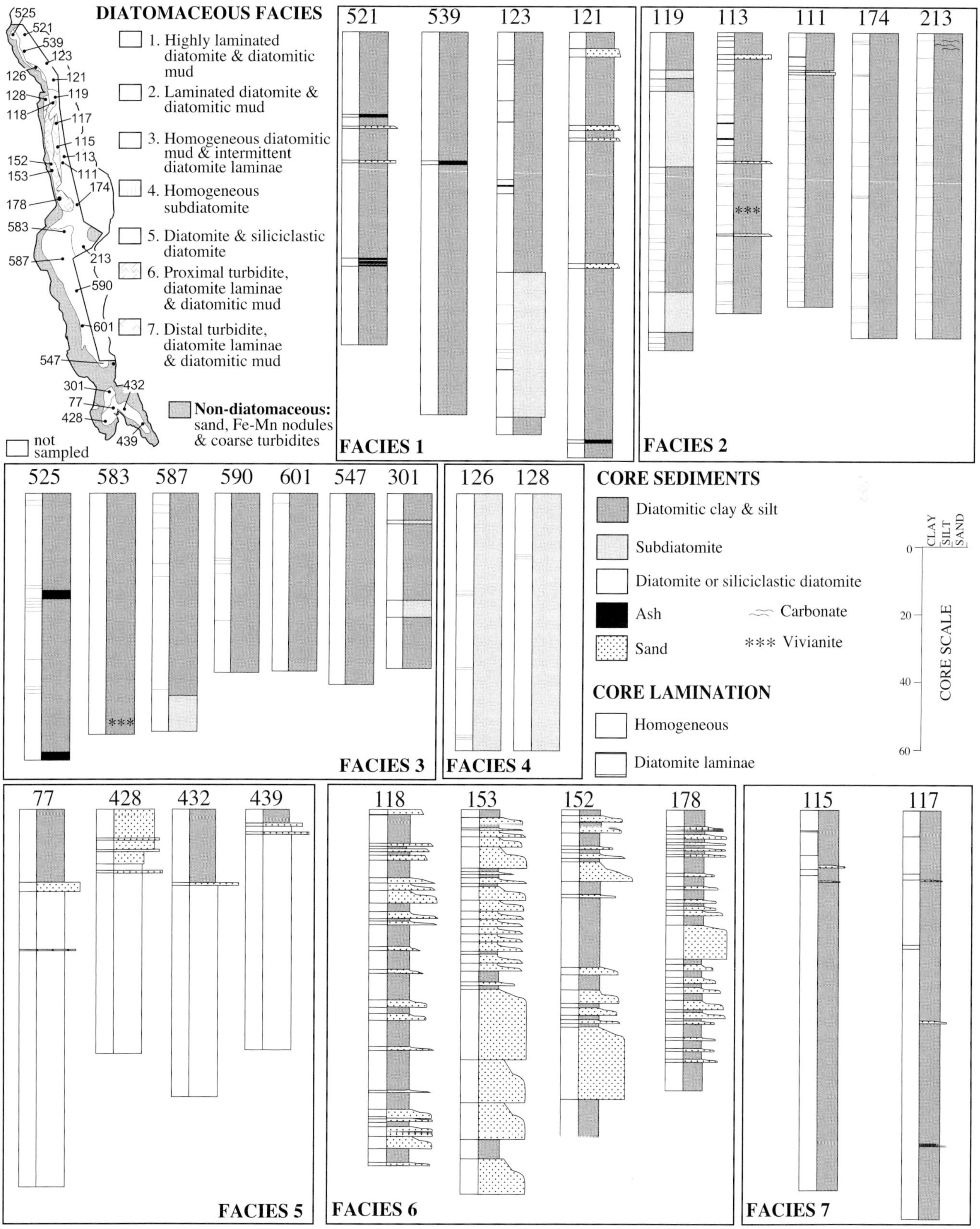


FIG. 7.—Spatial distribution of diatomaceous facies in Lake Malawi. Map shows simplified distribution of diatomaceous sediments (coarse-grained littoral facies are omitted), based on 242 short gravity cores and several hundred grab samples. Selected and simplified core logs are shown for each facies type.

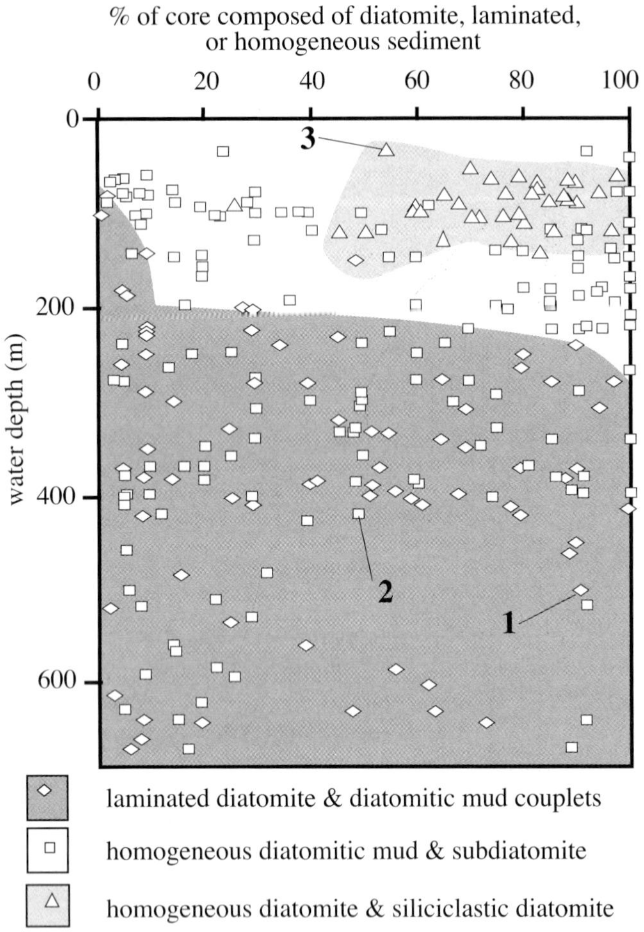


FIG. 8.—Dominance of major types of diatomaceous sediment in cores from Lake Malawi. Each point represents one core and indicates its depth and the percentage of the core composed of a particular facies. For example, Point 1 indicates a core with 90% of its length composed of laminated sediment. Point 2 indicates a core with 43% of its length occupied by homogeneous sediment. Point 3 indicates another core with 52% of its length composed of diatomite. See text for discussion.

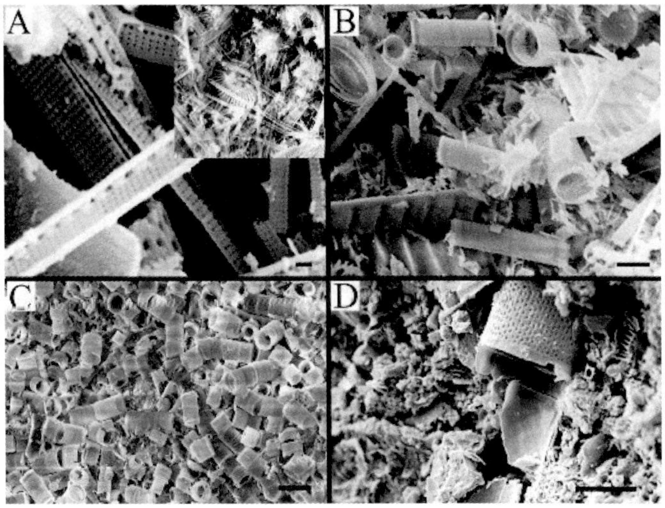


FIG. 9.—Diatomaceous sediments in Lake Malawi. **A)** *Nitzschia* diatomite lamina. **B)** Mixed *Aulacoseira*, *Surirella*, *Stephanodiscus*, and other taxa in a homogeneous diatomite. **C)** *Aulacoseira nyassensis* diatomite lamina. **D)** *A. nyassensis* in a diatomitic silty clay. Scale bars: A = 1 μm; B, C, and D = 10 μm.

ground sedimentation forms diatomitic mud. The occurrence of *Aulacoseira* and *Stephanodiscus* laminae may relate to the wind-induced mixing or riverine inputs noted above. However, *Nitzschia* laminae are also common in this facies, tending to occur close to submerged fault scarps both with and without turbidite deposits. This suggests two possibilities. In Model 3 it is suggested that turbidity currents may disrupt stratification, releasing nutrients. In Model 4 (Fig. 10E) nutrients are released by upwelling (e.g., near fault scarps).

In shallow areas, bioturbation is important. Model 5 (Fig. 10F) indicates deposition of homogeneous diatomitic mud, with full mixing of the water column to the lake floor. Nutrients are plentiful, but siliciclastic inputs are relatively high, diluting diatom deposition. This would be a typical situation for areas close to deltas. Model 6 (Fig. 10G) shows diatomite forming in areas with full mixing to the lake floor and low sediment inputs. This appears to be the situation in the southeast of Lake Malawi, close to its outlet river, though nutrient release here might also be related to upwelling caused by wind-induced seiches (Eccles, 1974).

### *Diatomaceous Facies and Geomorphology.—*

The diatomaceous sediments in Lake Malawi show a strong relationship with geomorphic, and tectonic, location. Homogeneous diatomaceous sediments generally occur in shallower locations such as lake-margin tilt blocks, horsts, and grabens (Fig. 11), where bioturbation can cause mixing. Diatomitic muds and diatom-poor sediments are probably also deposited as overbank sediments associated with deep-water turbidite channels (Scholz, 1995). Diatomite and siliciclastic diatomite are found in relatively shallow areas, with very low supplies of clastic sediment (Fig 11). Laminated diatomaceous facies occur on deep axial lake-floor plains, below the oxic–anoxic boundary. The proximal and distal turbidite, diatomite laminae, and diatomitic mud facies show a strong relationship with major border faults, occurring in deep-water settings (Fig.11).

## CONCLUSIONS

Diatomaceous sediments are common in later Tertiary and Quaternary rift deposits, yet they are often described without a clear definition. However, a simple classification can be carried out with reference to the proportion of diatoms to other components present in smear slides. Major terms suggested here include diatomite, siliciclastic or tuffaceous diatomite, subdiatomite, diatomitic, and diatom-poor. The common descriptor "diatomaceous" should continue to be used loosely for any sediment containing > 10% diatoms. If this set of terms, or similar alternatives, were to be adopted widely, then comparison of sedimentation in different basins would be improved considerably.

There are many factors that may alter living floras as they are transformed into fossil assemblages examined under a microscope. These include sedimentation factors, diagenetic alterations, and the effects of sampling. With a few exceptions, not enough attention has been paid to the taphonomy of diatom deposition. In general, diatom preservation becomes poorer as pH rises and biological activity (grazers and bioturbators) increases. Diatoms are often broken in coarse sediments laid down

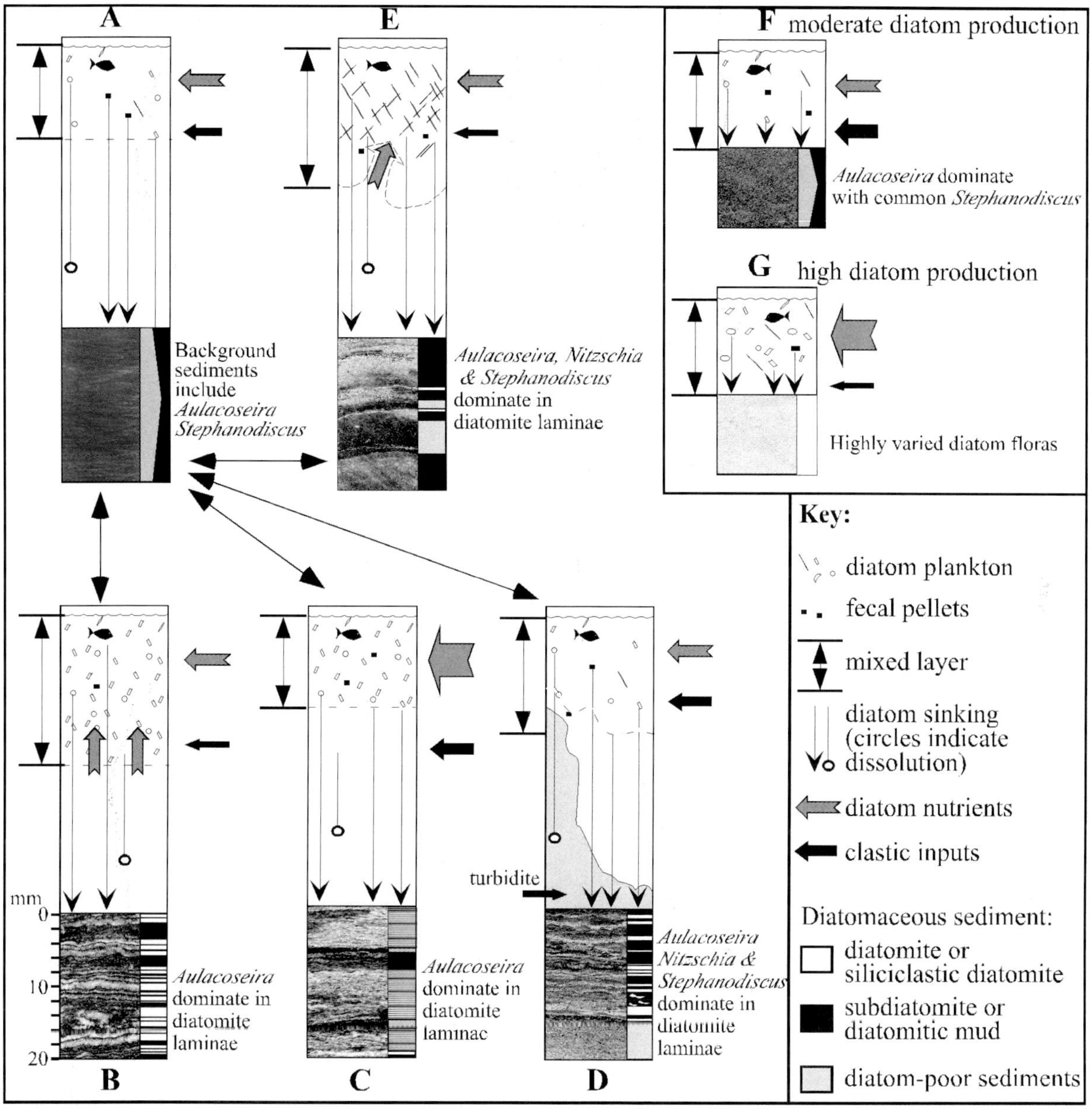


FIG. 10.—Sedimentation models for Lake Malawi. Diagrams A–E) represent deposition in deep stratified areas. Facies development involves alternations between (A) and (B–E). Both (F) and (G) are relevant for relatively shallow water (< 100 m in Lake Malawi). Diagram (A) gives scale for representative images of deposits shown at the bottom of each column. See text for discussion of sedimentation models.

under high-energy conditions. Fragmentation is common in sediment subject to wetting and drying (e.g., ephemeral lakes or lake margins). Diatoms tend to be lost as they sink to the bottom of lakes with silica-depleted water. Further loss may occur at the sediment–water interface, particularly where sedimentation rates are low. If living floras include robust and weakly silicified taxa, differential preservation may lead to false habitat conclusions. Current-induced lateral transport may introduce exotic species. Caution is clearly required in diatom paleoenvironmental interpretation.

In East Africa, diatomaceous sediments have been laid down in variety of lake types, ranging from small to large, deep to shallow, and fresh to saline–alkaline. Nevertheless, several diatomaceous facies can be recognized repeatedly through time, as in the Kenya rift, or spatially, as in the Malawi rift. In particular, diatomaceous sediments may occur as homogeneous facies or as laminated deposits. Further division of these two types depends on the proportion of diatoms in a sediment. Several major factors contribute to the development of these facies. These include seasonality (particularly with regard to rainfall and wind regimes), proximity to clastic sediment sources, and sedimentation rates. Drainage-basin geology exerts a vital control on nutrients such as silica, which is particularly important for the formation of diatomite. Tectonic history is also important, inasmuch as this determines basin morphology and potential water depths. Close relationships are evident, for example, between tectonic setting,

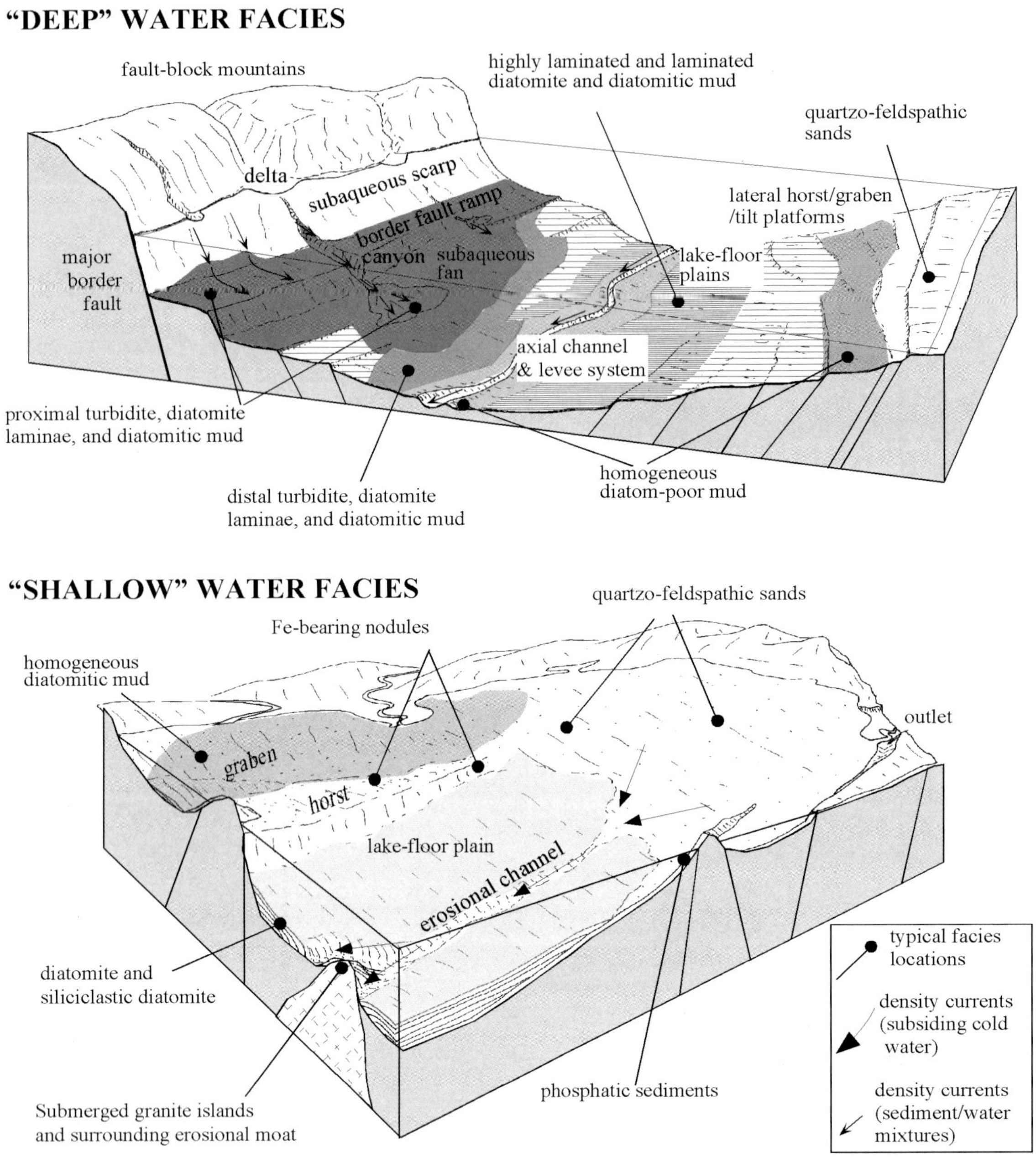


FIG. 11.—Geomorphic models and diatomaceous sedimentation in Lake Malawi. Deep regions are typically asymmetric with steep border faults opposed by gentler horsts and grabens. These show a close relationship with laminated and homogeneous diatomaceous sediments and turbidites. Shallow areas are characterized by homogeneous diatomaceous deposits, and coarse-grained facies that include Fe–Mn nodules, accumulations of phosphatic fish bones, and quartzo-feldspathic sand. The diatomaceous facies indicated are discussed in the text.

geomorphology, and diatomaceous sedimentary facies in Lake Malawi.

The Malawi and Kenya Rifts show several distinct contrasts in terms of diatom deposition through time. The Kenya Rift is occupied by many small to moderate-size lakes, with the larger Lake Turkana occurring in the extreme north. These water bodies are relatively shallow, and several occupy closed basins today. They have been particularly susceptible to climate change on decadal to millennial time scales and have experienced considerable variations in water chemistry, stratification, and sediment inputs. As a result, diatom floras frequently display major changes within sedimentary successions, showing variability in terms of species diversity, abundance, and preservation state. The relatively shallow-water conditions and proximity to sediment inputs and shoreline conditions also results in spatial changeability in many of the lake basins. In contrast, the

Malawi and Tanganyika Rifts are entirely flooded by very deep and very large water bodies. The diatomaceous sediments are laterally more extensive and tend to show, spatially, more gradual facies changes that reflect water depth. Diatoms are dominated by fewer species through time and generally vary less in their patterns of abundance, dissolution, and fragmentation. In turn, these broad contrasts in diatom deposition between the Kenya and Malawi Rifts reflect contrasts in basin types and ultimately the interplay of tectonism, volcanism, and climate through time.

## ACKNOWLEDGMENTS

Samples were prepared by Alison Lee. Dr. J. Aitchison made available SEM facilities at the Earth Science Department of Hong Kong University. Cores from Lake Malawi were collected together with Dr. R. Crossley. Prof. R. Renaut, and Dr. M. Pickford assisted with sampling in Kenya. This work was carried out while in receipt of a Faculty Research Grant from the Hong Kong Baptist University.

## REFERENCES

ALLDREDGE, A.L., AND GOTSCHALKE, C.C., 1989, Direct observations of the mass flocculation of diatom blooms: characteristics, settling velocities and formation of diatom aggregates: Deep-Sea Research, v. 36, p. 159–171.

ANDERSON, N.J., AND BATTARBEE, R.W., 1994, Aquatic ecology community persistence and variability: a paleolimnological perspective, *in* Giller, P.S., Hildrew, A.G., and Rafaelli, D., eds., Aquatic Ecology; Scale, Pattern and Process, Oxford, U.K., Blackwell, p. 233–259.

BARKER, P., 1992, Differential diatom dissolution in Late Quaternary sediments from Lake Manyara, Tanzania: an experimental approach: Journal of Paleolimnology, v. 7, p. 235–252.

BARKER, P., GASSE, F., ROBERTS, N., AND TAIEB, M., 1990, Taphonomy and diagenesis in diatom assemblages: a late Pleistocene palaeoecological study from Lake Magadi, Kenya: Hydrobiologia, v. 214, p. 267–272.

BARKER, P., FONTES, J-C., GASSE, F., AND DRUART, J-C., 1994, Experimental dissolution of diatom silica in concentrated salt solutions and implications for paleoenvironmental reconstruction: Limnology and Oceanography, v. 39, p. 99–110.

BODEN, P., AND BLACKMAN, J., 1996, A laminated sediment sequence from the northern North Atlantic Ocean and its climatic record: Geology, v. 24, p. 507–510.

BRADLEY, W.H., 1965, Vertical density currents: Science, v. 150, p. 1423–1428.

BRADLEY, W.H., 1969, Vertical density gradients: Limnology and Oceanography, v. 14, p. 1–3.

BREESE, R., 1994, Diatomite, *in* Carr, D.D., ed., Industrial Minerals and Rocks: Society for Mining, Metallurgy and Exploration Inc., Colorado, p. 397–412.

CHANG, A.S., GRIMM, K.A., AND WHITE, L.D., 1998, Diatomaceous sediments from the Miocene Monterey Formation, California: A lamina-scale investigation of biological, ecological, and sedimentary processes: Palaios, v. 13, p. 439–458.

COHEN, A.S., 1984, Effect of zoobenthic standing crop on laminae preservation in tropical lake sediment, Lake Turkana, East Africa: Journal of Paleontology, v. 58, p. 499–510.

CROSSLEY, R., AND OWEN, R.B., 1988, Sand turbidites and organic rich diatomaceous muds from Lake Malawi, Central Africa, *in* Fleet, A.J., Kelts, K., and Talbot, M.R., eds., Lacustrine Petroleum Source Rocks: Geological Society of London, Special Publication 40, p. 369–374.

DAMNATI, B., TAIEB, M., AND WILLIAMSON, D., 1992, Laminated deposits from Lake Magadi (Kenya). Climatic contrast effect during the maximum wet period between 12,000–10,000 years B.P.: Société Géologique de France, Bulletin, v. 163, p. 407–414.

DOLLEY, P.T., 1999, Diatomite: Minerals Yearbook: Volume 1—Metals and Minerals, U.S. Geological Survey, Chapter 24, p. 1–4.

ECCLES, D.H., 1974, An outline of the physical limnology of Lake Malawi (Lake Nyasa): Limnology and Oceanography, v. 19, p. 730–742.

FINNEY, B.P., AND JOHNSON, T.C., 1991, Sedimentation in Lake Malawi (East Africa) during the past 10,000 years: a continuous paleoclimate record from the southern tropics: Palaeogeography, Palaeoclimatology, Palaeoecology, v. 85, p. 351–366.

FLOWER, R.J., 1993, Diatom preservation: experiments and observations on the dissolution and breakage in modern and fossil material: Hydrobologia, v. 269/270, p. 473–84.

FOURTANIER, E., AND GASSE, F., 1988, Premiers jalons d'une biostratigraphie et évolution des diatomées lacustres d'Afrique depuis 11 Ma: Académie des Sciences [Paris], Comptes Rendus, v. 306 (Série 2), p. 1401–1408.

FRANCOIS, R., PILSKALN, C.H., AND ALTABET, M.A., 1996, Seasonal variation in the nitrogen isotopic composition of sediment trap materials collected in Lake Malawi, *in* Johnson, T.C., and Odada, E.O., eds., The Limnology, Climatology and Paleoclimatology of the East African Lakes: Amsterdam, Gordon & Breach, p. 241–250.

GASSE, F., 1974, Les diatomées des sédiments Holocènes du bassin du Lac Afrera (Giulietti) (Afar Septentrional, Ethipoie). Essai de reconsitition de l'évolution du milieu: Internationale Revue der gesamten Hydrobiologie, v. 59 (1), p. 95–122.

GASSE, F., 1986, East African diatoms. Taxonomy, ecological distribution: Bibliotheca Diatomologica, v. 11, Stuttgart, Cramer, 201 p.

GASSE, F., 1987, Diatoms for reconstructing palaeoenvironments and palaeohydrology in tropical semiarid zones: Hydrobiologia, v. 154, p. 127–163.

GASSE, F., AND FONTES, J-C., 1989, Palaeoenvironments and palaeohydrology of a tropical closed lake (Lake Asal, Djibouti) since 10,000 yr. B.P.: Palaeogeography, Palaeoclimatology, Palaeoecology, v. 69, p. 67–102.

GASSE, F., AND STREET, F.A., 1978, Late Quaternary lake-level fluctuations and environments of the Northern Rift Valley and Afar Region (Ethiopia and Djibouti): Palaeogeography, Palaeoclimatology, Palaeoecology, v. 24, p. 279–325.

GASSE, F., LEDEE, V., MASSAULT, M., AND FONTES, J-C., 1989, Water-level fluctuations of Lake Tanganyika in phase with oceanic changes during the last glaciation and deglaciation: Nature, v. 342, p. 57–59.

GASSE, F., JUGGINS, S., AND KHELIFA, L.B., 1995, Diatom-based transfer functions for inferring past hydrochemical characteristics of African lakes: Palaeogeography, Palaeolimnology, Palaeoecology, v. 117, p. 31–54.

GASSE, F., BARKER, P., GELL, P.A., FRITZ, S.C., AND CHALIÉ, F., 1997, Diatom-inferred salinity in palaeolakes: an indirect tracer of climate change: Quaternary Science Reviews, v. 16, p. 547–563.

GRIMM, K.A., LANGE, C.B., AND AMARPAL, A.S., 1997, Self-sedimentation of phytoplankton blooms in the geologic record: Sedimentary Geology, v. 110, p. 151–161.

HABERYAN, K.A., 1985, The role of copepod fecal pellets in the deposition of diatoms in Lake Tanganyika: Limnology and Oceanography, v. 30, p. 1010–1023.

HABERYAN, K.A., 1987, Fossil diatoms and the history of Lake Rukwa, Tanzania: Freshwater Biology, v. 17, p. 429–336.

HABERYAN, K.A., 1990, The misrepresentation of the planktonic diatom assemblage in traps and sediments: southern Lake Malawi, Africa: Journal of Paleolimnology, v. 3, p. 35–44.

HABERYAN, K.A., 1992, Algal communities near Cape Maclear, southern Lake Malawi, Africa: Hydrobiologia, v. 215, p. 175–188.

HABERYAN, K.A., 1998, The effect of volcanic ash influx on the diatom community of Lake Tanganyika, East Africa: Missouri Academy of Sciences, Transactions, v. 32, p. 102–105.

HABERYAN, K.A., AND HECKY, R.E., 1987, The late Pleistocene and Holocene stratigraphy and paleolimnology of lakes Kivu and Tanganyika: Palaeogeography, Palaeoclimatology, Palaeoecology, v. 61, p. 169–197.

Halfman, J.D., Jacobson, D.F., Cannella, C.M., Haberyan, K.A., and Finey, B.P., 1992, Fossil diatoms and the mid to late Holocene paleolimnology of Lake Turkana, Kenya: a reconnaissance study: Journal of Paleolimnology, v. 7, p. 23–35.

Harben, P.W., and Kuzvart, M., 1997, Diatomite, *in* Harben, P.W., and Kuzvart, M., eds., Industrial Minerals: A Global Geology: Industrial Minerals Information Ltd., Metal Bulletin Plc., London, p. 161–167.

Harvey, T.J., 1976, The Paleolimnology of Lake Mobuto Sese Seko, Uganda–Zaire: the last 28,000 years: Unpublished Ph.D. dissertation, Duke University, North Carolina, 104 p.

Harwood, D.M., and Gersonde, R.E., 1990, Lower Cretaceous diatoms from ODP leg 113 Site 693 (Weddell Sea); Part 2, Resting Spores, chrysophycean cysts, an endoskeletal dinoflagellate, and notes on the origin of diatoms: Proceedings of the Ocean Drilling Program, Scientific Results, v. 113, p. 403–425.

Hecky, R.E., Bootsma, R., Mugidde, R., and Bugenyi, F.W.B., 1996, Phosphorus pumps, nitrogen sinks and silicon drains: plumbing nutrients in the African Great Lakes, *in* Johnson, T.C., and Odada, E.O., eds., The Limnology, Climatology and Paleoclimatology of the East African Lakes: Amsterdam, Gordon & Breach, p. 205–224.

Hilton, J., 1985, A conceptual framework for predicting the occurrence of sediment focusing and sediment redistribution in small lakes: Limnology and Oceanography, v. 30, p. 1131–1147.

Holdship, S.A., 1976, The paleolimnology of Lake Manyara, Tanzania: a diatom analysis of a 56 m core: Unpublished Ph.D. dissertation, Duke University, 220 p.

Inglethorpe, S.D.J., Utha-Aroon, C., and Chanyavanich, C., 1997, An inventory of diatomite deposits of the Lampang basin, northern Thailand, *in* Dheeradilock, P., Hinthong, C., Chaodumrong, P., Putthapiban, W., and Utha-Aroon, C., eds., Proceedings of the International Conference on Stratigraphy and Evolution of Southeast Asia and the South Pacific, 1997: Department of Mineral Resources, Bangkok, p. 669–685.

Johnson, T.C., 1996, sedimentary processes and signals of past climatic change in the large lakes of the East African Rift Valley, *in* Johnson, T.C., and Odada, E.O., eds., The Limnology, Climatology and Paleoclimatology of the East African Lakes: Amsterdam, Gordon & Breach, p. 367–413.

Johnson, T.C., and Ng'ang'a, P., 1990, Reflections on a rift lake, *in* Katz, B.J., ed., Lacustrine Basin Exploration: American Association of Petroleum Geologists, Memoir 50, p. 113–135.

Kilham, P., Kilham, S.S., and Hecky, R.E., 1986, Hypothesized resource relationships among African planktonic diatoms: Limnology and Oceanography, v. 31, p. 1169–1181.

Kilham, S.S., and Kilham, P., 1975, *Melosira granulata* (Ehr.) Ralfs: morphology and ecology of a cosmopolitan freshwater diatom: Internationale Vereinigunge für Theoretische und Angewandte Limnologie, Verhandlungen, v. 19, p. 2716–2721.

Krebs, W.N., Bradbury, J.P., and Theriot, E., 1987, Neogene and Quaternary lacustrine diatom biochronology, western U.S.A.: Palaios, v. 2, p. 505–513.

Liutkus, C.M., 2000, Sedimentology, stratigraphy, and stable isotopes of the lowermost Bed II freshwater wetlands, Olduvai Gorge, Tanzania: Unpublished M.Sc. thesis, Rutgers University, 159 p.

Owen, R.B., 1981, Quaternary diatomaceous sediments and the geological evolution of Lakes Turkana, Baringo and Bogoria, Kenya Rift Valley: Unpublished Ph.D. dissertation, University of London, 465 p.

Owen, R.B., and Crossley, R., 1992, Spatial and temporal diatom variability in Lake Malawi and ecological implications: Journal of Paleolimnology, v. 7, p. 55–71.

Owen, R.B., and Renaut, R.W., 1981, Palaeoenvironments and sedimentology of the middle Pleistocene Olorgesailie Formation, southern Kenya Rift Valley: Palaeoecology of Africa, v. 13, p. 147–174.

Owen, R.B., and Renaut, R.W., 1986, Sedimentology, stratigraphy and palaeoenvironments of the Holocene Galana Boi Formation, northeast Lake Turkana, Kenya, *in* Frostick, L.E., Renaut, R.W., Reid, I., and Tiercelin, J.J., eds., Sedimentation in the African Rifts: Geological Society of London, Special Publication 25, p. 311–322.

Owen, R.B., and Renaut, R.W., 2000, Miocene and Pliocene diatomaceous lacustrine sediments of the Tugen Hills, Baringo District, Central Kenya Rift, *in* Gierlowski-Kordesch, E.H., and Kelts, K.K., eds., Lake Basins Through Space And Time: American Association of Petroleum Geologists, Studies in Geology, no. 46, p. 399–404.

Owen, R.B., and Utha-Aroon, C., 1999, Diatomaceous sedimentation in the Tertiary Lampang Basin, northern Thailand: Journal of Paleolimnology, v. 22, p. 81–95.

Patterson, G., and Kachinjika, O., 1993, Effect of wind induced mixing on vertical distribution of nutrients and phytoplankton in Lake Malawi: Internationale Vereinigunge für Theoretische und Angewandte Limnologie, Verhandlungen, v. 25, p. 872–876.

Pilskaln, C.H., and Johnson, T.C., 1991, Seasonal signals in Lake Malawi sediments: Limnology and Oceanography, v. 36, p. 544–557.

Renaut, R.W., and Owen, R.B., 1988, Opaline cherts associated with sublacustrine hydrothermal springs at Lake Bogoria, Kenya Rift Valley: Geology, v. 16, p. 699–702.

Richardson, J.L., and Richardson, A.E., 1972, History of an African rift lake and its climatic implications: Ecological Monographs, v. 42, p. 499–534.

Roberts, N., Taieb, M., Barker, P., Damnati, B., Icole, M., and Williamson, D., 1993, Timing of the Younger Dryas event in East Africa from lake level changes: Nature, v. 366, p. 146–148.

Round, F.E., 1964, The diatom sequence in lake deposits: some problems of interpretation. Internationale Vereinigunge für Theoretische und Angewandte Limnologie, Verhandlungen, v. 15, p. 1012–1026.

Scholz, C.A., 1995, Deltas of the Lake Malawi Rift, East Africa: seismic expression and exploration implications: American Association Petroleum Geologists, Bulletin, v. 79, p. 1679–1697.

Scholz, C.A., Johnson, T.C., and McGill, J.W., 1993, Deltaic sedimentation in a rift valley lake: new seismic reflection data from Lake Malawi (Nyasa), East Africa: Geology, v. 21, p. 395–398.

Schrader, H.J., 1971, Fecal pellets: role in the sedimentation of pelagic diatoms: Science, v. 174, p. 55–57.

Smayda, T.J., 1970, The suspension and sinking of phytoplankton in the sea: Annual Review of Oceanography and Marine Biology, v. 8, p. 353–414.

Smayda, T.J., 1971, Normal and accelerated sinking of phytoplankton in the sea: Marine Geology, v. 11, p. 105–122.

Stoffers, P., and Hecky, R.E., 1978, Late Pleistocene and Holocene evolution of the Kivu–Tanganyika basin, *in* Matter, A., and Tucker, M.E., eds., Modern and Ancient Lake Sediments: International Association of Sedimentologists, Special Publication 2, p. 43–55.

Trauth, M.H., and Strecker, M.R., 1996, Late Pleistocene lake-level fluctuations in the Naivasha Basin, Kenya, *in* Johnson, T.C., and Odada, E.O., eds., The Limnology, Climatology and Paleoclimatology of the East African Lakes: Amsterdam, Gordon & Breach, p. 549–557.

Tribble, J.S., Arvidson, R.S., Lane, M., and Mackenzie, F.T., 1995, Crystal chemistry, and thermodynamic and kinetic properties of calcite, dolomite, apatite and biogenic silica: applications to petrologic problems: Sedimentary Geology, v. 95, p. 11–37.

Volosov, A.G., Khodarovskiy, I.L., and Ryzhenko, B.N., 1972, Equilibria in the system $SiO_2$–$H_2O$ at elevated temperatures along the three-phase curve: Geochemica International, v. 9, p. 362–277.

Walther, J.V., and Helgeson, H.C., 1977, Calculation of the thermodynamic properties of aqueous silica and the solubility of quartz and its polymorphs at high pressures and temperatures: American Journal of Science, v. 277, p. 1315–1351.

Wang, F., Zhang, H., Feng, H., Chen, G., Wang, D., and He, H., 1997, Spectroscopic study of diatomite in Leizhou Peninsula, China: 30th International Geological Congress, Proceedings, v. 16, p. 159–175.

# BIOGENIC SILICA PROFILES IN THE SEDIMENTS OF LARGE TROPICAL LAKES: EXAMPLES FROM EAST AFRICA

THOMAS C. JOHNSON

*Large Lakes Observatory, University of Minnesota Duluth, Duluth, Minnesota 55812, U.S.A.*

ABSTRACT: Profiles of the abundance or mass accumulation rate of biogenic silica in sediment cores can provide important insight into the past productivity of diatoms and the climatic influence on a lake basin through time. While primary production of diatoms in the overlying waters can influence the burial rate of diatoms, other factors need to be considered as well. The geochemical mass balance of dissolved silica in the lake system is one such factor that plays an important role in controlling the rate at which diatoms can be buried in the lake sediments. Examples are provided from Lake Malawi and Lake Victoria, East Africa, that demonstrate how the rates of input and outflow of dissolved silica are related to biogenic silica profiles in the sediments, and how conclusions were drawn on past environmental conditions in these large tropical lakes.

## INTRODUCTION

The sedimentary record of past environmental conditions in lakes and their drainage basins is rich and elusive. Numerous proxy signals are to be found in the biogenic and authigenic components of lacustrine deposits, but their interpretation is often hindered by an incomplete understanding of what drives the signals, or by too many unknowns plugged into too few equations. The field of limnogeology is in its infancy compared to its more mature cousin, marine geology, especially in the realm of paleoenvironmental interpretation of the sedimentary record. Paleoceanography has benefited significantly from decades of well supported studies of modern ocean processes—the circulation dynamics of the world ocean, the chemistry of seawater, primary production and food web dynamics, and modern sedimentary processes. These investigations have all led to a much clearer understanding of sedimentary signals than would have been possible without such a rich and sustained interdisciplinary study of the sea. We paleolimnologists who work in the large lakes of the world are not blessed with such a fertile legacy of modern process studies in these globally important water bodies. Moreover, the chemistry of each lake basin is unique, unlike the well mixed oceans. Consequently we have no Universal Truths in paleolimnology such as the marine oxygen isotope curve, the alkenone paleothermometer, or the strontium isotope geochronometer.

However, we do recover a variety of reproducible signals from lacustrine sediment, such as pollen diagrams, diatom assemblage profiles, and carbonate oxygen isotope curves. One of the relatively easy parameters to measure in sediment cores is the abundance of biogenic silica. Biogenic silica (the opaline silica found in diatom frustules, sponge spicules, and phytoliths) is present in most lake sediments and can be quantified by a fairly simple chemical digestion. This involves dissolution of the biogenic silica in a warm base solution, followed by a spectrometric analysis for the dissolved silica that is released (Conley, 1988; DeMaster, 1979). Biogenic silica can be one of the most useful signals to be recovered by relatively simple and rapid analytical techniques, particularly in low-alkalinity lakes lacking calcareous sediments.

Unfortunately, profiles of biogenic silica are not easy to interpret in terms of their paleoclimatic significance. Paleoceanographers interpret such profiles in marine sediments simply in terms of upwelling intensity; more intense upwelling of nutrient-rich, deep waters leads to higher primary production and results in a higher burial rate of diatoms and radiolarians in the underlying sediments. In lakes, the reservoir of dissolved silica is far more limited than in the oceans, and a prolonged increase in upwelling intensity results in only a temporary rise in the accumulation rate of diatoms before the reservoir of dissolved silica is drawn down. Profiles of biogenic silica are best interpreted after consideration of the geochemical mass balance of silica in the lake basin, and an appraisal of what part of the mass balance has been perturbed to yield a shift in the burial rate of diatoms.

Here I present examples of records of biogenic silica from Lake Malawi and Lake Victoria in East Africa (Fig. 1). The examples provide evidence of major shifts in the burial rates of diatoms on time scales ranging from decades to millennia. The causes for these shifts are driven in most cases by climatic forcing, but anthropogenic influences are apparent as well in one of the examples that is provided.

## THE GEOCHEMICAL MASS BALANCE OF SILICA IN LAKES

Ultimately the mass accumulation rate of biogenic silica in lake sediments is limited by the net input of dissolved silica to the lake. Inputs can be dissolved silica from river and groundwater inflow, the silica released from "soluble particulate silica" (e.g., fluvial diatoms) introduced by rivers or wind, and the dissolved silica from submerged hot springs. In a large lake, groundwater inflow and outflow are likely to be relatively small components of the hydrological budget, and the input of wind-blown phytoliths is also likely to be relatively small. Soluble particulate silica from rivers may be a significant contributor to the silica budget (Bootsma and Hecky, 1998), although Treguer et al. (1995) did not consider this to be the case for the world ocean. In addition to the burial of diatoms, the authigenic formation of new silicate minerals can be an important sink for dissolved silica in large lakes. It is extremely difficult to quantify, however, because the newly formed silicates are likely to be poorly crystallized or difficult to distinguish from detrital silicates. In Lake Superior, diatom burial accounts for only 10% of the net input of dissolved silica, leaving smectite authigenesis as the most likely (but unsubstantiated) major sink (Johnson and Eisenreich, 1979). This apparently is unlike the oceans, where the annual input of dissolved silica from the continents is nearly balanced by the burial of diatoms in marine basins (Treguer et al., 1995).

Sedimentation in Continental Rifts
SEPM Special Publication No. 73, Copyright © 2002
SEPM (Society for Sedimentary Geology), ISBN 1-56576-082-4, p. 247–255.

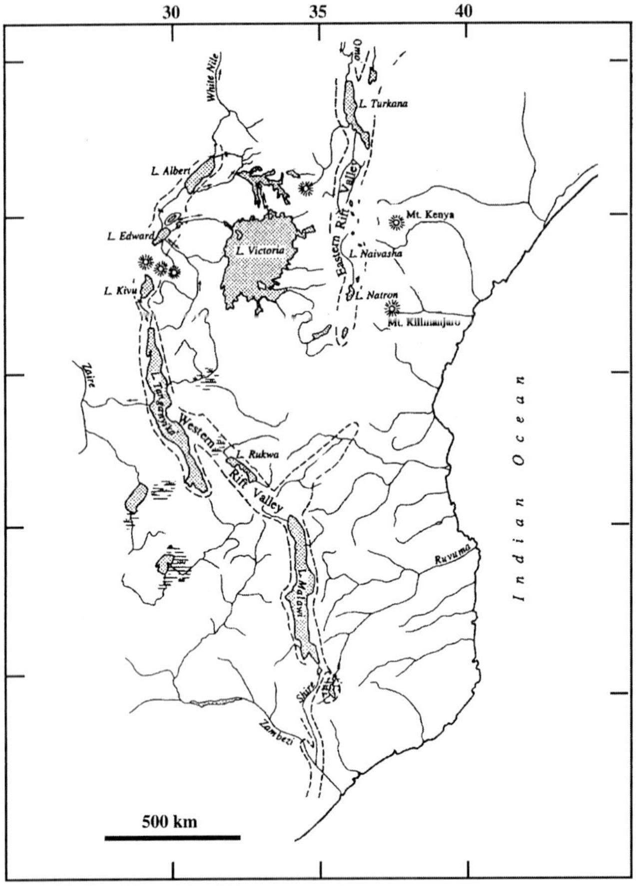


FIG. 1.—The large lakes of East Africa, showing the locations of Lake Malawi at the south end of the Rift Valley and Lake Victoria, situated on the equator between the western and eastern arms of the Rift Valley. (Modified from Beadle, 1981.)

The geochemical mass balance equation for dissolved silica is:

$$\frac{\partial S}{\partial t} = Q_r S_r - Q_0 S_0 - MAR_{BSi} A_s - MAR_{AS} A_s \qquad (1)$$

where:

$S$ = total mass of dissolved silica in the water column (g $SiO_2$)
$Q_r$ = annual input of river water ($cm^3/yr$)
$S_r$ = average concentration of dissolved silica in river water (g $SiO_2/cm^3$)
$Q_o$ = annual river outflow from the lake ($cm^3/yr$)
$S_o$ = average concentration of dissolved silica in the outflowing water (g $SiO_2/cm^3$)
$MAR_{BSi}$ = mass accumulation rate of biogenic silica in the sediments (g $SiO_2/cm^2yr$)
$A_s$ = the area of lake floor in which biogenic silica and authigenic silicates are accumulating ($cm^2$)
$MAR_{AS}$ = mass accumulation rate of $SiO_2$ in authigenic silicate minerals (g $SiO_2/cm^2yr$)

This equation does not address internal cycling of silica within the water column or the upper few centimeters of the sediments. Silica is taken up by diatoms in the water column, and most of the diatom frustules redissolve before they can be preserved in the sediments. In Lake Superior, for example, the total input of dissolved silica is about $5 \times 10^8$ kg $SiO_2/y$, and the annual uptake of silica by diatoms is estimated to be about an order of magnitude greater (Johnson and Eisenreich, 1979). Thus about 90% or more of the silica incorporated into diatom frustules each year is redissolved. A similar preservation factor was found for Lake Michigan (Parker and Edgington, 1976). Clearly, the magnitude of this biological uptake factor imposes seasonal or even interannual fluctuations on the concentration of dissolved silica in the water column that are not addressed in Equation 1, which concerns the long-term geochemical mass balance for silica. Equation 1 states that at steady state, the amount of biogenic and authigenic silica that can accumulate in the sediments equals the net input to the lake. If the burial rate exceeds the input rate, the dissolved silica concentration goes down, resulting in an acceleration of the rate of dissolution of biogenic silica in the hypolimnion or on the lake floor, approximated by the equation (Hurd, 1972):

$$\frac{\partial C}{\partial t} = -kA(C_{sat} - C) \qquad (2)$$

where:

$C$ = concentration of dissolved silica in the bottom water (μM)
$k$ = first-order rate constant for dissolution (μM/$cm^2$•s)
$A$ = specific surface area of biogenic opal exposed to dissolution ($cm^2$/μg $SiO_2$)
$C_{sat}$ = the saturation concentration for biogenic silica (μM). This value is about 2000 μM at 20°C and about half that value at 0°C (Hurd, 1972; Kato and Kitano, 1968).

Conversely, if the input rate exceeds the burial rate, the dissolved silica concentration in the lake rises, resulting in retardation of the dissolution rate and an increase in the burial rate.

So the system is self-correcting, leading to a geochemical mass balance in most instances. However, the response of the mass balance mechanisms may require decades to adjust.

## THE NONLINEAR RESPONSE OF PROFILES OF BIOGENIC SILICA TO CHANGES IN PRIMARY PRODUCTION

The concentration of biogenic silica varies down core in response to environmental conditions that ultimately, in most instances, are linked to climatic forcing. In both lakes and oceans, variations in the mass accumulation rates of biogenic silica are interpreted as indications of past primary production of diatoms (or, at places in the oceans, secondary production of radiolarians) (Charles et al., 1991; Colman et al., 1995; Ragueneau et al., 2000). This interpretation carries many assumptions that, at most, may be valid only in sediments where siliceous microfossils are sufficiently abundant to elevate pore-water concentrations of dissolved silica to saturation values with respect to biogenic opal. Otherwise, postdepositional dissolution of the microfossils may be substantial, with diagenetic imprint on the abundance of siliceous microfossils overwhelming any signature of past productivity or climatic forcing. Diatom productivity does not necessarily reflect total primary productivity, and the ratio Si:C in phytoplankton populations can vary substantially in response to the availability of light, dissolved silica, and trace metals (reviewed in Ragueneau et al., 2000).

In the marine environment, changing productivity is attributed primarily to a shift in the intensity of upwelling. At one time, volcanic eruptions were considered important, but the classic study by S.E. Calvert in the Gulf of California dispelled the idea. He carried out a mass-balance study of silica in the volcano-prone Gulf and demonstrated that the diatomaceous ooze found there could be explained solely by the net input of dissolved silica from offshore deep ocean water. It required no additional input of silica from nearby volcanoes (Calvert, 1966). Upwelling in response to wind-driven circulation brings nutrient-rich deep waters to the photic zone, promoting high production of diatoms. This results in a localized high mass accumulation rate of biogenic silica in the underlying sediments. Interpreting the curve of biogenic silica in marine sediments in terms of upwelling alone is viable on a local scale because the source of dissolved silica in the deep waters of the ocean can be considered as infinite relative to the area of upwelling and deposition.

In lakes, even large lakes, primary production may change as a result of variations in upwelling intensity or the influx of limiting nutrients (e.g., nitrate or phosphate) from the drainage basin. In addition, the reservoir of dissolved silica in the deep waters is finite. To illustrate this, consider Lake Malawi, the second largest lake in the East African Rift Valley. It is about 600 km long, 35 km wide, and 700 m deep. H. Bootsma and his colleagues have been working on aspects of the lake's silica budget (Bootsma and Hecky, 2000). River input of dissolved and soluble particulate silica to the lake, normalized to lake area, is in the range of 650–1110 mmol/m$^2$•yr. (Bootsma and Hecky define "soluble particulate silica" as the amount of silica dissolved when fluvial suspended sediment is digested in 0.1 N $NaHCO_3$ at 105°C for two hours. This may overestimate the amount of silica released when the suspended matter reaches a lake, because our research has found that a substantial amount of silica is released from silicate minerals after about 40 minutes of digestion in a warm base solution.) Outflow down the Shire River is just 13 mmol/m$^2$•yr, resulting in a net input of 635–1100 mmol/m$^2$•yr that can accumulate in the bottom sediments as biogenic or authigenic silica. We typically find 5–15% biogenic silica in Lake Malawi sediments, accumulating at a rate of 1 mm/yr. When the spatial variability in biogenic silica content in surface sediments is coupled with high-resoution seismic reflection profiles to better estimate the real extent of deposition, this translates into an overall burial rate of about 6 x 10$^8$ kg SiO$^2$/yr, or 360 mmol/m$^2$•yr. This is only one-third to one-half of the annual net input of dissolved silica and soluble particulate silica to the lake, and it suggests that silicate authigenesis plays a major role in Lake Malawi, just as it does in Lake Superior. The iron silicate nontronite appears to be at least one of the primary authigenic silicates formed in Lake Malawi sediments (Müller and Förstner, 1973). However, if soluble particulate silica input to Malawi is negligible, as Treguer et al. (1995) claim for the oceans, then burial of biogenic silica in the sediments nearly equals the river input of dissolved silica to the lake, and there is no need for silicate authigenesis to balance the silica budget.

The record of biogenic silica in lake sediments will not track a permanent rise in productivity indefinitely, because of the finite supply of dissolved silica available in the water column. An abrupt rise in diatom productivity, due to either increased upwelling or phosphate input, would initially be expected to cause a comparable rise in diatom burial rate in the underlying sediments. However, a silica deficit would result, creating conditions less favorable for both diatom production and preservation. Johnson and Eisenreich (1979) modeled the response of silica concentration in Lake Superior to eutrophication caused by phosphorus loading, assuming that the system adjustment was due solely to the mechanism of diatom preservation, and derived the following equation:

$$C = C_f + (C_0 - C_f)\exp\left[\frac{-k'P}{V}(1+y)t\right] \qquad (3)$$

where:

$C$ = the concentration of silica in the water column at time $t$ after the abrupt rise in diatom productivity (µM).
$C_0$ = the concentration of silica in the water column just prior to the change in productivity (µM)
$C_f$ = the new steady-state concentration of silica in the water column required to balance silica inputs and outputs (µM)
$k'$ = dissolution rate constant, similar to $k$ in Equation 2, but in this case related to diatom productivity rather than to diatom surface area.
$P$ = diatom frustule production (kg $SiO_2$/y)
$y$ = the proportional increase in diatom productivity
$V$ = volume of water in the lake basin

Equation 3 is applied to Lake Malawi to determine how long it would take for this tropical lake system to respond to a change in productivity. I assume that the silica deficit is ultimately compensated by a drawdown of dissolved silica concentration in the hypolimnion, causing the rate of diatom dissolution on the lake floor to increase and the diatom burial rate to return to its original state. The concentration of dissolved silica in the hypolimnion of Lake Malawi is about 225 µM (Bootsma and Hecky, 1998) and its volume (below 400 m) is 634 km$^3$. These are the values assigned to $C_0$ and $V$, respectively, in Equation 3. $C_f$ is determined from:

$$\Delta C_f = \frac{k'\Delta C_0 + y}{k'(1+y)}$$

$$\Delta Cf = C_{sat} - C_f \qquad (4)$$

$$\Delta C_0 = C_{sat} - C_0$$

(Johnson and Eisenreich, 1979) where $C_{sat}$ is the saturation concentration for biogenic opal at Malawi bottom water temperature, approximately 2000 µM (Hurd, 1972). The value of $k'$ is determined from a variation on Equation 2:

$$D = k'\Delta C_0 P \qquad (5)$$

where $D$ is amount of diatom frustule dissolution in kg $SiO_2$/y. Primary productivity in Lake Malawi has been measured on only a few occasions, so it is not well known. For this calculation I assume that 90% of the silica incorporated into diatom frustules redissolves, and only 10% of the annual diatom production is buried (Parker and Edgington, 1976). Given the burial rate of 6 x 10$^8$ kg $SiO_2$/y (above), the value for $P$ becomes 6 x 10$^9$ kg $SiO_2$/y.

A large increase in diatom production in Lake Malawi, caused by either increased phosphate loading or a change in upwelling intensity, would not result in a biogenic-silica record in the sediments that linearly tracked this history of events. Application of the equations above indicate that if upwelling increased abruptly, and remained high for a few millennia, the sedimentary record would show an equally abrupt rise in the burial rate of diatoms, but the rise would drop off exponentially over just a few decades (Fig. 2). The assumptions made in the above equations are certainly subject to refinement, but the decadal time scale of the lake response is not likely to change significantly as the assumptions are adjusted.

Despite this nonlinear response of only a few decades duration to a prolonged shift in the intensity of upwelling, dramatic and long-lived (millennial) changes are observed in the biogenic silica profiles of Lake Malawi sediments. This can happen only by changing the rate of input or outflow of dissolved silica to or from

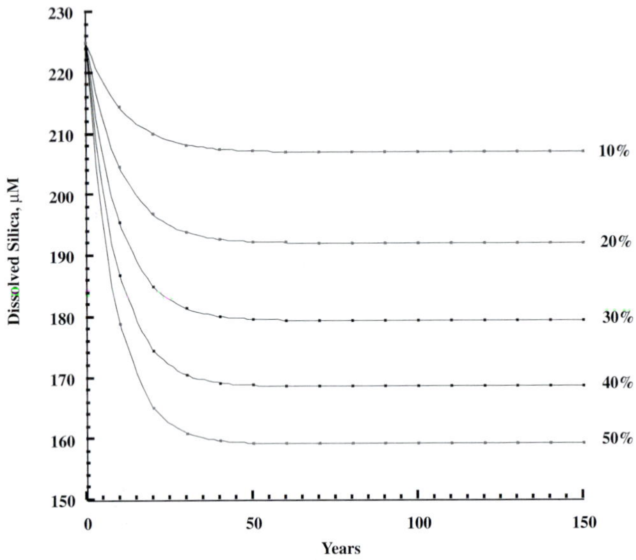


FIG. 2.—Model results for Lake Malawi based on Equations 3–5. The curves represent how the concentration of dissolved silica in the hypolimnion of Lake Malawi would change in time, after a sudden increase in the rate of diatom productivity at *t* = 0. Curves represent productivity increases of 10, 20, 30, 40 ,and 50%. In each case the concentration of dissolved silica drops within five decades to values that will increase the rate of diatom dissolution required to maintain a geochemical mass balance of dissolved silica in the system.

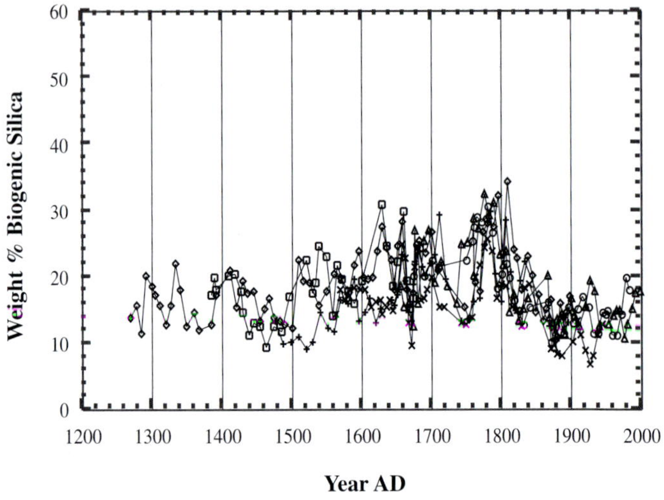


FIG. 3.—Biogenic silica abundance in six short cores recovered from the northern basin of Lake Malawi. The chronology is based on analyses of $^{210}Pb$ and varve counts. (Modified from Johnson et al., 2001.)

the lake, or by shifting the location of diatom deposition within the lake, or by changing the magnitude of formation of authigenic silicate in the sediments. Silicate authigenesis is difficult to quantify, and it is unlikely to cause the observed variations in biogenic silica profiles. Authigenesis of new silicate minerals is likely to be enhanced when diatoms are abundant in sediments, because under such circumstances the concentration of dissolved silica in pore waters is relatively high. Thus when diatom burial is retarded, silicate authigenesis should also be retarded—one sink for dissolved silica is not likely to replace the other. Thus we are left with variations in the inflow or outflow of dissolved silica or a shift in the location of major diatom deposition within the lake to explain observed shifts in biogenic silica profiles. We see evidence for both having happened in Lake Malawi.

## LAKE MALAWI

High-resolution profiles of biogenic silica obtained from six cores in northern Lake Malawi, spanning the last 600 years, show a strong signal of enhanced biogenic silica concentrations during the Little Ice Age (LIA) (Johnson et al., 2001) (Fig. 3). There is independent evidence for lake level having been about 100 m lower than present during much of this time because of drier conditions in the basin (Owen et al., 1990). With lake level this much lower than present, the very highly productive southwest and southeast arms of the lake would have dried up completely. Analyses of surface sediments recovered by Project PROBE in 1986 show that the richest accumulations of diatom ooze (approximately 50% BSi) are found in these southern extremeties of the lake. If these high-productivity regions disappeared (i.e., dried out) as a consequence of lower lake level, the mass balance of biogenic silica could be maintained only by shifting the locus of high rates of diatom burial to deeper sites, including the north basin.

While a lower lake level is one explanation for this case, it is not the only means by which enhanced diatom burial could be accomplished in the north basin of the lake. A shift in the prevailing direction of the winds might change the lake circulation patterns, with a consequent shift in regions of high productivity within the lake. On a much longer time scale, we see pronounced structure in biogenic silica profiles spanning the last 25,000 years in five piston cores from the north basin of the lake (Fig. 4). The profiles correlate well with one another, indicating that the cores record a basinwide phenomenon. However, this is not seen near the center of the lake, where a core exhibits little variation in biogenic silica over the past several thousand years (Fig. 4). This probably reflects the upwelling dynamics of a long rift lake, which are strongly influenced by internal seiches. Plisnier and Coenen (2001) documented this clearly for Lake Tanganyika, where upwelling is well established at the south end of the lake during the season of intense south winds. At the end of the windy season, the thermocline is depressed in the north basin because of prevailing winds out of the south. As the winds subside, the thermocline rebounds, causing an internal wave that oscillates back and forth through the basin with a period of about a month. The crest of this internal wave breaks into the photic zone at the north end of the lake periodically, creating a pulse of high productivity (Plisnier and Coenen, 2001). The same happens at the south end of the lake as well, but not in the middle of the lake, where the internal seiche is much subdued in amplitude. So primary production dynamics appear to be amplified at the ends of Lake Tanganyka and be much less variable near the center of the lake. This is probably the case in most long, narrow, rift lakes, including Malawi. This would explain the differences observed in profiles of biogenic silica from the center of the lake versus the north basin.

The profiles of biogenic silica in cores M98-1P and M98-2P are replotted in terms of mass accumulation rate (*MAR*) of biogenic silica, on the basis of the following equation:

$$MAR_{BSi} = \%BSiLSR(1 - \Phi)\rho \qquad (6)$$

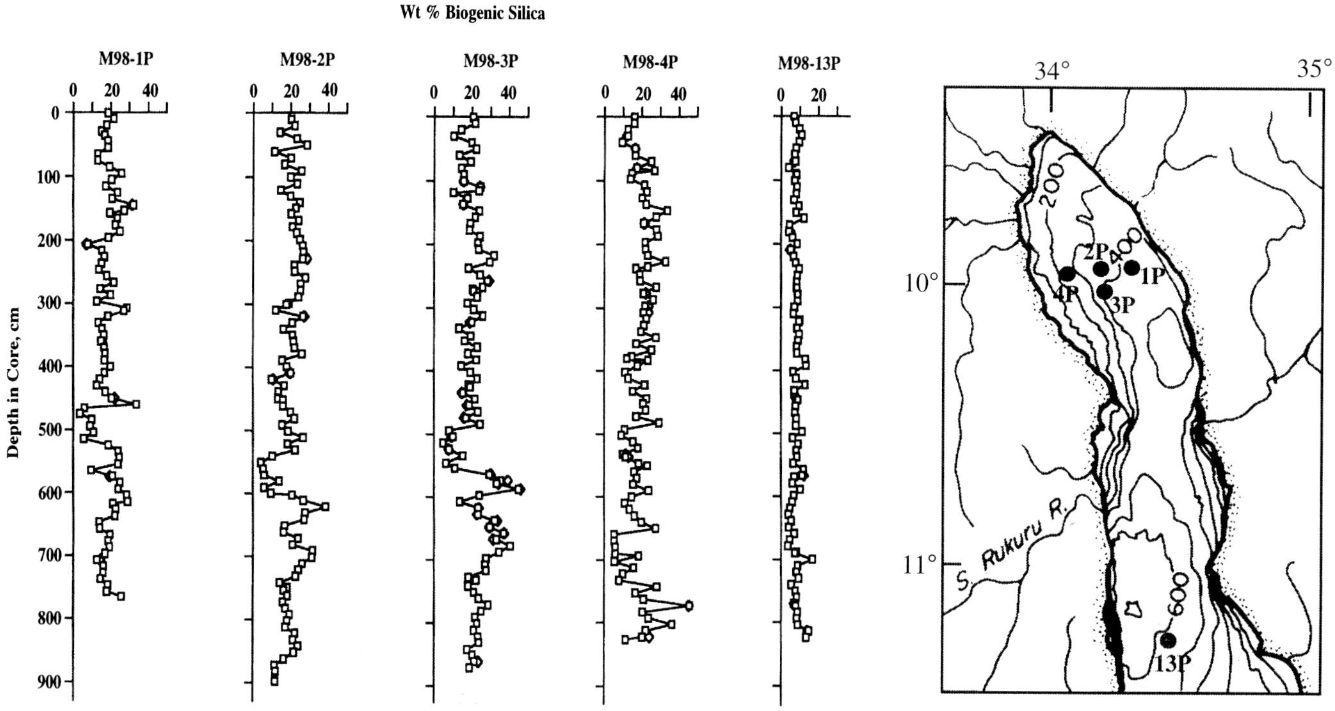


FIG. 4.—Profiles of biogenic silica in piston cores recovered from the northern and central basins of Lake Malawi. Note the lower abundance and less variability in core M98-13P from the central basin, reflecting the less dynamic and less productive environment there. Core locations are indicated on the bathymetric map (contours in meters).

where:
$MAR_{BSi}$ is the mass accumulation rate of biogenic silica
$LSR$ = the linear sedimentation rate
$\Phi$ = sediment porosity, derived from water content
$\rho$ = dry sediment density, assumed to equal 2.54 g/cm$^3$

The linear sedimentation rates in core M98-1P is based on six radiocarbon dates and in M98-2P is based on eight radiocarbon dates (Barry et al., in press; Johnson et al., 2002).

The $MAR_{BSi}$ profiles of the two cores show millennial-scale similarities with the Greenland ice core records of $\delta^{18}O$ (e.g., the GRIP record) (Fig. 5) (Johnson et al., 2002). The $MAR_{BSi}$ profiles have lower mean values in the Pleistocene than in the Holocene, and the transition from Pleistocene to Holocene values occurs in two abrupt steps, with an intervening interval of Pleistocene-like conditions. The timing of the glacial–interglacial transition, however, is different from the GRIP record. The initial rise to Holocene-like values at 13 ka occurred as the Northern Hemisphere plunged back into glacial-like conditions at the beginning of the Younger Dryas. In fact, all of the peaks and valleys of the Pleistocene portion of the $MAR_{BSi}$ profile are anti-phased with the GRIP record (Fig. 5). When $MAR_{BSi}$ values rise (i.e., towards Holocene values), the GRIP record drops to colder temperatures. The African tropical climate thus appears to have been linked to the ice-sheet and climate dynamics of the Northern Hemisphere in the late Pleistocene.

The relatively low $MAR_{BSi}$ during the Late Pleistocene is attributed to the dry, cool conditions that characterize much, if not all, of tropical Africa during the last ice age. Temperatures were likely 5–8°C cooler than present (Gasse, 2000; Livingstone, 1993), and recent evidence from the diatom record in core M98-2P suggests that Lake Malawi was at a low stand (Gasse et al., in press). Under these circumstances, the rate of weathering would have been lower than at present, resulting in a lower fluvial influx of dissolved silica to the lake basin.

The millennial-scale variations in $MAR_{BSi}$ (e.g., the peaks of about 1000–2000 years duration centered on 22 ka, 15 ka, and 13 ka) do not track lake-level changes that are estimated by Gasse et al. (in press) based on the relative abundance of periphytic diatoms. Johnson et al. (2002) hypothesize that these peaks in $MAR_{BSi}$ are the result of more frequent or stronger winds out of the north, which would promote upwelling in the north basin.

Measurements of the ratio of the niobium to the titanium concentrations in the sediments of core M98-2P support this hypothesis. Volcanic ash layers in northern Lake Malawi sediments are enriched in certain incompatible elements (i.e., resistant to incorporation into mineral phases, thus enriched in volcanic glass), including niobium. Consequently the relative abundance of volcanically derived material in the sediments is reflected in a profile of Nb/Ti. The intervals of high $MAR_{BSi}$ prior to 11,000 years ago all coincide with times of high Nb/Ti ratios (Johnson et al., 2002).

The volcanic ash in Lake Malawi comes from the Rungwe volcanoes to the north of the lake. Two discrete ash layers derived from these volcanoes are observed in the piston cores and date at 4.4 and 9.1 ka. Aside from the sharp peaks associated with these layers, the Nb/Ti ratio appears to have higher average values in the Pleistocene than in the Holocene sediments. North winds blowing over the dry volcanic landscape would have transported ash to the lake, resulting in higher Nb/Ti in lake sediments. Such winds, which would have caused upwelling and increased diatom productivity at the north end of the lake, may be a regional response to global shifts in late Pleistocene climate. For example,

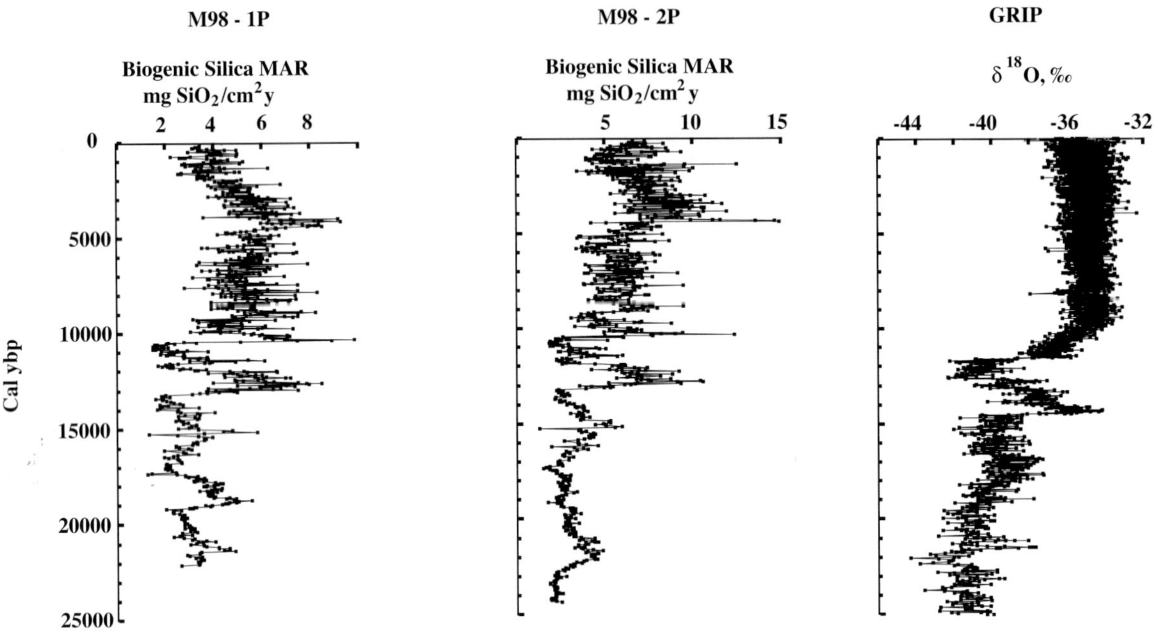


FIG. 5.—The mass accumulation rate of biogenic silica in cores M98-1P and 2P, based on analyses for BSi at 1 cm intervals. The geochronology of these cores is based on six radiocarbon dates in M98-1P and eight radiocarbon dates in M98-2P (Barry et al., in press). The profiles exhibit a lower diatom burial rate during the last glacial maximum than in the Holocene and a distinct, two-step deglaciation signal similar to the GRIP $\delta^{18}O$ record from Greenland (Johnson et al., 2002).

a more southward excursion of the Intertropical Convergence Zone would expose northern Malawi and southwestern Tanzania to the northeasterly winds of the northern side of the ITCZ.

While the Malawi $MAR_{BSi}$ profiles deviate significantly from the GRIP oxygen isotope record after 11 ka, subtle century-scale links remain between the climate of the Northern Hemisphere and the Lake Malawi basin. The $MAR_{BSi}$ in northern Lake Malawi was elevated during the Little Ice Age (Fig. 3) and it was low at 1 ka, the time of the Medieval Warm Period in the Northern Hemisphere. So perhaps it was not the low stand of the lake that accounts for the higher $MAR_{BSi}$ during the Little Ice Age, but rather it was the recurrence of more frequent or stronger north winds coinciding with a cooler Northern Hemisphere and a southward shift of the ITCZ (Johnson et al., 2002).

Although Calvert (1966) mostly dispelled the myth that volcanism plays a major role in the abundance of diatoms in marine sediments, his study does not necessarily apply to lakes in volcanic terrain. The Rungwe volcanic field lies just north of Lake Malawi in southern Tanzania. Rungwe and Keijo volcanoes appear to be very young, with an eruption from the latter estimated to have occurred around the end of the Seventeenth Century (Harkin, 1960). The piston cores from northern Lake Malawi contain two major ash layers with ages of about 4.4 and 9.1 ka, as well as several other ash horizons just millimeters thick. There appears to be a positive but short-lived enhancement of diatom productivity associated with some of the ash layers (Fig. 6). The ash layers are overlain by a spike in biogenic silica, but they do not mark the onset of prolonged high burial rates of diatoms.

## LAKE VICTORIA

Lake Victoria is substantially different than Lake Malawi in its morphology and origin. Malawi is a narrow, deep, rift lake. Victoria is much larger in surface area, nearly equidimensional in plan, and only 68 m deep (Johnson et al., 1996). Victoria formed around 400,000 years ago as a result of uplift on the western branch of the East African Rift, causing back flow of the major river drainage through the area, which had previously been east to west (Bishop and Trendall, 1967). Back flow resulted in ponding and the formation of this broad, shallow tropical lake.

Several piston cores were recovered from offshore Lake Victoria in 1995 and 1996 during the first two years of field work by the International Decade for the East African Lakes (IDEAL). Profiles of biogenic silica in these cores can be correlated across the entire lake basin (Fig. 7; Johnson et al., 1998). Lake Victoria dried out completely in the Late Pleistocene and began to refill around 14,600 yr BP (Johnson et al., 1996; Johnson et al., 2000). The concentration of biogenic silica rose rapidly as the lake basin filled, to about 20 wt.%, then dropped off dramatically around 9,800 yr BP. The following two thousand years were a time of very low diatom abundance and preservation in the offshore sediments of Lake Victoria. It apparently was an unfavorable time for diatoms to compete successfully with other phytoplankton. This was a relatively wet and warm period in equatorial Africa, and it is also believed to have been a period of low wind intensity and enhanced stratification in the water column (Stager and Johnson, 1999). Under such conditions, diatoms may have been outcompeted by blue-green algae. This would have resulted in less uptake of dissolved silica in the surface waters, allowing it to leave the lake via the Nile River outflow.

While it may seem unlikely that diatoms would not utilize nearly all dissolved silica available in the photic zone, two lines of evidence indicate that river outflow can indeed be an important sink for dissolved silica, even in a eutrophic lake. First, the surface-water concentration of dissolved silica in Lake Victoria in 1960 was quite high, at about 75 μM (Talling, 1966). Multiplying this concentration by the Nile outflow from the lake shows that

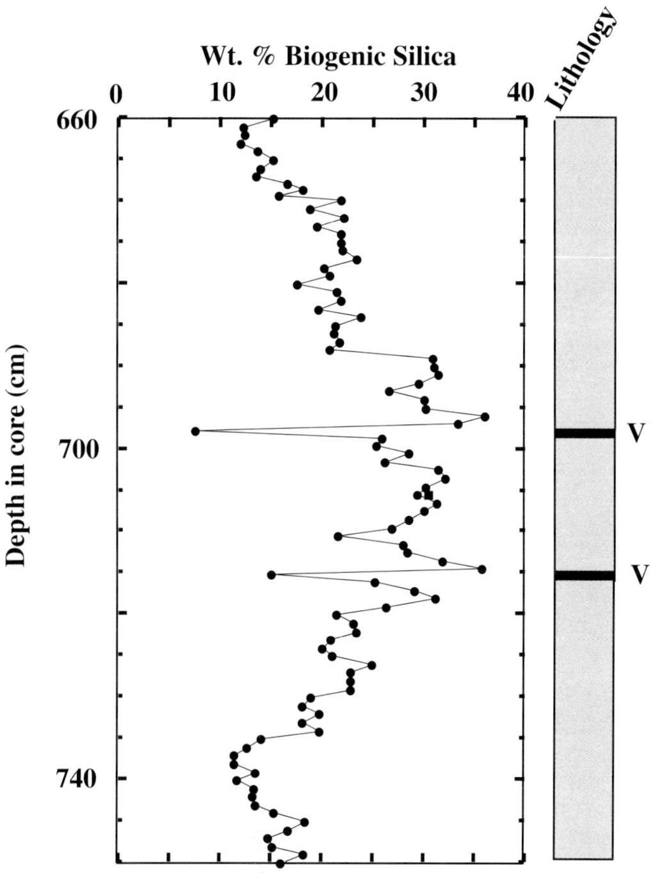


FIG. 6.—High-resolution data from core M98-2P, showing the short-lived effect of volcanism on diatom productivity in Lake Malawi. Volcanic ash layers are indicated by the black bands at 698 cm and 715 cm depths sandwiched between intervals of varved diatomaceous mud (light gray). The ash layers have relatively low percentages of biogenic silica in them. Unusually high abundances of biogenic silica are found in the 1–2 cm overlying the ash layers, corresponding to 15–20 years duration.

nearly half of the dissolved silica flowing into the lake at that time was lost to outflow (Johnson et al., 1998). The surface-water concentration of dissolved silica is even higher in Lake Edward, where it was measured at 150 μM in 1995 and even higher in the 1960s (Lehman et al., 1998). Second, mesocosm experiments have shown that when diatoms are outcompeted by blue-green algae, they may not take up dissolved silica available in the photic zone, even if silica is added to spur diatom productivity (Klemer and Barko, 1991).

The abrupt drop in diatom abundance in Lake Victoria at 9800 yr BP was accompanied by a substantial shift in the oxygen isotope composition of the lake waters, from $\delta^{18}O$ values of about +8 per mil to about 0 per mil (Beuning et al., 1997). This was interpreted to mark the reinitiation of permanent Nile outflow from Lake Victoria after the reestablishment of the lake in the early Holocene (Beuning et al., 1997; Johnson et al., 2000). Recent evidence based on the strontium isotope composition of shell and fish remains in the upper Nile, however, indicate that Nile overflow began much earlier, at about 13,500 yr BP (Talbot et al., 2000). So the precipitous drop in the apparent isotopic composition of Lake Victoria waters at about 9800 yr BP remains unexpained. Nevertheless, Nile outflow was occurring at this time, and here is a case where a major shift in biogenic silica resulted not from a change in input to the lake but from a change of outflow from the basin due to plankton dynamics tied to water-column stratification.

Lake Victoria has undergone severe eutrophication and degradation of the lake ecosystem in the past few decades, primarily due to increased human activity on the lake periphery (Hecky, 1993). A profile of biogenic silica in a short multi-core recovered from Lake Victoria records the anthropogenic impact quite clearly (Fig. 8) (Verschuren et al., 1998). The profile shows preindustrial concentrations of biogenic silica in the offshore sediments to be around 6–7%. Around 1930 the silica concentration began to rise, and it continued to rise until the 1980s. The abrupt decline in concentration of biogenic silica after the mid-1980s coincides with the replacement of diatoms by cyanobacteria as the dominant phytoplankton in the lake (Verschuren et al., 1998). The significant rise in concentration of biogenic silica in the uppermost 2 cm of the core is an artifact of early diagenesis and not an indication of a recent rise in diatom productivity (Verschuren et al., 1998).

## CONCLUSIONS

The abundance of biogenic silica in the sedments of large tropical lakes can shift significantly through time. In contrast to marine systems, such variations cannot be attributed solely to changes in upwelling intensity. The reservoir of dissolved silica in the water column of large lakes is finite, and it can be depleted within a few decades if upwelling intensifies because of climate forcing. Consequently such a phenomenon would be accompanied by a brief, not prolonged, increase in diatom burial rate.

Many factors associated with climate change can affect profiles of biogenic silica in lake sediments. In Lake Malawi, a shift in the locus of high productivity due to changing winds caused significant millennial-scale peaks in concentration of biogenic silica in sediments deposited in the north basin of the lake during the Late Pleistocene. North winds that promoted upwelling in the north basin of the lake appear to have been linked to relatively cool periods in the Northern Hemisphere. A more subtle but consistent relationship apparently existed between upwelling and diatom productivity in northern Lake Malawi and the climate of the Northern Hemisphere during the Little Ice Age and the Medieval Warm Period. Variations in lake level may also have played a role in this relationship, particularly during the Little Ice Age, with the locus of high diatom burial rates shifted into deeper basins as shallower high-productivity zones dried out. Such shifts in diatom productivity are likely to be most pronounced at the ends of rift lakes, not near their centers. Upwelling zones are most pronounced at the ends of rift lakes, in response to winds that are usually channeled up or down a rift valley. The oscillatory behavior of internal waves in the water column is also likely to be amplified at the ends of rift lakes as well, where they can impart pulses of high productivity in the euphotic zone.

A major reduction in the input rate of dissolved silica to a lake basin may result from dry conditions in the drainage basin, as in Lake Malawi during the Late Pleistocene. The burial rate of biogenic silica in a basin cannot exceed the net input rate of dissolved silica to the system. If the input rate is substantially reduced, then so too will be the diatom mass accumulation rate.

Another side of the geochemical mass balance of silica in a lake system is in the outflow term. Lake Victoria experienced a major loss of dissolved silica in Nile outflow, around 9800 years ago. Cyanobacteria apparently outcompeted diatoms for about

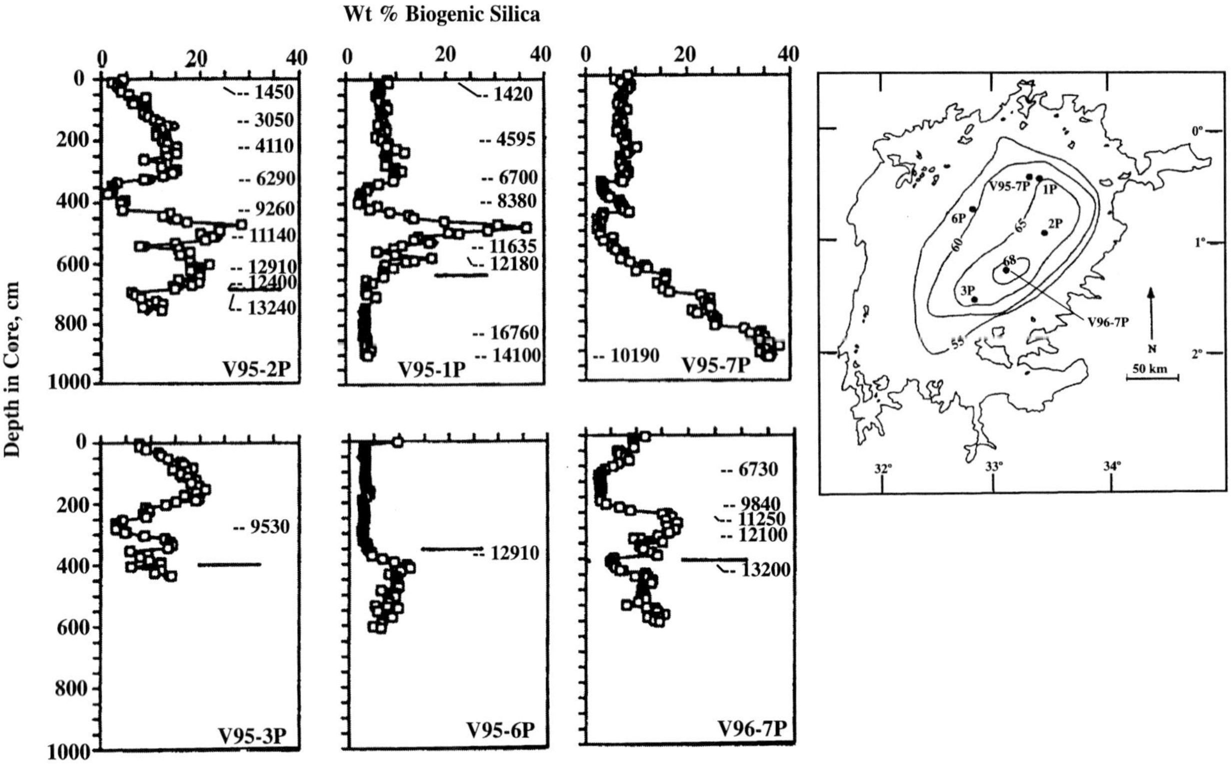


FIG. 7.—Profiles of biogenic silica in six piston cores from Lake Victoria, spanning about the last 17,000 years. Uncorrected radiocarbon dates are indicated on each profile and demonstrate that the peaks and valleys in these curves correlate in time. The horizontal line in five of the profiles indicates the Late Pleistocene desiccation surface. (Modified from Johnson et al., 1998.)

2000 years in this moist, highly stratified period, allowing concentrations of dissolved silica in the photic zone and river outflow to remain relatively high.

Finally, anthropogenic activities can affect the geochemical mass balance of silica to the point of leaving a measurable signal in the sediments of even very large tropical lakes. In Lake Victoria, the concentration of biogenic silica in sediments began to rise in the 1930s when human population and agricultural activities began to increase substantially in the lake basin. The resultant eutrophication of the lake system first resulted in an elevated burial rate of diatoms in the sediments, followed by a drop in the concentration of biogenic silica as cyanobacteria progressively outcompeted diatoms as the major primary producers in the lake system.

## ACKNOWLEDGMENTS

I thank Yvonne Chan of the Large Lakes Observatory for her extensive analyses of biogenic silica in sediment cores that we have recovered from the East African lakes. These cores could not have been obtained without the assistance of the able captains and crews of the R/V Usipa on Lake Malawi and the R/V Ibis on Lake Victoria. I greatly appreciate the help and enthusiasm of Mike Talbot, Dirk Verschuren, and other members of the IDEAL science teams in the field. This manuscript benefited from careful reviews by R.E. Hecky and H. Bootsma. This research was funded through grants from the U.S. National Science Foundation, with additonal ship time and logistical support in Malawi provided by the SADC/GEF Lake Malawi Biodiversity Project. This is publication number 130 of the International Decade for the East African Lakes (IDEAL).

## REFERENCES

BARRY, S.L., FILIPPI, M.L., TALBOT, M.R., AND JOHNSON, T.C., in press, A 20,000-yr sedimentological record from Lake Malawi, East Africa: the Late-Pleistocene / Holocene transition in the southern tropics, *in* Odada, E.O., and Olago, D.O., eds., The East African Great Lakes: Limnology, Palaeoclimatology and Biodiversity: Dordrecht, The Netherlands, Kluwer Academic Publishers.

BEADLE, L.C., 1981, The Inland Waters of Tropical Africa; An Introduction to Tropical Limnology: London, Longman, 475 p.

BEUNING, K., KELTS, K., ITO, E., AND JOHNSON, T.C., 1997, Paleohydrology of Lake Victoria, East Africa, inferred from $^{18}O/^{16}O$ ratios in sediment cellulose: Geology, v. 25, p. 1083–1086.

BISHOP, W.W., AND TRENDALL, A.F., 1967, Erosion surfaces, tectonics and volcanic activity in Uganda: Geological Society of London, Quarterly Journal, v. 122, p. 385–420.

BOOTSMA, H.A., AND HECKY, R.E., 1998, Water Quality Report: Draft Document: SADC/GEF Lake Malawi Nyasa Biodiversity Conservation Project.

BOOTSMA, H.A., AND HECKY, R.E., 2000, Inputs, outputs and internal cycling of silica in Lake Malawi, *in* IDEAL 2nd International Conference Symposium, Club Makokola, Malawi, p. 21–22.

CALVERT, S.E., 1966, Accumulation of diatomaceous sediments in the Gulf of California: Geological Society of America, Bulletin, v. 77, p. 569–596.

CHARLES, C.D., FROELICH, P.N., ZIBELLO, M.A., MORTLOCK, R.A., AND MORLEY, J.J., 1991, Biogenic opal in southern ocean sediments over the last 450,000 years: implications for surface water chemistry and circulation: Paleoceanography, v. 6, p. 697–728.

COLMAN, S.M., PECK, J.A., KARABANOV, E.B., CARTER, S.J., BRADBURY, J.P., KING, J.W., AND WILLIAMS, D.F., 1995, Contiinental climate response to

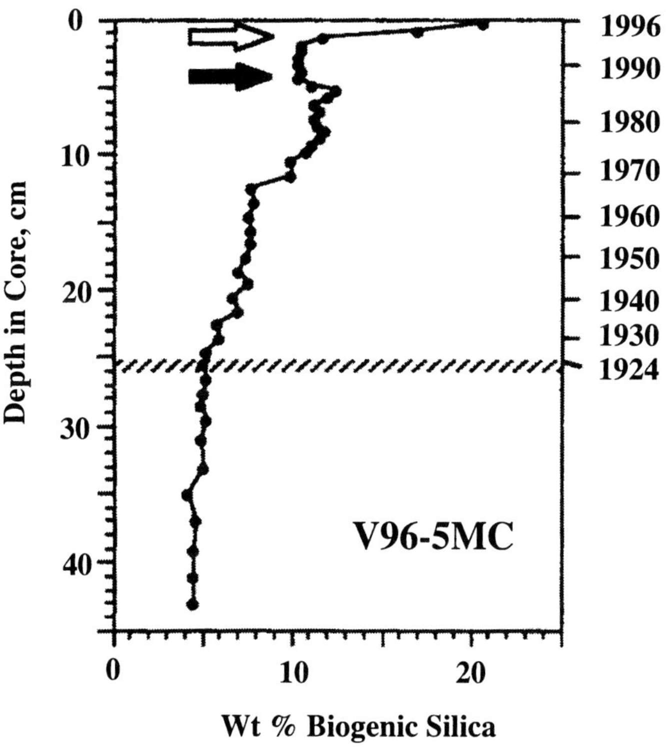


FIG. 8.—A profile of the abundance of biogenic silica in a short core from Lake Victoria, reflecting anthropogenic impact on the lake ecosystem. The line corresponding to 1924 indicates the beginning of the rise in biogenic silica that coincides with rising human population around the lake shore. The black arrow indicates the timing of cyanobacteria overtaking diatoms as the main primary producers in the lake, and the light arrow indicates a rise in biogenic silica in the upper 2 cm of the core associated with early diagenesis. (Modified from Verschuren et al., 1998.)

orbital forcing from biogenic silica records in Lake Baikal: Nature, v. 378, p. 769–771.

CONLEY, D.J., 1988, Biogenic silica as an estimate of siliceous microfossil abundance in Great Lakes sediments: Biogeochemistry, v. 6, p. 161–179.

DeMaster, D.J., 1979, The marine budgets of silica and $Si^{32}$: Ph.D. dissertation, Yale University, 308 p.

GASSE, F., 2000, Hydrological changes in the African tropics since the Last Glacial Maximum: Quaternary Science Reviews, v. 19, p. 189–211.

GASSE, F., BARKER, P., AND JOHNSON, T.C., in press, A 24,600 yr diatom record from the northern basin of Lake Malawi, *in* Odada, E.O., and Olago, D.O., eds., The East African Great Lakes: Limnology, Palaeoclimatology and Biodiversity: Dordrecht, The Netherlands, Kluwer Academic Publishers.

HARKIN, D.A., 1960, The Rungwe volcanics at the north end of Lake Nyasa: Dar Es Salaam, Tanzania, Tanganyika Geological Survey, Memoir II, 172 p.

HECKY, R.E., 1993, The eutrophication of Lake Victoria: Internationale Vereinigung für Theoretische und Angewandte Limnologie, Verhandlungen, v. 25, p. 39–48.

HURD, D.C., 1972, Factors affecting dissolution rate of biogenic opal in seawater: Earth and Planetary Science Letters, v. 15, p. 411–417.

JOHNSON, T.C., AND EISENREICH, S.J., 1979, Silica in Lake Superior: mass balance considerations and a model for dynamic response to eutrophication: Geochimica et Cosmochimica Acta, v. 43, p. 77–91.

JOHNSON, T.C., SCHOLZ, C.A., TALBOT, M.R., KELTS, K., RICKETTS, R.D., NGOBI, G., BEUNING, K.R.M., SSEMMANDA, I., AND MCGILL, J.W., 1996, Late Pleistocene desiccation of Lake Victoria and rapid evolution of cichlid fishes: Science, v. 273, p. 1091–1093.

JOHNSON, T.C., CHAN, Y., BEUNING, K.R.M., KELTS, K., NGOBI, G., AND VERSCHUREN, D., 1998, Biogenic silica profiles in Holocene cores from Lake Victoria: implications for lake level history and initation of the Victoria Nile, *in* Lehman, J.T., ed., Environmental Change and Response in East African Lakes: Dordrecht, The Netherlands, Kluwer Academic Publishers, p. 75–88.

JOHNSON, T.C., KELTS, K., AND ODADA, E.O., 2000, The Holocene history of Lake Victoria: Ambio, v. 29, p. 2–11.

JOHNSON, T.C., BARRY, S.L., CHAN, Y., AND WILKINSON, P., 2001, Decadal record of climate variability spanning the last 700 years in the southern tropics of East Africa: Geology, v. 29, p. 83–86.

JOHNSON, T.C., BROWN, E.T., MCMANUS, J., BARRY, S.L., AND BARKER, P., 2002, A high resolution paleoclimate record spanning the past 25,000 years in southern East Africa: Science, v. 296, p. 113–114, 131–132.

KATO, K., AND KITANO, Y., 1968, Solubility and dissolution rate of amorphous silica in distilled and sea water at 20°C: Oceanographic Society of Japan, Journal, v. 24, p. 147–152.

KLEMER, A., AND BARKO, J., 1991, Effects of mixing and silica enrichment on phytoplankton seasonal succession: Hydrobiologia, v. 210, p. 171–181.

LEHMAN, J.T., LITT, A.H., MUGIDDE, R., AND LEHMAN, D.A., 1998, Nutrients and plankton biomass in the rift lake sources of the White Nile: Lakes Albert and Edward, *in* Lehman, J.T., ed., Environmental Change and Response in East African Lakes: Dordrecht, The Netherlands, Kluwer Academic Publishers, p. 157–172.

LIVINGSTONE, D.A., 1993, Evolution of African climate, *in* Goldblatt, P., ed., Biological Relationships Between Africa and South America: New Haven, Connecticutt, Yale University Press, p. 455–472.

MÜLLER, G., AND FÖRSTNER, U., 1973, Recent iron ore formation in Lake Malawi, Africa: Mineralium Deposita, v. 8, p. 278–290.

OWEN, R.B., CROSSLEY, R., JOHNSON, T.C., TWEDDLE, D., KORNFIELD, I., DAVIDSON, S., ECCLES, D.H., AND ENGSTROM, D.E., 1990, Major low levels of Lake Malawi and implications for speciation rates in cichlid fishes: Royal Society [London], Proceedings, Series B, v. 240, p. 519–553.

PARKER, J.I., AND EDGINGTON, D.N., 1976, Concentration of diatom frustules in Lake Michigan sediment cores: Limnology and Oceanography, v. 21, p. 887–893.

PLISNIER, P.-D., AND COENEN, E.J., 2001, Pulsed and dampened annual limnological fluctuations in Lake Tanganyika, *in* Munawar, M., and Hecky, R.E., eds., The Great Lakes of the World (GLOW): Food-Web, Health and Integrity: Leiden, The Netherlands, Backhuys Publishers, p. 83–96.

RAGUENEAU, O., TREGUER, P., LEYNAERT, A., ANDERSON, R.F., BRZEZINSKI, M.A., DEMASTER, D.J., DUGDALE, R.C., DYMOND, J., FISCHER, G., FRANCOIS, R., HEINZE, C., MAIER-REIMER, E., MARTIN-JEZEQUEL, V., NELSON, D.M., AND QUEGUINER, B., 2000, A review of the Si cycle in the modern ocean: recent progress and missing gaps in the application of biogenic opal as a paleoproductivity proxy: Global and Planetary Change, v. 26, p. 317–365.

STAGER, J.C., AND JOHNSON, T.C., 1999, A 12,400 $^{14}C$ year offshore diatom record from east central Lake Victoria, East Africa: Journal of Paleolimnology, v. 23, p. 373–383.

TALBOT, M.R., WILLIAMSON, M.A.J., AND ADAMSON, D.A., 2000, Strontium isotope evidence for late Pleistocene reestablishment of an integrated Nile drainage network: Geology, v. 28, p. 343–346.

TALLING, J.F., 1966, The annual cycle of stratification and phytoplankton growth in Lake Victoria (East Africa): Internationale Revue der Gesamten Hydrobiologie, v. 51, p. 545–621.

TREGUER, P., NELSON, D.M., VAN BENNEKOM, A.J., DEMASTER, D.J., LEYNAERT, A., AND QUEGUINER, B., 1995, The silica balance in the world ocean: a reestimate: Science, v. 268, p. 375–379.

VERSCHUREN, D., EDGINGTON, D.N., KLING, H., AND JOHNSON, T.C., 1998, Silica depletion in Lake Victoria: sedimentary signals at offshore stations: Journal of Great Lakes Research, v. 24, p. 118–130.

# MAGADIITE AND MAGADI CHERT: A CRITICAL ANALYSIS OF THE SILICA SEDIMENTS IN THE LAKE MAGADI BASIN, KENYA

HANS-JÜRGEN BEHR

*Göttinger Zentrum Geowissenschaften, Georg-August University, Goldschmidtstrasse 3, 37077 Göttingen, Germany*

ABSTRACT: Sedimentological and microbiological studies have been undertaken of the cherts of the Magadi basin in the deepest part of the Kenya Rift. The results of field studies, supported by microscopic and SEM analyses, have produced new ideas on the genesis of the cherts. Formation of the cherts has previously been explained abiotically by the alteration of Na silicates. However, high concentrations of coccoid and filamentous cyanobacteria and other organisms were found in the cherts and the precursor sediments (calcified *Pleurocapsa* in colonies, blooms, mats, and bacterial stromatolites; *Microcoleus*; *Microcystis*).

Two main chert horizons with different precursor sediments have been distinguished: (1) a lower chert horizon of Pleistocene age that contains amorphous silica, minor Na-silicates, erionite, and calcite (Green Beds), and (2) a thinner upper chert horizon with preserved magadiite laminae (High Magadi Beds). The precursor sediments of both cherts were deposited on trachytic pyroclastic detritus in a muddy faulted depression. The pyroclastic deposits were decomposed by alkaline groundwater, which discharged from silica-enriched springs along fault lines and formed colloidal silica sols. The precipitation of silica phases, Na-silicates, and calcite was controlled by the production of biogenic $CO_2$ and its resulting effects on the pH.

The precursor sediments had an unusual viscous, putty-like consistency. The sediments acquired thixotropic properties when mixed with silica–erionite suspensions and became liquefied during earthquake activity. Numerous synsedimentary dike swarms and diapirs of pillow-shaped chert were formed during rift faulting. The precursor sediments lithified by the crystallization of quartzine spherules when they came in contact with hypersaline brines. The strong microbial metabolism was a catalyst for intense chemical sedimentation, which resulted in voluminous chert formation in the Magadi Basin.

## INTRODUCTION

"Magadi-type chert" was first described from Lake Magadi by Eugster (1967) and Hay (1968), and is recognized as a distinct variety of inorganic chert that forms by the diagenetic transformation of Na silicates. Precursor sediments that preserve the original magadiite, however, are very rare (Surdam et al., 1972; Eugster and Chou, 1973; Sheppard and Gude, 1974; Surdam and Eugster, 1976; Parnell, 1986, 1987; Krainer and Spötl, 1998). The processes of chert formation, therefore, remain speculative without relics of the precursor (Hattori et al., 1996).

In this paper the origins of the cherts at Magadi are reconsidered. Detailed mapping has shown that two main chert units are present in the Magadi basin. The older (Green Beds) chert contains abundant microbial remains but provides little evidence for a magadiite precursor. A revised model for its formation is presented, based on the formation of stable silica sols and gels. In contrast, magadiite was a precursor of the younger (High Magadi Beds) chert. Many of cherts show complex deformation structures that formed when the chert precursor underwent permanent strain during compression, and these were later preserved by chertification (Behr and Röhricht, 2000). On the basis of microscopic studies, we show that silica sediments derived from stable sols and gels may have the same rheological properties as magadiite.

## GEOLOGICAL SETTING

The Magadi Basin, located 120 km southwest of Nairobi, is a closed depression situated in the lowest part of the southern Kenya Rift, between Lake Naivasha to the north and Lake Natron to the south. The rift basin is about 70 km wide at Lake Magadi, and has a surface area of 81 $km^2$. The western rift escarpment is 2550 m high; the eastern escarpment is 1600 m high. The vertical displacements along the boundary faults are estimated at 3–4 km (Baker and Wohlenberg, 1971). The morphology of the basin, the lake shoreline, and the lacustrine sedimentation are strongly controlled by three narrow NNE-striking fault zones that are 40 km long and 2 km wide (Fig. 1). Lake "Little Magadi" (Nasikie Engida) is situated directly north and 30 m above Lake Magadi, which lies at an elevation of ~ 580 m. Lake Little Magadi, which covers an area of 9 $km^2$, is 0.7–120 cm deep, and is fed by hot (83°C) springs.

The region is tectonically and seismically active. More than 40 m of trona ($Na_2CO_3 \cdot NaHCO_3 \cdot 2H_2O$), which fills the central part of the graben, has accumulated during the last 10,000 years. Garrett (1992) summarized details about the origin, mining, and processing of the trona deposit. About 75% of the lake flats is covered by trona. Small lagoons and pools with depths from a few centimeters up to 2 m are found along the shoreline (Fig. 1A) and contain alkaline brines with a pH of 9.0–12.5, up to 430,000 mg/l total dissolved salts (TDS: measured in 1996), and a salinity of 5–35 wt %. The annual precipitation in the area is ~ 430 mm, compared to 1,800 mm of annual evaporation. More than 200 springs with a pH of 8–10.5, up to 39,000 mg/l TDS, a salinity of 1–39 wt %, and a temperature of 23–83°C are present locally along fault lines.

Systematic scientific research in the Magadi area began with detailed geological mapping by Baker (1958, 1963). The area became famous as one of the most alkaline lakes in the world, and as one of the richest trona deposits. The sodium silicate minerals and chert at Magadi were first described by Eugster (1967) and Hay (1968). They interpreted the chert as a special variety of inorganic chert that formed by the transformation of Na silicates. The Pleistocene sediments that host the massive and bedded cherts cover an area of ~100 $km^2$. The thickness of the chert sequence typically varies from 1 to 5 m, but may exceed 20 m in the central parts of the basin (Fig. 1B). Detailed hydrochemical data for the Magadi Basin were published by Eugster (1967, 1970, 1980, 1986), Jones et al. (1967, 1977), and Hay (1968). Age dating was carried out by Butzer et al. (1972), Fairhead et al. (1972), Goetz and Hillaire-Marcel (1992), Tichy and Seegers (1999), Röhricht (1999), and Behr and Röhricht (2000). The seismicity of the Magadi Basin was studied by Hollnack and Stangel (1998). Geothermal data were documented by Crane (1981) and Allen et al. (1989). Gloaguen (2000) summarized and discussed the main tectonic aspects of the Magadi–Natron Basin.

Sedimentation in Continental Rifts
SEPM Special Publication No. 73, Copyright © 2002
SEPM (Society for Sedimentary Geology), ISBN 1-56576-082-4, p. 257–273.

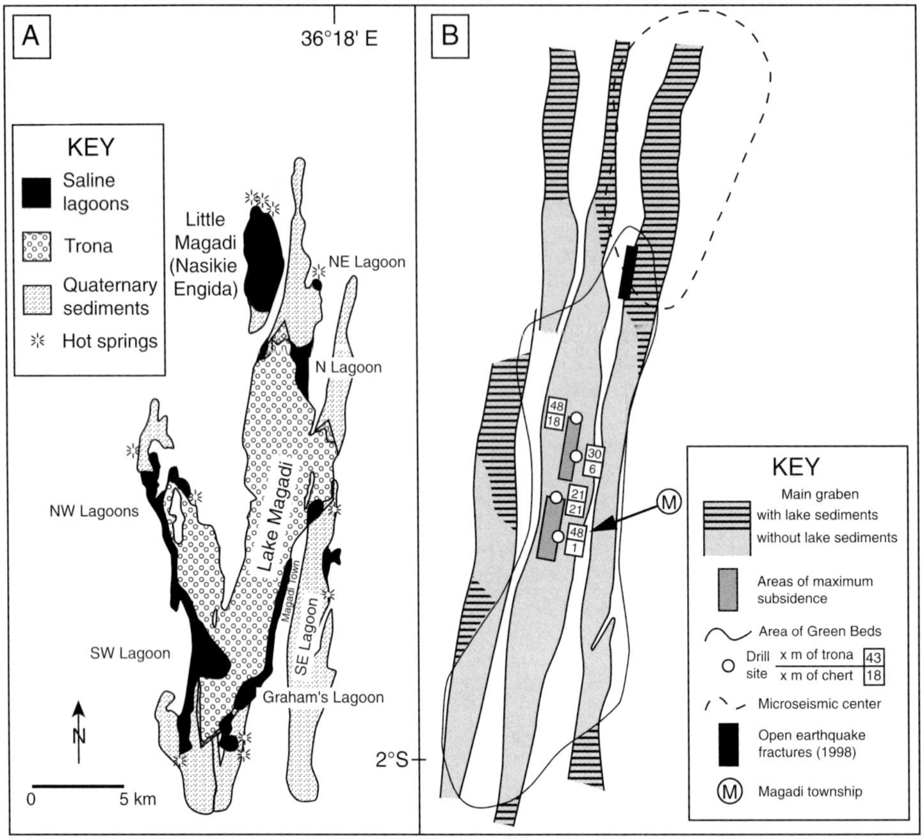


FIG. 1.—The Lake Magadi basin. **A)** General setting. (For a more detailed map of the sediment distribution see fig. 1 in Behr and Röhricht, 2000). **B)** Lake Magadi fault trench system. Boxes denote the zone of maximum rift subsidence as defined by the lower limits of the Green Beds chert at ~ 500 m above MSL.

The present study includes detailed geological mapping, which has led to modification of the stratigraphic sequence (Fig. 2). Behr and Röhricht (2000) showed that fault tectonics has played a larger role in the stratigraphic development of the basin than was formerly thought. Over time, the extensional tectonics shifted from the rift boundaries toward the central parts of the Magadi Basin. The deepest part of the rift depression lies at ~ 500 m elevation and is covered by a thick sediment pile. A complete stratigraphic section for the basin is not exposed. Up to 100 m of Pleistocene and Holocene sediments have been deposited on the Plateau Trachytes, which were dated at 1.4–0.7 Ma (Crossley, 1979). Caliche (calcrete) horizons that mark climatic hiati are present in some but not all sections. It is likely that a link between Lake Magadi and Lake Natron must have existed repeatedly in the Pleistocene, judging by the distribution of lacustrine stromatolites (Casanova, 1987; Hillaire-Marcel and Casanova, 1987; Williamson et al., 1993).

The oldest sediments, which lie upon the Plateau Trachytes, are the Oloronga Beds, which are a series of airfall tuffs, lacustrine carbonates, and diatomaceous cherts (Baker, 1958; Eugster, 1980; Behr and Röhricht, 2000). The stratigraphic sequence directly above the Oloronga Beds (Fig. 2) is referred to here as the Green Beds. These sediments are 30 m thick and contain about 90% of the chert at Magadi in a greenish matrix of Eolithic (erionite) tuffs and pyroclastic silts. Baker (1958, 1963) had already considered the Green Beds as a separate chert unit (his "Chert Series"). The cherts of the Green Beds cover an area of 100 km$^2$. Outcrops of similar chert beyond the Magadi area and up to 60 m above the present lake show that the original area of the sediments may have been much larger. The Green Beds were faulted by syndepositional extensional tectonic movements.

The younger sediments, termed the High Magadi Beds (Fig. 2), are only 0–3 m thick, but they represent the key stratigraphic unit in the discussion on the genesis of laminated chert. The High Magadi Beds consist of brown silts, erionite tuffs, laminites, magadiite, and chert. They cover the timespan between ~ 23,700 yr BP and the Holocene. The lack of absolute age dating led Eugster (1980, p. 217) to infer that the massive and widespread Green Beds chert and the local High Magadi Beds chert were the same stratigraphic unit with lateral transitions. His genetic model is based on the combined mineralogical and structural features of *both* types of chert. Radiometric dating, however, showed that the Green Beds chert, with an age of 98,000 yr BP, is much older than the High Magadi Beds (Goetz and Hillaire-Marcel, 1992). The two units are separated by an erosional disconformity. The upper part of the High Magadi Beds grades into laminites that contain only traces of magadiite and chert (Fig. 2). These, in turn, are covered by Holocene lake sediments, loessic soils, and hillwash deposits. The trona formed after ~ 10,000 yr BP (Butzer et al., 1972). Unlike the massive trona in the main lake, which is > 40 m thick, the trona in Little Magadi (dated ~ 9,000 yr BP) developed as one continuous layer only 8 cm thick.

Eugster (1969, 1980) proposed that the lakewide chert developed from a sodium silicate precursor that precipitated at

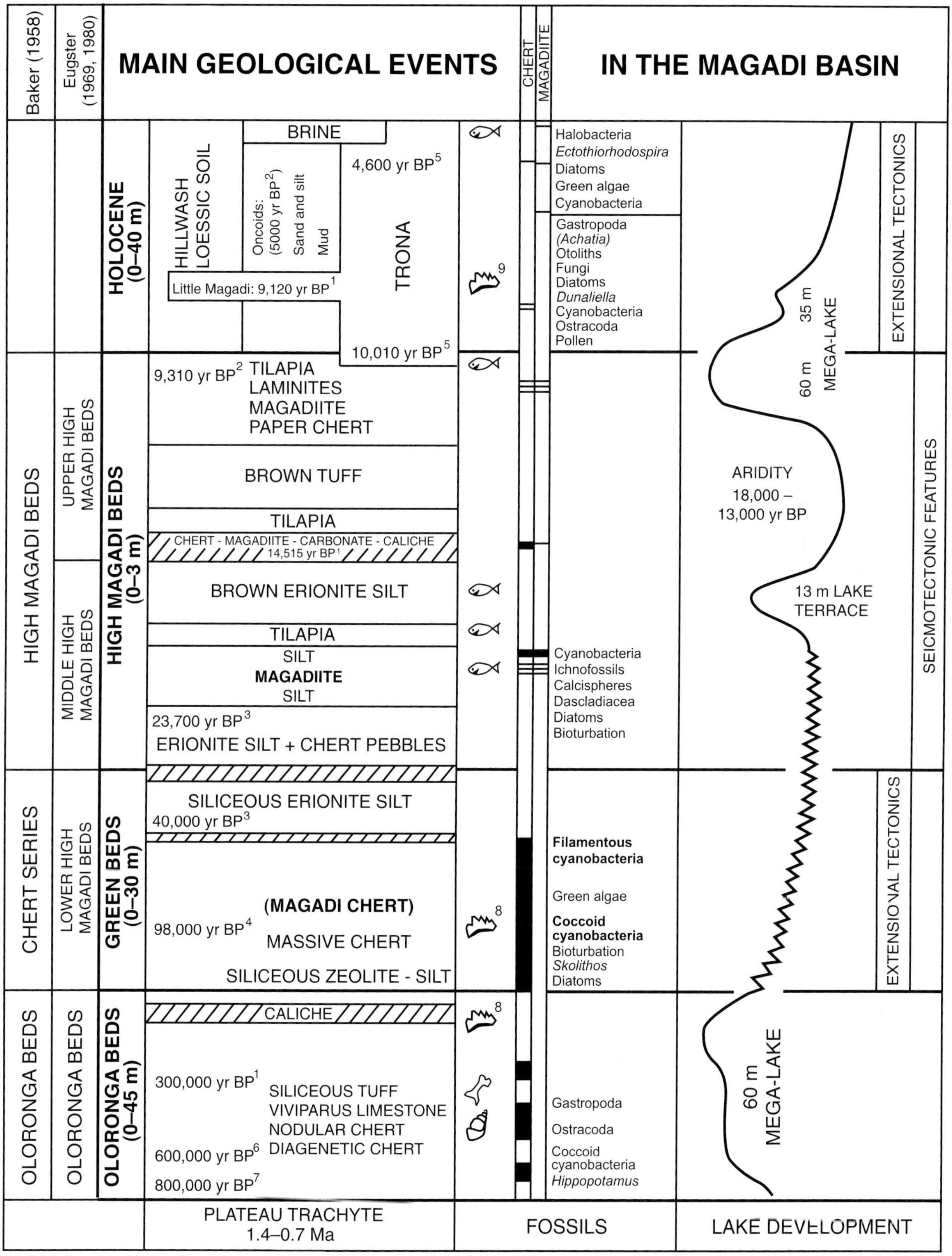


FIG. 2.—Stratigraphic record of the lake sediments in the Magadi Basin, showing the position of the chert and magadiite beds. Sources: (1) this paper; (2) Tichy and Seegers (1999); (3) Williamson et al. (1993); (4) Goetz and Hillaire-Marcel (1992); (5) Butzer et al. (1972); (6) Kaufmann et al. (1990); (7) Fairhead et al. (1972). Other information: (8) Oncoids and stromatolites with cyanobacteria; (9) Maximum of stromatolites.

the chemocline of a large stratified lake that underwent evaporation. Only in this way could the huge amounts of silica necessary to produce the magadiite and the chert deposits be explained. Using stromatolite carbonate ages, Casanova (1986) traced former water levels across the basin and distinguished two pluvial episodes, each corresponding to lakes 30–60 m deep that were connected with Lake Natron. The only convincing time marker, however, is a 13-m-high terrace in the middle part of the High Magadi Beds (White, 1953; Butzer et al., 1972; Baker, 1958). Sediments, shorelines, relics of lagoons and inlets for these large extensive paleolakes could not be demonstrated.

## DEFORMATION STRUCTURES IN THE PRECURSOR SEDIMENTS OF THE CHERTS AT MAGADI

### *Cherts of the Green Beds*

The thick chert horizon in the Green Beds consists of a lower unit with massive chert and a thinner upper unit with bedded chert (Fig. 2). Unique structures developed in the precursor sediments of the lower unit. They are seen in the field as black and brown hummocks and ridge-shaped structures that are aligned, especially at the rims of uplifted fault-blocks (Behr and Röhricht, 2000). The upper bedded chert is strongly fragmented into plates < 1 $m^2$ that mantle extensive plains, particularly in the northern part of the basin.

*Dike Chert.*—

Morphological features that are 2–50 m long, 10 cm to 2 m high, and up to 30 cm wide were described by Eugster (1980) as "chert ridges", and are found mainly along the lakeshores parallel to horsts (Fig. 3). Eugster (1969) interpreted these structures as the relics of desiccation polygons (< 50 m in diameter) that formed at the surface of former magadiite-bearing sediments. "Toothpaste" magadiite was squeezed out of desiccation cracks to form wall-like structures. When extruded, this material consisted of a highly viscous paste mixed with intraclasts. The viscous conditions resulted in rollovers, slickensides, ball-and-pillow structures, and other structures. According to Eugster (1980), sodium was then leached from the ridges by rainwater to form chert from the residue.

FIG. 3.—Vertical chert dike crosscutting bedded chert in foreground.

The dike orientations measured from aerial photographs and field observations reveal a 90° grid pattern with a dominant north–south direction and a maximum deviation of 20°. These dikes are cut discordantly by east–west trending dikes. True polygonal desiccation structures have not been recognized. Some dike zones are up to 3 km long and locally show an *en echelon* pattern. They can be correlated with faults with a mean length of 3.5 km. Neotectonic joints in the younger High Magadi Beds are up to 30 m long and show similar *en echelon* arrangements. These, however, have a predominant 10–15° (NNE) orientation with a subsidiary trend perpendicular to that direction (Atmaoui, 1998). Both systems are considered to be seismotectonic tension joints and signify a rotation of the stress field over time. Arched dikes with radial elements in the central parts of the basin may be explained by the locally higher subsidence rates.

*Pillow Chert.*—

The bottom of the SE Lagoon is covered with more than 200 ring-shaped diapirs that are ~ 5 m in diameter (Fig. 4). These, in turn, are composed of pillow-shaped chert bodies < 1 m in diameter. Some of these bodies are linked by tubes of chert. The pillows are cemented at their contacts by a younger generation of

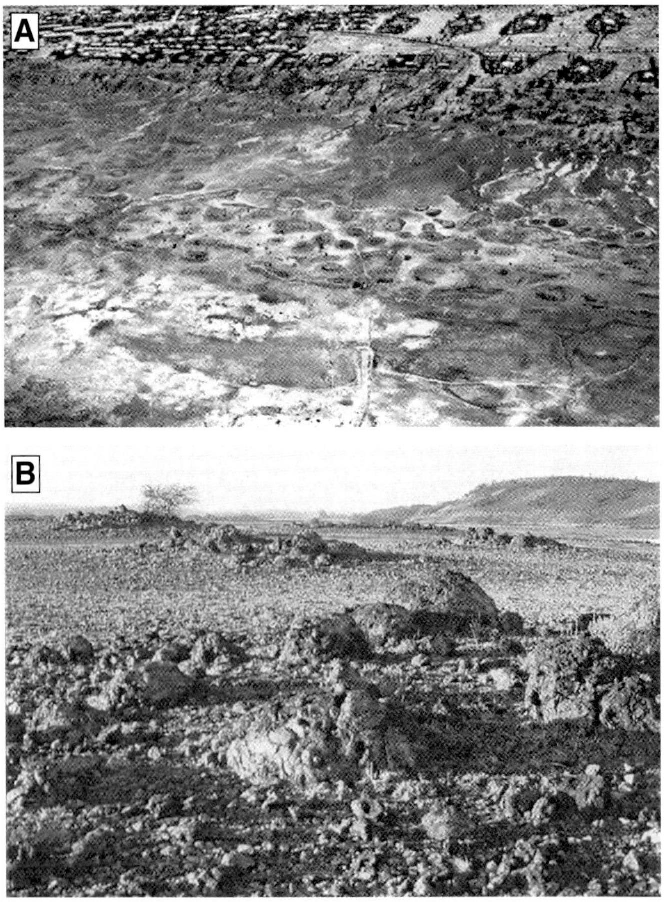

FIG. 4.—**A)** Chert diapirs (3–10 m in diameter) covering a horst in the SE Lagoon. Chert ridges have been squeezed out along a fault line. The Magadi-fault horst and the township are visible in the background. **B)** Diapiric mounds composed of chert pillows up to 1 m in diameter.

chert. In cross section, the pillows show complex turbulent flow, and "knead" and fragmentation textures that are healed by a younger translucent chert. The pillow chert formed from the extrusion of a more mobile precursor than the linear dikes.

*Slumped Chert.—*

Several meters of synsedimentary subsidence led to slumping along fault ramps before the precursor sediments had lithified. The chert precursor slipped and crept downward in a viscous state along the flanks of the ridges. This slumping produced a mash of brecciated crusts that were cemented by later generations of chert (Fig. 5). Evidence for active fault tectonics during the formation of the precursor can be found in many places.

*Chert Ridges.—*

Anticlinal diapiric ridges and bulge structures (10–30 m long, ~ 1 m high) formed along the fault ramps, which are 2–4 m high (Fig. 6). These structures are commonly broken near the crest, and show radial cracks and cavities in the bedding. The chert horizon may form flexures at fault ramps with a throw of up to 1 m. This confirms both the prolonged high viscosity of the precursor and synsedimentary fault tectonics.

*Bedded Chert.—*

The bedded chert consists of flat, lenticular plates up to 2 cm thick that are interlayered with erionite silt (Fig. 7). Casts and molds of evaporite minerals, mostly gaylussite and calcite (partly in "chains"), are abundant on the bedding planes. Water-escape structures, tepee structures, vague streamers, ripples, and "petee" structures (cf. Gerdes and Krumbein, 1987) are common (Fig. 8). "Petee" structures are characterized by wrinkled blisters and fold patterns that formed by sliding of thin layers on a mucilaginous substrate. Sliding was caused by strong winds blowing over the mudflats or by seismic tremors. The mucilaginous layers were coatings of bacterial films and silica–erionite suspensions (see below). Thin desiccated fragments that formed on the dried-out surface could migrate in similar manner. Shallow slickensides with a dip of about 2–5° are present in the bedded chert.

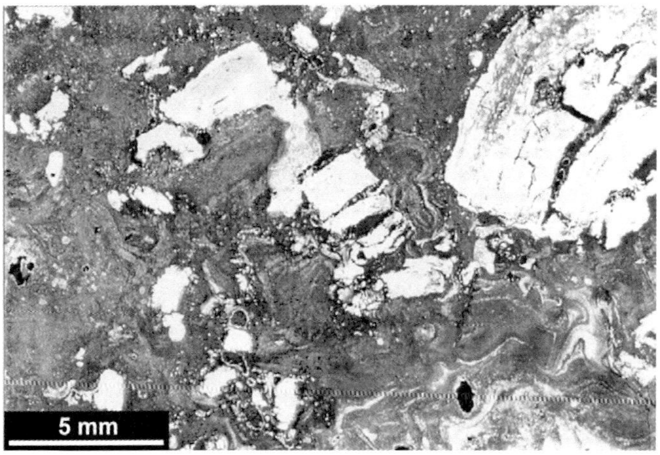


FIG. 5.—Slumped chert in thin section (plane-polarized light). Crushed fragments of calcified and silicified cyanobacterial bioherms are cemented by younger gray chert with flow structures.

FIG. 6.—The NE Lagoon, showing chert ridges along fault lines and the saline pan of trona. The volcano Oloolkisailie (Mt. Olorgesailie) is visible in the background.

*Reticulate Structures.—*

Reticular cracks are caused by volume changes. They cover the surfaces of many chert plates and nodules. The width of the reticulations and the depth of their wedge-shaped cracks vary from a few millimeters to 1 cm. Eugster (1980) attributed these structures to shrinkage during the magadiite to chert conversion. Hay (1968) inferred the same process for chert nodules in the Moinik Beds at Lake Natron. The irregular lobate and spinose shapes of these head-size nodules indicate contraction and distortion of the original more lenticular forms (Fig. 9).

## *Cherts of the High Magadi Beds*

The SE Lagoon (Fig. 1) is the best place to examine the High Magadi Beds. The deformation features of the thinner and partly silicified magadiite layers are different and less spectacular than the structures described for the Green Beds. Horizontal, millimeter-thick and paper-thin sheets of chert are characteristic. The "paper chert" is irregularly deformed, crumpled, and perforated, and shows reticulate textures. This paper chert is also present in the volcanic talus that covers the surfaces of local horsts. The magadiite layers (the inferred

FIG. 7.—Bedded chert with extrusive chert dikes parallel to a N–S fault ramp.

FIG. 8.—Petee structures on bedded chert that originated by sliding of the upper sediment layer on slippery bacterial films.

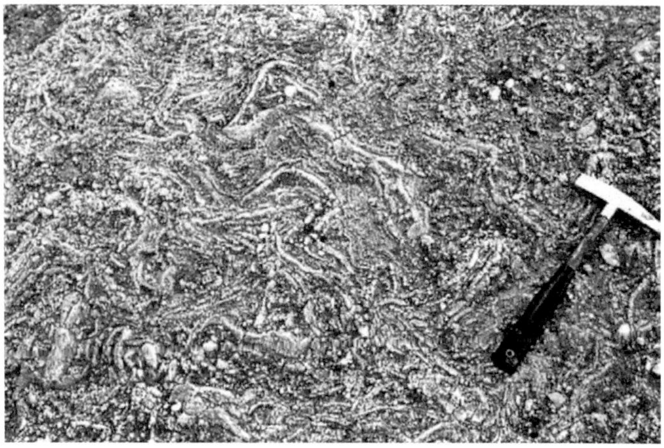

FIG. 10.—Chaotic slump folds in the High Magadi Beds. The 0.5-cm-thick chert layers are silicified magadiite.

chert precursor) were deformed by chaotic slumping before silicification (Fig. 10). During silicification, horizontal gliding was induced by liquefaction in a magadiite layer 3 cm thick over an area of 1 $km^2$. Flat-dipping slickensides in this material are covered with striations with random orientations (Fig. 11). This shows that seismotectonic activity occurred during sedimentation.

## ORGANIC REMAINS IN THE MAGADI CHERTS

Information on the fossil record of the Pleistocene and Holocene sediments of the Magadi Basin is limited to studies of the fossil fish populations (Tichy and Seegers, 1999; Seegers and Tichy, 1999), stromatolites (Casanova 1986, 1987), and diatoms (Damnati et al., 1992). In their description of a laminated sediment core taken from the High Magadi Beds, Damnati et al. (1992, p. 410) mentioned a "microlayer of microspherules (probably bacteria)", but microbial fossils have not previously been discussed. The following results are based on detailed field observations and the microscopic analysis of about 300 sediment samples.

### *Unicellular Cyanobacteria*

Unicellular (coccoid) cyanobacteria (5–20 μm) are present throughout the stratigraphic section (Fig. 2) and in the recent lake, but they are most extensive (or best preserved) in the massive chert of the Green Beds. Initially it was difficult to identify the microorganisms because of their resemblance to inorganic particles. The Green Beds chert consists of up to 90% quartzine (length-slow), chalcedony (length-fast) and opal-CT, most of which form aggregates of microspheres (10–30 μm). Furthermore, magadiite forms lepispheres of bladed crystals (10–20 μm). Dispersed globular calcite and pyrite fall within the same range of diameters. It is reasonable to assume that some silica phases represent pseudomorphs of magadiite. Under crossed nicols, aggregates of quartzine microspheres may show 90° extinction patterns. Schubel and Simonson (1990) suggested that these textures characterize chert that developed from magadiite. The microspheres, however, may mask coccoid cells.

The discovery of a horizon with bacterial stromatolites (Fig. 12) in the massive chert sequence of the Green Beds has played a key role in our new interpretation. The outcrop is located near the road junction at Magadi Township, along the eastern fault

FIG. 9.—Lobate chert nodule (15 cm diameter), with reticulation formed by shrinking. Moinik Beds, Lake Natron.

FIG. 11.—Porcellaneous chert formed by silicification of magadiite intercalated with horizontal chert formed by liquefaction. Coin 2 cm.

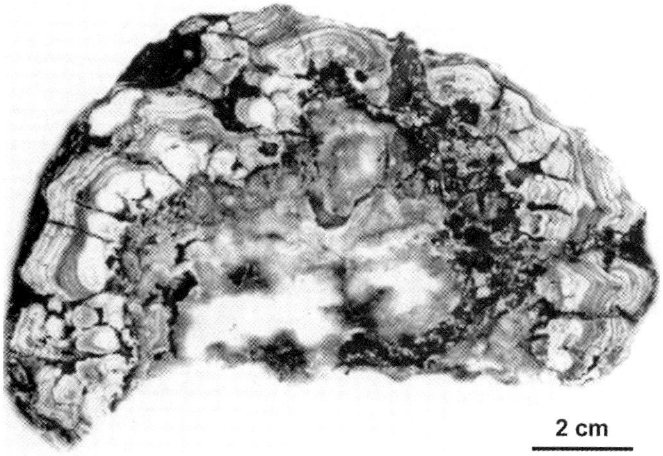


FIG. 12.—Cross section of a laminated and calcified bacterial stromatolite on clump of slumped chert.

escarpment (Behr and Röhricht, fig. 3B). There, a chert terrace has formed on lithic debris and trachytic pyroclastics. The horizon contains abundant calcareous nodules, conical to dome-shaped bodies up to 30 cm x 10 cm, and encrustations with stromatolites that show microlamellae (0.1–1 mm). These structures are composed of ovoid aggregates of coalescent unicellular cyanobacteria. Up to five concentric, calcified envelopes have formed around individual cells. Pores between the onion-skin layers are partly silicified. Colonies of *Pleurocapsa* sp. are revealed after etching with HCl (Fig. 13), showing silicified mucilage and calcified cell envelopes (Figs. 14, 15). Radial textures indicate that the slime in the voids between the envelopes formed at different stages. With successive division, the volume of the cell chambers was reduced and deformed until only a punctuate inclusion remains. Degradation processes similar to those that affected the coccoid cyanobacteria were described by Golubic and Hofmann (1976).

Large populations of minute (1 μm) spherical reproductive cells surround the stromatolites (Figs. 16A, 17). Calcite needles that have formed on the cell capsule can produce spherulitic structures that suggest larger cell diameters. Some reproductive cells may have left the colony, migrated, grown up to 10 μm, and formed new colonies. Flakes of secondary calcite commonly crystallized in the cell chambers after death. All these features are phenotypic for baeocytes. These are the reproductive cells of genera in the order Pleurocapsales, which have gliding motility.

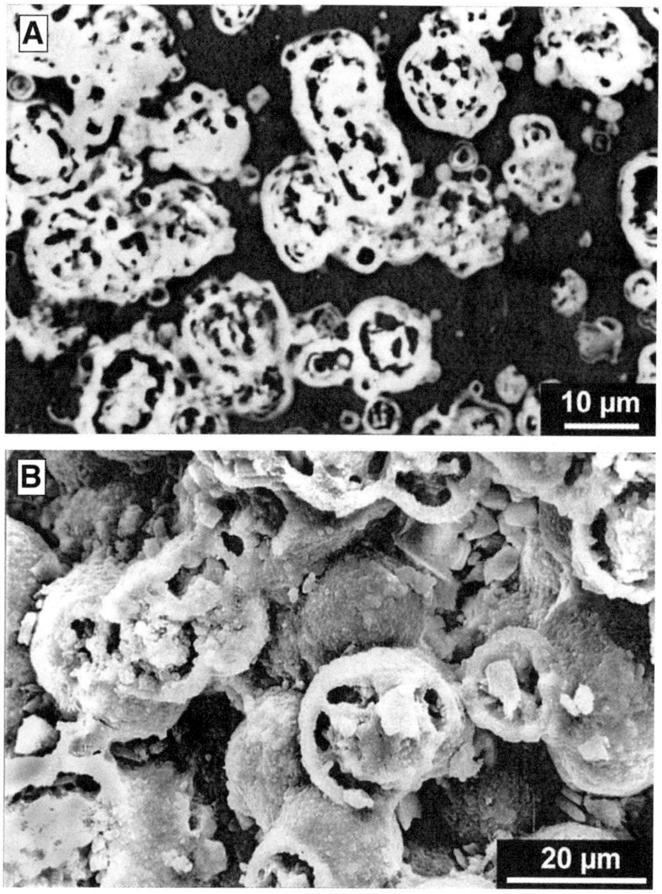


FIG. 14.—SEM images of *Pleurocapsa*. **A)** Aggregates of *Pleurocapsa* cells with thick calcified mucilaginous capsules. Cell chambers (black) are 1 μm. **B)** Partly silicified cells of *Pleurocapsa* with polydermate calcareous envelopes.

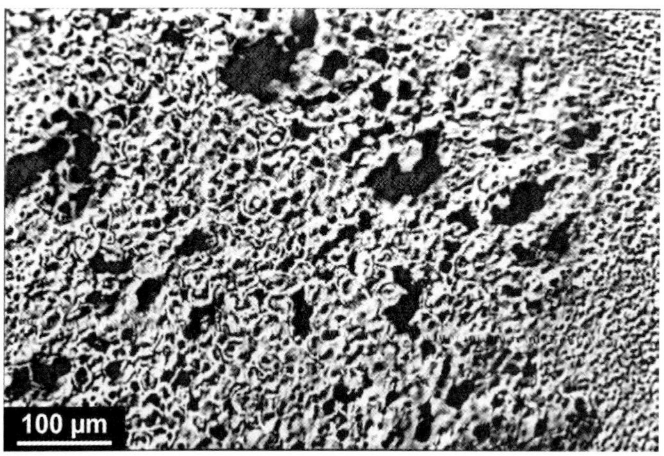


FIG. 13.—Pattern of silicified cyanobacterial colony visible after etching with HCl. SEM image.

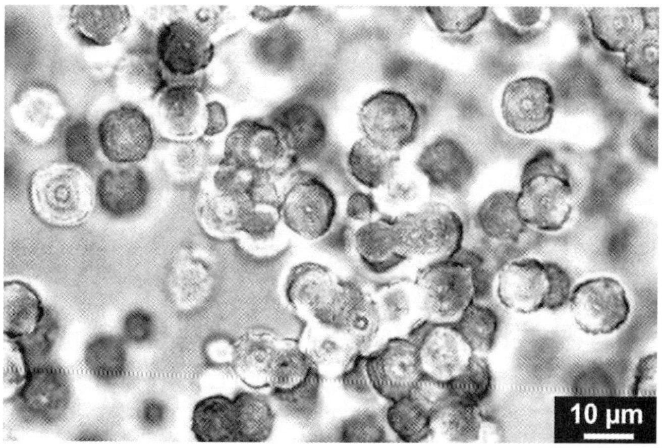


FIG. 15.—Calcified and partly silicified *Pleurocapsa* with polydermate capsules and cell chambers in the center. Thin section (plane-polarized light).

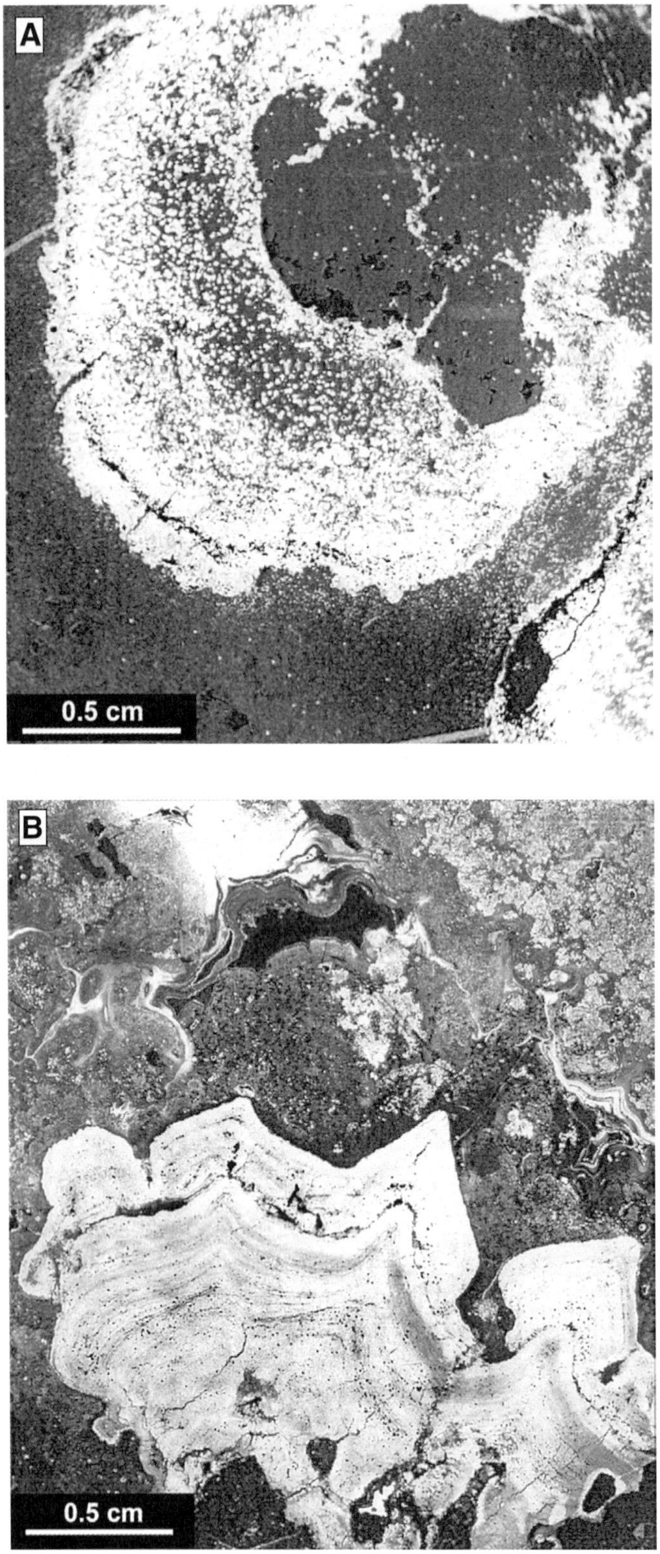


FIG. 16.—Thin sections (plane-polarized light) of **A)** calcified encrustation of coccoid cyanobacteria around a chert fragment, with dense concentrations of baeocytes, and **B)** calcified thinly laminated colony of filamentous bacteria stratified with coccoid cyanobacteria. Small cluster colonies of coccoid cyanobacteria (white) are visible at the upper right.

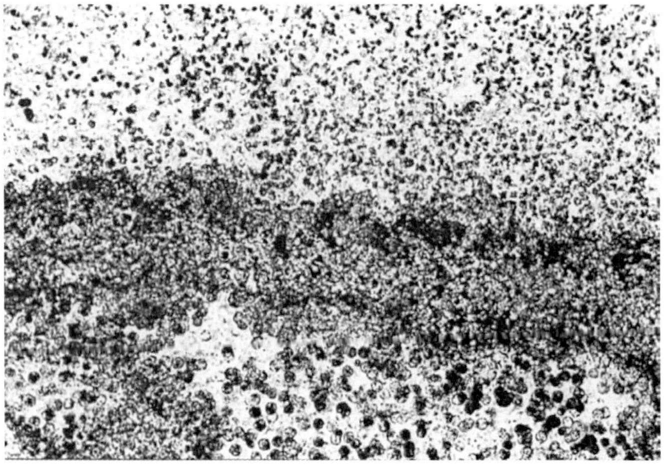

FIG. 17.—*Pleurocapsa* crust with swarms of baeocytes of different sizes. The larger cells are 10 μm in diameter. Thin section (plane-polarized light).

These genera may have multicellular filaments but still produce baeocytes. On top of the coccoid cells, layers with microlamination were formed by filamentous microbes (Fig. 16B). The latter consist of threads 0.5 μm in diameter, but they lack calcareous sheaths and consequently are poorly preserved. These microbes are assigned to the green non-sulfuric bacterium *Chloroflexus*. This genus includes filamentous, thermophilic, and gliding bacteria that commonly live in neutral to alkaline hot springs. Similar bacterial communities have been described from hot springs in Yellowstone (Doemel and Brock, 1974). This type of cyanobacterial–bacterial stromatolite preserved in chert differs from the calcareous stromatolites and oncoids that are widespread in the Magadi–Natron basin (cf. Casanova (1986, 1987).

Single cells that lie outside the massive calcareous colonies commonly are completely replaced by silica and coated with quartzine microspheres. Without calcification, however, it is difficult to recognize coccoid cyanobacteria. Cells with calcite remains can easily be recognized under crossed nicols and in the back-scattered mode of the scanning electron microscope (SEM). The outer zones of *Pleurocapsa* colonies are strongly calcified (Fig. 16A), whereas the older cells are commonly replaced by silica. Calcite is redeposited as encrustations in pores or precipitated as millimeter-size zoned crystals. Unlike the cyanobacterial cells, the morphologically similar magadiite lepispheres were never calcified.

Another unicellular species of *Pleurocapsa* with significantly smaller cell diameters of 1–5 μm, including the capsule, is also abundant (Fig. 18). Their capsules are less developed, but calcification is common. Typically, a minute inclusion is the only remnant of the central cell chamber. The cells are coccoid with disk-shape tendencies and show binary division. This species preferentially covers the walls of larger pores or grows on the surfaces of *Pleurocapsa* stromatolites. They also form pellet-shaped colonies (Fig. 19) and spongy thrombolitic crusts. A high production of mucilaginous substances (extracellular organic slime) and the large capsule protect against desiccation, UV radiation, and extreme salinities. In experiments, Thompson and Ferris (1990) found that the capsule size of *Synechococcus* increased from 6 to 15 μm when transferred to a highly saline medium. It can be inferred, therefore, that the larger *Pleurocapsa* species grew under conditions of high evaporation, high salinities, periodic desiccation, and intense UV radiation. In contrast, the smaller species lived in deeper water or in protected cavities.

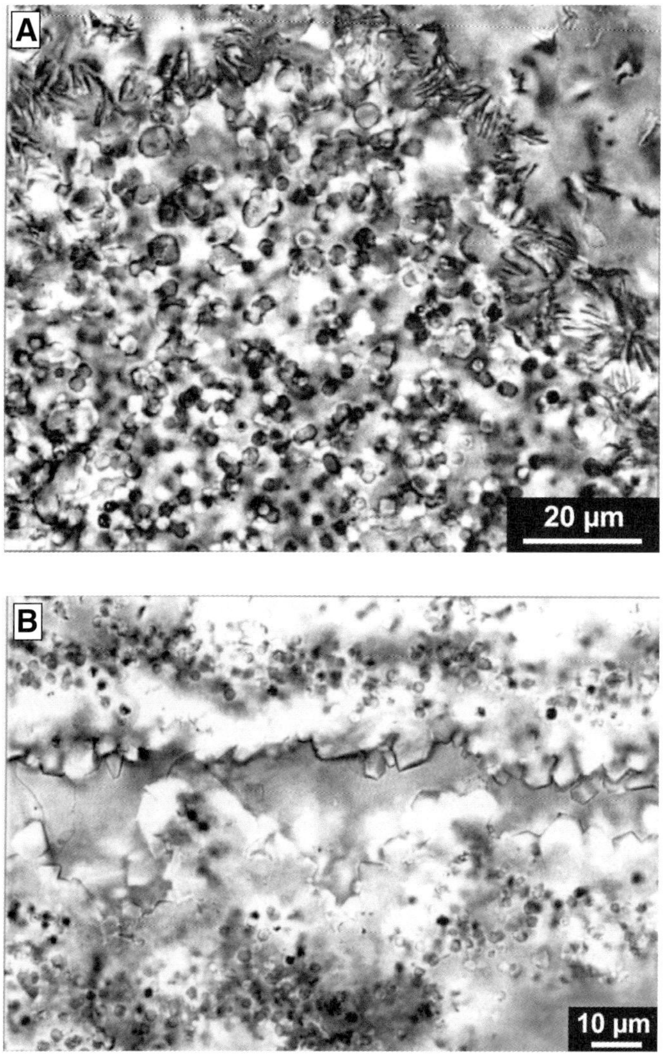


FIG. 18.—**A)** Small calcified species of *Pleurocapsa*. Secondary calcite flakes formed at the rim of the colony. **B)** The same species present as inclusions in microquartz that has coated a fenestral cavity. Thin sections (plane-polarized light).

A third species of *Pleurocapsa* forms irregular and large globular colonies, several hundred microns in diameter, which developed by the aggregation of large quantities of organic slime. The size of the cells is 1–2 µm, and the ratio of slime to cell volume is ~ 3–5. The individual cell chambers form icicle-shaped structures on a mat-like substrate (Fig. 20). An unidentified species forms strings of spherical spores enclosed in filaments that are approximately 100 µm long. These filaments grew on mats of dead and decomposed cell chambers.

Observations of the microorganisms allow reconstruction of the original biogenic structures (biofilms, mats, carpets, crusts) in the sediments that were precursors of the massive chert. Normally, unicellular colonies form loose mucilaginous benthic mats. However, these mats may be lithified by extensive precipitation of calcite in the extracellular slime. Bundles of filaments with carbonaceous sheaths (3 µm in diameter) of the cyanobacterium *Microcoleus* were responsible for the stabilization of the fabric. Fenestrae and other cavities are relics of the primary porosity, and are filled with volcanic fragments, sand, silt, and zeolites,

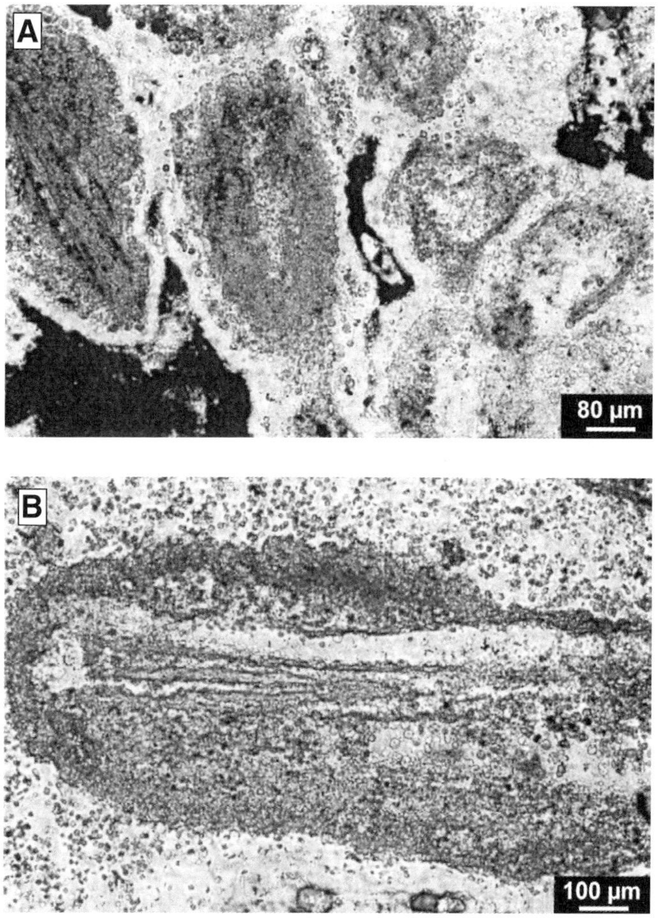


FIG. 19.—**A, B)** Different types of pelletal colonies of *Pleurocapsa*. Thin section (cross-polarized light).

including analcime. These components are now cemented with silicified organic slime. The walls of larger pores are coated with clear quartz crystals, which incorporate different species of coccoid microbes. The free-floating genus *Microcystis* is also very common in the precursor sediment. Blooms of *Microcystis* were stimulated by mineral enrichment due to evaporation. Biostromes of the unicellular cyanobacteria *Aphanothece* (*Synechococcus*) in Great Salt Lake, Utah (Carozzi, 1961), may serve as a model of the precursor sediment before its transformation into chert.

Calcification of cyanobacteria is a much-discussed problem. The Magadi samples show several stages of calcification. The calcified capsules are composed of calcite particles on a nanometer scale and are inferred to have been precipitated by bacterial activity. The textures are characteristic of calcified *Synechococcus* (Thompson and Ferris, 1990). Aggregates of calcite crystallites (< 0.5 µm) are common in the chert matrix. In general, photosynthetic bacteria have a Ca-enriched microenvironment surrounding the cells. Metabolic processes are responsible for alkalinization of this aquatic environment by $HCO_3$–OH exchange through the cell membrane. Cyanobacteria, therefore, are commonly entombed in calcite.

### *Filamentous Cyanobacteria*

Filamentous cyanobacteria are predominant in the bedded cherts, whereas unicellular species decline. Mats are com-

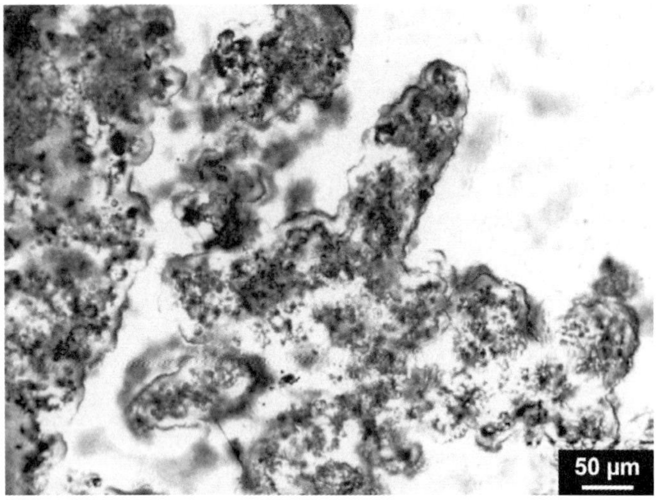


FIG. 20.—Thrombolitic *Pleurocapsa* colonies with large production of (now silicified) extracellular mucilaginous substances. Thin section (plane-polarized light).

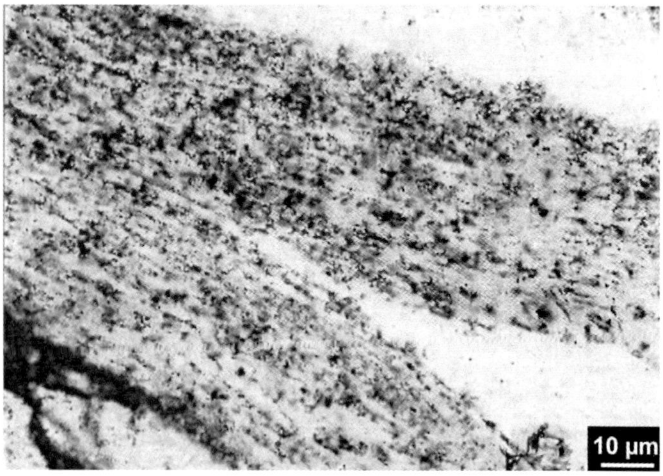


FIG. 22.—Mycelium mats of Actinomycetes. Thin section (plane-polarized light).

posed of filaments (3 µm in diameter) belonging to the genus *Microcoleus* (Fig. 21), which are bundled together in a calcified sheath. Extensive carpets of filamentous and unicellular cyanobacteria induced gliding and slumping at the sediment surface. The resulting "petee structures" are widespread along the northern margins of the Green Beds outcrop. In bedded calcite-bearing chert with ripple marks, isolated filaments of *Schizothrix* and *Phormidium* show parallel alignment caused by the current flow.

### *Actinomycetes*

Actinomycetes are present in some samples and can be identified by the morphology of the mycelium. The spores (0.5 µm) are locally aggregated in clusters. The short hyphae bear only few sporophores. The mycelia may form garlands that hang between chert fragments (Fig. 22). Actinomycetes are normally aerobic soil bacteria, but they also live in lake muds. Basophile species have been recognized around saline lakes, but they do not tolerate high $CO_2$ concentrations.

FIG. 21.—Colony of filamentous *Microcoleus*.

### *Diatoms*

Diatoms are widespread in thinly laminated horizons of the High Magadi Beds (Damnati et al., 1992). Although commonly corroded, six species have been identified. In contrast, diatoms are rare in the Green Beds, but they are locally concentrated in small pockets. Only *Navicula* sp. could be identified. Abundant diatoms with high diversity have been found in the chert of the Oloronga Beds, which lie stratigraphically below the Green Beds.

### *Other Fossil Remains*

Low concentrations of ostracods are present in the High Magadi Beds; ostracods are, however, more common in the Oloronga Beds. High concentrations of calcispheres from the silicified magadiite horizon of the NE Lagoon are useful indicators of the paleoenvironment of the High Magadi Beds and the Green Beds. Calcispheres are radial calcite capsules up to 5 mm in diameter with a hollow core of 50–200 µm. They are partly silicified and are the reproductive cysts of green algae that live in shallow water with restricted circulation. Standing Dascladiacea stems, 10–15 cm high, are found at the southern limits of the southeastern lagoons of Lake Magadi and Little Magadi (Fig. 23). The stems are in upright (growth) position and were embedded during different chert-forming phases. The cellular structure and sporangia in the stems are very well preserved.

Ichnofossils are common in both cherts. Dwelling burrows (1 cm long) with excavation pellets, similar to those of tiger beetle larva and salt beetles, are common (Fig. 24). The excavation pellets are surrounded by concentric structures, which consist of colonies of bacteria that grew radially. The circular burrows have a diameter of 1 mm. At the surface, the cherts show desiccation cracks and feeding structures. The burrows and cracks indicate subaerial exposure. Today these structures cover the extensive mudflats of the lagoons. Bioturbation by worms is widespread in both the High Magadi Beds and the Green Beds sediments. *Skolithos* and rhicospheres are present in the lower part of the Green Beds. Polymorphous fungal spores and hyphae, as well as pollen, are abundant in the High Magadi Beds. A *Viviparus* limestone with high concentrations of gastropods is present in the Oloronga Beds beneath the massive chert. These indicators of high freshwater inflow are absent in both the Green Beds and High Magadi Beds. Local freshwater supply on the Holocene

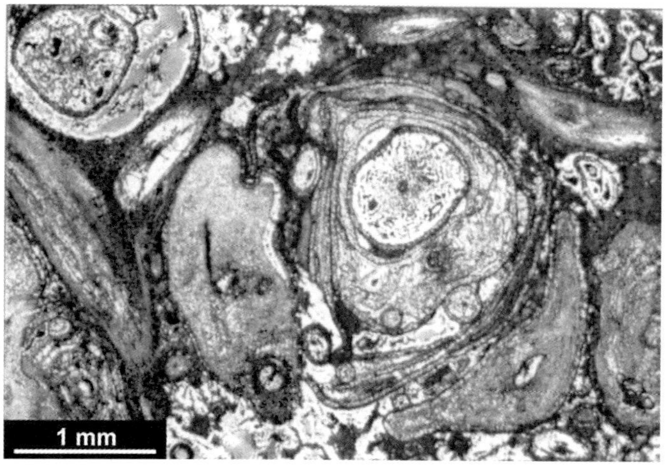


FIG. 23.—Cross section of Dascladiacea stems in chert. Thin section (plane-polarized light).

littoral plains of Lake Magadi was responsible for the large production of calcareous oncoids (up to 3 cm).

### *Tilapia*

Fishes of the group *Oreochromis alcalicus* (soda-tolerant *Tilapia*) are present in the recent Lake Natron–Magadi system. Baker (1958) classified the fossil fishes in the High Magadi Beds as *Tilapia nilotica*, a freshwater species that is widespread in Africa. The evolution, diversification, and radiation of the *Oreochromis* group during the last 7000 years since the two lakes separated is summarized by Tichy and Seegers (1999) and Seegers and Tichy (1999). Detailed studies of the fossil fishes have not yet been carried out. Nevertheless, it is clear that *Oreochromis alcalicus* lived in the Magadi–Natron paleolake(s). Complete fossil *Tilapia* skeletons are up to 18 cm long. Sagittal otoliths (2–3 mm), which are preserved as calcium carbonate, are very common. In addition, fish teeth (< 0.5 mm) and scales replaced by apatite are present. The *Tilapia* were deposited in brown volcanic silt as scattered specimens or in mass concentrations. Cyanobacterial blooms that lowered the oxygen concentration probably caused the mass extinctions. Unlike the High Magadi Beds, only a single fish scale was found in the Green Beds. The amount of freshwater during the time of Green Beds sedimentation must have been too small and the environment too toxic (as a result of continuous cyanobacterial blooms?) for the development of fishes.

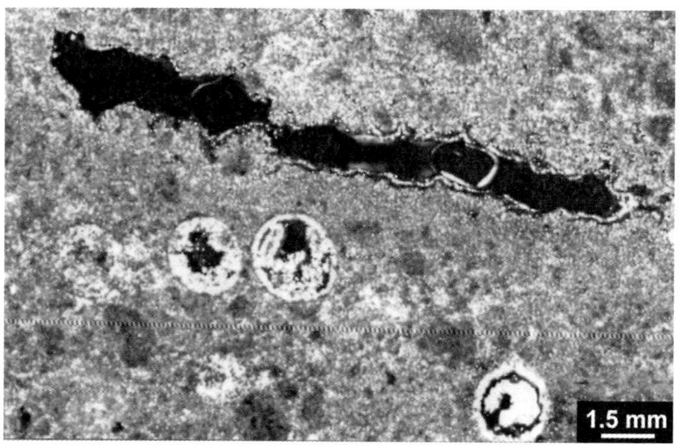


FIG. 24.—Long and basal sections of ichnofossils in chert. Thin section (cross-polarized light).

A comparison of the fossils in the Green Beds and in the other lake sediments confirms the unusual paleoenvironmental conditions of the Green Beds cherts. Many of the quartzine spherules in the Green Beds represent silicified calcareous cells of cyanobacteria, which formed in addition to primary quartzine spherules. Silicified magadiite lepispheres are subordinate. In contrast, silicified magadiite lepispheres predominate in the High Magadi Beds, whereas colonies of calcified coccoid microbes are subordinate. Commonly the microbes were thickly coated by magadiite, before silicification. Ichnofossils are widespread in both the Green Beds and the High Magadi Beds. Fishes are abundant in the High Magadi Beds but are absent in the Green Beds. Additionally, isolated cells and aggregates of calcareous unicellular cyanobacteria are found in the older Oloronga cherts and in lobate chert nodules in the Moinik Series (Peninj Beds, Lake Natron). Noncalcified *Pleurocapsa* are present in the recent lake brines and in the green Holocene lake muds. Diatoms are abundant in the Oloronga Beds, in the Holocene sediments, in the recent lakes, and in the upper parts of the High Magadi Beds, but they are very rare in the Green Beds. Green algae are predominant in Lake Little Magadi and in the springs of the Lake Magadi, but they are absent in the Green Beds (except for the calcareous alga Dascladiacea).

## CHERT FORMATION

Six types of chert can be distinguished in the Magadi area (Behr and Röhricht, 2000). The most important types are the massive chert in the Green Beds and the thin-bedded, papery magadiite chert in the High Magadi Beds. Until now, both cherts have been interpreted as abiotic sediments. In the established model, their formation was controlled by evaporation and the pH. Two stages of chert formation have been recognized: the precipitation of magadiite in saline, alkaline lakes, followed by leaching of $Na^+$ by surface water, which in turn leaves chert as the final product.

The large volume of the chert in the Magadi Basin can be explained by the deposition of large volumes of trachytic pyroclasts with volcanic glass, and their subsequent alteration by alkaline water according to the reaction (Eugster and Jones, 1968):

$$\text{Trachytic pyroclastics} + Na^+ + H_2O \rightarrow \text{Na–Al–Si gel} + SiO_2$$

The products of this reaction can be observed on a small scale today at the northern margins of Lake Little Magadi. The sodic brines that flow near the surface in porous volcanic debris are an important factor in the alteration. The silica concentrations in these brines may reach up to 1,760 mg/l $SiO_2$, and the brine pH is typically 9.5–12. In contrast, spring waters from deeper levels contain only 0.3% of those concentrations because of earlier silica precipitation and the formation of amorphous silica cements (Eugster, 1980). The alteration products consist of brown and green silts that are composed of volcanic particles, authigenic erionite, and analcime, with dispersed micron-size particles of magadiite, quartzine, and opal-A.

All of the cherts consist of up to 90% quartzine (length-slow) and chalcedony (length-fast). The rest consists of a late, green, amorphous silica-rich substance containing variable amounts of Al, Ca, K, Ba, Fe, microquartz, erionite, clinoptilolite, calcite, and volcanic fragments (Fig. 25). Opal-A (50–750 nm) was observed in the chert matrix, in silicified microbes and in micropores (Fig. 26).

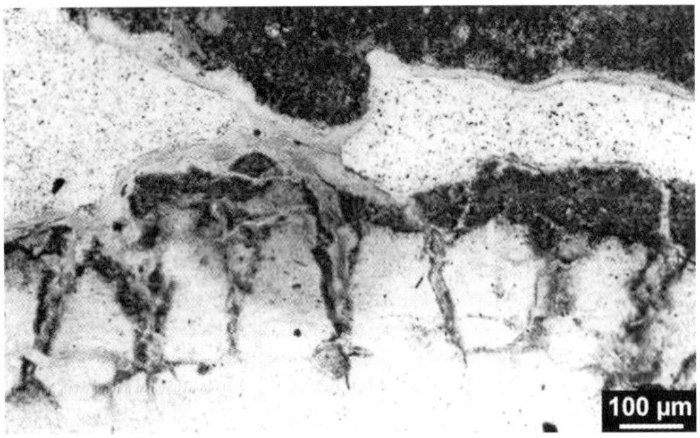


FIG. 25.—Reticular shrinkage structures in chert clumps (white), filled with green amorphous silica. Thin section (cross-polarized light).

Opal-CT spherules are subordinate here. Idiomorphic quartz crystals (< 10 µm) formed locally in the pores. The chert matrix consists of isotropic particles < 1 µm across. Birefringence can be observed in the coarser-grained parts of the matrix (~ 30%). Spherulites of quartzine up to 10–30 µm in diameter formed in the matrix and form a grid pattern under crossed nicols. This pattern allows a better contrast between the different structures within the chaotic polyphase chert. Various growth structures, viscous and brittle deformation structures, and cementation with quartzine and chalcedony can be differentiated. Schubel and Simonson (1990) previously considered that the spherulites represent replaced magadiite lepispheres.

## *"Magadiite Chert" in the High Magadi Beds*

Eugster (1967, 1969) and Hay (1968) assumed that the Na silicates represent the precursor of the chert, on the basis of the following arguments:

- All chert horizons represent the same sedimentary level and must have formed during one event.

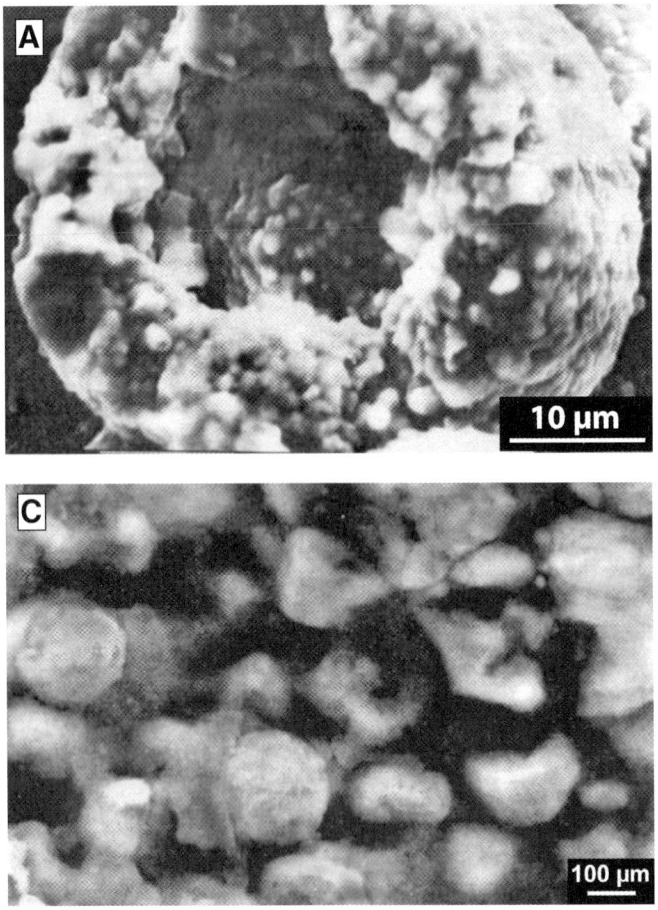


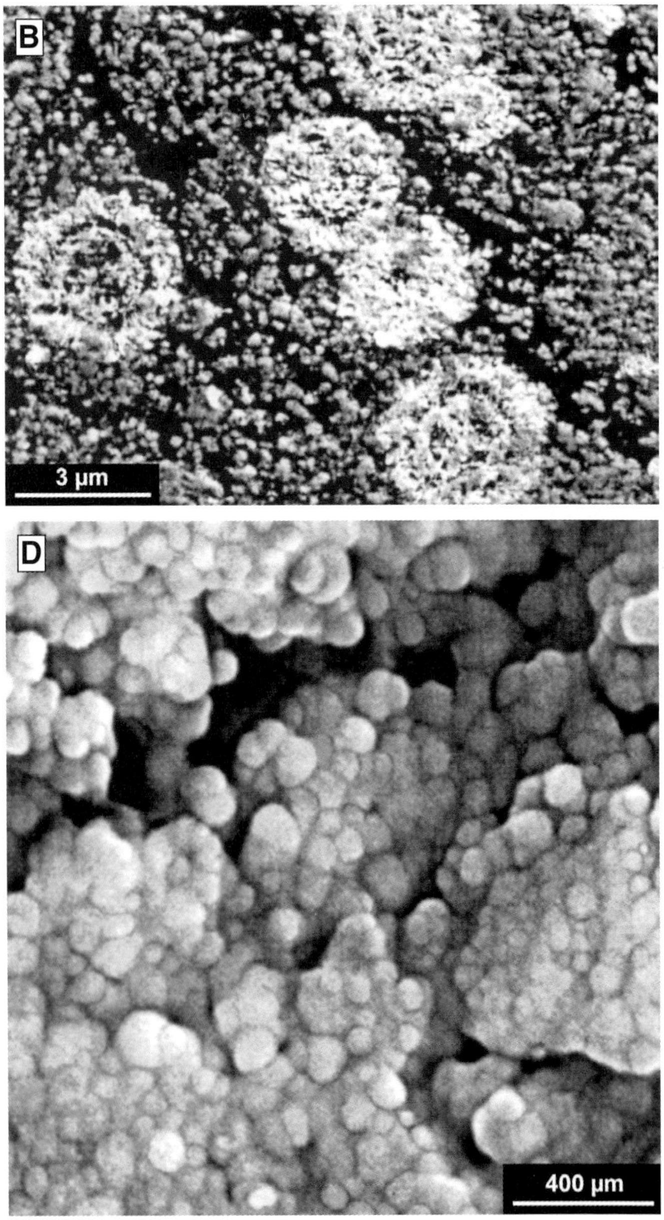


FIG. 26.—**A)** Cell of *Pleurocapsa* silicified by nanospheres of silica (opal-A). **B)** Calcified polydermate *Pleurocapsa* cells in a matrix of silica nanospheres. **C)** Opal-A spheres cemented by an amorphous silica phase. **D)** Spheres of chalcedony in matrix pores. SEM images.

- Magadiite layers show lateral transformation into chert.
- Because soft silica-phase sediments are unknown, the "toothpaste" deformation implies a soft precursor with properties like those of magadiite.

The transformation into chert was explained by the gradual removal of Na during the percolation of seasonal rainwater, runoff, and groundwater, or spontaneous crystallization of quartz from magadiite (Hay, 1968). This conversion should have resulted in shrinkage of at least 25 vol % (Eugster, 1980).

At present, the best outcrops of pure magadiite horizons (up to 20 cm thick) are exposed at the surface in erosion channels that become filled during rainfall or seasonal flooding. Silicification of Na silicates, characterized by volume-for-volume replacement, begins at levels deeper than 20 cm. Figure 27A shows fresh lepispheres of magadiite; the silicification took place without destruction of the primary crystals (Fig. 27B). However, the transformation is possible only if sufficient $SiO_2$-rich pore water is available. Since its formation ~ 20,000 years ago, the magadiite has remained soft and slightly plastic, despite its frequent contact with surface waters. Therefore, simple leaching of Na and transformation of the magadiite into quartzine is unlikely to be the main chert-forming process.

Chert that formed from magadiite builds up paper-thin, friable layers deformed by slumping before silicification, or homogeneous porcelain-like white to yellowish layers 1–3 cm thick with varnish-like striations. Was magadiite precipitated at the chemocline of a large water mass? The lake terrace situated 13 m above the High Magadi Beds, which confirms a former high water level, supports this interpretation. However, precipitation of the magadiite must have coincided with cyanobacterial activity and the formation of ichnofossils. Microbial mats, Al–Si gels, suspensions of erionite silt, and Na carbonate foam with calcite casts of gaylussite are found in fossilized form in the High Magadi Beds, as well as along the recent shoreline of the lake. We suggest, therefore, that the formation of magadiite and the silicification could have taken place at the same time at the surface of muddy plains that were temporarily covered with brines.

A recent example of this process was demonstrated in a drill core of Holocene sediments taken near hot springs at the south end of Lake Magadi. This 3-m-long core consists of soft, green, erionite silt with detrital volcanic minerals. Several millimeter-thick layers in the core are rich in equant microspherules of magadiite, kenyaite, chalcedony, and globular opal-A. The mineral phases are randomly mixed, and a clear genetic sequence cannot be recognized. This is in accord with the stability fields of these phases, which lie close together (Eugster, 1969). Larger aggregations of microspherules form flat and hard micronodules that consist entirely of chalcedony. This suggests that Na silicates and chert formed in part by early diagenesis in the lake mud. The Na-silicate-bearing horizons of the High Magadi Beds may have been silicified mainly by the supply of silica from surface brines and not by simple leaching and sodium removal. Eugster (1980) recognized that authigenic magadiite could form by the interaction of volcanic debris with alkaline brines. The common slumping structures in the chert formed during the stage of plasticity, before the final crystallization of quartzine.

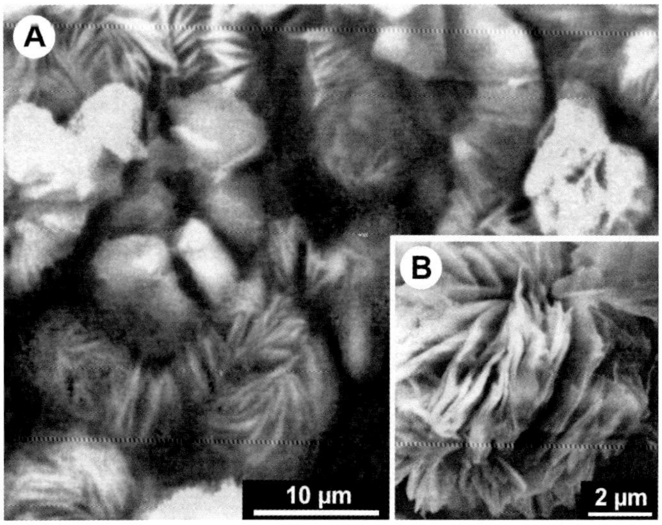


FIG. 27.—**A)** Lepispheres of magadiite. **B)** Bladed magadiite after silicification without disturbance of the crystal structure. SEM images.

### *Cherts of the Green Beds*

Unlike the thinly layered cherts in the High Magadi Beds, the cherts in the Green Beds are compact. The boundary between both types of chert is disconformable. This accords with the age difference revealed by radiometric dating (cf. Goetz and Hillaire-Marcel, 1992). The base of the High Magadi Beds consists of green silt with pebbles of the older (Green Beds) chert. Two units can be distinguished in the compact Green Beds:

(1) *A lower unit of massive chert.* Typical sedimentary features are microbial colonies and bioherms, bacterial stromatolite horizons, spring terraces, clumps of Si–Al gel and soft silica-phase nodules, slumping, fenestrae that include volcanic fragments, and flaser structures of former erionite silt suspensions. Diagenetic structures include "toothpaste" deformation features (pillows, dikes, flow structures, brecciation), slumping, striations, reticulation, dewatering channels, and gas vesicles.

(2) *An upper unit of bedded cherts.* Sedimentary structures are characterized by lenticular bedding (up to 2 cm), interlayering of former erionite silt suspensions, microbial mats, tepee and petee structures, ripples, shrinkage cracks, and small flood scours. Diagenetic features include casts of gaylussite, thermonatrite, and calcite on bedding planes, flat shear planes with striations, and distortion of bedding planes before crystallization of quartzine.

The lower unit of the chert shows chaotic fragmentation from a microscale to macroscale. The mineralogy, including the silica phases, is the same as described for the High Magadi Beds chert. Silicified magadiite is suspected in some samples, but identification is difficult. Late secondary magadiite formed in cavities from younger brines associated with the High Magadi Beds.

### *Processes of Silica Precipitation and the Biogenetic Catalyst*

The microbial communities of the stromatolite horizon show evidence of warm spring water that was discharged at the surface from volcanic debris. Silicification of the carbonaceous microbial structures and mucilaginous substances took place concurrently with precipitation of the silica phases. This is supported by the coalescence of silicified microbes and chert. The silica precipitation in the Green Beds may have taken place by the coagulation of colloidal silica sols, resulting in opaline substances, as shown by the nanometer-size silica spheres and abundant green and amorphous silica phases. Stable silica sols may contain > 50 wt %

silica and can be conserved for industrial purposes for many years when stored under favorable conditions. Electrostatic effects in the sols are the most likely stabilizing mechanism: the negative residual surface charge of the silica particles is compensated by $Na^+$ ions. The latter form diffuse ion layers around the silica particles. The repulsive forces between the ion layers are responsible for the stabilization of the silica sol. However, even minor variations in pH may change the surface charge and thereby destabilize the silica sol and already coagulated suspensions (Iler, 1973, 1979; Williams et al., 1985; Lagaly et al., 1998). The stability of silica particles increases with higher pH, because of the higher concentration of negative ions. In contrast, the sols become destabilized with increasing salt concentration. Silica particles coagulate to form a three-dimensional gel network at conditions of pH > 6. This can be explained by charge inversion of the silica particles from negative to positive, caused by ion adsorption.

Several authors have described examples of long-term stability of viscous silica sediments. Petersen and von der Borch (1965) described silica gel formation at pH 9.5–12 in the Coorong Lagoon, South Australia. Later chertification took place at pH 7.0–6.5. Colvinaux and Goodman (1971) described a silica gel 10 cm thick from a salt lake in the Galapagos Islands. Wheeler and Textoris (1978) attributed formation of Triassic chert in playas to crystallization of earlier silica gel. Renaut and Owen (1988) inferred a gelatinous siliceous precursor for opaline cherts around sublacustrine hot springs that discharged along fault lines. Behr and Watkins (2000) found silica gel as an intermediate phase in the formation of the Cretaceous opal in Australia. In that example, soil bacteria controlled the pH.

Coagulation is particularly effective in solutions with high concentrations of organic impurities and at sites near cyanobacteria. Decreasing pH is caused by higher concentrations of $CO_2$ and organic acids, which are produced by decay of organic material, notably during periods of cyanobacterial blooms. This process results in segregation of clumps of silica gel. The first gel clumps are of micron size, but with progressive accretion they may grow to centimeter-scale dimensions. At this stage, the viscous material is ductile like fresh magadiite and forms a heterogeneous lumpy paste like putty. In the cherts, larger gel clumps commonly contain numerous spindle-shaped and undeformed casts of gaylussite < 1 cm in size. These indicate temporary evaporitic stages during the main period of chert formation.

Lithification to chert begins along the rims of the clumps of gel. At first, an isotropic crust of fine-grained silica particles forms around the clumps. During this process, the water content decreases by several percent. Subsequent shrinkage results in concentric cracks and reticulate textures with wedge-shaped cross sections in the crust (Fig. 25). These shrinkage cracks also function as dewatering pathways. The remains of the silica sols generate an amorphous green substance in fenestrae, shrinkage cracks, and channels. Micro-scale spring mounds formed at the ends of fluid-escape channels and consist of flaky crystallites (probably silhydrite, $3SiO_2 \cdot H_2O$) with a lower refractive index than chalcedony (Fig. 28). The reticulated surfaces are commonly covered with craters about 50 μm in diameter (Fig. 29). Chalcedony fibers up to 20 μm long in the crater axes show parallel orientation and build palisade-like walls. The immediate surroundings of the craters are covered with bipyramidal and rhombohedral quartz crystals (~ 2.5 μm), which appear macroscopically as a white crust.

The Green Beds chert precursor showed thixotropic properties, which may explain the seismotectonic deformation textures. Thixotropy is based on the formation and breakdown of frame complexes. These frameworks are built up of suspended erionite needles that fill in the network of fenestrae and larger cavities. The suspension was subsequently thickened by the formation of hydrogen bridges between the silica particles and the molecules in the dispersion agent. Finally, the silica particles and erionite aggregates form one gelatinous mass. Mechanical effects may destabilize the cohesion of the suspension. Consequently, the viscosity is decreased and the yield point is lowered. Cohesion normally increases when the liquefied dispersion is undisturbed. Thixotropy of the precursor was especially influenced by calcium (cf. Lagaly et al., 1998), which can now be found in the calcified cyanobacteria. Gels may expand or shrink by a factor of a hundred owing to changing temperature. Volume changes also may have been caused by variations in pH, the ion valance, and resulting attractive forces between the particles. In this way, the unusual lobate morphology of the chert nodules and the reticulate textures can be explained.

Rise of gas bubbles is possible even in Newtonian liquids with a high viscosity. However, the mobility of gas bubbles in suspen-

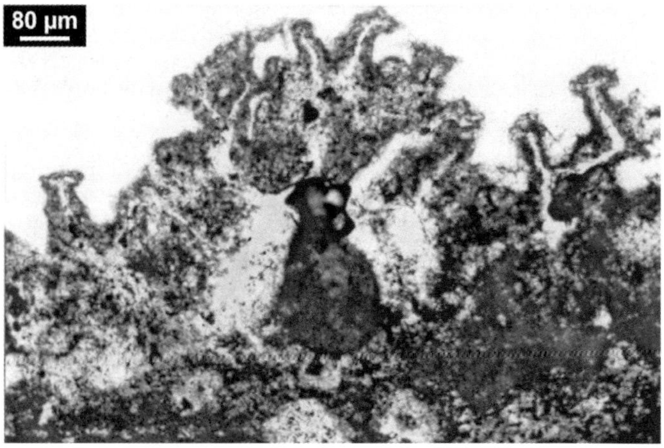


FIG. 28.—Dewatering microchannel system at the edge of a silica clump. Thin section (plane-polarized light).

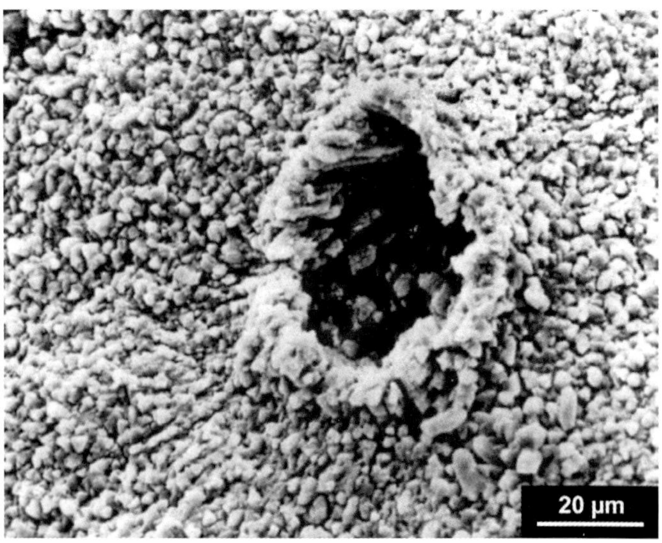


FIG. 29.—Dewatering craters on reticulated chert covered with rhombohedral quartz crystals. SEM image.

sion is possible until the yield point is reached and the gas bubbles become stuck. The deformed and dumbbell-shaped pores in many chert samples are characteristic of this process. Dewatering must have been strong in both the magadiite and silica gel precursors, which contained 10–15 wt % and ~ 50 wt % $H_2O$, respectively. Water-escape tubes are up to 20 cm in diameter (see also Eugster, 1980, p. 15). The high porosity of the Green Beds chert and the many genetically different pore generations are the result of polyphase development. This is in contrast to the denser cherts in the Moinik and Oloronga Beds, which do not show effective porosities. The distinguishing features of the two porosity types have been examined and can be summarized as follows:

Where there is effective porosity (typically 2.5–5 vol %), reticulate networks form an open pore system, partly filled with green amorphous silica. Concentric shrinkage cracks develop around gel clumps. Dewatering tubes are not interconnected, but have openings at the surface. Syneresis cracks are widespread in the chert matrix and are partly mineralized with chalcedony. Bioturbation can produce curled open worm tubes.

Where porosity is ineffective, the pores in the chert are isolated. Chert from the Moinik Beds and Oloronga Beds has a density similar to that of chalcedony (2.58–2.63 g/cm$^3$). Cherts with isolated pores at Magadi have a density of 2.4–2.5 g/cm$^3$. These cherts show globular pores (< 0.5–20 μm) that were created by microbes. Oblate fenestrae (up to several centimeters) formed as cavities between colonies of bacteria, volcanic debris, and gel clumps. Spherical and dumbbell-shaped, micron- to millimeter-size gas cavities ($CO_2$, $N_2$, $O_2$) may constitute up to 50 vol % of the chert layers, and are deformed by volume changes in the matrix. They show parallel alignment due to flow of the gel and are concentrated in the upper parts of the bioherms and crusts. Casts of gaylussite were common on the rims of gel clumps. Fluid inclusions were not found. The fluids (water and brine) must have escaped through syneresis cracks.

The pH is the most important parameter for the precipitation of Na silicates and/or amorphous silica at normal temperatures. The production of microbial $CO_2$ may have caused significant shifts of the pH levels. Therefore, the microbial activity likely plays the role of a catalyst for the chert genesis.

## DISCUSSION AND CONCLUSIONS

The main data with implications for the genesis of Magadi-type chert at the type locality in the Magadi Basin are the following:

1. Radiometric age dating and detailed stratigraphical observations show two genetically different chert types, which were considered as one unit in earlier studies.

2. Deposition of the chert precursor was strongly influenced by cyanobacteria and other microbes that lived in shallow water. Deformation textures in the precursor demonstrate thixotropic properties of the material before lithification.

3. Opal-A and amorphous silica point to an important role of colloidal silica sols and silica gel.

4. The Na-silicate-bearing horizons of the High Magadi Beds have been silicified mainly by the supply of silica from surface brines and not by simple leaching and sodium removal.

The cyanobacteria might easily have been overlooked because these cells, 5–20 μm in size, are similar to the microspherules of magadiite, kenyaite, chalcedony, calcispheres, pellets, and coated grains. Their recognition is even more difficult after silicification. The calcified cells are only partly preserved, because of dissolution and partial replacement of the calcite. The recent spring water (< 1 mg/l Ca) and the Holocene sediments and High Magadi Beds have low Ca contents (0.6–2 wt % CaO), which are assumed to originate in the Ca-poor trachytes (1.67 wt % CaO) of the Magadi Basin. The Oloronga Beds have significantly higher Ca contents (4.5 wt %) because these sediments formed by the alteration of pyroclastic fragments (6–18 wt % CaO) that originate from the Lake Natron Basin. Consequently, the Oloronga Beds are rich in limestone with gastropods and ostracods, Ca-bearing zeolites, and calcite. Sediments from the Oloronga Beds were redeposited in the Green Beds, so that CaO became available for calcification of the cyanobacteria and the formation of Ca-rich minerals (calcite, clinoptilolite, gaylussite). In the High Magadi Beds, the Ca content was low because substantial pyroclastics from the Lake Natron region are missing.

The geochemistry of silica-rich brines and the formation of Na silicates have been the object of several studies (Eugster, 1970; Jones et al., 1977; Fritz et al., 1987). Our observations agree with the findings of Dietzel and Usdowski (1995) and Dietzel and Böhme (1997), who studied the correlation between the stabilities of magadiite, quartz, and amorphous silica, and the ratio of polymeric to monomeric silica. At high pH, the monomers dissociate to form networks of polymers, which make up 40 mol % of the total silica. As a consequence, the $Si(OH)_4$ activity is reduced. The decisive factor for the destabilization of the polymers is the reduction of the pH due to dissolution of $CO_2$. This again results in higher $Si(OH)_4$ activities. Depending on the $Na^+/H^+$ ratio, Na silicates and (or) amorphous $SiO_2$ (with similar stability fields) are formed. Decreasing temperatures play only a limited role in the precipitation of the silica.

Cyanobacteria are obligate oxygenic, phototrophic organisms, which produce molecular oxygen by photosynthetic photolysis of water. They are able to bind hydrogen for the reduction of $CO_2$ to form carbohydrates. Bacterial decay was accepted by Bradley and Eugster (1969) and Eugster and Hardie (1978) as a mechanism to recharge brines with $CO_2$, resulting in the precipitation of the Holocene trona deposits at Magadi. The role of volcanic and atmospheric $CO_2$ was discussed by Hillaire-Marcel (1987). However, a biological $CO_2$ source has the most efficient carbonating effect and is most plausible here. The high growth rate of cyanobacteria in the Green Beds indicates an environment that was rich in dissolved organic carbon, oversaturated with respect to $CO_2$, and exposed to high light intensity. Additional stimulating factors are the high viscosity of the silica sols and the low salinity of the water, which hinder degassing of $CO_2$. In the Green Beds, sediments typical of alkaline lakes such as trona may not have been deposited. With progressive growth of the biomass, the $CO_2$ cycle became increasingly important for the dissolution of calcite and precipitation of silica, besides photosynthesis and respiration. The CaO content in the chert precursor of ~ 3–7 wt % in the stromatolite horizon locally decreased to < 1 % in the chert.

The $SiO_2$ content in the volcanic rocks of the Lake Natron Basin is 37–53 wt %, compared to 59–64 wt % in the trachytes of the Magadi Basin. Therefore, larger quantities of $SiO_2$ and chert were mobilized at Magadi than in the Natron Basin. The silica-bearing mother brines were generated by the reaction between percolating alkaline water and volcanic detritus. The brines were ponded in flat faulted depressions, evaporated, and became covered with carpets of cyanobacteria and stromatolites. These "bioreactors" controlled the pH levels and thereby the precipitation of Na silicates and amorphous silica. The following sequence of mineralization for the "Magadi chert" (Green Beds) can be inferred:

silica sol → silica gel → amorphous silica → quartzine

For the Green Beds chert, the role of Na silicates is subordinate. The precursor of the "magadiite chert" (High Magadi Beds) consisted of Na silicates with intercalations of microbial crusts and ichnofossils. During diagenesis, percolating brines eventually silicified the precursor. The chert-forming times were characterized by evaporation. Unlike previous explanations, we do not need a deeper lake with a chemocline in our model. In the Oloronga Beds, deposited in a large fresh-water lake, diatoms are the main silica source for a biogenic chert.

The $\delta^{18}O$ data of O'Neil and Hay (1973) on Magadi-type chert are not tied well to the stratigraphic record. In supplementary studies, Behr and Röhricht (2000) differentiated between the two types of chert using stable oxygen isotopes (Fig. 30). The $\delta^{18}O$ values of the Green Beds chert range from 41 to 44‰, and represent the highest values known for any geological material. The $\delta^{18}O$ of the High Magadi Beds chert is significantly lower (32.8–37.7‰) and is in the same range as the $\delta^{18}O$ of magadiite and kenyaite. Values of 31.8‰ in laminated chert can be explained by mixing with meteoric water. High $\delta^{18}O$ values are generally explained by very high evaporation rates. The inflow of concentrated saline brines may have induced the precipitation of silica phases from sols. The higher salinities may cause sudden gelling (Iler, 1973, 1979; Lagaly et al., 1998) and stiffening of the chert precursor. At a late stage (i.e., after the diapiric extrusion of the silica mash) quartzine fibers may spontaneously crystallize and the silica precursor is transformed into chert.

## ACKNOWLEDGMENTS

The field work in the period 1995–1999 was funded by the DFG (German Science Foundation) with logistical support of the Department of Geology, University of Nairobi. I am grateful to C. Röhricht, A. Arnold, A.M. van den Kerkhof (University of Göttingen), J. Ogola and O. Ochieng (Nairobi), and the Magadi Soda Co. (Kenya) for analytical and technical support, and to R. Renaut for his stimulation in continuing the studies.

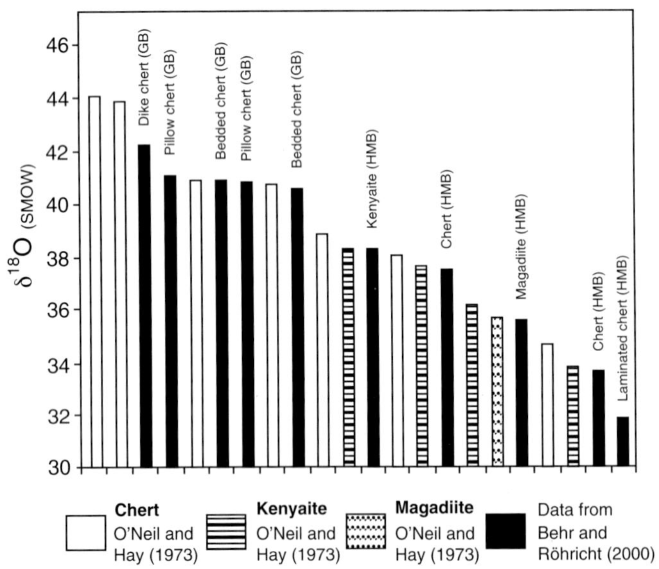


FIG. 30.—The range of $\delta^{18}O$ values for cherts from Magadi and Tanzania.

## REFERENCES

ALLEN, D.J., DARLING, W.G., AND BURGESS, W.G., 1989, Geothermics and hydrogeology of the southern part of the Kenya Rift Valley with emphasis on the Magadi–Nakuru area: British Geological Survey, Research Report (Hydrogeology Series) SD/89/1.

BAKER, B.H., 1958, Geology of the Magadi area: Geological Survey of Kenya, Report 42, 81 p.

BAKER, B.H., 1963, Geology of the area south of Magadi: Geological Survey of Kenya, Report 61, 27 p.

BAKER, B.H., AND WOHLENBERG, J., 1971, Structure and evolution of the Kenya Rift Valley: Nature, v. 229, p. 538–542.

BEHR, H.-J., AND RÖHRICHT, C., 2000, Record of seismotectonic events in siliceous cyanobacteria sediments (Magadi cherts), Lake Magadi, Kenya: International Journal of Earth Sciences, v. 89, p. 268–283.

BEHR, H.-J., AND WATKINS, J.J., 2000, Cretaceous microbes—producer of black opal at Lightning Ridge, NSW, Australia (abstact): Geological Society of Australia, 15th Convention, Abstracts, p. 28.

BRADLEY, W.H., AND EUGSTER, H.P., 1969, Geochemistry and paleolimnology of the trona deposits and associated authigenic minerals of the Green River Formation of Wyoming: U.S. Geological Survey, Professional Paper 496-B, 71 p.

BUTZER, K.W., ISAAC, G.L., RICHARDSON, J.L., AND WASHBURN-KAMAU, C., 1972, Radiocarbon dating of East African lake levels: Science, v. 175, p. 1069–1076.

CAROZZI, A.V., 1961, Observations on algal biostromes in the Great Salt Lake, Utah: Journal of Geology, v. 70, p. 246–253.

CASANOVA, J., 1986, East African rift stromatolites, *in* Frostick, L.E., Renaut, R.W., Reid., I., and Tiercelin, J.J., eds., Sedimentation in the African Rifts: Geological Society of London, Special Publication 25, p. 201–210.

CASANOVA, J., 1987, Stromatolites et hautes niveaux lacustrine pléistocènes du basin Natron–Magadi (Tanzanie–Kenya): Sciences Géologiques Bulletin, v. 40, p. 135–153.

COLINVAUX, P.A., AND GOODMAN, G.D., 1971, Recent silica gel from saline lake in Galapagos Islands (abstract): American Association of Petroleum Geologists, Bulletin, v. 55, p. 333–334.

CRANE, K., 1981, Thermal variations in the Gregory Rift of southern Kenya: Tectonophysics, v. 74, p. 239–262.

CROSSLEY, R., 1979, The Cenozoic stratigraphy and structure of the western part of the Rift Valley in southern Kenya: Geological Society of London, Journal, v. 136, p. 393–405.

DAMNATI, B., TAIEB, M., AND WILLIAMSON, D., 1992, Laminated deposits from Lake Magadi (Kenya). Climatic contrast effect during the maximum wet period between 12,000–10,000 yrs B.P.: Société Géologique de France, Bulletin, v. 163, p. 407–414.

DIETZEL, M., AND USDOWSKI, E., 1995, Depolymerisation of soluble silicate in dilute aqueous solutions: Colloid and Polymer Science, v. 273, p. 590–597.

DIETZEL, M., AND BÖHME, G., 1997, Adsorption und Stabilität von polymerer Kieselsäure: Chemie der Erde, v. 57, p. 189–203.

DOEMEL, W.N., AND BROCK, T.D., 1974, Bacterial stromatolites: origin of laminations: Science, v. 184, p. 1083–1085.

EUGSTER, H.P., 1967, Hydrous sodium silicates from Lake Magadi, Kenya: precursors of bedded chert: Science, v. 157, p. 1177–1180.

EUGSTER, H.P., 1969, Inorganic bedded cherts from the Magadi area, Kenya: Contributions to Mineralogy and Petrology, v. 22, p. 1–31.

EUGSTER, H.P., 1970, Chemistry and origin of the brines of Lake Magadi, Kenya: Mineralogical Society of America, Special Paper no. 3, p. 215–235.

EUGSTER, H.P., 1980, Lake Magadi, Kenya, and its precursors, *in* Nissenbaum, A., ed., Hypersaline Brines and Evaporitic Environments: Amsterdam, Elsevier, Developments in Sedimentology, no. 28, p. 195–232.

EUGSTER, H.P., 1986, Lake Magadi Kenya: a model for rift valley hydrochemistry and sedimentation, *in* Frostick, L.E., Renaut, R.W., Reid, I., and Tiercelin, J.J., eds., Sedimentation in the African Rifts: Geological Society of London, Special Publication 25, p. 177–189.

EUGSTER, H.P., AND CHOU, I.M., 1973, The depositional environments of Precambrian banded-iron formations: Economic Geology, v. 68, p. 1144–1168.

EUGSTER, H.P., AND JONES, B.F., 1968, Gels composed of sodium–aluminium-silicate, Lake Magadi, Kenya: Science, v. 161, p. 160–164.

EUGSTER, H.P. AND HARDIE, L.A., 1978, Saline lakes, *in* Lerman, A., ed., Lakes: Chemistry, Geology, Physics, p. 237–293.

FAIRHEAD, J.D., MITCHELL, L.G., AND WILLIAMS, L.A.J., 1972, New K/Ar determinations on rift volcanics of S. Kenya and their bearing on age of rift faulting: Nature, Physical Science, v. 238, p. 66–69.

FRITZ, B., ZINS-PAWLAS, M.-P., AND GUEDDARI, M., 1987, Geochemistry of silica-rich brines from Lake Natron (Tanzania): Université Louis Pasteur de Strasbourg, Sciences Géologiques, Bulletin, v. 40, p. 97–110.

GARRETT, D.E., 1992, Natural Soda Ash: Occurrences, Processing, and Use: New York, Van Nostrand Reinhold, 336 p.

GERDES, G., AND KRUMBEIN, W.E., 1987, Biolaminated Deposits: Berlin, Springer, 183 p.

GLOAGUEN, R., 2000, Analyse quantitative de l'extension continentale par imagerie satellitale optique et radar. Application au rift sud-kenyan: unpublished Ph.D. dissertation, Université de Bretagne Occidentale, Brest, France, 277 p.

GOETZ, C., AND HILLAIRE-MARCEL, C., 1992, U-series disequilibria in early diagenetic minerals from Lake Magadi sediments, Kenya: dating potential: Geochimica et Cosmochimica Acta, v. 56, p. 1331–1341.

GOLUBIC, S., AND HOFFMANN, H.J., 1976, Comparison of Holocene and mid-Precambrian Entophysalidaceae (Cyanophyta) in stromatolitic algal mats: cell division and degradation: Journal of Paleontology, v. 50, p. 1074–1082.

HATTORI, L, UMEDA, M., NAKAGAWA, T., AND YAMAMOTO, H., 1996, From chalcedonic chert to quartz chert: Diagenesis of chert hosted in a Miocene volcanic–sedimentary succession, Central Japan: Journal of Sedimentary Research, v. 66, p. 163–174.

HAY, R.L., 1968, Chert and its sodium-silicate precursors in sodium-carbonate lakes of east Africa: Contributions to Mineralogy and Petrology, v. 17, p. 255–274.

HILLAIRE-MARCEL, C., 1987, Hydrologie isotopique des lacs Magadi (Kenya) et Natron (Tanzanie): Université Louis Pasteur de Strasbourg, Sciences Géologiques, Bulletin, v. 40, p. 111–120.

HILLAIRE-MARCEL, C., AND CASANOVA, J., 1987, Isotopic hydrology and paleohydrology of the Magadi (Kenya)–Natron (Tanzania) Basin during the late Quaternary: Palaeogeography, Palaeoclimatology, Palaeoecology, v. 58, p. 155–181.

HOLLNACK, D., AND STANGEL, R., 1998, The seismicity related to the southern part of the Kenya Rift: Journal of African Earth Sciences, v. 26, p. 477–495.

ILER, R.K., 1973, Colloidal silica, *in* Matijevic, E., ed., Surface and Colloid Science: New York, Wiley, p. 11–51.

ILER, R.K., 1979, The Chemistry of Silica; Solubility, Polymerization, Colloid and Surface Properties, and Biochemistry: New York, Wiley, 866 p.

JONES, B.J., RETTIG, S.L., AND EUGSTER, H.P., 1967, Silica in alkaline brines: Science, v. 158, p. 1310–1314.

JONES, B.J., EUGSTER, H.P., AND RETTIG, S.L., 1977 Hydrochemistry of the Lake Magadi basin, Kenya: Geochimica et Cosmochimica Acta, v. 41, p. 53–72.

KAUFMANN, A., MARAGARITZ, M.P.M., HILLAIRE-MARCEL, C., HOLLOS, G., BOARETTO, E., AND TAIEB, M., 1990, The $^{36}$C1 ages of the brines in the Magadi–Natron Basin, East Africa: Geochimica et Cosmochimica Acta, v. 54, 2827–2833.

KRAINER, K., AND SPÖTL, C., 1998, Abiogenic silica layers within a fluvio-lacustrine succession, Bolzano volcanic complex, northern Italy: a Permian analogue for Magadi-type cherts?: Sedimentology, v. 45, p. 489–505.

LAGALY, G., SCHULZ, O., AND ZIMEHL, R., 1998, Dispersionen und Emulsionen: Darmstadt, Steinkopff, 560 p.

O'NEIL, J.R., AND HAY, R.L., 1973, $^{18}O/^{16}O$ ratios in cherts associated with the saline lake deposits of East Africa: Earth and Planetary Science Letters, v. 19, p. 257–266.

PARNELL, J., 1986, Devonian Magadi-type cherts in the Orcadian basin, Scotland: Journal of Sedimentary Petrology, v. 56, p. 495–500.

PARNELL, L., 1987, Mineralized Magadi-type chert in the Devonian of Scotland: Support for a model of sulfide deposition in low temperature alkaline conditions: Economic Geology, v. 28, p. 1053–1056.

PETERSON, M.N.A., AND VON DER BORCH, C.C., 1965, Chert: modern inorganic deposition in a carbonate-precipitating locality: Science, v. 149, p. 1501–1503.

RENAUT, R.W., AND OWEN, R.B., 1988, Opaline cherts associated with sublacustrine hydrothermal springs at Lake Bogoria, Kenya Rift Valley: Geology, v. 16, p. 699–702.

RÖHRICHT, C., 1999, Lithologie und Genese der Chertserien des Magadi Beckens, Lake Magadi, Kenia: unpublished Ph.D. dissertation, University of Göttingen, 108 p.

SCHUBEL, K.A., AND SIMONSON, B.M., 1990, Petrography and diagenesis of cherts from Lake Magadi, Kenya: Journal of Sedimentary Petrology, v. 60, p. 761–776.

SEEGERS, L., AND TICHY, H., 1999, The *Oreochromis alcalicus* flock (Teleostei: Cichlidae) from lakes Natron and Magadi, Tanzania and Kenya, with descriptions of two new species: Ichthyological Exploration of Freshwaters, v. 10, p. 97–146.

SHEPPARD, R.A., AND GUDE, L., 1974, Chert derived from magadiite in a lacustrine deposit near Rome, Malheur County, Oregon: U.S. Geological Survey, Journal of Research, v. 2, p. 625–630.

SURDAM, R.C., AND EUGSTER, H.P., 1976, Mineral reactions in the sedimentary deposits of the Lake Magadi region, Kenya: Geological Society of America, Bulletin, v. 87, p. 1739–1752.

SURDAM, R.C., EUGSTER, H.P., AND MARINER, R.H., 1972, Magadi-type chert in Jurassic and Eocene to Pleistocene rocks, Wyoming: Geological Society of America, Bulletin, v. 83, p. 2261–2266.

THOMPSON, J.B., AND FERRIS, F.G., 1990, Cyanobacterial precipitation of gypsum, calcite, and magnesite from natural alkaline lake water: Geology, v. 18, p. 995–998.

TICHY, H., AND SEEGERS, L., 1999, The *Oreochromis alcalicus* flock (Teleostei: Cichlidae) from lakes Natron and Magadi, Tanzania and Kenya: a model for the evolution of 'new' species flocks in historical times: Ichthyological Exploration of Freshwaters, v. 10, p. 147–174.

WHEELER, W.H., AND TEXTORIS, D.A., 1978, Triassic limestone and chert of playa origin in North Carolina: Journal of Sedimentary Petrology, v. 48, p. 765–776.

WHITE, T.H., 1953, Some speculations on the sudden occurrence of floods in the history of Lake Magadi: East African Natural History Society, Journal, v. 22, p. 69–71.

WILLIAMS, L.A., PARKS, O., AND CRERAR, D.A., 1985, Silica diagenesis, 1. Solubility controls: Journal of Sedimentary Petrology, v. 55, p. 301–311.

WILLIAMSON, D., TAIEB, M., DAMNATI, B., ICOLE, M., AND THOUVENY, N, 1993, Equatorial extension of the younger Dryas event: rock magnetic evidence from Lake Magadi (Kenya): Global and Planetary Change, v. 7, p. 235–242.

# TRACE-ELEMENT COMPOSITION OF CHERTS FROM ALKALINE LAKES IN THE EAST AFRICAN RIFT: A PROBE FOR ANCIENT COUNTERPARTS

ROBERT KERRICH, ROBIN W. RENAUT, AND THOMAS BONLI
*Department of Geological Sciences, University of Saskatchewan, Saskatoon, SK S7N 5E2, Canada*

ABSTRACT: Magadiite and cherts from the Magadi basin in southern Kenya and four other localities in the East African Rift all share distinctive compositional systematics: (1) negatively fractionated REE patterns; (2) high absolute concentrations of U, Nb, and Zr (up to 1,500 ppm versus a crustal average of 190 ppm); (3) normalized enrichments of U, Nb, and Zr relative to REE; (4) extreme fractionations of U–Th, Nb–Ta, and Zr–Hf; (5) positive Ce but negative Eu anomalies; and (6) normalized peaks at Mo, Ag, and Sb. Alkaline lake cherts composed of secondary quartz retain the trace-element patterns of the precursor, albeit at lower absolute element contents. Sublacustrine sinters at Lake Bogoria share some of the compositional features of the cherts from Magadi but lack the Ce and Eu anomalies and U–Th, Nb–Ta, and Zr–Hf fractionations. Cherts from ocean-ridge, pelagic, and continental-shelf settings are characterized by progressively larger Zr contents (≤ 170 ppm), but all are distinctly lower than in the alkaline-lake cherts, where micron-scale Zr-rich phases were identified as possible authigenic zircons. Both marine and alkaline-lake cherts share positive Ce anomalies. These result from scavenging of Ce by Fe–Mn oxyhydroxides in seawater, but from the solubility of Ce (IV) in oxidized alkaline brines for the lacustrine cherts. Europium anomalies range from small negative to positive in marine cherts, whereas alkaline-lake cherts feature large negative anomalies. These compositional systematics, which reflect the aqueous environment in which the chert precursor formed, confer a tool for interpreting the paleoenvironment of cherts in the geological record.

## INTRODUCTION

Lacustrine cherts are common in continental sediments. These cherts potentially have many origins, including primary precipitation from siliceous gels (Peterson and von der Borch, 1965; Wheeler and Textoris, 1978; Shaw et al., 1990; Chough et al., 1996); diagenetic alteration of diatomite (biogenic opal-A to microcrystalline quartz); silicification of other sediments, notably carbonates (Birney de Wet and Hubert, 1989; Behr, this volume); sublacustrine hydrothermal precipitation (Renaut and Owen, 1988; Khandikar et al., 1999; Renaut et al., 2002); and early diagenetic alteration of sodium-silicate precursor minerals in saline, alkaline lakes (Sheppard and Gude, 1986). Most lacustrine cherts are composed of microcrystalline or cryptocrystalline silica and are found in beds, nodules, or as irregular replacements of other sediments. Distinguishing the origin(s) of these cherts is not always easy. Morphology and petrographic features are commonly used, but like marine cherts, interpretations based on these criteria alone can be ambiguous (cf. Hattori et al., 1996; Bustillo, 2001; Behr, this volume).

One of the characteristic types of chert in continental rift basins is Magadi-type chert, first described from Lake Magadi in the southern Kenya Rift (Eugster, 1967, 1969). Magadi-type cherts form in highly alkaline lakes from hydrous sodium-silicate precursor minerals, such as magadiite ($NaSi_7O_{13}(OH)_3 \cdot 3H_2O$) and kenyaite ($NaSi_{22}O_{45} \cdot 10H_2O$) (Hay, 1968; Rooney et al., 1969; Surdam et al., 1972; Maglione, 1980; Sheppard and Gude, 1986; Sebag et al., 2001). The sodium-silicate precursor minerals transform to quartz by leaching of sodium (Eugster, 1967, 1969), recrystallization in contact with trapped lake water (Hay, 1968, 1970), or other mechanisms (Sheppard and Gude, 1986; Sebag et al., 2001). Heaney (1995) proposed that magadiite might transform to quartz through intermediate moganite. Given that the pH of the fluids in which the precursor minerals formed is typically > 10.5, the trace-element distribution inherited by such cherts might be highly distinctive, providing an additional tool to help the recognition of alkaline lake cherts in the sedimentary record.

To test this hypothesis, the trace-element compositions of magadiite, Magadi-type chert, and other cherts from the Magadi basin were determined. These alkaline-lake cherts possess distinctive rare earth element (REE) patterns unlike those reported for other types of chert. They also have anomalously high concentrations of U, Nb, Zr, Ag, and other trace elements. This composition is consistent with the mobility of these elements in the highly alkaline fluids from which the precursor minerals precipitated.

To confirm the retention of these distinctive trace-element patterns in the geological record, older lacustrine cherts from the East African Rift were analyzed. These included Pleistocene cherts from Magadi basin, late Pliocene–early Pleistocene cherts at Olduvai Gorge in Tanzania, and Pliocene cherts from the Central Kenya Rift. Lake and hot-spring waters from saline, alkaline Lake Bogoria, Kenya, were analyzed as a potential analog for Magadi lake waters during magadiite precipitation. Hydrothermal lacustrine chert (sublacustrine sinter) at Lake Bogoria was analyzed for comparison.

In this paper, we present results for REE, high-field-strength elements (HFSE: Th, Nb, Ta, Zr, Hf, Ti) and other trace elements in these samples, and discuss their application to the geological record of chert from continental rifts and other continental environments. The compositional systematics of the cherts at Magadi are then compared with marine cherts from several depositional settings. We show that the trace-element geochemistry is potentially a powerful tool for recognizing cherts that have formed in saline, alkaline lakes of continental rifts and elsewhere.

## GEOLOGICAL SETTING OF SAMPLING SITES

### *Lake Magadi Basin, Kenya*

The Lake Magadi basin lies in the semiarid southern Kenya Rift just north of the border with Tanzania (Fig. 1A–C). Lake Magadi is a hydrologically closed, saline pan containing up to 40 m of bedded sodium carbonate evaporites, principally trona ($NaHCO_3 \cdot Na_2CO_3 \cdot 2H_2O$) and nahcolite ($NaHCO_3$) (Baker, 1958; Eugster, 1980; Behr and Röhricht, 2000). The saline, alkaline lake occupies the lowest point in the southern Kenya Rift, which is up

Sedimentation in Continental Rifts
SEPM Special Publication No. 73, Copyright © 2002
SEPM (Society for Sedimentary Geology), ISBN 1-56576-082-4, p. 275–294.

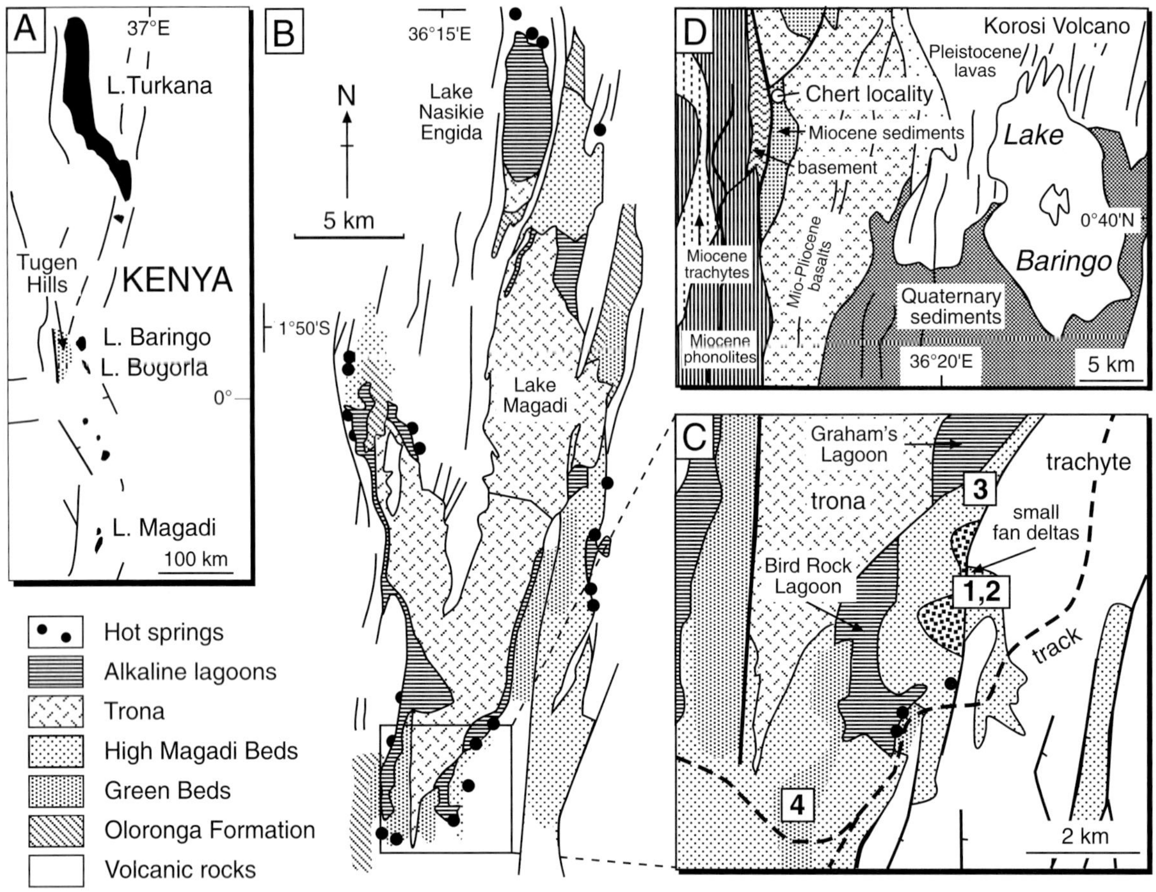


FIG. 1.—Location of chert sampling sites. **A)** Kenya Rift Valley, showing location of Lake Bogoria, Lake Magadi, and Tugen Hills. **B)** Simplified geological map of Lake Magadi (modified from Allen et al., 1989, and Behr and Röhricht, 2000). **C)** Inset of Part B, showing sampling locations. **D)** Simplified geological map of the central Tugen Hills (after Chapman et al., 1978), showing location of the Lukeino Formation cherts.

to ~ 70 km wide at this latitude. The Nguruman Escarpment, a large boundary fault, lies to the west. The eastern edge of the rift is less well defined but is marked by a series of smaller fault scarps. Most of the rift floor is characterized by subparallel grid faults that include the Lake Magadi trough (Crossley, 1979; Le Turdu et al., 1999). The rift floor is composed mainly of Pliocene and Pleistocene trachytes, alkali basalts, phonolites, and volcaniclastic deposits, with locally intercalated colluvial, fluviatile, and lacustrine sedimentary rocks.

Lake Magadi is the successor to a series of older, larger lakes that at different times have extended laterally across much of the rift floor. The oldest exposed sedimentary rocks, many of which are lacustrine, form the mid-Pleistocene Oloronga Formation (Baker, 1958; Eugster, 1980). This sequence, which is about 45 m thick, consists of tuffs, siltstones, bioclastic limestones, and concretionary cherts. The Oloronga Formation contains a trachyte flow dated at < 0.8 Ma (Fairhead et al., 1972), and a *Hippopotamus* tooth yielded an age of ~ 0.3 Ma (Behr and Röhricht, 2000). Following deposition of the Oloronga Formation, there was a major phase of faulting and basin subsidence. During this phase, the "Green Beds" were deposited (Behr and Röhricht, 2000; Behr, this volume). These consist of greenish tuffs, volcaniclastic silts, and at least three units of chert. The Green Beds have an aggregate total thickness of about 30 m (Behr, this volume). The Green Beds, which are dated at about 0.1 Ma, were formerly considered part of the Oloronga Formation (Behr and Röhricht, 2000) and are in part equivalent to the Chert Series of Baker (1958).

Younger phases of lacustrine sedimentation have been recorded from sediment outcrops around the lake margins and in adjacent graben, as well as in superimposed sequences of stromatolites up to 60 m above present lake level, the latest of which are dated at 130 ka and 12–10 ka, respectively (Hillaire-Marcel et al., 1986; Hillaire-Marcel and Casanova, 1987).

The phase of lacustrine sedimentation that preceded and includes the modern saline pan produced the High Magadi Beds, a series of clay- and zeolite-rich tuffaceous mudstones and siltstones (Baker, 1958; Butzer et al., 1972; Surdam and Eugster, 1976). These are preserved in small outcrops around the present shoreline and in adjacent graben. Radiometric age determinations on short cores have shown that three main lacustrine phases were present at ~ 40 ka, 24 ka, and 12–10 ka, with intervening phases of low lake level (Taieb et al., 1991; Roberts et al., 1993).

Lake Magadi is recharged by about 200 alkaline hot springs, located on fault lines along the margins of the lake basin. Supplementary recharge comes from dilute runoff and direct rainfall. With no perennial dilute inflow, the hydrothermal waters undergo intense evaporation to produce a highly saline, alkaline brine with > 300 g $l^{-1}$ total dissolved solids (TDS) and pH > 10.5 (Jones et al., 1977; Eugster, 1980; Allen et al., 1989).

### *The Lukeino Formation, Tugen Hills, Kenya*

Ancient saline, alkaline lakes are known to have existed in several parts of the Kenya Rift. The best evidence is seen in the Tugen Hills, a large uplifted fault block west of Lake Baringo in the central Kenya Rift (Fig. 1D). There, a Neogene succession of intercalated sedimentary and volcanic formations is well exposed, and includes several hominid-bearing sequences (Bishop et al., 1971; Chapman et al., 1978; Hill, 1999; Owen and Renaut, 2000).

Several of the sedimentary formations contain authigenic zeolites that indicate highly alkaline conditions during sedimentation and diagenesis (Renaut et al., 1999). We found small (1–5 cm) nodules of chert in lacustrine shales that were mapped as Pliocene Lukeino Formation by Pickford (1975, 1978), approximately 3 km from Kamego, directly below the Saimo Escarpment (Fig. 1D). These pale cream quartzose cherts have reticulate surface cracks similar to those illustrated by Eugster (1969) from Magadi. Lakes, both fresh and alkaline, existed in small fault-bounded basins during the period of Lukeino sedimentation (Pickford, 1978; Owen and Renaut, 2000).

### *Olduvai Gorge, Tanzania*

A suite of nodular chert samples was collected from the lacustrine sediments of Bed I at Olduvai Gorge, Tanzania (Fig. 2A). Lake Olduvai occupied a small, hydrologically closed basin at the western margin of the East African Rift in northern Tanzania during the late Pliocene and early Pleistocene (Hay, 1970, 1976; Ashley and Hay, this volume). During its history, Lake Olduvai fluctuated between being fresh to highly saline and alkaline. Two distinctive chert beds are exposed at location 80 of Hay (1976) (Fig. 2B). These nodular cherts have the typical characteristics of Magadi-type cherts, including soft-sediment deformation features and reticulation. The cherts may have formed from a sodium silicate precursor mineral that was precipitated during a period when Lake Olduvai was meromictic with a stable stratification, or through the interaction of alkaline brine with fresh water (Hay 1968, 1976; O'Neil and Hay, 1973).

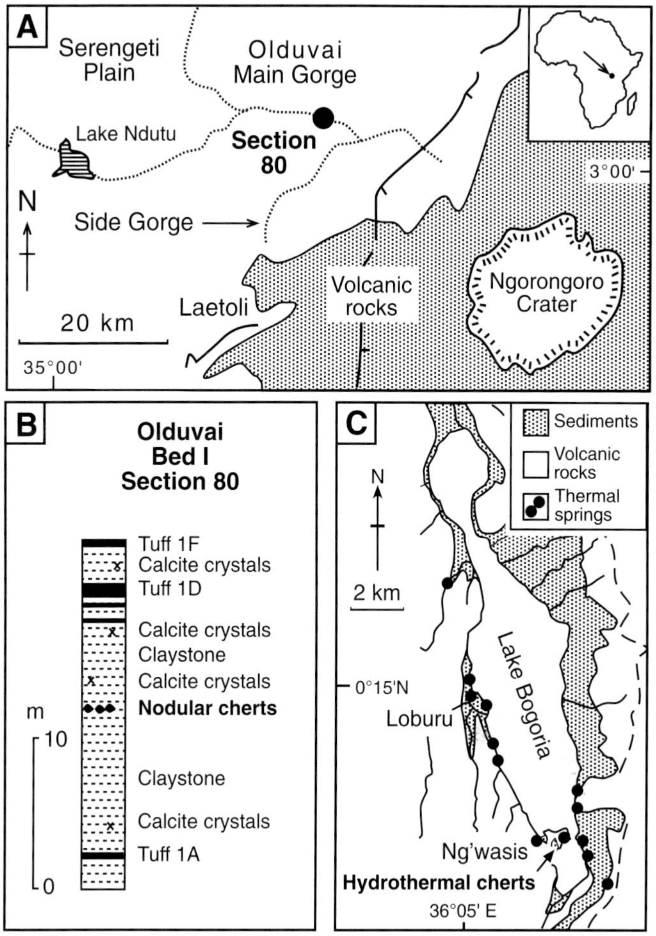


FIG. 2.—Location of chert sampling sites. **A)** Location of Section 80, Olduvai Gorge, Tanzania. **B)** Simplified stratigraphy of section 80, showing location of chert horizons (modified from Hay, 1976). **C)** Location of sublacustrine cherts at Ng'wasis, Lake Bogoria.

### *Lake Bogoria, Kenya*

Lake Bogoria is a saline, alkaline, meromictic lake located in a volcanic catchment in the central Kenya Rift (Fig. 2C). It is fed by ephemeral streams, precipitation on the lake surface, and about 200 thermal springs with Na–$HCO_3$–Cl waters (Renaut and Tiercelin, 1994). The lake waters, which are of Na–$CO_3$–$HCO_3$–Cl composition, are permanently stratified with a chemocline at about 6 m depth. The mixolimnion waters have a salinity of ~ 55–65 g $l^{-1}$ TDS, in contrast to the anoxic monimolimnion, with waters of ~ 90 g $l^{-1}$ TDS.

Sediment cores that span the Holocene taken from the deepest central part of the lake include cumulate crystal beds of trona and organic-rich oozes and laminites that locally contain magadiite and kanemite ($NaHSi_2O_5 \cdot 3H_2O$) (Tiercelin, 1981; Perinet et al., 1982; Renaut et al., 1986; Renaut and Tiercelin, 1994). The magadiite, which is present at between 7 and 10 m depth in the cores, appears nodular rather than bedded, but its lateral extent has not yet been determined.

The present hydrochemistry of Lake Bogoria may be similar to that of Lake Magadi during the period(s) of magadiite precipitation. Both lakes are strongly saline and alkaline with a pH > 10; they are partly fed by alkaline thermal springs and have catchments composed of intermediate to alkaline volcanic rocks. The surface waters of Lake Bogoria are diluted by runoff during unusually wet years. The aqueous trace-element concentrations in Lake Bogoria, therefore, are probably similar to those of a more dilute precursor of modern Lake Magadi.

Along the southern shoreline of Ng'wasis (formerly called Mwanasis), a peninsula at the southern end of Lake Bogoria, opaline cherts cement littoral gravels near a site of modern hydrothermal activity. These cherts contain planktonic diatoms and have been interpreted as precipitates that formed around the vents of sublacustrine hot springs at a time when Lake Bogoria was deeper and fresher than today (Renaut and Owen, 1988; Renaut et al., 2002). Samples of these hydrothermally influenced cherts, some of which resemble Magadi-type cherts in external morphology, were included in this study for comparison with those known to have had a sodium silicate precursor.

## SAMPLING AND ANALYTICAL METHODS

Samples of magadiite were collected from outcrops of High Magadi Beds east of a small fan delta on the southwestern shoreline of the lake (Fig. 1C). The magadiite is present as several discontinuous, irregular beds about 1–20 cm thick in brown

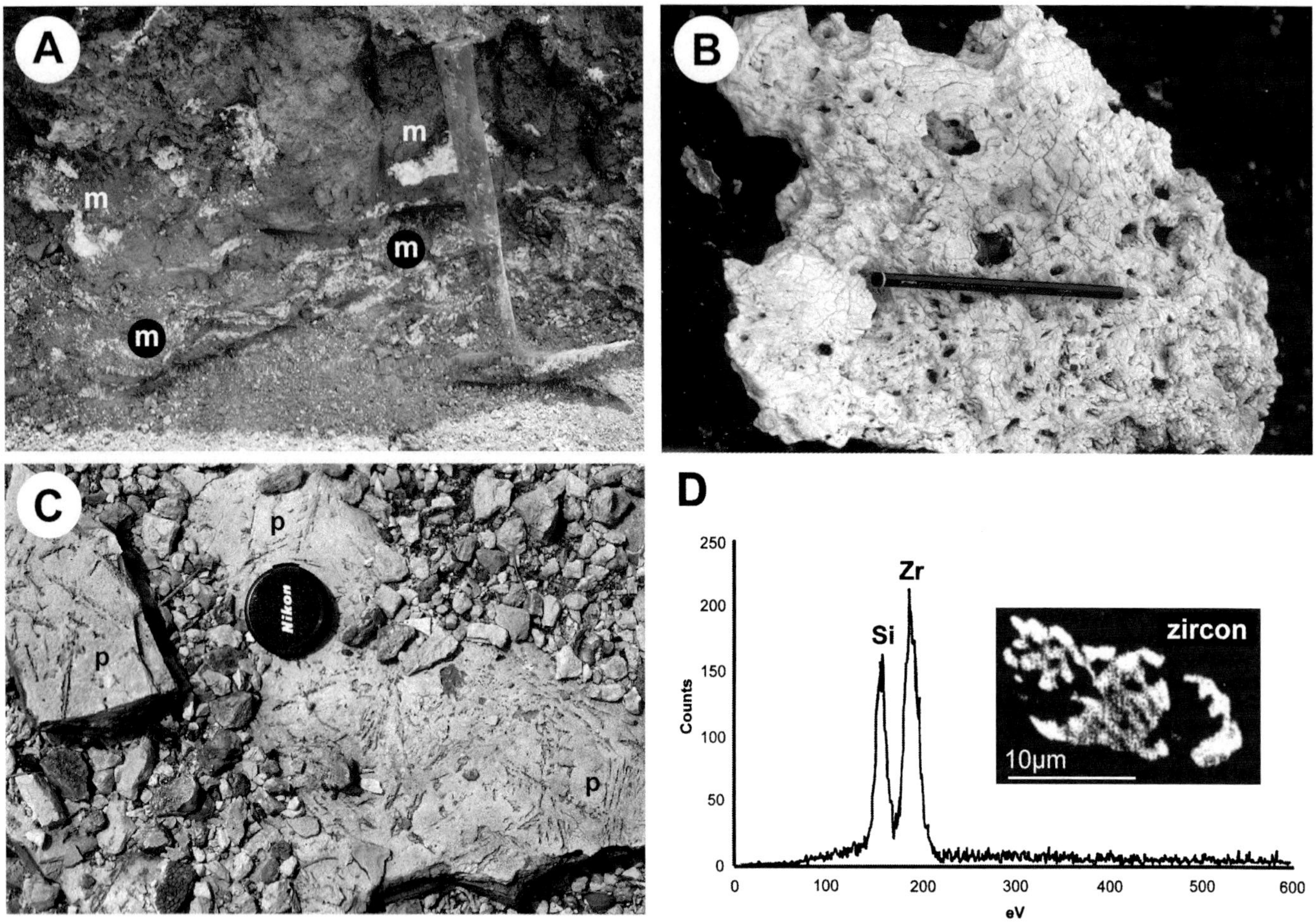


FIG. 3.—**A)** *In situ* magadiite nodules and stringers (m) in zeolite lacustrine tuffs from the High Magadi Beds, southwestern margin of Lake Magadi (Site 1 in Fig. 1C). **B)** Eroded block of Magadi-type chert (quartz, minor magadiite) from the High Magadi Beds, southwestern margin of Lake Magadi (Site 3 in Fig. 1C). Reticulate surface cracks and open porosity are well seen. Pencil length: 14.5 cm. **C)** Chert with evaporite pseudomorphs from the Green Beds near Bird Rock Lagoon, south of Lake Magadi (Site 4 in Fig. 1C). Lens cap diameter: 51 mm. **D)** X-ray spectrum of high-atomic-number particle shown in the backscattered electron image (inset). Chert from High Magadi Beds, Lake Magadi.

tuffaceous siltstones. Fresh samples (Fig. 3A) were obtained by digging up to 50 cm into the face of the outcrop because samples at the surface had already undergone significant diagenetic alteration to quartz. Samples of late Pleistocene–Holocene chert (Fig. 3B) were obtained from the same bed as the magadiite, and in a small exposure of High Magadi Beds along the base of a fault scarp, ~ 500 m north of that locality.

Samples of older (Pleistocene), pale green chert were obtained from outcrops of the Green Beds directly south of Lake Magadi (Fig. 1; see also Eugster, 1980, figs. 15–18). These chert samples are laminated with pseudomorphs of trona and gaylussite on their upper surface (Fig. 3C). Chert nodules from the Lukeino Formation were collected directly from the outcrop and in small gullies below the outcrop, where they had been concentrated by runoff. Samples from Bed 1 at Olduvai were collected from step trenches at section 80 that were newly excavated in June 1999. The analyzed cherts come from a horizon that lies ~ 10 m below Tuff 1F of the Olduvai sequence (see Hay, 1976, for stratigraphic details). The cherts from Ng'wasis at Lake Bogoria were collected from fossil beach exposures (see Renaut and Owen, 1988, for location).

Samples of chert and magadiite were analyzed by inductively coupled plasma mass spectrometer (ICP MS) at the University of Saskatchewan using the protocol of Fan and Kerrich (1997). Filtered acidified water samples from Lake Bogoria were analyzed using the same instrument and the method of McLaren (1992).

Qualitative chemical analyses were performed on a JEOL 8600 Super Probe electron microprobe analyzer in energy-dispersive spectrometry (EDS) mode. The instrument was operating with an accelerating voltage of 15 kV, probe current of 10 nA, and spot beam with a spatial resolution estimated at 2.0 μm. The magadiite samples were mounted as a conventional polished thin section and sputter coated with carbon before analysis. The samples were first scanned manually in a grid pattern using backscattered electron imaging (BEI). The detector contrast was set to high and brightness to low so that only phases composed of heavy elements relative to the rock matrix were visible on the screen. The magnification was set such that the field of view on the screen was 150 μm.

A Zr-bearing silicate (Fig. 3D), which is discussed below, was found embedded in a pit slightly below the surface of one sample, which prevented complete polishing of the grain. This and the small (< 10 μm) size of the zirconium silicate grain precluded any attempt to analyze the phase quantitatively using wavelength-dispersive spectrometry (WDS).

## RESULTS

The analyzed samples range from nearly pure (> 90%) magadiite to cherts composed almost entirely of secondary microcrystalline quartz. Zeolites (mainly erionite) and K-feldspar (both detrital and authigenic) are present in the samples from Magadi and Olduvai but constitute < 5 weight percent. Microbial organic matter is present in some samples of magadiite from the High Magadi Beds and Green Beds. Trace-element compositions are presented as REE, multi-element, and transition-metal diagrams, normalized to upper continental crust (Figs. 4, 5, 6; Appendix Tables 1a–c; Taylor and McLennan, 1985). Such diagrams permit inspection of a large array of elements of differing absolute concentration and allow an assessment of the degree of coherence of trace-element patterns in suites of samples (cf. Sun and McDonough, 1989).

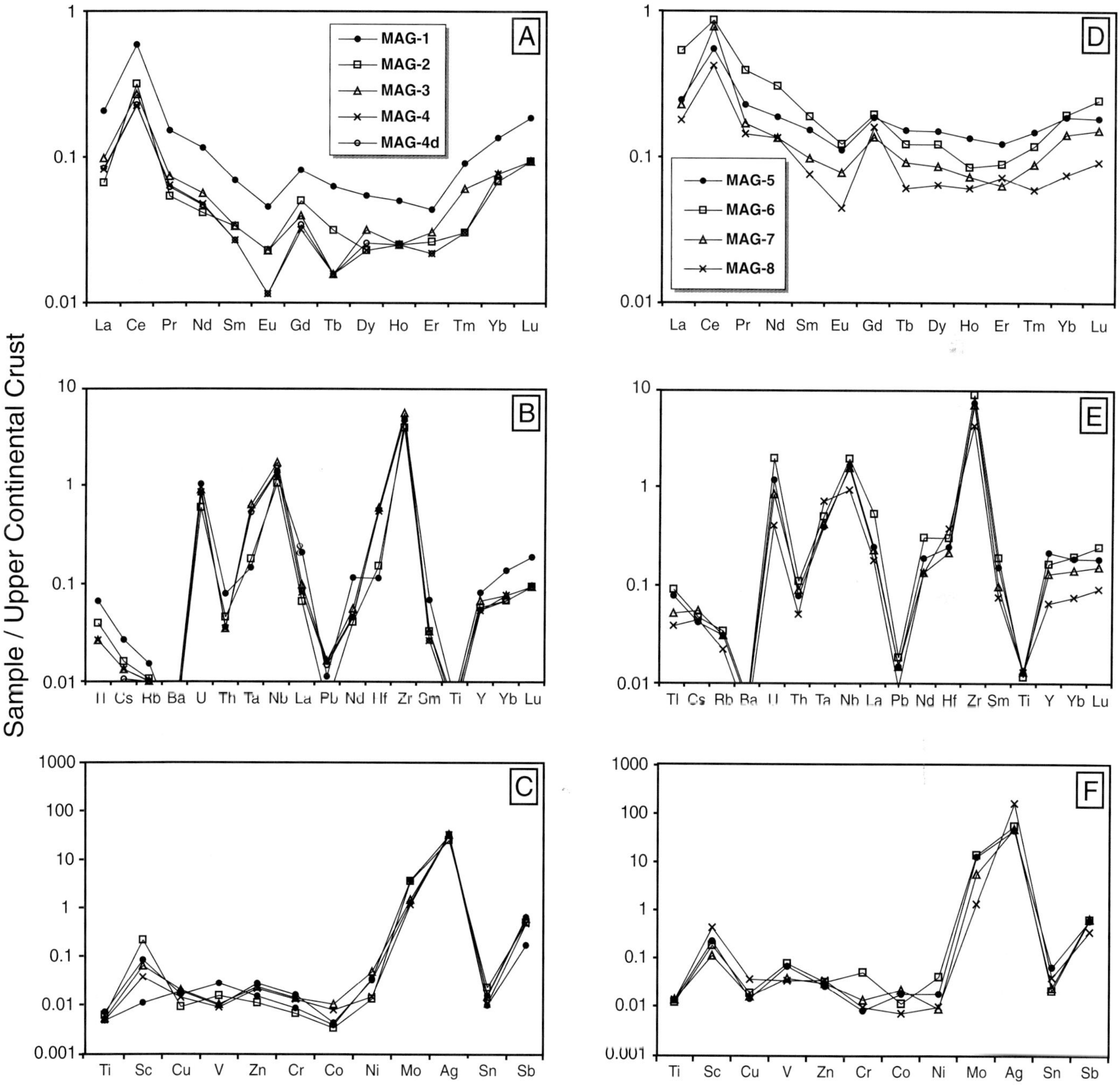


FIG. 4.—**A, D)** REE, **B, E)** multielement, and **C, F),** transition-metal diagrams normalized to upper-continental-crust composition, for magadiite and Magadi-type cherts from type locality (High Magadi Beds) at Lake Magadi (Sites 1 and 2 in Fig. 1C). See Appendix 1a for details of the proportion of magadiite and quartz in the analyzed samples.

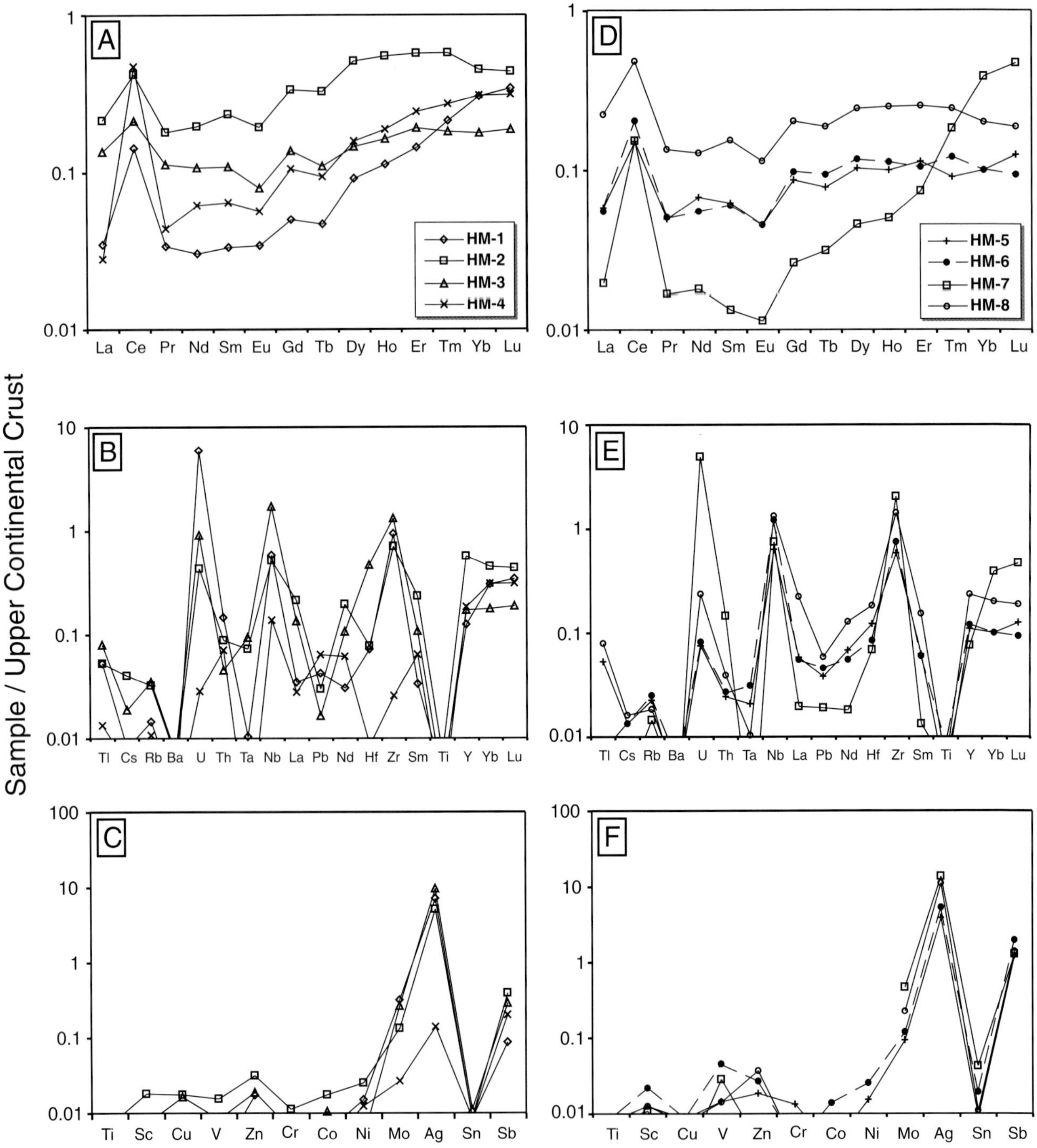


FIG. 5.—High Magadi beds chert, Lake Magadi. **A, D)** REE, **B, E)** multielement, and **C, F)** transition-metal diagrams, normalized to upper-continental-crust composition, for Magadi-type cherts from Site 3 (Fig. 1C) at Lake Magadi.

### *Magadiite and Cherts*

The magadiite and associated cherts from the High Magadi Beds have total REE abundances of 19–86 ppm. Most samples have coherent REE patterns, with nearly flat HREE, negative fractionations from Ho to Pr, and variably positive Ce but negative Eu anomalies (Figs. 4, 5). In addition, the magadiite and cherts are characterized by: (1) high U (1.2–5.5 ppm), Nb (13–27 ppm), and Zr (795–1656 ppm) contents, which are similar to, or greater than, average continental crustal abundances; (2) relatively low concentrations of Th, Ta, and Hf; (3) correspondingly high U/Th, Nb/Ta, and Zr/Hf ratios; (4) troughs at the alkali and alkali earth metals Tl, Cs, Rb, Ba, and Pb and Ti relative to neighboring elements; and (5) low concentrations of most transition metals (Sc to Ni), but relatively high contents of Mo, Ag, and Sb (Figs. 4, 5).

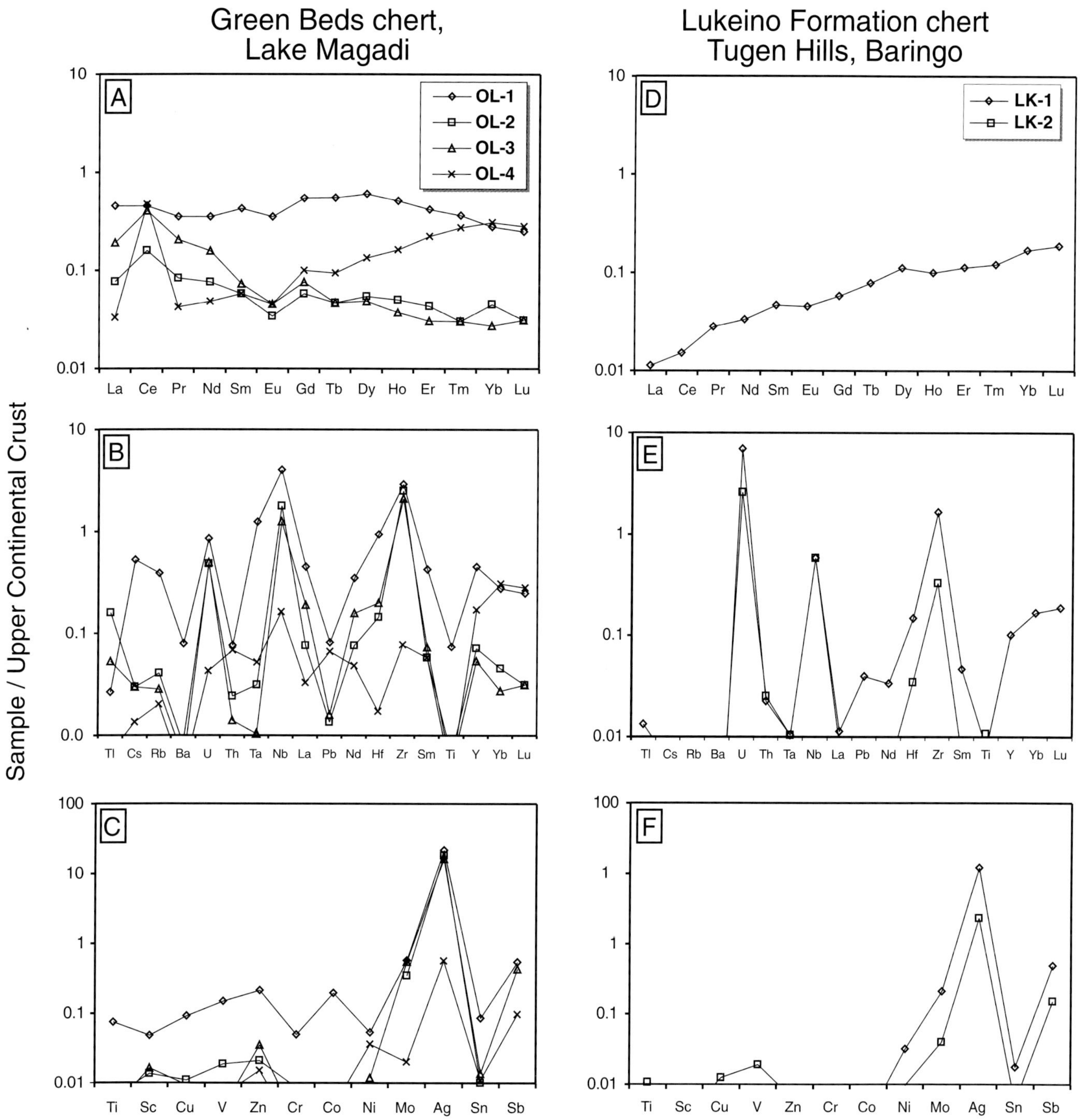


FIG. 6.—**A, D)** REE, **B, E)** multi-element, and **C, F)** transition-metal diagrams, normalized to upper-continental-crust composition, for cherts from the Pleistocene Green Beds at Magadi and Pliocene Lukeino Formation in the Tugen Hills, Kenya.

The older cherts from the Green Beds and the Lukeino Formation also show prominent maxima at U, Nb, Zr, and Ag, but possess lower overall concentrations of REE and other trace elements, and subdued Ce and Eu anomalies (Fig. 6). Cherts from Olduvai have the same maxima except for Ag, and like the Lukeino samples, tend to lower overall concentrations of trace elements (Fig. 7A–C).

## *Hydrothermal Cherts*

The two samples of hydrothermal chert (sublacustrine sinter) from Ng'wasis at Lake Bogoria show coherent trace-element patterns (Fig. 7D–F). These cherts are characterized by: (1) nearly flat REE patterns; (2) minor Ce and Eu anomalies; (3) minor enrichments of Th–U relative to Ba, and Nb–Ta and Zr–Hf rela-

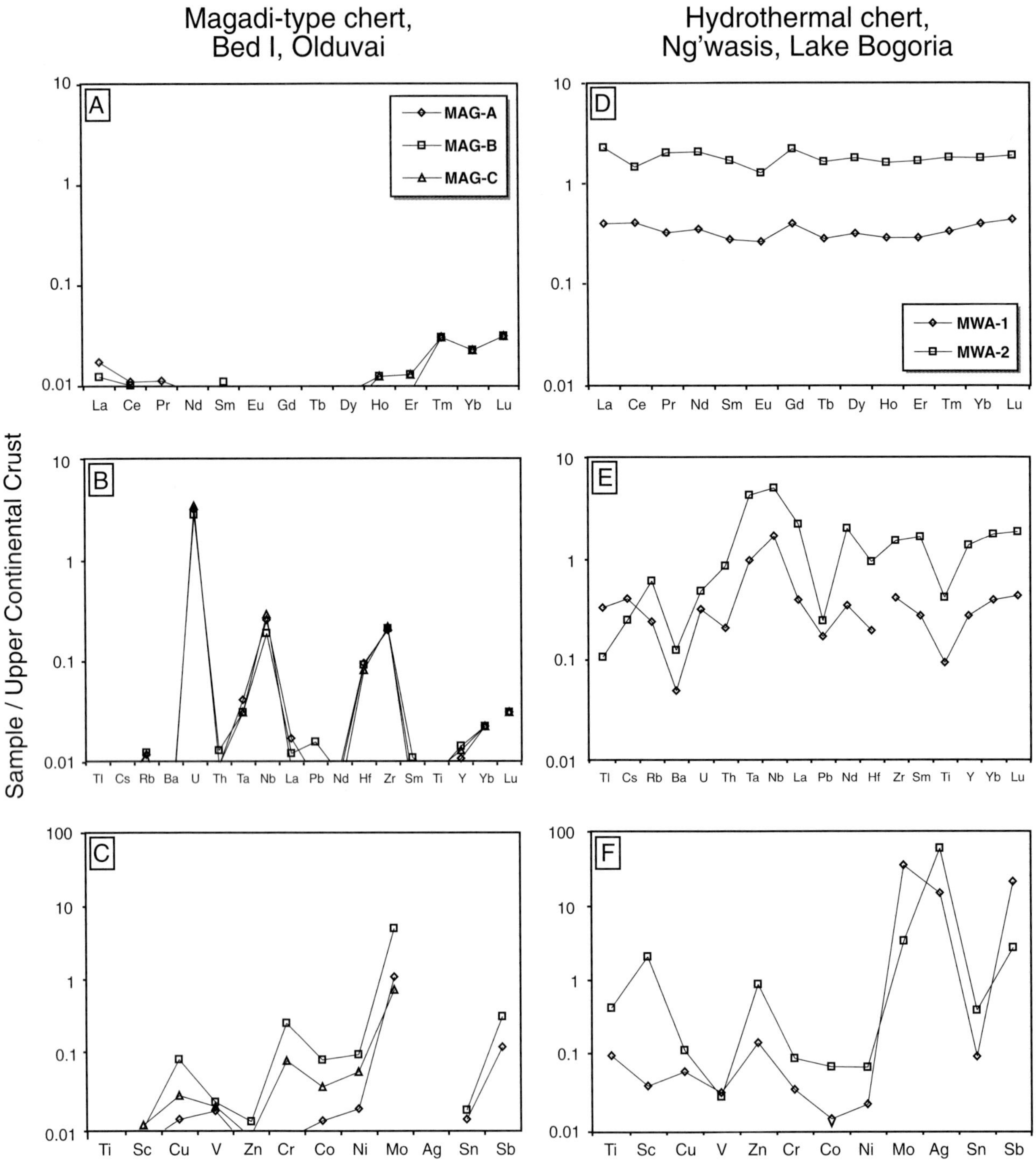


FIG. 7.—**A, D)** REE, **B, E)** multi-element, and **C, F)** transition-metal diagrams, normalized to upper-continental-crust composition, for Magadi-type cherts from Olduvai Gorge, Tanzania, and sublacustrine hydrothermal sinters at Lake Bogoria, Kenya.

tive to REE; (4) commensurately less fractionated U/Th, Nb/Ta, and Zr/Hf ratios, (5) minima at Ba, Pb, and Ti; and (6) enrichments of Mo, Ag, and Sb (Fig. 7D, E, F).

### *Waters*

Surface and bottom waters in Lake Bogoria have comparable fractionated REE patterns with minor negative Eu anomalies. Most other trace elements are close to or below detection limits, except a maximum at Mo (Fig. 8; Appendix Tables 1d, e ). Only Mo, Sn, and Sb were above detection limits in the hydrothermal fluids from Lake Bogoria.

### *Backscattered Electron Imagery*

Given the high bulk Zr contents in most cherts from the type locality at Magadi, a search for a Zr-bearing phase was made using backscattered electron imagery. Several high-atomic-num-

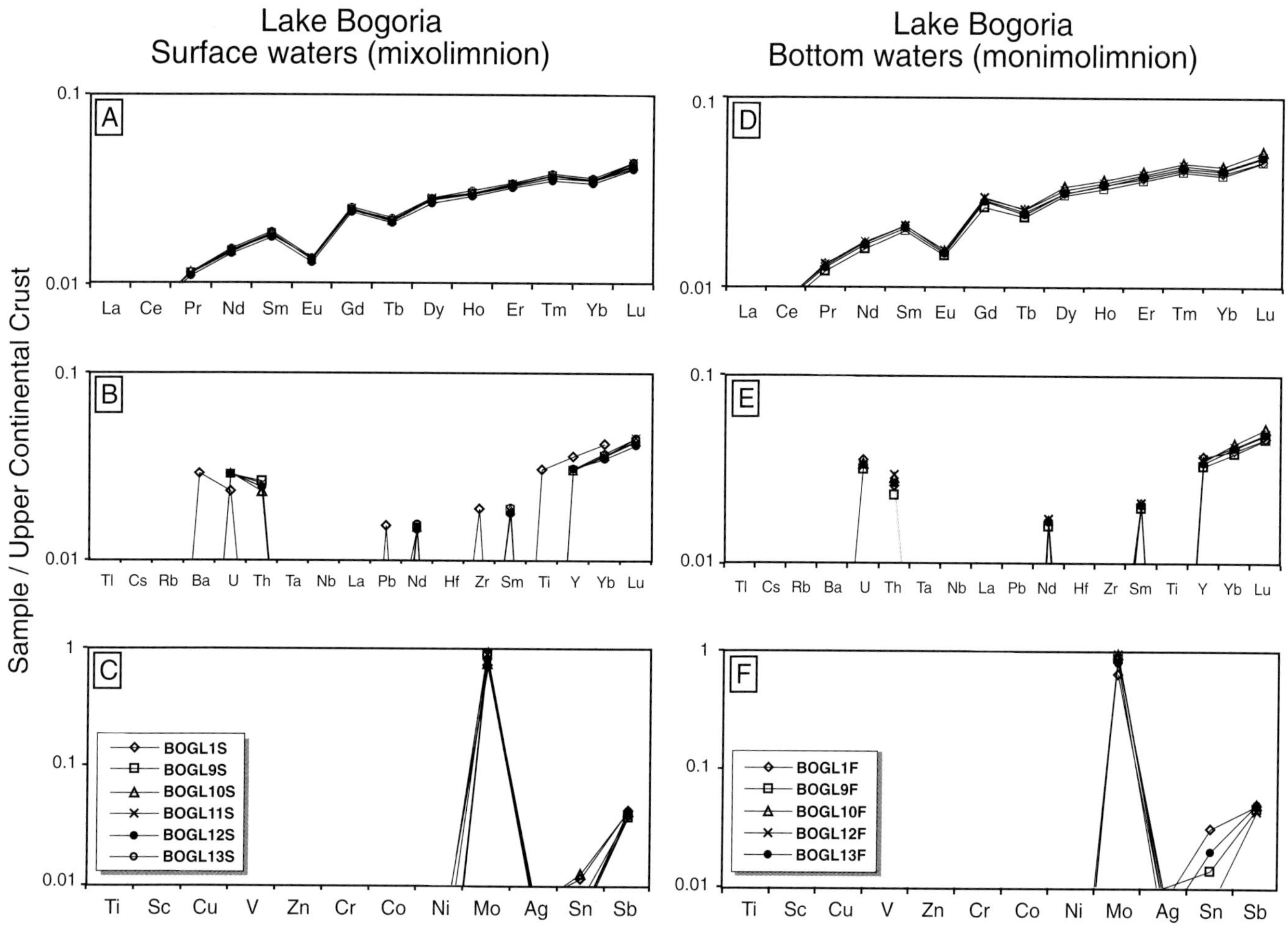


FIG. 8.—**A, D)** REE, **B, E)** multi-element, and **C, D)** transition-metal diagrams, normalized to upper-continental-crust composition, for surface and bottom waters, Lake Bogoria, Kenya.

ber spots were found, mainly in the submicrometer-to-micrometer size range, that yielded Si and Zr X-ray spectra. From the morphology of the larger particles and the X-ray spectra, these are tentatively interpreted as low-temperature hydrothermal or diagenetic zircons (Fig. 3D; cf. Saxena, 1966).

## DISCUSSION

### *Cherts and Sublacustrine Sinter*

All the lacustrine cherts analyzed in this study share distinctive trace-element characteristics with the magadiite and Magadi-type cherts from the type locality of the High Magadi Beds. Notable are: (1) flat to negatively fractionated REE; (2) variable positive Ce but negative Eu anomalies; (3) enrichment of U, Nb, and Zr above neighboring REEs; (4) extreme U/Th, Nb/Ta, and Zr/Hf fractionations; and (5) enrichments of Mo, Sb (± Ag) relative to neighboring transition metals. Accordingly, these compositional features appear to be characteristic of magadiite and alkaline-lake cherts from the East African Rift.

Behr and Röhricht (2000) and Behr (this volume) have recently undertaken detailed sedimentological, petrographic, and scanning electron microscopy studies of the cherts at Magadi, and have shown that most (> 90%?) of the cherts that are *older* than the High Magadi Beds probably formed by replacement of calcified microbial mats and bioherms by silica rather than by abiotic replacement of magadiite as originally proposed by Eugster (1967, 1969). They suggested that some of the cherts formed in shallow hypersaline alkaline mudflats and lagoons that underwent intense evaporation rather than in a basinwide meromictic lake (Behr and Röhricht, 2000). Silica precipitation, either as soft gels or by replacement of $CaCO_3$, would have been controlled by pH variations induced by microbial activity (bacterial degradation and photosynthesis) and evaporative concentration. Our samples from the Green Beds, therefore, may not have had a magadiite precursor. Nonetheless, a consistent alkaline-lake chert signature is still preserved in the trace-element characteristics of the Green Beds cherts and those from Olduvai and Lukeino. The similar patterns exhibited by magadiite, Magadi-type chert, and Green Beds cherts imply that they have all inherited the composition of the highly alkaline fluids.

Trace-element compositional studies have also been conducted on saline, alkaline lakes and associated sediments in the Sasykkul depression of the Pamirs in central Asia. There, low atmospheric precipitation, combined with intense evaporation, generates Na–Cl–$CO_3$–$SO_4$ and Na–Cl–$CO_3$ brines that have 40–130 g $l^{-1}$ TDS and pH of 9–10. Thermal springs discharge around the lakes (Volkova, 1998). The lake-margin sediments are composed of trona, halite, and burkeite ($Na_2CO_3 \cdot 2Na_2SO_4$). The Sasykkul sediments share many of the compositional features of alkaline-lake cherts from the Kenya rift, including: (1) enrichment of U over Th; (2) low Ta contents; (3) variable Zr and Hf contents and ratios; (4) relatively low Rb and Cs contents; and (5) variably positive Ce but negative Eu anomalies. Consequently, these compositional features may be common to most alkaline-lake cherts, reflecting the distinctive properties of the alkaline brines from which they precipitate.

The hydrothermal cherts (sublacustrine sinters) at Lake Bogoria have some of the compositional features of the alkaline-lake cherts, including enrichment of Nb and Zr relative to neighboring REE, and Mo, Ag, and Sb maxima. However, they lack the Ce and Eu anomalies or the extreme U–Th, Nb–Ta, and Zr– Hf fractionations of the Magadi cherts and are relatively enriched in Ti and transition metals (cf. Figs. 5D–F and 7D–F).

### *Aqueous Transport*

Most cherts from Magadi and the waters from Lake Bogoria have negatively fractionated REE patterns. Alkaline waters typically show relative enrichment of HREE, proportional to increasing pH (Johannesson et al., 1994; Johannesson and Zhou, 1999). According to the speciation modeling of Johannesson et al. (1994), the REE form dicarbonate complexes in alkaline aqueous fluids, with preferential complexing of HREE, consistent with the results of this study.

The systematic but variable Ce anomalies reflect redox conditions. The Ce(III)/Ce(IV) redox potential lies between that of Fe(II)/Fe(III) and Mn(II)/Mn(IV) (Elderfield and Sholkovitz, 1987; Byrne and Sholkovitz, 1996). Fe and Mn are two of the most geochemically active elements in redox cycles or at interfaces within waters and sediments. The formation and dissolution of Mn and Fe oxides control the cycling of many trace elements without multiple redox states; the strictly trivalent lanthanides, Ce and Eu excepted, are a good example of this coupling. Cerium, independently of the other lanthanides, also exhibits an active redox-driven geochemistry in natural waters and sediments given its two oxidation states (Byrne and Sholkovitz, 1996, and references therein).

Processes invoked to explain the development of Ce anomalies include: (1) biologically mediated oxidation of Ce(III) to Ce(IV) in coastal waters and the open ocean (Moffett, 1990, 1994a, 1994b; Sholkovitz and Schneider, 1991; Sholkovitz et al., 1994); (2) Ce(III)–Ce(IV) redox shift across anoxic–oxic interfaces in the water columns of marine basins (De Baar et al., 1988; German and Elderfield, 1989; Sholkovitz et al., 1992); and (3) the development of the largest Ce anomalies in rivers with the highest pH (Elderfield et al., 1990; Goldstein and Jacobsen, 1998a, 1998b). Surface oxidation of Ce(III) to Ce(IV) results in highly insoluble $CeO_2$ coatings on Mn nodules or oxyhydroxide coatings on precipitating mineral phases with a compensating negative Ce anomaly characteristic of seawater (Byrne and Sholkovitz, 1996).

In contrast to observations in seawater, positive Ce anomalies reported for the alkaline waters from Lake Van, Turkey, indicate an enhanced tendency of Ce to remain in solution compared to its trivalent REE neighbors (Möller and Bau, 1993). Positive Ce anomalies have been explained by oxidation of Ce(III) to Ce(IV) and stabilization of carbonato–Ce(IV) complexes in solution. Thus, positive Ce anomalies might be a feature typical of oxic, alkaline, carbonate-rich, low-temperature waters in general, given that at high pH and high carbonate concentrations, formation of carbonato–Ce(IV) complexes in solution can prevent scavenging of Ce, which counteracts the process that leads to Ce depletion in seawater (Möller and Bau, 1993). Similarly, the enrichment of U accords with the high aqueous solubility of U in oxidized fluids. Consequently, the positive Ce anomalies of the Magadi cherts and some marine cherts may arise from distinct processes: redox-controlled solubility of Ce(IV) in the former, but for marine cherts, scavenging of Ce by Fe and Mn oxides and oxyhydroxides. Variably developed negative Eu anomalies are consistent with the low solubility of $Eu^{2+}$ in oxidized thermal waters (Humphries, 1984). Collectively, the Magadi cherts plot to variably positive Ce anomalies but systematic negative Eu anomalies. Lake Bogoria waters have negative Eu anomalies acquired by the cherts but no excursion of Ce/Ce* (Fig. 9). Consequently, Lake Bogoria waters may be only an approximate proxy for lake water at Magadi when magadiite was precipitating.

The REEs, Al, Ti, Y, and HFSE have low solubilities in most hydrothermal fluids evolved from marine, meteoric, or metamorphic fluids, and they behave isochemically in rocks influenced by high water–rock ratios in most hydrothermal systems (Humphries, 1984; Kerrich and King, 1993; Terakado and Fujitani, 1998, and references therein). The mobility of Zr and other elements that are generally immobile in hydrothermal systems has been described from a variety of late-stage magmatic fluid events, specifically in alkaline complexes (Cerny, 1982; Rubin et al., 1988). The mobility of Ti, Zr, and REE has been demonstrated where there is intense hydrothermal tourmaline replacement, and precipitation of hydrothermal zircons at ~300°C (Alderton et al., 1980; Ludden et al., 1984; Kerrich and King, 1993). Gieré (1990) reviewed the mobility of Ti, Zr, and REE, summarizing evidence for aqueous transport of these elements under a number of conditions where alkaline magmatism was inactive. Their mobility was attributed to high aqueous activities of $CO_2$, $B^-$, $F^-$, or $PO_4^{3}$. Of specific relevance to this study is a report of high concentrations of Zr and REE in saline, highly alkaline (pH 12) groundwaters (Kraynov et al., 1969). As shown by Eugster and Jones (1979), weathering of volcanic rocks commonly yields alkaline sodic surface waters and ground waters; these solutions favor leaching and transport of elements such as Zr and REE that have low solubility in near-neutral surface waters or hydrothermal fluids. Zirconium is mobile during near-surface tropical weathering of soils developed on volcanic rocks (Kurtz et al., 2000).

Alkaline magmas tend to intrinsically greater concentrations of incompatible elements, including LREE, Th, U, Nb, Ta, Zr, and Hf, than tholeiitic counterparts, especially for melts derived from metasomatized mantle lithosphere as in continental rifts (Menzies et al., 1987; Sun and McDonough, 1989). The East African Rift is the site of within-plate alkaline and carbonatitic magmatism (Baker, 1987; MacDonald et al., 1994). Accordingly, the conjunction of incompatible element–rich alkaline magmas with oxidizing alkaline hydrothermal fluids creates conditions that favor leaching, aqueous transport, and concentration of REE and HFSE, and the precipitation of hydrous sodium silicates or siliceous gels enriched in this suite of elements under intense evaporation. In alkaline igneous rocks, Th, U, Zr, and Hf may not be present in refractory minerals such as zircon or baddelyite that are resistant to weathering or hydrothermal alteration, but rather in silicates that weather readily, whereas Ti is present as a refractory oxide. These features of alkaline rocks may account for the enrichment of HFSE, but not Ti, in the Magadi cherts (Fig. 4). It is possible that there is a contribution to the REE and HFSE budget from volcanic ash. In volcanic rocks, however, U/Th (0.15–0.4), Nb/Ta (10–36), and Zr/Hf (28–60) ratios have restricted ranges (Sun and McDonough, 1989), whereas these element pairs may be strongly fractionated in hydrothermal systems (Kerrich and King, 1993), as evidenced by Nb/Ta of 46–1872, and Zr/Hf ratios spanning 275–1315 in the Magadi cherts (Appendix 1a–c; Figs. 4, 10).

### *Ancient Alkaline Lake Cherts and Comparison with Marine Cherts*

Fossil Magadi-type cherts have commonly been recognized by their distinctive surface textures, notably reticulation (Fig. 3B),

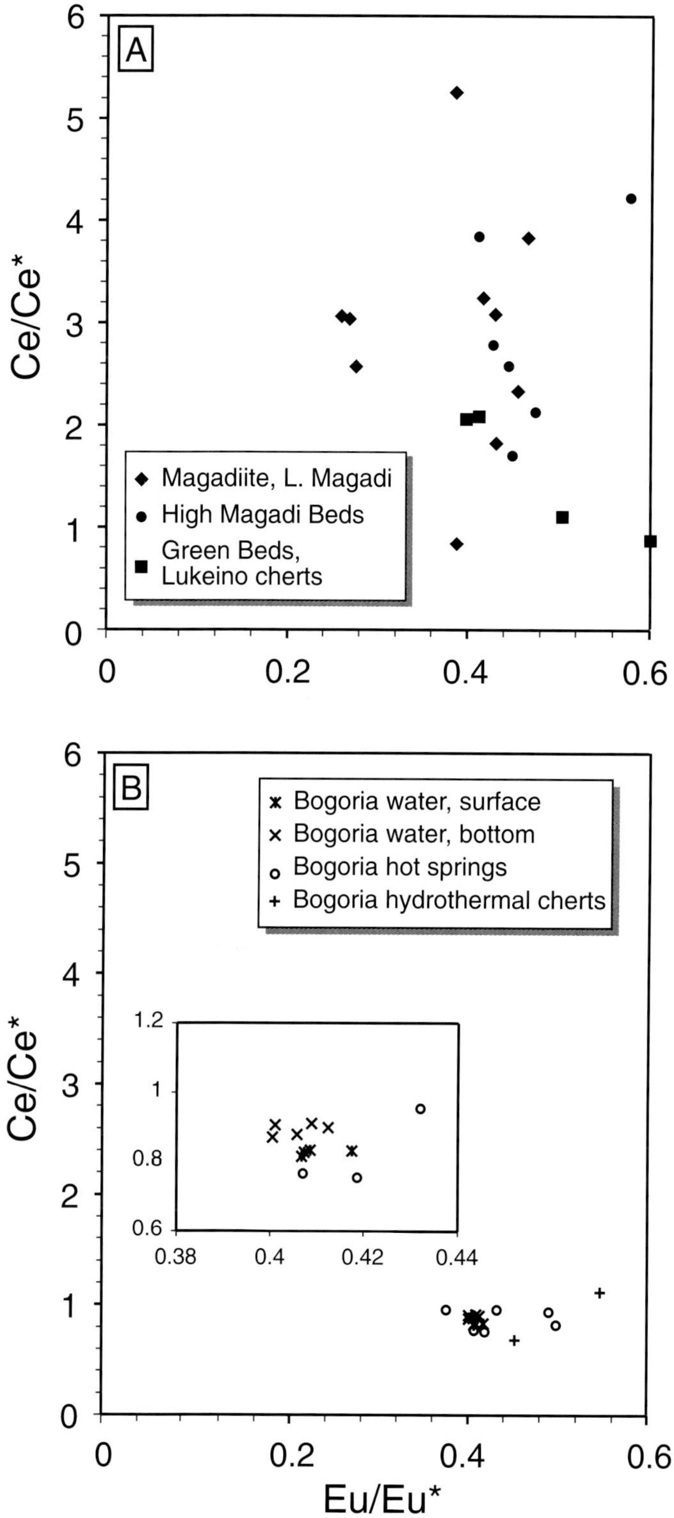


FIG. 9.—Plots of Eu/Eu* vs. Ce/Ce* for **A)** alkaline lake cherts from the East African Rift, and **B)** lake and hydrothermal waters, and sublacustrine sinters from Lake Bogoria.

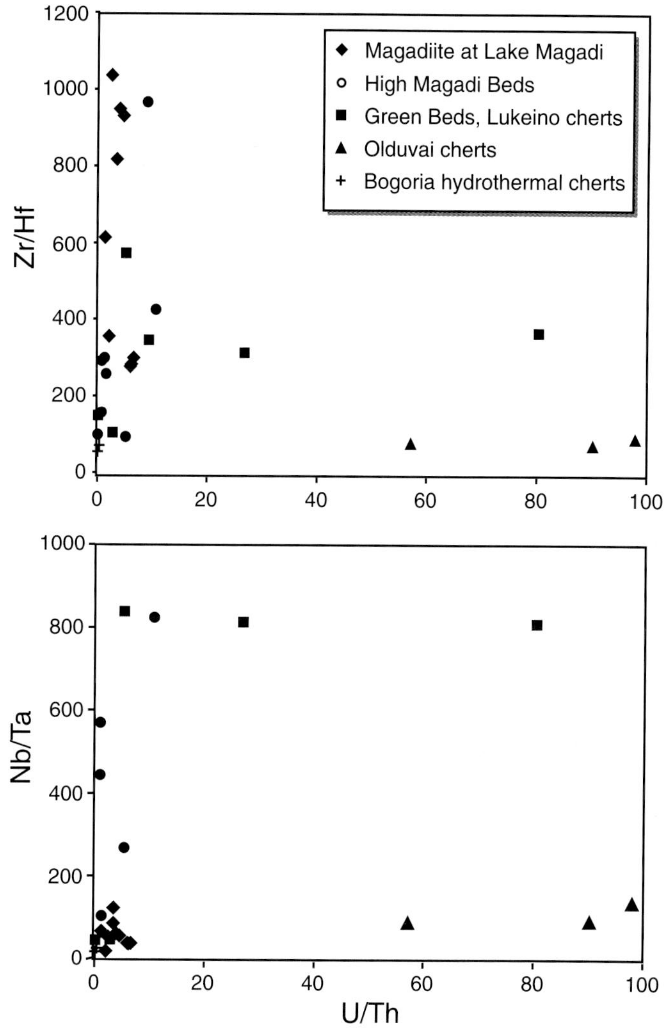


FIG. 10.—Plots of Zr/Hf and Nb/Ta ratios versus U/Th for specified cherts and sublacustrine sinters from the East African Rift.

soft-sediment deformation structures, and crystal pseudomorphs (Fig. 3C), or by an association with closed-basin lacustrine sedimentary rocks (e.g., Hay, 1968; Parnell, 1986). Where those lake sediments contain zeolites, authigenic K-feldspar, or other phases indicative of high alkalinity, the confidence level in the identification of the chert as of Magadi-type is high. The petrographic fabrics of Magadi-type cherts were once considered distinctive (Schubel and Simonson, 1990), but similar textures have since been illustrated other types of chert (e.g., Hattori et al., 1996; Bustillo, 2001; Behr, this volume). Our observations at Magadi, and those of Behr and Röhricht (2000), show that not all alkaline lake cherts possess these features. Some alkaline lake cherts lack reticulation, and in isolation would be difficult to distinguish from cherts of other origins. The results of this study confirm that trace-element systematics can provide an additional tool to distinguish the origin of ancient cherts in the geological record.

The pre-Pleistocene alkaline lake cherts that have been replaced by quartz retain trace-element patterns similar to those of the younger cherts at Magadi but have lower absolute element contents. The most straightforward interpretation of the overall decrease is volume change accompanying diagenesis of the precursor (e.g., magadiite or silica gel) to quartz, through intermediate phases such as moganite, opal-CT, and opal-C.

Many, but not all, examples of modern magadiite formation are present in regions with active or recent volcanism. The parent alkaline fluids that evolve from weathering and hydrothermal alteration of volcanic rocks have similar compositions, although geographically dispersed. Records of recent hydrous sodium silicates include Lake Chad (Maglione, 1980; Sebag et al., 2001); Lake Bogoria (Perinet et al., 1982; Icole and Perinet, 1984; Renaut and Tiercelin, 1994); Lake Natron (Hay, 1968; Gueddari, 1984); and several alkaline lakes in the northwestern United States (e.g., Eugster, 1967; Rooney et al., 1969). Fossil examples, identified as Magadi-type cherts, are also commonly found in regions of contemporaneous volcanism (e.g., Houser, 1985; Parnell, 1986; Krainer and Spötl, 1998; Dunagan, 2000).

Hydrothermal cherts at Lake Bogoria share some of the compositional features with Magadi-type cherts, but they differ in other elemental concentrations and ratios. This is probably due to mixing of the hydrothermal fluids with lake waters of variable salinity and alkalinity (cf. Renaut et al., 2002).

Trace-element geochemistry has been used to distinguish the depositional and geodynamic setting of marine cherts (Murray et al., 1990, 1992a, 1992b; Murray, 1994; Owen et al., 1999). Rare earth elements, Zr contents, and $Fe_2O_3/TiO_2$ and $Al_2O_3/SiO_2$ ratios are distinct in cherts from Phanerozoic ocean-ridge, deep-ocean, and passive-margin settings, the first having a dominant hydrothermal input, whereas the third has a significant terrigenous siliciclastic component. Kato et al. (1998), for example, measured up-section variations of REE and Eu/Eu* values in 3.3 to 3.2 Ga banded iron formations from the Pilbara craton, Western Australia, which they interpreted in terms of oceanic plate translation from a ridge environment to a convergent-margin environment.

There are few systematic datasets for all REE and HFSE, and other elements in marine cherts from different depositional settings (cf. Murray et al., 1992a). Accordingly, the Magadi cherts and sublacustrine sinters from the East African Rift are compared with marine cherts using Ce/Ce*, Eu/Eu*, and La/Yb (Fig. 11). All cherts show positive Ce anomalies, but the Magadi cherts extend to more extreme values of Ce/Ce* and La/Yb. Marine cherts are distinctive in their field for Eu/Eu* (Fig. 11).

The budgets of REE in deep-ocean cherts and in passive-margin and convergent-margin cherts (Eu and Ce excepted) include a siliciclastic input, and this is likely the case for Zr–Hf and Nb–Ta. Moreover, siliciclastic sediments, reflecting upper continental crust, are characterized by normalized Nb–Ta depletions relative to REE and unfractionated Nb/Ta (~ 12) and Zr/Hf (~ 36) ratios (Taylor and McLennan, 1985). Consequently, Magadi-type cherts are highly distinctive relative to marine counterparts in terms of relatively high Nb and Zr concentrations (Fig. 12), Nb and Zr enrichment relative to REE, and extreme Nb/Ta and Zr/Hf fractionations.

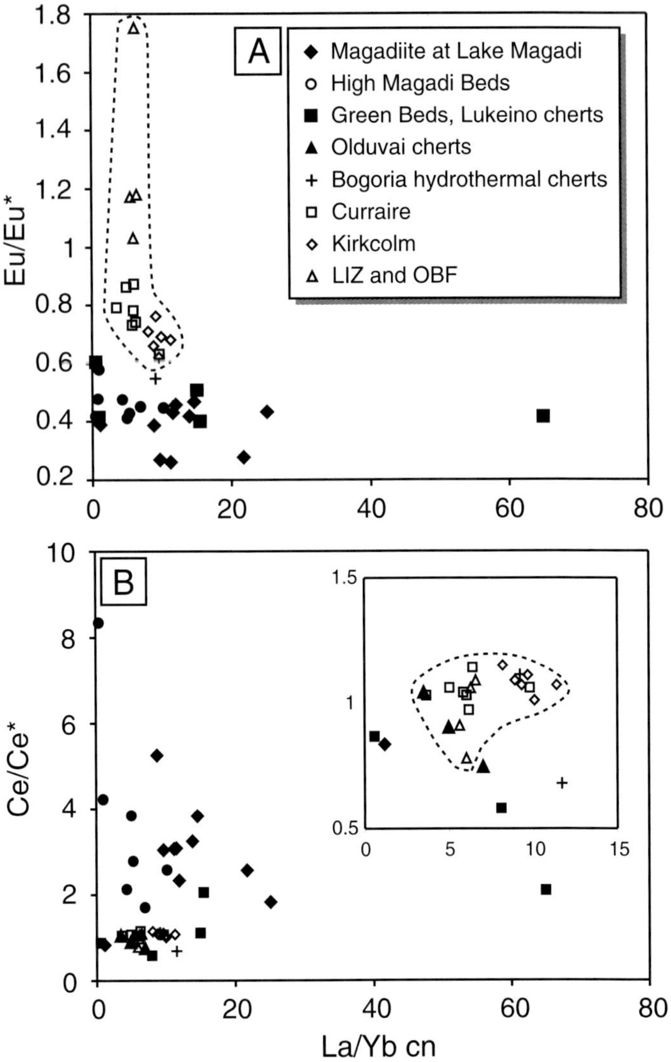


FIG. 11.—Plots of Eu/Eu* and Ce/Ce* versus La/Yb for cherts and sinters from the East African Rift and marine cherts (shale normalized values). Data for Curraire, Kirkcolm, LIZ, and OBF are from Owen et al. (1999).

## CONCLUSIONS

The trace-element (including REE) composition of cherts and hydrous sodium silicates that have formed in alkaline lakes is highly distinctive and differs from that found in marine and hydrothermal cherts. Alkaline-lake cherts and magadiite can be distinguished by: their negatively fractionated REE patterns; high absolute concentrations of U, Nb, and Zr; normalized enrichments of U, Nb, Zr relative to REE; extreme fractionations of U–Th, Nb–Ta, and Zr–Hf; positive Ce but negative Eu anomalies; and peaks at Mo, Ag, and Sb. Ancient alkaline-lake cherts that have been diagenetically replaced by quartz retain these features but have lower absolute element contents. The distinctive trace-element geochemical signatures of alkaline-lake cherts provide a valuable new tool to discriminate them from other types of chert in continental rifts and in other depositional environments.

## ACKNOWLEDGMENTS

We thank Brian Pratt for critique of an initial draft, Lily Song for word-processing, and Brian Morgan for the ICP MS analyses. R.K. and R.W.R acknowledge NSERC research grants, and an NSERC MFA to the departmental geochemical facility. Fieldwork and analyses were partly supported by a grant from Elf-Norge. R.K. also acknowledges the George McLeod endowment to the University of Saskatchewan. Permission to work at Magadi and in Lake Bogoria National Reserve (LBNR) was granted by the Office of the President, Government of Kenya, with logistical help from the National Oil Corporation of Kenya. We thank William Kimosop, Senior Warden of LBNR, for his continued support of this research, and Jean-Jacques Tiercelin, Corinne Tarits, Gail Ashley, Robert Blumenschine (OLAPP) and Richard Hay for their help with this study. The original manuscript benefited from careful reviews by Drs. Angeles Bustillo and Richard W. Murray.

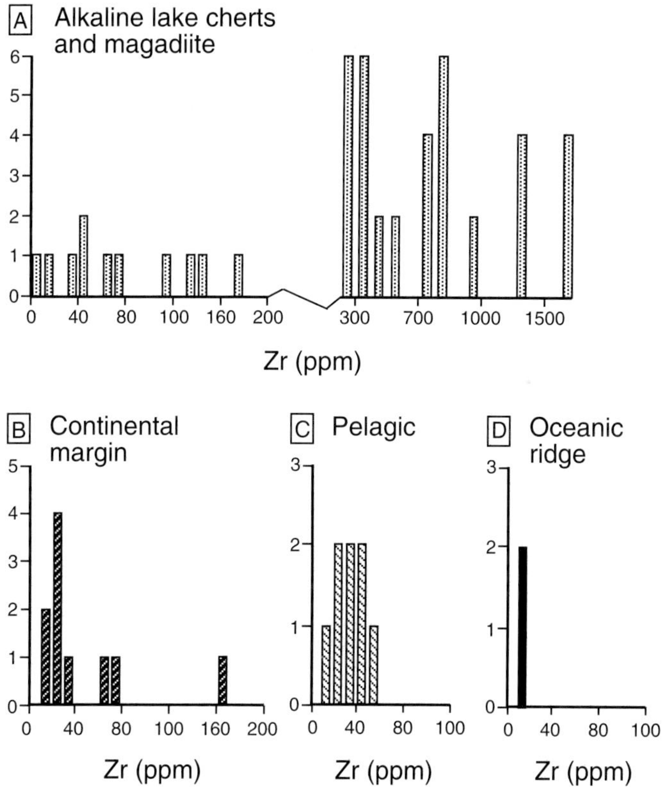


FIG. 12.—Histograms of Zr contents for Magadi-type cherts, and marine cherts from specified geodynamic settings (marine data from Murray, 1994).

## REFERENCES


ALDERTON, D.H.M., PEARCE J.A., AND POTTS P.J., 1980, Rare earth element mobility during granite alteration: Evidence from southeast England: Earth and Planetary Science Letters, v. 49, p. 149–165.

ALLEN, D.J., DARLING, W.G., AND BURGESS, W.G., 1989, Geothermics and hydrogeology of the southern part of the Kenya Rift Valley with emphasis on the Magadi–Nakuru area: British Geological Survey, Research Report (Hydrogeology Series) SD/89/1.

BAKER, B.H., 1958, Geology of the Magadi area: Geological Survey of Kenya, Report no. 42, 81 p.

BAKER, B.H., 1987, Outline of the petrology of the Kenya rift alkaline province, *in* Fitton, J.G., and Upton B.G.J., eds., Alkaline Igneous Rocks: Geological Society of London, Special Publication 30, p. 293–311.

BEHR, H.-J., AND RÖHRICHT, C., 2000, Record of seismotectonic events in siliceous cyanobacteria sediments (Magadi cherts), Lake Magadi, Kenya: International Journal of Earth Sciences, v. 89, p. 268–283.

BIRNEY DE WET, C.C., AND HUBERT, J.F., 1989, The Scots Bay Formation, Nova Scotia, Canada, a Jurassic carbonate lake with silica-rich hydrothermal springs: Sedimentology, v. 36, p. 857–874.

BISHOP, W.W., HILL, A., AND MILLER, J.A., 1971, Succession of Cainozoic vertebrate assemblages from the northern Kenya Rift Valley: Nature, v. 233, p. 389–394.

BUSTILLO, M.A., 2001, Cherts with moganite in continental Miocene Mg-clay deposits: an example of "false" Magadi-type cherts, Madrid Basin, Spain: Journal of Sedimentary Research, v. 71, p. 436–444.

BUTZER, K.W., ISAAC, G.L., RICHARDSON, J.L., AND WASHBURN-KAMAU, C., 1972, Radiocarbon dating of East African lake levels: Science, v. 175, p. 1069–1076.

BYRNE, R.H., AND SHOLKOVITZ, E.R., 1996, Marine chemistry and geochemistry of the lanthanides, *in* Gschneidner, K.A., and Eyring, L., eds., Handbook on the Physics and Chemistry of Rare Earths, vol. 23: Amsterdam, Elsevier, p. 497–593.

CERNY, P., 1982, Anatomy and classification of granitic pegmatites, *in* Cerny, P., ed., Short Course in Granitic Pegmatites in Science and Industry: Mineralogical Association of Canada, Winnipeg, p. 1–39.

CHAPMAN, G.R., LIPPARD, S.J., AND MARTYN, J.E., 1978, The stratigraphy and structure of the Kamasia Range, Kenya Rift Valley: Geological Society of London, Journal, v. 153, p. 265–281.

CHOUGH, S.K., KIM, S.B., AND CHUN, S.S., 1996, Sandstone/chert and laminated black shale couplets, Cretaceous Uhangri Formation (southwest Korea): depositional events in alkaline lake environments: Sedimentary Geology, v. 104, p. 227–242.

CROSSLEY, R., 1979, The Cenozoic stratigraphy and structure of the western part of the rift valley in southern Kenya: Geological Society of London, Journal, v. 136, p. 393–405.

DE BAAR, H.J.W., GERMAN, C.R., ELDERFIELD, H., AND VAN GAANS, P., 1988, Rare-earth element distributions in anoxic waters of the Cariaco trench: Geochimica et Cosmochimica Acta, v. 52, p. 1203–1219.

DUNAGAN, S.P., 2000, Lacustrine carbonates of the Morrison Formation (Upper Jurassic, Western Interior), East Central Colorado, U.S.A., *in* Gierlowski-Kordesch, E.H., and Kelts, K., eds., Lake Basins through Space and Time: American Association of Petroleum Geologists, Studies in Geology no. 44, p. 181–188.

ELDERFIELD, H., AND SHOLKOVITZ, E.R., 1987, Rare-earth elements in the pore waters of reducing nearshore sediments: Earth and Planetary Science Letters, v. 82, p. 280–288.

ELDERFIELD, H., UPSTILL-GODDARD, R.C., AND SHOLKOVITZ, E.R., 1990, The rare-earth elements in rivers, estuaries, and coastal seas and their significance to the composition of ocean waters: Geochimica et Cosmochimica Acta, v. 54, p. 971–991.

EUGSTER, H.P., 1967, Hydrous sodium silicates from Lake Magadi, Kenya: precursors of bedded chert: Science, v. 157, p. 1177–1180.

EUGSTER, H.P., 1969, Inorganic bedded cherts from the Magadi area, Kenya: Contributions to Mineralogy and Petrology, v. 22, p. 1–31.

EUGSTER H.P., 1980, Lake Magadi, Kenya, and its precursors, *in* Nissenbaum, A., ed.. Hypersaline Brines and Evaporitic Environments: Amsterdam, Elsevier, Developments in Sedimentology no. 28, p. 195–232.

EUGSTER H.P., AND JONES B.F., 1979, Behavior of major solutions during closed-basin brine evolution: American Journal of Science, v. 279, p. 609–631.

FAIRHEAD, J.D., MITCHELL, J.G., AND WILLIAMS, L.A.J., 1972, New K/Ar determinations on rift volcanics of S. Kenya and their bearing on the age of rift faulting: Nature (Physical Sciences), v. 238, p. 66–69.

FAN, J., AND KERRICH, R., 1997, Geochemical characteristics of aluminum depleted and undepleted komatiites and HREE-enriched low-Ti tholeiites, western Abitibi greenstone belt: A heterogeneous mantle plume–convergent margin environment: Geochimica et Cosmochimica Acta, v. 61, p. 4723–4744.

GERMAN, C.R., AND ELDERFIELD, H., 1989, Rare-earth elements in Saanich inlet, British Columbia, a seasonally anoxic basin: Geochimica et Cosmochimica Acta, v. 53, p. 2561–2571.

GIERÉ, R., 1990, Hydrothermal mobility of Ti, Zr, and REE: Examples from the Bergall and Adamello contact aureoles (Italy): Terra Nova, v. 2, p. 60–67.

GOLDSTEIN, S.J., AND JACOBSEN, S.B., 1988a, Rare-earth elements in river waters: Earth and Planetary Science Letters, v. 87, p. 35–47.

GOLDSTEIN, S.J., AND JACOBSEN, S.B., 1988b, Nd and Sr isotopic systematics of river water suspended material—implications for crustal evolution: Earth and Planetary Science Letters, v. 89, p. 249–265.

GUEDDARI, M., 1984, Géochimie et thermodynamique des evaporites continentales. Etude du lac Natron en Tanzanie et du Chott el Jerid en Tunisie: Sciences Géologiques Mémoire, Institut de Gólogie, Université Louis Pasteur de Strasbourg no. 76, 143 p.

HATTORI, I., UMEDA, M., NAGAGAWA, T., AND YAMAMOTO, H., 1996, From chalcedonic chert to quartz chert: diagenesis of chert hosted in a Miocene volcanic–sedimentary succession, Central Japan: Journal of Sedimentary Research, v. 66, p. 163–174.

HAY, R.L., 1968, Chert and its sodium-silicate precursors in sodium-carbonate lakes of east Africa: Contributions to Mineralogy and Petrology, v. 17, p. 255–274.

HAY, R.L., 1970, Silicate reactions in three lithofacies of a semiarid basin, Olduvai Gorge, Tanzania, *in* Morgan, B.A., ed., Mineralogy and Petrology of the Upper Mantle; Sulfides; Mineralogy and Geochemistry of Non-Marine Evaporites: Mineralogical Society of America, Special Paper no. 3, p. 237–255.

HAY, R.L., 1976, Geology of the Olduvai Gorge: Berkeley, California, University of California Press, 203 p.

HEANEY, P.J., 1995, Moganite as an indicator for vanished evaporites: a testament reborn?: Journal of Sedimentary Research, v. 65, p. 633–638.

HILL, A., 1999, The Baringo Basin: from Bill Bishop to BPRP, *in* Banham, P., and Andrews, P., eds., Late Cenozoic Environments and Hominid Evolution: a Tribute to Bill Bishop: Geological Society of London, p. 85–97.

HILLAIRE-MARCEL, C., AND CASANOVA, J., 1987, Isotopic hydrology and paleohydrology of the Magadi (Kenya)-Natron (Tanzania) Basin during the late Quaternary: Palaeogeography, Palaeoclimatology, Palaeoecology, v. 58, p. 155–181.

HILLAIRE-MARCEL, C., CARRO, O., AND CASANOVA, J., 1986, $^{14}C$ and Th/U dating of Pleistocene and Holocene stromatolites from East African paleolakes: Quaternary Research, v. 25, p. 312–329.

HOUSER, B.B., 1985, Magadi-type chert, indicator of a lacustrine environment in the middle Eocene McBean Formation, South Carolina: Southeastern Geology, v. 25, p. 185–197.

HUMPHRIES, S.E., 1984, The mobility of the rare earth elements in the crust, *in* Henderson, P., ed., Rare Earth Element Geochemistry: Amsterdam, Elsevier, p. 317–342.

ICOLE, M., AND PERINET, G., 1984, Les silicates sodiques et les milieux évaporitiques carbonaté bicarbonatés sodiques: une revue: Revue de Géologie Dynamique et de Géographie Physique, v. 25, p. 167–176.

JONES, B.J., EUGSTER, H.P. AND RETTIG, S.L., 1977, Hydrochemistry of the Lake Magadi basin, Kenya: Geochimica et Cosmochimica Acta, v. 41, p. 53–72.

JOHANNESSON, K.H., LYONS, W.B., AND BIRD, D.A., 1994, Rare earth element concentrations and speciation in alkaline lakes from the Western U.S.A: Geophysical Research Letters, v. 21, p. 773–776.

JOHANNESSON, K.H., AND ZHOU, X., 1999, Origin of middle rare earth element enrichments in acid waters of a Canadian High Arctic lake: Geochimica et Cosmochimica Acta, v. 63, p. 153–165.

KATO, Y., OHTA, I., TSUNEMATSU, T., WATANABE, Y., ISOZAKI, Y., AND MARUYAMA, S., 1998, Rare earth element variations in mid-Archean banded iron formations: Implications for the chemistry of ocean and continent and plate tectonics: Geochimica et Cosmochimica Acta, v. 62, p. 3475–3497.

KERRICH, R., AND KING, R., 1993, Hydrothermal zircon and baddeleyite in Val d'Or Archean mesothermal gold deposits: Characteristics, compositions, and fluid inclusion properties: Canadian Journal of Earth Sciences, v. 30, p. 2334–2351.

KHANDIKAR, A.S., SANT, D.A., GOGTE, V., AND KARANTH, R.V., 1999, The influence of Deccan volcanism on climate: insights from lacustrine Intertrappean deposits, Anjar, western India: Palaeogeography, Palaeoclimatology, Palaeoecology, v. 147, p. 141–149.

KRAINER, K., AND SPÖTL, C., 1998, Abiogenic silica layers within a fluvio-lacustrine succession, Bolzano volcanic complex, northern Italy: a Permian analogue for Magadi-type cherts? Sedimentology, v. 45, p. 489–505.

KRAYNOV, S.T., MES'KOV, A.N., PETROVA, N.G., BATURINSKAYA, I.V., AND ZHARIKOVA, V.M., 1969, Highly alkaline (pH 12) fluorsilicate waters in the deeper zones of the Lavozeso massif: Geochemistry International, v. 614, p. 635–640.

KURTZ, A., DERRY, L.A., CHADWICK, O.A., AND ALFANO, M.J., 2000, Refractory element mobility in volcanic soils: Geology, v. 28, p. 683–686.

LE TURDU, C., TIERCELIN, J.-J., RICHERT, J.P., ROLET, J., XAVIER, J.-P., RENAUT, R.W., LEZZAR, K.E., AND COUSSEMENT, C., 1999, Influence of preexisting oblique discontinuities on the geometry and evolution of extensional fault patterns: evidence from the Kenya Rift using SPOT imagery, *in* Morley, C.K., ed., Geoscience of Rift Systems—Evolution of East Africa: American Association of Petroleum Geologists, Studies in Geology no. 44, p. 173–191.

LUDDEN J.N., DAIGNEAULT, R., ROBERT, F., AND TAYLOR, R.P., 1984, Trace element mobility in alteration zones associated with Archean Au lode deposits: Economic Geology, v. 79, p. 1131–1141.

MACDONALD, R., WILLIAMS, L.A.J., AND GASS, I.G., 1994, Tectonomagmatic evolution of the Kenya rift valley: some geological perspectives: Geological Society of London, Journal, v. 151, p. 879–888.

MAGLIONE, G., 1980, An example of recent continental evaporitic sedimentation, the Chadian Basin (Africa), *in* Chambre Syndicale de la Recherche et de la Production du Petrole et du Gaz Naturel, ed., Evaporite Deposits: Editions Technip, Paris, p. 5–9.

MCLAREN, J.W., 1992, The analysis of natural waters by ICP-MS, *in* Jarvis, K.E., Gray, A.L., and Houk, R.S., eds., Inductively Coupled Plasma Mass Spectrometry: New York, Blackie, p. 265–278.

MENZIES, M., ROGERS, N., TINDLE, A., AND HAWKESWORTH, C., 1987, Metasomatic and enrichment processes in lithospheric peridotites, an effect of asthenosphere–lithosphere interaction, *in* Menzies, M., and Hawkesworth, C., eds., Mantle Metasomatism: London, Academic Press, p. 313–361.

MOFFETT, J.W., 1990, Microbially mediated cerium oxidation in sea-water: Nature, v. 345, 421–423.

MOFFETT, J.W., 1994a, A radiotracer study of cerium and manganese uptake onto suspended particles in Chesapeake Bay: Geochimica et Cosmochimica Acta, v. 58, p. 695–703.

MOFFETT, J.W., 1994b, The relationship between cerium and manganese oxidation in the marine-environment: Limnology and Oceanography, v. 39, p. 1309–1318.

MÖLLER P., AND BAU, M., 1993, Rare-earth patterns with positive cerium anomaly in alkaline waters from Lake Van, Turkey: Earth and Planetary Science Letters, v. 117, p. 671–676.

MURRAY, R.W., 1994, Chemical criteria to identify the depositional environment of chert: general principles and applications: Sedimentary Geology, v. 90, p. 213–232.

MURRAY, R.W., BUCHHOLTZ TEN BRINK, M.R., JONES, D.L., GERLACH, D.C., AND RUSS, G.P., 1990, Rare earth elements as indicators of different marine depositional environments in chert and shale: Geology, v. 18, p. 268–271.

MURRAY, R.W., BUCHHOLTZ TEN BRINK, M.R., GERLACH, D.C., RUSS, G.P., AND JONES, D.L., 1992a, Interoceanic variation in the rare earth, major, and trace element depositional chemistry of chert: perspectives gained from the DSDP and ODP record: Geochimica et Cosmochimica Acta, v. 56, p. 1897–1913.

MURRAY, R.W., BUCHHOLTZ TEN BRINK, M.R., GERLACH, D.C., RUSS, G.P., AND JONES, D.L., 1992b, Rare earth, major, and trace element composition of Monterey and DSDP chert and associated host sediment: assessing the influence of chemical fractionation during diagenesis: Geochimica et Cosmochimica Acta, v. 56, p. 2657–2671.

O'NEIL, J.R., AND HAY, R.L., 1973, $^{18}O/^{16}O$ ratios in cherts associated with the saline lake deposits of East Africa: Earth and Planetary Science Letters, v. 19, p. 257–266.

OWEN, R.B., AND RENAUT, R.W., 2000, Miocene and Pliocene diatomaceous lacustrine sediments of the Tugen Hills, Baringo District, Central Kenya Rift, *in* Gierlowski-Kordesch, E.H., and Kelts, K., eds., Lake Basins through Space and Time: American Association of Petroleum Geologists, Studies in Geology no. 44, p. 465–472.

OWEN, A.W., ARMSTRONG, H.A., AND FLOYD, J.D., 1999, Rare earth element geochemistry of Upper Ordovician cherts from the Southern Upland of Scotland: Geological Society of London, Journal, v. 156, p. 191–204.

PARNELL, J., 1986, Devonian Magadi-type cherts in the Orcadian basin, Scotland: Journal of Sedimentary Petrology, v. 56, p. 495–500.

PERINET, G., TIERCELIN J.J., AND BARTON, C.E., 1982, Présence de kanemite dans les sédiments récents du lac Bogoria, Rift Gregory, Kenya: Bulletin Minéralogique, v. 105, p. 633–639.

PETERSON, M.N.A., AND VON DER BORCH, C.C., 1965, Chert: modern inorganic deposition in a carbonate-precipitating locality: Science, v. 149, p. 1501–1503.

PICKFORD, M.H.L., 1975, Late Miocene sediments and fossils from the northern Kenya Rift Valley: Nature, v. 256, p. 279–284.

PICKFORD, M.H.L., 1978, Stratigraphy and mammalian palaeontology of the late-Miocene Lukeino Formation, Kenya, *in* Bishop, W.W., ed., Geological Background to Fossil Man: Edinburgh, Scottish Academic Press, p. 237–262.

RENAUT, R.W., AND OWEN, R.B., 1988, Opaline cherts associated with sublacustrine hydrothermal springs at Lake Bogoria, Kenya Rift Valley: Geology, v. 16, p. 699–702.

RENAUT, R.W., AND TIERCELIN, J.-J., 1994, Lake Bogoria, Kenya Rift Valley—a sedimentological overview, *in* Renaut, R.W., and Last, W.M., eds., Sedimentology and Geochemistry of Modern and Ancient Saline Lakes: SEPM, Special Publication 50, p. 101–123.

RENAUT, R.W., TIERCELIN, J.J., AND OWEN, R.B., 1986, Mineral precipitation and diagenesis in the sediments of the Lake Bogoria basin, Kenya Rift Valley, *in* Frostick, L.E., Renaut, R.W., Reid., I., and Tiercelin, J.J., eds., Sedimentation in the African Rifts: Geological Society of London, Special Publication 25, p. 159–175.

RENAUT, R.W., EGO, J., TIERCELIN, J.J., LE TURDU, C., AND OWEN, R.B., 1999, Saline, alkaline palaeolakes of the Tugen Hills–Kerio Valley region, Kenya Rift Valley, *in* Banham, P., and Andrews, P., eds., Late Cenozoic Environments and Hominid Evolution: a Tribute to Bill Bishop: Geological Society of London, p. 41–58.

RENAUT, R.W., JONES, B., TIERCELIN, J.-J., AND TARITS, C., 2002, Sublacustrine precipitation of hydrothermal silica in rift lakes: evidence from Lake Baringo, central Kenya Rift Valley: Sedimentary Geology, v. 148, p. 235–257.

ROBERTS, N., TAIEB, M., BARKER, P., DAMNATI, B., ICOLE, M., AND WILLIAMSON, D., 1993, Timing of the younger Dryas event in East Africa from lake-level changes: Nature, v. 366, p. 146–148.

ROONEY, T.P., JONES, B.F., AND NEAL, J.T., 1969, Magadiite from Alkali Lake, Oregon: American Mineralogist, v. 54, p. 1034–1043.

RUBIN, J.M., HENRY C.D., AND PRIE, J.C., 1988, Hydrothermal zircons and zircon overgrowths, Sierra Blanca Peaks, Texas: American Mineralogist, v. 74, p. 865–869.

SAXENA, S.K., 1966, Evolution of zircons in sedimentary and metamorphic rocks: Sedimentology, v. 6, p. 1–33.

SCHUBEL, K.A., AND SIMONSON, B.M., 1990, Petrography and diagenesis of cherts from Lake Magadi, Kenya: Journal of Sedimentary Petrology, v. 60, p. 761–776.

SEBAG, D., VERRECCHIA, E.P., SEONG-JOO, L., AND DURAND, A., 2001, The natural hydrous sodium silicates from the northern bank of Lake Chad: occurrence, petrology and genesis: Sedimentary Geology, v. 139. p. 15–31.

SHAW, P.A., COOKE, H.J., AND PERRY, C.C., 1990, Microbialitic silcretes in highly alkaline environments: some observations from Sua Pan, Botswana: South African Journal of Geology, v. 93, p. 803–808.

SHEPPARD, R.A., AND GUDE, A.J., 1986, Magadi-type chert: a distinctive diagenetic variety from lacustrine deposits, *in* Mumpton, F.A., ed., Studies in Diagenesis: U.S. Geological Survey, Bulletin 1578, p. 335–345.

SHOLKOVITZ, E.R., AND SCHNEIDER, D.L., 1991, Cerium redox cycles and rare-earth elements in the Sargasso Sea: Geochimica et Cosmochimica Acta, v. 55, p. 2737–2743.

SHOLKOVITZ, E.R., SHAW, T.J., AND SCHNEIDER, D.L., 1992, The geochemistry of rare-earth elements in the seasonally anoxic water column and porewaters of Chesapeake Bay: Geochimica et Cosmochimica Acta, v. 56, p. 3389–3402.

SHOLKOVITZ, E.R., LANDING, W.M., AND LEWIS, B.L., 1994, Ocean particle chemistry—the fractionation of rare-earth elements between suspended particles and seawater: Geochimica et Cosmochimica Acta, v. 58, p. 1567–1579.

SUN, S.S., AND MCDONOUGH, W.F., 1989, Chemical and isotopic systematics of oceanic basalts; implications for mantle composition and processes, *in* Saunders, A.D., and Norry, M.J., eds., Magmatism in the Ocean Basins: Geological Society of London, Special Publication 42, p. 313–345.

SURDAM, R.C., AND EUGSTER, H.P., 1976, Mineral reactions in the sedimentary deposits of the Lake Magadi region, Kenya: Geological Society of America, Bulletin, v. 87, p. 1739–1752.

SURDAM, R.C., EUGSTER, H.P., AND MARINER, R.H., 1972, Magadi-type chert in Jurassic and Eocene to Pleistocene rocks, Wyoming: Geological Society of America, Bulletin, v. 83, p. 2261–2266.

TAIEB, M., BARKER, P., BONNEFILLE, R., DAMNATI, B., GASSE, F., GOETZ, C., HILLAIRE, C., ICOLE M., MASAULT, M., ROBERTS, N., VINCENS, A., AND WILLAMSON, D., 1991, Histoire paléohydrologique du lac Magadi (Kenya) au Pleistocène superieur: Academie des Sciences [Paris], Comptes Rendues, Série 2, v. 313, p. 339–346.

TAYLOR, S.R., AND MCLENNAN, S.M., 1985, The Continental Crust; Its Composition and Evolution: Oxford, U.K., Blackwell, 312 p.

TERAKADO, Y., AND FUJITANI, T., 1998, Behavior of the rare earth elements and trace elements during interactions between acidic hydrothermal solutions and silicic volcanic rocks, southwestern Japan: Geochimica et Cosmochimica Acta, v. 62, p. 1903–1917.

TIERCELIN, J.J., 1981, Rifts continentaux, tectonique, climats, sédiments. Exemples: la sédimentation dans le Nord du Rift Gregory, Kenya, et dans le rift de l'Afar, Ethiopie, depuis le Miocène: Ph.D. thesis, Université Aix.Marseille II, 260 p.

VOLKOVA, N.L., 1998, Geochemistry of rare elements in waters and sediments of alkaline lakes in the Sasykkul depression, East Pamirs: Chemical Geology, v. 147, p. 265–277.

WHEELER, W.H., AND TEXTORIS, D.A., 1978, Triassic limestone and chert of playa origin in North Carolina: Journal of Sedimentary Petrology, v. 48, p. 765–776.

APPENDIX TABLE 1A.—Trace-element compositions (ppm) of magadiite and Magadi-type chert from the High Magadi Beds, Lake Magadi.

| | MAG-1 | MAG-2 | MAG-3 | MAG-4 | MAG-4d | MAG-5 | MAG-6 | MAG-7 | MAG-8 |
|---|---|---|---|---|---|---|---|---|---|
| Ti | 32 | 27 | 23 | 22 | 21 | 61 | 55 | 62 | 61 |
| P | 0 | 0 | 13 | 4 | 17 | 19 | 13 | 16 | 16 |
| Cr | 0.3 | 0.2 | 0.5 | 0.5 | 0.6 | 0.3 | 1.8 | 0.5 | 0.3 |
| Co | 0.04 | 0.03 | 0.1 | 0.08 | 0.04 | 0.18 | 0.11 | 0.22 | 0.07 |
| Ni | 0.69 | 0.26 | 0.95 | 0.29 | 0.64 | 0.36 | 0.82 | 0.18 | 0.19 |
| Rb | 1.73 | 1.22 | 1.15 | 1.14 | 1.14 | 3.57 | 3.95 | 3.58 | 2.55 |
| Sr | 27.3 | 14.3 | 36.9 | 28 | 29 | 31.2 | 43.6 | 43.7 | 31.2 |
| Cs | 0.1 | 0.06 | 0.05 | 0.05 | 0.04 | 0.16 | 0.18 | 0.21 | 0.17 |
| Ba | 11.1 | 5.8 | 9.2 | 7.2 | 7.4 | 14.1 | 21.4 | 17.9 | 12.9 |
| Sc | 0.91 | 2.38 | 0.7 | 0.4 | 0.12 | 2.44 | 2 | 1.24 | 4.72 |
| V | 1.63 | 0.91 | 0.61 | 0.53 | 0.57 | 4.08 | 4.68 | 2.26 | 2.03 |
| Ta | 0.14 | 0.17 | 0.61 | 0.55 | 0.51 | 0.38 | 0.49 | 0.41 | 0.7 |
| Nb | 17.2 | 14.7 | 23.7 | 20 | 19.4 | 23.3 | 27.4 | 22.1 | 13.3 |
| Zr | 868 | 718 | 1018 | 898 | 887 | 1365 | 1656 | 1315 | 795 |
| Hf | 0.66 | 0.88 | 3.42 | 3.18 | 3.22 | 1.44 | 1.78 | 1.27 | 2.24 |
| Th | 0.85 | 0.49 | 0.38 | 0.38 | 0.39 | 0.85 | 1.2 | 0.98 | 0.5 |
| U | 2.86 | 1.66 | 2.5 | 2.37 | 2.32 | 3.32 | 5.5 | 2.38 | 1.16 |
| Y | 1.79 | 1.27 | 1.49 | 1.19 | 1.2 | 4.81 | 3.71 | 2.93 | 1.48 |
| La | 6.21 | 1.98 | 2.93 | 2.46 | 2.51 | 7.5 | 16.4 | 6.96 | 5.47 |
| Ce | 38.1 | 20.4 | 17.2 | 14.4 | 14.6 | 35.8 | 56.3 | 50.8 | 27.3 |
| Pr | 1.08 | 0.38 | 0.52 | 0.45 | 0.44 | 1.65 | 2.83 | 1.23 | 1.04 |
| Nd | 3 | 1.07 | 1.46 | 1.23 | 1.2 | 4.97 | 8.06 | 3.58 | 3.58 |
| Sm | 0.31 | 0.15 | 0.15 | 0.12 | 0.12 | 0.7 | 0.87 | 0.45 | 0.35 |
| Eu | 0.04 | 0.02 | 0.02 | 0.01 | 0.01 | 0.1 | 0.11 | 0.07 | 0.04 |
| Gd | 0.31 | 0.19 | 0.15 | 0.12 | 0.13 | 0.72 | 0.76 | 0.53 | 0.62 |
| Tb | 0.04 | 0.02 | 0.01 | 0.01 | 0.01 | 0.1 | 0.08 | 0.06 | 0.04 |
| Dy | 0.19 | 0.08 | 0.11 | 0.08 | 0.09 | 0.54 | 0.44 | 0.31 | 0.23 |
| Ho | 0.04 | 0.02 | 0.02 | 0.02 | 0.02 | 0.11 | 0.07 | 0.06 | 0.05 |
| Er | 0.1 | 0.06 | 0.07 | 0.05 | 0.05 | 0.29 | 0.21 | 0.15 | 0.17 |
| Tm | 0.03 | 0.01 | 0.02 | 0.01 | 0.01 | 0.05 | 0.04 | 0.03 | 0.02 |
| Yb | 0.3 | 0.15 | 0.17 | 0.17 | 0.15 | 0.42 | 0.44 | 0.32 | 0.17 |
| Lu | 0.06 | 0.03 | 0.03 | 0.03 | 0.03 | 0.06 | 0.08 | 0.05 | 0.03 |
| Cu | 0.41 | 0.22 | 0.49 | 0.35 | 0.47 | 0.36 | 0.47 | 0.4 | 0.9 |
| Zn | 1.06 | 0.76 | 1.63 | 1.51 | 1.95 | 1.86 | 2.25 | 2.12 | 2.44 |
| Mo | 5.49 | 5.3 | 2.17 | 1.75 | 1.9 | 19 | 20.7 | 8.37 | 1.97 |
| Ag | 1.56 | 1.24 | 1.65 | 1.61 | 1.69 | 2.22 | 2.76 | 2.37 | 8.16 |
| Tl | 0.05 | 0.03 | 0.02 | 0.02 | | 0.06 | 0.07 | 0.04 | 0.03 |
| Pb | 0.23 | 0.12 | 0.33 | 0.32 | 0.3 | 0.31 | 0.38 | 0.31 | 0.19 |
| Sn | 0.05 | 0.12 | 0.08 | 0.06 | 0.08 | 0.36 | 0.11 | 0.13 | 0.22 |
| Sb | 0.03 | 0.1 | 0.12 | 0.1 | 0.13 | 0.12 | 0.13 | 0.13 | 0.07 |
| La/Yb cn | 14.3 | 9.1 | 11.9 | 10 | 11.6 | 12.4 | 25.7 | 15.1 | 22.3 |
| La/Sm cn | 12.6 | 8.3 | 12.2 | 12.8 | 13.1 | 6.7 | 11.8 | 9.7 | 9.8 |
| Gd/Yb cn | 0.85 | 1.05 | 0.73 | 0.58 | 0.72 | 1.42 | 1.43 | 1.37 | 3.01 |
| Eu/Eu* | 0.39 | 0.36 | 0.4 | 0.25 | 0.24 | 0.43 | 0.41 | 0.44 | 0.26 |
| Ce/Ce* | 3.52 | 5.61 | 3.34 | 3.27 | 3.33 | 2.44 | 1.98 | 4.16 | 2.75 |
| Total REE | 49.8 | 24.5 | 22.9 | 19.1 | 19.4 | 53 | 86.6 | 64.6 | 39.2 |
| Zr/Hf | 1315 | 816 | 298 | 282 | 275 | 948 | 930 | 1035 | 355 |
| Nb/Ta | 122.9 | 86.5 | 38.9 | 36.4 | 38 | 61.3 | 55.9 | 53.9 | 19 |
| La/Nb | 0.36 | 0.14 | 0.12 | 0.12 | 0.13 | 0.32 | 0.6 | 0.32 | 0.41 |
| Th/Nb | 0.05 | 0.03 | 0.02 | 0.02 | 0.02 | 0.04 | 0.04 | 0.04 | 0.04 |
| Th/La | 0.14 | 0.25 | 0.13 | 0.15 | 0.16 | 0.11 | 0.07 | 0.14 | 0.1 |
| Zr/Y | 485 | 566 | 683 | 754 | 739 | 284 | 446 | 449 | 537 |
| Th/U | 0.3 | 0.3 | 0.15 | 0.16 | 0.17 | 0.26 | 0.22 | 0.41 | 0.48 |
| **% | 70 | 80 | 70 | 90 | 90 | 50 | 50 | 90 | 60 |

**Percentage of magadiite in sample as determined by XRD.

Appendix Table 1b.—Trace-element compositions (ppm) of cherts from the High Magadi Beds, Lake Magadi.

| | HM-1 | HM-2 | HM-3 | HM-4 | HM-5 | HM-6 | HM-7 | HM-8 |
|---|---|---|---|---|---|---|---|---|
| Ti | 19 | 32 | 17 | 10 | 25 | 40 | 35 | 36 |
| P | 13 | 119 | 78 | 7 | | 15 | 48 | |
| Cr | 0.0 | 0.4 | 0.2 | 0.1 | 0.5 | 0.1 | 0.1 | 0.2 |
| Co | 0.06 | 0.18 | 0.11 | 0.04 | 0.06 | 0.14 | 0.04 | 0.06 |
| Ni | 0.30 | 0.50 | 0.10 | 0.25 | 0.31 | 0.51 | 0.00 | 0.00 |
| Rb | 1.63 | 3.65 | 3.93 | 1.20 | 2.53 | 2.82 | 1.63 | 2.07 |
| Sr | 13.4 | 55.0 | 32.4 | 19.7 | 33.0 | 57.8 | 19.7 | 36.3 |
| Cs | 0.03 | 0.15 | 0.07 | < 0.04 | 0.05 | 0.05 | 0.02 | 0.06 |
| Ba | 1.63 | 3.65 | 3.93 | 1.20 | 2.53 | 2.82 | 1.63 | 2.07 |
| Sc | 0.10 | 0.20 | 0.08 | 0.08 | 0.14 | 0.24 | 0.12 | 0.14 |
| V | 0.16 | 0.94 | 0.50 | 0.29 | 0.87 | 2.75 | 1.72 | 0.86 |
| Ta | 0.01 | 0.07 | 0.09 | < 0.002 | 0.02 | 0.03 | < 0.002 | 0.01 |
| Nb | 8.2 | 7.3 | 24.0 | 1.9 | 8.9 | 17.1 | 10.5 | 18.7 |
| Zr | 178 | 134 | 251 | 5 | 110 | 142 | 386 | 272 |
| Hf | 0.42 | 0.45 | 2.74 | 0.05 | 0.71 | 0.49 | 0.40 | 1.06 |
| Th | 1.57 | 0.95 | 0.49 | 0.76 | 0.26 | 0.29 | 1.55 | 0.42 |
| U | 16.6 | 1.21 | 2.56 | 0.08 | 0.21 | 0.23 | 13.8 | 0.66 |
| Y | 2.75 | 12.56 | 3.78 | 4.06 | 2.41 | 2.63 | 1.66 | 5.13 |
| La | 1.04 | 6.41 | 4.06 | 0.84 | 1.75 | 1.66 | 0.59 | 6.70 |
| Ce | 9.16 | 27.0 | 13.6 | 30.2 | 9.64 | 13.0 | 9.81 | 30.9 |
| Pr | 0.24 | 1.28 | 0.80 | 0.31 | 0.35 | 0.36 | 0.12 | 0.96 |
| Nd | 0.79 | 5.09 | 2.78 | 1.61 | 1.76 | 1.44 | 0.47 | 3.34 |
| Sm | 0.15 | 1.06 | 0.49 | 0.29 | 0.28 | 0.27 | 0.06 | 0.69 |
| Eu | 0.03 | 0.17 | 0.07 | 0.05 | 0.04 | 0.04 | 0.01 | 0.10 |
| Gd | 0.19 | 1.28 | 0.52 | 0.40 | 0.33 | 0.37 | 0.10 | 0.77 |
| Tb | 0.03 | 0.21 | 0.07 | 0.06 | 0.05 | 0.06 | 0.02 | 0.12 |
| Dy | 0.32 | 1.79 | 0.51 | 0.55 | 0.36 | 0.41 | 0.16 | 0.85 |
| Ho | 0.09 | 0.44 | 0.13 | 0.15 | 0.08 | 0.09 | 0.04 | 0.20 |
| Er | 0.33 | 1.32 | 0.44 | 0.56 | 0.26 | 0.24 | 0.17 | 0.58 |
| Tm | 0.07 | 0.19 | 0.06 | 0.09 | 0.03 | 0.04 | 0.06 | 0.08 |
| Yb | 0.67 | 0.99 | 0.39 | 0.68 | 0.22 | 0.22 | 0.85 | 0.44 |
| Lu | 0.11 | 0.14 | 0.06 | 0.10 | 0.04 | 0.03 | 0.15 | 0.06 |
| Cu | 0.20 | 0.44 | 0.41 | 0.15 | 0.22 | 0.22 | 0.10 | 0.19 |
| Zn | 1.24 | 2.25 | 1.34 | 0.60 | 1.33 | 1.90 | 0.53 | 2.62 |
| Mo | 0.48 | 0.20 | 0.40 | 0.04 | 0.14 | 0.18 | 0.70 | 0.34 |
| Ag | 0.36 | 0.26 | 0.49 | 0.01 | 0.19 | 0.26 | 0.68 | 0.56 |
| Tl | 0.04 | 0.04 | 0.06 | 0.01 | 0.04 | < 0.01 | < 0.01 | 0.06 |
| Pb | 0.85 | 0.60 | 0.33 | 1.29 | 0.76 | 0.92 | 0.38 | 1.17 |
| Sn | 0.05 | 0.05 | 0.06 | 0.05 | 0.05 | 0.11 | 0.23 | 0.06 |
| Sb | 0.02 | 0.08 | 0.06 | 0.04 | 0.25 | 0.39 | 0.26 | 0.28 |
| La/Yb cn | 1.11 | 4.64 | 7.46 | 0.89 | 5.70 | 5.41 | 0.50 | 10.9 |
| La/Sm cn | 4.48 | 3.91 | 5.35 | 1.87 | 4.04 | 3.97 | 6.35 | 6.27 |
| Gd/Yb cn | 0.23 | 1.07 | 1.10 | 0.49 | 1.24 | 1.39 | 0.10 | 1.45 |
| Eu/Eu* | 0.54 | 0.45 | 0.42 | 0.45 | 0.40 | 0.39 | 0.39 | 0.42 |
| Ce/Ce* | 4.33 | 2.18 | 1.74 | 14.5 | 2.85 | 3.94 | 8.54 | 2.63 |
| Total REE | 16.5 | 51.3 | 27.1 | 38.0 | 18.2 | 22.2 | 15.5 | 51.0 |
| Zr/Hf | 424 | 299 | 91 | 98 | 155 | 290 | 966 | 257 |
| Nb/Ta | 823 | 104 | 267 | | 444 | 569 | | 1872 |
| La/Nb | 0.13 | 0.88 | 0.17 | 0.43 | 0.20 | 0.10 | 0.06 | 0.36 |
| Th/Nb | 0.19 | 0.13 | 0.02 | 0.39 | 0.03 | 0.02 | 0.15 | 0.02 |
| Th/La | 1.51 | 0.15 | 0.12 | 0.90 | 0.15 | 0.17 | 2.63 | 0.06 |
| Zr/Y | 65 | 11 | 66 | 1 | 46 | 54 | 233 | 53 |
| Th/U | 0.09 | 0.79 | 0.19 | 9.50 | 1.24 | 1.26 | 0.11 | 0.64 |

Note: These cherts contain < 10% magadiite and are predominantly quartz.

APPENDIX TABLE 1C.—Trace-element compositions (ppm) of Pliocene and Pleistocene cherts, and hydrothermal cherts at Lake Bogoria.

| | Green Beds cherts (Magadi) | | | | Olduvai Bed I cherts | | | Lukeino Fm cherts | | Hydrothermal cherts | |
|---|---|---|---|---|---|---|---|---|---|---|---|
| | OL-1 | OL-2 | OL-3 | OL-4 | MAG-A | MAG-B | MAG-C | LK-1 | LK-2 | MWA-1 | MWA-2 |
| Ti | 339 | 30 | 22 | 19 | 34 | 36 | 39 | 34 | 49 | 426 | 1921 |
| P | 59 | 73 | 23 | 16 | 10 | 15 | 16 | 8 | 2 | 39 | 206 |
| Cr | 1.8 | 0.3 | 0.1 | 0.1 | 0.2 | 9.1 | 2.7 | 0.1 | 0.0 | 1.2 | 3.1 |
| Co | 1.93 | 0.08 | 0.06 | 0.05 | 0.11 | 0.78 | 0.33 | 0.03 | 0.07 | 0.13 | 0.66 |
| Ni | 1.06 | n.d. | 0.24 | 0.72 | 0.34 | 1.84 | 1.08 | 0.64 | 0.19 | 0.41 | 1.31 |
| Rb | 43.7 | 4.59 | 3.19 | 2.26 | 1.34 | 1.40 | 1.07 | < 0.154 | 0.12 | 27.0 | 69.1 |
| Sr | 75.2 | 9.2 | 55.3 | 25.5 | 3.3 | 7.4 | 3.4 | 2.4 | 3.4 | 11.3 | 39.5 |
| Cs | 1.96 | 0.11 | 0.11 | 0.05 | < 0.04 | < 0.04 | < 0.04 | < 0.04 | < 0.04 | 1.51 | 0.93 |
| Ba | 43.7 | 4.59 | 3.19 | 2.26 | 1.34 | 1.40 | 1.07 | < 0.09 | 0.12 | 27.0 | 69.1 |
| Sc | 0.53 | 0.15 | 0.18 | 0.06 | 0.06 | 0.10 | 0.11 | 0.08 | 0.05 | 0.40 | 22.8 |
| V | 8.86 | 1.13 | 0.27 | 0.41 | 0.94 | 1.24 | 1.07 | 0.43 | 1.15 | 1.77 | 1.57 |
| Ta | 1.20 | 0.03 | 0.01 | 0.05 | 0.04 | 0.03 | 0.03 | 0.01 | 0.01 | 0.94 | 4.22 |
| Nb | 56.4 | 25.1 | 17.7 | 2.3 | 3.6 | 2.7 | 4.1 | 8.1 | 8.1 | 24.3 | 72.9 |
| Zr | 557 | 480 | 400 | 15 | 39 | 41 | 42 | 313 | 62 | 79 | 295 |
| Hf | 5.43 | 0.84 | 1.16 | 0.10 | 0.56 | 0.54 | 0.48 | 0.86 | 0.20 | 1.14 | 5.55 |
| Th | 0.82 | 0.26 | 0.15 | 0.73 | 0.10 | 0.14 | 0.10 | 0.24 | 0.27 | 2.22 | 9.25 |
| U | 2.39 | 1.35 | 1.40 | 0.12 | 9.03 | 8.00 | 9.80 | 19.3 | 7.24 | 0.89 | 1.36 |
| Y | 9.95 | 1.58 | 1.18 | 3.77 | 0.24 | 0.32 | 0.29 | 2.22 | 0.13 | 6.09 | 30.8 |
| La | 13.61 | 2.30 | 5.76 | 0.99 | 0.52 | 0.37 | 0.26 | 0.34 | 0.24 | 11.9 | 68.4 |
| Ce | 29.0 | 10.2 | 25.9 | 30.3 | 0.71 | 0.65 | 0.53 | 0.97 | 0.24 | 25.9 | 93.8 |
| Pr | 2.50 | 0.59 | 1.46 | 0.30 | 0.08 | 0.07 | 0.05 | 0.20 | 0.03 | 2.30 | 14.4 |
| Nd | 9.10 | 1.98 | 4.12 | 1.25 | 0.23 | 0.20 | 0.13 | 0.87 | 0.12 | 9.09 | 53.6 |
| Sm | 1.92 | 0.26 | 0.33 | 0.26 | 0.03 | 0.05 | 0.02 | 0.21 | 0.03 | 1.24 | 7.64 |
| Eu | 0.31 | 0.03 | 0.04 | 0.04 | < 0.01 | < 0.01 | < 0.01 | 0.04 | < 0.01 | 0.23 | 1.12 |
| Gd | 2.06 | 0.22 | 0.29 | 0.38 | 0.02 | 0.02 | 0.01 | 0.22 | 0.02 | 1.51 | 8.45 |
| Tb | 0.35 | 0.03 | 0.03 | 0.06 | < 0.002 | < 0.002 | < 0.002 | 0.05 | < 0.002 | 0.18 | 1.06 |
| Dy | 2.11 | 0.19 | 0.17 | 0.47 | 0.03 | 0.03 | 0.02 | 0.39 | 0.02 | 1.11 | 6.29 |
| Ho | 0.41 | 0.04 | 0.03 | 0.13 | < 0.003 | 0.01 | 0.01 | 0.08 | < 0.003 | 0.23 | 1.29 |
| Er | 0.96 | 0.10 | 0.07 | 0.51 | 0.02 | 0.03 | 0.03 | 0.26 | 0.01 | 0.66 | 3.87 |
| Tm | 0.12 | 0.01 | 0.01 | 0.09 | 0.01 | 0.01 | 0.01 | 0.04 | < 0.004 | 0.11 | 0.60 |
| Yb | 0.61 | 0.10 | 0.06 | 0.68 | 0.05 | 0.05 | 0.05 | 0.37 | 0.02 | 0.87 | 3.96 |
| Lu | 0.08 | 0.01 | 0.01 | 0.09 | 0.01 | 0.01 | 0.01 | 0.06 | < 0.005 | 0.14 | 0.61 |
| Cu | 2.28 | 0.28 | 0.24 | 0.24 | 0.30 | 1.99 | 0.63 | 0.13 | 0.31 | 1.42 | 2.78 |
| Zn | 15.1 | 1.49 | 2.50 | 1.07 | 0.30 | 0.80 | 0.49 | 0.25 | 0.44 | 9.92 | 61.9 |
| Mo | 0.86 | 0.52 | 0.83 | 0.03 | 1.61 | 7.44 | 1.08 | 0.32 | 0.06 | 54.71 | 5.08 |
| Ag | 1.09 | 0.91 | 0.81 | 0.03 | | | 0.61 | 0.12 | 0.76 | 3.08 | |
| Tl | 0.02 | 0.12 | 0.04 | < 0.01 | < 0.01 | < 0.01 | < 0.01 | 0.01 | < 0.01 | 0.25 | 0.08 |
| Pb | 1.64 | 0.27 | 0.32 | 1.33 | 0.14 | 0.32 | 0.13 | 0.79 | 0.18 | 3.42 | 4.91 |
| Sn | 0.47 | 0.06 | 0.08 | 0.06 | 0.07 | 0.09 | 0.04 | 0.10 | 0.03 | 0.50 | 2.13 |
| Sb | 0.11 | 0.00 | 0.09 | 0.02 | 0.02 | 0.06 | 0.00 | 0.10 | 0.03 | 4.32 | 0.55 |
| La/Yb cn | 16.0 | 16.5 | 68.8 | 1.04 | 7.46 | 5.31 | 3.73 | 0.66 | 8.60 | 9.77 | 12.4 |
| La/Sm cn | 4.58 | 5.71 | 11.28 | 2.46 | 11.20 | 4.78 | 8.40 | 1.05 | 5.17 | 6.18 | 5.78 |
| Gd/Yb cn | 2.79 | 1.82 | 4.00 | 0.46 | 0.33 | 0.33 | 0.17 | 0.49 | 0.83 | 1.44 | 1.76 |
| Eu/Eu* | 0.47 | 0.37 | 0.39 | 0.39 | | | 0.56 | | 0.51 | 0.42 | |
| Ce/Ce* | 1.13 | 2.10 | 2.13 | 13.5 | 0.76 | 0.92 | 1.07 | 0.90 | 0.59 | 1.14 | 0.70 |
| Total REE | 84.6 | 19.7 | 43.2 | 38.3 | 4.2 | 12.2 | 3.5 | 6.4 | 1.9 | 131 | 346 |
| Zr/Hf | 102 | 571 | 345 | 147 | 69 | 75 | 88 | 364 | 311 | 69 | 53 |
| Nb/Ta | 47 | 836 | 1772 | 46 | 91 | 89 | 137 | 808 | 812 | 26 | 17 |
| La/Nb | 0.24 | 0.09 | 0.33 | 0.43 | 0.14 | 0.14 | 0.06 | 0.04 | 0.03 | 0.49 | 0.94 |
| Th/Nb | 0.01 | 0.01 | 0.01 | 0.32 | 0.03 | 0.05 | 0.02 | 0.03 | 0.03 | 0.09 | 0.13 |
| Th/La | 0.06 | 0.11 | 0.03 | 0.74 | 0.19 | 0.38 | 0.38 | 0.71 | 1.13 | 0.19 | 0.14 |
| Zr/Y | 56 | 304 | 339 | 4 | 162 | 127 | 145 | 141 | 479 | 13 | 10 |
| Th/U | 0.34 | 0.19 | 0.11 | 6.08 | 0.01 | 0.02 | 0.01 | 0.01 | 0.04 | 2.49 | 6.80 |

Appendix Table 1d.— Trace element compositions (ppb) of surface and bottom waters from Lake Bogoria.

| | Surface waters | | | | | | Bottom waters | | | | |
|---|---|---|---|---|---|---|---|---|---|---|---|
| | BOG L1S | BOG L9S | BOG L10S | BOG L11S | BOG L12S | BOG L13S | BOG L1F | BOG L9F | BOG L10F | BOG L12F | BOG L13F |
| Ti | 315 | 264 | 298 | 254 | 282 | 299 | 399 | 288 | 351 | 416 | 340 |
| P | | | | | | | | | | | |
| Cr | 37.3 | 14.2 | 34.0 | 12.7 | 23.6 | 25.9 | 68.9 | 29.2 | 30.1 | 33.5 | 48.7 |
| Co | 0.49 | 0.66 | 0.42 | 0.72 | 0.52 | 0.60 | 0.26 | 0.37 | 0.13 | 0.33 | 0.24 |
| Ni | n.d. | 1.27 | n.d. | 1.92 | 1.39 | 1.77 | n.d. | 0.92 | 1.54 | 0.63 | n.d. |
| Rb | 593 | 612 | 579 | 618 | 613 | 592 | 785 | 723 | 745 | 768 | 738 |
| Sr | 445 | 437 | 438 | 440 | 439 | 443 | 553 | 478 | 506 | 519 | 507 |
| Cs | 1.73 | 1.67 | 1.62 | 1.81 | 1.72 | 1.73 | 2.35 | 2.43 | 2.26 | 2.34 | 2.39 |
| Ba | 123 | 120 | 121 | 118 | 118 | 122 | 166 | 152 | 169 | 167 | 167 |
| Sc | 18.1 | 16.7 | 16.9 | 14.4 | 18.0 | 17.2 | 20.2 | 15.2 | 17.6 | 21.6 | 18.1 |
| V | 35.4 | 28.3 | 32.3 | 26.9 | 31.4 | 32.6 | 33.4 | 24.2 | 28.4 | 33.3 | 29.1 |
| Ta | 0.45 | 0.42 | 0.38 | 0.42 | 0.41 | 0.45 | 0.50 | 0.43 | 0.52 | 0.52 | 0.46 |
| Nb | 4.26 | 4.78 | 4.28 | 4.90 | 4.41 | 4.54 | 4.05 | 5.32 | 5.73 | 5.45 | 4.12 |
| Zr | 131 | 162 | 132 | 160 | 151 | 127 | 96 | 176 | 180 | 124 | 79 |
| Hf | 0.44 | 0.39 | 0.39 | 0.39 | 0.32 | 0.38 | 0.38 | 0.37 | 0.93 | 0.34 | 0.37 |
| Th | 246 | 277 | 245 | 270 | 257 | 257 | 280 | 249 | 308 | 321 | 289 |
| U | 80 | 79 | 80 | 79 | 80 | 80 | 101 | 90 | 96 | 93 | 95 |
| Y | 661 | 657 | 659 | 660 | 671 | 666 | 827 | 737 | 784 | 805 | 764 |
| La | 227 | 231 | 228 | 232 | 227 | 229 | 245 | 227 | 241 | 248 | 236 |
| Ce | 456 | 459 | 457 | 465 | 454 | 463 | 538 | 509 | 551 | 545 | 538 |
| Pr | 82.1 | 81.1 | 82.5 | 81.0 | 78.5 | 82.2 | 93.8 | 85.5 | 91.6 | 93.2 | 90.0 |
| Nd | 393 | 383 | 387 | 392 | 376 | 401 | 448 | 413 | 446 | 453 | 434 |
| Sm | 83.6 | 83.1 | 82.9 | 81.6 | 79.5 | 85.3 | 95.6 | 89.9 | 95.4 | 95.1 | 93.0 |
| Eu | 11.8 | 11.8 | 11.8 | 12.1 | 11.4 | 12.2 | 13.5 | 12.9 | 13.9 | 13.8 | 13.2 |
| Gd | 95.2 | 93.9 | 94.7 | 95.3 | 91.4 | 97.2 | 110 | 101 | 112 | 114 | 109 |
| Tb | 13.9 | 13.7 | 14.1 | 13.8 | 13.5 | 14.4 | 16.0 | 15.0 | 16.7 | 16.5 | 15.6 |
| Dy | 97.8 | 98.1 | 99.4 | 100 | 93.9 | 101 | 111 | 108 | 120 | 115 | 111 |
| Ho | 24.0 | 24.2 | 24.4 | 24.3 | 23.5 | 25.2 | 27.9 | 26.7 | 29.7 | 28.7 | 27.8 |
| Er | 76.1 | 77.1 | 77.7 | 78.3 | 74.6 | 79.2 | 87.2 | 84.8 | 94.3 | 91.6 | 89.3 |
| Tm | 12.1 | 12.4 | 12.5 | 12.4 | 11.7 | 12.7 | 13.9 | 13.6 | 15.1 | 14.7 | 14.2 |
| Yb | 77.7 | 78.0 | 78.9 | 78.3 | 75.4 | 80.7 | 88.6 | 86.1 | 96.6 | 92.8 | 91.7 |
| Lu | 13.2 | 13.6 | 13.7 | 14.1 | 13.0 | 14.1 | 14.9 | 14.8 | 16.9 | 15.9 | 15.7 |
| Cu | 2.38 | 4.10 | 2.83 | 7.60 | 2.61 | 3.38 | 1.51 | 3.00 | 1.27 | 1.90 | 1.53 |
| Zn | n.d. | 7.48 | n.d. | 20.5 | 5.39 | 6.06 | 9.96 | 2.95 | 2.47 | 1.04 | n.d. |
| Mo | 1032 | 1250 | 1062 | 1263 | 1168 | 1133 | 952 | 1288 | 1401 | 1406 | 1182 |
| Ag | 0.36 | 0.34 | 0.33 | 0.39 | 0.44 | 0.33 | 0.35 | 0.50 | 0.36 | 0.43 | 0.24 |
| Tl | | | | | | | | | | | |
| Pb | 1.09 | 3.93 | 2.24 | 6.55 | 44.7 | 3.19 | 0.65 | 1.30 | 0.62 | 0.72 | <0.05 |
| Sn | 62.8 | 28.2 | 68.9 | 30.4 | 30.7 | 40.5 | 171 | 75.3 | 51.8 | 36.1 | 110.1 |
| Sb | 8.52 | 7.52 | 8.43 | 7.79 | 8.54 | 7.73 | 9.86 | 9.20 | 10.04 | 8.78 | 10.06 |
| La/Yb cn | 2.10 | 2.12 | 2.07 | 2.12 | 2.16 | 2.04 | 1.98 | 1.89 | 1.79 | 1.92 | 1.85 |
| La/Sm cn | 1.75 | 1.79 | 1.78 | 1.84 | 1.84 | 1.74 | 1.66 | 1.63 | 1.63 | 1.69 | 1.64 |
| Gd/Yb cn | 1.01 | 1.00 | 0.99 | 1.01 | 1.00 | 1.00 | 1.03 | 0.97 | 0.96 | 1.02 | 0.98 |
| Eu/Eu* | 0.40 | 0.41 | 0.41 | 0.42 | 0.41 | 0.41 | 0.40 | 0.41 | 0.41 | 0.41 | 0.40 |
| Ce/Ce* | 0.82 | 0.82 | 0.82 | 0.83 | 0.83 | 0.83 | 0.87 | 0.90 | 0.91 | 0.88 | 0.90 |
| Total REE | 2771 | 2961 | 2810 | 3017 | 2884 | 2892 | 3049 | 3168 | 3409 | 3392 | 3182 |
| Zr/Hf | 296 | 413 | 340 | 413 | 473 | 332 | 255 | 481 | 193 | 365 | 211 |
| Nb/Ta | 9 | 11 | 11 | 12 | 11 | 10 | 8 | 12 | 11 | 10 | 9 |
| La/Nb | 53.3 | 48.3 | 53.3 | 47.4 | 51.5 | 50.5 | 60.5 | 42.7 | 42.1 | 45.5 | 57.2 |
| Th/Nb | 57.7 | 57.9 | 57.1 | 55.1 | 58.3 | 56.6 | 69.2 | 46.9 | 53.9 | 58.9 | 70.0 |
| Th/La | 1.08 | 1.20 | 1.07 | 1.16 | 1.13 | 1.12 | 1.14 | 1.10 | 1.28 | 1.29 | 1.22 |
| Zr/Y | 0.20 | 0.25 | 0.20 | 0.24 | 0.23 | 0.19 | 0.12 | 0.24 | 0.23 | 0.15 | 0.10 |
| Th/U | 3.07 | 3.51 | 3.06 | 3.41 | 3.23 | 3.22 | 2.78 | 2.78 | 3.21 | 3.45 | 3.05 |

APPENDIX TABLE 1E.—Trace element compositions (ppb) of hot spring waters, south end of Lake Bogoria.

| | MW3 | MW4 | MW2 | KIB1 | KIB2 | LOS1 |
|---|---|---|---|---|---|---|
| Ti | 8.0 | 13.1 | 12.6 | 10.3 | 9.2 | 13.7 |
| P | | | | | | |
| Cr | 4.23 | 6.12 | 6.14 | 10.3 | 11.9 | 3.32 |
| Co | 0.18 | 0.10 | 0.11 | 0.02 | 0.20 | 0.16 |
| Ni | 0.09 | n.d. | n.d. | 0.24 | 0.85 | n.d. |
| Rb | 582 | 590 | 605 | 546 | 573 | 527 |
| Sr | 90 | 91 | 92 | 83 | 82 | 81 |
| Cs | 11.5 | 11.8 | 11.0 | 11.4 | 12.1 | 10.4 |
| Ba | 38.1 | 46.6 | 20.2 | 24.0 | 28.6 | 24.3 |
| Sc | 8.36 | 10.4 | 9.2 | 12.4 | 11.9 | 8.73 |
| V | 2.63 | 3.32 | 4.29 | 3.71 | 3.21 | 5.34 |
| Ta | 0.08 | 0.10 | 0.10 | 0.06 | 0.07 | 0.08 |
| Nb | 0.67 | 1.12 | 1.12 | 0.49 | 0.56 | 0.68 |
| Zr | 7.92 | 10.9 | 8.7 | 3.42 | 25.4 | 5.92 |
| Hf | 0.07 | 0.06 | 0.05 | 0.04 | 0.10 | 0.07 |
| Th | 0.14 | 0.34 | 0.92 | 0.13 | 0.44 | 0.20 |
| U | 7.15 | 8.56 | 11.8 | 2.13 | 5.54 | 11.3 |
| Y | 9.97 | 15.9 | 19.2 | 2.84 | 11.7 | 4.76 |
| La | 0.26 | 0.56 | 0.99 | 0.26 | 0.30 | 0.25 |
| Ce | 0.59 | 1.17 | 2.19 | 0.50 | 0.71 | 0.52 |
| Pr | 0.09 | 0.25 | 0.50 | 0.08 | 0.11 | 0.07 |
| Nd | 0.45 | 1.25 | 2.69 | 0.41 | 0.73 | 0.51 |
| Sm | 0.24 | 0.51 | 1.05 | 0.11 | 0.31 | 0.23 |
| Eu | 0.05 | 0.09 | 0.17 | 0.02 | 0.06 | 0.05 |
| Gd | 0.54 | 0.93 | 1.44 | 0.20 | 0.61 | 0.37 |
| Tb | 0.12 | 0.22 | 0.30 | 0.04 | 0.13 | 0.07 |
| Dy | 1.24 | 1.83 | 2.49 | 0.30 | 1.14 | 0.61 |
| Ho | 0.32 | 0.51 | 0.67 | 0.08 | 0.34 | 0.14 |
| Er | 0.98 | 1.56 | 2.10 | 0.25 | 1.19 | 0.51 |
| Tm | 0.13 | 0.22 | 0.33 | 0.04 | 0.21 | 0.07 |
| Yb | 0.86 | 1.28 | 1.95 | 0.23 | 1.28 | 0.42 |
| Lu | 0.11 | 0.17 | 0.28 | 0.03 | 0.19 | 0.06 |
| Cu | 0.54 | 0.59 | 0.20 | 1.31 | 0.87 | n.d. |
| Zn | n.d. | 3.11 | n.d. | 67.5 | n.d. | 11.7 |
| Mo | 177 | 184 | 202 | 150 | 169 | 157 |
| Ag | 0.02 | 0.08 | 0.01 | 0.03 | 0.07 | 0.00 |
| Tl | | | | | | |
| Pb | 1.92 | 0.80 | 1.06 | 0.67 | 0.73 | 1.36 |
| Sn | 54.0 | 55.9 | 66.7 | 72.7 | 77.0 | 49.6 |
| Sb | 12.9 | 12.7 | 12.0 | 14.9 | 15.3 | 6.2 |
| La/Yb cn | 0.21 | 0.31 | 0.37 | 0.82 | 0.17 | 0.42 |
| La/Sm cn | 0.69 | 0.71 | 0.61 | 1.54 | 0.61 | 0.70 |
| Gd/Yb cn | 0.51 | 0.60 | 0.61 | 0.72 | 0.39 | 0.73 |
| Eu/Eu* | 0.38 | 0.41 | 0.42 | 0.50 | 0.43 | 0.49 |
| Ce/Ce* | 0.95 | 0.77 | 0.75 | 0.82 | 0.95 | 0.94 |
| Total REE | 253 | 268 | 299 | 310 | 270 | 230 |
| Zr/Hf | 112 | 179 | 160 | 80 | 265 | 90 |
| Nb/Ta | 8 | 12 | 12 | 8 | 9 | 9 |
| La/Nb | 0.38 | 0.50 | 0.89 | 0.54 | 0.52 | 0.36 |
| Th/Nb | 0.20 | 0.30 | 0.82 | 0.27 | 0.78 | 0.29 |
| Th/La | 0.53 | 0.61 | 0.92 | 0.50 | 1.50 | 0.80 |
| Zr/Y | 0.79 | 0.68 | 0.45 | 1.20 | 2.16 | 1.24 |
| Th/U | 0.02 | 0.04 | 0.08 | 0.06 | 0.08 | 0.02 |

# SEDIMENTARY PROCESSES AND LITHOFACIES IN LAKE-MARGIN GROUNDWATER-FED WETLANDS IN EAST AFRICA

DANIEL M. DEOCAMPO

*Department of Geological Sciences, Rutgers University, Piscataway, New Jersey 08854-8066, U.S.A.*
*Present address: Department of Mineral Sciences, National Museum of Natural History,*
*Smithsonian Institution, MRC 119, Washington, DC 20560-0119, U.S.A.*
*e-mail: kerolite@earthlink.net*

ABSTRACT: Groundwater-fed wetlands are common features of many lake margins, and they may leave characteristic deposits that record hydrologic conditions at their time of formation. Geomorphic and sedimentologic observations of wetland complexes on the margins of Lakes Eyasi, Makat (Ngorongoro Crater), and Natron, Tanzania, show that clastic, chemical, and biological processes in the wetlands produce discernible lithofacies variation. Five distinct sets of lithofacies are identified: sands and gravelly sands (S and Sg), silty to sandy muds (Ms), bioturbated muds (Mb), alkaline muds (Ma), and organic-rich rootmats, peat, and clay (R, P, Co). Several wetland depositional subenvironments can be defined on the basis of lithofacies associations, including drainage channels (facies S, Sg, Mb, Ms), marshes (R, P, Co, Mb), hippopotamus flats (Co, Mb, Ms), pools (S, Ms, Mb), and alkaline flats (Ma, Co). Sediment supply, aqueous geochemistry, activity of large mammals, and composition and distribution of vegetation are the primary controls on sedimentation. Sediments in these subenvironments are generally thin (< 1 m), but they are thicker where large mammals like *Hippopotamus amphibius* generate deep (> 1.5 m) bioturbation zones. Recognizable lithofacies are found up to ~ 1 km basinward from the sources of spring waters, with down-gradient sediment fining. Very strong evaporatively driven geochemical gradients are seen as well, with increasing salinity and alkalinity downstream and laterally. Wetland sediments are laterally restricted to only a few tens of meters, and they abruptly interfinger with sediments from adjacent fluviolacustrine environments. In all three basins, these Recent deposits overlie a regressive disconformity or hiatus that signifies a drop in lake level, recording the onset of regional aridity during the late Holocene. Such wetland sediments can be important paleohydrological indicators, generally indicating a fluctuation that led to subaerial exposure and colonization of lake flats by wetland vegetation.

## INTRODUCTION

Wetlands associated with groundwater discharge are common features of closed-basin lakes in the East African Rift, especially in drylands where minerotrophic mires develop on exposed mudflats (Hardie et al., 1978). Laterally extensive sheets of fine-grained sediment may accumulate in these settings, although they have not been widely recognized in the sedimentary record, and there are few reports of modern facies (Hamilton and Taylor, 1986; Quade et al., 1995). These deposits typically comprise multiple lithofacies, and are found within larger sedimentary systems. The depositional context in which the wetland sediments form may be perennial lacustrine (Hardie et al., 1978; Renaut et al., 1986), playa (Hay et al., 1986; Quade et al., 1995), or fluvial (Haynes, 1985; Holliday, 1995).

Wetland deposits have been difficult to recognize in the sedimentary record for several reasons. First, wetlands are usually small features relative to their sedimentary basins. Their deposits, therefore, can be expected to be limited in areal extent (Quade et al., 1995). Second, because many biotic, chemical, and hydrologic processes are active in spring systems, many lithologies can form. This complexity is due in part to their close proximity to environments, like alluvial fans or lakes, that may overprint or destroy the sedimentary record of a spring. Third, the fine grain size and organic-rich composition of wetland sediments can be similar to those of lacustrine deposits. Fourth, the familiar image of spring deposits as travertine mounds or siliceous sinters has focused attention on laterally restricted chemical precipitates. Some styles of groundwater discharge do indeed favor chemical precipitation, especially those with thermal ground waters, and substantial evaporative concentration of waters can induce authigenic mineralization even at ambient temperatures. However, a volumetrically more significant component of sedimentation can occur over a much larger area around springs, and in pools, marshes, and other wetlands (Hay et al., 1986; Holliday, 1995; Quade et al., 1995; Deocampo, 1997).

Significant relief in wetland systems is usually found only adjacent to spring discharge that is associated with fault conduits (Haynes, 1985; Quade et al., 1995), mounds formed of spring deposits (e.g., Roberts and Mitchell, 1987), or spring-sapping features (Baker et al., 1990). The preservation of wetland sediments is therefore enhanced in low-gradient lake-margin settings.

In the playa-margin groundwater-fed wetlands of the Great Basin, U.S.A., Quade et al. (1995) recognized four main depositional subenvironments: (1) an upslope alluvial fan or plain, (2) a sand / phreatophyte flat, (3) a wet meadow, and (4) marshes with spring pools. These late Quaternary deposits were useful in reconstructing paleo–water table fluctuations, and demonstrated the importance of groundwater-fed wetlands as a depositional environment (Quade et al., 1995; Quade et al., 1998).

Chemical precipitates can also constitute a significant volume of spring deposits, and these reflect groundwater and surface-water chemistries. Chemical sedimentation is dominant at spring heads (Hardie et al., 1978) and in playa lakes (Rosen, 1994), and also may occur in associated marshes (Hay et al., 1986; Deocampo and Ashley, 1999a). Precipitation at spring heads, typically carbonate travertine and tufa, is perhaps the most diagnostic deposit of a spring, but spring mounds form only where calcium carbonate supersaturation is achieved.

Wetlands associated with ground-water or surface-water discharge are an important component of hydrologically closed basins, such as those in the East African Rift. In these settings, stratigraphic relationships between marginal lacustrine wetland deposits and lake sediments could provide high-resolution, potentially datable evidence for paleoenvironmental and paleo-

Sedimentation in Continental Rifts
SEPM Special Publication No. 73, Copyright © 2002
SEPM (Society for Sedimentary Geology), ISBN 1-56576-082-4, p. 295–308.

climate reconstruction. Description of modern facies in some of these environments is an important step in understanding these dynamic systems.

The purpose of this paper is to determine the major controls on sedimentation in lake-margin groundwater wetlands in closed-basin East African lakes, and to construct a generalized facies model. This is accomplished by characterizing the clastic and chemical lithofacies and major geomorphic processes in several modern lake-margin spring-fed wetlands in the basins of northern Tanzania.

## GEOLOGIC SETTING

The study was carried out in the Eyasi, Ngorongoro Crater, and Natron basins in northern Tanzania (Fig. 1). This part of the Gregory Rift is dominated by large-scale extensional block faulting that began about 30 Ma, with most of the basin subsidence occurring in the last 4 Ma (Baker, 1986). Volcanic rocks in the region range from basaltic and trachytic lavas to trachyandesitic volcaniclastics and, more recently, carbonatitic ashes from Oldoinyo Lengai (Hay, 1976; Williams, 1978; Dawson et al., 1994). The basins of the region were produced by a combination of rift-controlled subsidence and basin segmentation due to spatially distributed volcanic centers, such as the Ngorongoro Volcanic Highlands (Fig. 1; Grove, 1986).

Lake Eyasi lies in the graben basin on the east side of the Hamwawa Escarpment in northern Tanzania, southwest of the Ngorongoro Volcanic Highlands. Input into the lake is dominated by fluvial systems draining the Serengeti Plains to the northwest and the uplifted metamorphic highlands to the southeast. The northeast lake margin abuts the fluvial drainage eroding Oldeani Mountain, a deeply incised Neogene trachyandesitic volcano. Ephemeral streams, sheetwash, and rare debris flows carry poorly sorted conglomeratic material downslope. The eastern lake margin is dominated by deltaic sands from the southern part of the basin that have been redistributed along the lake shore by littoral processes.

Ngorongoro Crater occupies the caldera depression of a giant trachytic volcano. Since the caldera collapsed approximately 2 million years ago, the internally drained basin, ~ 15 km in diameter, has contained a lake that has varied in size and chemistry since at least the late Pleistocene (Pickering, 1965; Gromme et al., 1970; Hay, 1976; Deocampo, 1997). Most of the caldera floor is covered with Pleistocene lacustrine deposits, ranging from zeolitic claystones to massive silty beds of ostracodal sand. There are many active springs on the basin floor, and several springs support extensive wetlands. Most of these are adjacent to the caldera wall, although some discharge nearer the center of the basin. In addition to feeding large wetlands, these zones of groundwater discharge support tens of thousands of large mammals during the dry seasons (Estes and Small, 1981; Deocampo et al., 1998). This stands in contrast to the Eyasi and Natron Basins, which have fewer mammals because of human interference, and essentially no water-loving species such as elephant, hippopotamus, and buffalo.

The Natron (Magadi) basin sits in the deepest part of the Gregory Rift (Fig. 1) and has been occupied by lakes of varying size throughout the Plio-Pleistocene (Baker, 1963; Isaac, 1977; Baker and Mitchell, 1976). The western flank of Lake Natron is bordered by the Nguruman Escarpment, which exposes Plio-Pleistocene volcaniclastic and lacustrine deposits, and the southern and eastern margins are adjacent to the distal volcaniclastic slopes of Oldoinyo Lengai and Gelai volcanoes. Clastic sediment input to the lake system is dominated by the alluvial plains at the footslopes of Oldoinyo Lengai and Gelai and by perennial fluvial input from the Engaresero and Moinik rivers, which incise the Rift escarpment (Vincens and Casanova, 1987).

## METHODS

Geomorphic features, surface water distribution, and sampling points were mapped by a combination of electronic total station, differentially corrected global positioning system, and tape and compass. Surface sediments were collected by trowel and shallow hand-dug pits, and selected locations were probed by core auger to examine and sample sediment up to a depth of ~ 2 m. Grain-size distribution was estimated by wet-sieve and pipette methods, and organic carbon content was estimated by loss on ignition (LOI, Lewis and McConchie, 1994). Bulk mineralogy of selected samples was determined by x-ray diffraction of randomly oriented powders (Moore and Reynolds, 1997) and for coarse sediments, petrography of resin-impregnated samples (Murphy, 1986). Water samples were analyzed *in situ* for temperature, pH, and conductivity by electronic meters, and carbonate alkalinity was estimated by field titration. Samples were drawn through a 0.45 μm filter in the field and analyzed by atomic absorption spectroscopy, direct-current plasma atomic emission spectroscopy, and ion chromatography. C and O stable-isotope analyses of selected carbonates were determined on an Optima

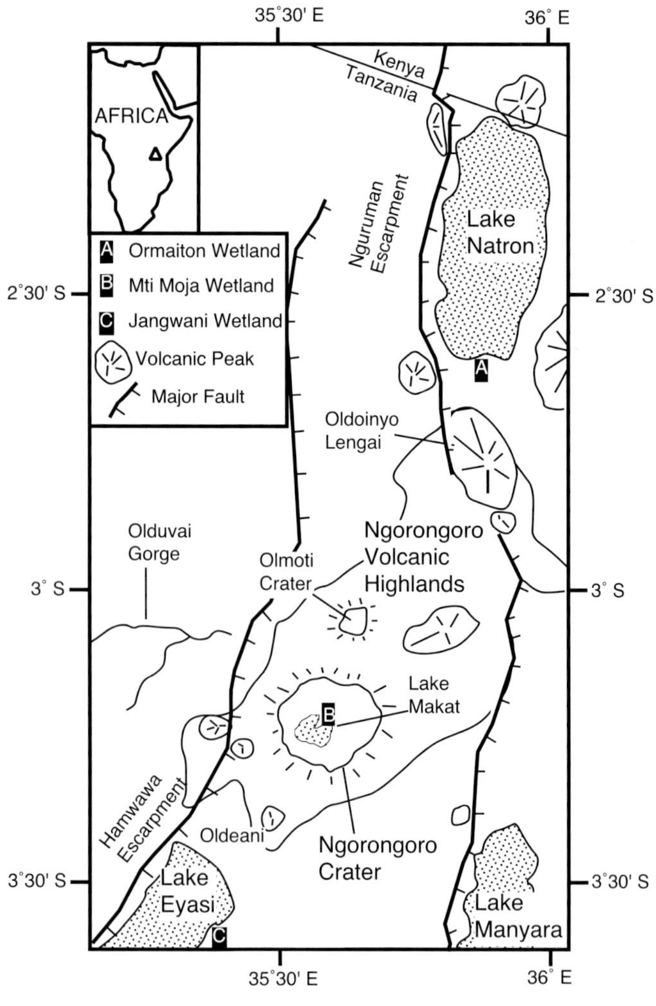


FIG. 1.—Location of the field area, with study sites indicated.

mass spectrometer, and are reported versus V-PDB (Coplen et al., 1983).

## RESULTS

### *Ormaiton, Lake Natron Basin*

Groundwater erupts at Ormaiton through two outcrops of inclined Plio-Pleistocene coarse-grained tuffs and conglomeratic, reworked volcaniclastics that overlie a folded vesicular lava (Fig. 2A–C). These waters appear to be discharging from a long flowpath, judging by the high total dissolved solids and depletion of magnesium and calcium (Table 1). Marshes are developed around the two major point sources of discharge, and cover the area between the two outcrops (Fig. 3). Drainage from these wetlands has incised the adjacent lake-margin deposits (i.e., recent sands and gravels), isolating the marsh from the distal alluvial plain to the east and delivering coarse material to the lake flat. The Plio-Pleistocene outcrops are partially covered by Late Pleistocene to Recent accumulations of coarse-grained reworked volcaniclastics that originate in the distal alluvial plain between the southern shore of Lake Natron and Oldoinyo Lengai (Fig. 1). These gravelly sands, which are constrained on the west by the Ormaiton outcrops and on the east by additional lava outcrops, were water saturated at a depth of < 0.5 m, with local saturation at the surface indicated by < 1-m-diameter patches of sedge growth. The elevated water table in the area implies that diffuse seepage may contribute to the wetland, although most of the water in the drainage channels appears to originate from the point sources of discharge next to the rock outcrops.

Clastic sediments at Ormaiton consist of (1) pebbly sands carried into the wetland environment from upslope by alluvial processes; (2) clasts of basaltic lava and coarse-grained tuff from the adjacent outcrops; and (3) reworked lacustrine muds. Marsh deposits are restricted to the area between the two main discharge points, and they consist of up to 110 cm of rootmat and fibrous peat (> 70% LOI, $n = 6$) composed of a mixture of *Cyperus immensus* and other unidentified sedges. Sediment to the east of the marsh is much coarser, with negligable organic content, including material both on the sand flats to the east and in the drainage channel along the edge of the marsh, which prevents coarse material from entering the wetland (Fig. 3). Clastic sediment in the wetland is thus much finer than that of the surrounding alluvial environments, despite its close proximity to sources of coarse sediment.

Although the waters discharging are not supersaturated with respect to any mineral phases that are known to precipitate in the area (cf. Jones et al., 1977), $CO_2$ degassing and evaporative concentration bring the spring waters to supersaturation with

FIG. 2.—**A)** View south toward Ormaiton spring, point B in Fig. 3. Groundwater discharge at the base of the outcrops feeds the marsh and drainage channel. A small herd of white goats can be seen at the eastern (right) edge of the contact between the marsh and rock outcrop. **B)** View north from the Ormaiton spring, where discharge is eroding coarse materal and carrying it as bedload in the drainage channel. Thick peat is accumulating on the far side of the drainage channel. **C)** View from the northern edge of Ormaiton marsh towards Lake Natron, following the heavy 1998 rains. Mudflats beyond the edge of the marsh are encrusted with trona and calcite.

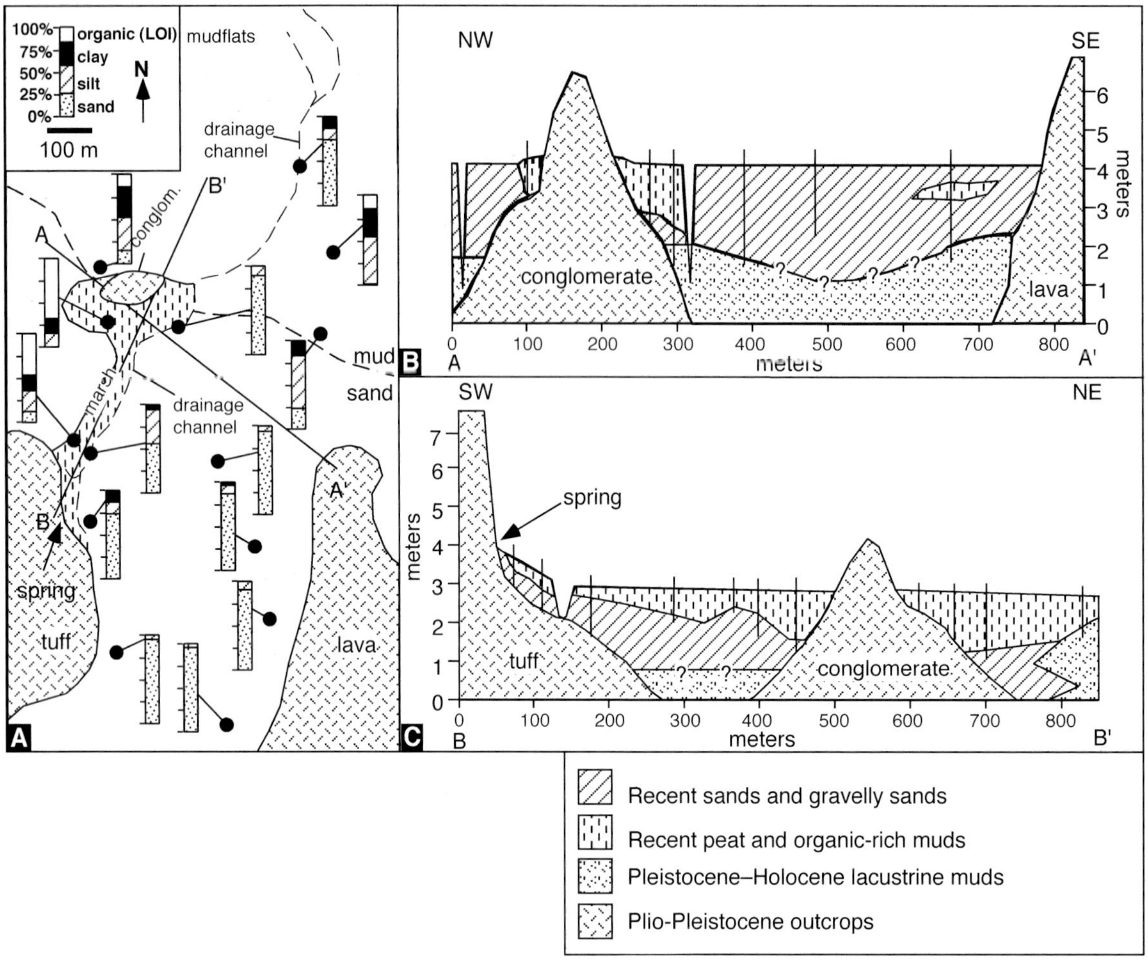


FIG. 3.—**A)** Major surface features at Ormaiton groundwater wetland, Lake Natron basin, with selected surface-sample grain size and loss on ignition (LOI) analyses. **B)** Cross section along the A–A' transect, with core-auger probes indicated by vertical lines. **C)** Cross section along the B–B' transect, with core-auger probes indicated.

respect to magnesium silicates in distal areas of the wetland (Tables 1, 2). Calcium levels, however, remain below detection limits, implying that only trivial amounts of calcite can form at the surface. Concentrated brines are commonly found in evaporating puddles, and efflorescent crusts covering distal mudflats are composed of halite and trona ($Na_3(HCO_3)(CO_3)\bullet 2H_2O$). Some lakeflat muds contain up to 20 weight % fine sand composed of euhedral crystals of calcite, of the same morphology as those reported by Hay (1976) in the Pliestocene lake sediments of the Olduvai Basin, but finer grain size. Although the origin of these crystals remains unclear, they are likely reworked diagenetic alteration products of lacustrine evaporite minerals such as gaylussite (Hay, 1976; Cerling and Hay, 1986). Relatively light $\delta^{18}O$ values (average $\delta^{18}O$ = -3.68‰ PDB; s.d. = 0.74; $n$ = 4) and $\delta^{13}C$ (average $\delta^{13}C$ = 0.05‰; s.d. = 0.24; $n$ = 4), however, suggest that they do not record a strictly evaporitic, sublacustrine diagenetic history.

Shallow auger-core holes and hand-dug soil pits show that the recent wetland sediments range from < 20 cm to > 100 cm in thickness, and are limited to the western part of the channel, adjacent to groundwater discharge (Fig. 3B). These deposits appear to overlie lacustrine clays and alluvial sands that are intercalated at the lake edge.

## *Mti Moja, Ngorongoro Crater*

The Mti Moja wetland is fed by a point source of groundwater adjacent to a small rise (3–4 m) on a peninsula on the northeast lake margin of the crater lake Makat (Fig. 1). Fresh groundwater erupts at the base of an eroding outcrop of carbonate that has been interpreted as a Pleistocene lacustrine carbonate deposit (Deocampo, 1997) that underlies the nearby section described by Hay (1976), and flows toward the northwest (Figs. 4, 5A–E). The resistant carbonate outcrop appears to have locally halted headward erosion along the lakeshore, because Holocene erosion has eroded the adjacent cliffs of weakly cemented lacustrine silts and clays back > 1 km.

Clastic sediments are composed of reworked Pleistocene lake clays and sandy silts, volcanic ash, and material derived from the adjacent eroding Pleistocene lacustrine deposits. Marsh deposits, found within 200 m of the discharge point, are typically silty clay with somewhat high organic content (15–20% LOI; $n$ = 3). The vegetation in this marsh was dominated by *Cyperus immensus*, but after the heavy rains of 1998, which inundated the wetland with freshwater, a stand of *Typha* grew, suggesting that a rootstock had been dormant. Beyond 200 m, the deposits are dominated by tuffaceous sandy muds, mostly

TABLE 1.—Selected chemical analyses of water samples.

| samp no. | Locality | $SiO_2$ (mM) | $Na^+$ (mM) | $K^+$ (mM) | $Ca^{2+}$ (mM) | $Mg^{2+}$ (mM) | $Cl^-$ (mM) | $SO_4^{2-}$ (mM) | pH | conductivity (μS cm$^{-1}$) | temp (°C) | alkalinity[a] (meq/L) | log $PCO_2$ |
|---|---|---|---|---|---|---|---|---|---|---|---|---|---|
| ***Lake Natron*** | | | | | | | | | | | | | |
| 99-46 | Ormaiton Proximal | 0.36 | 57.42 | 0.16 | n.d. | n.d. | 7.77 | 3.71 | 9.51 | 5020 | 26 | 35.17 | -3.59 |
| 99-48 | Ormaiton Distal | 0.88 | 133.10 | 0.38 | n.d. | 0.04 | 31.03 | 5.41 | 9.86 | 10,300 | 22 | 55.96 | -3.85 |
| ***Ngorongoro Crater*** | | | | | | | | | | | | | |
| 96-3 | Mti Moja Proximal | 0.96 | 2.80 | 0.39 | 0.15 | 0.28 | 0.63 | 0.03 | 8.00 | 550 | 28 | 4.80 | -2.82 |
| 96-13 | Mti Moja Distal | 1.08 | 13.35 | 2.33 | 0.10 | 0.02 | 2.65 | 0.63 | 9.37 | 3000 | 22 | 15.99 | -3.56 |
| ***Lake Eyasi*** | | | | | | | | | | | | | |
| 99-13 | Jangwani Proximal | 0.55 | 45.32 | n.d. | 0.03 | n.d. | 16.81 | 0.92 | 9.02 | 4540 | 36 | 22.38 | -3.22 |
| 99-16 | Jangwani Distal | 0.17 | 137.02 | 0.02 | 0.05 | 0.68 | 63.75 | 3.04 | 9.52 | 11,100 | 20 | 43.96 | -3.50 |

a Alkalinity determined by titration in the field.
n.d. Not detected.

reworked from the underlying Pleistocene lake beds. Winnowing of mud has produced a downstream fining in the spring drainage, and fine to medium sands in the channel locally have centimeter-scale ripple marks.

Large mammals play an important geomorphic role in the wetland system, especially *Hippopotamus amphibius* (Deocampo and Ashley, 1997; Ashley et al., 2000). In addition to producing a > 1.5 m deep bioturbation zone in the muds, they control the surface hydrology on the mudflats because water flow tends to follow the 1–2 m wide, > 1 m deep trails plowed as the hippos move across and through the soft sediment substrate. They also add a considerable amount of fecal organic matter to the sediment, largely in the form of partially decayed vegetation. This is evident in the large difference in LOI between lakeflat clays (LOI = 5–10%; $n = 4$) and clays associated with hippo pools and trails (LOI = 10–20%; $n = 4$), even though these environments are nearly barren of vegetation, except for minor *Cyperus laevigatus*.

The waters erupting at Mti Moja are undersaturated with respect to calcite (Tables 1, 2), and no significant mineral precipitation was observed at the discharge site. Efflorescent crusts of trona and halite, and pore-filling micritic calcite, are abundant on the distal mudflats > 250 m away from the point of discharge. Sand-size euhedral calcite grains are also found in these lake-margin mudflats. They are somewhat coarser grained than the Ormaiton calcites at Lake Natron and are apparently reworked from the underlying Pleistocene lake beds. They have very high $\delta^{18}O$ values (average $\delta^{18}O$ = 2.14‰ PDB; s.d. = 0.3; $n = 4$) and $\delta^{13}C$ (average $\delta^{13}C$ = 4.74‰; s.d. = 0.26; $n = 4$), more clearly reflecting evaporitic, sublacustrine conditions than those from Natron, and possibly a methanogenic origin for the high $\delta^{13}C$ (Cerling and Hay, 1986).

Auguring shows that Recent sediments are generally thin (Fig. 4). They range from the ~ 10 cm muddy sands that overlie a resistant ledge of indurated claystone near the springhead to > 1.5 m of bioturbated, organic-rich muds associated with hippopotamus activity (Deocampo and Ashley, 1997). Beyond ~ 600 m from the springhead, no lithologic distinction can be made between wetland-related sediment and modern lake-flat sediments.

### *Jangwani, Lake Eyasi Basin*

The Jangwani spring wetlands are fed by a series of seeps discharging slightly warm water (36°C) into the lake margin. Groundwater discharges from unconsolidated late Pleistecene to Recent lake-margin sands that form shoreface dunes. These waters are likely from a deep thermal source, perhaps related to nearby basin-bounding deep faults. Standing water and marsh vegetation accumulate in the depression behind the dunes. Rootmat and peat over 80 cm in thickness are found overlying lake-margin sands.

The Jangwani wetland is surrounded by deposits of well-sorted medium- to coarse-grained quartz sand that form shoreface dunes on the lake (east) side of the wetland (Fig. 6). These deposits appear to represent lacustrine reworking of deltaic sands that enter the lake several kilometers south of Jangwani.

TABLE 2.—Mineral saturation indices for water samples calculated with the NETPATH thermodynamic database (Plummer et al., 1994).

| Sample No. | Sample Name | Calcite Saturation | Kerolite Saturation | Amorphous Silica Saturation |
|---|---|---|---|---|
| 99-46 | Ormaiton Proximal | –0.267 | –1.317 | –0.940 |
| 99-48 | Ormaiton Distal | –0.201 | 4.099 | –0.659 |
| 96-3 | Mti Moja Proximal | –0.156 | –1.014 | –0.337 |
| 96-13 | Mti Moja Distal | 1.120 | 5.831 | –0.891 |
| 99-13 | Jangwani Proximal | 0.319 | –1.641 | –0.732 |
| 99-16 | Jangwani Distal | 0.699 | 3.961 | –1.185 |

Positive indices indicate supersaturation, negative indices indicate undersaturation.

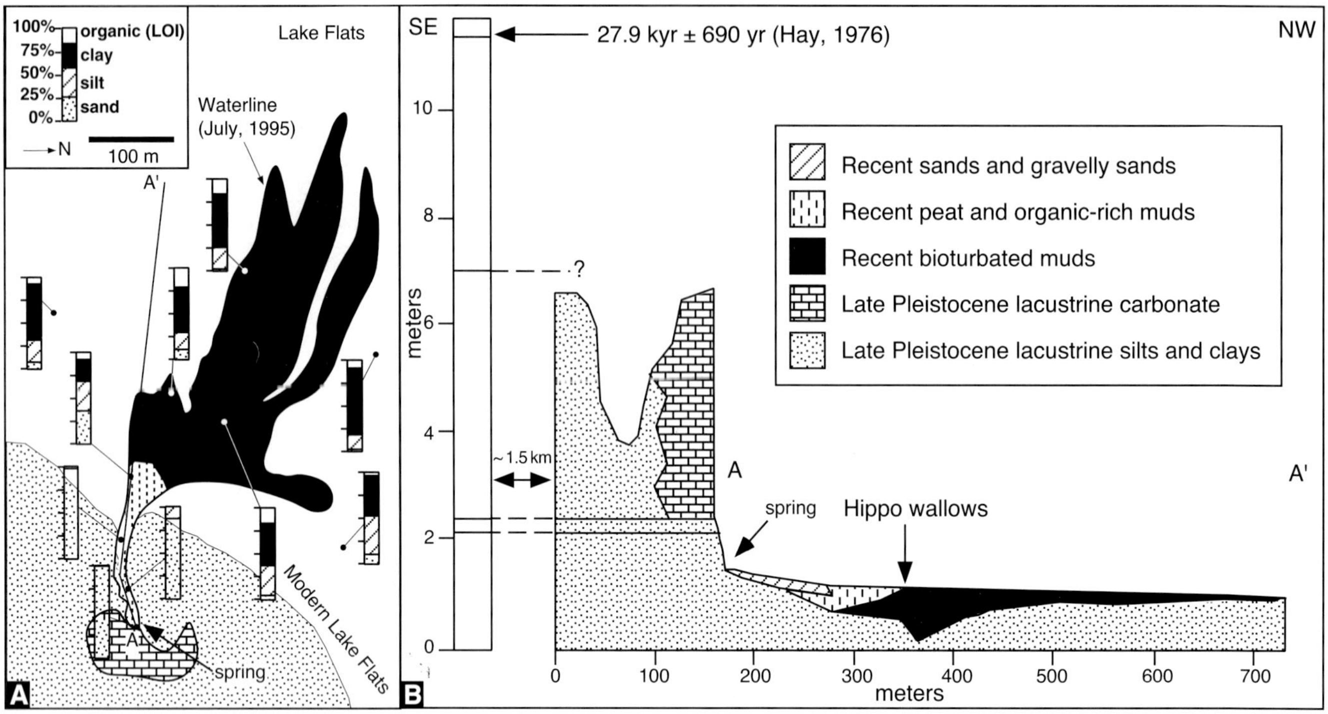


FIG. 4.—**A)** Major surface features at the Mti Moja groundwater wetland, Lake Makat, Ngorongoro Crater, with selected surface-sample grain size and loss on ignition (LOI) analyses. **B)** Cross section along the A–A' transect, with correlation to the section reported by Hay (1976) ~ 1.5 km southeast of Mti Moja. The date of 27.9 kyr is an uncalibrated radiocarbon date on several ostracode tests (Hay, 1976).

These quartz sands are the only significant detrital source for the wetland, and the sediment in the center of the marsh is nearly all organics, mostly a heavy fibrous unidentified sedge (> 90% LOI, $n = 7$).

Upon $CO_2$ degassing at the source, erupting Jangwani waters are supersaturated with respect to calcite (Tables 1, 2). Some minor encrustation of vegetation and the substrate by calcite was observed at one spring orifice. Evaporative concentration of these waters maintains supersaturation with respect to calcite and magnesium silicates, but biogenic $CO_2$ in the marsh buffers pH, apparently preventing carbonate precipitation (Deocampo and Ashley, 1999a). Mineral precipitation is instead concentrated in peripheral areas of the marsh, especially where the marsh drains onto the Eyasi lake flat, where degassing continues, and efflorescent crusts of trona and halite, and pore-filling calcite, are precipitated.

Sediments associated with the Jangwani groundwater discharge are limited to ~ 1 m of peaty organic deposits that overlie the lake-margin sands. A small amount of detrital material has been washed or blown into the marsh from around the periphery, but the deposits are otherwise dominated by organic material. A single AMS radiocarbon date, determined on a plant fragment from the base of the peat, gave a modern age.

## DISCUSSION

### *Clastic Facies*

Although the local source lithologies are a major control on clastic grain sizes, five sets of lithofacies can be readily distinguished in the wetland environments studied (Table 3): (1) Sands and gravelly sands (S and Sg) are associated with flowing drainage from the wetlands, and so usually form a ribbon-like geometry. They are generally thin, up to about 10 cm thick, and constitute the coarsest material in these depositional systems. Muds are common in areas where the water is not flowing, and these occur as three facies: (2) silty to sandy muds (Ms) representing reworked lake-margin or drainage-channel muds; (3) bioturbated muds (Mb), which are largely produced through bioturbation, aeration, and admixture of decaying organic matter and plant fragments by hippopotamus motion up to 1.5 m into the substrate, and include deep (> 30 cm) vertebrate footprints and hippopotamus trails; and (4) alkaline muds (Ma), which contain alkaline earth carbonates and evaporites like trona and halite, as well as shallow vertebrate footprints. The muds can occur in various geometries, ranging from a ribbon geometry, as in a drainage channel, to a sheet geometry, as in expansive mudflats. (5) The last group are deposits dominated by decaying vegetation, which include rootmats (R, LOI > 85%) to peat (P, LOI > 45%) and organic-rich clay (Co, LOI > 10%). These organic-rich deposits can be thick, > 100 cm thick, and they may form sheets that reflect the geometry of the marsh that produces them, either elongate along a channel margin or laterally extensive as on mudflats.

Deposits in wetlands dominated by clastic sediments are nearly devoid of sedimentary structures because of the extreme bioturbation in all depositional subenvironments. Structures that may be preserved are ripple-marked sands and channeling in the drainage channels, vertebrate tracks and trails in the muds, and desiccation cracks in the clayey lithofacies.

FIG. 5.—**A)** View northwest from the top of the carbonate outcrop at the Mti Moja wetland, Ngorongoro Crater (Fig. 4). Water erupts at the base of the carbonate (out of view) and feeds a small linear stream and marsh before spreading out onto the mudflats of Lake Makat. **B)** Sediment profile sampled in marsh about 20 m downstream of the spring discharge site. **C)** Occupied hippo pool, about 200 m downstream of the spring discharge site. **D)** Dried hippo flats, with the white carbonate outcrop visible about 350 m distant, behind and to the left of the vehicle. The linear features visible in the organic-rich muds are trails created by hippo movement. **E)** Alkaline flats at the distal end of the groundwater-fed wetland system. Waters in this environment are highly saline and alkaline, promoting precipitation of carbonates and other minerals.

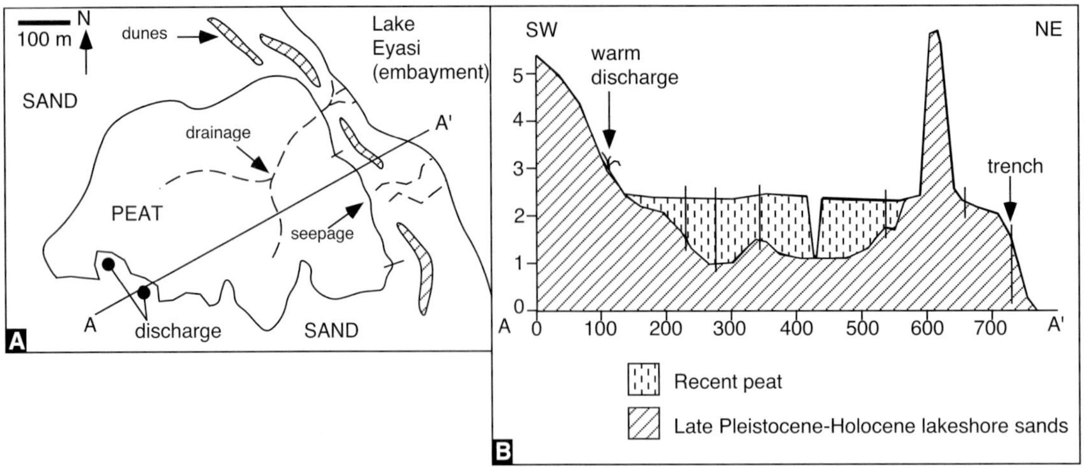


FIG. 6.—**A)** Major surface features at Jangwani groundwater wetland, Lake Eyasi basin. **B)** Cross section along the A–A' transect, with core-auger probes indicated by vertical lines.

## *Chemical Sedimentation*

Although aqueous analyses show solute losses, and therefore mineral precipitation, in these environments, precipitates are volumetrically insignificant compared to the clastic components of the sediment. Moreover, lacustrine minerals like evaporitic calcite crystals appear to be derived from older deposits and be reworked into the wetland sediments. It seems inappropriate, therefore, to assign chemifacies designations to the various deposits. Several important processes appear to affect the chemical sedimentary record, however, which may be discernible if reworking and diagenesis have been limited.

The wetlands studied are all fed by near-ambient-temperature ground waters, and so mineral precipitation due to thermal processes are not observed. Such precipitation of carbonate or silica under high-temperature conditions would provide an added aspect of chemical sedimentation that indicates a zone of groundwater discharge (Renaut et al., 1986; Renaut and Tiercelin, 1994; Renaut and Jones, 1997). Spring waters with high temperatures and total dissolved solids would also act to discourage colonization by hippos, possibly producing thinner wetland sediments as a result.

In these dryland environments, marshes are likely to develop in freshwater settings that are surrounded by evaporative, saline, alkaline environments. Therefore, the fresh to saline, alkaline geochemical gradients that are observed in modern lake-margin wetlands seem to be characteristic features of these environments (Figs. 2C, 5A). In all the wetland settings described here, such gradients produced a zonation in precipitation of calcite and magnesium silicates, with increased precipitation with distance from the freshwater source (Deocampo and Ashley, 1999a; Deocampo et al., 2002). Geochemical gradients may be difficult to detect in the sedimentary record, particularly where sediment reworking and diagenesis obscure the geochemical signature. In low-gradient environments, however, it may be possible to constrain these complexities.

Second, the marshes in the Jangwani and Ormaiton wetlands are large and productive enough to maintain elevated levels of $PCO_2$, which have been shown to suppress pore-water pH and precipitation of calcite and magnesium silicates (Deocampo and Ashley, 1999a, 1999b; Deocampo et al., 2000). The centers of these wetlands, therefore, are characterized by enhanced production of amorphous silica by diatoms (frustules) and marsh vegetation (phytoliths), with calcite and magnesium silicates precipitating in evaporative mudflats or in peripheral areas. This pattern may be obscured by seasonal (or longer-term) drying and oxidation, but it may be preserved if the wetland remains a long-term feature of the landscape.

## *Depositional Subenvironments*

Several depositional subenvironments can be defined on the basis of facies associations (Table 4). The coarsest-grained deposits in lake-margin wetlands are likely to be found in the *channels* (facies S, Sg, Mb, Ms) that drain these wetlands, if a source of clastic sediment is available (Figs. 2A, 5A). In the three study sites described, wetland drainage channels incise the substrate by > 1 m, and can carry a high sediment load. The composition of the sands and gravel reflects the composition of the locally eroding outcrops, like the Pleistocene lacustrine beds exposed next to Mti Moja, Ngorongoro (Fig. 4), or the Plio-Pleistocene volcanics at Ormaiton, Lake Natron (Fig. 3). In such a setting, much of the coarse sediment is derived from outcrop lateral to the drainage. Marsh vegetation growing on the fringes of the channel may also become reworked into the sediment, providing a source of organic-rich facies in the deposits.

*Marsh* development in these wetlands is represented by accumulation of organic-rich sediment (Figs. 2B, 5B; facies R, P, Co, Mb). Coarse-grained sediment is generally excluded from these subenvironments, because vegetation limits the competence of sheetwash and floodwaters. This is seen clearly by the very fine grain size of marsh sediment in the Ormaiton and Jangwani wetlands, despite their location within meters of coarse-grained sources. This effect of vegetation on grain size is similar to that observed for wetlands of similar spatial scale in distal alluvial fan settings of the Great Basin, U.S.A. (Quade et al., 1995).

The *hippopotamus flat* (facies Co, Mb, Ms) is distinguished by the dominant geomorphic role of *Hippopotamus amphibius* (Fig. 5C–D). Sedimentation in this subenvironment is controlled by hippos in two ways. First, they actively control the movement of surface and groundwater by plowing trails with their bodies, dramatically changing the local surface-water drainage daily and over seasonal timescales (Deocampo and Ashley, 1997; McCarthy et al., 1998). By modifying the flow of surface water, hippos potentially control the distribution of clastic material carried by the water and minerals precipitated as the spring waters evapo-

TABLE 3.—Wetland lithofacies identified in this study, with associated characteristics.

| Facies Code | Facies | LOI | GS | Mineral Precipitates | Sedimentary Structures | Geometry | Basal contact |
|---|---|---|---|---|---|---|---|
| **Sg** | Gravelly sands | < 5% | sand and gravel | none | minor ripples | ribbon | sharp, erosional |
| **S** | Sand | < 5% | sand | none | minor ripples | ribbon | sharp, erosional |
| **Ms** | Sandy muds | 5–10% | sandy mud | none | desiccation cracks, mammal tracks | ribbon to sheet | sharp to gradational |
| **Mb** | Bioturbated muds | 10–20% | silty to sandy clay | biogenic amorphous silica | desiccation cracks, mammal tracks and trails, esp. hippo | elongate sheet | gradational |
| **Ma** | Alkaline muds | 5–10% | silty to sandy clay | calcite, trona, halite | desiccation cracks, animal traces | sheet | gradational |
| **R** | Rootmat | > 90% | minor clastics, clay | biogenic amorphous silica | massive | ribbon to sheet | sharp |
| **P** | Peat | 35–90% | silty to sandy clay | biogenic amorphous silica | massive | ribbon to sheet | gradational |
| **Co** | Organic-rich clay | 15–35% | silty to sandy clay | biogenic amorphous silica | massive | ribbon to sheet | gradational |

rate. Second, they create a bioturbation zone > 1.5 m thick that reworks the underlying substrate, significantly increasing the organic carbon content. In effect, this creates a basal unconformity at the bottom of the mixing zone.

*Pools* (facies Sd, Ms, Mb) in the wetlands can range in size from meter-scale puddles to water bodies several tens of meters across. Pools often are created and maintained by groups of hippopotami (Fig. 5C), and are transitional environments between the drainage channels that feed the pools and the more distal hippopotamus flats.

In the very low-gradient mudflats distal to freshwater discharge, evaporation produces high solute concentrations, and authigenic mineral precipitation is widespread. Because of the high salinity and alkalinity, vegetation coverage is very low to absent, and animal activity is limited to animals in transit, often to or from the wetland sources of potable water (Estes and Small, 1981; Deocampo et al., 1998). These *alkaline flats* (Fig. 5E; facies Ma, minor Co) are characterized by fine-grained sedimentation, presumably during flood stages or as eolian input, followed by desiccation and mineral precipitation of efflorescent crusts and pore-filling precipitates of halite, trona, and calcite.

The remaining areas surrounding these depositional subenvironments can be grouped together informally as *ephemeral wetlands* (Deocampo, 1997), representing the seasonal nature of their water saturation. These settings are commonly components of the larger depositional environment within which the wetland exists, as, for example, the lacustrine mudflats of Mti Moja, Ngorongoro or the distal alluvial fan channels of Ormaiton, Lake Natron. Therefore, the lithofacies associated with these settings depends entirely on the surrounding depositional environment.

## *Composite Facies Model*

On the basis of the three examples of lake-margin wetlands, a schematic facies model can be constructed for sedimentation in marginal wetlands of arid or semiarid, closed-basin lakes (Fig. 7). An important aspect of such systems is that a regressive disconformity overlies lacustrine sediments, representing erosion or hiatus due to exposure of the land surface, which allows subsequent colonization by wetland vegetation. In the Jangwani and Ormaiton systems, this disconformity is then overlain by coarse-grained deposits recording the lacustrine regression. At Jangwani, the erosion and sands are attributable to littoral longshore currents and shoreface erosion, whereas at Ormaiton, these features

TABLE 4.—Wetland depositional subenvironments with lithofacies associations.

| SUBENVIRONMENT | LITHOFACIES |
|---|---|
| Drainage channel | Sg, S, Mb, Ms |
| Pool | S, Ms, Mb |
| Marsh | R, P, Co, Mb |
| Hippopotamus flats | Co, Mb, Ms |
| Alkaline flats | Ma, Co |

are associated with alluvial processes. In contrast, coarse-grained material is found overlying the basal disconformity at Mti Moja only in proximal reaches of the drainage channel, and muds overlie the disconformity elsewhere (at the base of the hippo bioturbation zone). This appears to be because the primary geomorphic agent responsible for the disconformity is not related to flowing water, but instead is due to the wallowing of hippos. Channel sands can then predicted to be interbedded with facies from the other wetland subenvironments. Because these are low-gradient environments where sediment is continually reworked, any package of sediment preserved by a lacustrine transgression would record only conditions immediately prior to flooding by the lake.

Although the wetland sediments are produced by a complex depositional system, a predictable sedimentary geometry may be produced. As exemplified by the Ormaiton (Fig. 3) and Mti Moja (Fig. 4) systems, wetland sediments may be as much as an order of magnitude longer in the down-gradient axis (hundreds of meters) than along-gradient (meters to tens of meters). Lateral facies changes along-gradient tend to be abrupt, such as alluvial sands deposited adjacent to marsh sediments at Ormaiton, or

Lake-flat muds adjacent to hippo pools at Mti Moja. Down-gradient facies changes are more gradual, however, because of the combined effects of downstream fining and a more gentle ecotone, such as that in the gradient from drainage channel to hippo flats to alkaline flats at Mti Moja.

## *Controls on Sedimentation*

The major controls on sedimentation in lake-margin wetlands are summarized by the flowchart in Figure 8. Tectonic setting provides two critical controls on the system: (1) the steep regional topographic gradients common in rift-valley systems that influence sediment supply and incision rates; and (2) volcanism, which is the primary control on water chemistry throughout the region. The impacts of climate are seen (1) geochemically, through the evaporative concentration of surface waters and the relative proportions of hydrothermal versus meteoric waters in the wetlands (Renaut and Tiercelin, 1994); and (2) biologically, through controls on the composition and distribution of vegetation. The aqueous geochemistry of the lake-margin wetlands is a critical component of the system not only in terms of mineral precipitation and diagenesis but also because it controls the distribution of biota, especially *Hippopotamus amphibius* (Deocampo et al., 1998).

Bioturbation and organic enrichment of the sediment by large mammals can be an important process in natural wetlands, and it appears to be a major difference between the Ngorongoro wetlands and those in Natron and Eyasi. In effect, the large mammals can increase the thickness of sediment associated with the groundwater wetland at Ngorongoro through their interaction with the substrate. The more intense hippo activity at Mti Moja may be due to the far fresher waters available there as opposed to the more saline and alkaline Ormaiton and Jangwani wetlands (Table 1). Alternatively, human activities may have

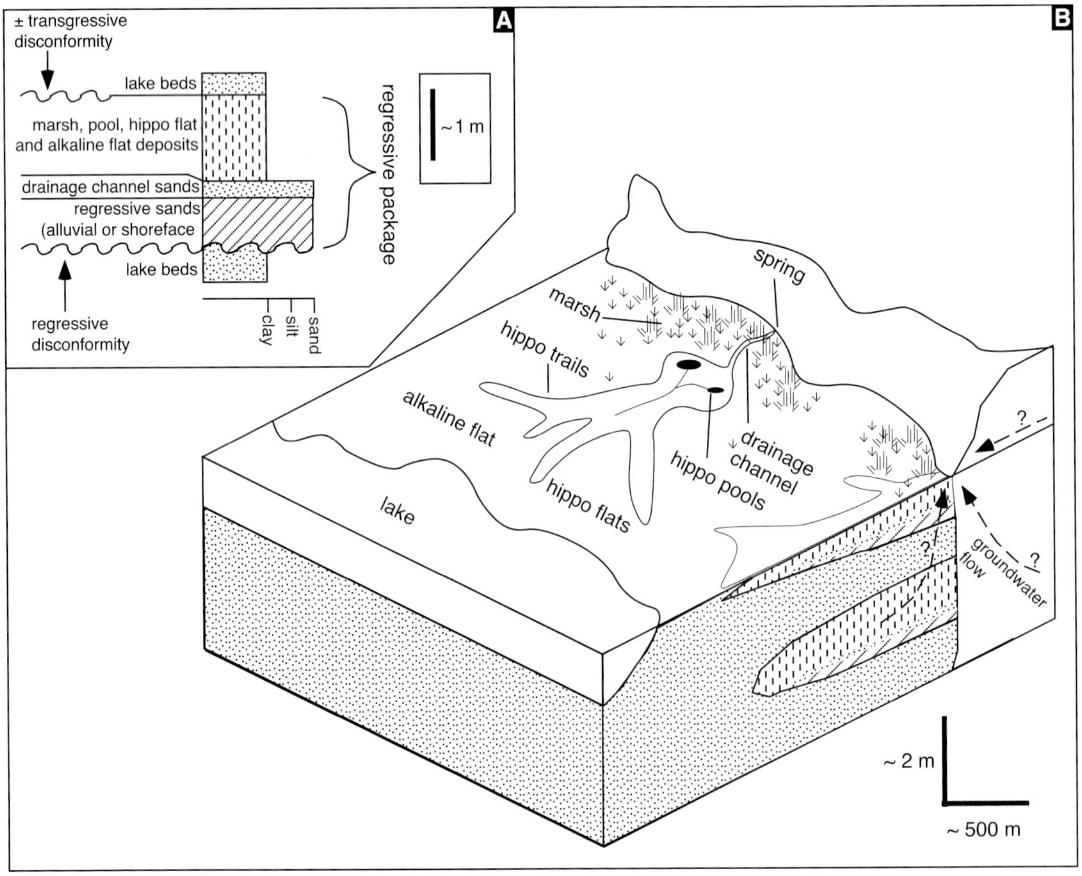


FIG. 7.—**A)** Schematic stratigraphic facies model based on observations of modern wetland sedimentation. **B)** Block model of a lake-margin groundwater-fed wetland system and schematic stratigraphy under conditions of fluctuating lake level.

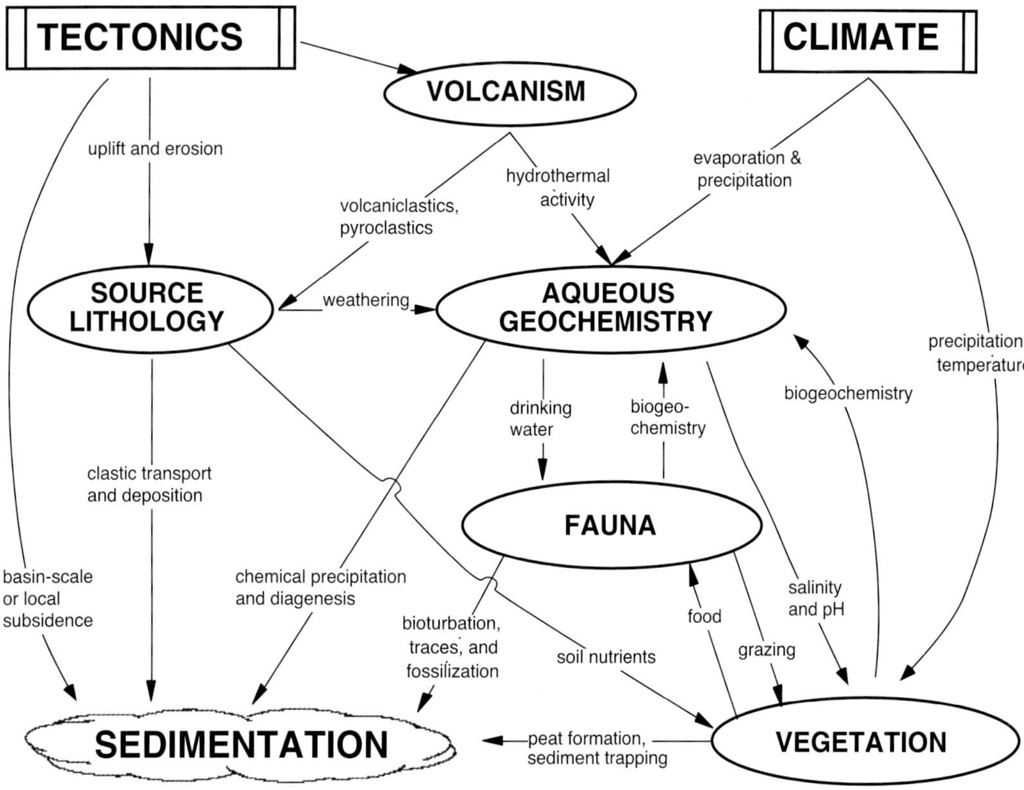


FIG. 8.—Flowchart summarizing the major controls and feedbacks on sedimentation in East African lake-margin wetlands, modified from Deocampo (1997).

prevented occupancy of the latter wetlands by hippos, but no evidence of prior hippo presence, such as deep sediment profiles of hippo pools, trails, and other geomorphic features, was evident (Deocampo and Ashley, 1997; McCarthy et al., 1998).

Diagenesis of sediments from these environments may alter lithologies significantly because of the great postdepositional shifts in environmental conditions that are possible. With burial and minimal diagenesis, the organic-rich facies from these wetland settings will likely produce very thin coal seams that grade laterally into the more clastic-rich facies. Extended periods of subaerial exposure, however, may lead to oxidation and minimal preservation of the organic material. The mineralogical and geochemical signal of local freshwater conditions seems likely to persist, unless conditions are extreme, because it is kinetically easier to dissolve indicators of salinity and alkalinity than it is to precipitate new minerals (Deocampo et al., 2002). Postdepositional hydrothermal activity, however, may obscure an older geochemical signature because of significant mineralization.

### *Wetland Sediments and Paleohydrology*

The modern wetlands discussed here developed in subaerially exposed lake-margin environments. The relatively low modern lake levels seen throughout northern Tanzania are due to the regional drying since about 7,000 years ago (Kendall, 1969; Stager et al., 1997). Humidity is thought to have peaked at that time, as suggested by high lake levels in the region (Street-Perrott and Roberts, 1983), a peak in woodland cover around Lake Tanganyika (Vincens, 1991), minimal dust flux to the Arabian Sea (Clemens and Prell, 1990, 1991), and maximal monsoon intensity (Zonneveld et al., 1997). Subsequent late Holocene drying led to the subaerial exposure of lake flats throughout the region, and the development of wetlands around zones of groundwater discharge such as those discussed in this study. Deposits associated with these wetland environments, therefore, indicate a drop in regional water levels over the past few thousand years. The modern radiocarbon date at the base of the Jangwani peat suggests, however, that the deposits may not be strongly aggradational, and that they are continually reworked. This was also indicated by observations of allochthonous peat carried into a nearby wetland by flooding lake waters after the 1998 rains. If buried by a future lake transgression, such as hypothesized in the facies model of Figure 7, these sediments would therefore represent a "snapshot" of only the final stage of wetland history, perhaps overlain by a transgressive unconformity.

In hydrologically closed basins such as those in East Africa, climate change over thousands of years can produce large lake-level fluctuations (Kutzbach and Street-Perrott, 1985), and in a basin with low-gradient lake margins, a regression can freshly expose many square kilometers of lake flats. As in the wetlands described above, hydrologic contexts such as aquifer outcrops or barriers (e.g., faults and dikes) that force water to the surface can lead to wetland development (Quade et al., 1995; Deocampo, 1997; Deocampo and Ashley, 1999b). Even without such hydrogeological conditions, unconfined groundwater flow can create discharge zones in playa margins because of an emergent water table (Barnes et al., 1990; Rosen, 1994). If the deposits produced by such wetlands are then buried by subsequent lake-level rise, a sedimentary record of one complete lake-level fluctuation would be produced (Figs. 7, 9). An important component of such a record would be geochemical and mineralogical evidence for paleochemical changes (Deocampo and Ashley, 1999a; Deocampo et al., 2000), with saline–alkaline indicators forming during lacustrine conditions (evaporites, carbonates, Mg-clays) and freshwater indicators forming under wetland conditions (biogenic amorphous silica, alteration–dissolution of Mg-clays). Sedimentary packages such as this could provide important evidence for studies in closed-basin paleohydrology.

Although this model has some similarities to other lacustrine facies models incorporating nearshore or littoral marsh-deposited facies such as peat and gyttja (Fouch and Dean, 1982; Cabrera and Saez, 1987; Platt, 1989; Kasinski, 1991), it differs in two main respects. First, it recognizes a basal disconformity as a common feature signifying regression. This is likely more typical in low-gradient lake basins that have broad lake flats that are episodically inundated and dessiccated, and may or may not exhibit erosional features. Deposits in several hydrologically closed or restricted basins in East Africa show that such a response is common (Hay, 1976; Renaut and Tiercelin, 1994). Second, lake-marginal wetlands tend to be more stable, and hence produce a more distinct sedimentary record, where they form because of discharge of confined groundwaters, as along a fault conduit or where a confined aquifer crops out (Quade et al., 1995; Deocampo and Ashley, 1999b). These same localities may then be areas of repeated wetland development if the hydrogeologic framework is relatively stable, and these deposits may predictably interfinger laterally with adjacent alluvial or lacustrine sediments.

Although paleohydrological studies in East Africa have generally been limited to lacustrine sediment (e.g., Roberts et al., 1993) or lake-margin stromatolites (Casanova and Hillaire-Marcel, 1992), several studies in drylands of the southwestern U.S. have shown the applicability of wetland sedimentology. Holliday (1995) found that wetland sediments, manifested as clayey, organic-rich deposits and marl, are important components of the southern High Plains record of regional drying in the late Pleistocene and early Holocene. An initial period of widespread lacustrine deposition ended about 20,000 years ago, and was followed by a period of incision and alluviation. Continued lowering in regional moisture contributed to a drop in stream competence, an end to the alluviation, and the development of wetlands and small pools of standing water. These wetland sediments tend to be localized and discontinuous, and they overlie alluvial sands with a basal erosional unconformity (Holliday, 1995). The outcrop-scale geometry of lithofacies is similar to the depositional model of the Ormaiton system described in this study (Fig. 3), where wetland organic-rich sediments overlie alluvial sands and lacustrine muds. In both cases, the wetland sediments are indicative of a drop in hydraulic competence within the alluvial channels, consitituing a record of dropping water table.

A similar record of drying is also evident in wetland sedimentation in the southern Great Basin, Nevada, although the underlying stratigraphy is not as well known because of poor exposure (Quade, 1986; Quade and Pratt, 1989; Quade et al., 1995; Quade et al., 1998). In that playa setting, low-gradient valley-margin seeps generated marshy "wet meadow" environments and phreatophyte flats upslope that created down-gradient clastic fining due to vegetation trapping, similar to the process reported at Ormaiton, Lake Natron. Repeated episodes of groundwater discharge in these environments have produced a complex sedimentary record spanning much of the time since the last glacial maximum (Quade et al., 1995; Quade et al., 1998). Over late Quaternary timescales, it can be seen that the wetland sediments record a relative drop in regional water table, inasmuch as they overlie coarse gravels indicative of much higher flow regimes, such as those exposed in the Pahrump Valley, Nevada (Quade et al., 1998).

In both these cases from the U.S., the wetland sediments record a relative drop in regional humidity analogous to the late Holocene aridity that led to the development of the lake-margin wetlands of this study. The application of freshwater wetland sediments as an indicator of reduced water availability for a given basin may seem counterintuitive, particularly when the deposits are interbedded with lacustrine sediments with saline–alkaline indicators. In contrast to lacustrine deposits, however, wetland sediments record only highly localized subaqueous conditions, and in a fluviolacustrine system they require regression and loss of fluvial competence to form.

## CONCLUSIONS

Sedimentation in lake-margin wetlands in the Natron, Ngorongoro, and Eyasi Basins of northern Tanzania is controlled by a complex system involving clastic sedimentation, mineral precipitation and diagenesis, bioturbation, and peat formation and vegetative trapping of sediments. Tectonic and climatic setting produce a set of controls on sedimentation, including sediment supply, aqueous geochemistry, large-mammal activity, and patterns of vegetation. Complex feedbacks among these controls are characteristic of these dynamic environments.

In the dryland basins of East Africa, with ambient-temperature groundwater discharge, these environments produce five distinct sets of lithofacies: sands and gravelly sands (S and Sg), silty to sandy muds (Ms), bioturbated muds (Mb), alkaline muds (Ma), and organic-rich deposits such as rootmats (R), peat (P), and organic-rich clay (Co). Several depositional subenvironments have characteristic lithofacies associations in these lake-margin wetlands, including drainage channels (facies S, Sg, Mb, Ms), marshes (R, P, Co, Mb), hippopotamus flats (Co, Mb, Ms), pools (S, Ms, Mb), and alkaline flats (Ma, Co). Other peripheral environ-

| - Lake Level + | Depositional Environment | Lithology | Primary Authigenic/Biogenic Minerals | Important Biological Factors |
|---|---|---|---|---|
| 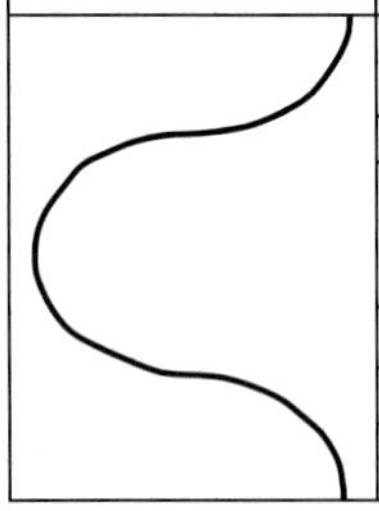 | Lacustrine | fine-grained | evaporites, carbonates, Mg-clays | invertebrate |
| | Erosion?-Lacustrine? | coarse-grained | absent | invertebrate |
| | Wetland | fine-grained, organic-rich | biogenic amorphous silica (from marsh plants and diatoms) | hippo and other vertebrates, plants |
| | Erosion?-Alluvial? | coarse-grained | absent | hippo |
| | Mudflats | fine-grained | evaporites, carbonates, Mg-clays | invertebrate |
| | Lacustrine | | | |

FIG. 9.—Hypothetical generalized sedimentary record of a freshwater wetland on the margin of a hydrologically closed, saline, and alkaline lake during hydrological fluctuation.

ments may be ephemeral wetlands that are flooded during wet times, which are commonly components of the surrounding lacustrine or alluvial depositional environment.

Sediments in these lake-margin environments are thin, generally < 1 m, but greater thicknesses are found where large mammals such as *Hippopotamus amphibius* are present to generate deep (> 1.5 m) bioturbation profiles. Wetland sedimentary bodies are > 1 km elongate perpendicular to the topographic gradient, and interfinger over tens of meters laterally with adjacent fluviolacustrine deposits.

Because the occurrence of lake-margin wetlands is tightly linked to regional hydrology, these deposits are important indicators of paleohydrology within hydrologically closed basins. They occur only when water tables are lowered sufficiently to allow subaerial exposure of the lake margin, hence they are indicators of increased ratios of evaporation to precipitation. In all three basins studied here, modern wetland sediments overlie a regressive disconformity, thereby recording the onset of late Holocene aridity in East Africa.

## ACKNOWLEDGMENTS

Thanks to the Tanzania Commission on Science and Technology and the Ngorongoro Conservation Area Authority for allowing this research. This project was supported by student research grants from the Geological Society of America, and Sigma Xi—the Scientific Research Society. Major field support was provided by the Olduvai Landscape Paleoanthropology Project, with support from the National Science Foundation (Archaeology BNS-9000099, SBR-9601065, SBR-9602478, and Geoscience-9903258), the L.S.B. Leakey Foundation, the Wenner-Grenn Foundation, the National Geographic Society, the Sokkia Corporation, and Rutgers University. Thanks to Cynthia Liutkus, John Mosi, and Zacharia, for field assistance, and James Ebert , Victoria Hover, John Hutch, IMO Industries, Gavin Lao, Egon Weber III, and James Wright for technical and laboratory assistance. Thanks also to Gail Ashley, Richard Hay, Charles Peters, and Robin Renaut for helpful conversations.

## REFERENCES


Ashley, G.M., Deocampo, D.M., and Liutkus, C.M., 2000, Track records of large mammals in lake-margin wetlands, L.M. Bed II, Olduvai Gorge, Tanzania (abstract): International Association of Sedimentologists, Dublin, Ireland, Regional Meeting Abstracts.

Baker, B.H., 1963, Geology of the area south of Magadi: Geological Survey of Kenya, Reports, v. 61, 27 p.

Baker, B.H., 1986, Tectonics and volcanism of the southern Kenya Rift Valley and its influence on rift sedimentation, *in* Frostick, L.E., Renaut, R.W., Reid, I., and Tiercelin, J.J., eds., Sedimentation in the African Rifts: Geological Society of London, Special Publication 25, p. 45–57.

Baker, B.H., and Mitchell, J.G., 1976, Volcanic stratigraphy and geochronology of the Kedong–Olorgesailie area and the evolution of the south Kenya rift valley: Geological Society of London, Journal, v. 132, p. 467–484.

Baker, V.R., Kochel, R.C., Laity, J.E., and Howard, A.D., 1990, Spring sapping and valley network development, *in* Higgins, C.G., and Coates, D.R., eds., Groundwater Geomorphology: Geological Society of America, Special Paper 252, p. 235–265.

Barnes, C.J., Chambers, L.A., Herczeg, A.L., Jacobson, G., Williams, B.G., and Wooding, R.A., 1990, Mixing processes between saline groundwater and evaporation brines in groundwater discharge zones, *in* Proceedings of the International Conference on Groundwater in Large Sedimentary Basins: Australian Water Resources Council, Conference Series no. 20, p. 369–378.

Cabrera, L., and Saez, A., 1987, Coal deposition in carbonate-rich shallow lacustrine systems: the Calaf and Mequinenza sequences (Oligocene, eastern Ebro Basin, NE Spain): Geological Society of London, Journal, v. 144, p. 451–461.

Casanova J., and Hillaire-Marcel, C., 1992, Chronology and paleohydrology of Late Quaternary high lake levels in the Manyara Basin (Tanzania) from isotopic data ($^{18}O$, $^{13}C$, $^{14}C$, Th/U) on fossil stromatolites: Quaternary Research, v. 38, p. 205–226.

Cerling, T.E., and Hay, R.L., 1986, An isotopic study of paleosol carbonates from Olduvai Gorge: Quaternary Research, v. 25, p. 63–78.

Clemens, S.C., and Prell, W.L., 1990, Late Pleistocene variability of Arabian Sea summer-monsoon winds and dust source-area aridity: A record from the lithogenic component of deep-sea sediments: Paleoceanography, v. 5, p. 109–145.

Clemens, S.C., and Prell, W.L., 1991, One million year record of summer monsoon winds and continental aridity from the Owen Ridge (Site 722), Northwest Arabian Sea: Proceedings of the Ocean Drilling Program, Scientific Results, v. 117, p. 365–388.

Coplen, T.B., Kendall, C., and Hopple, J., 1983, Comparison of stable isotope reference samples: Nature, v. 302, p. 236–238.

Dawson J.B., Pinkerton H., and Nyameru, C., 1994, June, 1993 eruption of Oldoinyo Lengai, Tanzania: Exceptionally viscous and large carbonatite lava flows and evidence for coexisting silicate and carbonate magmas: Geology, v. 22, p. 799–802.

Deocampo, D.M., 1997, Modern sedimentation and geochemistry of freshwater springs: Ngorongoro Crater, Tanzania: unpublished M.Sc. thesis, Rutgers University, 140 p.

Deocampo, D.M., and Ashley, G.M., 1997, Hippo trails: A large-scale freshwater biogenic structure in arid East Africa (abstract): Geological Society of America, Abstracts with Programs, v. 29, p. 40–41.

Deocampo, D.M., and Ashley, G.M., 1999a, Siliceous islands in a carbonate sea: Modern and Pleistocene records of spring-fed wetlands in Ngorongoro Crater and Oldupai Gorge, Tanzania: Journal of Sedimentary Research, v. 69, p. 974–979.

Deocampo, D.M., and Ashley, G.M., 1999b, Sedimentology, geochemistry, and paleohydrologic potential of groundwater wetlands in East African lacustrine basins: International Union for Quaternary Research, XV Congress, Durban, South Africa, p. 52.

Deocampo, D.M., Ashley, G.M., and Copeland, S.R., 1998, Springs and streams in an East African closed basin: Controls on biological communities in Ngorongoro Crater, Tanzania (abstract): Geological Society of America, Abstracts with Programs, v. 30, p. 120.

Deocampo, D.M., Hay, R.L., Ashley, G.M., Kyser, T.K., and Liutkus, C.M., 2000, Lacustrine clay diagenesis in northern Tanzania, with paleoenvironmental application at Olduvai Gorge (abstract): Geological Society of America, Abstracts with Programs, v. 32, p. A366.

Deocampo, D.M., Blumenschine, R.J., and Ashley, G.M., 2002, Wetland diagenesis and traces of early hominids, Olduvai Gorge, Tanzania: Quaternary Research, v. 57, p. 271–281.

Estes, R.D., and Small, R., 1981, The large herbivore populations of Ngorongoro Crater: African Journal of Ecology, v. 19, p. 175–185.

Fouch, T.D., and Dean, W.E., 1982, Lacustrine and associated clastic depositional environments, *in* Scholle, P.A., and Spearing, D., eds., Sandstone Depositional Environments: American Association of Petroleum Geologists, Memoir 31, p. 87–114.

Gromme, C.S., Reilly, T.A., Mussett, A.E., and Hay, R.L., 1970, Palaeomagnetism and potassium–argon ages of volcanic rocks of Ngorongoro Caldera, Tanzania: Geophysical Journal of the Royal Astronomical Society, v. 22, p. 101–115.

Grove, A.T., 1986, Geomorphology of the African Rift System, *in* Frostick, L.E., Renaut, R.W., Reid, I., and Tiercelin, J.J., eds., Sedimentation in the African Rifts: Geological Society of London, Special Publication 25, p. 9–18.

Hamilton, A., and Taylor, D., 1986, Mire sediments in East Africa, *in* Frostick, L.E., Renaut, R.W., Reid, I., and Tiercelin, J.J., eds., Sedimentation in the African Rifts: Geological Society of London, Special Publication 25, p. 211–217.

Hardie, L.A., Smoot, J.P., and Eugster, H.P., 1978, Saline lakes and their deposits: a sedimentological approach, *in* Matter, A., and Tucker, M.E., eds., Modern and Ancient Lake Sediments: International Association of Sedimentologists, Special Publication 2, p. 7–41.

Hay, R.L., 1976, Geology of the Olduvai Gorge: Berkeley, California, University of California Press, 203 p.

Hay, R., Pexton, R.E., Teague, T.T., and Kyser, T.K., 1986, Spring-related carbonate rocks, Mg-clays, and associated minerals in Pliocene deposits of the Amargosa Desert, Nevada and California: Geological Society of America, Bulletin, v. 97, p. 1488–1503.

Haynes, C.V., 1985, Mastodon-Bearing Springs and Late Quaternary Geochronology of the Lower Pomme de Terre Valley, Missouri: Geological Society of America, Special Paper 204, 35 p.

Holliday, V.T., 1995, Stratigraphy and Paleoenvironments of Late Quaternary Valley Fills on the Southern High Plains: Geological Society of America, Memoir 186, 136 p.

Isaac, G.L., 1977, Olorgesailie: Archaeological Studies of a Middle Pleistocene Lake Basin in Kenya: Chicago, University of Chicago Press, 272 p.

Jones B.F., Eugster H.P., and Rettig, S.L., 1977, Hydrochemistry of the Lake Magadi Basin, Kenya: Geochimica et Cosmochimica Acta, v. 41, p. 53–72.

Kasinski, J.R., 1991, Tertiary lignite-bearing lacustrine facies of the Zittau Basin: Ohre rift system (Poland, Germany and Czechoslovakia), *in* Anadon, P., Cabrera, L., and Kelts, K., eds., Lacustrine Facies Analysis: International Association of Sedimentologists, Special Publication 13, p. 93–107.

Kendall, R.L., 1969, An ecological history of the Lake Victoria basin: Ecological Monographs, v. 39, p. 121–176.

Kutzbach, J.E., and Street-Perrott, F.A., 1985, Milankovitch forcing of fluctuations in the level of tropical lakes from 18 to 0 kyr BP: Nature, v. 317, p. 130–134.

Lewis, D.W., and McConchie, D., 1994, Analytical Sedimentology: New York, Chapman & Hall, 197 p.

McCarthy, T.S., Ellery, W.N., and Bloem, A., 1998, Some observations on the geomorphological impact of hippopotamus (*Hippopotamus amphibius L.*) in the Okavango Delta, Botswana: African Journal of Ecology, v. 36, p. 44–56.

Moore, D.M., and Reynolds, R.C., 1997, X-Ray Diffraction and the Identification and Analysis of Clay Minerals: Oxford, U.K., Oxford University Press, 378 p.

Murphy, C.P., 1986, Thin Section Preparation of Soils and Sediments: Berkhamsted, U.K., A B Academic Publishers, 149 p.

Pickering, R., 1965, Ngorongoro: Quarter Degree Sheet 53: Geological Survey of Tanzania.

Platt, N.H., 1989, Lacustrine carbonates and pedogenesis: sedimentology and origin of palustrine deposits from the Early Cretaceous Rupelo Formation, W Cameros Basin, N Spain: Sedimentology, v. 36, p. 665–684.

Plummer, L.N., Prestemon, E.C., and Parkhurst, D.L., 1994, An interactive code (NETPATH) for modeling NET geochemical reactions along a flow PATH—version 2.0. U.S. Geological Survey, Water-Resources Investigations Report 94-4169.

Quade, J., 1986, Late Quaternary Environmental Changes in the Upper Las Vegas Valley, Nevada: Quaternary Research, v. 26, p. 340–357.

Quade, J., and Pratt, W.L., 1989, Late Wisconsin groundwater discharge environments of the southwestern Indian Springs Valley, southern Nevada: Quaternary Research, v. 31, p. 351–370.

Quade, J., Mifflin, M.D., Pratt, W.L., McCoy, W., and Burckle, L., 1995, Fossil spring deposits in the southern Great Basin and their implications for changes in water-table levels near Yucca Mountain, Nevada, during Quaternary time: Geological Society of America, Bulletin, v. 107, p. 213–230.

Quade, J., Forester, R.M., Pratt, W.L., and Carter, C., 1998, Black mats, spring-fed streams, and late-glacial-age recharge in the southern Great Basin: Quaternary Research, v. 49, p. 129–148.

Renaut, R.W., and Jones, B., 1997, Controls on aragonite and calcite precipitation in hot spring travertines at Chemurkeu, Lake Bogoria, Kenya: Canadian Journal of Earth Sciences, v. 34, p. 801–818.

Renaut, R.W., and Tiercelin, J.J., 1994, Lake Bogoria, Kenya Rift Valley—a sedimentological overview, *in* Renaut, R.W., and Last, W.M., eds., Sedimentology and Geochemistry of Modern and Ancient Saline Lakes: SEPM, Special Publication 50, p. 101–123.

Renaut, R.W., Tiercelin, J.J., and Owen, R.B., 1986, Mineral precipitation and diagenesis in the sediments of the Lake Bogoria basin, Kenya Rift Valley, *in* Frostick, L.E., Renaut, R.W., Reid, J., and Tiercelin, J.-J., eds., Sedimentation in the African Rifts: Geological Society of London, Special Publication 25, p. 159–175.

Roberts, C.R., and Mitchell, C.W., 1987, Spring mounds in southern Tunisia, *in* Frostick, L., and Reid, I., eds., Desert Sediments: Ancient and Modern: Geological Society of London, Special Publication 35, p. 321–334.

Roberts, N., Taieb, M., Barker, P., Damnati, B., Icole, M., and Williamson, D., 1993, Timing of the Younger Dryas event in East Africa from lake-level changes: Nature, v. 366, p. 146–148.

Rosen, M.R., 1994, The importance of groundwater in playas: A review of playa classifications and the sedimentology and hydrology of playas, *in* Rosen, M.R., ed., Paleoclimate and Basin Evolution of Playa Systems: Geological Society of America, Special Paper 289, p. 1–18.

Stager, J.C., Cuming, B., and Meeker, L, 1997, A high-resolution 11,400-Yr diatom record from Lake Victoria, East Africa: Quaternary Research, v. 47, p. 81–89.

Street-Perrott, F.A., and Roberts, N., 1983, Fluctuations in closed-basin lakes as an indicator of past atmospheric circulation patterns, *in* Street-Perrott, F.A., Beran, M., and Ratcliffe, R., eds., Variations in the Global Water Budget: Dordrecht, The Netherlands, Reidel, p. 331–346.

Vincens, A., 1991, Late Quaternary vegetation history of the South-Tanganyika Basin. Climatic implications in South Central Africa: Palaeogeography, Palaeoclimatology, Palaeoecology, v. 86, p. 207–226.

Vincens, A., and Casanova, J., 1987, Modern background of Natron–Magadi Basin (Tanzania–Kenya): Physiography, climate, hydrology and vegetation, *in* Lac Natron: Géologie, Géochimie et Paléontologie d'un Bassin Evaporitique du Rift Est-Africain: Sciences Géologiques, v. 40, p. 9–21.

Williams, L.A.J., 1978, Character of Quaternary volcanism in the Gregory Rift Valley, *in* Bishop, W.W., ed., Geological Background to Fossil Man: London, Scottish Academic Press, p. 55–70.

Zonneveld, K.A.F., Ganssen, G., Troelstra, S., Versteegh, G.J.M., and Visscher, H., 1997, Mechanisms forcing abrupt fluctuations of the Indian Ocean summer monsoon during the last deglaciation: Quaternary Science Reviews, v. 16, p. 187–201.

# DEPOSITION AND GEOCHEMISTRY OF LACUSTRINE AND SPRING CARBONATES IN MESOZOIC RIFT BASINS, EASTERN NORTH AMERICA

CAROL B. DE WET

*Department of Geosciences, Franklin and Marshall College, Lancaster, Pennsylvania 17604, U.S.A.*

CLAUDIA I. MORA

*Department of Geological Sciences, University of Tennessee, Knoxville, Tennessee 37996, U.S.A.*

PAMELA J.W. GORE

*Department of Geology, Georgia Perimeter College, Clarkston, Georgia 30021, U.S.A.*

ELIZABETH GIERLOWSKI-KORDESCH

*Department of Geological Sciences, Ohio University, Athens, Ohio 45701, U.S.A.*

AND

STEVEN J. CUCOLO

*Department of Geosciences, Franklin and Marshall College, Lancaster, Pennsylvania 17604, U.S.A.*

ABSTRACT: We recognize a causal association between volcanism and lacustrine limestone precipitation in Newark Supergroup strata. Two other associations are apparent: limestone with vegetation, and limestone interlaminated with siliciclastic beds. These associations reflect the different origins of the limestone and relate to basinal hydrological parameters. Hettangian-age limestone deposits reflect an increase in carbonate production and increased rates of groundwater circulation due to the high geothermal heat flow that accompanied volcanism and rifting. Most meter-scale limestone deposits stratigraphically overlie basalt and have covarying stable-isotope compositions, indicating closed-basin conditions. Limestones with spongy to framework texture, interpreted as tufa deposits, reveal sites of sublacustrine or nearshore springs or seeps. Tufa $\delta^{18}O$ values are consistently depleted relative to meter-scale carbonates from the same basin, indicating higher temperature waters at debouching localities.

Strong climatic seasonality is inferred from carbonate turbidites and euhedral calcite laminae interbedded with siliciclastic deposits. The turbidites formed when rains flushed littoral-zone carbonates into profundal depths, and euhedral calcites reflect reducing conditions in stratified lake waters associated with dry periods. Paleoclimatic differences north to south along the rift basin trend are confirmed by a 5‰ difference in $\delta^{18}O$ values of meter-scale limestones. The northern Hartford and Fundy Basins have enriched values, indicating increasing aridity northward, with a resulting increase in evaporation from those rift lakes. Littoral-zone carbonates associated with plants and stromatolites contain $^{13}C$-depleted carbon isotope signatures, reflecting organic mediation in carbonate precipitation. In contrast, profundal-zone carbonates have more uniform $\delta^{13}C$ values, reflecting a greater degree of equilibration between the atmosphere and the lake water. Newark Supergroup rift-basin limestones reflect a combination of environmental and climatic factors that increased the dissolved-load and/or bedload transportation of calcite, coupled with mechanisms such as basalt mantling, turbidite deposition, *in situ* precipitation, and vegetative associations, which enhanced carbonate deposition. Stable-isotope compositions help define these varied hydrogeochemical and paleoenvironmental conditions.

## INTRODUCTION

The early Mesozoic rift basins of the Newark Supergroup, eastern North America (Fig. 1), contain a complex succession of alluvial, fluvial, lacustrine, palustrine, hot-spring, eolian, and igneous deposits. Lacustrine strata are predominantly siliciclastic mudstones and siltstones, interbedded with mudflat and fluvial sheetflood deposits and alluvial-fan deposits. Relatively thin lacustrine limestones are present in many of the rift basins. Lake sediments are interpreted as deposits in shallow playa lakes (Wheeler and Textoris, 1978; Hubert et al., 1982; Demicco and Gierlowski-Kordesch, 1986; Gierlowski-Kordesch and Rust, 1994), floodplain lakes (Gore, 1986a, 1986b), and deep, stratified lakes (Olsen, 1990a, 1990b, Manspeizer and Olsen, 1981; Hentz, 1985; LeTourneau and McDonald, 1985). Localized carbonate deposits, interpreted as having formed in hot springs or seeps (Wheeler and Textoris, 1978; Steinen et al., 1987; de Wet and Hubert, 1989), are associated with lacustrine, fluvial, and alluvial-fan deposits.

Carbonate deposits in Mesozoic rift basin deposits are volumetrically less significant than siliciclastic rocks; however, the different mechanisms responsible for carbonate precipitation and deposition provide important paleoenvironmental information such as biological, climatic, and source-terrane changes. Carbonate precipitation can occur in freshwater lakes when evaporation and increased salinity alter ion activities, or when pH is affected by plants and algae. When older carbonate terrane is weathered, clasts and ions are transported into the catchment area and new carbonate beds can accumulate. Weathering of metamorphic and/or basaltic rocks can add considerable $Ca^{2+}$ to surface water and groundwater, enhancing deposition of carbonate at springs and seeps. Carbonate can also precipitate within soils where soil moisture is low.

This paper describes a combined study of the sedimentology and the trace-element and stable-isotope compositions of lacustrine carbonate deposits in the Newark Supergroup. The goals of this study are to characterize the occurrence of lacustrine limestones in the rift basins, to constrain the general mechanisms of carbonate precipitation, and to develop a substantive picture of the surface-water and ground-water systems, and the climatic and environmental factors contributing to limestone precipitation.

## GEOLOGICAL SETTING AND REGIONAL STRATIGRAPHY

The Newark Rift System (Fig. 1) is a series of northeast-trending, early Mesozoic fault-bounded basins found in eastern North America from Nova Scotia through South Carolina. The basins formed during the earliest phase of rifting associated with

Sedimentation in Continental Rifts
SEPM Special Publication No. 73, Copyright © 2002
SEPM (Society for Sedimentary Geology), ISBN 1-56576-082-4, p. 309–325.

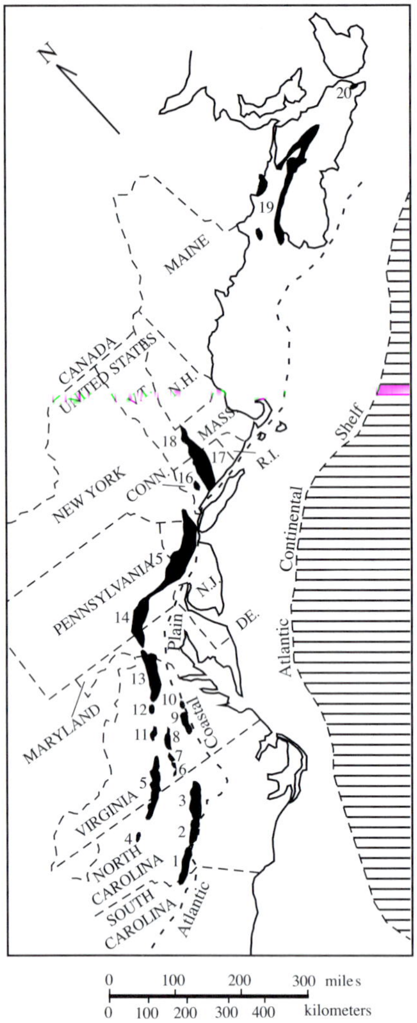


FIG. 1.—Eastern North America and the distribution of onshore Mesozoic rift basins. Basins discussed in this study are in bold italics. After Froelich and Olsen (1984).

the breakup of Pangea and formation of the Atlantic Ocean. The onset of the rifting initiated movement along a series of Alleghenian thrust faults that were reactivated as extensional, listric normal faults (Manspeizer and Olsen, 1981); crustal extension along these faults formed basins that were filled with continental deposits. These rocks consist principally of alluvial-fan, fluvial, and lacustrine sedimentary strata; some basins also contain coal, eolian sandstone, carbonate or siliceous spring deposits, and evaporites. The sedimentary rocks, together with interstratified Lower Jurassic basalts and associated intrusive sills and dikes, constitute the Newark Supergroup (Froelich and Olsen, 1984; Olsen et al., 1989; Weems and Olsen, 1997). Weems and Olsen's (1997) refinement of the Newark Supergroup stratigraphy divides the succession into three groups, on the basis of the nearly synchronous volcanism and plutonism throughout the rift basins (Fig. 2). The limestones described in this study occur in the Chatham and Meridan Groups.

The pre-Mesozoic bedrock of the catchment area varies from basin to basin, but all of the rocks are potential contributors to the rift-lake limestones. For example, bedrock around the Hartford and Pomperaug Basins consists of schist, granite, marble, amphibolite, and gneiss. Clasts of the marble occur in Mesozoic strata in the Pomperaug Basin, indicating that the source area was nearby and that carbonate entered the rift sediments both as bedload and, presumably, dissolved-load constituents Gierlowski-Kordesch, personal communication). In the Gettysburg Basin, clasts of local Paleozoic limestone and dolomite are common in alluvial-fan conglomerates stratigraphically below limestone lake deposits (Cloos and Pettijohn, 1973; de Wet et al., 1998). Paleozoic carbonates dominate the stratigraphy of the region and must have made significant contributions to the rift lakes as both dissolved-load and bedload components. Lithologies adjacent to the Culpeper Basin are mostly limestone and dolostone along the northern edge of the rift basin, and mostly greenstone epidosite, granite, and arkose conglomerate along the southern edge (Lindholm et al., 1979). Groundwater and surface-water processes leached carbonate components from the bedrock of the catchment area and supplied it to the rift lakes.

## METHODS

Thin sections were stained (Dickson, 1966) and examined in transmitted light, using cathodoluminescence, and thin sections and rock chips were examined using scanning electron microscopy (SEM). Ferroan and nonferroan calcite were the only carbonate minerals present, and they were sampled separately to distinguish diagenetic phases. Carbonate material was sampled for minor-element and stable-isotope analyses using a modified Dremel drill. Powder for elemental analysis was dissolved in a dilute HCl and Cd solution and then filtered to separate out the fine-grained clastic component. Carbonate samples were then analyzed for Mg, Fe, Sr, and Mn content, using a Thermo-Jarrell Ash ICAP-61 inductively coupled argon spectrometer (ICP) and results are reported in parts per million (mg/kg); all results were well above detection limits of a few mg/kg for these elements. Samples drilled for isotopic analysis were roasted for one hour to remove volatile organics and reacted with phosphoric acid at 25°C after McCrea (1950). Isotopic ratios were measured on a Finnigan MAT Delta Plus mass spectrometer at the University of Tennessee. Results are reported in standard $\delta$-permil notation relative to V-PDB with a precision of ± 0.10‰ for oxygen and ± 0.05‰ for carbon.

## NEWARK SUPERGROUP LIMESTONE CATEGORIES

Limestone and argillaceous limestone are present in most of the exposed, onshore rift basins, and numerous shale beds are also highly calcareous. We recognize three categories of limestone in the Newark Supergroup strata, on the basis of lateral and vertical field relations: (1) meter-scale micritic limestone beds; (2) millimeter- to centimeter-scale limestone beds interstratified with siliciclastic beds; and (3) limestone with a distinctive spongy or framework texture and localized extent. The categories are significant because they represent different depositional environments and mechanisms of carbonate precipitation.

Not every basin contains every limestone category. For example, the Fundy Basin contains both meter-scale and spongy textured carbonates but lacks millimeter- to centimeter-scale beds. The Gettysburg Basin has thick meter-scale limestone in two locations but lacks millimeter- to centimeter-scale and spongy textured to framework-textured deposits. The Newark, Pomperaug, and Culpeper Basins contains predominantly millimeter- to centimeter-scale limestone beds interlayered with thick siliciclastic strata. The Durham sub-basin contains pre-

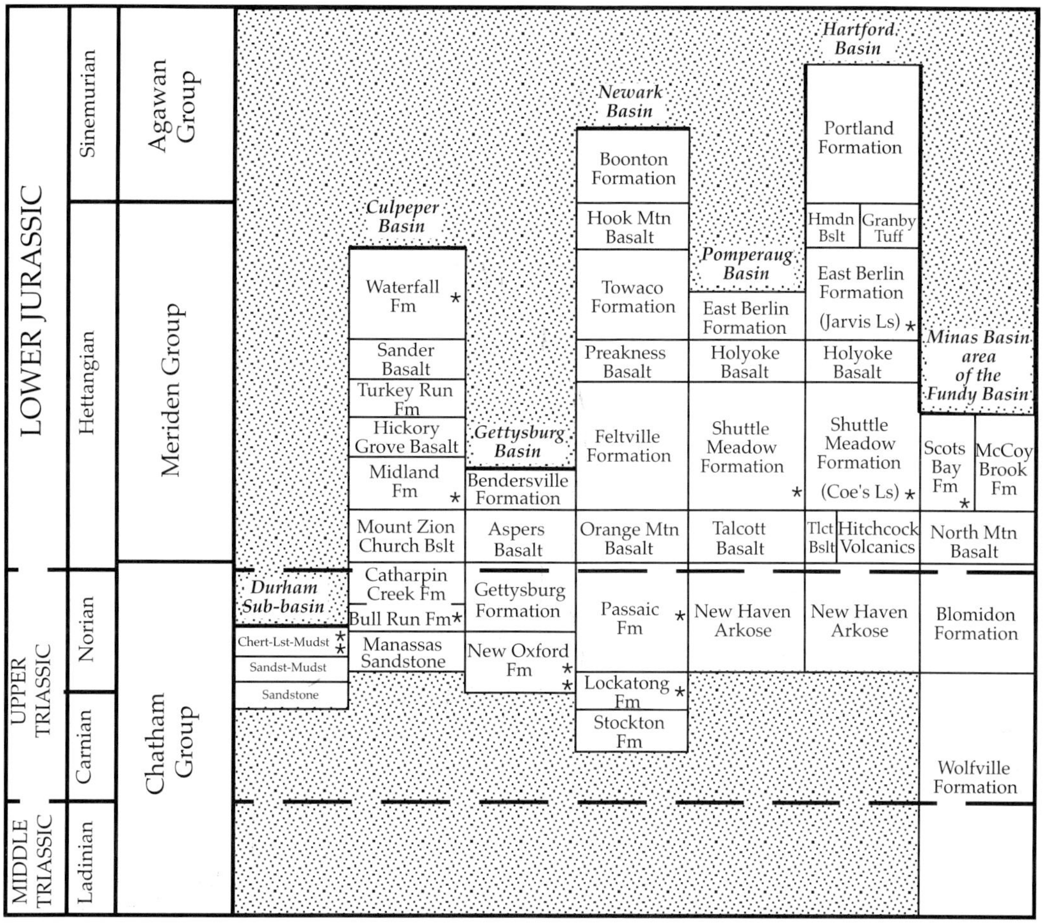


FIG. 2.—Newark Supergroup stratigraphy divided into Weems and Olsen's (1997) three groups and formations. The star symbol marks the approximate stratigraphic positions of limestone deposits discussed in the text. After Weems and Olsen (1997).

dominantly centimeter-scale beds, but one location consists of spongy textured to framework-textured limestone. Only the Hartford Basin contains all three categories of limestone. These occurrences may reflect broad depositional trends, as discussed below.

### *Category 1: Meter-Scale Limestone Beds*

Limestone occurs as beds, 1–23 m thick, that are traceable throughout an outcrop and have either gradational or abrupt lower and upper contacts (Fig. 3). The beds may contain shaly partings or sandy lenses. The limestone varies from mudstone to wackestone, and rarely, packstone. It is white to green to gray to almost black in color, depending on the amount of disseminated organic matter. Features common to all of these deposits include: a dense, peloidal, micritic matrix with a variable terrigenous clay component; stromatolites or microbial relicts; and ostracode and/or bivalve shell fragments. The limestone deposits rarely exhibit grading, but they can be highly fossiliferous, containing plant remains (Figs. 3, 4), abundant ostracodes, fish bone or scale fragments, gastropods, small bivalves, and other microfauna. In the Hartford and Fundy Basins, meter-scale limestone directly overlies basalt; other examples are associated with sandstone, siltstone, or mudstone interpreted as fluvial or alluvial-plain deposits.

### *Category 2: Millimeter- to Centimeter-Scale Limestone Beds*

Millimeter- to centimeter-scale beds of limestone and/or calcareous mudstone are the most common type of carbonate in the rift basins (Fig. 5). They occur only in lacustrine deposits, interstratified with siliciclastic beds. Beds range from < 1 cm to 80 cm thick and can be traced laterally throughout individual outcrops. They typically occur as gray or tan, graded beds, consisting predominantly of carbonate clasts in a carbonate matrix, or as layers of micronodules and carbonate stringers. Sand-size carbonate clasts and scour surfaces delineate the bases of graded limestone beds, and the beds fine upward into micrite and argillaceous micrite (Fig. 5). The beds may be fossiliferous or oolitic, forming thin wackestone to packstone beds, and, locally, ostracodal packestones. A subset of this category consists of euhedral micrite to microspar crystals in an argillaceous matrix (Fig. 6). Upper and lower contacts with terrigenous siliciclastic beds are sharp, and no shelly faunal remains are present in the carbonate layers.

### *Category 3: Spongy to Framework-Textured Limestone*

Relatively pure limestone, with a spongy or framework texture (Fig. 7A, B) and localized extent, has been described from the Fundy Basin (de Wet and Hubert, 1989), the Hartford Basin

FIG. 3.—Field photograph of the 5-m-thick Scot's Bay Formation. Green marly limestone underlies pure white packstone deposits containing large silicified tree remains (arrow). Hammer on right for scale (arrow). Fundy Basin.

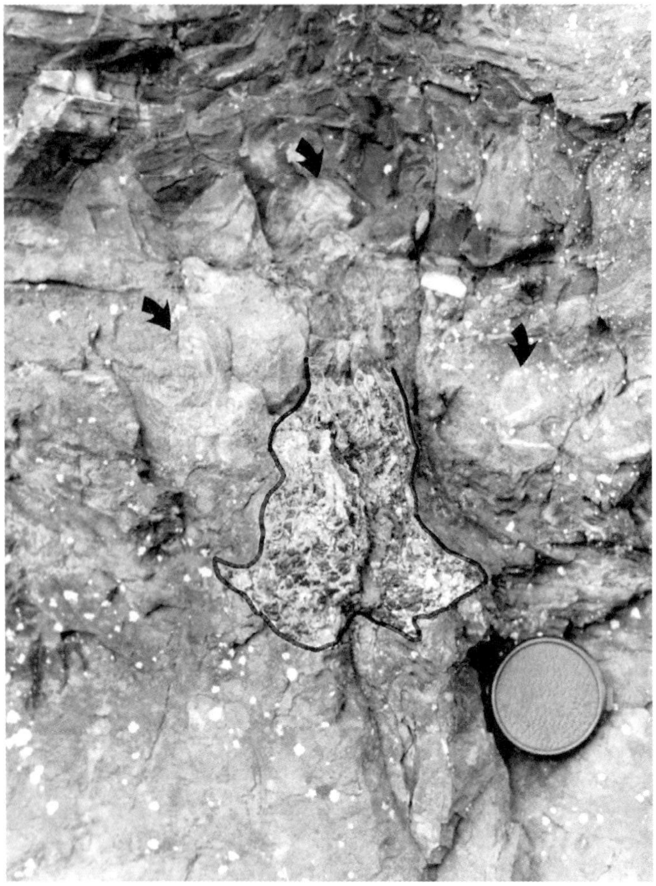

FIG. 4.—Field photograph of carbonaceous tree remains (outlined) and stromatolites (arrows) in an oolitic packstone in the Passaic Formation. Stromatolites encrust the plant material and suggest that most of the micrite at this site formed through biogenic modification of lake water chemistry. Bright spots are Recent lichen. Western Newark Basin.

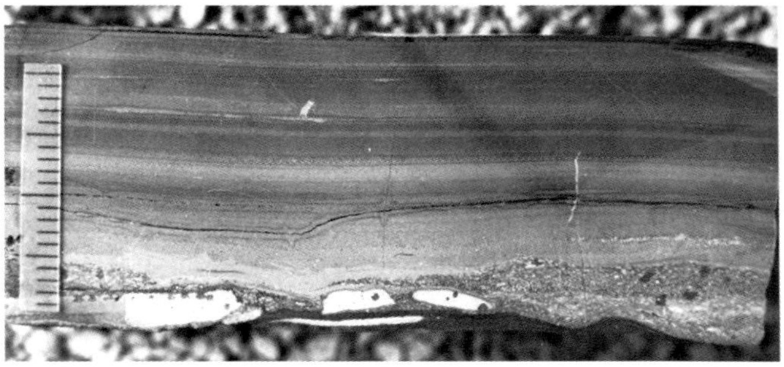

FIG. 5.—Hand specimen of limestone illustrating typical millimeter- to centimeter-scale limestone layering. This graded bed is from the Bull Run Formation, Catlett, Virginia. The light-colored clasts at the base are composed of locally derived micrite, and sand-size carbonate clasts fine upward into argillaceous micrite. Scale is in centimeters. Culpeper Basin.

(Steinen et al., 1987), and the Durham sub-basin (Wheeler and Textoris, 1978). The rocks range in color from gray to greenish but have striking red, black, and gray colors where they contain silica. Faunal remains are rare in these limestones, except for microbial fabrics, which represent algal or bacterial growth (Fig. 7B). Charophytes and other algae have been described from limestone directly adjacent to the spongy deposits (Birney–de Wet, 1984).

### *Other Carbonates*

Carbonate also occurs as finely disseminated diagenetic dolomite within lake-basin mudstones, described chiefly from the Newark (Van Houten, 1965) and Hartford (Hubert et al., 1976; Hubert et al., 1992) Basins. Pedogenic carbonate, precipitated in soil horizons and overbank muds, is common in all of the basins. Paleozoic carbonate clasts are contained in alluvial-fan conglomerates. These types of carbonate are beyond the scope of this paper and will not be addressed further.

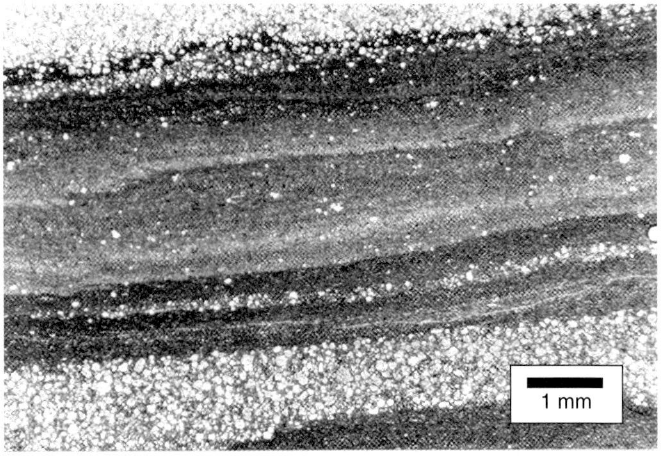


FIG. 6.—Photomicrograph of laminite consisting of dense, euhedral micrite and microspar interlayered with silty lacustrine mudstone from the Passaic Formation near Pottstown, Pennsylvania. The carbonate is nonferroan by staining. Western Newark Basin.

## RESULTS

### *Limestone Distribution by Basin and Category*

*Fundy Basin.—*

The only occurrence of limestone (other than caliche and alluvial-fan clasts) in the Fundy Basin is in the Lower Jurassic (Hettangian) Scots Bay Formation, which crops out in small coves along the Bay of Fundy in the Minas Basin (de Wet and Hubert, 1989). The limestone consists of a 5-m-thick, category one deposit, containing within it a localized bed of spongy textured to framework-textured, category three limestone.

The thick limestone directly overlies the North Mountain Basalt. Carbonate drapes the undulating basalt surface for over 5.5 km of exposure. It ranges from a pure, white packstone (Fig. 8) to gray-green argillaceous lime mudstone containing abundant mudcrack horizons. Stromatolites are present locally, and one horizon contains silicified remains of tree trunks up to 0.5 m in diameter (Fig. 3) (de Wet and Hubert, 1989).

The localized deposit consists of a 1.2-m-thick, open-framework, porous-textured carbonate containing geothite spherulites, chalcedony, drusy quartz, and chert (de Wet and Hubert, 1989) (Fig. 7A, B). Chalcedony and chert have replaced calcite shell fragments, charophyte oogonia, and lime mud. Drusy quartz fills final pores. Locally patches of carbonate remain unsilicified. An anastomosing network of silica and mudstone-filled fractures leads down below the originally porous framework horizon to feeder pipes and fissures in the North Mountain Basalt. Overlying wackestone and packstone beds drape over the deposit (Birney–de Wet, 1984).

*Hartford Basin.—*

All three limestone categories occur within the Hartford Basin. Meter-thick beds occur in basal Shuttle Meadow Formation strata previously exposed at Coe's Quarry, near North Branford, Connecticut, and in the lower East Berlin Formation near Holyoke, Massachusetts (the "Jarvis Limestone") (Gierlowski-Kordesch and Huber, 1995). Both occurrences are Lower Jurassic (Hettangian). The Coe's Quarry limestone was 17.6 m thick with a lateral extent of at least 25 m (the exposure is now gone; Starquist, 1943; Krynine, 1950; Steinen et al., 1987); the "Jarvis Limestone" is a minimum of 5 m thick (Starquist, 1943). Both units are fossiliferous, containing abundant fragments of ostracodes, bivalves, and fish bones in a peloidal micrite matrix.

At Coe's Quarry, category 3, spongy textured to framework-textured limestone occurs in the meter-scale carbonate strata. The spongy textured beds rest upon carbonatized agglomerate on top of the Talcott Basalt. The basalt is highly altered. Laterally discontinuous, spongy textured, boxwork-textured, to framework-textured limestones occur as micritic and banded travertine, pustular tufa, and siliceous tufa (Fig. 9). Microbial textures occur locally.

The Shuttle Meadow and East Berlin Formations also contain graded, thin-bedded (category 2) limestone. At Spruce Brook, Connecticut, (in the East Berlin Fm.) a layer of wackestone 4–10 cm thick is interstratified with ~ 8 m of massive to rippled sandstone, siltstone, and sandy mudstone. Ostracod and bivalve fragments and angular quartz grains are present in the wackestone. Similar horizons occur in the Shuttle Meadow Formation in exposures in Plainville and Durham, Connecticut, where limestone beds 2–40 cm thick are interstratified with massive to laminated mudstone, siltstone, and sandstone (Gierlowski-Kordesch and Huber, 1995; Gierlowski-Kordesch, 1998).

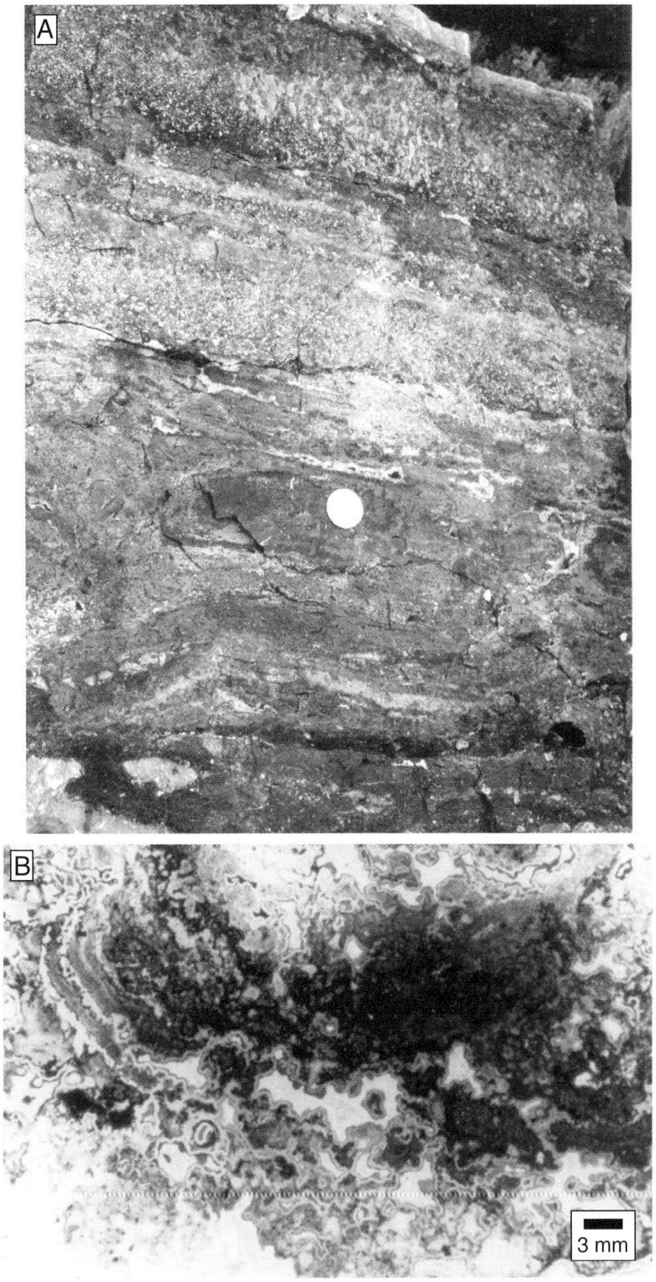


FIG. 7.—Silicified tufa deposits of the Scot's Bay Formation. **A)** Field photograph of part of a silicified tufa deposit within the meter-scale limestone succession. The original tufa was calcareous and highly porous, and a few large cavities remain (note one just above the coin). Coin is 2 cm in diameter. **B)** Photomicrograph of the silicified tufa in Part A. Iron-rich (dark) and iron-poor (light) banding highlights the original porous texture and clotted microbial fabric. Fundy Basin.

*Pomperaug Basin.—*

At Red Spring, Connecticut, 30–40 cm of millimeter- to centimeter-scale Lower Jurassic (Hettangian) limestone are exposed. The slightly argillaceous mudstone to packstone contains radial ooids, fish-bone fragments, and shelly fragments. The limestone

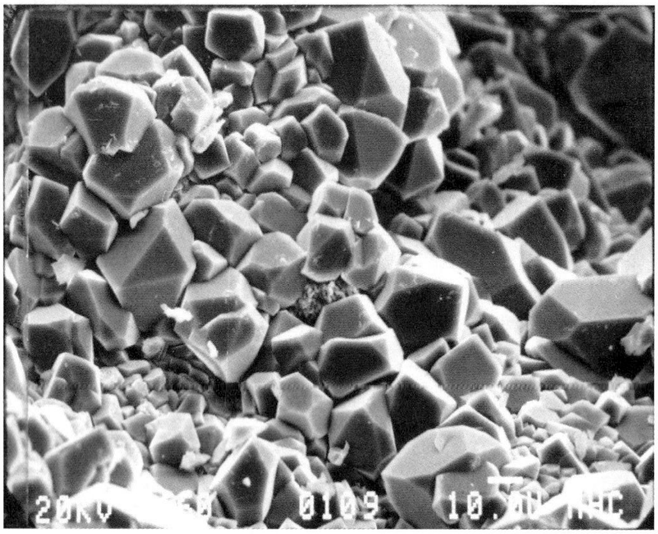

FIG. 8.—Scanning electron image (SEM) of low-Mg calcite crystals from the Scot's Bay Formation. This carbonate contains little argillaceous material, and is the purest limestone reported from Newark Supergroup sedimentary rocks. Fundy Basin.

has been correlated to two centimeter-scale limestone beds in the Shuttle Meadow Formation at Plainville, Connecticut (Huber, personal communication, 1997).

*Newark Basin.—*

The Newark Basin contains only millimeter- to centimeter-scale limestone beds. Lacustrine strata are dominated by siliciclastic rocks, but in the western Newark Basin, near Pottstown, Pennsylvania, a number of carbonate beds are interspersed in siltstone and sandstone horizons. In the Upper Triassic (Norian) Passaic Formation a carbonate bed 0.9 m thick contains microbial tufas and stromatolites (Fig. 4). Cross-bedded ooid sands, fish scales, ostracodes, and quartz sand grains are abundant in this unit.

FIG. 9.—SEM image of irregular, spongy texture common to category 3 deposits. This sample is from Shuttle Meadow Formation, Coe's Quarry, Connecticut. Hartford Basin.

Upsection, three thinner micritic limestone beds (maximum thickness 0.8 m) are interbedded with 200 m of massive red siltstone and mudstone. The limestone horizons consist of graded beds of quartz silt and micrite intraclasts (Fig. 5) or unusually pure, nonferroan, euhedral, low-Mg calcite (Fig. 6 ).

In New Jersey, the Carnian Lockatong Formation contains microlaminations of carbonate and millimeter- to centimeter-scale limestone beds. The carbonate contains abundant fish scales and ostracodes. These carbonate horizons have been correlated between cores and outcrop (Olsen et al., 1996) and occur on a Milankovitch time-scale frequency (Olsen et al., 1996; Weems and Olsen, 1997).

*Gettysburg Basin.—*

Two occurrences of meter-scale limestone have been described from the Upper Triassic (late Carnian to Norian) New Oxford Formation in the Gettysburg Basin (de Wet et al., 1998). The limestones are roughly time equivalent. A 6-m-thick limestone crops out in Rheems, Pennsylvania, within 300 m of fluvial sandstone and overbank mudstone. The limestone consists of carbonate mudstone to packstone, locally overprinted by pedogenic processes (de Wet et al., 1998). Micrite is argillaceous and contains thin sandstone lenses and shale stringers. Fossils include ostracodes, bivalves, rare gastropods, and plant fragments.

West of Rheems, near Thomasville, Pennsylvania, 245 m of New Oxford Formation alluvial-fan conglomerate, fluvial sandstone, overbank mudstone, and carbonate were cored at the Rife Farm site (first described by Cloos and Pettijohn, 1973). The core contains 23 m of limestone and marlstone. Light gray, meter-scale limestone beds contain abundant rip-up clasts and ostracode and bivalve fragments. Darker, more organic-rich layers are laminated and argillaceous. Overlying sandstone and mudstone rocks mark the return of siliciclastic deposition.

*Culpeper Basin and Durham Sub-Basin.—*

The southern basins are characterized by category two, millimeter- to centimeter-scale limestone deposits. Centimeter-scale limestones occur at several localities in the Upper Triassic (Carnian) Bull Run Formation in the Culpeper Basin (Fig. 10) (Gore, 1988a, 1988b). The thin limestones occur as black, laminated, stromatolitic micrites with abundant sulfide minerals (Culpeper Crushed Stone Quarry, Stevensburg, Virginia; Young and Edmundson, 1954; Smoot and Olsen, 1989) or gray, graded, laminated limestone containing intraclasts, gray-coated carbonate grains (0.1 mm in diameter), and fish scales. Wurtzite partially replaces some of the larger intraclasts and the surrounding matrix (locality near Calverton, Virginia; Gore, 1989). Limestones are overlain by gray or red, fossiliferous lacustrine mudstones or clayshales with conchostracans, ostracodes, notostracans, plant fragments, fish scales, and insects.

Millimeter- to centimeter-scale limestone layers are also found in Upper Triassic rocks in the Durham (North Carolina) sub-basin, exposed in the Triangle Brick Quarry and in the Borden Brick and Tile Quarry. In both quarries, thin (1–4 cm) limestones occur at the base of 1–2 m thick unit of gray to pale brown, noncalcareous mudshale to clayshale containing conchostracans, pelecypods, ostracodes, and scarce crayfish and insect fragments. The limestone-bearing shale unit is truncated by an irregular fluvial scour surface and abruptly overlain by trough-cross-stratified sandstone interbedded with red massive siltstone. The limestones overlie paleosol-bearing siltstone and mudstone of the Upper Triassic (Carnian) Sanford Formation (Gore, 1986a; Driese and

FIG. 10.—Hand-sample photograph showing black stromatolitic limestone associated with calcite pseudomorphs after evaporite crystals (arrow) from the Bull Run Formation, Groveton section. This sample is from a 7-m-thick lacustrine sequence that documents the deepening and shallowing of a perennial lake. Scale = 5 mm. For a detailed interpretive sketch of this sample, see Gore (1988b). Culpeper Basin.

Mora, this volume). In the Triangle Brick Quarry, the limestone is a massive, ostracodal biomicrite with fish scales, fish bones and fish teeth, phosphatized coprolites, and *Scoyenia* burrows. Ostracode valves are both disarticulated and articulated, and valves are oriented both concave up and concave down (Fig. 11). In the Borden Brick and Tile Quarry, the beds are discontinuous silty limestone to very calcareous siltstone. Wheeler and Textoris (1978) describe highly porous and microlaminated limestones near the center of the Durham sub-basin; the limestones are patchily silicified and contain few fossil remains.

In the Jurassic section of the Culpeper Basin at Thoroughfare Gap, Virginia, the (Hettangian) Waterfall Formation consists of ~ 35 m of laminated black shale (Hentz, 1985). Interlaminated with the shale are highly calcareous layers composed of alternating laminae of light-colored silty calcilutite or microspar, and dark, calcareous, organic-rich clayshale. Locally, the Waterfall Formation contains millimeter- to centimeter-scale graded limestone beds, similar to those described from the Hartford Basin. Other samples from the Culpeper Basin come from the Midland and Bull Run Formations.

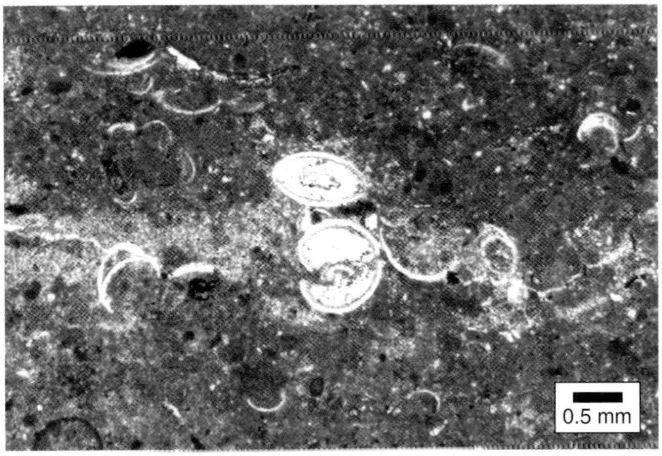


FIG. 11.—Photomicrograph showing ostracodal limestone from the Chatham Group lake sequence exposed in the Triangle Brick Quarry. Ostracodes are both articulated and disarticulated, and valves are oriented both concave up and concave down. Durham sub-basin.

### *Stable Isotopes and Minor Elements by Basin and Category*

Isotopic and minor-element compositions for limestones are summarized by basin and category in Figures 12, 13, and 14. Few trends emerge. Carbon isotope $\delta^{13}C$ values between –0.7 and –5.8‰ typify most meter-scale and millimeter- to centimeter-scale limestones of the Newark Supergroup (Fig. 12), with exceptions in the Gettysburg Basin and millimeter-to-centimeter-scale samples in the Culpeper Basin. Meter-scale limestones from the Gettysburg Basin are depleted in $^{13}C$ compared to similar rocks in the other basins. Spongy textured to framework-textured limestones have a much wider range of $\delta^{13}C$ values, to –15.1‰ in Rife Farm samples, Gettysburg Basin. Minor-element (Mg, Fe, Sr, and Mn) contents vary between basins and limestone types (Fig. 14), with mean concentrations generally lower and less variable in meter-scale limestones than in other limestone types.

Some modern lake limestones have $\delta^{13}C$ values that fall in the 0 to –5‰ PDB range (Komor, 1994; Benson et al., 1996), suggesting that these are fairly common values for alkaline lake carbonates. Arenas et al. (1997) report $\delta^{13}C$ values of between +0.4 and –6.4‰ from Miocene carbonate lake strata in the Ebro Basin, Spain. They suggest that these values reflect overall groundwater compositions, modified by photosynthetic $CO_2$ extraction, which shifts the precipitating carbonate towards heavier $\delta^{13}C$ values. Pratt et al.'s (1985) whole-rock carbonate data exhibit a wider range in $\delta^{13}C$ values, but because the data do not distinguish between types of limestone, the range reflects a variety of diagenetic and depositional parameters.

Samples from the Gettysburg Basin and some millimeter- to centimeter-scale and spongy textured to framework-textured limestones from the Fundy and Culpeper/Durham basins exhibit $\delta^{13}C$ values in the –5 to –12‰ range. This more $^{13}C$-depleted field (Fig. 12F) is more typical of lake carbonate produced from groundwaters and surface waters that have interacted with volcanic bedrock terrain, as in northern regions of the Tanzanian Manyara Basin, where the average $\delta^{13}C$ value is –9‰ PDB (Casanova and Hillaire-Marcel, 1992b).

#### *Fundy Basin.—*

Isotopic compositions of meter-scale limestones show a positive covariance, with a limited range of $\delta^{13}C$ values, –2.1 to –4.6‰, and $\delta^{18}O$ values from –2.2 to –8.0 ‰ (Fig. 12A). Spongy textured to framework-textured limestones have lower and less variable $\delta^{18}O$ values (–6.8 to –8.3‰), but $^{13}C$-depleted carbon isotope compositions, with $\delta^{13}C$ values varying from –3.3 to –15.9‰. Meter-scale limestones from the Fundy Basin exhibit Fe (mean concentration of 1061 mg/kg), Mg (1309 mg/kg), and Sr (204 mg/kg) contents that are low relative to similar rocks in other Supergroup basins (Fig. 14).

#### *Hartford Basin.—*

Stable-isotope compositions of meter-scale limestones in the Hartford Basin are similar to the Fundy Basin samples, with $\delta^{13}C$ = +1.4 to –5.3‰ and $\delta^{18}O$ = –2.5 to –5.5 ‰, except for a single sample of micritic limestone from the Shuttle Meadow Formation

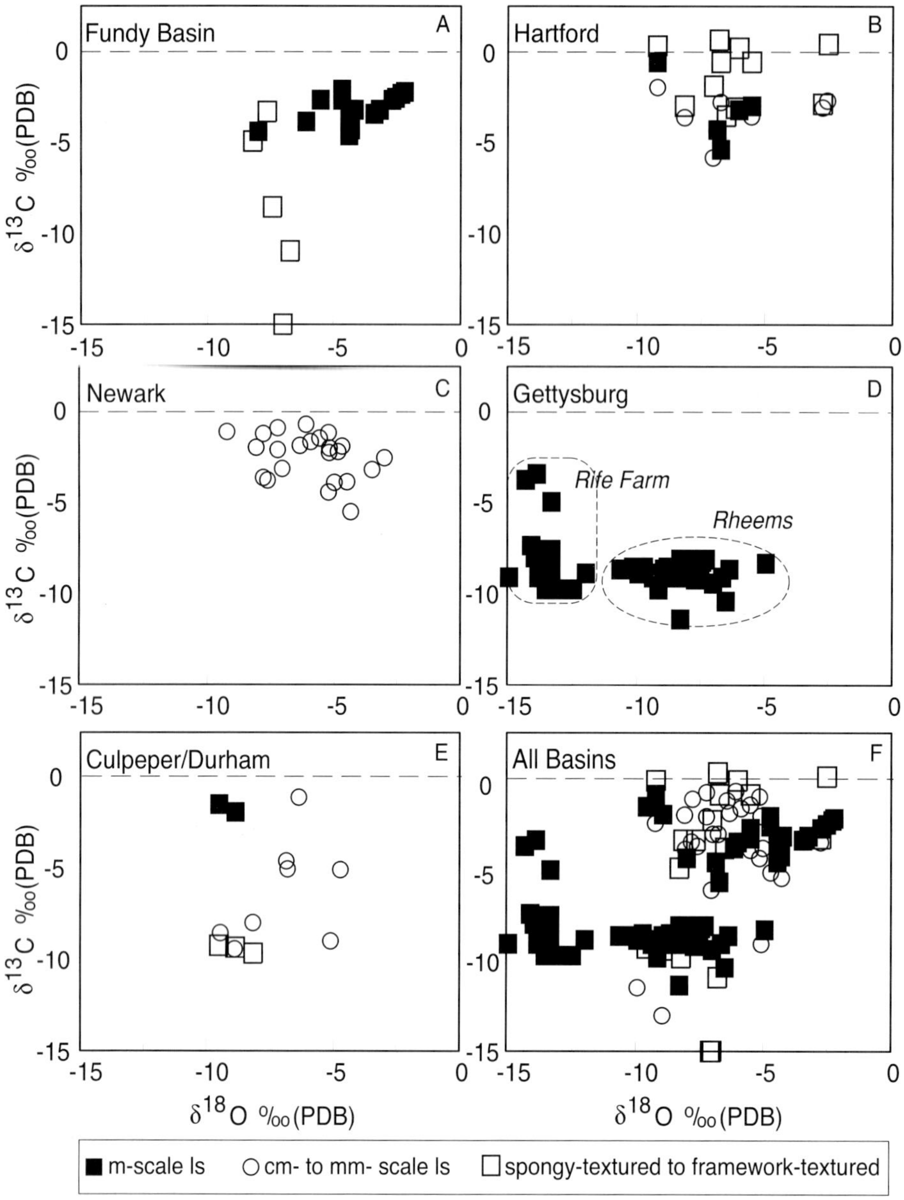


FIG. 12.— Carbon and oxygen isotope cross plots for samples from all limestone categories.

that is in direct contact with the underlying Talcott Basalt (–0.5 and –9.2‰, respectively (Fig. 12B). Thin, millimeter- to centimeter-scale limestones have similar oxygen isotope composition, but a more restricted range of $\delta^{13}C$ (–2.0 to –5.8‰). Spongy textured to framework-textured limestones also have similar $\delta^{18}O$ values but relatively $^{13}C$-enriched $\delta^{13}C$ values, +0.2 to –3.5‰, (Fig. 12B). Meter-scale limestones in both the Shuttle Meadow and East Berlin Formations have similar Mg and Fe contents, but Shuttle Meadow limestones have less Sr (average 511 vs. 142 mg/kg) and more Mn (2381 vs. 218 mg/kg) (Fig. 14). Limestone in direct contact with basalt (see comment above) has markedly higher Fe (15160 mg/kg), lower Sr (94 mg/kg), and lower Mn (1149 mg/kg). Millimeter- to centimeter-scale and spongy textured to framework-textured limestones have similar compositions, with greatest variation in Mg (up to 7.7 wt%) and, to a lesser extent, Mn contents.

*Pomperaug Basin.—*

Isotopic analyses were not made for carbonate from the Pomperaug Basin. Compared to millimeter- to centimeter-scale Shuttle Meadow Formation limestone from the nearby Hartford Basin, these samples have significantly lower Mg and Fe concentrations (Mg 2779 mg/kg, Fe 961 mg/kg) (Fig. 14).

*Newark Basin.—*

Stable-isotope and minor-element compositions in millimeter- to centimeter-scale limestones of the Newark Basin are similar to values in other basins (Fig. 12C). The $\delta^{13}C$ values range from –0.7 to –5.5‰, and the $\delta^{18}O$ values range from –0.4 to –8.8‰. Ooid-rich beds contain many of the most Mg- and Fe-rich samples (Fig. 14).

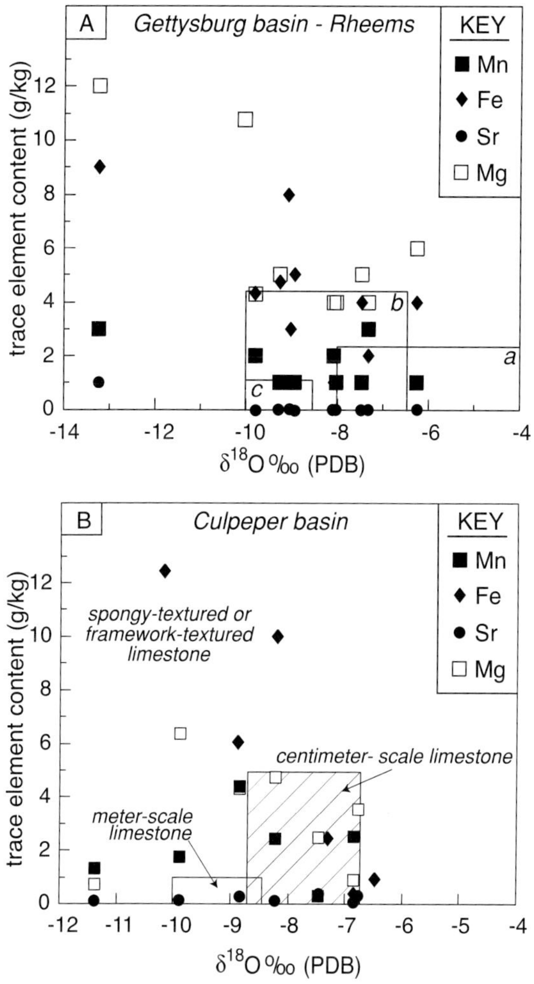


FIG. 13.—Trace-element composition of meter-scale limestone contrasted with oxygen isotope compositions. **A)** Comparison of values at Rheems locality (Gettysburg Basin) with all other basins. Box [a] delimits compositions in the Fundy Basin; box [b] shows values from the Hartford Basin; box [c] from the Culpeper Basin. Meter-scale limestone in the Gettysburg Basin has a much wider field of values, suggesting significant postdepositional (pedogenic and/or diagenetic) alteration. **B)** A comparison of trace-element and oxygen isotope compositions across limestone types in the Culpeper Basin. Spongy textured to framework-textured limestones (data) exhibit significantly greater geochemical variability than other limestone types (delimited fields), reflecting greater local temperature and source controls on their composition. Meter-scale limestones reflect more constant environmental conditions during carbonate precipitation.

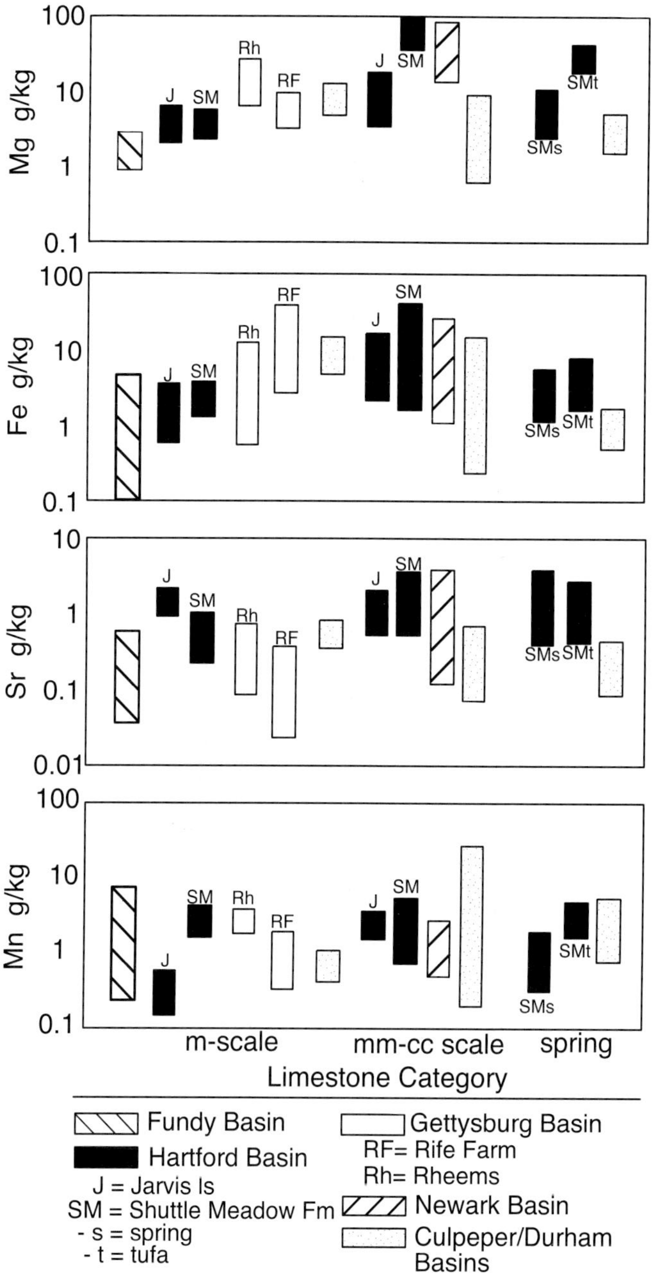


FIG. 14.—Trace-element geochemical plots showing the relative abundances of Mg, Fe, Sr, and Mn in all samples from the basins studied.

*Gettysburg Basin.—*

Isotopic compositions in meter-scale limestones form two distinct arrays (Fig. 12D). Samples from the Rheems locality have a relatively small range of $\delta^{13}C$ values (–7.6 to –11.4‰) and large range of $\delta^{18}O$ values (–4.9 to –10.6‰), whereas the Rife Farm samples show the opposite, with variable $\delta^{13}C$ (–3.4 to –9.7‰) and a small range of $\delta^{18}O$ (–12.0 to –15.1‰). These are the lowest $\delta^{18}O$ values for any of the Newark Supergroup limestones. The rocks contain slightly elevated concentrations of Mg and Fe (Fig. 14) relative to similar rocks in other basins, and there are compositional differences, particularly in Mn, between limestone from the two Gettysburg Basin sites (Fig. 14).

*Culpeper Basin.—*

The millimeter- to centimeter-scale limestones have $\delta^{13}C$ = –1.1 to –9.4‰, and include the most negative values for this category of limestone in the Newark Supergroup. Carbon isotope compositions of meter-scale limestones are $^{13}C$-enriched relative to the thin beds (Fig. 12E). Spongy textured to framework-textured limestone has the most negative isotopic compositions in the basin, with $\delta^{13}C = -9.2$ to $-9.7‰$ and $\delta^{18}O = -7.0$ to $-8.5‰$ (Fig. 12E). The millimeter- to centimeter-scale limestones exhibit marked variability in elemental compositions (Fig. 14), particularly in Mn content. The Mg, Fe, and Sr content of millimeter- to centimeter-scale and spongy textured to framework-textured carbonates are significantly lower than values from similar rocks in other basins (Fig. 14).

## INTERPRETATION OF ISOTOPIC AND GEOCHEMICAL COMPOSITIONS

Carbon isotope compositions of lacustrine limestones reflect the spectrum of sources and processes contributing to dissolved bicarbonate, including organic productivity, exchange with atmospheric $CO_2$, catchment-area bedrock, oxidation of aquatic or marginal vegetation, and incorporation of isotopically light soil $CO_2$ during lake lowstands (Komar, 1994; Abu-Jaber, 1998). Oxygen isotope compositions of lacustrine limestones are largely controlled by the temperature of carbonate precipitation and the composition of lake water, itself dependent on water source, residence time, and modifications controlled by the ratio of precipitation to evaporation (P:E) in the basin (Talbot and Kelts, 1990; Casanova and Hillaire-Marcel, 1992a). As such, wide variations in both carbon and oxygen isotope compositions are typical of lacustrine limestones (Talbot and Kelts, 1990; Casanova and Hillaire-Marcel, 1993) and interpretation of the isotopic data must consider lithologic, paleogeographical, paleoenvironmental, and diagenetic factors that may have influenced the compositions. Minor-element (Mg, Fe, Sr, Mn) chemistry of the carbonates may further constrain carbonate origin or alteration. Positive covariance between $\delta^{13}C$ and $\delta^{18}O$ values is commonly seen in perennial lakes from hydrologically closed basins (Talbot and Kelts, 1990). Other trends may result from pedogenic alteration (Hillaire-Marcel and Casanova, 1987) or ephemeral-lake carbonate deposition (Camoin et al., 1997). In concert with other observations, isotopic analyses may help constrain paleohydrology and paleoclimate in rift-basin lakes (Hillaire-Marcel and Casanova, 1987; Casanova and Hillaire-Marcel, 1993; Casanova, 1994; Arenas et al., 1997; Camoin et al., 1997).

### *Meter-Scale Limestones*

Meter-scale limestones of the Fundy and Hartford Basins have relatively heavy values of $\delta^{13}C$ (Fig. 12), a wide range of $\delta^{18}O$ values, and the isotopic compositions covary. These results support closed-basin hydrology (Talbot and Kelts, 1990). Temperature variation during carbonate precipitation cannot account for the almost 10‰ range in $\delta^{18}O$ values within a given basin. Enriched oxygen isotope compositions more likely reflect increased residence time or decreased P:E ratios. The most negative $\delta^{18}O$ values can be interpreted as the baseline value of incoming waters, with more positive values reflecting evaporation (Casanova and Hillaire-Marcel, 1992a, 1992b). Carbon isotopes in these limestones most likely reflect lake systems with total dissolved inorganic carbon (TDIC) dominated by exchange with atmospheric $CO_2$, moderated by regional source rocks or biological activity affecting the $\delta^{13}C$ of TDIC (Komor, 1994; Benson et al., 1996).

Meter-scale limestones from the more northerly Fundy and Hartford basins are 5‰ enriched in $^{18}O$ compared to similar limestones from the more southerly Gettysburg and Culpeper basins. Isotopic enrichment is the opposite of what would be expected from a latitude effect on precipitation (Dansgaard, 1964; Swart et al., 1993). Sedimentological indicators suggest that the northern basins were more arid than those to the south (Weems and Olsen, 1997), and the heavier oxygen values can be accounted for by decreased P:E towards the north. Assuming a temperature of 25°C and using the fractionation factor of Friedman and O'Neil (1977), oxygen isotope compositions of open lacustrine micrite at Rheems (Gettysburg Basin) suggests Triassic lake water with $\delta^{18}O = -7.8‰$ V-SMOW (de Wet et al., 1998). Compositions in meter-scale limestones from other basins suggest lake waters that were slightly enriched, with $\delta^{18}O$ values of about $-4.2‰$ V-SMOW.

The lithologies present at the Rheems locality (Gettysburg Basin; Fig. 12D) indicate well-mixed and well-oxygenated lake waters (de Wet et al., 1998). Low $\delta^{13}C$ values of Rheems limestone ($-7.6$ to $-11.4‰$) are consistent with petrographic evidence of subaerial exposure and pedogenic modification. In contrast, the large range of $\delta^{13}C$ values from the Rife Farm locality (Gettysburg Basin; Fig. 12D) show positive covariance with the total organic content of the limestone, and reflect variable levels of lake productivity (de Wet et al., 1998). This is consistent with petrographic evidence for lake deepening and subsequent shallowing as recorded in the change from fauna-rich, bioturbated gray limestone to laminated, organic-rich, black limestone, and back to gray, mottled limestone.

The 23-m-thick Rife Farm limestone is the oldest and thickest of the meter-scale Newark Supergroup limestones and represents a long-lived lacustrine system dominated by carbonate deposition (de Wet et al., 1998). Oxygen isotope values derived from meter-scale carbonates from the Rife Farm site are, on average, 5‰ lower than limestones at Rheems, located only 30 km away and within the same basin. It is unlikely that the lakes were sourced by significantly different water or deposited under significantly different climatic conditions. Rather, the very low oxygen isotope values at Rife Farm reflect a localized burial history accompanied by reequilibration with warm diagenetic fluids (de Wet et al., 1998).

Figure 14 compares the minor-element content of Rheems limestone with meter-scale limestones in other basins, delimited by boxes. Fairly uniform Sr contents between 100 and 50 mg/kg, but more variable Mn compositions, suggest slight, and variable, modification of the original lake-water chemistry. Much higher Fe and Mg contents in a few Rheems samples correlate well with more negative $\delta^{18}O$ values. The Sr-rich but Fe- and Mn-poor "Jarvis Limestone" (Hartford Basin) may reflect the fossiliferous nature of the rock, rich in ostracodes and fish-bone fragments, and a very low degree of diagenetic alteration, because Sr is strongly partitioned into most fluids relative to calcite. High Mn and variable Sr contents in Shuttle Meadow limestones (Hartford Basin) likewise suggest only slight diagenetic modification. Thus, interbasinal variations in minor-element and isotopic compositions are likely due to local differences in both the depositional and diagenetic environments.

Local bedrock may also affect lacustrine isotopic compositions. For example, where bedrock consists of crystalline rock, the average $\delta^{13}C$ of lake water may be $+5.0‰$, but within the same catchment area, where the groundwater and surface water drain volcanic bedrock, the average $\delta^{13}C$ value of lake water is $-9.0‰$ (Casanova and Hillaire-Marcel, 1992b). Lake water may represent a mix of these two isotopically distinct waters, having a general value of $\delta^{18}O = -1.9$ to $-2.4‰$ (Casanova and Hillaire-Marcel, 1992b). In the Fundy, Hartford, and Culpeper Basins, Newark Supergroup limestones have a restricted range of $\delta^{18}O$ values, 0 to $-5.0‰$. Carbonates from Rife Farm have a broader range, $\delta^{18}O = -3.0$ to $-10.0‰$, and limestone from Rheems has a narrow range of only 3‰ ($-7.0$ to $-10.0‰$). The depleted limestones are more typical of lacustrine deposits precipitated from lake waters derived from volcanic terrane, such as the northern regions of the Tanzanian Manyara Basin where the average lake water has a $\delta^{18}O$ value of –9.0‰ (Casanova and Hillaire-Marcel, 1992b). Note that in the Fundy and Hartford Basins the thickest limestone directly stratigraphically overlies basalt. The more positive Newark Supergroup limestones may represent lake carbonate produced from groundwaters and surface waters that have interacted with crystalline bedrock (Blum et al., 1998).

### *Millimeter- to Centimeter-Scale Limestones*

The millimeter- to centimeter-scale limestones in the Newark and Hartford Basins have relatively restricted ranges of $\delta^{13}C$ values that are similar to compositions of meter-scale limestones in the same basins. Carbon isotope compositions are more variable in the Culpeper Basin. We interpret carbon isotope compositions from millimeter- to centimeter-scale beds of the Newark Supergroup to largely reflect an approach to atmosphere–water equilibration, with lighter values attributed to an increased component of organically derived carbon. Light carbon can be contributed to lake carbonate by pedogenesis or oxidation of marginal vegetation.

The range of carbon and oxygen isotopic values in millimeter- to centimeter-scale samples reflects the diverse origins of these beds. Their stratigraphic and sedimentological features indicate that they generally formed as turbidites, but their isotopic compositions indicate that the source carbonate may have had either a shallow, littoral source or a more profundal origin. For example, the small range of carbon isotope compositions in the Hartford and Newark Basins indicates a relatively stable source for the carbonate material, probably from deeper parts of the lake, where pedogenic or lake-margin processes would have had little effect on the carbon isotope signature. In contrast, negative and variable $\delta^{13}C$ values in the Culpeper Basin may reflect marginal or pedogenic processes and the incorporation of vegetation-derived carbon, suggesting that these carbonate clasts were sourced in the shallow, littoral zone.

Millimeter- to centimeter-scale limestone composed of low-Mg micrite and interbedded with shale or siltstone in the Newark and Culpeper Basins do not contain shelly faunal remains and have very fine-grained clay matrices (revealed by scanning electron microscopy) suggesting microbially mediated carbonate precipitation, directly from lake water. Relatively $^{13}C$-enriched carbon isotope compositions in pure micrite beds from the Newark Basin have tightly clustered $\delta^{13}C$ values (–0.7 to –0.9‰; Fig. 12C), suggesting a similar origin for each bed. One possible interpretation is that the fine-grained micrite formed by seasonal events, as documented in Thompson et al. (1990). Thompson et al. (1990) attributed precipitation of fine-grained micrite in Green Lake, New York, to coccoid cyanobacterial epicellular biomineralization. Bio-induced calcite precipitation of micrite has been reported in other lakes as well (Hsü and Kelts, 1978; Donovan, 1994). Isotopic evidence for carbonate origin in the Newark Basin is ambiguous, however, inasmuch as the measured isotopic compositions are neither significantly light nor heavy, making it, therefore, unclear to what extent, and in what way, the limestones were affected by biological processes. Marine whiting events, or the precipitation of calcite directly from seawater, is controversial (Shinn et al., 1989), but lake whiting events have been documented (Muller and Wagner, 1978; Thompson et al., 1990). This process reflects seasonal changes in the P:E ratio, with whitings being associated with evaporative periods.

Very thin, laminated, organic-rich limestones in the Lockatong and Waterfall Formations (Newark and Culpeper Basins, respectively) are interpreted as having formed in deep lakes in which water-column stratification and low dissolved oxygen created conditions conducive to micrite precipitation (Gore, 1986c; Thompson et al., 1990; Platt and Wright, 1991). Conditions of low dissolved oxygen may explain the relatively high values of Mn and Fe (up to 2,000 mg/kg Mn and 18,000 mg/kg Fe), inasmuch as reducing conditions are necessary to yield divalent cations, promoting their subsequent incorporation into calcite (Fig. 14). The presence of calcareous micronodules, stringers, and wavy laminae of calcite within mudstone deposits indicates that calcite precipitation occurred at the sediment-water interface, or within the upper few centimeters of sediment.

Van Houten (1987) suggested that some of these types of layers, particularly in the Newark Basin, were "varves" in that they may record an annual cycle of lacustrine stratification and deposition. Our interpretation that these very pure micrite laminae represent precipitation events, either biogenic or abiogenic, strengthens the idea that the climate of the Mesozoic rift basins was strongly seasonal, or monsoonal (Olsen and Kent, 1996). Month-long dry periods resulted in lake stratification and carbonate supersaturation, producing the thin, sharply bounded limestone laminae and beds.

### *Spongy Textured to Framework-Textured Limestone*

Recent sediments in Lake Bogoria (Kenya rift valley; Renaut et al., 1986; Renaut and Tiercelin, 1994), the Ethiopian rift valley (Le Turdu et al., 1999), and Lake Tanganyika (Tiercelin, 1991) show that shallow rift-basin lakes and shorelines may be sites of discharge of heated groundwater. Spongy textured to framework-textured carbonates in the Newark Supergroup sediments are interpreted similarly as spring or seep carbonates. Previous workers have interpreted tufa textures and wurzite-bearing stromatolite fabrics in the Culpeper, Fundy, and Hartford Basins, and the Durham sub-basin, as spring deposits (Carozzi, 1964, Steinen et al., 1987; Wheeler and Textoris, 1978; Gore, 1988a; de Wet and Hubert, 1989).

Within the Fundy and Culpeper–Durham Basins, the isotopic composition of spongy textured to framework-textured limestones are systematically lower than associated limestones (Fig. 12A, E). Extremely variable carbon isotope compositions range by more than 16‰, to values as low as –15.9‰. In contrast, spongy textured to framework-textured carbonate from the Hartford Basin has more $^{13}C$-enriched $\delta^{13}C$ values (~ 0 to –5.5‰). The more depleted isotopic values may indicate maximum organic influence, whereas the more enriched values possibly result from abiotic precipitation and $CO_2$ degassing (Farmer, 2000; Fouke et al., 2000).

Spongy textured to framework-textured carbonates in the Fundy Basin are depleted in $^{18}O$ relative to meter-scale limestones from the same basin (Fig. 12A). $\delta^{18}O$ values in the Hartford and Culpeper Basins are more similar to those of meter-scale limestones. $\delta^{18}O$ values in all basins are in the range –5.0 to –9.0‰, consistent with carbonate precipitation from warm (20–50°C) meteoric water (de Wet and Hubert, 1989). Thus, spongy textured to framework-textured deposits are attributed to precipitation from fluids debouching at slightly elevated temperatures as springs on the lake floor or shoreline (de Wet and Hubert, 1989). The variation in $\delta^{18}O$ in the Hartford and Culpeper Basins reflects more variable spring sources or precipitation temperatures.

In general, minor-element chemistry of spongy textured to framework-textured limestone deposits are much more variable than millimeter- to centimeter-scale and meter-scale carbonates from the same basin (e.g., Culpeper Basin; Figs. 13, 14).

More variable Fe and Mn contents may reflect groundwater interaction with bedrock and/or different oxidation states of circulating groundwater. The relatively high Sr content of Shuttle Meadow tufas (Hartford Basin; Fig. 14) may be a result of the interaction of circulating groundwaters with basalt prior to spring debouching.

## ORIGIN OF RIFT-BASIN LIMESTONES

The presence of limestone in continental lakes provides insight into tectonic and climatic factors that affect sediment accumulation (Carroll and Bohacs, 1999; May et al., 1999; Bohacs et al., 2000). Lacustrine carbonate precipitation is readily overwhelmed by siliciclastic sedimentation, and, therefore, deposition of Newark Supergroup limestone reflects periods when siliciclastic sedimentation was significantly reduced. The origin of these Mesozoic rift-basin limestones is linked to volcanic processes and seasonal (?) climatic variations, with different sedimentological products for each process. We recognize three limestone associations common to multiple Newark Supergroup basins. These are: (1) a carbonate–basalt association; (2) carbonate–profundal sediment interlayering; and (3) a carbonate–plant and microbialite association. Each association is the result of different sedimentological processes and results in one or more of the limestone categories described earlier. For example, meter-scale limestone may be associated with basalt or with vegetation and microbialite, but tufa is associated with basalt. Millimeter- to centimeter-scale limestone beds are interlayered with fine-grained siliciclastic material, indicating a distal littoral or profundal depositional environment. Each association and its carbonate products are described in the following three sections.

Although carbonate production may be only a part of the evolution of a lake, as described by Carroll and Bohacs (1999), it is likely to yield detailed information about the balance between source-area dynamics and climate. The results of this study, which focused only on Newark Supergroup *carbonate* strata, concurs with Olsen's (1990b) tripartite division of all Newark Supergroup lake deposits. His "Richmond-type", or the fluvial–lacustrine facies of Carroll and Bohacs (1999), typifies carbonates from the southern basins but also includes the Rheems location in the mid-Atlantic Gettysburg Basin. The "Newark-type", or fluctuating profundal facies, is characteristic of millimeter- to centimeter-scale turbidite deposition and whiting precipitation, as documented from the Culpeper, Gettysburg, Newark, and Hartford Basins. The abundance of mudcracks in meter-scale limestone in the Fundy Basin typifies the "Fundy-type" depositional style of Olson (1990b), or the evaporative facies of Carroll and Bohacs (1999).

The Olsen (1990b) and Carroll and Bohacs (1999) schemes do not, however, include tufa facies. Tufas are potential indicators of another important process in lacustrine carbonate deposition, those related to igneous processes. Volcanism may exert a primary control on whether carbonate deposition ensues, because a significant amount of volcanism changes drainage patterns, ponding surface waters (Yuretich, 1982). Cooling igneous rocks provide heat to groundwater, increasing circulation rates and ion exchange processes, facilitating carbonate ion accumulation. Hydrothermal plumbing systems associated with igneous activity generate springs and seeps, potentially resulting in deposition of calcareous tufa, and basalt mantling may inhibit siliciclastic weathering long enough for carbonates to accumulate in significant quantity to be preserved in the rock record. We suggest that there is a causal relationship between the association of limestone with volcanic rocks in some rift-basin settings.

### *Basalt Mantling and the Role of Volcanism*

The presence of meter-scale limestone deposits in the Newark Supergroup poses the interesting paradox that increasing lake water depths (Olsen et al., 1993) and the formation of thick limestone successions occurred coevally with major changes in rift-basin dynamics and depositional style (Weems and Olsen, 1997). These changes correlate with the Triassic–Jurassic boundary and are characterized by increased regional tectonic extension, outpouring of flood basalt, and increased siliciclastic sedimentation rates (Olsen et al., 1993). Yet meter-scale limestones directly overlie Hettangian Basalts in three localities, one in the Fundy Basin (Scots Bay Formation), and two in the Hartford Basin ("Jarvis" and Coe's Quarry Formations). The depositional style of the Scots Bay, "Jarvis", and Coe's Quarry limestones are very similar, but probably only the Scots Bay and Coe's Quarry rocks are coeval. Only the thick carbonates of the Gettysburg Basin appear to be unrelated to basalt flows. Extrusive volcanism is documented as having a significant influence on lake characteristics and styles of sedimentation in East African rift lakes (Yuretich, 1982; Renaut and Owen, 1991; Cohen et al., 1997). In Miocene rift-basin deposits in Idaho, limestone distribution follows the contours of the basalt plain (although alluvial-plain sediments separate the limestone from the basalt) (Straccia et al., 1990), suggesting a correlation between presence of basalt and deposition of limestone. Our study suggests that mantling of siliciclastic rift sediments by basalts might have created a barrier to siliciclastic input into lacustrine depositional sites, allowing carbonate to accumulate without significant dilution by siliciclastic sediments. For example, in the Newark Basin, the Orange Mountain, Preakness, and Hook Mountain Basalts reached a minimum area of several thousand square kilometers and a thickness of over 130 m (Puffer and Student, 1992; Tollo and Gottfried, 1992). The volcanic rocks would effectively have mantled all underlying siliciclastic rocks and inhibited local siliciclastic erosion for some time. Rainwater leaching through fresh basalt can yield large quantities of calcium ions (Lavoie, 1995; Renaut, 1999, personal communication), and this process may be rapid compared with the formation of clays by basalt weathering. This may give the carbonates a head start in accumulation, allowing limestone to form until being superseded by siliciclastic deposition (Gierlowski-Kordesch, 1998).

Weathering rates of basaltic rocks might have been accelerated by an increase in humidity postulated for Hettangian climate relative to earlier Triassic times (Parrish et al., 1982). Circulation of heated groundwater associated with basaltic volcanism might have increased chemical weathering of calcium-rich minerals, chiefly plagioclase, in the basalt. Heated groundwater may also increase chemical breakdown of catchment-area rocks, resulting in an increased dissolved load, yielding carbonate precipitation at discharge sites such as spring and seeps. Tufa, and its associated plumbing system in underlying basalt, is exposed in the Scots Bay Formation (Fundy Basin) (de Wet and Hubert, 1989). Spongy textured to framework-textured limestones from the Hartford Basin ("Jarvis" and Coe's Quarry limestones) are interpreted to have formed as tufas associated with sublacustrine or littoral-zone springs.

### *Turbidites and Whitings*

The thin limestone deposits preserved in distal littoral or proximal profundal strata in Newark Supergroup lacustrine rocks formed either as turbidites or as whitings. Coarse-grained carbonate layers with normally graded bedding, interlayered with finer-grained, laminated lacustrine sediments, are interpreted as turbidite or tempestite deposits (Sturm and Matter, 1978; Platt and Wright, 1991; Harris et al., 1994). Scoured bases and gradational upper surfaces indicate deposition from high-energy currents into an otherwise relatively quiet-water setting; broken and abraded lacustrine faunal and floral remains in these beds further support intermittent storm or turbidity current deposition, potentially associated with seasonal monsoons that would flush nearshore carbonates into deeper water. Articulated shelly fauna, particularly ostracodes, in intervening fine-grained layers suggest background low-energy deposition in the epilimnion. Many millimeter- to centimeter-scale deposits in the Newark Supergroup contain intraclasts composed of micritic limestone. Intraclasts likely represent erosion of littoral-zone semiconsolidated lime mud and its redeposition as profundal sediments. The presence of intraclasts indicates that transportation distances were relatively short, with limited reworking of the sediment *en route*. Thus, bedload transport of carbonate clasts into profundal sediments was probably associated with storms that reworked the clasts and/or fractured carbonate-cemented underwater crusts and redistributed them as downslope intraclastic turbidites. In addition to storm-generated turbulence, earthquakes may have triggered slope failure and turbidite deposition. For example, Lake Biwa, Japan, contains 20 turbidite beds within 3 m of sediment that correlate with the historical record of earthquakes over the past 1300 years (Inouchi et al., 1996).

Homogeneous micritic laminae with sharp upper and lower contacts were precipitated directly from supersaturated lake water, triggered by bio-mediated coccoid algal blooms or evaporative concentration and abiotic whiting formation. These laminae are always interlayered with quiet-water siliciclastic strata, lack shelly faunal remains, and consist of euhedral calcite crystals that are relatively enriched in Fe and Mn. Enrichment in iron and manganese indicates that porewaters were dysaerobic, enabling cations to be reduced to their divalent state to then be incorporated into precipitating calcite. This sedimentological and geochemical evidence, coupled with documentation that seasonal whiting events occur in modern lakes (Muller and Wagner, 1978; Thompson et al., 1990; Platt and Wright, 1991), suggests that the Newark Supergroup rift lakes also experienced seasonal precipitation events. Isotopic evidence is equivocal, however. The carbon isotope values are not significantly $^{18}C$ depleted, which would have suggested a bio-mediated origin, and the oxygen isotope values are not significantly $^{18}O$ enriched, indicating an evaporative origin. Therefore, we conclude, judging from sedimentological evidence and minor-element data, that the millimeter- to centimeter-scale micrite laminae probably formed because of seasonal factors such as algal blooms or evaporative concentration, but further isotopic analyses are needed to substantiate this interpretation.

### *Associated Organic Matter*

A number of the Newark Supergroup limestones are associated with fossilized plant and microbial material. In particular, centimeter-thick limestones from the western Newark Basin (Passaic Formation) contain fossil tree remains and root traces (see Fig. 4). Stromatolites and microbial encrustations around the tree remains indicate that rising lake levels drowned palustrine vegetation (Olsen, 1989). Carbonate precipitation continued in the form of ooids, oncolites, and micrite. The abundance of microbial fabrics and stromatolites suggests that micrite precipitation was associated with biologically induced pH changes, possibly enhanced by seasonal evaporation (Platt and Wright, 1991).

Similar features such as fossil tree remains, stromatolites, and oncolites occur in basal limestones in the Fundy Basin (Scots Bay Formation) (see Figures 3 and 4). These features are interpreted as indicators of a lacustrine transgression over a previously vegetated setting (de Wet and Hubert, 1989). The Rheems locality (Gettysburg Basin) also contains evidence for vegetation as root traces and patches of amorphous organic matter (de Wet et al., 1998). The Rheems Limestone is interpreted as a shallow, alluvial-plain palustrine to lacustrine deposit (de Wet et al., 1998). In all of these situations the close association of vegetation and limestone suggests a causal relationship. It indicates that vegetation might have served as a baffle to incoming siliciclastic material, or fluvial transport of siliciclastic grains might have been largely confined to an anastomosing river system, allowing carbonates to form in adjacent floodplain low areas (e.g., Flores and Hanley, 1984; Platt and Wright, 1991; Valero Garces et al., 1997). Plant remains and mottles in the rift limestones suggest that most of the calcium carbonate was transported to the lake site in solution rather than as bedload, because carbonate clasts might also have been caught in a vegetative buffer zone (Dungan and Driese, 1999) or simply dissolved after long transport distances (Gierlowski-Kordesch, 1998). Because many of the catchment areas of the Gettysburg, western Newark, and Fundy rift Basins contain older limestone or dolomite, there may have been relatively minor amounts of siliciclastic material in the erosional system, and supersaturated surface and groundwater readily resulted in carbonate precipitation (Donovan, 1994; Gierlowski-Kordesch, 1998).

The mechanisms proposed for limestone deposition in Newark Supergroup rift lakes take into account the observed carbonate–basalt association, carbonate–profundal sediment interlayering, and carbonate–plant and microbialite association. In each association there are multiple limestone examples from widely spaced, independent basins, indicating that each depositional mechanism transcends local factors and represents significant depositional processes.

## CONCLUSIONS

Distinct patterns in limestone deposition can be recognized in siliciclastic-dominated strata, and reveal information on past volcanic and climatic events. Our documentation of three types of carbonate deposition, linked to three distinct associations and depositional processes, provides new insight into lacustrine rift sedimentation. The relevance of these depositional styles is in what they reveal about the response of an individual basin to external factors, as well as processes common to all of the basins. Such patterns may be recognizable in other ancient lacustrine successions, and further study may strengthen the causal relationships between the three associations documented here.

We note the significant impact that volcanism may have on carbonate deposition in lakes. Basalt mantling alters drainage patterns and inhibits siliciclastic deposition. Volcanism accelerates groundwater circulation and chemical alteration of aquifer rocks. This results in prolonged periods dominated by carbonate deposition, resulting in thick (meter-scale) limestone strata. Sites of spring or seep discharge associated with heated groundwater produced Newark Supergroup tufa deposits. Variable $\delta^{13}C$ and $\delta^{18}O$ values, and minor-element content in the tufas reflect local conditions such as type of bedrock associated with the plumbing of spring or seeps, temperature of spring or seep waters, and whether there was any biological mediation in carbonate precipitation. Covarying $\delta^{13}C$ and $\delta^{18}O$ values in meter-scale limestones situated stratigraphically above basalts indicate closed-basin hydrology, and $^{18}O$-enriched oxygen compositions reflect increased residence time of the lake waters. Depleted carbon isotope compositions are due to migration of source waters through volcanic bedrock.

Thinner limestones, such as the millimeter- to centimeter-scale limestone beds described here, are consistently interbedded with fine-grained siliciclastic deposits. This indicates that limestone deposition was episodic, related either to storm-generated or earthquake-generated substrate instability, or to whiting events. Newark Supergroup carbonate turbidites derived from littoral zones are depleted in $^{13}C$ and have more variable minor-element and $\delta^{18}O$ values relative to turbidites sourced from distal littoral or profundal areas, reflecting incorporation of pedogenic or vegetation-related calcite precipitation. We conclude that storm-related turbulence, particularly if linked to monsoonal conditions of high rainfall following long dry periods, resulted in flushing of proximal littoral-zone carbonates into distal settings.

Particular laminae in the Newark Supergroup lake strata are composed of uniform, euhedral calcite crystals. These laminates represent inorganic whitings, or bio-induced precipitation related to algal blooms. Stable-isotope compositions of the calcite crystals are inconclusive as to their origin, but their enriched $Fe^{2+}$ and $Mn^{2+}$ concentrations suggest that water-column stratification and low-oxygen conditions were important influences on precipitation. Such conditions may have been the result of long dry periods associated with a monsoonal climate. Therefore, the presence of thin limestone beds can be used as a paleoclimatic indicator suggestive of episodic deposition linked to short-term, seasonal (?) climate fluctuations. Similarly, the 5‰ oxygen isotope enrichment in the limestones of the northern basins, relative to those of the southern basins, indicates increased evaporation and a more arid climate to the north.

Although carbonate rocks are susceptible to diagenetic alteration, we conclude that that all of the Newark Supergroup limestones, except those from the Rife Farm site (Gettysburg Basin), reflect little diagenetic alteration, and preserve their depositional carbon isotope signatures and discernible oxygen isotopic trends. With the exception of Rife Farm samples (Gettysburg Basin), oxygen isotope compositions vary systematically according to limestone category, associated sedimentary features, and origin, indicating that they are a useful tool in reconstructing paleolacustrine conditions.

## ACKNOWLEDGMENTS

We are grateful to R. Steinen, who provided samples of the carbonates from Coe's Quarry, D. Textoris, who gave samples of the Wheeler and Textoris site rocks near Durham, as well as oolites from the Culpeper Basin, P. Huber, who provided samples and information from the little-known Pomperaug Basin, and N. McDonald, who provided some of the Hartford Basin samples. All other samples were collected by the authors. This study was supported by National Science Foundation EAR-9512288 grant to C. Mora. V. Pedone and J. Casanova provided helpful reviews.

## REFERENCES

Abu-Jaber, N.S., 1998, A new look at the chemical and hydrological evolution of the Dead Sea: Geochimica et Cosmochimica Acta, v. 62, p. 1471–1479.

Arenas, C., Casanova, J., and Pardo, G., 1997, Stable-isotope characterization of the Miocene lacustrine systems of Los Monegros (Ebro Basin, Spain): Palaeogeographic and palaeoclimatic implications: Palaeogeography, Palaeoclimatology, Palaeoecology, v. 128, p. 133–155.

Benson, L., White, L.D., and Rye, R., 1996, Carbonate deposition, Pyramid Lake Subbasin, Nevada: 4. Comparison of the stable isotope values of carbonate deposits (tufas) and the Lahontan lake-level record: Palaeogeography, Palaeoclimatology, Palaeoecology, v. 122, p. 45–76.

Birney–de Wet, C.C., 1984, Sedimentology and Petrology of the Scots Bay Formation (Lower Jurassic), Nova Scotia, Canada: Unpublished M.S. Thesis, University of Massachusetts, Amherst, 325 p.

Blum, J.D., Gazis, C.A., Jacobson, A.D., and Chamberlain, C.P., 1998, Carbonate versus silicate weathering in the Raikhot watershed within the High Himalayan Crystalline Series: Geology, v. 26, p. 411–414.

Bohacs, K.M., Carroll, A.R., Mankiewicz, P.J., and Neal, J.E., 2000, Lake basin type, source potential, and hyrdrocarbon character: an integrated sequence stratigraphic geochemical framework, *in* Gierlowski-Kordesch, E.H., and Kelts, K.R., eds., Lake Basins Through Space and Time: American Association of Petroleum Geologists, Studies in Geology 46, p. 3 31.

Camoin, G., Casanova, J., Rouchy, J.-M., Blanc-Velleron, M.-M., and Deconinck, J.-F., 1997, Environmental controls on perennial and ephemeral carbonate lakes: the central palaeo–Andean Basin of Bolivia during late Cretaceous to early Tertiary times: Sedimentary Geology, v. 113, p. 1–26.

Carozzi, A.V., 1964, Complex ooids from Triassic lake deposit, Virginia: American Journal of Science, v. 262, p. 231–241.

Carroll, A.R., and Bohacs, K.M., 1999, Stratigraphic classification of ancient lakes: Balancing tectonic and climatic controls: Geology, v. 27, p. 99–102.

Casanova, J., 1994, Stromatolites from the East African rift: A synopsis, *in* Bertrand-Sarfati, J., and Monty, C., eds., Phanerozoic Stromatolites II: Dordrecht, The Netherlands, Kluwer Academic Publishers, p. 193–226.

Casanova, J., and Hillaire-Marcel, C., 1992a, Late Holocene hydrological history of Lake Tanganyika, East Africa, from isotopic data on fossil stromatolites: Palaeogeography, Palaeoclimatology, Palaeoecology, v. 91, p. 35–48.

Casanova, J., and Hillaire-Marcel, C., 1992b, Chronology and paleohydrology of Late Quaternary high lake levels in the Manyara Basin (Tanzania) from isotopic data ($^{18}O$, $^{13}C$, $^{14}C$, Th/U) on fossil stromatolites: Quaternary Research, v. 38, p. 205–226.

Casanova, J., and Hillaire-Marcel, C., 1993, Carbon and oxygen isotopes in African lacustrine stromatolites: Palaeohydrological interpretation, *in* Swart, P.K., Lohmann, K.C., McKenzie, J., and Savin, S., eds., Climate Change in Continental Isotopic Records: American Geophysical Union, Geophysical Monograph 78, p. 123–133.

Cloos, E., and Pettijohn, F.J., 1973, Southern border of the Triassic Basin, west of York, Pennsylvania: fault or overlap?: Geological Society of America, Bulletin, v. 84, p. 523– 536.

Cohen, A.S., Talbot, M.R., Awramik, S.M., Dettman, D.L., and Abell, P., 1997, Lake level and paleoenvironmental history of Lake Tanganyika, Africa, as inferred from late Holocene and modern stromatolites: Geological Society of America, Bulletin, v. 109, p. 444–460.

Dansgaard, W., 1964, Stable isotopes in precipitation: Tellus, v. 16, p. 436–468.

de Wet, C.B., and Hubert, J.F., 1989, The Scots Bay Formation, Nova Scotia, Canada, a Jurassic carbonate lake with silica-rich hydrothermal springs: Sedimentology, v. 36, p. 857– 873.

de Wet, C.B., Yocum, D.A., and Mora, C.I., 1998, Carbonate lakes in closed basins: sensitive indicators of climate and tectonics; an example from the Gettysburg Basin (Triassic), Pennsylvania, USA, *in* Shanley, K., and McCabe, P., eds., Relative Role of Eustasy, Climate, and Tectonism in Continental Rocks: SEPM, Special Publication 59, p. 191–209.

Demicco, R.V., and Gierlowski-Kordesch, E.H., 1986, Facies sequences of a semi-arid closed basin: the lower Jurassic East Berlin formation of the Hartford Basin, New England: Sedimentology, v. 33, p. 107–118.

Dickson, J.A.D., 1966, Carbonate identification and genesis as revealed by staining: Journal of Sedimentary Petrology, v. 36, p. 491–505.

DONOVAN, J.J., 1994, On the measurement of reactive mass fluxes in evaporative groundwater-source lakes, *in* Renaut, R.W., and Last, W.M., eds., Sedimentology and Geochemistry of Modern and Ancient Saline Lakes: SEPM, Special Publication 50, p. 33–50.

DUNGAN, S.P., AND DRIESE, S.G., 1999, Control of terrestrial stablization on Late Devonian palustrine carbonate deposition: Catskill Magnafacies, New York, USA: Journal of Sedimentary Research, v. 69, p. 772–783.

FARMER, J.D., 2000, Hydrothermal systems: Doorways to early biosphere evolution: GSA Today, v. 10, no. 7, p. 1–9.

FLORES, R.M., AND HANLEY, J.H., 1984, Anastomosed and associated coal-bearing fluvial deposits: Upper Tongue River Member, Palaeocene Fort Union Formation, northern Powder River Basin, Wyoming, U.S.A., *in* Rahmani, R.A., and Flores, R.M., eds., Sedimentology of Coal and Coal-Bearing Sequences: International Association of Sedimentologists, Special Publication 7, p. 85–103.

FOUKE, B.W., FARMER, J.D., DES MARAIS, D.J., PRATT, L., STURCHIO, N.C., BURNS, P.C., AND DISCIPULO, M.K., 2000, Depositional facies and aqueous–solid geochemistry of travertine-depositing hot springs (Angel Terrace, Mammoth Hot Springs, Yellowstone National Park, USA): Journal of Sedimentary Research, v. 70, p. 565–585.

FRIEDMAN, I., AND O'NEIL, J.R., 1977, Compilation of isotopic fractionation factors of geochemical interest: U.S. Geological Survey, Professional Paper 440-KK, 96 p.

FROELICH, A.J., AND OLSEN, P.E., 1984, Newark Supergroup, a revision of the Newark Group in eastern North America: U.S. Geological Survey, Bulletin 1537-A, p. A55–A58.

GIERLOWSKI-KORDESCH, E.H., 1998, Carbonate deposition in an ephemeral siliciclastic alluvial system: Jurassic Shuttle Meadow Formation, Newark Supergroup, Hartford Basin, USA: Palaeogeography, Palaeoclimatology, Palaeoecology, v. 140, p. 161–184.

GIERLOWSKI-KORDESCH, E.H., AND HUBER, P., 1995, Lake sequences of the Central Hartford Basin, Newark Supergroup: State Geological and Natural History Survey of Connecticut, Guidebook no. 7, p. B1–B39.

GIERLOWSKI-KORDESCH, E.H., AND RUST, B.R., 1994, The Jurassic East Berlin Formation, Hartford Basin, Newark Supergroup (Connecticut and Massachusetts): a saline lake–playa–alluvial plain system, *in* Renaut, R., and Last, W.M., eds., Sedimentology and Geochemistry of Modern and Ancient Saline Lakes: SEPM, Special Publication 50, p. 249–265.

GORE, P.J.W., 1986a, Depositional framework of a Triassic rift basin: The Durham and Sanford sub-basins of the Deep River Basin, North Carolina, *in* Textoris, D.A., ed., SEPM, Southeastern United States: Third Annual Midyear Meeting, Raleigh, North Carolina, Field Guidebook, Field Trip no. 3, p. 55–115.

GORE, P.J.W., 1986b, Late Triassic and Early Jurassic lacustrine sedimentation in the Culpeper Basin, Virginia, *in* Manspeizer, W., ed., Triassic–Jurassic Rifting: Continental Breakup and the Origin of the Atlantic Ocean and Passive Margins, Part A: Amsterdam, Elsevier, Developments in Geotectonics, ser. 22, p. 369–400.

GORE, P.J.W., 1986c, Early diagenetic nodules, compaction and secondary lamination in Early Jurassic lacustrine black shale, Culpeper Basin, Virginia (abstract): American Association of Petroleum Geologists, Bulletin, v. 70, p. 596.

GORE, P.J.W., 1988a, Lacustrine sequences in an early Mesozoic rift basin: Culpeper Basin, Virginia, USA, *in* Fleet, A.J., Kelts, K., and Talbot, M.R., eds., Lacustrine Source Rocks: Geological Society of London, Special Publication 40, p. 247–278.

GORE, P.J.W., 1988b, Paleoecology and sedimentology of a Late Triassic lake, Culpeper Basin, Virginia, USA: Palaeogeography, Palaeoclimatology, Palaeoecology, v. 62, p. 593–608.

GORE, P.J.W., 1989, Toward a model for open- and closed-basin deposition in ancient lacustrine sequences: The Newark Supergroup (Triassic–Jurassic), Eastern North America: Palaeogeography, Palaeoclimatology, Palaeoecology, v. 70, p. 29–51.

HARRIS, N.B., SORRIAUX, P., AND TOOMEY, D.F., 1994, Geology of the Lower Cretaceous Viodo carbonate, Congo Basin: A lacustrine carbonate in the South Atlantic Rift, *in* Lomando, A.J., Schreiber, B.C., and Harris, P.M., eds., Lacustrine Reservoirs and Depositional Systems: SEPM, Core Workshop no. 19, p. 143–172.

HENTZ, T.F., 1985, Early Jurassic sedimentation of a rift-valley lake: Culpeper Basin, northern Virginia: Geological Society of America, Bulletin, v. 96, p. 92–107.

HILLAIRE-MARCEL, C., AND CASANOVA, J., 1987, Isotopic hydrology and paleohydrology of the Magadi (Kenya)–Natron (Tanzania) Basin during the Late Quaternary: Palaeogeography, Palaeoclimatology, Palaeoecology, v. 58, p. 155–181.

HSÜ, K.J., AND KELTS, K., 1978, Late Neogene sedimentation in the Black Sea, *in* Matter, A., and Tucker, M.E., eds., Modern and Ancient Lake Sediments: International Association of Sedimentologists, Special Publication 2, p. 129–145.

HUBERT, J.F., REED, A.A., AND CAREY, P.J., 1976, Paleogeography of the East Berlin Formation, Newark Group, Connecticut Valley: American Journal of Science, v. 276, p. 1183–1207.

HUBERT, J.F., GILCHRIST, J.M., AND REED, A.A., 1982, Jurassic redbeds of the Connecticut Valley: (1) Brownstones of the Portland Formation; and (2) Playa–Playa Lake–Oligomictic Lake model for parts of the East Berlin Formation, Shuttle Meadow, and Portand Formations: Hartford, State Geology and Natural History Survey of Connecticut, Guidebook 5, p. 103–141.

HUBERT, J.F., FESHBACH-MERINEY, P.E., AND SMITH, M.A., 1992, The Triassic–Jurassic Hartford Rift Basin, Connecticut and Massachusetts: Evolution, sandstone diagenesis, and hydrocarbon history: American Association of Petroleum Geologists, Bulletin, v. 76, p. 1710–1734.

INOUCHI, Y., KINUGASA, Y., KUMON, F., NAKANO, S., YASUMATSU, S., AND SHIKI, T., 1996, Turbidites as records of intense palaeoearthquakes in Lake Biwa, Japan: Sedimentary Geology, v. 104, p. 117–125.

KOMOR, S.C., 1994, Bottom-sediment chemistry in Devil's Lake, northeast North Dakota, *in* Renaut, R.W., and Last, W.M., eds., Sedimentology and Geochemistry of Modern and Ancient Saline Lakes: SEPM, Special Publication 50, p. 21–32.

KRYNINE, P.D., 1950, Petrology, stratigraphy, and the origin of the Triassic sedimentary rocks of Connecticut: Hartford, Connecticut State Geology and Natural History Survey, Bulletin 73, p. 1–247.

LAVOIE, D., 1995, Carbonate botryoids in Lower Devonian amygdaloidal basalts: evidence for precipitation of high-Mg calcite from heated and volcanic $CO_2$-buffered marine waters: Journal of Sedimentary Research, v. A65, p. 541–546.

LETOURNEAU, P., AND MCDONALD, N.G., 1985, The sedimentology, stratigraphy, and paleontology of the Lower Jurassic Portland Formation, Hartford Basin, central Connecticut: State Geology and Natural History Survey of Connecticut, Guidebook 6, p. 353–391.

LE TURDU, C., TIERCELIN, J.-J., GIBERT, E., TRAVIC, Y., LEZZAR, K.-E., RICHERT, J.-P., MASSAULT, M., GASSE, F., BONNEFILLE, R., DECOBERT, M., GENSOUS, B., JEUDY, V., TAMRAT, E., MOHAMMED, M.U., MARTENS, K., ATNAFU, B., CHERNET, T., WILLIAMSON, D., AND TAIEB, M., 1999, The Ziway–Shala lake basin system, Main Ethiopian Rift: Influence of volcanism, tectonics and climatic forcing on basin formation and sedimentation: Palaeogeography, Palaeoclimatology, Palaeoecology, v. 150, p. 135–177.

LINDHOLM, R.C., HAZLETT, J.M., AND FAGIN, S.W., 1979, Petrology of Triassic–Jurassic conglomerates in the Culpeper Basin, Virginia: Journal of Sedimentary Petrology, v. 49, p. 1245–1262.

MANSPEIZER, W., AND OLSEN, P.E., 1981, Rift basins of the passive margin: tectonics, organic-rich lacustrine sediments, basin analyses, *in* Hobbs, G.W., ed., Field Guide to the Geology of the Paleozoic, Mesozoic, and Tertiary Rocks of New Jersey and the Central Hudson Valley, Atlantic Margin Energy Symposium: American Association of Petroleum Geologists, Eastern Section, p. 25–103.

MAY, G., HARTLEY, A.J., STUART, F.M., AND CHONG, G., 1999, Tectonic signatures in arid continental basins: an example from the Upper

Miocene–Pleistocene, Calama Basin, Andean forearc, northern Chile: Palaeogeography, Palaeoclimatology, Palaeoecology, v. 151, p. 55–77.

McCrea, J.M., 1950, On the isotopic chemistry of carbonates and a paleotemperature scale: Journal of Chemical Physics, v. 18, p. 849–857.

Müller, G., and Wagner, F., 1978, Holocene carbonate evolution in Lake Balaton (Hungary): a response to climate and impact of man, *in* Matter, A., and Tucker, M.E., eds., Modern and Ancient Lake Sediments: International Association of Sedimentologists, Special Publication 2, p. 57–81.

Olsen, P.E., 1989, Geology of the Newark Basin, *in* Olsen, P.E., and Gore, P.J.W., eds., Tectonic, Depositional, and Paleoecological History of Early Mesozoic Rift Basins, Eastern North America: 28th International Geological Congress, Fieldtrip Guidebook T351, p. 69–90.

Olsen, P.E., 1990a, The latest Triassic and Early Jurassic formations of the Newark Basin (Eastern North America Newark Supergroup): Stratigraphy, structure, and correlation: New Jersey Academy of Science, Bulletin 25, p. 25–51.

Olsen, P.E., 1990b, Tectonic, climatic and biotic modulation of lacustrine systems—Examples from Newark Supergroup of eastern North America, *in* Katz, B.J., ed., Lacustrine Basin Exploration: American Society of Petroleum Geologists, Memoir 50, p.209–224.

Olsen, P.E., and Kent, D.V., 1996, Milankovitch climate forcing in the tropics of Pangaea during the Late Triassic: Palaeogeography, Palaeoclimatology, Palaeoecology, v. 122, p. 1–26.

Olsen, P.E., Smoot, J.P., and Gore, P.J.W., 1989, Triangle Brick Quarry, Durham, NC, *in* Olsen, P.E., and Gore, P.J.W., eds., Tectonic, Depositional, and Paleoecological History of Early Mesozoic Rift Basins, Eastern North America: 28th International Geological Congress, Fieldtrip Guidebook T351, p. 29–31.

Olsen, P.E., Kent, D.V., and Cornet, B., 1993, Early Mesozoic lacustrine record of cyclical climate change from core and outcrops of the Newark Basin: SEPM, Lamont Doherty Geological Observatory of Columbia University, Field Guide, p. 1–21.

Olsen, P.E., Kent, D.V., Cornet, B., Witte, W.K., and Schlische, R.W., 1996, High-resolution stratigraphy of the Newark rift basin (Early Mesozoic, eastern North America): Geological Society of America, Bulletin, v. 108, p. 40–77.

Parrish, J.T., Ziegler, A.M., and Scotese, C.R., 1982, Rainfall patterns and the distribution of coals and evaporites in the Mesozoic and Cenozoic: Palaeogeography, Palaeoclimatology, Palaeoecology, v. 40, p. 67–101.

Platt, N.H., and Wright, V.P., 1991, Lacustrine carbonates: facies models, facies distributions and hydrocarbon aspects, *in* Anadon, P., Cabrera, Ll., and Kelts, K., eds., Lacustrine Facies Analysis: International Association of Sedimentologists, Special Publication 13, p. 57–74.

Pratt, L.M., Vuletich, A.K., and Davis, T.A., 1985, Geochemical and isotopic characterization of organic matter in rocks of the Newark Supergroup, *in* Robinson, G.R., and Froelich, A., eds., Proceedings of the Second United States Geological Survey Workshop on the Early Mesozoic Basins of the Eastern United States: U.S. Geological Survey, Circular 946, Alexandria Virginia, p. 74–78.

Puffer, J.H., and Student, J.J., 1992, Volcanic structures, eruptive styles, and posteruptive deformation and chemical alteration of the Watchung flood basalts, New Jersey, *in* Puffer, J.H., and Ragland, P.C., eds., Eastern North American Mesozoic Magmatism: Geochemistry, Structure, and Stratigraphy: Geological Society of America, Special Paper 268, p. 261–277.

Renaut, R.W., and Owen, R.B., 1991, Shore-zone sedimentation and facies in a closed rift lake: the Holocene beach deposits of Lake Bogoria, Kenya, *in* Anadon, P., Cabrera, Ll., and Kelts, K., eds., Lacustrine Facies Analysis: International Association of Sedimentologists, Special Publication 13, p. 175–195.

Renaut, R.W., and Tiercelin, J.J., 1994, Lake Bogoria, Kenya Rift Valley—A sedimentological overview, *in* Renaut, R.W., and Last, W.M., eds., Sedimentology and Geochemistry of Modern and Ancient Saline Lakes: SEPM, Special Publication 50, p. 100–123.

Renaut, R.W., Tiercelin, J.-J., and Owen, R.B., 1986, Mineral precipitation and diagenesis in the sediments of the Lake Bogoria Basin, Kenya Rift Valley, *in* Frostick, L.E., Renaut, R.W., Reid, I., and Tiercelin, J.-J., eds., Sedimentation in the African Rifts: Geological Society of London, Special Publication 25, p. 159–176.

Shinn, E.A., Steinen, R.P., Lidz, B.H., and Swart, P.K., 1989, Whitings, a sedimentological dilemma: Journal of Sedimentary Petrology, v. 59, p. 147–161.

Smoot, J.P., and Olsen, P.E., 1989, Culpeper Crushed Stone Quarry, Culpeper (Stevensburg), VA, *in* Olsen, P.E., and Gore, P.J.W., eds, Tectonic, Depositional, and Paleoecological history of Early Mesozoic Rift Basins, Eastern North America: 28th International Geological Congress, Field Trip Guidebook T351, p. 60–63.

Starquist, V.L., 1943, The stratigraphy and structural geology of the central portion of the Mount Tom and East Mountain ridges: Unpublished M.S. thesis, Smith College, Northampton, Massachusetts, 48 p.

Steinen, R.P., Gray, N.H., and Mooney, J., 1987, A Mesozoic carbonate hotspring deposit in the Hartford Basin of Connecticut: Journal of Sedimentary Petrology, v. 57, p. 319–326.

Straccia, F.G., Wilkinson, B.H., and Smith, G.R., 1990, Miocene lacustrine algal reefs—southwestern Snake River Plain, Idaho: Sedimentary Geology, v. 67, p. 7–23.

Sturm, M., and Matter, A., 1978, Turbidites and varves in Lake Brienz (Switzerland): deposition of clastic detritus by density currents, *in* Matter, A., and Tucker, M.E., eds., Modern and Ancient Lake Sediments: International Association of Sedimentologists, Special Publication 2, p. 147–168.

Swart, P.K., McKenzie, J.A., and Lohmann, K.C., 1993, Introduction, *in* Swart, P.K., Lohmann, K.C., McKenzie, J.A., and Savin, S., eds., Climate Change in Continental Isotopic Records: American Geophysical Union, Geophysical Monograph 78, p. xi–xiii.

Talbot, M.R., and Kelts, K., 1990, Paleolimnological signatures from carbon and oxygen isotopic ratios in carbonates from organic carbon-rich lacustrine sediments, *in* Katz, B., ed., Lacustrine Basin Exploration: American Association of Petroleum Geologists, Memoir 50, p. 99–112.

Thompson, J.B., Ferris, F.G., and Smith, D.A., 1990, Geomicrobiology and sedimentology of the mixolimnion and chemocline in Fayetteville Green Lake, New York: Palaios, v. 5, p. 52–75.

Tiercelin, J.-J., 1991, Natural resources in the lacustrine facies of the Cenozoic rift basins of East Africa, *in* Anadon, P., Cabrera, Ll., and Kelts, K., eds., Lacustrine Facies Analysis: International Association of Sedimentologists, Special Publication 13, p. 3–37.

Tollo, R.P., and Gottfried, D., 1992, Petrochemistry of Jurassic basalt from eight drill cores, Newark Basin, New Jersey: Implications for the volcanic petrogenesis of the Newark Supergroup, *in* Puffer, J.H., and Ragland, P.C., eds., Eastern North American Mesozoic Magmatism: Geological Society of America, Special Paper 268, p. 233–259.

Valero Garces, B.L., Gierlowski-Kordesch, E., and Bragonier, W.A., 1997, Pennsylvanian continental cyclothem development: no evidence of direct climatic control in the Upper Freeport Formation (Alleghenian) of Pennsylvania: Sedimentary Geology, v. 109, p. 305–320.

Van Houten, F.B., 1965, Crystal casts in Upper Triassic Lockatong and Brunswick Formations: Sedimentology, v. 4, p. 301– 313.

Van Houten, F.B., 1987, Late Triassic cyclic sedimentation: Upper Lockatong and Lower Passaic Formations (Newark Supergroup), Delaware Valley, west-central New Jersey: Geological Society of America, Centennial Field Guide, Northeastern Section, p. 81– 86.

Weems, R.E., and Olsen, P.E., 1997, Synthesis and revision of groups within the Newark Supergroup, eastern North America: Geological Society of America, Bulletin, v. 109, p. 195–209.

Wheeler, W.H., and Textoris, D.A., 1978, Triassic limestone and chert of playa origin in North Carolina: Journal of Sedimentary Petrology, v. 48, p. 765–776.

Yuretich, R.F., 1982, Possible influences upon lake development in the East African Rift Valleys: Journal of Geology, v. 90, p. 329–337.

Young, R.S., and Edmundson, R.S., 1954, Oolitic limestone in the Triassic of Virginia: Journal of Sedimentary Petrology, v. 24, p. 275–279.

# Index

## A



## B



## C



## D



## E



## F



## G



## H



## I



## J



## K

# L



# M



# N



# O



# P



# Q



# R



# S

# T



# V



# W